To
Walter
With love from
Judy.
Christmas 1971

BOTANY

An Introduction to Plant Biology

One of the five stigmas of the cranesbill flower (*Erodium chamaedroyides*) with pollen. Taken with the Stereoscan Electron microscope. ×300. (Courtesy of R. H. Falk and E. M. Gifford.)

BOTANY
An Introduction to Plant Biology

Fourth Edition

T. ELLIOT WEIER

C. RALPH STOCKING

MICHAEL G. BARBOUR

John Wiley & Sons, Inc., New York · London · Sydney · Toronto

Cover. Stereoscan view of the cuticle on a petal of a flower of Shasta Daisy (*Chrysanthemum maximum*), prepared and taken by R. H. Falk. Magnification ×2,600

Line drawings by Alice B. Addicott with the assistance of Jean A. McKinnon.

Library of Congress Catalogue Card Number: 79-105385

ISBN   0-471-92467-9

Printed in the United States of America

10   9   8   7   6   5   4   3   2   1

Preface

During the past decade, advances in biology and chemistry have introduced important concepts that the student must study before taking the introductory course in botany. The new material poses questions for all botanists, but the problem of selecting the subject matter for the introductory course is a particularly vexing one. Although this fourth edition of *Botany, An Introduction to Plant Biology* is still an introductory textbook, the student reading it must know more about biological concepts than those who studied earlier editions. We still assume, however, that students have little or no knowledge about plants as living entities. The objective of the text remains unchanged: *to acquaint students with the great range of plant life that inhabits our biosphere.* With concepts of energy transfer, electron flow, and DNA replication already established, we can proceed directly to the exciting aspects of plant form—its development, its relation to function, its similarities over wide ranges of plants and, in seeming contradiction, the great diversity that allows for unique individuals within species.

To help maintain breadth of approach and to foster the ideas of younger botanists currently making the decisions on modern curricula, Professor Michael G. Barbour, who recently joined the staff of the Department of Botany of the University of California at Davis, was asked to become a third author. His interest in plants, particularly their ecology and taxonomy, his dedication to teaching and to his students, and his ability to write and to criticize have already made him an appreciated and valued co-worker.

The quality of illustrations has been greatly enhanced. Alice B. Addicott, a biologist with a master's degree in biology from Stanford University and a botanist by association, has acted as a collaborator, not only in the preparation of new illustrations that bear her signature (ABA), but also in redrawing, coloring, and rearranging figures that have been retained from earlier editions. She has been assisted in her work by Jean A. McKinnon (original figures signed JAM). Color has been added to this edition; most line drawings are shaded in greens and grays, and there are fourteen plates in full color. The majority of the halftones are new; they range from exciting micrographs from recent research papers to greatly improved views of subjects presented in earlier editions.

The choice of new subject matter and the deletion of older material reflect the advances in knowledge and the changes in thinking that have occurred over the past decade. For instance, the electron microscope has supplied significant information about the cell wall, sieve elements, food transport, the chloroplast and mitochrondria, and other aspects of cell structure. The electron-scanning microscope has revealed details of leaf surfaces and the design of pollen cell walls. The chapters on cells and cell division have been completely revised to convey this new information about the ultrastructure of plant cells.

The increased importance of ecology during the decade is indicated in the new chapters on ecology and taxonomy. The chapter on ecology now includes an analysis of many living and nonliving factors in the environment; the ecotype, community, and ecosystem concepts; a review of selected world vegetation types; and a discussion of problems related to pollution. In the taxonomy chapter, we have tried to show the great diversity of taxonomic research methods, such as biosystematics and numerical taxonomy, and have included considerable material on the history of taxonomy.

The policy in earlier editions of *Botany* has been to preceed from the known to the unknown. Although we still believe that this is the best possible approach, in the present edition we have revised the opening paragraphs of many chapters to emphasize the traits that are common to all of the plants discussed in that particular chapter. Thus we now also proceed from uniformity to diversity.

In earlier editions, problems of plant development were confined for the most part to a single chapter but, in the fourth edition, development is a continuous theme running through many chapters. The algae are used to illustrate principles in the development of form. Additional aspects of development, particularly information gained through experimentation, are brought together and discussed in Chapter 20, "Plant Growth and Development."

Thus, the chapters on morphology are more complete throughout the text because the processes of development are given parallel treatment in each of them. This approach permits a great many complementary illustrations. The stem and root chapters have complementary diagrams on development, and the illustrations of leaf and flower development relate directly to the illustrations of the shoot apex. Matching figures illustrate cross sections of the stipe of the brown algae and cross sections of the stems of the vascular plants.

The discussion of algae, here, is frankly experimental. Our aim is to tell a connected story about plant development that will be meaningful without naming particular species. Chapter 22 is a study of the *biology of the algae.* The traits that group algae together are described first, and then the diversity that separates them into phyla is considered. In the second part of the chapter we compare the plant bodies of algae by describing three lines of evolutionary development. Although the principal concepts of this story can be told without referring to particular species, examples are given, and each example is illustrated. Reproductive cycles are then discussed in terms of gametic, zygotic, and sporic life cycles. Again, the concepts involved do not require our mentioning specific algae, but each life cycle is accompanied by a diagram and examples. Because the electron microscope has so greatly increased our knowledge of algal cytology, the presentation of the structure of algae cells in the fourth and last section concentrates chiefly on ultrastructure. The material on classification and life cycles (in the first and third sections of Chapter 22) corresponds closely to similar subject matter in most introductory botany textbooks, but the treatment of the evolution of algal form and the fine structure of the algal cell is somewhat new to an elementary text.

The problem of classification has not been resolved. Our decision for this edition has been to follow, with one exception, the system adopted in a current textbook on plant morphology, *An Evolutionary Survey of the Plant Kingdom,* by R. F. Scagel, R. J. Bandoni, G. E. Rouse, W. B. Schofield, J. R. Stein, and T. M. C. Taylor. Our exception is the retention of Sphenophyta in preference to Arthrophyta. Unquestionably the term Arthrophyta, meaning "jointed plants," better describes the horsetails than does the phrase "wedge plants," which Sphenophyta designates. But we have found the word Arthrophyta to be confusing; it is too similar to the Arthropoda of the zoologist and to Anthophyta used by Robert F. Seagel and his co-workers for the flowering plants. In adopting this classification we do not mean to express an opinion on the evolutionary development of plant phyla, although we are skeptical about the theory of a monophyletic origin of the vascular plants.

The chapter on inheritance has been revised to point out that modern plant breeding is no longer based on simple Mendelian genetics, but is complicated by polygenes, multiple alleles, heterosis, and polyploidy. Much thought was given as to whether to restate the genetic code, but we decided to mention it only briefly in the account of DNA replication, on the premise that all biology students will be subjected to it several times before obtaining a degree.

The close and frequently exciting interchanges with colleagues have been a constant source of stimulation and pleasure during the preparation of this edition. Their response to requests for illustrative material, for advice on problems such as that posed by the phylogeny of the plant kingdom, and for reviews of the manuscript has been phenomenal. The graduate students in our own department have been particularly helpful in contributing information, illustrations, and criticism. David Brown and Lewis Feldman gave willingly of their time and experience in the preparation of several excellent figures. Professors E. M. Gifford, Elizabeth Cutter, and Norma Lang provided unfailing stimulation and help. We owe a special debt to Professor Daniel Branton, who contributed very materially with a criticism of the third edition and with important suggestions for changing the fourth edition. Harold Drever, Bijan Dehgan, and Walter Russell constantly supplied fresh material for the pen or camera. Since we have received the important and willing assistance of many colleagues in the preparation of this edition, the book in a sense expresses the points of view of a wide range of botanists. However, we are responsible for its accuracy of statement, its organization, and its philosophy. The writing of this volume has been difficult and exciting—we hope it will be a genuine contribution to plant biology.

T. ELLIOT WEIER
C. RALPH STOCKING
MICHAEL G. BARBOUR

Davis, California, 1970

Contents

BOTANY

An Introduction to Plant Biology

1

Introduction

An educated person is one who has a fair knowledge and an appreciative understanding of the environment in which he lives, and of the part man plays in modifying it. Our environment is most complex. It embraces everything that surrounds us, both living and nonliving. It includes the air we breathe, the food we eat, the water we drink; it includes temperature, rainfall, light, humidity of the atmosphere, wind, soil, and atmospheric pressure; it involves the varied plant and animal life, both wild and domesticated; and it **includes, as well as plants and animals, the human beings with whom we are associated, their racial characteristics,** and **their social** and **economic systems.** All elements of the environment affect our lives, our health, our well-being, our happiness, our freedom, our philosophy, and religious views.

This environment, in its totality, is called the biosphere. Our current interest lies in plant life, and this first chapter briefly discusses some of the roles that plants play in helping to stabilize the biosphere. It also points out ways in which botanists study plants and their interrelationships with other components of the biosphere. The second chapter sketches some of the similarities and diversities among plants, as well as briefly relating the botanists' efforts to arrange the similarities into a system that will bring order to the diversity. This is classification and it draws upon knowledge from many fields of botany: anatomy, plant biochemistry, heredity, and morphology. This effort, which is the central theme of the second chapter, results in a more or less satisfying separation of plants into groups that give evidences of a genetic relationship through the sharing of a common ancestor. All details of anatomy, cytology, and physiology used here will be discussed in more depth in subsequent chapters, and are of importance here only insofar as they serve to clarify the groupings of plants. Do not try to memorize them all at this point.

THE BIOSPHERE

Our environment constitutes the biosphere of the planet Earth. It is a thin layer of water, soil, and atmosphere which supports life as we know it. The soil and water teem with microorganisms, but relatively few large organisms inhabit the soil. On the other hand, many species, including some of the largest forms, inhabit the water. But it is the narrow band of atmosphere about the land which supports the greatest number and variety of organisms.

The temperature range within our solar system is from nearly absolute zero to $10^7\,°C$, and the wave lengths of radiant energy range from 10^{-9} to 10^{-10} cm for the extremely short gamma rays to 10^5 cm for the long radio waves. To the best of our knowledge, life requires a temperature between $0\,°C$ and $50\,°C$, with the optimum around $30\,°C$. The human eye is sensitive to radiant energy from the sun between 5000 and 7000 A in wave length (5 to 7×10^{-5} cm). Plants, the sole converters of radiant energy, use mainly blue and red light for this purpose. Their growth responds to blue light and other wave lengths in a variety of ways (page 412). Thus, life is tolerant to a very narrow temperature band and responds to an equally narrow band of radiant energy.

The solar radiation reaching the earth's surface is modified by the composition of the atmosphere, which is a mixture of 21 percent oxygen, 78 percent nitrogen, 0.03 percent carbon dioxide and less than 0.006 percent of 3 other gases. At sea level, the column of atmosphere above one square inch weighs 15 pounds and will support a column of water 34 feet high (Fig. 5.13). There is abundant life in the biosphere from sea level to an elevation varying from about 2.5 miles at the equator to about one mile in the northern Canadian Rockies. In the oceans, abundant life is found from sea level to a depth of 100 feet.

But the biosphere has not always been constituted as described. Life probably originated in an atmosphere devoid of oxygen. Thus, the first living organisms were anaerobic forms and probably never advanced much beyond simple filaments or masses of cells (Fig. 2.5A). They may have been subjected to large amounts of ultraviolet radiation. With the advent of photosynthesis (Chap. 13), there arose organisms able to utilize the radiant energy of sunlight and to split water into hydrogen and oxygen (Fig. 13.9). The hydrogen was used to reduce carbon dioxide to a carbohydrate (Chap. 13), thus permitting storage of radiant energy as chemical energy for future use. The oxygen escaped to the atmosphere (Fig. 13.9), and ozone (O_3), formed in the atmosphere from oxygen, filtered out much of the ultraviolet radiation. Early primitive plants played a dominant part in establishing oxygen in the atmosphere, and present-day plants are essential for its maintenance there.

It may not forever remain so. Man's ability to utilize the enormous amounts of fossil energy stored in eons past, together with his utilization of nuclear energy, is bringing about changes in the biosphere. We already know that these changes are presently harmful to some forms of life, including man himself. We do not know the permanent long-term changes that may benefit or harm man, and man's ability to bring about such changes in the biosphere poses a very important problem for contemporary biologists.

CERTAIN BIOLOGICAL UNIVERSALS

If we assume that the original anaerobic organisms were similar in appearance and function to those of today (Fig. 21.9), and if the first aerobic and photosynthetic forms are also similar to those of today (Fig. 2.5A), then certain general principles about life emerge. These principles may be called **biological universals.** First, *all life is cellular* (Fig. 2.5). Since viruses cannot live without host cells, we shall not consider them true living organisms. Second, the use and transfer of energy within all cells involves the same compound, known as **adenosine triphosphate,** or **ATP** for short (Chap. 5). Third, some steps in respiration are identical in all aerobic organisms and in some anaerobic forms (Chap. 14). Fourth, some steps in respiration in all organisms and in all cells are associated with membranes (Chap. 4). Fifth, the green pigment chlorophyll *a* is required in the conversion of the radiant energy of sunlight to food (photosynthesis) in all green plants; also, ATP is involved in respiration and photosynthesis, and the light-trapping steps in photosynthesis are always associated with membranes (Chap. 4). And sixth, genetic information which carries inheritance from one generation to the next is stored in the same compound in all organisms. This is deoxyribose nucleic acid, or simply DNA (Chap. 5). In other words, if the present day bacteria and single-celled algae (both anaerobic and aerobic forms), which we consider to be primitive organisms, are similar to their ancestors of 400 million years ago, certain important basic processes of life become apparent: respiration, photosynthesis, and inheritance are basically the same in all organisms and have been so from the beginning. If this assumption is false, we have no starting point and an understanding of past life will be much more difficult to attain.

DIFFERENCES BETWEEN ORGANISMS OF ANAEROBIC AND AEROBIC TIME

We have just assumed that present-day anaerobic forms closely resemble the forms living in anaerobic time. Are there any major differences between the organisms of anaerobic and aerobic time? There seem to be. In present day aerobic organisms, the processes of respiration, photosynthesis, and inheritance are confined to their own special compartments within the cell; mitochondria, chloroplasts, and nuclei respectively (Figs. 2.5B, 4.4). But in the primitive forms, these three activities, as well as others, occupy communal space in a single non-compartmented cell (Fig. 2.5A). Since the more primitive forms lack a nucleus, they are known as the **prokaryotes.** The forms with a true nucleus are the **eukaryotes.** From another point of view, there are really only two kinds of organisms, consumers of energy and producers of energy. This is a less scientific but more descriptive way of saying that there are two kinds of kingdoms, Animal (consumers) and Plant (producers). *Man is the greatest consumer of all.*

PLANTS AND PLANT SCIENTISTS

In the plant kingdom only the blue green algae are prokaryotes and only the fungi, together with a few isolated species, lack chlorophyll. The metabolic processes of all plants, with the possible exception of the fungi, are similar. They are all producers!

In all of life on Earth, plants are the only producers. All consumers, particularly man, are dependent upon plants for food, for fiber, for wood, for energy, and for oxygen. A knowledge of plants, their habitats, their structure, their metabolism, and their inheritance is thus the basic foundation for man's survival.

What are the problems that concern the plant scientists in their efforts to accumulate significant knowledge of the plant life of the biosphere? Plants must first be named. Early botanists, such as the Greek, Theophrastus (400 B.C.), were largely concerned with naming plants. In fact, many current scientific names for plants are the same as, or similar to those, given to the same plants by Greek or Roman botanists. Concomitant with naming plants came an increase in the numbers of known plants. Plant explorers accompanied the expansion of empires. As a result, botanic gardens and herbaria, storehouses of botanic information, were established in Sweden, England, and France. In order to retrieve information there must exist some system of classification. Modern methods of classifying plants date from the Swedish botanist, Linnaeus (1707–1778). New species of plants are still being described, but the standing of a plant scientist no longer depends upon the numbers of plants he can recognize.

Some other important problems claiming the attention of plant scientists are as follows. (1) What are the principles underlying the great diversity of plants? (2) What is the mechanism that controls the precise patterns of growth? (3) Since plants are the only producers, what kind of plants, growing under what conditions of temperature, soil, moisture, and nutrients will produce most abundantly? (4) How are plants related to their environment, and will the changes in the biosphere brought about by man eventually prove harmful to man? These problems can only be studied with experiments.

Experimentation

There are many kinds of plants growing in many different places. Just how are plant form and plant metabolism interrelated so that in spite of the great array of plants and plant habitats, a fairly constant relationship has developed between plants and their environment? There is no complete answer to this question. Attempts to answer it still require much experimentation and study.

It is through experimentation that we hope to derive answers. Experimentation asks questions about plants that can be answered by manipulating plants under specific conditions designed to answer the question. In 1770, the Dutch physician van Ingenhousz asked the question, "Will a glowing sliver burst into flame in an atmosphere in which plants have been confined in the dark?" The answer was, "No". Will it burst into flame in an atmosphere in which plants have been confined in the light? Now the answer is, "Yes". What is the difference between the two atmospheres?

Asking the right question is frequently the most difficult step. Generally it should be a simple question, hopefully with not more than two possible answers. The procedure must be within the realm of available experimental technique. The experimental system set up for study of the question must be under reasonably good control. The fewer the variables there are in the system, the better. And, of considerable significance, the answer to the right questions must lead to a greater understanding of the whole problem and, in addition, to other questions.

Plant Taxonomists

Taxonomists wish to complete an inventory of the earth's plant resources; they want to categorize, for easy reference, the sum total of existing plant variation. As the inventory becomes complete (and it is far from complete today), they hope it will be possible to organize this diversity into some evolutionary scheme. What sort of patterns exist, and do these patterns mean genetic relationship and evolution from common ancestors? Which plants or plant characteristics are the most primitive —imitating early, now-extinct species—and which characteristics are advanced—indicating a recent time of evolution? Taxonomists use morphology, anatomy, biochemistry, hybridization, and chromosome studies to arrive at tentative conclusions about the *phylogenetic* (evolutionary) relationships between species.

Taxonomy is basic to botany, for all botanists use taxonomic information in their research. The physiologist, for example, must know the particular species he works with. Since species differ in their physiology, his results would be meaningless if the name of his study organism were not known; other workers could neither corroborate nor extend his findings. The anatomist, developmental morphologist, and cytologist all face the same necessity of knowing the species with which they work.

Plant Ecologists

Plant ecologists want to understand how plants are adapted to their environment. What are the environmental forces that regulate plant growth and what kinds of responses can plants make to them? Living organisms and their environment are interdependent in complicated, often unexpected ways. Can man learn to imitate this natural balance or to manipulate it for his own ends? Is it possible, with the vegetation we have, to meet several pressing needs at once: those of agriculture, industry, recreation, conservation? What is the wisest, most efficient way to use our plant resources? Have we changed our environment dangerously by pollution? What level of pollution can be safely tolerated? Are plants adapting to pollution?

The ecologist also wishes to inventory the earth's plant resources. Unlike the taxonomist, he is interested in categorizing groupings of species (communities) rather than individual species. How is it that certain species repeatedly group together to the exclusion of all other species? What is the geologic history of these communities, and what are their paths of migration through time?

Ecologists are often generalists, utilizing information from taxonomy, physiology, anatomy, geology, and meteorology to search for answers to questions like those posed above.

Plant Morphologists

The great diversity of plant forms is readily apparent. There are sea weeds, mosses, ferns, pine trees, corn plants, palms, and dandelions. Obviously, corn bears seeds (Chap. 16), but do ferns, mosses, and sea weeds also bear seeds? If they do not, how do they propagate (Chaps. 22, 26, 27)? Both pines and palms have single straight trunks, but the palm has large divided leaves, found at the top only, and the pine has many lateral branches clad with needles. Why? (Chap. 17). Botanists interested in plant form are known as morphologists. Their interests lie in the diversity of plant form and the manner in which cells are arranged to give this form. All plants require water and food; are water and food conducted and transported in cells of similar form in plants of the ocean and plants of the desert? How do the reproductive structures of plants of the ocean, marsh, and desert differ (Chaps. 22, 27)? Are the cellular details of the stems of mosses, ferns, pine trees, and palms alike (Chaps. 7, 27)?

The morphologists are also interested in knowing how form develops in such a precise fashion. When a fertilized egg cell in the flower of a pea plant starts to grow, it follows a pathway similar to that followed by fertilized eggs of all other seed plants. Eventually a pea seed develops identical to all other pea seeds but different from every other kind of seed. When the seed is planted, it germinates and grows just as do the seeds of all other species of plants, but the pea seed always produces pea plants, and only pea plants. Cells divide in precise planes and enlarge in precise locations within the developing seed at regular rates. Differentiation of cells to form tissues also occurs in a regular fashion. The morphologist has learned how to modify some of these steps. He would like to know more of the manner in which they are controlled and directed within the plant itself. A better knowledge of these things will give us a deeper insight into the developmental process of all cells.

Plant Physiologists

While it is true that the basic processes of respiration and photosynthesis are similar in all plants, the end products of the synthetic processes in plants are as variable as there are different species of plants. Why? There are hard wheats and soft wheats and the flour made from them is different. Some wheat seeds need to be held at a low temperature before they will germinate. Some lettuce seeds won't germinate until they have been briefly exposed to red light. Again, why? The answer to these and similar questions lie in the realm of plant physiologists.

Morphologist and physiologist frequently unite in searching for aspects of form that have a chemical basis. A vegetative shoot of *Chrysanthemum* will not flower as long as it receives more than 10 hours of daylight. Soon after the vegetative shoot receives a day-night cycle of only 9 hours of daylight and 15 hours of darkness, flower buds form. What are the morphological changes that take place in the *Chrysanthemum* shoot as it changes from the vegetative to the flowering condition? What chemical substance, i.e. hormone, is induced by the change in light period and how does it bring about the change in the growth pattern?

One of the major goals of the plant physiologist is to understand all of the biochemical and physical processes that are the basis of life, and that are responsible for the great diversity of life form and activities. The more the physiologist knows about life reactions, the more he is able to regulate and to control these activities and so shape plant growth and development. Is it possible to predict or even regulate when a desirable plant will flower? Can fruit without seeds be produced? Why are some plants bushy-shaped dwarfs and others of the same species tall and vine-like?

Cytologists

Cytologists, or cellular biologists, as they are now more popularly known, are concerned with many problems at the cellular level. It might seem that since the basic phenomena of inheritance, photosynthesis, and respiration are essentially identical in all organisms and these three processes are confined to specific organelles within the cell, that understanding them would be easy. We do know that the genetic code present in the nucleus directs the synthesis of specific proteins. But we don't know how protein is actually synthesized in the cell. We are not sure of the physical arrangement of DNA in a chromosome. We would like to know the precise relationship between the steps in photosynthesis and respiration and the molecules of the membranes associated with these processes. Respiration and photosynthesis supply energy for cellular synthesis. Protein synthesis probably takes place in all parts of the cell, even in the mitochondria and chloroplasts which supply energy for synthesis elsewhere in the cell.

Since protein synthesis is coded by DNA (Chap. 6), and mitochondria and chloroplasts have their own DNA (page 47), what is the relationship between the nuclear or chromosomal DNA (Chap. 4) and mitochondrial and chloroplastic DNA in the synthesis of the important array of enzymes characteristic of these organelles? What other organelles may be involved? How? Questions are being answered slowly. The cell physiologist and the biochemist find themselves seeking solutions to similar questions but seeking answers by different approaches.

CONCLUSIONS

Since plants are the producers and animals (including man) are the consumers, we must know about plants. Universities must continue to educate people knowledgeable in the ways of plants, just as they train specialists in space, in medicine, in law and in social theory. For without plants there would be no life.

2

Classification—the Relationship of Plants

Plants are (a) those organisms supplied with chlorophyll, capable of trapping the radiant energy of sunlight, transforming it into stored potential energy while at the same time releasing oxygen to the atmosphere; and (b) an assemblage of organisms lacking chlorophyll but morphologically very similar to certain groups of the lower plants supplied with chlorophyll.

Over 550,000 species of plants clothe the earth and inhabit most of the waters to a depth of approximately 200 feet. What is a **species?** How are species arranged or classified, to bring order to this vast assemblage of plants? A classification serves one practical purpose: the storage of information in a retrievable manner. Most classifications have a common basis; similar things are grouped together. Retrievable information about Americans is kept in computer centers in Washington. Books are kept in libraries. Plants are kept in herbaria and botanic gardens. *A species is a convenient unit of information.*

Retrieval of information is made easy in a library by placing like information together on shelves and giving it a numerical designation. All botanical literature in a library using the Library of Congress classification is found under the letters QK. Medical literature is shelved under R. This is accompanied by an elaborate card index system in which extensive cross references allow the sought-after information to be found in a number of ways. One type of information about kinds of plants may be retrieved by the use of books called floras. In these floras, plant names and descriptions are arranged so that with enough information about a given plant specimen, unknown plants may be named and placed in the proper group by the skilled person.

A NATURAL CLASSIFICATION

Any classification then, is the placing of similar things together. The problem in biological classification is the meaning of the word "similar." Are plants living in the same habitat similar? (Fig. 2.1, see color plate 1). They must certainly have similar nutritional requirements. The Swedish botanist Linnaeus (Chap. 19) divided all flowering plants into 10 groups, or families, based entirely on the number of stamens in each flower (Chap. 15). This

Figure 2.2. There are 35 different families having flowers with many stamens. Three are shown, A, Rosa; B, Hibiscus; C, Camellia.

sometimes worked: all plants currently grouped together as members of the pea family have ten stamens (Fig. 2.4). All buttercups and their relatives have many stamens, but so do some roses (Fig. 2.2*A*), mallows (Fig. 2.2*B*), and camellias (Fig. 2.2*C*). Consideration of other characteristics today allows us to separate plants bearing flowers with many stamens into at least 35 different categories or families. What do modern taxonomists mean by similar?

Plants are placed in categories thought to have a close genetic relationship. Thus, in the most homogeneous populations, the plants are not only very similar in many aspects of their morphology but they *interbreed freely* (Fig. 2.3 *A, B*; see color plate 1). Their genetic constitutions are similar, even identical; they are very closely related. Such a population of closely related interbreeding individuals constitutes a **species.** The most basic requirement of a biological classification is that it shows a true genetic relationship. Several different species may have some characteristics in common and an occasional mating may take place between members of two different but similar species (Fig. 2.3*C*). Such similar species are grouped together in **genera** (*s.* genus), which are thus groups of genetically related species. A comparison of some genera will show that they have traits in common. (Fig. 2.3*D*). Genera with similar traits are grouped in **families,** which

are therefore also genetically related but less closely than genera. Plants in the same family have common characteristics that indicate they may all have evolved from the same ancestor. For instance, members of the pea family have flower parts in fives, pods like pea or bean pods, and dissected leaves like pea or locust leaves. Many members of the family have flowers resembling the sweet pea, with ten stamens, nine of them grown together (Fig. 2.4).

Families are grouped together in **orders,** and orders may be grouped in **classes.** At the top level it is customary to divide all plants into **phyla.** Differences in photosynthetic pigments (Chap. 22), manner of leaf development (Chap. 27), structure of the conducting, or vascular, tissues (Chap. 27), and methods of reproduction (Chap. 22) are the most important elements in separating plants into phyla.

A classification depicting genetic relationships is said to be a **natural** or **phylogenetic classification.** Grouping plants by habitats (such as desert or marsh plants), or by the number of the stamens or color of their flowers has limited usefulness and is said to be an **artifical classification.** A true natural classification is impossible to arrive at and there have been several major attempts which are discussed in detail in Chapter 29.

Thus the word similar, in a natural classification, means organisms having genetic constitutions indi-

2 Classification—the Relationship of Plants

Color Plate 1

Figure 2.1. Cranberry, club moss, a mushroom living in close association. They have similar environmental requirements. Are they similar?

A

B

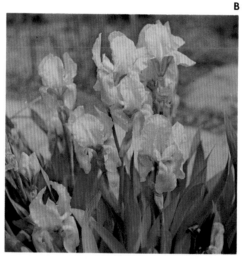

Figure 2.3. *A, B,* two clones of irises (*Iris germanica*). These two clones freely interbreed and thus constitute a species. Notice, however, that the flower colors of the two clones are different. *C,* a single population of Dutch iris (*Iris Xiphium*). These plants interbreed freely, but only very rarely do they cross with members of the populations shown in *A* and *B.* *C* is a second species in the genus *Iris. D,* a population of *Freesia.* These plants have some characteristics similar to those of *Iris.* Both *Iris* and *Freesia* are genera in the family Iridaceae.

C

D

A

B

C

D

Color Plate 2

Figure 2.7. Pigments, hence color, are of importance in distinguishing phyla in the thallophytes. *A, Amanita muscaria,* or the death angel, is a Basidiomycete; it lacks chlorophyll but does have a bright red pigment in the cap. *B, Ulva,* or sea lettuce, of the phylum Chlorophyta has both chlorophylls *a* and *b*. Because its thallus is only two cells in thickness, the green color shows best through overlapping folds of the blade. *C,* the Phaeophyta, represented by *Postelsia,* or the sea palm, have chlorophylls *a* and *c* and a brown pigment. *D,* the Rhodophyta have only chlorophyll *a* and a red pigment which shows well in the blades of *Iridophycus. E,* the blades of *Porphyra* are two cells thick; because of the thinness of the blades and a frequently poor development of the red pigment, this species of Rhodophyta is not as red as other members of the phylum. *F,* a narrow strip of intertidal zone shows the green alga *Ulva,* a brown alga resembling *Fucus,* and the red alga *Gigartina.* (All with the assistance of David Brown.)

E

F

cating relationship. This system is exemplified in the following complete classification of the sweet pea. Note the characteristic ending for each category.

Sweet Pea

	Name	Ending
Species designation	*odoratus*	—
Genus	*Lathyrus*	—
Family	Leguminosae	aceae*
Order	Rosales	ales
Subclass	Dicotyldeonae	ae
Class	Angiospermae	ae
Phylum	Anthophyta	phyta

*Taxonomists have agreed that all names of families should end in *aceae* (from the Latin suffix *aceae, aceous* meaning "of the nature of"). But in this case, they have retained an old name which is very popular. (Fabaceae is an alternative name for the pea family which conforms with the rule.) However, see other plant families such as Rosaceae, "of the nature of a rose"; Liliaceae, "of the nature of a lily"; and the like.

SOME CONVENIENT GROUPS OF PLANTS:

Broad divisions of plants may be made that include several phyla. While these categories are convenient,

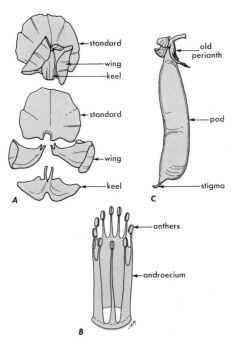

Figure 2.4. Diagram showing traits that are important in the classification of the sweet pea (*Lathyrus odoratus* L.). A, flower, exploded view; B, 10 stamens, 9 attached together, 1 free; C, pod with seeds.

they do not have taxonomic importance. For instance, all plants can be divided in two groups: those with true nuclei (the **eukaryotes**) (Fig. 2.5*B*) and those without true muclei (the **prokaryotes**) (Fig. 2.5*A*). A large group of primitive plants, mainly aquatic and filamentous, may be set apart from other plants by a very distinctive characteristic: the sexual cells, eggs, sperms, or gametes are not protected by a jacket of vegetative or sterile cells (Fig. 2.5*C*). These plants may be referred to collectively as the **thallophytes.** The thallophytes constitute the 8 algal (Fig. 2.5*E*) and 2 fungal phyla. The only unifying characteristic of all the remaining plants is that their reproductive cells are protected by a jacket of sterile cells (Fig. 2.5*D*). A small group of these plants lack vascular or specialized conducting tissue. They are the **liverworts** and **mosses** (bryophytes) (Fig. 2.8). All other plants have developed a specialized conducting tissue, and are known as **vascular plants.** The more advanced vascular plants **bear seeds** and are frequently referred to as the **seed plants** (Fig. 2.5*F*). These are rather precisely separated into **gymnosperms** (Fig. 2.13), whose seeds are borne naked on a cone scale, and **angiosperms** (Fig. 2.15), whose seeds are protected within an ovary.

PLANT PHYLA

We have seen that plants constituting a species are very closely related. And we have said that all members of a family arose, in the primitive past, from a common ancestor. Do all members of a phylum have a common ancestor? Are phyla related? Do all plants have a common ancestor? A partial answer comes from the fossil record.

Plants of past eons have left their remains, or marks, on geologic strata. These remains or marks are **fossils** (Fig. 2.6). Fossils are found in sedimentary rocks, in volcanic ash, and in quantity in coal. By comparing the morphology of present day plants with the morphology of fossils from past geologic periods, it is possible to establish broad lines of relationship between the phyla. Unfortunately, fossils are few in number, do not reveal the entire plant, and are preserved only in certain habitats. Therefore, the fossil record must be used with caution. This is why agreement on the evolutionary relationship (**phylogyny**) of plant phyla is difficult to achieve.

Figure 2.5. Diagrams to show differences between groups of plants; A, a prokaryotic cell (blue-green alga); B, a eukaryotic cell (green alga, *Chlorella*); C, a group of reproductive cells lacking a protective jacket (the brown alga, *Ectocarpus*); D, a group of reproductive cells provided with a protective jacket (the liverwort, *Riccia*); E, a thallophyte (the singled-celled green alga, *Acetabularia*); F, a seed-bearing vascular plant (bean, *Phaseolus vulgaris*).

2 Classification—the Relationship of Plants

Figure 2.6. An imprint of a leaf of *Pecopteris,* a common plant growing during the Pennsylvanian geologic period about 300 million years ago. (Courtesy of Field Museum of Natural History.)

Phyla of Thallophytes

Thallophytes, a very large group of primitive plants, do not have a jacket of sterile protecting cells around their reproductive cells (Fig. 2.5C). Those without chlorophyll may be conveniently called **fungi** (Fig. 2.7A; see color plate 2); those with chlorophyll are **algae** (Figs. 2.5E, 2.7B, C, D).

Algal Phyla

The algae may be separated further into groups on the basis of photosynthetic pigments. Some of these groups have phylum rank. A large number of algae have only chlorophyll *a* (Chap. 22), plus two accessory water-soluble pigments, one blue and the other red (Chap. 22). This assemblage of forms is divided in two phyla. One phylum, the blue-green algae (Cyanophyta) (Fig. 2.5A), are prokaryotes and hence lack true nuclei. The other phylum, the red algae (Rhodophyta) (Fig. 2.7D), is composed of eukaryotes that inhabit the coastal waters of the tropic and temperate zone oceans. While both phyla contain only chlorophyll *a* and similar accessory pigments, their structure is so different that any relationship between the two phyla must be very distant (Chap. 22).

A third group of algae also has only chlorophyll *a*. It is separated from the Cyanophyta and Rhodophyta because it lacks the water soluble pigments, and it possesses another chloroplast pigment, xanthophyll. This phylum is known as the Xanthophyta (Chap. 22).

A large group of algae contain chlorophylla *a* and *b*. These algae are easily divided in two phyla. One of these, the Euglenophyta, consists of single-celled forms (**unicells**) that lack a cell wall. The other phylum is a large one, the green algae (Chlorophyta) (Fig. 2.7B). Because higher plants also contain chlorophyll *a* and *b*, it is thought that the Chlorophyta are ancestors of the higher plants (Chap. 22).

Other algae contain chlorophylls *a* and *c*. One phylum comprises the sea weeds—kelps—of ocean coasts. These are brown algae known as the Phaeophyta (Fig. 2.7C). Members of a second phylum, the diatoms (Bacillariophyta), are quite small, but anyone who has looked at a fresh slide of sea water under the microscope has probably seen

Table 2.1 *Summary of thallophyte phyla*

NO.	PHYLUM	COMMON NAME	CHLOROPHYLLS	DESCRIPTION
1	Cyanophyta	blue-green algae	a	Prokaryotic cells
2	Pyrrophyta		a, c	Mostly unicells
3	Chrysophyta	golden algae	a, c	Many unicells, some with silica in cell wall
4	Phaeophyta	brown algae	a, c	Marine forms, some very large
5	Rhodophyta	red algae	a	Mostly marine forms, many with a feathery plant body
6	Xanthophyta	golden-green algae	a	Many unicells, some simple filaments
7	Euglenophyta	Euglena	a, b	Motile unicells without cell walls
8	Chlorophyta	green algae	a, b	Mostly rather small forms of great diversity
9	Schizomycophyta	bacteria	—	Prokaryotic cells, a few with bacterial chlorophyll
10	Myxomycota	slime molds	—	Amorphous masses of protoplasm in vegetative stage
11	Eumycota	fungi	—	Mostly filamentous forms of considerable diversity

them, for they are extremely common unicells or filaments. Their cell walls are impregnated with silica, and are so constructed that the two halves fit together like top and bottom of a petri dish. The third phylum having chlorophylls *a* and *c* is comprised of small cells, many of which lack cell walls. It is known as the Pyrrophyta. These algae are not common.

Fungal Phyla

Two thallophyte phyla lack chlorophyll; thus, they are unable to make their own food and are therefore called heterotrophs. A primitive group of fungi, the Mxyomycota, are the **slime molds** which in their vegetative stage, are nothing but naked masses of protoplasm. An advanced group, the

Eumycota, are **true fungi** (Fig. 2.7*A*). Some of them very closely resemble the algae, and one group is known as the algal fungi. The thallophyta phyla are summarized in Table 2.1.

Classification of Land Plants

There is general agreement that the land plants take their origins from the green algae, and a particular genus, *Fritschiella* (Fig. 22.5) has been selected as a possible starting point. One group of land plants lacking a vascular system forms a natural phylum of liverworts and mosses, the Bryophyta. All other land plants have conducting, or vascular, tissue. General agreement is lacking as to their natural relationship.

All vascular plants have a system for conducting

water and nutrients throughout the plant. They all have roots, stems, and leaves (Fig. 2.5F), except for some primitive living genera which lack roots (Fig. 2.9). Also, they have all developed a cuticle on their aerial portions which serves as protection against desiccation. The classification problem facing us is: Did all these characteristics, which distinguish vascular plants from all other plants, arise just once in the long evolutionary history of plants? If so, all extant plants (plants living today) have evolved from the same vascular ancestral form. This theory of plant evolution asumes a **monophyletic origin** of vascular plants; all such plants would then belong to a single phylum—Tracheophyta.

It is equally conceivable that there were numerous steps in the evolution of the characteristics of the vascular plants, and that the cuticle, vascular tissue, roots, stems, and leaves may have arisen in several different periods and localities. The plants living today, then, would not have a common vascular ancestral form. There would be several common ancestral forms, at least one for each suggested phylum. This is known as a **polyphyletic origin.**

It therefore becomes difficult to present a system for the classification of the vascular plants in an introductory botany course. Three such systems are compared in Table 2.2. The first column summarizes a recently proposed monophyletic system. Column 2 shows a polyphyletic system used in a recent textbook of plant morphology, which we shall follow in this book. The classification which seemed likely seven years ago and was followed in the third edition of this text appears in column 3. There is fairly general agreement among botanists on the main subgroups of vascular plants; the differences in these 3 columns lie in the number of phyla. According to column 1 there is a single phylum for all vascular plants; thus implying a single ancestor. According to column 2, there are ten vascular plant phyla, implying at least ten ancestors. But according to column 3, there are four phyla of vascular plants, implying fewer ancestors.

The fourth column gives the common names for these groups of plants. While these names refer as precisely to a group of plants as do the scientific names, the common names do not suggest relation-

Table 2.2 *Comparison of systems of classifying the major divisions of the land plants*

MONOPHYLETIC (H. P. BANKS, 1968)	POLYPHYLETIC (SCAGEL ET AL., 1967)	POLYPHYLETIC (ROBBINS, WEIER, STOCKING, 1963)	COMMON NAMES
Bryophyta	Bryophyta	Bryophyta	Bryophytes; mosses and liverworts
Tracheophyta			Vascular plants
	Psilophyta	Psilophyta	Psilophytes
Rhyniophytina	*Rhynia*	*Rhynia*	*Rhynia*
Zosterophyllophytina			
Psilophytina	*Psilotum*	*Psilotum*	*Psilotum*
Lycophytina	Lycophyta	Lycophyta	Lycopods, club mosses
Sphenophytina	Arthrophyta	Sphenophyta	*Equisetum*, horsetails
Trimerophytina			
Pterophytina		Pterophyta	
Cladoxylopsida			
Filicopsida	Pterophyta	Filicinae	Ferns
		Gymnospermae	Gymnosperms
Coenopteropsida			
Progymnospermopsida	Pteridospermophyta	Pteridospermales	Seed ferns
Cycadopsida	Cycadophyta	Cycadales	Cycads
	Ginkgophyta	Ginkgoales	*Ginkgo*
Coniferopsida	Coniferophyta	Coniferales	Conifers
Gnetopsida	Gnetophyta	Gnetales	Gnetum
Angiospermopsida	Anthophyta	Angiospermae	Angiosperms

ships between the plant groups they designate. We shall use these common names in all situations except those having evolutionary implications.

Phyla of Land Plants

The mosses and liverworts (Bryophyta) (Fig. 2.8, Chap. 26) are easily separated from all other land plants. Instead of true roots, they have rhizoids that act as absorbing organs. They also have small leafy stalks which produce sperm and egg cells. Frequently, in mosses, a brown upright stalk bearing a capsule is seen growing from the low leafy stalk; this is the offspring of sperm and egg fusion. The mosses, which lack vascular tissue, are thought to have arisen from the Chlorophyta but not to have given rise to any other land plants. In other words, the mosses we see today may be much as they were when plants first invaded the land.

The most primitive extant vascular plants are known as the **psilophytes,** and there are only two living genera, *Psilotum* and *Tmesipteris*. *Psilotum* is easily cultivated and is fairly common as a potted plant in botany greenhouses (Fig. 2.9). The name *Psilotum* means naked, or smooth, and refers to the

Figure 2.8. The hair cap moss, *Polytrichum*, is a representative of the Bryophyta.

Figure 2.9. A potted plant of *Psilotum nudum*, a living species closely resembling a fossil form found in sedimentary rocks laid down 400 million years ago.

2 *Classification—the Relationship of Plants*

simple dichotomously branching stem, with small scale-like leaves and with rhizoids rather than roots (Chap. 27). *Psilotum* appears very similar in structure to a fossil (*Rhynia*) from Scotland found in sedimentary rocks that were deposited 400 million years ago. Two other very ancient groups of plants, the **club mosses** or **lycopods** (Lycophyta), and the **horsetails**, or *Equisetum* (Sphenophyta), once formed a dominant earth flora with many tree forms. Today there are only three genera of club mosses, and a single genus of horse tails, *Equisetum*. Both are ancient plants, being easily related to fossils found in rocks nearly 400 million years old.

Equisetum has hollow, jointed stems which have an encircling sheath of small leaves at each joint. Spores, or reproductive cells, are borne in specialized cones at the summit of some of the upright stalks (Fig. 2.10).

The club mosses (Lycophyta) more closely resemble the higher plants (Fig. 2.11). One genus is known as ground pine and has been used extensively for Christmas wreaths. The club mosses are separated from the other groups of vascular plants by their distinctive reproductive structures; their vascular system; the single veins in their small leaves; and the union of stem and leaf without any break in the vascular tissue of the stem (Chap. 27).

The living **ferns** (Pterophyta) are generally small plants and are characterized by their highly dissected, characteristic leaves, which unroll at the tip as they open. Seeds are not formed by any fern (Fig. 2.12).

The remaining six groups of plants (Table 2.2) are seed plants. Five of these groups have seeds borne upon scales in a cone (Fig. 2.13) or as naked seeds on short stems; these are the **gymnosperms.**

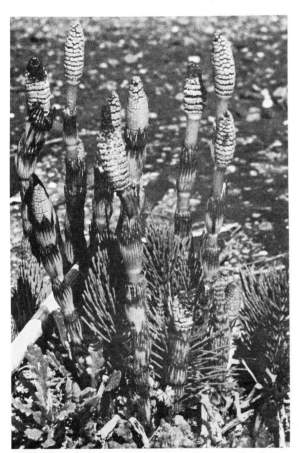

Figure 2.10. Horsetails, *Equisetum hyemale* L.; this is the only genus in the phylum Sphenophyta.

Figure 2.11. There are three genera in the Lycophyta. This is *Lycopodium clavatum* L., or running pine.

Figure 2.12. The fern, *Polypodium vulgare,* representing the Pterophyta.

The remaining group of seed plants has seeds enclosed in a seed case or ovary (Fig. 2.14); these are the **angiosperms** (Anthophyta) (Fig. 2.15).

Two of the groups of gymnosperms, *Ginkgo* (Ginkgophyta) and the cycads (Cycadophyta) are the most primitive of the living seed plants. *Ginkgo biloba* is the only species in its group. It grows to be a large tree, has been cultivated for many centuries in Chinese temple gardens, and is currently finding much favor in western countries. It is extinct in the wild. *Ginkgo* has fan-shaped leaves (Fig. 10.15*B*) with many veins radiating from the petiole and a break or gap in the vascular cylinder where the vascular tissue from the leaf joins the stem. Naked seeds are borne on the end of short stems.

The cycads (Cycadophyta) comprise eleven genera, two of which are extinct. Living cycads range in size from small forms to large trees resembling palms.

The most conspicuous group of gymnosperms are the **conifers** (Coniferophyta) (Fig. 2.16). Together with the angiosperms (Anthophyta), they dominate the present flora. The needle or scale leaves of the conifers have but one or two veins, and in some forms there does not appear to be a break in the

Figure 2.13 (*above*). An open cone of the pinon pine, *Pinus monophylla,* showing two seeds in place upon a scale. A gymnosperm.

Figure 2.14 (*right*). A fully matured ovary of a bell pepper, *Capsicum frutescens var grossum,* showing seeds within the matured ovary. An angiosperm.

2 Classification—the Relationship of Plants

vascular cylinder where the vascular tissue from the leaf joins with the vascular cylinder of the stem. In contrast, leaves of the angiosperms (Anthophyta) have many veins and there is always a gap at the junction of leaf vascular tissue with the stem vascular tissue. We may divide the angiosperms in two subclasses. The seeds of one have two seed-leaves, or **cotyledons,** as seen in the peanut or bean. These are the Dicotyledonae. The seeds of the other group (onions, lilies) have a single seed-leaf. These are the Monocotyledonae.

The association between the phyla is shown by the key on page 18.

THE BINOMIAL

The species name of sweet pea is *odoratus* and it belongs to the genus *Lathyrus* in the Leguminosae family. To call it just *odoratus* would not be very meaningful, for it does not show relationship with anything else. To call it simply *Lathyrus* would not separate it from other species in the genus *Lathryus.* The scientific name of every species is a binomial consisting of two words: the genus plus the species (*Lathyrus odoratus*).

Frequently the genus name refers to the ancient name of the same or a similar plant. *Lathyrus* is the Greek word for pea. An effort is usually made to make the species name descriptive. *Odoratus* is the Latin word designating that something is sweet smelling. So altogether the Greek genus and Latin species translate to sweet-smelling pea. There are also black peas, *L. niger;* cultivated peas, *L. sativus;* beach peas, *L. maritimus;* splendid peas, *L. splendens;* and so on. Since an effort is made to keep the scientific names of plants and animals in the correct Latin or Greek form, they are really foreign terms and as such are always written in *italics.* As the species designations are generally descriptive, they may be used in conjunction with other genera. We have, for instance, black pepper, *Piper nigrum;* black mustard, *Brassica nigra;* and black walnut, *Juglans nigra.* There is sweet alyssum, *Alyssum odoratum,* and cultivated oats, *Avena sativa.*

The Swedish botanist, Linnaeus (Fig. 2.17) was largely responsible for establishing the binomial system, not only for plants but for all living organisms. In fact, the full name of the sweet pea is *Lathryus ordoratus L.* The cultivated pea should likewise be *Lathryus sativus L.* The *L.* after these two names indicates that they were first described by Linnaeus. The large numbers of plant names followed by the letter *L.* are a living tribute to the greatness of Linnaeus' taxonomic contributions.

Figure 2.15. The angiosperms all bear flowers and fruit as shown here by orange blossoms and oranges (*citrus sinensis*).

Key to Plant Phyla		PHYLA	COMMON NAMES
A. Plants without roots, stems or leaves, and no protective jacket of vegetative cells around the reproductive cells.			thallophytes
B. Plants lacking chlorophyll			fungi
C. Vegetative stage of naked cytoplasm		Myxomycota	slime molds
C. Vegetative stage with cell wall, mostly filamentous		Eumycota	fungi
B. Plants with chlorophyll			algae
D. Chlorophyll *a* alone			
E. Prokaryotes		Cyanophyta	blue-green algae
E. Eukaryotes			
F. With water soluble blue and red pigments		Rhodophyta	red algae
F. With plastid pigment, xanthophyll		Xanthophyta	
D. Chlorophyll *a* and *c*			
G. Large sea weeds with a cell wall		Phaeophyta	brown algae
G. Small, mostly unicells, with a cell wall		Chrysophyta	golden algae
G. Small unicells generally lacking a cell wall		Pyrrophyta	
D. Chlorophyll *a* and *b*			
H. Unicells without a cell wall		Euglenophyta	*Euglena*
H. Greatly diversified forms with cell walls		Chlorophyta	green algae
A. Plants with a protective jacket of vegetative cells around reproductive cells (often have roots, stems, or leaves)			
I. Plants lacking vascular tissue		Bryophyta	liverworts, mosses
I. Plants with vascular tissue			vascular plants
J. Plants without seeds			
K. Roots absent		Psilophyta	*Psilotum*
K. Roots present			
L. Small leaves, no gap at union with stem, leaves do not unroll as they open.			
M. Stems not jointed		Lycophyta	club mosses
M. Stems jointed		Sphenophyta	horsetails
L. Leaves well-developed, gap at union with stem, leaves unroll as they open		Pterophyta	ferns
J. Plants with seeds			seed plants
N. Seeds not covered			gymnosperms
O. Seeds on short stems		Ginkgophyta	*Ginkgo*
O. Seeds in cones			
P. Trees palm-like		Cycadophyta	cycads
P. Trees conical		Coniferophyta	conifers
N. Seeds covered		Anthophyta	angiosperms
R. 2 seed leaves		Dicotyledonae	dicotyledons
R. 1 seed leaf		Monocotyledonae	monocotyledons

And today when the sweet pea binomial appears in the literature, every plant scientist, no matter what language he speaks, knows that reference is made to a specific pea with certain characteristics.*

It is not uncommon, particularly in cultivated

*Realizing that many species of plants have several common names, thus giving rise to confusion, the American Joint Committee on Horticultural Nomenclature has prepared a book entitled *Standardized Plant Names*, which in the second edition is described as "a revised and enlarged listing of approved scientific and common names of plants and plant products in American commerce or use." In the preface to the second edition the statement is made that "the purpose of *Standardized Plant Names* is to bring intelligent order out of the chaos in names of plants and plant products existing the world over. Such standardization, supported by adequate authority, will not only promote satisfactory understanding between those who sell and those who buy, but will also improve the multifarious relations, scientific, educational and social, into which the advancing plant consciousness of America has grown."

Figure 2.17. Linnaeus (1707–1778), the Swedish botanist who initiated the binomial system of nomenclature. (Courtesy of the Library of The New York Botanical Garden.)

Figure 2.16. Pinus *Jeffreyi* representing the Coniferophyta

plants, to have races develop within a species. Indeed, in the many years devoted to breeding sweet peas, such races have arisen. One has a condensed, bushy type of growth. Bailey gives this the name *Lathyrus odoratus nanellus Bailey,* or dwarf sweet pea. This is a trinomial.

Several thousand new species of plants are discovered every year. The plant specimens may have been collected in parts of the world previously unexplored by botanists, although new species are being discovered every year in areas traversed many times by botanists. These specimens ultimately fall into the hands of a systematic botanist who is interested in and has given special attention to the group (family or genus) to which the plant in question belongs. This botanist sets out to identify the species. In so doing, he compares it with allied forms, of which he has herbarium specimens or published descriptions. He may find that the plant has characters similar to those of a species heretofore described; or he may find that the plant has characters so different from those of any known species that he concludes it is a new species (**species novum,** sp. nov.). In other words, it is a species that has never been described. Accordingly, he describes this new species and gives it a name. The description is published in one of the many recognized botanical journals, and the specimen or specimens he used in making the description are properly labeled and placed in one or more of the herbaria of the world. This specimen is known as the **type specimen.**

HERBARIA AND FLORAS

There are about 550,000 species of plants, divided as we have just indicated. No attempt has been made to describe all of these plants in one treatise. It requires, for instance, 10 large volumes to describe the vascular plants in Engler and Prantl's current *Genera Plantarum.* For general convenience, books describing local floras are in common use. We have, for instance, for the United States:

Leroy Abrams. 1940–1960. Illustrated flora of the Pacific States. 4 Vol. Stanford University Press. Stanford, Calif.

L. H. Bailey. 1949. Manual of cultivated plants. MacMillan Co., New York, 1116 pp.

H. A. Gleason. 1952. Illustrated flora of the northeastern United States and adjacent Canada. Lancaster Press, Lancaster, Penn. 3 Vols.

J. K. Small. 1933. Manual of the southeastern flora. University of North Carolina Press, Chapel Hill, 1554 pp.

J. T. Howell. 1949. Marin flora. University of California Press, Berkeley, Calif., 322 pp.

J. T. Howell, P. H. Raven, P. Rubtzoff. 1958. A flora of San Francisco, Calif. University of San Francisco, San Francisco, Calif., 157 pp.

A. S. Hitchcock. 1950. Manual of the grasses of the United States. 2nd Edition. U.S. Government Printing Office, Washington D.C. 1051 pp.

We have noted that through the use of such manuals one could, with sufficient knowledge and skill, identify an unknown plant. This is sometimes difficult, even for an expert. For the agriculturist, cattle rancher, conservationist, tomato grower, or orchidist struggling with forage plants, poisonous plants, or weeds, the task of identifying plants is likely to be difficult and time-consuming.

Specimen plants, stored in herbaria, make identification of unknown plants easier and more definite. Such specimens have been identified by specialists, sometimes being sent long distances to these storehouses of information about plant classification.

Here, unknown plants may be accurately identified. Herbaria are generally associated with universities, botanic gardens, and certain governmental agencies. Several of the larger herbaria are:

Herbarium	No. of Specimens
Royal Botanic Gardens, Kew, England	6,500,000
Komarov Botanical Institute, Leningrad, USSR	4,500,000
Conservatoire et Jardin Botaniques de Geneve, Switzerland	4,000,000
U.S. National Herbarium, Washington, D.C.	3,000,000
New York Botanical Garden, New York	3,000,000
Missouri Botanical Garden, St. Louis	1,700,000
National Herbarium, Melbourne, Australia	1,500,000

BOTANIC GARDENS

Botanic Gardens are of particular interest because they store the plants themselves (Fig. 2.18). Furthermore, these gardens are not confined to the dissemination of knowledge about plants, but are also concerned with dissemination of the plants themselves. Some of the great botanic gardens of today, like the Kew Gardens in England, have a long history of distributing valuable crop and ornamental species to areas far from their native habitats. Sometimes the species actually grow better in their new habitats because insect or fungal parasites are absent. In other cases, the garden grows many individual plants but selects only the most robust ones for transplanting.

Collecting and distributing valuable plants has sometimes involved questionable activities such as smuggling. For example, by the mid-nineteenth century, rubber had become a valuable, desired plant product. The best rubber trees (*Hevea brasiliensis*) were native to Brazil, and the Brazilian government definitely wished to maintain its monopoly. Export of rubber tree cuttings or seeds was strictly forbidden.

The India Office of England, however, was determined to initiate rubber plantations in the English colony of Ceylon. In 1876 it secretly commissioned an English rubber worker in Brazil named H. A. Wickham to smuggle out rubber tree seeds, agreeing to pay him $50 per thousand. Wickham proved to

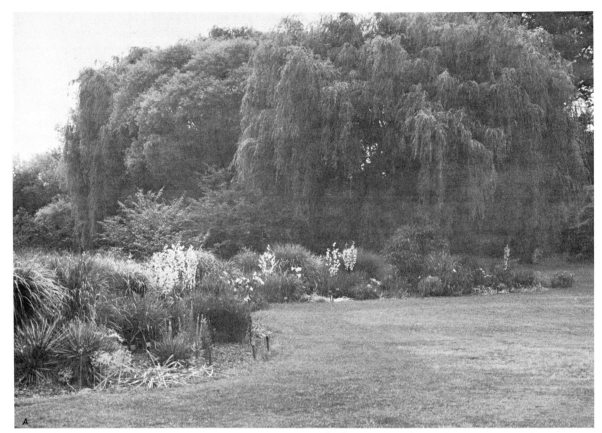

Figure 2.18. Botanic Gardens are storehouses and disseminators of information. *A*, the extensive collection of monocotyledons at the Brooklyn Botanic Garden is displayed in a large and attractive border planting; *B*, The Childrens' Garden at the Brooklyn Botanic Garden gives practical information about plants to the children of the neighborhood. (*A*, courtesy of Brooklyn Botanic Garden; *B*, courtesy of Mr. Bertram Wiener.)

be very efficient and soon collected some 70,000 seeds, packing them in covered baskets which he labeled simply, "Botanical specimens for Her Majesty's Gardens at Kew." He eventually found a willing ship captain with available space and slipped past the customs office at Belem without incident.

Once in England, he turned the seed over to botanists at the Kew Gardens who carefully germinated and nutured 2400 seedlings from the 70,000 seeds. Most of these plants were shipped within a year to plantations (Fig. 2.19) which had been made ready in Ceylon. Wickham had his picture taken in a Ceylon plantation in 1912, when it was producing a princely $18,000 worth of rubber per acre per year! Not surprisingly, he was knighted in 1920. Today, 95 percent of the world's rubber comes from the Far East.

Figure 2.19. Firestone rubber plantation near Robertsville, Liberia. (Courtesy of Dr. Georg Gerster—Rapho Guillumette.)

Botanic gardens serve as a definite link in the distribution of knowledge about plants to the general public. In addition to actual displays of growing plants, informative lectures and booklets dealing with plants are available to the interested and advanced amateur.

Among the oldest and best know botanic gardens in the world (with the dates of establishment in parentheses) are the following.

Royal Botanic Garden of Padua, Padua, Italy (1545).
National Museum of Natural History, Paris, France (1635).
Royal Botanic Garden, Edinburgh, Scotland (1670).
University Botanic Gardens, Cambridge, England (1762).
Botanic Garden and Botanical Institute of the University of Vienna, Vienna, Austria (1754).
Botanic Garden of Harvard University, Cambridge, Mass. (1807).
Botanical Garden of Rio de Janeiro (1808).
Botanic Garden, Munich, Germany (1809).
Botanic Gardens of New South Wales, Sydney, Australia (1816).
Government Botanic Gardens, Buitenzorg, Java (1817).
U.S. Botanic Garden, Washington, D.C. (1820).
Royal Botanic Garden, Kew, England (1841).
Missouri Botanical Garden, St. Louis, Mo. (1859).
Arnold Arboretum of Harvard University, Jamaica Plain, Mass. (1872).
The New York Botanical Garden, New York City (1895).
Brooklyn Botanic Garden, Brooklyn, N.Y. (1910) (Fig. 2.18).
Montreal Botanical Garden, Montreal, Canada (1936).

These gardens have plantings of species from all parts of the world, arranged by plant families or according to habitats. They contain outdoor plantings and often greenhouses controlled to grow tropical plants. Their reference libraries may contain thousands of volumes of books, pamphlets, and photographs, and the herbaria may be extensive. For example, the New York Botanical Garden, with 40 acres, has a reference library of 43,500 volumes and thousands of pamphlets; an herbarium of approximately 1,706,000 specimens; and a systematic plantation, arboretum, fruticetum, rose gardens, and rock gardens. The Missouri Botanical Garden, at St. Louis, with 75 acres, has a library of 50,000 volumes, an herbarium of 3,000,000 specimens, 6500 species growing under glass, and 7500 species growing out-of-doors. The famed Royal Botanic Garden at Kew, England, with 288 acres, has a reference library of 44,000 volumes, an herbarium of 6.5 million specimens, an arboretum and fruticetum together of 7000 species and varieties, 13,000 species growing under glass, and 8000 herbaceous species growing out-of-doors.

BOTANICAL JOURNALS

We have previously mentioned the storing of knowledge about plants in a retrievable form. But knowledge must be accumulated before it can be stored. Today the greatest increase in our knowledge of plants and animals comes from the biology departments of many universities. Since most of this work receives support from public funds, it must be available to all who desire to know about it. As a consequence, this new information is published in some 2000 scientific journals. A number of examples of such journals is given in Table 2.3.

Table 2.3 *Representative journals publishing the results of botanical research*

LIBRARY OF CONGRESS CALL NUMBER	JOURNAL	PUBLISHED BY	COUNTRY OF ORIGIN	DATE ESTABLISHED
		I. General		
QK 1 A55	American Journal of Botany	Botanical Society of America	U.S.	1914
QK 1 A65	Annals of Botany	Oxford University Press	England	1887
QK 1 B54	Botanical Gazette	The University of Chicago Press	U.S.	1875
QK 1 C35	Canadian Journal of Botany	National Research Council, Canada	Canada	1922
QK 1 D4	Deutsche botanische Gesellschaft, Berichte	Deutsche botanische Gesellschaft	Germany	1882
QK 1 F5	Flora	Gustav Fischer Verlag	E. Germany	1817
QK 1 J3	Japanese Journal of Botany	National Research Council of Japan	Japan	1923
QK 1 J5	Journal of Experimental Botany	Society for Experimental Biology	England	1950
QK 1 L45	Linnean Society of London, Journal of Botany	Linnean Society of London	England	1856
QK 1 N45	New York Botanic Garden Journal	New York Botanic Garden	U.S.	1951
QK 1 N3	New Phytologist	Blackwell Scientific Publications	England	1902
QK 1 P58	Planta	Springer-Verlag	Germany	1926
QK 1 R39	Revue Générale du Botanique	Centre National de la Recherche Scientifique	France	1889
QK 1 T6	Bulletin Torrey Botanical Club	Torrey Botanical Club	U.S.	1870
QH 5 A4	Annales des Sciences Naturelles, Botanique	Masson et Cie	France	1824
QK 1 E3	Economic Botany	Society for Economic Botany	U.S.	1947
QK 1 R8	Doklady Botanical Science	Academy of Sciences of the USSR	Russia	1921

LIBRARY OF CONGRESS CALL NUMBER	JOURNAL	PUBLISHED BY	COUNTRY OF ORIGIN	DATE ESTABLISHED
	II. Cells and Cellular Biology			
QK 1 R38	Revue de Cytologie et de Biologie Végétales	Centre National de la Recherche Scientific	France	1937
QH 201 J68	Journal de Microscopie	Société Française de Microscopie Electronique	France	1962
QH 301 C4	La Cellule	P. Debaiseux and P. Martens	Belgium	1884
QH 301 C5	Chromosoma	Springer-Verlag	Germany	1939
QH 581 A1J6	Journal of Cell Biology	Rockefeller University Press	U.S.	1955
QH 591 P6	Protoplasma	Springer-Verlag	Austria	1927
QH 581 A1 J7	Journal of Cell Science	Cambridge University Press	England	1966
QH 201 Q8	Quarterly Journal of Microscopical Science	Cambridge University Press	England	1900
QH 301 C9	Cytologia	Wada-Kunkokae Foundation	Japan	1929
	III. Taxonomy			
QK 1 B63	Botanische Jahrbücher	E. Schweizerbart'sche Verlagsbuch-handlung	Germany	1881
QK 1 R5	Rhodora	New England Botanical Club	U.S.	1899
QK 1 S3	Annals of the Missouri Botanical Garden	Missouri Botanical Garden	U.S.	1914
QK 1 B675	Brittonia	American Society of Plant Taxonomists	U.S.	1931
	IV. Ecology			
QH 301 J6	Journal of Ecology	British Ecological Society Blackwell Scientific Publications	England	1913
QH 301 E34	Ecology	Ecological Society of America	U.S.	1920

Table 2.3 *Representative journals publishing the results of botanical research (cont.)*

LIBRARY OF CONGRESS CALL NUMBER	JOURNAL	PUBLISHED BY	COUNTRY OF ORIGIN	DATE ESTABLISHED
QH 540 D5	Oikos	Acta Oecologica Scandinavica	Denmark	1949
QK 901 A1	Vegetatio	International Phytosociological Association	The Netherlands	1948
V. Inheritance and Evolution				
QH 301 G4	Genetics	Genetics Society of America	U.S.	1916
QH 301 H4	Hereditas	Nordick Genetiker-förening	Sweden	1920
QH 301 E9	Evolution	Society for the Study of Evolution	U.S.	1946
VI. Plant Physiology				
QK 1 P48	Plant and Cell Physiology	Japanese Society of Plant Physiologists	Japan	1959
QK 1 P5	Plant Physiology	American Society of Plant Physiologists	U.S.	1926
QK 1 P4	Physiologia Plantarum	Scandinavian Society of Plant Physiology	Sweden	1948
QK 1 P452	Physiologie Végétale	Gauthrie-Villars	France	1963
QK 1 P52	Soviet Plant Physiology	Academy of Sciences of the USSR	Russia	1954
QK 1 Z4	Zeitschrift für Pflanzenphysiologie	Gustav Fischer-Verlag	Germany	1908
VII. Morphology, Including Algae and Fungi				
QK 504 A7	Archiv für Mikrobiologie	Springer-Verlag	Germany	1934
QK 600 M9	Mycologia	Mycological Society of America	U.S.	1909
QK 564 P46	Phycologia	International Phycological Society	U.S.	1961
QK 534 B7	Bryologist	American Bryological Society	U.S.	1898

3

The Plant Body of Seed Plants

In this chapter we shall give attention to the general structures and functions of a familiar seed plant, the common garden bean. Any one of many thousands of seed plants would serve our purposes as well. It is recognized that seed plants are very complex organisms, as compared with unicellular, filamentous, and platelike plant bodies or even multicellular plants like the mushrooms. Other types of plants have different types of plant bodies. They will be discussed in some detail after the structure and function of the plant body of seed plants is understood. See Figs. 22.1, 24.43, 28.1 for other plant body types.

The bean seed is borne in a pod on a "mother plant." When removed from the pod under suitable growing conditions, the bean seed germinates and a young plant (seedling) emerges (Fig. 3.1*D*) having roots, stems, and leaves. The plant *grows,* extends its roots into the soil and its stems with their leaves into the atmosphere, and after a number of weeks attains adult size. Flowers are formed, then pods "set," and in the pods are seeds ("beans"). The bean plant has then completed a **life cycle.** Many of the physiological processes involved are complex and require a knowledge of chemistry and physics in order to be understood. A bean plant is an individual—it has a birth, a young life, an adult life, a period of reproduction, and then it dies. It has **organs,** just as does the human being, which perform various functions necessary to maintain its life. These organs—**roots, stems, leaves, flower parts**—are composed of **tissues,** groups of cells that carry on different activities (Figs. 3.1*E* and *F* and 3.2). The cells are the small compartments that make up tissues; they possess living material **(protoplasm)** and hence are the *living units of plant structure and function* (Fig. 4.2). The plant body is composed of cells and the products of cells. All the different kinds of work the plant does are really the work of its cells.

A plant body like that of the bean (Fig. 3.1*D*) possesses all the organs and tissues necessary to maintain life. The stem (1) supports the foliage leaves and flowers, (2) conducts water and mineral salts from the roots, (3) conducts plant foods from tissues where they are manufactured to tissues where they are needed for growth or where they are stored for future use, and (4) may store foods. The roots (1) anchor the plant in the soil, (2) absorb water and

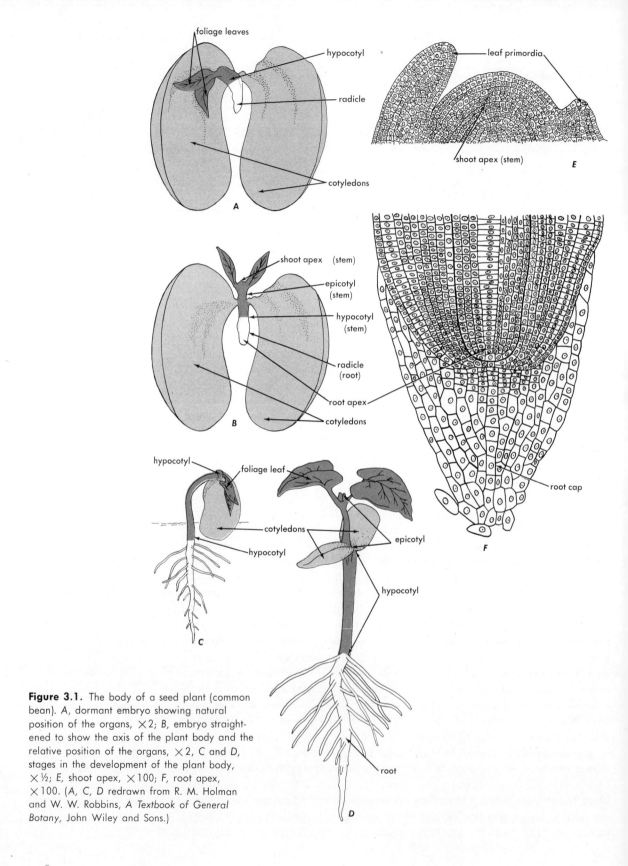

Figure 3.1. The body of a seed plant (common bean). *A*, dormant embryo showing natural position of the organs, ×2; *B*, embryo straightened to show the axis of the plant body and the relative position of the organs, ×2, *C* and *D*, stages in the development of the plant body, ×½; *E*, shoot apex, ×100; *F*, root apex, ×100. (*A, C, D* redrawn from R. M. Holman and W. W. Robbins, *A Textbook of General Botany,* John Wiley and Sons.)

mineral salts from the soil, and (3) may store food. The leaves (1) manufacture foods and (2) give off water in the form of water vapor. These activities involve both chemical and physical processes, which will be discussed in later chapters.

One cannot have a clear understanding of the structures and functions of a plant without a knowledge of its anatomy, such as is revealed by the use of a compound microscope. All-important at the very beginning is an understanding of cells— the units of plant structure and function. Chapters 4–6 deal with the cell.

The gross development of the seed plant body is shown in Fig. 3.1A to D. After a bean seed is soaked in water for several hours, the outer skin or seed coat is easily removed. The body exposed is the **embryo** (Fig. 3.1A, B). The term embryo as used in biology refers to a living organism in the early stages of development. In most plants and the higher animals, the embryo is more or less dependent upon the maternal parent for its food. The bean embryo is a very young plant. There are two large **cotyledons** or seed leaves, several young **foliage leaves,** a **stem** (composed of **epicotyl** and **hypocotyl**), and a **rudimentary root (radicle).** The two cotyledons contain stored food, chiefly starch. Although the young foliage leaves are small, their veins can be easily seen with a lens. Fig. 3.1A, B shows the bean embryo spread open so that the relationship of the different parts is more easily observed.

The embryo plant body has two regions of importance to future growth—the **shoot tip** and the **root tip** (Fig. 3.1E, F). The shoot tip includes: (1) at the very end of the young plant, between the rudimentary leaves, a dome-shaped mass of undifferentiated cells constituting the shoot apex, and (2) cells directly below the apex that are enlarging and in early stages of elongation and differentiation. The root tip, at the lower end of the young plant, is composed of similar tissues: (1) the root apex of unidifferentiated cells, and (2) the elongating and differentiating cells above the root apex, and (3) a root cap protecting the root apex. The undifferentiated cells that make up these two tips are capable of **division, differentiation,** and **growth. Meristematic cells** are cells capable of active cell division with a resulting increase in the number of cells. Until the death of the plant, there are always meristematic tissues at the tips of stems and roots.

As long as they are alive, all multicellular plants contain localized regions of meristematic tissue, enabling them to grow, and differentiate new stems and roots. Indeed, we shall see that a continual growth of new roots is essential for continued absorption of water and nutrients. Plants are thus said to have open growth, and it has been demonstrated that many fully differentiated individual plant cells contain a full genetic complement and are able to give rise, through mitosis, to new individuals. On the other hand, animals do not have meristematic tissue; only the undifferentiated cells of embryos are able to divide and to differentiate into specialized cells. With but few exceptions, mature animal tissue can only produce new cells similar to those of their own tissue type. Moreover, while well-fed animals may increase somewhat in girth, they do not *continually grow* in girth or in length as does a tree or vine.

During the early stages in the development of the seedling bean plant, the cells are nourished by food (chiefly starch) stored in the cotyledons. Before this food moves from the cotyledons to other parts of the plant, it is changed to sugar. Sugar, unlike starch, is soluble in the water of cells, and in solution it moves from the cotyledons to all other cells of the embryo, especially to those actively growing cells of the shoot tip and root tip. As soon as the cotyledons are raised into the light by the lengthening of the shoot, they become green, and then they begin the manufacture of food. By this time the roots are well established in the soil and are absorbing water and mineral salts. The foliage leaves that were present in the seed expand; they, too, engage in the process of food making. As food moves from the cotyledons, they shrivel and after a time disappear altogether. The seedling, with its roots absorbing water and mineral salts from the soil, and with its green leaves manufacturing food, is now an *independent plant.* The plant, as shown in Fig. 3.1D, is equipped with all the organs and tissues necessary to carry on the processes that are required to keep it alive and growing. At this stage the structural framework of the adult plant is established. It is an axis with meristematic tissue at both ends (Fig. 3.2). The lower portion of the axis anchors the plant in the soil and absorbs water and mineral nutrients. The upper end of the axis produces leaves for photosynthesis, and flowers and fruits for reproductive purposes. The central portion of the axis raises the

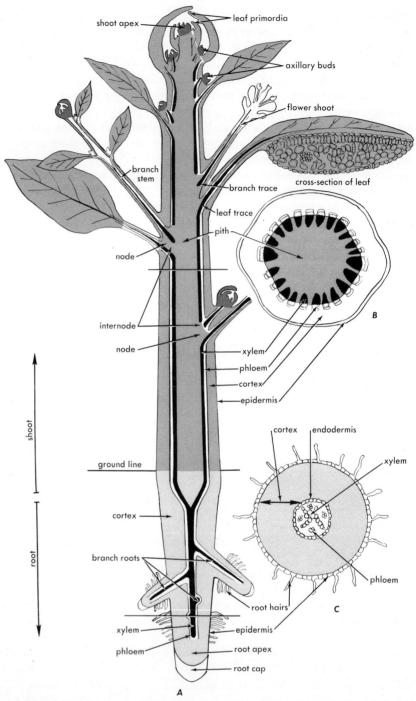

Figure 3.2. Diagram A, showing the principal organs and tissues of the body of a seed plant; B, cross section of stem; C, cross section of root. (A, redrawn from R. M. Holman and W. W. Robbins, A *Textbook of General Botany*, John Wiley and Sons.)

3 *The Plant Body of Seed Plants*

photosynthetic and reproductive part into the light and air of the biosphere and serves as a communication pathway between the absorbing and synthetic ends of the axis.

The axis may branch. Branch roots arise from certain meristematic cells inside the main root (Fig. 3.2A). Branch shoots originate in undifferentiated cells found in an angle formed by a young leaf at the point of its attachment to the stem. This is the leaf **axil** (Fig. 3.2A). Leaves originate from the undifferentiated cell or meristematic cells at the upper end of the shoot or shoot apex (Figs. 3.1, 3.2A). These undeveloped leaves are known as **leaf primordia.**

A diagram of a seed plant (Fig. 3.2) will serve to summarize the foregoing discussion and illustrate the structural organization of a seed plant. The plant body is made up of two parts: the shoot and the root. The shoot is composed of two kinds of organs—stems and leaves. In the seedling there are two kinds of leaves, the cotyledons or seed leaves, which are temporary in most species, and the foliage leaves, which are relatively longer-lasting. The stem, in contrast to the root, is divided into **nodes** and **internodes,** regions that alternate throughout the length of the stem. Nodes are the slightly enlarged portions where leaves and buds arise and where branches originate from the buds. An internode is the region between two successive nodes. As we shall see later, the anatomy of nodes and internodes differs. In Fig. 3.2, structures and tissues are shown which will be discussed in later chapters.

4

The Metabolic Plant Cell

Galen, the last of the great Greek doctors, who lived in Asia Minor during the second century A.D., thought that all tissues such as spleen, brain, and kidney of animals; leaves, stems and roots of plants were a "sensible element, of similar parts all through, simple and uncompounded." Others before him had thought that animal tissues were simply coagulated "juices" seeping through the walls of the intestine. No one dreamt that these tissues and their plant counterparts had a fabulous and complicated structure, a structure that could not be seen until the invention of a microscope which Zacharias Jansen, a spectacle maker of Holland, accomplished in 1590. This instrument was later improved by Robert Hooke, an Englishman, who was not only interested in optics but was also an architect and an experimenter with flying machines. Hooke, who lived from 1636 to 1703, examined all sorts of natural objects with his improved microscope. Among these were thin slices of cork (the dead outer bark of an oak). Figure 4.1 is an illustration of the cork tissue as Hooke saw it under his microscope. This figure was published in 1664 in an article entitled *Micrographia, or Some Physiological Descriptions of Minute Bodies Made by Magnifying Glasses.* The term **cell** was first used by Hooke to denote in cork "little boxes or cells distinct from one another . . . that perfectly enclosed air." He estimated that a cubic inch of cork would contain about 1259 million such cells.

Because of the prominence of cell walls in plant tissue, the cell early became the unit of structure and of life for botanists. However, during these early years, zoologists considered the tissues as the true centers of life of the body. There were supposed to be 21 different tissues, able to change within themselves depending upon the organ in which they were located. Life was thought to reside in these tissues. However, numerous observations upon protozoa and animal tissues led to a gradual accumulation of evidence that cells also existed in animals. In 1838 the German zoologist Theodor Schwann and the Belgian botanist Matthias Jacob Schleiden collaborated in a paper entitled, *Microscope investigations on the similarity of the structure and growth in animals and plants.*

This paper established on a firm basis *the theory* that the cell is a basic unit of structure *in both plants and animals.* However, another 30 years of research were necessary for the general acceptance of such a new idea and for the establishment of *the fact* that *all organisms are composed of cells,* that cells are indeed the basic units or building blocks of life. Cells are maintained by the division of cells. Thus meristematic cells of shoot apices (which we have just discussed) have a direct line of descent going back to the beginning and continuing into the future. In animals, such a continuity of cells occurs only in the germ plasm. Cells have existed as long as life and will exist as long as life exists.

CELL CONCEPT

We have already noted the existence of two types of cells, the prokaryotic cells which characterize the more primitive organisms, and the eukaryotic cells of all other animals and plants (Fig. 2.5*A*, *B*). A detailed discussion of the prokaryotic cell will be reserved for the chapters discussing the more primitive plants. The following discussion is concerned mainly with the structure of the eukaryotic plant cell (Fig. 4.2).

All eukaryotic cells separate many of the metabolic processes of the cell into discrete membrane-bounded, organized, protoplasmic particles called **organelles.** Each organelle has a characteristic structure and performs the same function wherever it occurs. The organelles are: the **nuclei** (Fig. 4.3), the **mitochondria** (Figs. 4.3, 4.8), the **endoplasmic reticulum** (Figs. 4.3, 4.7), the **dictyosomes** (Figs. 4.3, 4.11), the **microbodies** (Figs. 4.8*B*, 4.13) or similar particulates, and the **plastids** (Figs. 4.9, 4.10). While the **ribosome** is a small particle rather than a membrane-bounded organelle, it does have a specific niche in cell metabolism and it is convenient to include it in this list.

In the nuclei, nuclear division, and the associated ribosomes, cells have a means of storing genetic information and a system of distributing it to direct cellular development. In the mitochondria cells have a mechanism for delivering useful energy for cellular activities. In the endoplasmic reticulum, dictyosomes, and possibly the microbodies, there is a mechanism for the synthesis and probably degradation of cellular metabolites. The organelles are bathed in ground **cytoplasm** through which diffusion of materials occurs, where certain phases of metabolism take place, and where the activities of the organelles are coordinated.

Figure 4.1. Cork tissue as Robert Hooke observed it under his microscope. (From Hooke, *Micrographia* (1664), The Council of the Royal Society of London for Improving Natural Knowledge.)

Eukaryotic plant and animal cells differ in several basic ways: (1) Those plant cells specialized to carry out photosynthesis are provided with an organelle, the **chloroplast** (Figs. 4.2, 4.3), in which reside most of the enzymes, energy compounds, and intermediate substances required for photosynthesis; (2) with few exceptions, plant cells are provided with a cell wall composed of **cellulose** and secondarily deposited materials which gives greater rigidity and lessened mobility to plants; (3) most plant cells are provided with a large central aqueous **vacuole** (Figs. 4.2, 4.19); (4) synthesis of materials occurs at a much higher rate in some animal cells, for example gland cells, where the organelles involved in synthesis (endoplasmic reticulum and dictyosomes) are more numerous.

Considering for the moment only the morphology of plant cells, we may ask four questions. (1) What is the structure of the meristematic cell? (2) What changes occur during cell division? (3) What changes take place during differentiation? (4) How is the specialized cell particularly fitted to perform its part in the overall economy of the whole plant?

These questions will demand our attention in the pages that follow. We shall first consider the structure of a cell from a young bean leaf, first as it appears under the light microscope at about 1000 times magnification, and then as it appears under the electron microscope at from 3000 to 60,000 times magnification.

Cells become specialized to perform definite functions. This results in different specialized tissues and organs. In all plants and animals, cells have been organized into comparable tissues to perform comparable functions. For instance, upon reaching a certain state of complexity, all plants—whether thallophytes, mosses or vascular plants—develop flattened blades similar to leaves for carrying on photosynthesis (Figs. 22.7, 26.14, 27.7). All multicellular plants and animals may start life as a single cell, the fertilized egg cell (Fig. 16.21). Subsequently, through stages of differentiation, undifferentiated cells become cells specialized for particular purposes. Most plants retain, for relatively long periods, undifferentiated cells capable of becoming specialized cells. Indeed, it seems that most all plant cells carry full potency to return to the undifferentiated state and to again divide and give rise to a line of cells able to form a complete plant or a complete functioning tissue (page 85).

UNITS OF MEASUREMENT

At this point a consideration of a few measurements is in order (Table 4.1). We have already stated that Robert Hooke calculated that there would be 1259 million cells in a cubic inch of cork. This figure is based on his estimate of about 1100 cork cells along a line an inch long. One

Figure 4.2. Plant cells. A, diagram of a living cell from an *Elodea* leaf. When viewed on the light microscope, organelles in living cells, other than chloroplasts, lack contrast. The nucleus and nucleolus are generally visible. Sometimes one of the categories of microbodies may be seen. Mitochondria are seen only with perfect lighting conditions, ×700; B, a cell from a root tip of *Vicia faba* heavily stained with osmium to show the dictyosomes, ×500; C, a cell from a young leaf of Elodea, fixed to show plastids and mitochondria, ×500. (B after R. Bowen, *Zeitschr. Zelf. Mikr. Anat.,* **9,** 1.)

Figure 4.3. A cell from a leaf of a bean seedling (*Phaseolus vulgaris*). A is an electron micrograph; *B* is a diagram of A. Chloroplasts, nucleus, and vacuole are same size as in A; other organelles are slightly enlarged or otherwise accentuated. Ribosomes not drawn. Original magnification of A is ×10,000.

4 The Metabolic Plant Cell

plasmalemma

cell wall

microbody

intercellular space

nucleolus

nuclear envelope

nucleus

intercellular
space

mitochondrion

plasmodesma

vacuole

cytoplasm

nuclear pore

membranes

plastid

ribosomes

endoplasmic reticulum

dictyosome

B

Table 4.1 *Some dimensions*[a]

	MICRONS	ANGSTROMS
1 inch	25,000	25,000,000
1 millimeter	1,000	10,000,000
1 micron		10,000
Resolution of unaided eye	100–300	
Resolution of light microscope	0.25	2,500
Practical resolution of electron microscope for biological material		20
Wavelength of blue light	0.50	5,000
Wavelength of red light	0.70	7,000
Wavelength of electron beam 75 KV (KV is 1,000 volts)		0.005
Diameter of cork cell	25	
Volume of cork cell	15,625 μ^3	
Diameter of leaf cell	200 to 400	
Diameter of vessel in xylem	500	
Diameter of *Scenedesmus*	10	
Length of *Scenedesmus*	30	

[a]One foot magnified 100,000 times is nearly 19 miles; 1 inch magnified so each Angstrom in it measures 1 inch in length appears about 46,000 miles long.

thousand cells along an inch would give us about 40 cells for the length of a millimeter, so a single cork cell is $\frac{1}{40}$ of a millimeter in diameter. It now becomes convenient to use a smaller unit of measurement, the **micron,** symbolized by the Greek letter mu, μ. There are 1000 microns in a millimeter, and a cell $\frac{1}{40}$ of a millimeter in diameter would be 25 microns in diameter. This is about correct for a cork cell, although many plant cells are larger.

Under the best conditions one can detect particles 0.25 μ in diameter under the light microscope. This is about one-half the wavelength of blue light, and shorter wavelengths are needed to see smaller particles. Just as a large ocean wave is not modified as it passes by a small post, particles smaller than one-half the wavelength of light do not disturb the advancing wave front of light and therefore the particle cannot be seen.

An electron beam has a wave motion with a distance of 0.0000005 μ from crest to crest. Such small distances call for a still shorter unit of measurement. This is the **angstrom,** named after the Swedish physicist A. S. Ångström and abbreviated as **A.** There are 10,000 A in one micron. With the short wave lengths of a beam of electrons, a resolution of distances separating atoms is theoretically possible. But the difficulties in making perfect magnetic lenses and in the nature of biological material itself means that lines or particles in cells closer than 20 A can not be distinctly separated from each other regardless of the magnification obtained in cells. However, the best figure obtainable with the light microscope is 0.25 μ, or 2500 A, so the electron microscopes allow more than a 100-fold increase in magnification.

Since the unaided eye cannot see particles closer together than 0.2 mm or 200 microns, a magnification of ×100,000 would be needed to visualize a particle 20 A in diameter. Because of the complexity of the cell and the relative thickness of the tissue section (25 times the 20 A particle), direct magnifications of ×100,000 are not satisfactory. So for higher magnifications, photographic negatives are generally made at from 40,000 to 50,000 magnification and the additional magnification is obtained in an enlarged photographic print. Most biological work with the electron microscope is done at lower magnifications.

Technique

Living cells may be observed for many hours on the light microscope by maintaining them in a proper medium. Because the parts of the protoplast have an index of refraction close to that of water (the protoplasm bends light rays to the same degree that water does), many details of cell structure are not apparent in living cells. Enhancement of detail is attained by studying living cells with phase contrast optics or with polarized light. Still more detail is obtained by killing cells in various mixtures of alcohol, acetic acid, chromic acid, formaldehyde, or osmium tetroxide. The latter two compounds are of particular interest, because in killing cells they do not coagulate the proteins. The killed tissue is dehydrated, embedded in paraffin, and cut into sections from 5 to 20 microns in thickness. When this tissue is mounted on a glass slide and properly stained, much detail becomes apparent.

For electron microscopy, small pieces of leaf or other tissue are killed in glutaraldehyde, stained with osmium tetroxide and dehydrated and embedded in a plastic that must be held at 70°C for 24 hours to harden it properly. Sections of tissue and plastic about 500 A thick are then placed in the electron microscope in a high vacuum. A beam of electrons is passed through the sections and an image is formed on a fluorescent screen. Magnetic fields bend the beam of electrons just as glass bends light rays. Thus the lenses in the electron microscope are precisely formed magnetic fields. The dark masses seen on the screen of the electron microscope will not be images of organic matter but shadows of the high atomic-weight metal, osmium, which has become preferentially bound to some components of the protoplasm. It is not currently known why or how osmium is bound to cellular constituents. Actually, an electron microscope is a highly refined TV set tuned onto a closed circuit (Fig. 4.4). The contrast in the image on the electron microscope screen is low and details can be studied best on photographic prints prepared to enhance contrast.

A recent development in technique makes it possible to examine the ultrastructure of frozen cells without chemical fixation. Cells or isolated organelles are rapidly frozen at the temperature of liquid nitrogen. They are placed under a vacuum and the frozen suspension is kept at −147°C. The top of the frozen drop is chipped away and some of the top ice allowed to evaporate. This leaves several angstroms of frozen tissue extending up out of the frozen block. A very thin layer of carbon and platinum is now deposited over the tissue and ice surface. This layer of carbon and platinum, actually the impression of the exposed tissue, may now be placed on a grid and observed with the electron microscope (Fig. 4.18). The technique is referred to as **freeze-etch.** Bacteria may be quickly frozen to −147°C and upon thawing still survive, so bacterial cells examined by the freeze-etch technique are still alive. This is as close to living cells as the electron microscopist is likely to get.

THE METABOLIC PLANT CELL: STRUCTURE RELATED TO FUNCTION

Cells may be considered from the viewpoint of their (1) structure (Chapter 4), (2) activity (Chapter 6) or (3) chemical organization (Chapter 5). Knowledge of the interrelationships of all three approaches is necessary for an understanding of the cell as a whole. Within a cell, certain activities are confined to definite structures or organelles. Photosynthesis takes place in chloroplasts; respiration is confined to mitochondria; and the nuclei store the genetic information.

Cells are dynamic living units maintained in equilibrium with their surroundings only through the expenditure of energy. Cells are continually renewing their substance. Cells divide. To keep pace with the increasing number of cells, the mass of protoplasm must also increase. Cells are also accumulating materials to be used in their own metabolism. For instance, leaf cells synthesize sugar from simpler substances, and store it in large amounts as starch.

It is sometimes convenient to think of cells in terms of the structure and chemical nature of the organelles—such as chloroplasts—which lie within the cell. Among other things, chloroplasts are formed of proteins, lipids, and pigments. Some of the lipids are phospholipids containing phospho-

Figure 4.4. An electron microscope (Hitachi, HU-12) giving magnifications from 1000 to 500,000 diameters. (Courtesy of Perkin-Elmer Corporation.)

rous, while other lipids contain sulphur and, like Dreft, have a very high surface activity. The most common pigments are chlorophylls *a*, *b*, and *c*, and a variety of **carotenoids** (page 69) (substances resembling, or related to, carotene). We know least about the proteins. A great deal is known about the chemistry of plastids, mitochondria, and nuclei. Less is known about the other organelles.

The products, or accumulations, of protoplasmic activity (water, sugar, cellulose, hormones and similar items) are found in the dynamic cooperating unit of protoplasm, called the **protoplast** (Figs. 4.2*A*,

4.3). The protoplast secretes about itself a **cell wall** of cellulose, which frequently later receives a deposition of secondary products (Fig. 4.21). All of the protoplast, plus the cell wall, is the **plant cell.**

The activities of whole cells may be studied. Their ability, through the expenditure of energy, to accumulate materials in greater concentration than occurs in the surrounding media is a subject of considerable interest. The organelles may be isolated from cells, and their activities or chemical characteristics be studied in test tubes. Both isolated mitochondria and chloroplasts will carry on many of their normal cellular activities for a short period *in vitro*.

Whatever the approach to the study of cells, it is important to remember that energy is required to maintain a living cell as an individual entity. The maintenance of a steady living state, or equilibrium, through the expenditure of energy is known as **homeostasis.** Disruption of the source of the necessary energy must result in the death of the cell.

The parts of the cell are shown in Table 4.2. Also included are the dates when the parts were first observed by the light microscopist and by the electron microscopist, the approximate dimensions of the parts, and a notation as to their function. Dates for first observation with the light microscope are in **bold face type.**

LIGHT MICROSCOPY

The Living Cell

In a living leaf of *Elodea* or a young bean plant (*Phaseolus vulgaris*), the chloroplasts are the most prominent organelles. They are embedded in the clear hyaline cytoplasm which is confined to the periphery of the cell, or to thin strands crossing the clear central vacuole (Fig. 4.2*A*). The central vacuole (Fig. 4.2*A*) is largely water with dissolved salts, various organic solutes, and sometimes water-soluble pigments. The interfaces of the cytoplasm-vacuole and the cytoplasm-cell wall are characterized by special cytoplasmic membranes that exercise some control over the passage of substances into and out of the cytoplasm. These membranes are not

Table 4.2 *The parts of the cell* (date in **bold face type** refers to the year first reported with the light microscope; date in light face type indicates the year first seen with the electron microscope)

CELL PART	DIMENSION	FUNCTION
The plant cell, **1880**		
Primary cell wall, **1665**		Protection, strength
Cellulose microfibrils, 1948	5000 A	Mechanical
Cellulose molecules, 1922 (X-ray) 1949	100–250 A	Strength
Amorphous matrix	8.34 A × 7.9 A	Strength
Middle lamella	2 μ	Adhesion between cell walls
Protoplast, **1846**	0.025 to 2 mm	The homeostatic unit
Protoplasm, **1840**		Synthetic and developmental reactions under direction of nucleus
Cytoplasm, **1882,** 1957		
Chloroplast, **1702,** 1953	2 to 20 μ	Photosynthesis
Chromoplast, **1900,** 1958	2 to 10 μ	Accumulation of carotenes and similar pigments
Amyloplast, **1884,** 1955	2 to 25 μ	Starch storage
Leucoplast, **1883,** 1957	2 to 10 μ	?
Mitochondrion, **1897,** 1947	2 μ × 2–100 μ	Respiration
Dictyosomes, **1927,** 1956	3 μ	Enzyme synthesis
Endoplasmic reticulum, 1957	170 A	Protein synthesis
Ribosomes, 1955	200 A	Protein synthesis
Sphaerosomes, **1919,** 1967	2 μ	Lipid synthesis (?)
Microbodies, 1965	0.1–2.0μ	Compartmentalization of various enzymes
Microfibrils, 1959	280 A (indefinite length)	Various functions
Tonoplast, **1877,** 1960	80 A	Regulation of exchange between vacuole and cytoplasm
Plasmalemma, **1880**(?), 1954(a)	80 A	Regulation of exchange between cytoplasm and external solution
Crystals, 1963	10 μ	Unclear, possibly protein storage
Plasmodesmata, **1879,** 1957	2 μ	Protoplasmic bridge between cells

Table 4.2 *The parts of the cell* (date in **bold face type** refers to the year first reported with the light microscope; date in lightface type indicates the year first seen with the electron microscope) (cont.)

CELL PART	DIMENSION	FUNCTION
Nucleus, **1831**	100 to 15,000 μ	Contains genetic information necessary for normal cell development and activity
Nuclear envelope, **1907,** 1955	250 A	Separates nucleoplasm from cytoplasm
Nucleoplasm, **1879** Chromatin, **1880**		
Nucleoproteins **(1869)**		
Nucleic acids **(1889)** DNA (1924) DNA helix, 1953	18 A	Carries genetic code
Unit fibers, 1963	125 A, (indefinite length)	Encompasses DNA helix and nucleoproteins
Nucleolus, **1882,** 1958 (animal, 1952) RNA	10 to 50 μ	RNA synthesis Transferral of information from DNA to cytoplasm
Ribosomes, 1956	150 A	Protein synthesis
Mitosis, **1882,** 1960		Replication of DNA and distribution to daughter cells
Interphase, **1830** Chromosome, **1888,** 1955	2 to 200 μ	Replication of DNA Vehicle carrying DNA helix in replication and distribution
Kinetchore Centromere 1925		Region of chromosome to which spindle fibers are attached One-half of a chromosome
Chromatid, **1900** Prophase Metaphase, **1884** Anaphase, **1884** Telophase, **1894**	1 to 10 μ	Stages of mitosis
Spindle, **1883,** 1960		Cytoplasmic structure involved in moving chromosomes
Fibers, **1881,** 1960 Microtobules, 1960	various indefinite lengths 280 A (indefinite length)	During mitosis
Cytokinesis, **1891,** 1960		Division of cytoplasm to daughter cells, frequently unequal
Equatorial plate, **1875,** 1958	various	First stage in separation of daughter cells
Vacuole, **1835,** 1957		Various functions, important in water economy of cell
Water Inorganic salts Various organic solutes Crystals		

visible with the light microscope, but, because of their activity, their existence has long been recognized. The membrane separating cytoplasm and vacuole is known as the **tonoplast;** that bounding the outer surface of the cytoplasm is the **plasmalemma.** The cytoplasm and plasmalemma may be removed, leaving the tonoplast as a sac enclosing the vacuole. A single nucleus (Fig. 4.2A) is embedded in the cytoplasm and may be seen flattened against one of the walls. In the living cell it generally appears structureless, although with careful focusing, one to four small denser bodies, **the nucleoli,** (nucleolus s.) can be seen (Fig. 4.2A). There is also an envelope around the nucleus. This may be punctured and a clear fluid, the **nucleoplasm,** will escape through the puncture. In addition, small spherical particles about 0.54 μ in diameter may be seen in the living cytoplasm. They belong to a class of particles called **microbodies** (Fig. 4.2A).

The cytoplasm in a leaf cell, after a short sojourn under the microscope, will show active streaming, carrying both the chloroplasts and the microbodies with it. With time, the velocity increases up to 5 to 10 centimeters per minute. This rapid flowing of the cytoplasm is a dramatic indication of its plasticity, and should be recalled when studying microscope slides and electron micrographs, for it is sometimes difficult to reconcile the activity of flowing cytoplasm with the static and frequently complicated images seen in electron micrographs.

Fixed and Stained Cells

Cells fixed and stained show considerably more detail. The cell wall is evident and the cellulose wall of one protoplast is frequently seen separated from the cellulose wall of an adjacent protoplast by an intervening substance. This is the **middle lamella** (Fig. 4.21), which is largely pectin and formed as a final step in cell division (page 96). Chloroplasts, when present, are evident. After proper fixation and staining, numerous small filamentous bodies, perhaps 2 to 4 μ long and 0.54 μ in diameter, are seen in the background cytoplasm. These are mitochondria, which are typically seen in the light microscopes as simple darkly-staining granules (Fig. 4.2C). Microbodies and dictyosomes are apparent after special fixing and staining. The nuclear membrane and the nucleolus are both much more distinct after fixation and staining. Within the nu-

cleus, there are generally clumps of stained material called **chromatin** (Fig. 4.2C). The **nucleoplasm,** however does not stain.

ELECTRON MICROSCOPY

A cell from a young bean leaf (*Phaseolus vulgaris*) is shown in Fig. 4.3. Since the cells at this stage have not elongated and contain all of the organelles, they are satisfactory examples of plant cell structure as seen by the electron microscope. Figure 4.3A illustrates the relative size and abundance of the organelles just mentioned. Figure 4.3B is a diagram of this cell which should be used to definitely locate the organelles in the micrograph. The chloroplasts are the most abundant and conspicuous organelles, since only the nucleus is larger. Mitochondria and dictyosomes are small and the dictyosomes occur in relatively small numbers. There are not many cisternae of endoplasmic reticulum and they are masked by the abundant ribosomes. In the diagram, mitochondria, dictyosomes, and the endoplasmic reticulum are enlarged. The free ribosomes are not indicated. The tonoplast is accentuated.

Membranes

Perhaps the most notable thing about cells at this magnification is their **compartmentalization** into **membrane-bounded organelles.** The protoplasm itself is bounded externally by the plasmalemma (Figs. 4.3, 4.5, 4.18), and internally by the tonoplast (Figs. 4.3, 4.5). The plasmalemma separates the protoplast from the external cell wall or other nonliving systems. Within the protoplast, the tonoplast separates the protoplasm from the vacuole. In any organized living system, high energy regions must be separated from low energy regions by membranes.

Note that each of these membranes consists of a light line sandwiched between two dark lines. The approximate widths of the three membrane components are 20-35-20 A, respectively. Note that the plasmalemma is more heavily stained and the two darker components may not have identical widths. These membranes are known as **unit membranes.** They are thought to be composed of a central bimolecular leaflet of lipid sandwiched between two monomolecular layers of fully extended protein (Fig. 4.6B).

A drop of oil or other lipid may be spread on

Figure 4.5. A cell wall bounded on each side by a thin peripheral layer of cytoplasm. Bordering portions of the vacuoles of each protoplast are present. The tonoplasts separate cytoplasm and vacuole. The plasmalemma bounds the protoplast, separating it from the cell wall. The plasmalemma stains heavier than the tonoplast. The unit membrane structure is evident in portions of all three membranes, ×100,000.

water to form a monomolecular film. (The colors seen when oil is spread on water are indicative of very thin oil films, some of which may be only a single molecule in thickness.) A force of 9 dynes is required to break such a lipid film. Natural oils spread on water (Harvey and Danielli used natural oils from salmon eggs) are disrupted by forces as low as 2 dynes. Protein was found to be associated with the salmon egg oil and to account for the lessened strength of the film formed when this oil was floated on water. Later experiments demonstrated that a membrane consisting of a bimolecular layer of lipid sandwiched between monomolecular layers of protein could account for many of the properties of cellular membranes. The Davson-Danielli model of a cellular membrane is given in Fig. 4.6A. Note that these workers proposed the protein layers to be in the more natural form of globular proteins. The idea that the protein is fully extended (Fig. 4.6B) was proposed by Robertson because, in studying the membranes of

the rather inert myelin sheath protecting nerve fibers, he saw that there was not space enough for the globular proteins. The difference in the chemical activity of globular proteins and fully stretched proteins is significant and would greatly influence the activity of the membranes. It was proposed early in the initial days of electron microscopy, and generally accepted, that all cellular membranes were tripartite membranes composed of a bimolecular leaflet of lipid sandwiched between two monomolecular layers of fully stretched protein, similar to that of the myelin sheath of a nerve cell (Fig. 4.6B). While many prominent investigators are still of this opinion, we shall see (page 241) that evidence is accumulating to seriously question the reality of such a broad generalization. For instance, the myelin sheath is **chemically inert** and **unique.** The problem has great significance because many important cellular reactions are membrane-bound. Electron transport activity in both photosynthesis (page 241) and respiration (page 264) is bound to membranes. The binding of enzymes to membranes relegates the enzymes to a particular spot in a team of cooperating enzymes.

Ribosomes

The ground cytoplasm, after glutaraldehyde fixation and osmium staining, is moderately electron-transparent and has a soft gray granular appearance. In it always appears at least one type of densely-stained granule, roughly angular and from 170 A to 200 A in diameter (Fig. 4.3A). The enzyme **RNAase** specifically degrades RNA. Treating tissue with RNAase results in the disappearance of these granules, with no change in the other organelles. This is good evidence that RNA is present in these particulates, which are called **ribosomes.** Ribosomes are present in plastids and in mitochondria, as well as in the cytoplasm. They appear to be lacking in nuclei (Figs. 4.3A, 4.16). Some micrographs show them attached to the cytoplasmic side of the nuclear envelope. If they do occur within the nucleus, they are probably confined to the nucleolus (Fig. 4.16C). Ribosomes are frequently associated with the cytoplasmic membrane system, the endoplasmic reticulum (Figs. 4.3, 4.7A). Under certain circumstances they appear in groups, frequently as helical aggregations. These arrays are designated as **polyribosomes** or simply **polysomes** (Fig. 4.7B).

Organelles

Endoplasmic Reticulum

In cross section, the endoplasmic reticulum is represented by profiles of two parallel membranes separated by a narrow light space about 40 A wide (Figs. 4.7, 4.18). These profiles are actually cross sections through extensive flattened vesicles. Note that these membranes again form a closed system; their ends are never open to the ground cytoplasm. The endoplasmic reticulum frequently has ribosomes appressed to its outer, or cytoplasmic, surface. This is known as **rough endoplasmic reticulum. Smooth endoplasmic reticulum** lacks ribosomes and is not as common as rough endoplasmic reticulum (Fig. 4.7A). There is considerable evidence suggesting that protein synthesis may be associated with the rough endoplasmic reticulum. For example, endoplasmic reticulum is most highly developed in cells active in protein synthesis, such as the pancreas, and the glandular hairs of *Drosera* leaves (Fig.

10.16B, C). It is poorly developed in mature leaf cells (Fig. 4.3A). The endoplasmic reticulum is associated with **plasmodesmata,** or cytoplasmic connections from cell to cell across cell walls. It has proven difficult, however, to conclusively show that the endoplasmic reticulum, as such, passes through a plasmodesma. However, it seems clear in Fig. 4.7C that a strand of endoplasmic reticulum does pass through the plasmodesma, connecting a sieve tube element with its companion cell (page 125).

As exemplified by the endoplasmic reticulum, membrane systems within the protoplast, with the single exception of microtubules, form closed systems. Thus, membranes form sacs or vesicles. The

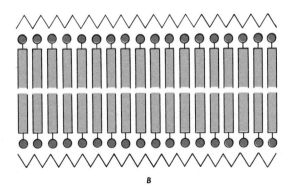

Figure 4.6. The unit membrane. A, the model proposed by Davson and Danielli in 1935. It shows a bimolecular layer of lipid bounded by a monomolecular layer of globular proteins; B, the model proposed by Robertson in 1959. Here the bounding monomolecular layer is fully stretched protein.

vesicles may be roughly spherical or they may be flattened. Flattened vesicular forms may be in stacks or may occur singly, irregularly, or regularly, distributed throughout the cytoplasm and organelles.

Mitochondria

The mitochondrial envelope separates internal mitochondrial space from the ground cytoplasm (Figs. 4.8, 4.18). The mitochondrial envelope is double and each component is thought by some to be a typical unit membrane. The outer component forms a continuous barrier around the mitochondrion. The internal component connects with a series of internal membranes, the **cristae** (Fig. 4.8). The cristae form a closed system, separating an internal space from the **mitochondrial matrix.** Note

Figure 4.7. Endoplasmic reticulum. A, endoplasmic reticulum in an apical cell of the alga *Chara.* Both rough and smooth endoplasmic reticulum are present and are continuous with each other. A microbody is also visible, ×15,000; B, endoplasmic reticulum in a meristematic cell from a root tip of wheat. Both rough and smooth ER may be seen. There are also free ribosomes and some in helical aggregations that form polyribosomes, ×30,000; C, endoplasmic reticulum in a sieve element and companion cell of curcurbit phloem. The endoplasmic reticulum may be seen in the plasmodesmata that pass through the cell walls connecting the two cells, ×60,000. (A, courtesy of J. B. Pantastico; B, courtesy of P. Bartels; C, courtesy of K. Esau and J. Cronshaw, *J. Ultrastruct. Res.,* **23,** 1. Copyright Academic Press.)

Figure 4.8. Mitochondria. A, mitochondrion from a young leaf of a safflower (*Carthamus tintorius*) seedling, ×60,000; B, mitochondrion and a microbody from a seven-day-old seedling of bean (*Phaseolus vulgaris*). The mitochrondrion is spherical; the association between a crista and the inner component of the envelope is distinctly seen. The light areas have fine fibrils in them, probably of DNA. The double envelope of the mitochondrion may be compared with the single membrane bounding the microbody, ×70,000; C, mitochondrion in a cortical cell of the brown alga *Egregia menziesii*; the double envelope, cristae, and DNA strands are apparent, ×75,000. (A, courtesy of A. H. P. Engelbrecht; C, courtesy of T. and A. A. Bisalputra, *J. Cell Biol.*, **33,** 511.)

that the cristae are continuous with the internal space of the envelope. The form and amount of the cristae varies. They are numerous in mitochondria from animal cells and generally are parallel to each other and at right angles to the long axis of the mitochondrion. Cristae in plant mitochondria frequently are sparse, irregularly arranged, and not always seemingly connected with the inner membrane of the envelope.

Plastids

The chloroplasts, the plastids which contain chlorophyll, are prominent organelles (Fig. 4.9). The chloroplasts are bounded by an envelope of two membranes, just as are the mitochondria. The plastids have a granular background material, the **stroma.** Particulates having the properties of cytoplasmic ribosomes may be seen in the stroma. They are from 170 to 200 A in diameter. Clear areas are also present, and in favorable sections, such clear areas containing an array of slender fibrils may be

seen. These fibrils are 50 A in diameter, which is the diameter of the DNA fibrils in other locations. These fibrils are absent from sections treated to remove DNA. Presumably, then, they represent chloroplastic DNA. They are more apparent in plastids from young leaves, as shown in Fig. 4.9.

Plastid membranes. Some plastid membranes occur in cylindrical stacks (Fig. 4.9), called **grana.** The grana are connected at irregular intervals by **frets** or **lamellae.** Note that where the frets enter the grana, adjacent membranes merge, back to back, to form an appressed membrane about twice as thick as the fret membranes. These double membranes in the grana are the **partitions.** (Fig. 4.9).

Various kinds of plastids. Plastids respond readily to changes in age or environmental conditions. A seedling of bean or barley just emerging from the ground may have yellow first leaves. The inner leaves of lettuce or cabbage are yellowish. Both leaves will become green in the light, yet the plastids in the two types of leaves have a very different appearance. The plastids in the leaves of the etiostroma frets   granum   envelope

Figure 4.9. Chloroplast from a corn (Zea mays) leaf, ×25,000. (Courtesy of L. K. Shumway.)

lated bean seedling have an elaborate crystalline-like structure generally referred to as a **prolamellar body** (Fig. 4.10*B*). The plastids in the lettuce leaf are small and have only a few single lamellae (Fig. 4.10*A*). These plastids, since they will become normal plastids in light, may be called **proplastids.** Other colorless plastids, **leucoplasts,** are found in the epidermal cells of leaves, in onion, in storage tissue of apples and in other white tissues. Some of these plastids store large amounts of starch and may be called **amyloplasts** (Fig. 4.10*C*).

As fruits ripen on trees and leaves prepare to fall in the end of summer, they change from green to red, orange, or yellow. This occurs because there is a destruction of chlorophyll, accompanied by an accumulation of yellow or red pigments known as the **carotenoids.** The plastids containing the red and yellow pigments are **chromoplasts.** A chromoplast from red pepper is shown in Fig. 4.10*D*. Plants grown in solutions deficient in mineral nutrients show characteristic changes in plastid structure.

Carotenoids. As just noted, there are two categories of plastid pigments, the chlorophylls and the carotenoids. The function of the chlorophylls is to trap light energy. The distribution of three of the chlorophylls (*a, b, c*) has already been considered. There are many types of carotenoids, and specific groupings may be constant throughout different phyla or genera. Their chemistry is briefly discussed in the following chapter.

Mutant plants unable to synthesize their own carotenes are known. When exposed to light and oxygen, these plants may be rapidly killed. This common destructive effect of light on cells is known as **photo-oxidation.** Such photosensitizing oxidations are widespread. Many pigments, including synthetic dyes, may be photosensitizers. They act upon a wide range of organisms, including man. Chlorophyll may have a very strong photosensitizing effect on a wide range of organisms, including the very plants that manufacture it. Chlorophyll photosensitizing does not occur in plants that have their natural supply of carotenoids. It does occur when the carotenes are absent. One function of the carotene pigments is to protect plants from photosensitization by their own chlorophyll.

        4 The Metabolic Plant Cell

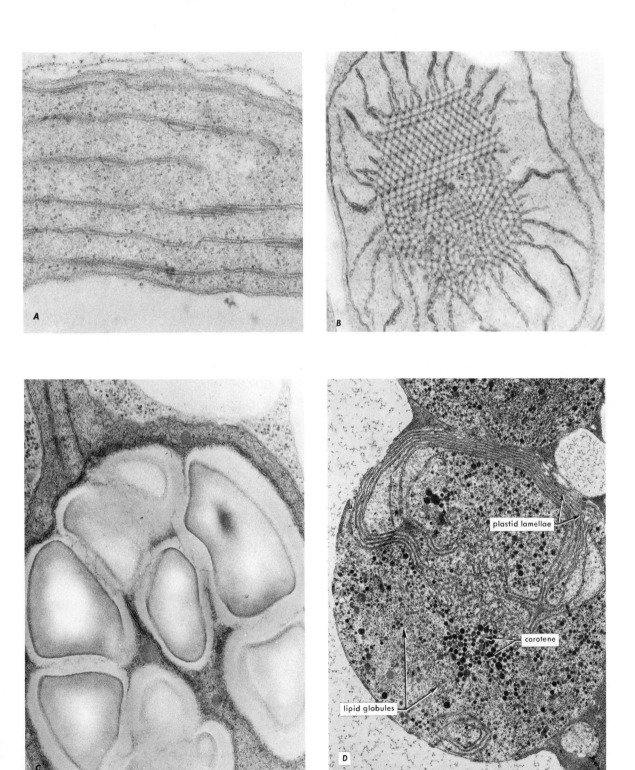

Figure 4.10. Four types of plastids. *A*, a small leucoplast from an inner white leaf of endive (*Cichorium endiva*), ×20,000; *B*, a proplastid from a yellow leaf of a dark-grown bean seedling (*Phaseolus vulgaris*), ×20,000; *C*, an amyoplast from a bean seedling (*Phaseolus vulgaris*), ×20,000; *D*, a chromoplast from a mature red pepper (*Capsicum*), ×21,000. (*D*, courtesy of A. R. Spurr and W. H. Harris, *Am. J. Bot.*, **44,** 1210).

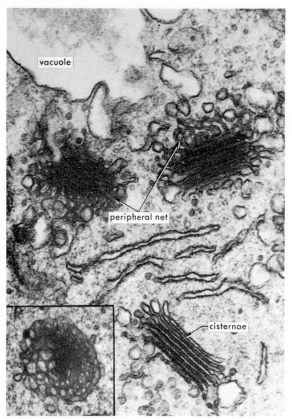

Figure 4.11. Four dictyosomes in a cell from the root tip of a water plant (*Hydrocharis*). One dictyosome has been sectioned at right angles to its cisternae, and another one almost parallel to its cisternae. Note the peripheral net around this second dictyosome. The two other dictyosomes have been sectioned obliquely to the cisternae and some material appears to be passing from them to the vacuole, ×15,000. (Courtesy of E. G. Cutter.)

Cytoplasmic inheritance. Both mitochondria (Fig. 4.8*C*) and chloroplasts (Fig. 4.9) have a supply of DNA and RNA. Does this mean that they have their own store of genetic information and that they are able to synthesize their own proteins? The case is not so clear for mitochondria, but evidence to be presented in the chapter on inheritance definitely demonstrates that chloroplast traits may be passed on from one generation to the next independently of the nucleus.

Dictyosomes

Dictyosomes are composed of from 5 to 15 circular flattened vesicles, or **cisternae,** aligned in stacks (Figs. 4.11, 4.18). In cross sections, many small vesicles appear at the margins of the cisternae. However, face views reveal that the margins of each cisterna form a course net with true vesicles only at the extreme outer regions (Fig. 4.11). The peripheral vesicles, in most cases, seem to be granules of secretory material elaborated in the dictyosome cisternae, possibly in collaboration with the endoplasmic reticulum. There is considerable evidence that the precursors of cell wall material are synthesized in the dictyosomes. The stack of cisternae appears to be polarized; that is, there is a forming face and a disappearing face. The latter is the concave face which is somehow used up in the formation of the vesicles.

There are many dictyosomes in plant cells, and collectively they may be referred to as the **golgi apparatus,** or golgi zone. This name is derived from the Italian neurologist, C. Golgi, who first described them in nerves cells in 1898. In animal cells, the golgi apparatus is generally associated with other cell constituents to form a complex region active in the elaboration of secretory products.

The dictyosomes are about 2 μ in diameter and about 0.5 μ in width. They are thus large enough to be seen with the light microscope. They probably were seen in 1925 by Robert Bowen, who called them osmiophilic platelets (Fig. 4.2*C*) because they combined with OsO_4. He was sure that they were the plant counterpart of animal dictyosomes.

Figure 4.12. Microtubules. A cell from a root of *Juniperus*. Note the microtubules at right angles to the cell wall, ×50,000. (Courtesy of M. C. Ledbetter and K. R. Porter, *J. Cell Biol.,* **19,** 139.)

Figure 4.13. Microbodies. Particulates from squash cotyledons, as well as other plant and animal parts, may be isolated and concentrated so that their ultrastructure may be studied with the electron microscope. Mitochondria are easily recognized. Also observed are a number of other particulates, bounded by a single membrane and varying in structural details. These are microbodies, designated by **mb** on the figure. The single membranes of some have been ruptured by the processing techniques, ×5000. (Courtesy of John Lott.)

Microtubules

There remains a membranous system which is not closed to the ground cytoplasm: the **microtubules** (Fig. 4.12). Microtubules are of indefinite length, are 280 A in diameter and are provided with a pore 80 A in diameter. The function of the microtubules is not definitely known. They occur in the spindle during mitosis (Fig. 6.11, page 91). They also occur in the cilia, or flagella, of motile cells of all eukaryotes. A function of motility has been ascribed to them. In some conducting cells they are wavy, and it is thought that they may be active in moving material through conducting cells. They may have a structural function and they may conduct stimuli. In many elongating cells the microtubules are parallel with each other and lie at right angles to the long dimension of the cell wall (Fig. 4.12).

Microbodies

Preparations of a number of plant and animal tissues show a variety of spherical organelles that vary from about mitochondrial size to very much smaller. These organelles can be distinguished morphologically from mitochondria, since they have a single outer membrane rather than a double membrane envelope that is characteristic of the mitochondria. Cisternae and cristae are absent from microbodies, and the central area frequently appears rather dense under the electron microscope and may contain a variety of crystals (Figs. 4.8*B*, 4.13, 4.18).

Microbodies not only are morphologically different from other cellular structures but also can be separated from them by centrifuging a mixture of ground tissue containing all organelles on a density gradient of sucrose solution.

Considerable research is being carried out at the present time to determine the exact chemical and physiological significance of these organelles. Evidence is accumulating that many different enzymes may be found in single membrane-bound spherical organelles in both plant and animal cells. To some extent, the types of enzymes found depend upon the tissue and kind of cells being studied. Thus, it appears that there is not just one kind of microbody but that the term microbody as defined above includes a fairly large number of different kinds of organelles that contain different enzymes and perform different functions in the cell.

Peroxisomes

The beginning student can appreciate some of the contradictions in terminology when he realizes that microbodies isolated from leaves and containing a complex of oxidative enzymes have been called peroxisomes; those isolated from castor bean seeds and containing other oxidative enzymes have been called glyoxysomes; while some microbodies observed in plant cells have been given the name sphaerosomes. Liver and various other animal tissues have yielded microbodies called lysosomes (high in hydrolytic enzymes) and peroxisomes (containing certain oxidative enzymes).

Crystals

Other particulates are found in the protoplasm. Among the more wide-spread of these are crystals, sometimes of protein. They have been observed in the nucleus, cytoplasm, mitochondria, and chloroplast. In isolated chloroplasts (Fig. 4.14), they are generally noted when the isolation medium has a high molar concentration.

Nucleus

Structure

In general, the complex details of nuclear structure and organization have been worked out with the light microscope, using fixatives of high acidity (lower than pH 6). Such fixatives, which coagulate proteins, are of little use in studying cellular detail at a resolution level of 50 A. Consequently, with but three exceptions—(1) the nuclear envelope (Fig. 4.15), (2) the nuclear **unit fibers** (Fig. 4.16B), and (3) a significant stage of meiosis (page 342)—the electron microscope has yielded a paucity of information about the nucleus in contrast to the im-

mense amount of information derived from electron microscope studies on the cytoplasm and cytoplasmic organelles. New information about the nucleus and nuclear activity has come mainly from two sources: (1) studies of the growth of organisms in which nuclear control of differentiation has been subjected to experimental techniques, and (2) biochemical studies of the nucleus and of inheritance. These investigations have furnished us with an insight into nuclear activity and structure that is far beyond the details of the cytoplasmic organization we have just discussed. Indeed, the biochemical results so far obtained are so spectacular that they form the basis for a large share of current beginning biology courses. Biochemistry of one sort or another usually continues the studies so initiated.

Electron micrographs of nuclei show few details not already recognized by the light microscopist (Figs. 4.3, 4.18). There are one or more nucleoli and some denser areas embedded in a granular nucleoplasm. The denser areas may represent sections of chromosomes, but such a relationship has not yet been definitely demonstrated. High magnification sometimes suggests fibrils in sectioned nuclei (Fig. 4.16A), and fibrils are certainly present in isolated nuclei (Fig. 4.16B) which have been freeze-dried.

Figure 4.14. Crystals within the cytoplasm. Crystals, probably of protein, in a chloroplast isolated from a leaf of bean (*Vicia faba*). Similar ones are frequently seen, not only in chloroplasts, but also in mitochondria and in the cytoplasm, ×60,000. (Courtesy of L. K. Shumway, *Protoplasma*, **76,** 182. Copyright Springer-Verlag.)

Figure 4.15. The nuclear envelope. *A*, a section of a nucleus from a meristematic cell of the shoot apex of *Chenopodium album*. The section is at right angles to the envelope, which is distinctly double and shows definite interruptions; *B*, a section parallel to the nuclear envelope from a cell in a squash cotyledon. The pores are very distinct, ×20,000; *C*, the surface of the nuclear envelope as it appears in a freeze-etch preparation, ×30,000. (A, courtesy of E. M. Gifford, Jr., and K. Stewart; *B*, courtesy of J. Lott, and C, courtesy of D. Branton).

The nuclear envelope is double and is provided with pores, which are shown in cross section in Fig. 4.15*A*. Figure 4.15*B* shows a tangential cut through a small portion of the envelope of a nucleus in a cell of a squash cotyledon. The pores, in face view, are very evident. They are also evident in the surface of the nuclei when the freeze-etch technique is used for preparation (Fig. 4.18). Although the pores appear as distinct holes, and there is some evidence that rather large molecules can pass through them, not all workers are agreed that they do serve as passageways.

Function

While it is well established that genetic information is stored within, and distributed from, the nucleus, the details of how this occurs are not all clear. Cells deprived of nuclei will continue normal activities for as long as three months. The green alga, *Acetabularia* (Figs. 2.5*E*, 4.17) is a single celled plant between 1 and 1½ inches tall. It shows considerable differentiation for a single cell. It has rhizoids for attachment, an upright stalk, and an expanded cap which differs from species to species.

Figure 4.16. The nucleoplasm. A, portion of a section of a nucleus from a leaf of tobacco (*Nicotina rustica*); fibers and aggregations of fibers are present. Under well-controlled processing procedures, the diameters of such fibers approximate 125 A; B, fibers, about 225 A in diameter, are a constant feature of the nucleoplasm obtained from isolated nuclei of barley Hordeum vulgare endosperm processed by the critical freezing point method; C, a view of a section of a nucleolus from the root tip of the water plant, *Hydrocharis*. Note the aggregation of polysome-like particles and the absence of a limiting membrane between the nucleolus and the nucleoplasm, all ×30,000. (B, courtesy of S. L. Wolfe; *J. Cell Biol.*, **37**, 610; C, courtesy of E. G. Cutter.)

The nucleus is situated at the bottom of one of the rhizoidal branches (Fig. 4.17A). It may easily be removed by cutting off the tip of the rhizoid containing it (Fig. 4.17B). Such an enucleated *Acetabularia* plant has been kept alive in the light for seven months, although *normal synthetic activities* ceased after two weeks. When the cap and a portion of the stalk are removed from an enucleated cell (Fig. 4.17B), a new cap will be regenerated. The shorter the piece of stalk that is removed, the better the regeneration. It appears that materials capable of inducing the formation of a new cap are released by the nucleus and accumulate in the upper portion of the stalk. Here they are readily available for the renewal of a cap, should the original cap be damaged or lost. These experiments would seem to indicate that the presence of a nucleus is not immediately necessary for the normal day-to-day metabolic activities of *Acetabularia*. There must have been sufficient genetic information present in one form or another, in the cytoplasm, to provide for the immediate regeneration of the new cap. It can be shown in *Acetabularia* that this material is used up. There must be a continual renewal of the cytoplasmic informational materials. The nucleus is also necessary for cell division. We shall see later on (page 85) that growth and differentiation do not occur without cell division. In the periods between cell divisions, the nucleus is actively preparing for a subsequent division.

This line of reasoning leads to an enigma—the lack of a nucleus in the food-conducting cells of the vascular plants, the sieve elements of the phloem (page 124). Mature and actively conducting sieve elements lack a nucleus. Such cells normally function for periods not longer than three years, but in palm trees they must function for 200 years! The only other enucleated functioning cell is the mammalian erythrocyte, and its life span is about three months.

Figure 4.18 shows the freeze-etch appearance of many of the organelles we have just discussed. The surface view of the nuclear envelope with the pore, both in face view and sectioned, is particularly striking. Three dictyosomes are shown: a face view, an oblique view, and a cross section. The peripheral net in the face view is very distinct. Two vesicles are adjacent to the peripheral net and one appears to be associated with the endoplasmic reticulum. A microbody appears close to the sectioned dictyosome. A cell wall separates the cells, and the upper cell contains a mitochondrion and vacuoles.

Nonprotoplasmic Portions of the Cell

All of the organelles and particulates so far discussed have been an integral part of the protoplasm. Energy has been expended within the protoplasm to maintain its state of homeostasis. The cell also contains regions outside of the protoplasm which are maintained in a state of equilibrium just as are organelles. The vacuoles and crystals within the protoplast, as well as the cell walls bounding the exterior of the protoplast, are such extra-protoplasmic regions.

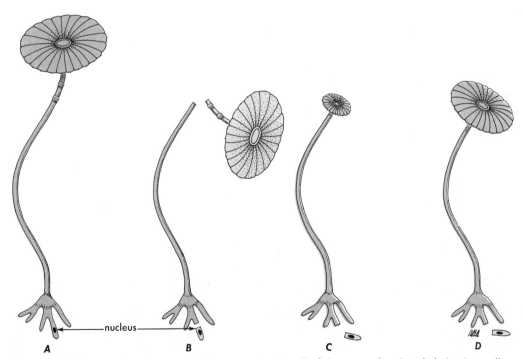

Figure 4.17. Regeneration of a new cap by an enucleated cell of the green alga *Acetabularia*. A, a cell of *Acetatularia*; note nucleus at tip of rhizoid; B, nucleus and cap have been cut off; C, new cap grows; D, growth of new cap completed; it may be smaller than the original cap.

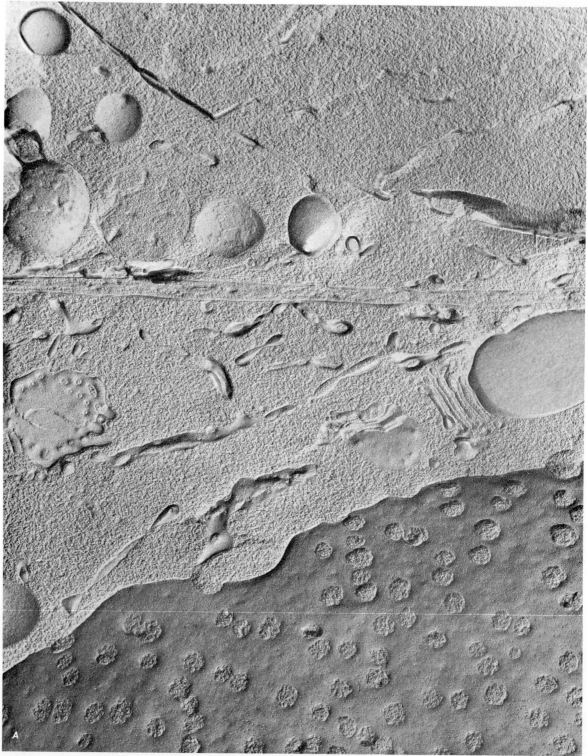

Figure 4.18. Freeze-etch preparation of an onion root tip cell, showing the organelles just discussed as they appear when the cells are examined after freezing, rather than after chemical fixation. Bacteria and other cells, after a similar cold treatment, are viable. The similarity between the chemical fixation image

and the freezing (living) image is striking. *A*, freeze etch preparation; *B*, diagram of *A*, with labels, ×10,000. (*A*, courtesy of D. Branton.)

Vacuoles

Vacuoles are definitely a part of the protoplast (Figs. 4.3, 4.19). We shall see later (page 74) that vacuoles have a higher energy potential (solute concentration) than the solutions outside the cells and that energy expended by the protoplasm is required to maintain this higher energy potential.

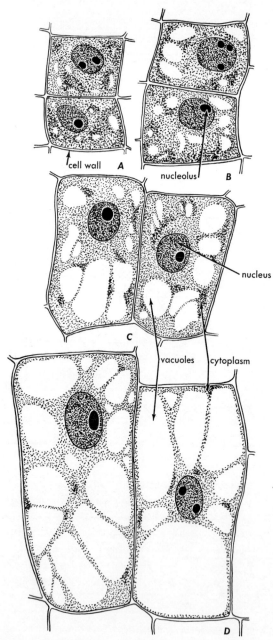

cell wall **A**

nucleolus **B**

nucleus

C

vacuoles cytoplasm

D

Figure 4.19. Stages in the growth of a cell. Progressively older cells shown from A through D. ×2000.

Certain of the materials of the vacuole may be absorbed passively from solutions surrounding cells. This absorption requires no energy. Other substances, like inorganic salts, are actively accumulated; energy must be expended to take them into the vacuole and to retain them within the vacuole in a higher concentration than that outside the protoplast. Other materials, like sugars and pigments, are synthesized within the cytoplasm and discharged into the vacuole. These materials present an interesting problem in that they easily pass through the tonoplast to reach the vacuole. They also pass easily from one cell to another, but they do not pass out of the cell into external solutions.

In many young cells, the cytoplasm occupies much of the space in the cell. Small vacuoles are, however, present (Fig. 4.19A). As the cell grows larger, the small vacuoles within the cytoplasm increase in size, coalesce, and become smaller in number (Fig. 4.19B, C). Finally, when the cell has attained its mature size, only a few large vacuoles, or even only one, may remain (Fig. 4.19D), and the protoplasmic contents (nucleus and cytoplasm) of the cell lie compressed against the cell wall. The nucleus may occupy a position near the center of the cell, in which event it is connected with the cytoplasm around the cell wall by strands of cytoplasm (Fig. 4.19D).

Vacuoles do not contain pure water, but rather a highly dilute solution of many substances. This aqueous solution in the cell is termed **cell sap.** Among the substances dissolved in the water, and thus constituents of cell sap, are the following: (1) atmospheric gases, including nitrogen, oxygen, and carbon dioxide; (2) inorganic salts, such as nitrates, sulfates, phosphates, and chlorides of potassium, sodium, calcium, iron, and magnesium; (3) organic acids, such as oxalic, citric, malic, tartaric; (4) salts of organic acids; (5) sugars, such as grape sugar (glucose) and cane sugar (sucrose); (6) water-soluble proteins, alkaloids, and certain pigments, such as anthocyanin. Generally the cell sap is slightly acid. The concentration of cell sap varies from cell to cell, and it may vary in the same cell during the course of the cell's life.

The most common pigments of the vacuolar sap are the **anthocyanins.** These pigments are responsible for the red color of the roots and leaves of garden beet and the red, purple, or blue of the petals of many flowers or of other parts of plants. The yellow coloration of poppy flowers is also due to yellow vacuolar anthocyanin-like pigments called anthoxanthins.

Inorganic Crystals

Cells with crystals can be found in almost all plants and in many different plant tissues. Crystals vary in chemical composition and in form (Fig. 4.20). The most common crystals are of calcium oxalate; it is generally held that they are an excretory product of the protoplast formed by the union of calcium and oxalic acid. This acid, a by-product of certain activities of the protoplast, is soluble in

Figure 4.20. Types of inorganic crystals found in the vacuoles of living cells and in the cell walls of older non-living cells. A, raphides; B, a cluster of crystals; C, a single crystal, ×2,000.

cell sap and is toxic to the protoplasm if it attains a high concentration in the cell. By its union with calcium, the soluble oxalic acid is converted into the highly insoluble calcium oxalate, which will not injure the protoplasm. In addition to calcium oxalate, crystals of calcium sulfate or of protein sometimes occur.

The Cell Wall

The protoplast is surrounded by a plasmolemma or cytoplasmic membrane. Outside this membrane, and surrounding the entire protoplast, is a rigid wall secreted by the protoplast that it encloses. When, as is usual, protoplasts are separated from each other by walls, the walls are cemented together by an intercellular substance, the **middle lamella** (Fig. 4.21), which is characterized by pectates and certain other substances. The first wall formed by the protoplast is the **primary wall** (Fig. 4.21) and it is composed mainly of **cellulose.** Aging of the cell may bring about the deposition by the protoplast of more wall material, which is laid down on the primary wall. Thus a **secondary wall** (Fig. 4.21) is formed, and the complete mature cell wall may finally come to have a thickness many times as great as the primary wall. In some tissues, the secondary wall is stratified and composed of several layers. In others the cells do not lay down secondary wall material, in which event the common wall between two adjacent protoplasts is composed of the middle lamella, with primary wall material on each side. The secondary wall may be of cellulose or of cellulose impregnated with other substances. Some of these substances, notably **lignin,** lend hardness to wood; others, like **suberin** and **cutin,** (Figs. 7.25, 8.14, 10.9) are waxy and protect leaves and stems against water loss. In addition, certain other materials may enter into the composition of the cell wall—gums, tannins, minerals, pigments, proteins, fats, and oils. It should be emphasized that in mature hard tissues, such as wood, lignin may be deposited not only in the secondary wall but also in the primary wall and middle lamella.

Although the walls of cells vary considerably in composition in different species, and from one part to another in the same individual plant, **cellulose** constitutes the greatest percentage of the material of which most cell walls are made. It is synthesized by the protoplast.

Other than cellulose, the commonest and most

widely found substance in the walls of plant cells is **lignin,** a material that adds to their hardness. By appropriate chemical treatments, cellulose and lignin can be separated. For example, when wood is treated with 72 percent sulfuric acid, cellulose is changed and becomes water-soluble, whereas lignin remains as an insoluble precipitate.

Suberin is a waxy substance associated with cellulose in the walls, particularly of cork cells. Since neither water nor gases can pass through suberin, cork tissue is an excellent protection against excessive loss of water.

Cutin is a fatty substance usually found as an external coating on the outer cellulose wall of the epidermal cells of leaves and stems.

Other substances, besides lignin, suberin, and cutin, that may be associated with cellulose in the walls of cells will be discussed at other points in the text.

SUMMARY

1. The cell may be considered the unit of life because it is the smallest unit known to be able to replicate itself, and to develop in an orderly fashion from a simple beginning to a very specialized end.

2. Cells may be divided into protoplast and cell wall.

3. Protoplasm is divided into many compartments by an array of particles and membranes or organelles.

4. The plasmalemma is the membrane separating the protoplast from the cell wall, and the tonoplast is the membrane separating the vacuole from the cytoplasm.

5. Ribosomes are small particulates having a high RNA content.

6. The endoplasmic reticulum consists of an array of extensive flattened vesicles. It may or may not have ribosomes associated with the outer surface of the vesicles. It appears to be involved in protein synthesis.

7. Mitochondria are small organelles about 0.5 μ in width and from 1 to 3 μ in length. They are bounded by a double envelope. The inner component invaginates to form cristae which are surrounded by a homogeneous matrix. Respiration is localized in the mitochondria.

8. Chloroplasts are approximately 5 by 10 μ in size. They are bounded by a double envelope. The internal chloroplast lamellae aggregate to form cylindrical grana which are connected by intergranal lamellae, the frets. The photochemical reactions of photosynthesis are membrane-bound; the enzymatic reactions of photosynthesis are located in the stroma.

9. In addition to chloroplasts, cells contain leucoplasts, amyloplasts, proplastids, and chromoplasts.

10. Chloroplasts contain fibrils of DNA and give evidence of having an inheritance pattern that is independent of nuclear control.

11. Dictyosomes are stacks of from 3 to 10 flattened cisternae. Each cisterna is surrounded by a peripheral net. Dictyosomes appear to be involved in the synthesis of various cellular products.

Figure 4.21. Development of cell walls during primary growth while stem is elongating. A, cell with cellulose walls; B, cutin may be laid down on the outside of an epidermal cell; C, in another location, suberin may be deposited within the cellulose primary cell wall to form a cork cell. Lignin is deposited within the primary wall and between the primary cell wall and the protoplast; D, during rapid elongation, it is laid down as rings or as a spiral band; E, when elongation stops (except for pits), the lignin forms a complete box—the secondary wall—about the cell. Each specialized cell is derived from its own antecedent meristematic cell and never gives rise to another specialized cell.

12. Microtubules are of indefinite length; they are about 280 A in diameter, and have an internal core about 80 A in diameter. Many functions are ascribed to them.

13. Microbodies are all bounded by a single membrane. They are variable in size and in morphology. They are closely associated with various types of intracellular enzyme activities.

14. Crystals, probably largely of protein, occur within the cytoplasm and cytoplasmic organelles. Their function is not clear.

15. The nucleus is bounded by a double envelope provided with pores. The nucleoplasm appears to be characterized by a closely packed array of unit fibrils about 225 A in diameter and of indefinite length. While cells may live and even differentiate for a short time without a nucleus, a nucleus is required for the continued life of a cell and for cell division.

16. Not all parts of the cell are located within the cytoplasm. Vacuoles are aqueous solutions within the protoplast, separated from the cytoplasm by the tonoplast. Inorganic crystals are located in vacuoles.

17. The cell wall, which bounds the protoplast, is formed of cellulose fibrils embedded in an amorphous matrix. Other compounds (suberin, cutin, and lignin) may become a part of it.

18. The type of cell just summarized is highly compartmentalized. It is known as a eukaryotic cell. More primitive cells are not compartmentalized; DNA, photosynthetic processes, and respiratory activity all share a common ground cytoplasm. These are prokaryotic cells.

5

Chemistry and Physiology of the Cell

CHEMISTRY OF THE CYTOPLASM

Plants receive radiant energy from the sun, and convert it into chemical energy in the form of food, from which it is released to be used in the multiple activities of the living cell. The capture of radiant energy, its transformation into chemical energy, its storage, and its release with the resulting growth, duplication and differentiation of the cell, occur within the protoplasm. The astonishing properties of the protoplasm in which all this occurs are inherent in its structure. The structure of protoplasm may be studied morphologically under microscopes or chemically in test tubes. Significantly, since the advent of the electron microscope, the distinction between chemistry and morphology is becoming less and less clear. In other words, the chemical elements (Table 5.1) group together in a little-understood manner to give the visible structure of protoplasm and the intricate seriation of energy transformations that occur within it. Therefore, in discussing the activities of protoplasm, we must first take into consideration the arrangements of atoms and of molecules that are at the bases of its chemistry and morphology.

We have within the living cell a wide variety of inorganic molecules, as well as carbon-containing molecules, making up the cell wall, protoplasm, and vacuolar materials. Although a discussion of all these compounds would involve a whole course in plant chemistry, we should have a general understanding of the diversity of these chemical compounds in the cell and of the major roles that the more important compounds play in the life of the cell. In general, we may divide the chemicals in a plant into two groups: (1) inorganic (substances taken into the cell from the outside) and (2) organic (carbon-containing compounds synthesized by the living cell from inorganic raw materials).

Inorganic Ions and Molecules in the Cell

The most abundant inorganic compound in the cell is water, which permeates the cell wall, makes up more than 90 percent of the fresh weight of protoplasm, and is the chief constituent of the cell vacuole. Many inorganic mineral salts are essential for the growth of plant cells (see page 213) and

Table 5.1 *Some of the common chemical elements found in plants*

ELEMENT	SYMBOL	APPROX-IMATE ATOMIC WEIGHT	COMMON VALENCE NUMBERS	COMMON IONS IN WHICH ELEMENTS ARE FOUND
Elements Essential for Plant Growth				
Carbon	C	12	$-4, +4$	CO_3^{--}, HCO_3^{---}
Hydrogen	H	1	$+1$	H^+, H_3O^+, OH^-
Oxygen	O	16	-2	OH^-, NO_3^-
Phosphorus	P	31	-5	$H_2PO_4^-$, HPO_4^{--}, PO_4^{---}
Potassium	K	39.1	$+1$	K^+
Nitrogen	N	14	$-3, +5$	NH_4^+, NO_3^-
Sulfur	S	32.1	$+6, -2$	SO_4^{--}
Calcium	Ca	40.1	$+2$	Ca^{++}
Iron	Fe	55.9	$+2, +3$	Fe^{++}, Fe^{+++}
Magnesium	Mg	24.3	$+2$	Mg^{++}
Boron	B	10.8	$+3$	BO_3^{---}
Zinc	Zn	65.4	$+2$	Zn^{++}
Manganese	Mn	54.9	$+2, +3$	Mn^{++}
Molybdenum	Mo	96	$+3, +5$	Mo^{+++}, MoO_2^+
Chlorine	Cl	35.5	-1	Cl^-
Copper	Cu	63.6	$+1, +2$	Cu^+, Cu^{++}
Some Other Elements				
Sodium	Na	23	$+1$	Na^+
Silicon	Si	28.1	$+4$	SiO_4^{---}
Aluminum	Al	27	$+3$	Al^{+++}

are obtained by the plant from its external environment, particularly from the soil or water in which it is growing. Some of these inorganic salts are found in the cell in low amounts or for a short time in an uncombined condition, since they serve as raw materials which, together with carbon dioxide and water, are used to synthesize the many complex organic materials in the cell. Other inorganic materials in the cell, such as potassium, may accumulate to relatively high levels as inorganic **ions** (charged atoms or groups of atoms) (page 181).

Organic Compounds in the Cell

Most of the dry weight of a plant cell is organic matter synthesized from inorganic raw materials by a cell itself or by other cells in the plant and moved to that cell. A wide variety of organic compounds from relatively simple sugars to extremely complex

proteins occurs in the cell. Some of these molecules are made up of hundreds of atoms. Organic compounds make up the food for the cell, the structural components of the wall and the protoplasm, and many of the more special compounds such as hormones, pigments, and enzymes. We will look at some of the principal types of organic compounds found in the cell.

Carbohydrates

Carbohydrates are organic compounds containing carbon, hydrogen, and oxygen atoms with the general formula $(CH_2O)_n$, where n may be any number. They make up the major part of the cell wall and are, in addition, one of the three general types of **food.** The major structural carbohydrate in the wall is cellulose whereas the principle carbohydrate foods are sugars and starches. A **food** is any organic material that serves the plant as a source of energy and as a source of carbon from which new compounds may be made. A food also supplies animals with the same source of energy and carbon.

Sugars. Let us consider briefly the structure of a sugar molecule. This will show us how the atoms in one relatively simple organic molecule are arranged. In one group of simple sugars, the **hexoses,** each molecule contains six carbon atoms and has the general formula $C_6H_{12}O_6$. When sugar is in solution, four or five of the carbon atoms (depending on the kind of sugar) and an oxygen atom form a closed ring structure (Fig. 5.1).

It is obvious that the OH and H groups of the sugar molecule may be arranged in different positions in the ring without changing the relative numbers of carbon, hydrogen and oxygen in the formula. In fact, a shifting of these groups results in sugars of different properties. Note that in fructose (Fig. 5.1), the carbon atoms at either end of the molecule lie outside of the ring and the hydrogen and OH (hydroxyl) groups are placed differently from those in glucose.

Actually, sixteen different hexoses are possible and all sixteen different sugars are known, although only a few occur naturally in plants. Chemicals such as these sugars with the same chemical composition but with different chemical properties are called **isomers.** The most common hexoses in plants are glucose and fructose. Sugars with three carbon atoms are called trioses; with four carbon atoms, tetroses; with five and even seven carbon atoms, pentoses and heptuloses. **Ribose,** a constituent of nucleic acids and other complex protoplasmic molecules, is a pentose sugar (Fig. 5.1).

Two molecules of simple sugars, **monosaccharides,** may combine with each other. Thus, if two molecules of glucose are united, with the loss of a water molecule, a **disaccharide** is formed. One such disaccharide is known as **maltose.** The union of the

Straight chain structures of various sugars

Ring structure of glucose and fructose

Figure 5.1. Structural formulas of various kinds of sugars.

monosaccharides, fructose and glucose, results in the commonest of all sugars, **sucrose.** Sucrose can easily be split into fructose and glucose. One water molecule is required to split a molecule of sucrose and the process is called **hydrolysis.** While many different kinds of disaccharides are possible, sucrose, our common table sugar, is by far the most abundant sugar in plants, where it serves as a soluble stored or moveable food. Maltose occurs in a free state only to a limited extent but may be readily obtained by the hydrolysis of starch.

Polysaccharides. Three or more monosaccharide molecules may join to form tri-, tetra-, or **polysaccharides;** the latter are composed of the union of many simple sugar molecules with the loss of a water molecule for each pair of simple sugar molecules united. Polysaccharides are not generally soluble in water, nor are they sweet. **Starch** and **cellulose** are the two most abundant polysaccharides in the plant. Each is composed of a long chain of many glucose molecules. In starch the chain may be coiled because of the way the glucose units are linked together and some chains are branched, while in cellulose the chains are unbranched and more or less straight. Cellulose is a major structural material in the wall, while starch is a reserve water-insoluble food that is stored in many cells. Some cells, instead of storing starch, store a polysaccharide called **inulin** that is composed of fructose units instead of glucose units.

The union of relatively simple molecules, like sugar, into long-chain gigantic molecules composed of the repetition of simple units is a common chemical process known as **polymerization.** We shall meet it again in our discussion of proteins and nucleic acids.

The structural formula of cellulose is given in Fig. 5.2.

Lignin

A cotton fiber is almost pure cellulose. Cellulose alone would not provide a stem with much strength for support. In cells destined for support, a secondary material is often laid down by the protoplast within the cellulose wall; this is lignin, and it constitutes a significant fraction of the secondary wall of many plant cells (Fig. 4.21). Between 20 and 30 percent of the cell walls in wood is lignin. The unit structure in lignin is not a sugar but a complex alcohol. It does not polymerize in straight chains as glucose does in cellulose but, instead, molecules of the alcohol join with each other in several directions to form a tight net.

Fats and Related Substances—Lipids

All living plant cells contain **lipids** (fats and fat-like substances) in their cytoplasm, and in some cells

Figure 5.2. Structural formula of glucose units in cellulose.

Figure 5.3. Structural formula for a fatty acid (lauric acid), glycerol, and a fat.

Side chain | Amino acid

Glycine

Tryptophane

Arginine

Cystine

Figure 5.4. Some amino acids.

long chains of carbon with only hydrogen atoms as side attachments. In lauric acid, the fatty acid shown in Figure 5.3, the carbon chain is composed of twelve carbon atoms.

Glycerol is a rather simple three-carbon-atom compound. Its formula is $C_3H_8O_3$ and its structural configuration is shown in Fig. 5.3. When glycerol unites with three fatty acids, three molecules of water are lost and a fat is produced. Notice that in the fats there are only six oxygen atoms while there may be many hydrogen atoms. This arrangement leaves many spots where oxygen may be introduced when fats are broken down in the cell. Fats are easily oxidized and as foods are very good sources of energy.

Proteins

Proteins are one of the most important classes of compounds found in protoplasm. Proteins are complex nitrogen containing compounds of high molecular weight. Egg albumin and gelatin are good examples of proteins. While there are many kinds of plant proteins, they do not occur in such large masses in plants as they do in animals; hence, plant proteins are not so commonly known.

When proteins are taken apart, it is found that they are all formed of about 25 different units, the **amino acids.** Each amino acid consists of at least two linked carbon atoms, one —NH_2 group, the **amino** group; and one —COOH

group, the **carboxyl** group (Fig. 5.4). There remains, in the example shown, a single hydrogen atom which we shall call the **side chain.** The simplest of these amino acids is **glycine,** shown in Fig. 5.4. All amino acids have the amino group, the carboxyl, and a side chain. The side chain is different in each of the amino acids. Four amino acids, glycine, tryptophane, arginine, and cysteine, are shown in Fig. 5.4, giving some idea of the variations met with in these side chains.

Polymerization of these amino acids takes place in the formation of protein and a long chain is formed in which the side chains of the amino acids

fatty compounds are present in the cell walls. These fats serve as food. Phospholipids (lipids containing phosphorus) and other complex lipids are found in protoplasmic membranes. **Cutin** and **suberin** are lipid materials that function in protecting the plant against excessive water loss. One common property of lipids is their relatively poor solubility in water.

Fats, the most abundant and simplest of plant lipids, are composed of fatty acids united with the three-carbon-atom alcohol, glycerol. Fatty acids have very little oxygen and are generally relatively

1st amino acid   2nd amino acid   3rd amino acid

Figure 5.5. A polypeptide chain is a part of a protein molecule. The backbone of the protein molecule is formed by many amino acids joined by the union of the amino group (NH$_2$) of one amino acid to the acid group (COOH) of another amino acid by the removal of a water molecule. The R groups represent side chains of the different amino acids.

extend outwardly from the backbone of the molecule (Fig. 5.5).

Consider that each of the amino acids is a symbol so that we can form an alphabet with them of more than 25 characters. From the 26 letters of our alphabet, all of the countless words of all of the known languages can be formed. Furthermore, the great majority of these words have no more than five letters and very few of them have more than 20 letters. Even the simplest proteins have 1000 or more amino acids, while the common number of amino acids per protein is of the order of 100,000. To calculate the possible number of different kinds of proteins formed from 100,000 different amino acids at a time in any combination is a job for a modern computer. One of the most exciting fields of modern biology is in the determination of the exact sequence of amino acids in protein molecules.

Proteins are the basic building block of protoplasm. They account for its great specificity and are the reason for our previous statement that the protoplasms of similar cells are always somewhat different.

The long protein molecules can fold, bend and coil in many ways. Particularly important is a folded or helical structure that many protein chains assume. In the cytoplasm of the cell, the long protein molecules tend to react with one another to form weak cross-linkages, and this increases the viscosity of the solution. The streaming of protoplasm in the cell must be viewed as involving the breaking and reforming of many weak cross-linking bonds among the protein molecules. The primary valence bonds in the side chains or backbones of protein molecules remain intact under these conditions of streaming.

Enzymes

While there are many ways in which proteins play a part in the economy of the cell, let us consider for the moment only the important part they take in controlling the chemical reactions of the cell. As an example, we may look at one step in the energy flow in the cells of plants and animals, in which the energy stored in foods is released to perform useful work. Burning gasoline in a flask releases uncontrolled and therefore wasted energy. Burning gasoline in a car also releases energy, but the design of the motor controls the release so that useful work is performed. Still, much of the energy is lost as heat, as the temperature of the motor illustrates. Sugar is burned in the cells of your body and in the cells of a plant, again with the release of energy and the perfomance of useful work, although some energy is lost as heat. A primary function of certain specific proteins is to control and to direct chemical reactions in the cell, limiting them to very specific pathways and bringing them about at relatively low temperatures.

Substances that increase the rate of chemical change without appearing to enter into the reaction are known as **catalysts.** There are both *inorganic catalysts* and *organic catalysts.* Organic catalysts produced by living organisms are called **enzymes.** As far as we know, all enzymes are proteins, although not all proteins are enzymes.

Enzymes are proteins of a complex nature. They are frequently very *specific* in their activity; that is, one enzyme activates only one or two chemical changes. The specificity of enzymes and their ability to speed the rate of reaction are in many instances due to their shape. Because of their shape and activity, enzyme molecules are able to combine with only one or two specific substances. As a result of this combination, the original compound or compounds react with each other or with surrounding substances. The resulting compounds are released

Figure 5.6. Chlorophyll *a* molecule.

occur in the plastids within the cytoplasm of the protoplast. Vacuolar pigments (anthocyanins and anthoxanthins) have already been discussed on page 58. Several kinds of other pigments occur in the plastids. The principal ones in the chloroplasts of higher plants are the green chlorophylls, chlorophyll *a* (Fig. 5.6) and chlorophyll *b*, and the yellow and red carotenoids, the carotenes and the xanthophylls.

We have noted that carotene is an effective protective agent against photo-oxidation. It also is a precursor of vitamin A and probably plays some part in cell nutrition. The carotenes and xanthophylls are responsible for much of the fall leaf coloring and the color of ripe fruits. This fact suggests that they have some functions in senescence.

There is an interesting group of water soluble pigments important in the trapping of light in the blue-green and red algae. These are known as the phycobilins. They are able to trap light of low intensity and pass the trapped energy on to chlorophyll for the normal steps of photosynthesis. The phycobilins are responsible for some of the color of the blue-green and red algae (Fig. 22.20).

Chlorophyll has a molecular structure of moderate complexity. It consists of a head and a tail region. At the center of the head is a single atom of magnesium which is surrounded by four complex rings of carbon and nitrogen. The fatty tail is a long carbon chain flanked by atoms of hydrogen. The structural formula of chlorophyll *a* is shown in Fig. 5.6.

Other very important pigments occur in plant cells but are inconspicuous. One such pigment is **phytochrome.** It is found in extremely small amounts in leaves and is involved in a number of physiological processes including the flowering of plants. Other pigments that occur in minute amounts in cells are riboflavin, which is important in the bending response of plants to light and the cytochromes, yellow pigments that are found in the cytoplasm and are concerned with certain steps in the processes of respiration and photosynthesis (page 241).

from the enzyme molecule, and the enzyme is now ready for further activity. Thus, an enzyme greatly speeds the rate of a reaction without being much changed itself by the reaction.

Enzymes have certain optimum conditions for their greatest activity. They may be destroyed by heat and by some metals and certain other substances. Many of the numerous steps involved in photosynthesis, respiration, and digestion are catalyzed by enzymes. Enzymes are active in practically all cellular processes.

Protoplasm may be thought of as a very complex colloidal system in which molecules of one sort are being constantly changed into molecules of another sort. Green plant cells store energy in one process and release it in another; both processes may go on within any green cell at the same time. Enzymes help to control the precision and the rate of the changes taking place in the protoplasm.

Pigments in the Cell

Pigments are colored compounds, and since they are colored they must somehow absorb certain wavelengths of light (Fig. 13.6). Plant pigments may be divided for convenience into water-soluble pigments that are usually found in the vacuoles of the protoplasts, and lipid-soluble pigments that

Hormones

It is a common observation that if the tip of a plant is pinched or cut off, new shoots develop below the injured area. Why do the buds below the tips grow only after the tip is removed? Are they stimulated in some way by the injury? Does the presence of the tip prevent growth? It can be shown that a chemical produced in very small amounts in the growing shoot tip moves down the stem and prevents the growth of the lateral buds. When the tip is removed, the inhibitory chemical decreases in concentration in the stem and the lateral buds begin to grow.

Thus it can be demonstrated that plants, like animals, produce chemical messengers that move in very low concentrations from one part of the plant to another and bring about many important changes in growth and development. Such naturally produced growth regulators are called **hormones.** The hormone that is involved in the dominance of the apical bud over the lateral buds, as described above, belongs to a class of plant growth regulators called **auxins.** Auxins were the first plant hormones to be studied, isolated, and chemically identified. The most common natural auxin is a complex organic compound known as indole acetic acid or **IAA.** Indole acetic acid in very low concentrations has a wide range of effects on plant growth and development (page 404) and is very important because it influences cell elongation.

Figure 5.7 shows that the indole acetic acid molecule is made up of two rings and an acid side chain. Although one of the rings contains nitrogen, other compounds that have a structure similar, in some respects, to the indole acetic acid molecule, but which lack nitrogen, are also effective growth regulators. Chemists are able to synthesize artifical growth regulators, such as naphthelene acetic acid (Fig. 5.8), which in many ways act like indole acetic acid in affecting plant growth and development.

Soon after auxins were discovered, it became apparent that both plant growth and development were regulated not by one hormone but by several, and that a shift in the balance of hormones in a cell could profoundly affect the way in which the cell would develop and grow. Two other classes of hormones that have been studied extensively are **gibberellins** and **phytokinins.**

Gibberellins, like auxins, affect cell elongation (page 405). However, gibberellins differ from auxins not only in their chemical nature but also in many of the other effects that they have on plant growth. Figure 5.9 shows the complex ring structure of a molecule of gibberellic acid. A number of other gibberellins have been identified in various

Figure 5.7. The molecule of the auxin, indole acetic acid (IAA), is made up of two rings and a side chain.

Figure 5.8. Naphthelene acetic acid is a synthetic growth regulator.

Figure 5.9. A molecule of gibberellic acid.

Figure 5.10. A molecule of kinetin.

plants and have molecular structures similar to gibberellic acid.

Not only are the elongation and differentiation of plant cells under hormonal influence and control, but cell division itself is also regulated by hormones. Phytokinins are particularly important in their influence on nuclear division; interestingly, they are chemically related to nucleic acid. The first phytokinin to be studied and identified was not a natural plant hormone but a chemical that, almost by chance, was found to be effective in regulating the growth and development of cells in a tissue culture. When small amounts of this chemical were added to a tobacco tissue culture, cell division was stimulated. The chemical was called **kinetin** (Fig. 5.10). A number of related compounds have been shown to be active in promoting cell division. Phytokinins have been shown to affect other growth phenomena such as cell elongation and development of buds.

In addition to the compounds that we have discussed, many other important substances are found in the cytoplasm, nucleus, and vacuoles of the cell. The chemistry of the nucleus will be considered in chapter 6.

THE PHYSIOLOGY OF THE CELL

We have studied the structure and some of the chemistry of the individual plant cell and we have seen that most plants are composed of countless numbers of these cells. A knowledge of the structure of a cell, a tissue, or an organ is essential to an understanding of their activities or functions. An automobile is composed of several thousand parts, each of which has a particular composition, shape, and location upon which depends its special role in the proper operation of the machine. The human eye is a complex organ in which the many individual parts are "put together" in such a way that the function of sight is possible. Likewise, the cell is a most complex structure–a *living structure*. It is the structural and functional unit of the plant. It has work to do; that is, it has functions or activities. **Physiology** is the study of the functions or activities of cells or organisms.

Our study of the physiology of cells will be simplified if we consider a chlorophyll-containing plant such as a unicellular alga (Fig. 22.4) floating freely in water. This one-celled plant carries on all the functions that are necessary to maintain its life.

Within this microscopic bit of living substance a number of physical and chemical processes are occurring simultaneously. Each cell is (1) absorbing materials such as water, mineral salts, and gases from outside its own body, and simultaneously losing materials to the external environment; (2) building foods from the materials and light energy that it absorbs; (3) digesting foods (changing complex foods to a simple form); (4) respiring, thus releasing energy for various activities; (5) building protoplasm from the foods; (6) growing; and (7) producing new cells.

We shall briefly discuss each of these physiological processes. However, it is not possible to elaborate on several of them without a knowledge of the structure of the plant as a whole. Therefore, the following discussion merely defines some of these processes, reserving a more complete discussion for later chapters.

Since the life activities of all cells are carried out in a water environment and often involve the movement of solutes throughout the individual cell, between cells, and between the cell and its external environment, it is appropriate that we first consider how a cell obtains water and solutes. Why does water move into some cells and out of others? Can the cell regulate the flow of water through it? Does the cell select the solutes that it absorbs? How does the structure of the cell influence water or salt absorption? Is energy necessary for either process? The answers to these and similar questions lie in the ultimate relation between cellular structure and function.

Permeability of Plant Membranes

Permeable Membranes

Let us now consider the green plant floating in water. Bathing the cell on all sides is a dilute but exceedingly complex solution of molecules and ions that are constantly bombarding the outer surface of the cell, the cell wall. This wall, though apparently continuous when viewed through a microscope, is in reality quite porous, with **microcapillary spaces** existing between the interwoven cellulose **microfibrils** (Fig. 5.11). Many of these spaces are filled with substances such as pectates or lignin, but many are filled instead with liquid. There is a strong attraction between the insoluble cell wall materials and water. Consequently, water

Figure 5.11. Electron micrograph of the cell wall of a cotton fiber. ×60,000. (Courtesy of P. A. Roelofsen.)

is found throughout the wall, forming a continuous pathway through which solute particles can freely diffuse. The process in which water is attracted to **(adsorbed on)** the surfaces of the cellulose microfibrils, causing them to move apart and the wall to swell, is called **imbibition.** Cellulose imbibes water and the cell wall thus allows the relatively free passage of water and dissolved materials. The cell wall is **permeable.** Cell walls found in the outer surface of higher plants may be impregnated with fatty materials such as cutin and suberin (Fig. 4.21). These walls do not imbibe water, and are impermeable to water and dissolved substances, though they may be permeable to substances soluble in fat.

Differentially Permeable Membranes

Lying in intimate contact with the wall of a living cell is the outer cytoplasmic or plasma membrane, the **plasmalemma,** which is the first important bar-rier to the free passage of the molecules and ions that bombard the cell from the outside (Figs. 4.3, 4.5). This limiting surface of the protoplast is characterized by a different physical and chemical composition from the rest of the protoplasm. It is not a rigid, unchanging surface but an ever-changing dynamic "guardian of life," for if it is destroyed, the cell dies. Although the exact nature of this membrane is not definitely established, it is thought to be a **unit membrane** (Figs. 4.5, 4.6), through which water can readily pass but which restricts or prevents the passage of many dissolved materials. It is **differentially permeable,** allowing some substances to pass freely, others to pass slowly, and still others to pass hardly at all. Although the cell cannot select those substances which are beneficial to it and prevent from entering those which are harmful or toxic, the presence of the cytoplasmic membrane does prevent the *free* diffusion of materials into and out of the cell.

The inner cytoplasmic membrane, the **tonoplast,** surrounding the vacuole, is also differentially permeable and is a unit membrane. It is more fatty in nature than the plasmalemma and consequently, certain substances diffuse through the cell wall and the plasmalemma into the cytoplasm but do not readily find their way into the vacuole. As we have already seen, the vacuolar sap frequently contains high concentrations of plant products (such as anthocyanins) that are not found in the cytoplasm. Destruction of the cytoplasmic membranes—by immersing the cell in alcohol, for instance—renders the membranes permeable, allowing the outward diffusion of vacuolar materials, and results in the death of the cell.

Within the cell itself, solutes are not free to diffuse at random throughout the protoplast. In addition to the tonoplast which restricts the movement of materials between the cytoplasm and the vacuole, there are the nuclear and plastid membranes, which restrict the free diffusion of materials into and out of these bodies. Moreover, the colloidal nature of the cytoplasm itself presents a delicate but ever-changing architecture within which the foods, vita-

mins, hormones, enzymes, and inorganic ions are so distributed that the intricate yet precise processes of life proceed.

Diffusion through Differentially Permeable Membranes

Nonliving Membranes

The differentially permeable membranes of the cell play very important roles in the maintenance of life activities. Certain features of their behavior can best be explained by reference to the simple apparatus called an **osmometer** (Fig. 5.12). The cylindrical sac of an osmometer is made of parchment paper, plastic, or other material that is differentially permeable. This sac is filled with a concentrated sugar solution and stoppered with a tight-fitting rubber stopper through which a glass tube is fitted; the sac is then immersed in distilled water (pure solvent). In following this discussion we should keep in mind that the sap of a living algal cell is usually more concentrated in total solutes (salts, sugars, and the like) than is the pond water in which it may be floating. The movement of water

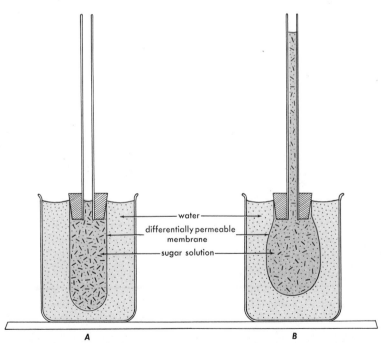

water
differentially permeable membrane
sugar solution

A   B

Figure 5.12. Diagram of osmometer and diffusion through a differentially permeable membrane. A, before osmosis has occurred; B, after osmosis.

into the plant cell obeys the same laws as those governing the movement of water into an osmometer.

Turgor Pressure

Outside of the osmometer are water molecules, and inside are water molecules and sugar molecules. The membrane, being differentially permeable, permits the free movement of water molecules inward or outward and retards or prevents the free movement of sugar molecules outward. After a short time the liquid rises in the tube, and the sac becomes distended, or **turgid.** The pressure of the solution against the sac, the **turgor pressure,** is a measurable pressure opposed by an equal and opposite pressure of the wall of the sac against its contents (the **wall pressure**). Evidently water molecules are moving inward more rapidly than outward. Under these conditions the **activity (chemical potential)** or **"free energy"** of water molecules outside the sac is greater than the activity or "free energy" of water molecules in the sac; thus, water is driven into the sac. As water continues to diffuse into the sac, turgor pressure is built up inside it. This pressure increases the activity of water molecules in the sac, and the tendency for water to diffuse back out of the sac gradually increases.

This outward diffusion of water finally equals the inward diffusion. At this point there is no further net movement of water into the sac, and the system is at equilibrium. Water molecules are passing through the membrane in both directions at the same rate. The activities of water on both sides of the sac are now equal. An important conclusion to be drawn from this demonstration is that (1) the activity of water is reduced by the presence of solute particles that reduce the concentration of the water, and (2) the activity of water is increased by the development of a physical turgor pressure in the solution. These two factors, solutes and turgor pressure, thus have opposite effects on the activity of water. In our example, equilibrium was attained when these two factors balanced each other and water within the sac had the same activity as the pure water outside.

The absolute value for the activity or chemical potential of water in a cell or in a solution is not easily measured. *Differences* in activity, however, can be determined, and pure water is the standard reference by which these differences can be conveniently measured. This difference between the activity of water molecules in pure water at atmospheric pressure (i.e., in an open container) and the activity of water molecules in any other system (i.e., water in a cell, in a solution, or in the sac) may be called the **water potential** of the cell or solution. Pure water has a water potential of zero under standard conditions of pressure and temperature. Since pure water is used as a standard and the activity of water in a cell or a solution is usually less than that of pure water, the water potential of a cell or a solution is usually a negative number.

The activity of water molecules, and thus the water potential in a system, is affected by several factors. The activity is decreased by the addition of solutes; it is increased by the physical pressure of cell walls on the aqueous cell contents and also by an increase in temperature. Water will diffuse from regions of high water potential to regions of lower water potential (i.e., from a dilute solution toward a more concentrated solution), when solute concentration is the only factor affecting the water potential. The water potential may be regarded as the driving force that causes a net movement of water in any system.

Osmosis

The diffusion of water (a solvent) through a differentially permeable membrane is called **osmosis.** Osmosis is not mysterious but is simply a special example of diffusion in which the differentially permeable membrane of a sac, or of a cell, separates the internal water that has a given water potential from the water of the surrounding environment that has a different water potential. Osmosis occurs (i.e., water diffuses across the membrane) along the water potential gradient from a region of higher to a region of lower potential of water. The cell gains or loses water until the potentials of water on both sides of the membrane are equal.

Water movement into living cells is usually controlled by two major factors that affect the water potential and hence the direction of diffusion of water. These are (1) the presence of solute particles in the water and (2) the existence of a turgor pressure. Other factors, such as a change in temperature or the presence of water-attracting materials like some colloids, also affect the potential of water in a cell. As we have seen, the presence of solute particles (i.e., molecules or ions) lowers the activity of

Figure 5.13. Diagram showing the relationship of one atmosphere pressure to the height of mercury and water columns that it will balance.

The osmotic pressure of a solution is usually expressed in units of atmospheric pressure. One atmosphere of pressure is approximately 15 pounds per square inch (Fig. 5.13). The osmotic pressure of a solution is not an actual physical pressure that exists in the solution. We might say that it is a power rating of the solution, just as an automobile engine has a rating of 150 horsepower. We do not mean that when the motor is not operating it is doing the work of 150 horses. It is only when the proper conditions exist (i.e., when the motor is operating), that the ability to do work—the power—actually results in the work being done. A solution having an osmotic pressure of five atmospheres and standing in a beaker has the capacity to develop a turgor pressure of five atmospheres only when placed in an ideal osmometer with pure water outside. This turgor pressure will develop because the water potential of pure water is five atmospheres higher than the water potential in the solution. If, however, the solution were placed in an osmometer and immersed in another solution having an osmotic pressure of three atmospheres, osmosis would occur only until a turgor pressure of two atmospheres had developed in the osmometer, because in this case the water potential in the solution outside the sac would only be two atmospheres higher than the water potential inside the sac.

Living Membranes

Water moves along a gradient from the external environment into the plant cell—whether it is the single cell of an alga, a root cell of a flowering plant, or a seed—along a gradient of water potential. Usually, the concentration of the osmotically active solutes (the osmotic pressure of the cell sap) and the turgor pressure in the cell largely determine the water potential in the cell. In certain instances, such as air-dried seeds, there is present a high percentage of colloidal material such as starch, protein, and cellulose, all of which have a great affinity for water and hence a large capacity for imbibing water. When these seeds are planted in moist soil, the water in the seeds has a very low potential compared with the water in the soil. Consequently, water diffuses into these seeds, causing them to swell. In this case, the colloidal materials largely contribute to the low activity of water in the seed.

Colloidal soil particles also lower soil water potential. The activity of water in a pond is pri-

the water in which they are dissolved. This effect on the water potential is proportional to the total number of dissolved particles (molecules and ions) in a given volume of water. Ten grams of sugar dissolved in 50 ml of water will lower the water potential approximately twice as much as will five grams of sugar dissolved in 50 ml of water.

This reduction of the activity of water caused by the presence of dissolved particles in it can be measured: if this solution is placed in an osmometer with pure water outside, osmosis will occur and turgor pressure will develop as long as there is any difference in the potential of water across the membrane. The theoretical maximum turgor pressure that could develop in this system equals the amount that the water potential in the solution is reduced by the solute particles in it.

Osmotic Pressure

The term **osmotic pressure,** sometimes called osmotic concentration or osmotic potential, is used to express these two properties of solutions, namely (1) the amount of reduction of the water potential in the solution caused by the presence of dissolved particles, and (2) the potential maximum turgor pressure that could develop as a result of osmosis.

Figure 5.14. Plasmolysis. *A,* photomicrograph of cells of garden beet in which a red pigment (anthocyanin) is dissolved in the cell sap; *B,* photomicrograph of leaf cells containing chloroplasts. Space between protoplast and wall occupied by plasmolyzing solution. ×640. (Courtesy of H. B. Currier.)

marily determined by the concentration of dissolved material, but in the case of water in the soil an additional factor is involved, namely, the presence of colloidal soil particles that imbibe water and thus lower its activity. Nevertheless, the direction of water movement will be from a region where the water potential is high (near zero) to a region where the water potential is lower (more negative).

We have seen that turgor pressures result because of differences in solute concentrations. Although the pressure outward is often considerable, the protoplast is prevented from bursting by the elastic cell wall, which resists stretching and exerts a pressure inward. The cell wall serves the same purpose that the casing of an automobile tire does in preventing the bursting of the inner tube.

It is obvious that in a cell the *turgor pressure outward* against the wall is equal to the *wall pressure inward* against the cell contents. A cell may have varying degrees of turgidity. The maximum turgor pressure that can be developed in a cell is equal to the osmotic pressure of the cell contents. That is, when a cell has absorbed all the water it can, osmotic pressure, turgor pressure, and wall pressure are equal. Normally, most living cells are in a condition in which turgor pressure is somewhat less than the maximum pressure possible. This means they are

capable of taking in more water by osmosis. However, if they lose water, cells may become *flaccid,* and have a very low turgor pressure.

The crispness of the leaves and the rigidity of the young parts of plants are due to the turgid condition of the individual cells. A young bean seedling stands erect chiefly because of the turgor pressure in all the cells of the stem. In such a young plant, strengthening tissue is not plentiful. If the cells of the seedling lose water rapidly, the whole plant becomes flaccid and droops, but it may recover if water is resupplied.

Plasmolysis

If a cell or group of cells is immersed in a solution that has a higher solute concentration than that of the cell sap, water diffuses outward, and the turgor pressure in the cell is reduced. The volume of the cell decreases somewhat, but more striking is the withdrawal of the protoplast from the cell wall and the decrease in the size of the vacuole. This phenomenon is called **plasmolysis.** As shown in Fig. 5.14, the space between the cytoplasm and the cell wall in a plasmolyzed cell is filled with the plasmolyzing solution in which the cells are immersed. If plasmolyzed cells are immersed in water or in a solution whose concentration is less than that of the cell sap, the cells regain their turgor; water molecules diffuse inward. If cells remain long in a state of pronounced plasmolysis, however, death ensues. A normal, healthy, and functioning cell is one in a turgid condition.

It is instructive to plasmolyze living cells that contain a pigment dissolved in the vacuolar sap. For this purpose, cells from the root of a red garden beet may be used. When they are immersed in a strong solution, such as one of table salt, plasmolysis soon follows. It will be seen, however, that the red pigment is retained in the vacuole (Fig. 5.14). Obviously, the tonoplast is impermeable to this pigment. If the cells are heated or are treated with various chemicals, such as chloroform, alcohol, or ether, the red pigment readily diffuses from the protoplast. Cytoplasmic membranes manifest differential permeability *only when the cell is alive;* when the cell is killed, cytoplasmic membranes lose their differential permeability. As we shall see later on, however, many nonliving, as well as living, membranes have differentially permeable properties.

A cell may die if plasmolysis is pronounced and prolonged. For example, if heavy applications of ordinary salt are placed on the soil where weeds are growing, so that the root cells are surrounded by a solution of high concentration, water diffuses from the cells, and they become severely plasmolyzed. If this state is prolonged, the roots die. The salt killed, not because it was toxic to root cells, but rather because severe plasmolysis was brought about by the high concentration of the soil solution. In parts of the western United States where rainfall is low and the evaporation rate high, salts of the soil may accumulate on the surface and form what are known as "alkali flats." In such soils, the concentration of the soil solution may be so high that ordinary crop plants cannot grow; only those species that are especially adapted to tolerate high salt concentration are able to survive. Plants of this type are known as **halophytes.**

Absorption of Dissolved Substances by Living Cells

In our discussion of absorption thus far, we have explained how water enters the living cell. In addition to water, the cell absorbs various inorganic ions such as potassium, calcium, nitrates, phosphates, and sulfates; and the gases, oxygen and carbon dioxide, both of which are soluble in water. Apparently they enter and leave the cell by simple diffusion.

Water and the different solute particles move into the cell *independently.* That is, if water molecules are diffusing through the cell wall and the cytoplasmic membranes and are entering the vacuole at a certain rate, it does not follow that any particular solute particle is entering the vacuole at the same rate. Moreover, in the event that the solution surrounding the cell contains $NaNO_3$, the entering of NO_3^- ions into the cell, does not necessarily mean that Na^+ ions are moving in at the same rate. Although loss of water in vapor form from a plant does affect the proportions of water and solutes absorbed, it appears that the kind and quantity of ions absorbed are chiefly determined not by the volume of water absorbed but by certain chemical and physical properties of root cells. Different solute particles diffuse independently of each other, and the direction of diffusion of any solute is independent of the direction of the diffusion of any other solute that may be a part of the same solution.

It has been found experimentally that the concentration of a solute particle (ion or molecule) may be greater in the vacuole than in the solution outside the cell (Table 5.2). How, then, do we harmonize *simple diffusion* with that in which solute particles move from a place where they are in low concentration to a place where they are in high concentration? It has been demonstrated that the **accumulation** of solute particles by plant cells is usually attended by high respiration rates. It may be assumed that the energy released by respiration is utilized by the cell to perform the labor of *forcing* the solute particles to move *against a concentration gradient,* and this same energy maintains the concentration difference. Thus the living cell performs work, the energy for which is derived from respiration. Ions that are accumulated by root cells are apparently "pumped" into the vacuole of the cell from the soil solution.

Summary of Absorption

1. All substances that enter the cell, gases included, are in solution in water.

2. The cell wall is normally permeable to all substances in true solution.

3. The cytoplasmic membranes are differentially permeable.

Table 5.2 *The concentration of various ions in the vacuolar sap of* Nitella clovata *and in the pond water in which it was growing* (*data from Hoagland and Davis, Protoplasma, 1929*)

ION	CELL SAP CONCENTRATION IN MILLEQUIVALENTS PER LITER	POND WATER CONCENTRATION IN MILLEQUIVALENTS PER LITER	ACCUMULATION RATIO, SAP CONCENTRATION EXTERNAL CONCENTRATION
Ca^{++}	1.3	1.3	10.0
Mg^{++}	10.3	3.0	3.6
Na^+	49.9	1.2	41.6
K^+	49.3	0.51	96.7
Cl^-	101.1	1.0	101.0
SO_4^{--}	13.0	0.67	19.0
$H_2PO_4^{---}$	1.7	0.008	212.5

4. Cell colloids imbibe water and swell.

5. Osmosis is the diffusion of water or other mobile molecules through a differentially permeable membrane. Water diffuses along a gradient of water potential.

6. Any particular solute may accumulate in the cell and thus maintain in the vacuole a higher concentration of that substance than exists outside of the cell. To accomplish and maintain this higher concentration, the cell must perform work. The energy for this work is furnished by respiration.

7. The diffusion of water and different solutes is independent.

8. Solute particles decrease the activity of water molecules; the solution then has a low water potential.

9. When water diffuses into a cell, the internal turgor pressure increases.

10. Turgor pressure increases the potential of water.

11. Water enters the cell when the water potential in the cell sap is less than that of the water in the solution outside the cell. Water leaves the cell when the water potential in the cell exceeds that in the solution outside the cell.

12. Cells immersed in a solution of high concentration become plasmolyzed. Water diffuses out of the protoplast and the protoplast withdraws from the wall.

METABOLISM

The sum total of all of the chemical reactions that go on in the plant body is called **metabolism.** Food manufacture, formation of cell walls, and synthesis of protein all involve many chemical reactions and result in the production of new materials. Such synthetic metabolic reactions, **anabolic reactions,** contrast with the breakdown reactions such as are involved in respiration or the oxidation of foods. The phase of metabolism that involves breakdown or destructive reactions is known as **catabolism.** Both aspects of metabolism are essential for normal cell activities, since anabolic reactions are necessary for the continued production of new materials and catabolic reactions release, in an available form, energy stored in foods.

Photosynthesis

Every day the combined oil wells of the world deliver some 2 billion barrels of oil for running the cars, boats, trains, and factories of our highly industrialized civilization. The energy this oil so conveniently stores came originally from the sun. It was "captured" by plants, and because of certain geological conditions was trapped and stored in the earth's crust for many millions of years. The process by which plants in the past captured, and today continue to capture, the radiant energy of the sun is known as **photosynthesis,** a term that literally means putting together (*synthesis*) by means of light (*photo*).

The principal features of this physiological process will be mentioned here, but a more complete discussion will be given in Chapter 13. Photosynthesis is a process that goes on only in cells that have *chloroplasts* and only when these cells are *illuminated*. Certain blue-green algae and purple bacteria are exceptional in carrying on photosynthesis in the absence of chloroplasts, but chlorophyll bound to membranes is present.

In photosynthesis, the simple compounds *water* and *carbon dioxide* are united to form *sugars* and *oxygen*. Glucose is one of the principal sugars produced during photosynthesis.

The oxygen may go into solution in the cell sap or diffuse out of the cell or be used in another cellular process, respiration. Carbon dioxide is dissolved in the water in which the plant is floating. As carbon dioxide is taken out of solution and used in photosynthesis, its concentration in the cell sap becomes less than that in the water outside the cell. As a result, diffusion of that gas into the cell goes on as long as active photosynthesis continues. In land plants, green cells obtain carbon dioxide from the atmosphere, but this gas must go into solution in the imbibed water of the cell wall and cytoplasm before it can diffuse to the chloroplasts.

Photosynthesis is an energy-storing process. Energy is required to bring about the synthesis of glucose from carbon dioxide and water. Light is the source of energy for photosynthesis and the glucose molecule contains transformed light energy holding the atoms together. If the glucose molecule is burned, oxygen is consumed and carbon dioxide and water are formed. The heat energy evolved in burning represents the release of the glucose molecule's chemical potential energy, which is equal to the light energy transformed and stored as chemical energy in the photosynthetic process.

Photosynthesis may be simply represented as follows:

$$6CO_2 + 6H_2O + 673 \rightarrow C_6H_{12}O_6 + 6O_2$$
carbon + water + kcal → sugar + oxygen
dioxide

Glucose may be utilized in a number of different ways: (1) broken down in the process of respiration, yielding energy; (2) converted into some closely related carbohydrate, such as sucrose (cane sugar, $C_{12}H_{22}O_{11}$), or cellulose (cell-wall building material), or starch (a reserve food supply); (3) converted into fatty substances; (4) united with the nitrogen, sulfur, and phosphorus derived from various inor-ganic salts absorbed from the water of the soil or the lake or stream in which the plant lives, thus forming proteins; and (5) employed as the chemical basis for a number of other substances that may be found in the cell, such as chlorophyll, or alkaloids, such as nicotine.

The conversion of glucose into other carbohydrates may be simply represented as follows:

$$nC_6H_{12}O_6 \rightarrow n(C_6H_{10}O_5) + nH_2O$$
glucose → carbohydrate + water

Digestion

Complex carbohydrates are formed by union of sugar molecules, fats are composed of fatty acids and glycerine, and proteins are polymers of amino acids. These materials are not only synthesized in cells but may be broken apart within cells and outside of them. Enzymes are involved in these processes of building and degradation. The breaking apart of complex carbohydrates, fats, and proteins into sugars, fatty acids, glycerine, and amino acids occurs easily in the presence of specific enzymes and without the release of large amounts of energy. This process whereby complex organic compounds are broken down into simpler compounds is known as **digestion.** Since water molecules are generally used in the digestion process, the process is chemically one of hydrolysis.

Plants as well as animals **digest** *foods;* that is, they break down complex foods into simple compounds. Plants have no special organs for digestion; it is carried on *in any cell that may store food,* even temporarily. In most green cells, photosynthetic activity may proceed at such a rate during the day that the food (glucose) accumulates faster than it is used in those cells. When this occurs, the glucose may be changed temporarily to starch, which appears as granules in the chloroplasts. Glucose is soluble in the water of cell sap; starch is insoluble. When the time comes for this temporary starch reserve to be used by the cell or transported out of the cell, it must be changed back into sugar. This chemical transformation of the insoluble starch into soluble glucose is one example of digestion. The equation for starch digestion is as follows:

$$(C_6H_{10}O_5)_n + nH_2O \rightarrow nC_6H_{12}O_6$$
starch + water → glucose

Any other foods that are insoluble in the cell sap, such as proteins and fats, must be digested (rendered soluble) before they can diffuse and nourish the cell. Enzymes *facilitate digestion.*

Digestion may occur not only in organs that store large quantities of foods but also in any living cell where food is stored, even in small quantities and temporarily.

Respiration

Many processes taking place in plants require energy. These include: the absorption of mineral salts against an osmotic gradient; the synthesis of complex compounds such as proteins; the maintenance of the protoplasm in a living state; cell division and cell growth; movement; photosynthesis; and translocation. Energy for photosynthesis is supplied by sunlight; energy for the translocation of water comes mainly from the evaporation of water from leaf surfaces. All other processes obtain energy from respiration.

Respiration may be defined as the **oxidation** *of organic substances, with the release of energy, within cells.* In its simplest form, oxidation involves chemical union of some substance with oxygen. However, in its broadest sense, oxidation covers the energy-releasing reactions, in which molecular oxygen itself may not be involved. In the burning of a match, oxygen of the air unites with wood, heat energy is released, and the wood is broken down into carbon dioxide and water. We have seen that photosynthesis is the reverse: carbon dioxide and water unite, oxygen is released, and energy is stored. Photosynthesis represents a type of reaction termed **reduction.** Actually, the two reaction types are much more complicated than expressed here and will be considered in more detail in later chapters.

Sugar is the usual organic substance oxidized in all plant cells. The energy stored in the sugar is, as we have learned, derived from sunlight. Sugar is a **food,** and all foods are energy-rich compounds whose molecules are so constructed that the energy they store may be released with comparative ease.

Respiration is a cellular process. All living cells respire. Although respiration is a complicated chemical process, the overall reaction occurring in cells may be described as follows: 1 gram-molecule of glucose sugar (180 grams) combines with 6 gram-molecules of oxygen (192 grams) to form 6

gram-molecules of water (108 grams) and 6 gram-molecules of carbon dioxide (264 grams), releasing 673 kilocalories of heat. A **kilocalorie** is the amount of heat required to raise the temperature of 1 liter of water (slightly more than a quart) 1° C.

This statement may be simplified as follows:

$$C_6H_{12}O_6 + 6O_2 \rightarrow 6CO_2 + 6H_2O + 673$$

sugar + oxy- → carbon + water + kcal
gen   dioxide

Energy Flow

When sugar is burned, carbon bonds are broken and heat is generated. The heat may be used to do work. If it were possible to use the rearrangement of the atoms of the sugar molecules directly to do work, little heat would be generated, and our machine would be highly efficient. This is exactly what happens in living cells during the process of respiration; reactions that release energy are **coupled** with reactions that absorb energy. Molecules are rearranged, new materials are formed, and work is done. These **coupled reactions** are key reactions in biosynthetic processes.

When carbon bonds are broken in the cell during the oxidative reactions of respiration, small amounts of energy may be made available for work. Some phosphate bonds, on the other hand, are able to store large amounts of energy, and when broken release it easily to drive other reactions along in **coupled processes.** Two most important compounds associated with energy transfer in cells are **adenosine diphosphate** or **ADP** and **adenosine triphosphate, ATP.** There is also an **adenosine monophosphate, AMP,** with only one phosphate group on the molecule. When a second phosphate is added to the single phosphate of AMP in the formation of ADP, a bond is formed that concentrates much more energy than is present in the phosphate bond of AMP. When a third phosphate is added to form ATP, an additional large amount of energy is concentrated in this bond. These two bonds that hold the two terminal phosphate groups of the ATP molecule are known as **high energy bonds** and are designated with the symbol $\sim P$. These bonds may be easily broken by hydrolysis, and each then releases about 8000 calories of energy per mole of ATP hydrolyzed. When the carbon oxygen bonds in a disaccharide or in a fat molecule are hydro-

lyzed, only about 1000 to 2000 calories are released.

ATP is the principal compound that appears to be universally formed during respiration in the cells of microorganisms, plants, and animals, and in which the energy from the oxidized food is temporarily stored. ATP has been called the energy currency of the cell. It may be formed from ADP and inorganic phosphate in the mitochondria during respiration, or it may be formed in the chloroplasts of green cells exposed to light during photosynthesis. From its site of synthesis, ATP may move to other locations where it gives up its energy in the performance of cellular work and it is converted to ADP and inorganic phosphate. ADP may now migrate back to the mitochondria where, in cooperation with respiration, it is again combined with inorganic phosphate to form ATP and the process is repeated.

Let us see how all of these topics are related. Our problem is to follow the radiant energy of the sun as it is captured by the cells of green plants, stored in the carbohydrate molecules, and finally used to synthesize animal and plant protoplasm which in turn utilize the energy of the carbohydrate molecule in various ways, ranging from simple leaf movements to the intellectual activities of man. Light may be thought of as small packages of energy, **photons** (page 242). The absorption of photons (light energy) by chlorophyll is thought to start a flow of electrons (electrochemical energy) through a series of compounds that are found in the chloroplast. This flow of electrons can be compared to the flow of electrons in a wire as an electric current passes through it. Just as work may be done by the passage of the current in the wire, so also work can be done as electrons are transported from one substance to another in the chloroplast under the influence of light energy. In the chloroplast, ATP is formed during this electron flow, but the energy that it temporarily holds is rapidly used during the synthesis of sugar. ATP is an excellent means of transferring cellular energy, but it is not a particularly good means of storing energy, since only a small amount of ATP is formed in the cell at any one time. The storage of chemical energy is a function of the sugar and other foods. Stored energy is unused energy; to be utilized it must be released in the cell, and this is done, as we have seen, during the process of respiration where there is again a flow of electrons from the food, this time to oxygen.

During this electron flow, some of the energy is again trapped as ATP is formed. The ATP may be transferred throughout the cell and its energy used to perform work such as the synthesis of more protoplasm and many other cellular constituents.

Assimilation

We have seen how the green plant manufactures glucose from carbon dioxide and water. This sugar may be transformed into related carbohydrates, such as sucrose or starch; or it may be changed to fats; or, with the addition of chemical elements (particularly nitrogen, sulfur, and phosphorus) from the soil, it may be used to form proteins. Carbohydrates, fats, proteins and simpler nitrogenous substances are the **foods** of plants, as well as of animals. They are *organic compounds.* The foods are manufactured from *raw materials,* such as water, carbon dioxide, nitrates, phosphates, and sulfates; these raw materials are *inorganic compounds.* The cell takes the foods (nonliving substances) and from them builds protoplasm (living substance). *The conversion of foods into the living material (protoplasm) of the cell is called* **assimilation.** The chemical changes involved in this conversion are not understood, but energy is probably used in the process. Except in a very general way, the chemical nature of protoplasm itself is unknown.

Animal cells and non-green plant cells can also convert foods into protoplasm. Also, any plant cell, with or without chlorophyll, can synthesize fats and proteins and can change glucose into starch and other carbohydrates. But only chlorophyll-bearing cells can manufacture glucose from carbon dioxide and water.

The steps leading up to the building of protoplasm in the free-floating water plant, or in any green cell, may be summarized as shown in the diagram on the following page.

Control of Metabolic Reactions

Although protoplasm is complex, the wide variety of chemical reactions which occur in it simultaneously during the life of the cell are not haphazard. There are intricate spatial and time interrelationships among them. The regulated sequential nature of cellular reactions is one of the

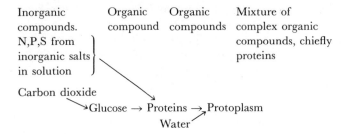

Inorganic compounds. N,P,S from inorganic salts in solution

Carbon dioxide

Organic compound

Organic compounds

Mixture of complex organic compounds, chiefly proteins

Glucose → Proteins → Protoplasm
Water

attributes of life. How are these reactions so precisely regulated and controlled in such a minute space as the volume of a cell? From the point of view of the size of an enzyme molecule, this volume is enormous; but our efforts toward miniaturization of electrical circuits in sophisticated electronic equipment can hardly be compared to the ultimate in miniaturization, the living cell.

Part of the regulation and order of metabolic reactions is dependent on cell structure. We have already seen (page 43) that there are many organelles and membranes within a cell and that certain enzymes are confined, during the life of the cell, to very specific locations. This compartmentalization of enzyme activities can be readily demonstrated by separating the cell into its various parts (page 51). Then it is found that specific metabolic reactions occur only in certain parts of the cell. For example, a sugar molecule synthesized in the chloroplast by photosynthesis may move into the cytoplasm, where it is changed and partially oxidized by the action of enzymes in the cytoplasm during a series of reactions called **glycolysis.** This oxidized product of glycolysis then moves into a mitochondrion where it is further oxidized. In this example, the sugar molecule has passed through three compartments (chloroplast, cytoplasm, mitochondrion) and has been treated uniquely in each. This beautifully precise compartmentalization in the cell is an important part of the mechanisms that control metabolic reactions.

Another aspect of metabolic control depends on genetic information in the nucleus that ultimately regulates the synthesis of specific enzymes in the cell.

A further form of metabolic control is dependent on the sensitivity of enzyme reactions to certain ions and molecules produced or used during metabolism.

For example, if a plant is growing on an inorganic medium containing ammonium ions as a source of nitrogen, the plant may be unable to utilize nitrate ions rapidly when they are placed in the solution. However, in a short time, a rapid increase in the plant's ability to use nitrate is often observed. It can be demonstrated that the absorption of nitrate from the culture medium has stimulated the synthesis of an enzyme, **nitrate reductase.** This enzyme is said to be "induced" in the cell by the **substrate** (nitrate).

Often the end product of a series of metabolic reactions can inhibit an enzyme catalyzing one of the early steps in the reaction sequence. This regulation of the course of a reaction through the action of one of the products of the reaction is known as **feedback.** It is a basic type of control in electronics. In cells, the concentrations of certain key metabolites play similar regulating roles in many phases of metabolism.

As an example of feedback control, we can look at some effects of ATP in the cell. During respiration, ATP is formed. The energy in ATP may be used to do useful work in the cell. The more work that is done, the more rapidly ATP is used. When little work is done, ATP tends to accumulate in the cell. It is known that a high level of ATP in the cell decreases the rate of at least one enzyme reaction in respiration. Consequently, an increase in ATP in the cell causes a decrease in the rate of respiration. Conversely, when a cell does more work, the level of ATP decreases, its inhibitory effect on the enzyme reaction involved in respiration decreases, and respiration increases. More ATP is generated. Thus, the rate of respiration is in part controlled by the level of cellular ATP which in turn is influenced by the rate at which work is being done.

5 Chemistry and Physiology of the Cell

SUMMARY OF CELLULAR PHYSIOLOGY

1. Metabolism is the sum total of all catabolic or breakdown reactions and all anabolic or synthetic reactions in the cell.

2. Cells absorb water and mineral salts from the external environment.

3. The numerous chemical reactions that go on in the cell are controlled by organic catalysts called enzymes.

4. Green plant cells are able to synthesize carbohydrates from water, carbon dioxide, and light energy, through the process of photosynthesis.

5. Carbohydrate foods can be transformed into fats and, with the addition of certain chemicals, into proteins.

6. Complex foods are rendered simple through the process of digestion.

7. Respiration is the oxidation of organic substances within living plant cells.

8. Protoplasm is built up through the process of assimilation of foods.

9. Energy flows from sunlight through chlorophyll into sugar molecules where it is stored as chemical energy during the process of photosynthesis.

10. Some chemical energy in sugar is transferred to ATP during respiration.

11. ATP is used by the cell to drive the energy-consuming reactions of metabolism.

12. The control of metabolism may depend in part on compartmentalization, enzyme synthesis, enzyme sensitivity to ions and molecules, and may involve a feedback mechanism.

6

The Dividing Cell

Why must cells divide? We shall see in the chapter on the algae that in a large order of algae, the Siphonales, cell division does not occur, yet these single-celled multinucleate plants become large and relatively complex (Fig. 22.5, page 449). Their habitat is the Mediterranean floor. No successful invasion of the land was ever made by autotrophic plants whose cells do not divide. It would appear that the mechanical strength of many cells and the increased flexibility of specialization made possible by a cellular pattern are vital survival factors in the evolution of land plants.

But the nuclei do divide in the Siphonales. In only a few genera do mature plants occur with a single nucleus, and these are all small plants. Perhaps the largest of them, *Acetabularia* (Fig. 2.5*E*) attains a height of about one inch. Even in *Acetabularia,* cell division is a prerequisite for sexual reproduction. Cell division enables an increase in size and complexity of the plant body.

Cell division is apparently required because a single nucleus with a full complement of genes can react with only a small amount of cytoplasm. Increase in size and complexity of the plant body requires cell division. The resulting increase in number of cells makes possible specialization for the different physiological activities including absorption, conduction, reproduction, photosynthesis, and support. Cell division always precedes active growth. Growth is the direct result of three activities: (1) cell division, (2) cell elongation, and (3) cell differentiation.

It is common knowledge that deoxyribose nucleic acid, found in chromosomes, is responsible for passing along the species characteristics from one generation to the next. The genetic information is stored in the DNA molecule in units known as genes, which occur in an orderly sequence along the length of the DNA molecule. The distribution of information from one cell generation to another is accomplished in meristematic cells through mitosis. Differentiation, however, is brought about only when the genetic information is transferred from the nucleus to the cytoplasm. While we are not sure just when this transfer takes place, differentiation occurs only during the interphase.

The process of differentiation is poorly understood. When tissues like the pith from a tobacco stem, or bits of cotyledon from a small *Streptanthus* seed, or phloem from a carrot root, are placed on agar under proper conditions, cells will divide without a definite polarity until an irregular callus results (Fig. 6.1*A*). With the proper hormones, normal chloroplasts will develop and the callus will become green (Fig. 6.1*B*). Growth eventually slows and subcultures must be made. These new cultures lose their chlorophyll, cell divisions take place, and the new culture grows. Only when growth slows down does the tissue become green. Examination of the chloroplasts shows that they reverted to a juvenile stage during the growth period (Fig. 6.1*C*).

When a callus from carrot phloem is transferred from a solid nutrient medium to an aqueous culture medium and gently shaken, individual cells will separate from the callus. These individual cells may be used to start new cultures. In the proper nutrient medium, individual cells, originally from phloem tissue, will behave as fertilized egg cells. They develop typical filamentous embryos and have been called **embryoids** (Fig. 6.2). Roots grow from the embryoids in liquid culture and, upon transferal to a solid agar culture, stems originate above the roots. These "artificial seedlings" may be brought to flower and produce seeds. These experiments indicate two very significant facts: (1) all plant cells carry the full complement of genetic information needed for the development of a complete mature plant; (2) before cells can divide they must return to the undifferentiated meristematic state.

CELL DIVISION

Cell division seems to have been an important factor for survival in the evolution of land plants. It is not surprising, then, that cell division is the first step in growth and differentiation. The formation of new tissue cannot take place without cell division.

During the period of division, both nucleus and cytoplasm divide. It is customary to designate the period of nuclear division as **mitosis** and to divide it into four phases: (1) **prophase;** (2) **metaphase;** (3) **anaphase;** and (4) **telophase.** The division of the cytoplasm is known as **cytokinesis.** Mitosis gives rise to daughter nuclei having identical gene complements. Cytokinesis gives rise to two new parcels of cytoplasm which are probably never identical (Fig. 6.3). The period of preparation for division is known

as **interphase** and we shall see that the sequence of events taking place during interphase is equal in importance, in the complete process of cell division, to that occuring during mitosis and cytokinesis. We may summarize the complete process of cell division as follows:

Figure 6.1. Cell division is necessary for growth. A, at start of the experiment, the green callus, grown in sterile tissue culture from a few cells taken from the cotyledon of a *Streptanthus* seed, had stopped growing; 14 days after being transferred to fresh culture medium, the chlorophyll disappeared from the tissue. A week later, 21 days from the start of the experiment, the callus had doubled in size and some chlorophyll was being synthesized; after another three weeks, 42 days from the start of the experiment, the callus had increased greatly in size and was very green; B, chloroplasts in the green thallus resemble chloroplasts in mature leaf tissue where cell divisions are not taking place, ×10,000. C, plastids in the subcultured colorless callus resemble the proplastids found in meristematic cells, ×10,000. (Courtesy of R. Sjolund.)

Figure 6.2. Embryonic development in free cell suspensions of carrot roots. *A*, a suspension of carrot cells from carrot phloem; *B*, a globular cell mass forms through unpolarized cell divisions; *C*, polarized cell divisions give rise to a torpedo-shaped embryoid; *D*, typical plantlet; *E*, carrot plant grown from a freely suspended cell from a carrot root. (Courtesy of F. C. Stewart, M. O. Mapes, A. E. Kent, and R. D. Holsten, *Science,* **143:**(3601), 20–27. Copyright 1964 by American Association for the Advancement of Science.)

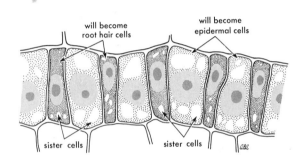

Figure 6.3. Sister cells frequently develop into different kinds of specialized cells. Unequal cell divisions occur in the protoderm, the meristematic cells of a root, which will differentiate into the epidermis: the small darker cells will become root hairs and the larger, less dense sister cells will remain epidermal cells, ×7000. (Courtesy of E. G. Cutter and L. Feldman.)

Cell division
  Preparation for division, interphase
  Period of division
    Mitosis → identical daughter nuclei
      prophase
      metaphase
      anaphase
      telophase
    Cytokinesis → different parcels of daughter cytoplasm

The Interphase Cell

An interphase cell in meristematic tissue has characteristic differences from an interphase cell of mature tissue. Portions of two interphase cells taken from the shoot apex of *Chenopodium album* are shown in Figs. 6.4*A* and *B*. It may be compared with the young leaf cell of Fig. 4.4. Meristematic cells of apical regions are almost spherical, and their shapes may be compared with a compressed mass of bubbles or lead BB shot. There is no large vacuole and the nucleus occupies a major proportion of the cell volume. Mitochondria, dictyosomes, endoplasmic reticulum, and ribosomes are similar to those found in the leaf cell. The shoot apex of *Chenopodium* was killed in potassium permanganate, so the ribosomes do not show. However, microbodies (sphaerosomes and multivesicular bodies, MVB) are present. The plastids, which lack chlorophyll in the apical cells, differ from those found in the leaf cell, which do contain chlorophyll. In meristematic cells the young plastids, or proplastids, are slightly larger than mitochondria and possess only a few paired membranes (Fig. 6.4).

The microtubules may be present and appear, at least just before the onset of mitosis, to form a

Figure 6.4. Meristematic cells in the shoot apex of *Chenopodium album*. *A* is an electron micrograph. *B* is a drawing of A. Note large nucleus, mitochondria, dictyosomes, proplastids, microbodies and absence of vacuoles. Compare with Fig. 4.3 ×10,000. (Courtesy of E. M. Gifford and K. Stewart.)

dictyosome

plasmodesma

old cell wall

new cell wall

microbody

ribosomes

endoplasmic reticulum

multivesicular body

nuclear pore

nucleus

nucleolus

cytoplasm

nucleus

nuclear envelope

membranes

proplastid

new cell wall

mitochondrion

intracellular space

B

Figure 6.5. Meristematic cells in a young leaf of tobacco (*Nicotiana tabacum*). *A*, section at right angles to the plane of the future cell plate shows cross section of microtubules, ×1,000; *B*, section in plane of future cell plate shows microtubules encircling nucleus, ×1,000; *C*, Three-dimensional interpretation of *A* and *B*. (Redrawn after K. Esau and J. Cronshaw, *Protoplasma*, **65,** 1. Copyright Springer-Verlag.)

narrow band encircling the nucleus in the plane of the forthcoming equatorial plate (Fig. 6.5).

Thin sections of interphase nuclei frequently show the nucleoplasm to be evenly and finely granular (Fig. 6.4). A coarser, more irregular clumping of granules is also seen (Figs. 4.4, 4.16*A*). In addition, denser aggregations of granular or fibrilar material may be observed (Figs. 4.4*A*, 6.4*A*). DNA may be localized in quantity in these areas and they may represent sections of chromosomes. Higher magnifications and special techniques reveal short coiled fibrils in the nucleoplasm. Some fibrils are visible in the nucleoplasm of the nucleus in Fig. 4.16*A*. Measurements of well-preserved fibrils in the nuclei of grasshopper sperm show them to be about 100 A in diameter.

Considerable detail is shown in the nucleolus from a root tip of the water plant *Hydrocharis* (Fig. 4.16*C*). Ribosome-like particles in spiral aggregations appear to form the basic nucleolar structure. Note the absence of a barrier between the nucleolus and nucleoplasm.

Nuclei isolated from young barley root tips and spread on distilled water will burst. The resulting film may be picked up, carefully dried, and observed on the electron microscope. The nuclei appear (Fig. 4.16*B*) to consist of loosely coiled or folded fibers about 200 A in diameter. Both thin sections and isolated nuclei indicate that the basic structure in interphase nuclei is an array of fibers called **unit fibers.**

MITOSIS

Prophase

The unit fibers in the interphase nucleus are long, slender, and seem tangled. We do not know their exact disposition in the interphase nucleus nor their precise relationship with chromosomes. But the onset of mitosis, as seen in the light microscope, is heralded by the presence of definite chromatin threads (Figs. 6.6*A*, 6.7). These threads gradually shorten and thicken and become easier to see. They stain more heavily with certain dyes. For this reason they are called colored (*chromo*) bodies (*soma*) or **chromosomes.** Each chromosome is derived from an individual thread. The nucleolus slowly decreases in size and finally disappears during this stage (Figs. 6.6A, 6.7).

It becomes apparent that at late prophase each chromosome is composed not of one but of two threads coiled about each other (Fig. 6.6B, C, D). The nuclear membrane disappears toward the end of the prophase, but there is reason to believe that the nucleoplasm may not mix with the general cytoplasm of the cell and that the chromosomes may remain embedded within it.

Metaphase

Forces active within the cell now arrange the chromosomes, or at least a specialized portion of each chromosome (the **kinetochore**), in the equa-torial plane of the cell (Figs. 6.7, 6.8H). The nucleoplasm appears to elongate slightly, and **spindle fibers** appear attached to the kinetochores. This structure, composed of chromosomes, fibers, and nucleoplasm (and possibly some cytoplasm), is the **spindle** (Figs. 6.6E, 6.7, 6.8).

As the nucleoplasm elongates, the chromosomes, or at least the kinetochore, is moved to the equa-torial plane of the cell (Figs. 6.6F, 6.7E, 6.8). Spindle fibers extend from the chromosomes to the op-posite poles of the cells. Other fibers apparently reach to the poles but are not attached to the chro-mosomes. The electron microscope demonstrates that these spindle fibers have the characteristics of microtubules (Fig. 6.8). This structure, chromo-

Prophase

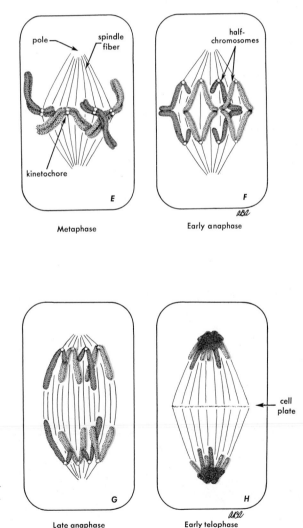

Metaphase   Early anaphase

Late anaphase   Early telophase

Figure 6.6. Mitosis, diagrammatic representation; A to D, stages in prophase. Two pairs of chromosomes are repre-sented; the green pair have a median kinetochore, while in the gray pair, the kinetochore is close to one end. The chro-mosomes shorten and thicken, and chromatids become appar-ent; E, metaphase, F, early anaphase, G, late anaphase; H, telophase.

Figure 6.7. Mitosis as followed in living unflattened cells of the endosperm of seeds of the blood lily (*Haemanthus katherinae*). Photographs in Nomarski system optics. *A*, early prophase; the nuclear membrane is still present. Note the clear zone of cytoplasm surrounding the nucleus; *B*, the nuclear envelope has disappeared, and the clear zone still surrounds the chromosomes. A few short spindle fibers are already present; *C*, full metaphase; the coiled chromatids are distinct. The two chromatids moving to the upper pole have separated, and a few spindle fibers may be seen; *D*, early anaphase; the chromatids are moving to opposite poles of the cell; *E*, mid-anaphase; spindle fibers are in evidence between the chromosomes and the poles of the cell; *F*, telophase; the chromatids have aggregated at the opposite poles of the cell. A nuclear envelope has not yet formed. Time intervals after A, B, 14 minutes; C, 64 minutes; D, 74 minutes; E, 93 minutes; F, 107 minutes. ×2000. (Courtesy of A. Bajer, *Chromosoma*, **25**, 249. Copyright Springer-Verlag.)

Figure 6.9. Isolated metaphase chromosomes. *Vicia faba* consists of unit fibers similar to those seen in the interphase nucleus, ×30,000. (Courtesy of S. Wolfe and R. G. Martin, *J. Exptl. Res.,* **50,** 140. Copyright Academic Press.)

Figure 6.8. Electron micrographs of thin sections of metaphase chromosomes in the endosperm of *Haemanthus katherinae. A,* metaphase plate showing the aligned chromosomes with microtubules (spindle fibers) attached to the kinetochore; *B,* several metaphase chromosomes showing the kinetochores, the two sister chromatids as well as subchromatids. Unit fibers appear to be distributed in two arrays in clumps and in connecting fibers between the clumps, ×40,000. (Courtesy of A. Bajer.)

somes, microtubules or spindle fibers, plus any adherent nucleoplasm is called the **mitotic spindle.**

Electron micrographs of isolated metaphase chromosomes shows them to be constructed as was the interphase nuclei of unit fibers. But here the fibrils are more tightly folded than in the interphase nuclei (Fig. 6.9).

The chromosomes are now distinct bodies of two closely associated halves, each half being known as a **chromatid** (Figs. 6.6*E,* 6.7). In those plants, such as corn, which have been intensively studied, each chromosome can be recognized and numbered. There are twenty chromosomes in corn, but only ten different types that can be distinguished by their size and form. There are thus two chromosomes of each type. The twenty chromosomes of corn may be arranged in ten pairs. Maps have been prepared

Figure 6.10. Influence of polarity of cell divisions on development of form. *A,* non-polarized cell divisions result in an irregular mass of cells (Fig. 6.1A); *B,* cell division parallel with the axis of the root (gray arrow and cells) results in growth in length; cell division parallel with the circumference of the root (dark green arrow and cells) increases the circumference of the root; cell division at right angles to the circumference of the root (light green arrows and cells) increases the diameter of the root.

of corn chromosomes showing the relative positions of the genes along them. Since the chromosomes split longitudinally, each gene is divided and each half-chromosome contains a full set of genes.

Anaphase

The chromosomes do not remain long in the equatorial plane. The chromatids soon separate from each other and move to opposite poles of the cell. This period of separation of half-chromosomes is called the **anaphase** (Figs. 6.6*F, G,* 6.7*D, E*).

Telophase

When the divided chromosomes have reached the opposite poles of the cell, they group together and spin out into long, thin threads of the reticulum. The nuclear membrane and the nucleolus again become apparent. This period of transition from chromosomes to interphase nucleus is known as the **telophase.** The half-chromosomes that have gone into the constitution of the new daughter nuclei grow, forming full-sized chromosomes once again (Figs. 6.6*F,* 6.7*H*).

Interphase

Even in the most rapidly dividing cells, the telophase nucleus gradually returns to the interphase

state. Either a new division must be prepared for, or the genetic information must be transmitted to the cytoplasm for the synthesis of new proteins, wall materials, and structure that accomplish differentiation in the new cell.

POLARITY OF CELL DIVISION

If the spindles in a meristematic tissue were oriented at random, an irregular mass of tissue would result. This does occur when a single cell or small group of cells from a carrot root in tissue culture produces an irregular mass of cells called a **callus** (Figs. 6.1, 6.10*A*). For orderly growth there must be a precise orientation of the mitotic spindles. A polarity is established so that, in general, the axes of spindles are parallel with each other and with the axis of the shoot or root (Fig. 6.10*B*). When the spindle axes are oriented parallel to the root-shoot axis, the divisions are called **anticlinal.** When the axes are perpendicular to the root-shoot axis, the divisions are called **periclinal** divisions (Fig. 6.10*B*).

What is responsible for polarity? We do not know. Polarity seems to be inherent in cells and in the tissue of which it is a part (however see *Fucus,* page 471.) Changes in polarity may be induced by hormones (Fig. 20.6). Occasionally, the orienta-

Figure 6.11. The formation of the cell plate in the endosperm of *Haemanthus katherinae*. A, B, living unflattened cells; photography by Nomarski system optics. A, telophase; the chromosomes are tightly clustered at opposite poles of the cell, and the cell plate has started to form; compare with C, ×4000. B, late telophase; the cell plate is almost continuous, and fibrils may be seen extending poleward a short distance from the plate; compare with D, ×4000. C, D, comparable stages of cell plate formation as seen with the electron microscope. C, early stage in plate formation; the microtubules are clustered into groups which are associated with a denser material, apparently in the plane of the future cell plate, ×26,000. D, late stage; the microtubules are fewer in number and are not clustered; vesicles are present and apparently fusing to form a continuous separation phase between the sister cells, ×26,000. (A and B, courtesy of A. Bajer, *Chromosoma*, **25**, 249. Copyright Springer-Verlag. C and D, courtesy of P. K. Hepler and W. T. Jackson, *J. Cell Biol.*, **38**, 437.)

tion to be assumed by the spindle can be detected in plants at the cellular level before metaphase. In some instances the cellular organelles will pass largely to one end of the interphase cell. Some recent observations report that microtubules form a circular band at the equatorial plane of the cell (Fig. 6.5) during interphase, just before the onset of mitosis.

In animal cells, polarity is generally indicated before the onset of mitosis by the presence and activity of the centrioles.

CYTOKINESIS

In the great majority of cases, the division of the nucleus is followed by the division of the cytoplasm.

Light microscopy demonstrated that a cell plate forms at the equatorial plane of the cell at right angles to the spindle fibers (Figs. 6.7, 6.11*A, B*). Electron microscopy has shown that the spindle fibers are microtubules. At the equatorial region, the spindle fibers or microtubules are gathered together in bundles and each bundle is surrounded by an amorphous material (Fig. 6.11*C*). Vesicles appear in this region and eventually fuse to form the first membrane dividing the daughter protoplasts (Fig. 6.11*D*). This membrane, formed of pectin, is known as the **middle lamella.** It has not yet been decided whether the vesicles are elaborated by the dictyosomes or by the endoplasmic reticulum, both of which are present in the vicinity. The new cell wall is formed by the deposition of cellulose by each

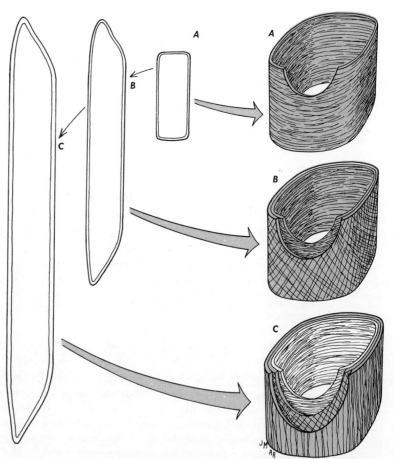

Figure 6.12. Diagram to show deposition and change in orientation of the cellulose fibrils in an elongating primary cell wall. A, B, and C represent increasing age and length of the same cell. The cellulose fibrils are first deposited parallel with the circumference of the protoplast. They are then pulled out of this orientation as the cell wall elongates. (In the actual wall, the different stages of fibril orientation would grade into each other.) Green represents the earliest fibrils deposited, and, light gray represents those most recently deposited.

Figure 6.13. Microtubules in the root tip of *Arabidopsis thaliana*. The microtubules are oriented parallel with the circumference of the cell wall and thus parallel with the cellulose fibrils, ×50,000. (Courtesy of M. C. Ledbetter.)

protoplast on its side of the middle lamella. This is the **primary cell wall.**

The formation of the primary wall, and subsequently secondarily deposited materials poses an interesting problem of genetic control. The cellulose is laid down outside the protoplast, apparently resulting from the polymerization of many sugar molecules to form long unbranched molecules of cellulose. We shall see in Chapter 7 that cell walls have definite characteristics. In the case of the pollen grain, the wall has very intricate and species-specific designs (Fig. 15.27). How can genes determine a pattern laid down by inert cellulose outside of the protoplast?

The molecules of cellulose, which may be several microns long, are held together in a crystalline array by hydrogen bonds to form long unbranched fibrils. These fibrils are of indefinite length, somewhat flattened, from 100–300 A in breadth and about 75 A in thickness. A cross section of a fibril having an area of 100 A^2 would show, on the average, the cross section of three molecules of cellulose. Between 20 and 30 percent of the primary cell wall consists of fibrils of cellulose. The fibrils, embedded in an amorphous matrix, give support to the cell wall as do the steel reinforcement rods in a concrete wall. In a *Valonia* cell (an alga with a cell diameter of 10 mm), it has been calculated that at one atmosphere of pressure one square centimeter of wall is subjected to an internal stress of 20,000 kilograms.

After removal of the amorphous materials from the wall, the array of fibrils may be studied with the electron microscope. When first deposited by the protoplast, in cells destined to lengthen, the fibrils are in parallel array and form a band around the protoplast (Figs. 6.12*A*, 6.13). The mechanism bringing about this precise arrangement is not understood. However, the microtubules in the cytoplasm are also present in a similar parallel array. (Figs. 6.13, 6.14). It is logical to postulate that the cytoplasmic microtubules may be involved in the orientation of the cellulose fibrils in the developing cell wall.

As the cell lengthens, the deposition of fibrils by the protoplast continues, with no change in the orientation of the newly deposited fibrils. However, elongations of the wall must change the orientation of older fibrils. They first separate to form a net more or less parallel with their original position (Fig. 6.12). As elongation continues, the fibrils, now further removed from the protoplast, assume a position more in line with the axis of the elongating cell. At a terminal position, the oldest cellulose fibers are all parallel with the elongating axis of the cell (Figs. 6.12, 6.14). This type of wall growth is known as **multinet growth.** A section through a primary cell wall after completion of elongation shows the oldest cellulose fibrils to be parallel to the long axis of the cell; the most recently deposited cellulose fibrils encircle the cell at right angles to the long axis (Fig. 6.14). Intermediate fibrils have intermediate positions.

SUMMARY OF CELL DIVISION

Cell division is now completed and its general aspects may be summarized.

1. Cell division increases the numbers of cells.
2. All cells have the potentiality to divide but are normally blocked from doing so.
3. Preparation for division takes place during interphase.
4. Nuclear division is called mitosis, cytoplasmic division is called cytokinesis.
5. The daughter nuclei are genetically identical.
6. The daughter parcels of cytoplasm may be unlike each other.
7. In general, mitotic spindles are oriented parallel with each other and with the axis of the shoot. A definite predetermined polarity brings this about.
8. The stages of mitosis are prophase, metaphase, anaphase and telophase.
9. The spindle, which is the mechanism for chromo-

Figure 6.14. The wall of a hair cell from the base of a sedge (*Juncus*) leaf. Note that the innermost fibrils are parallel with the girth of the cell and that the outermost cellulose fibrils are parallel with the long axis of the cell, × 50,000. (Courtesy of A. L. Houwink and P. A. Roelofsen, *Act. Bot. Neerland*, **3,** 385.)

some movement is thought to be constructed from many microtubules.

10. The middle lamella is the first membrane to divide the two daughter protoplasts.

11. Vesicles collect around the microtubules in the equatorial plane of the cell. They fuse to form the middle lamella.

12. Cellulose formed by the two daughter protoplasts and deposited on the middle lamella forms the primary wall.

DNA AND ITS REPLICATION

Deoxyribose nucleic acid, DNA, is the relatively simple molecule in which, it appears, is stored all of the information needed for the development of the several hundred thousand species, past and present. DNA is a prime biological universal. As far as we can tell, it has existed as long as life has existed. Although there may be variations in the type and distribution of the nitrogen bases in DNA from different species, there is no reason to suppose that there has ever been any fundamentally different kind of carrier of genetic information than the DNA known today.

As early as 1887, Weismann suggested that "qualities" responsible for inheritance had a linear arrangement in the chromosomes, which were just becoming known at that time. Between 1900 and 1904 the combined results of a number of investigations demonstrated the relationship between chromosomes, meiosis, and Mendel's newly rediscovered laws of inheritance. These observations ushered in a very exciting quarter-century of research on inheritance and chromosomes.

Miescher in 1869 was able to separate nuclear material from the other constituents of pus cells. The characterization of the acidic component of this nuclear material as nucleic acid led to many studies aimed at determining the occurrence, nature, and chemistry of the nucleic acids. These nucleic acids have three parts: a phosphate group, a five-carbon-atom (pentose) sugar, and organic nitrogen containing bases known as **pyrimidines** and **purines** (page 103). It was first believed that there were two different nucleic acids, one derived from animals and the other from plants. Later work showed that animal and plant cells each contain both types of nucleic acid.

Isolated nuclei always yielded nucleic acid, but did not demonstrate its location in the nuclei. Some sort of histochemical test was required to demonstrate the location of the nucleic acids during interphase and in dividing nuclei.

The search for color reactions that would demonstrate nucleic acids led Feulgen in 1924 to the discovery of the reaction that now bears his name. It had been known for some time that aldehydes such as the sugar aldehyde ribose, will reconstitute the color of a red dye, basic fuchsin, which has been decolorized by sulfurous acid. The basic fuchsin changes from colorless to violet. Now, both kinds of nucleic acids have pentose sugars (deoxyribose in DNA, ribose in RNA) that are aldehydes, but these potentially reactive groups are blocked, in the nucleic acids, by their association with the purine and pyrimidine bases. Feulgen wondered if the aldehyde group of the pentose sugar in a nucleic acid could be made available to bring about the color reaction with the decolored basic fuchsin. The answer was yes, and more. A mild hydrolysis with dilute hydrochloric acid broke the bond between the purines and the deoxypentose in DNA but did not readily break a similar bond between the nitrogen bases and the pentose of RNA. So the Feulgen technique of subjecting tissue or cells to mild acid hydrolysis followed by treatment with colorless fuchsin became, under properly controlled conditions, a specific and quantitative method of measuring the deoxypentose nucleic acid, DNA. The histochemical demonstration of DNA in nuclei was now a reality.

What is the relationship between DNA and the fibrils we have seen in the interphase nuclei and metaphase chromosomes? How is DNA replicated? What is the mechanism for the equal distribution of DNA in the dividing chromosomes?

In 1953, Watson and Crick in a short, highly significant paper in *Nature* presented the evidence for the molecular structure of DNA. The structure shown in Fig. 6.18 shows the now famous Watson-Crick model for the double helix of DNA. While a complete DNA molecule might be several million angstroms long it is only 25 A in diameter.

The fibrils found in isolated interphase nuclei and metaphase chromosomes are 125 A in diameter. We do not know their length. The diameter and length of chromosomes varies with the stage, so a direct comparison of their diameters with those of the unit fiber is not profitable. However, a number of unit fibers could fit easily into an extended chromosome, and several unit fibers have been seen in chromosomes of Vicia faba (Fig. 6.9). It is not known whether these strands represent one unit fiber folded back on itself or several different strands. Is a chromosome multistranded or not? A similar situation exists for the state of the DNA molecule within the unit fiber. This hierarchy of structure complicates

level of chromosome study. How, for instance, are the many 125 A unit fibers, with their enclosed DNA double helix, divided equally between the two chromatids?

We shall now consider the manner in which DNA itself is replicated. From each double helix must come two new double helices without, at the same time, disturbing the precise ordering of the base pairs. The dividing double helix may be several million angstrom units long, and since a single base pair occupies the space of only a few angstroms, there are many base pairs along a double helix.

DNA is localized in the chromosomes during mitosis. It ranges in amount from 2×10^{-9} to 10.36×10^{-9} milligrams per nucleus per cell. It is constant in amount for a given tissue and is definitely related to the number of chromosomes a tissue contains. Measurements of the DNA content in relation to stages of cell division are shown in Fig. 6.15. It should be noted that the late interphase and prophase nuclei have a value for DNA constant within the limits of the experiment. When sets of separating anaphase and telophase chromosomes are measured, each set now has half the amount of DNA found in the interphase, prophase, and metaphase nuclei. It follows that DNA must be synthesized during interphase. There is excellent evidence, not only that this is actually so, but that mitosis will not normally start until the DNA of the nucleus has been doubled.

Precise information telling us the exact time and location of the synthesis of new DNA may be obtained by using radioactive precursors of nucleic acid. A commonly used substance of this type is radioactive thymidine. It seems to enter cells and to be directly incorporated into nuclei of dividing cells of both plants and animals. Root tips of *Tradescantia* grown in a culture solution containing radioactive thymidine show that the interphase may be divided into three periods. During the first of these stages the nucleus increases in volume but there seems to be no incorporation of thymidine-H^3. Later, radioactive material is incorporated into newly-synthesized DNA. A period follows in which the nucleus maintains a constant size without any further incorporation of labeled thymidine. One investigator calculated that it took about 20 hours for a complete cell division in *Tradescantia* root tips. The time intervals for each stage are as follows: the pre-synthetic interphase, (a period during which the nucleus is increasing in size but no DNA

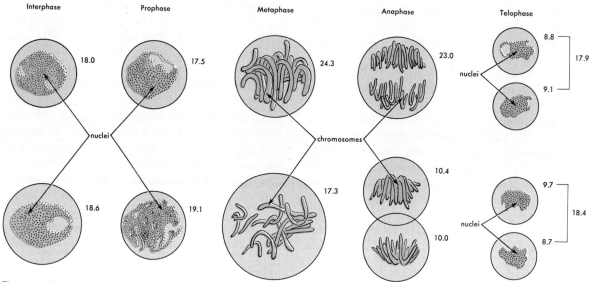

Figure 6.15. DNA content of nuclei at different stages of mitosis. Nuclei are shown for each of five stages. All pictures were made with the same magnification, so the circles show the relative sizes of the nuclei and chromosomes during mitosis. The numbers to the right of each circle indicate the amount of DNA present. (Redrawn after Patau and Swift, *Chromosoma*, **6**, 149. Copyright Springer-Verlag.)

Figure 6.16. Autoradiograph showing presence of DNA synthesized with radioactive thymidine. *A*, first mitotic division after incorporation of thymidine-H^3; *B*, second division after incorporation of thymidine-H^3. (Courtesy of J. Herbert Taylor.)

is being incorporated), 4 hours; the period of DNA synthesis, 10.8 hours; the postsynthetic interphase (a period during which the nuclear volume is constant and there is no DNA synthesis), 2.7 hours; prophase, 1.6 hours; metaphase, 0.3 hours; and anaphase—telophase, 0.6 hours.

It is possible to pinpoint the replication of new DNA molecules fairly well by growing seedlings in nutrient solutions containing radioactive thymidine. The following experiment was carried out by Taylor with seedlings of *Crepis capillaris*. Roots were placed in a nutrient solution containing radioactive thymidine for six hours and were then transferred to a similar nutrient solution with unlabeled thymidine for another six hours. The root tips were squashed on a slide and then carefully pressed out onto a very thin photographic film where they were left for about a month. Radiation sensitized silver

grains just as light rays would. The film was then developed and the sensitized silver grains were changed to small visible grains of metallic silver. The root tips themselves were stained in a dye to emphasize chromosome structure. Examination under the microscope showed an equal concentration of silver grains over both chromatids of the late prophase chromosomes (Fig. 6.16).

If the root tips are allowed to remain in the nutrient solution with unlabeled thymidine until they undergo a second mitotic division, then, portions of each chromatid pair will be labeled, but the label never occurs at the same level on sister chromatids. This situation can be explained by assuming that a double helix of DNA is present in early interphase (Fig. 6.17). This helix may untwist, or in some fashion become so arranged that components of the helix separate and two new strands

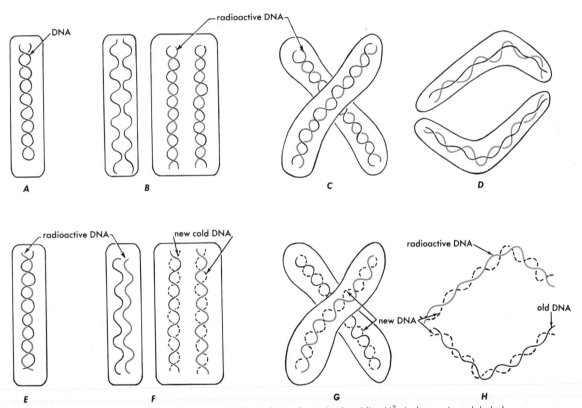

Figure 6.17. Diagram of postulated steps in DNA synthesis. Roots in thymidine-H³ six hours, in unlabeled nutrient solution during second mitotic division. Double helix represents DNA molecule, solid line indicates original molecule, broken line indicates DNA formed in presence of thymidine-H³, and dotted line shows DNA formed after incorporation of thymidine-H³. A to D, first mitosis: A, chromosome in early interphase before replication; B, mid-interphase replication of DNA; C and D, late prophase and anaphase respectively of first mitosis after thymidine-H³ incorporation; E to H, second mitosis: E, early interphase before incorporation of unlabeled thymidine; F, mid-interphase after incorporation of unlabeled thymidine; G and H, late prophase and anaphase respectively of second division.

incorporating the labeled thymidine are formed as new pairs for each old strand (Fig. 6.17*B*, *C*). The new helices become incorporated into chromatids of the late prophase chromosome (Fig. 6.17*C*), where they are made evident by the presence of the silver grains, as shown in Fig. 6.16*A*. Each of the separating daughter chromosomes of this division will carry labeled thymidine (Fig. 6.17*D*). Following a nucleus formed from the chromosomes of Fig. 6.17 through a second division in

nutrient solution lacking thymidine-H³, we find that the double helix of DNA (Fig. 6.17*E*) untwists and the two newly formed DNA helices incorporate unlabeled material (Fig. 6.17*F*). Thus one DNA double helix is entirely without a label, and the other DNA double helix has the older half of the molecule-bearing label, and a new half without a label (Fig. 6.17*F*). Now at late prophase, one chromatid will show the label and the other will lack it (Figs. 6.16*B* and 6.17*G*). One anaphase chromosome will be marked and the other will not (Fig. 6.17*H*).

While the evidence is conclusive that the double helix of the DNA molecule is an important part of a chromosome, it is difficult to think that a single DNA molecule coiled upon itself extends for the full length of a chromosome, as suggested in Fig. 6.17. The physical act of untwisting and separating such a long molecule in a chromosome during its replication presents a formidable mechanical problem.

The DNA Molecule

The molecular structure of the DNA double helix is shown in Fig. 6.18. Each strand forming part of the backbone of the helix is made up of alternating sugar (deoxypentose) (**S**) and phosphate (**P**) groups. These backbone strands are held together by hydrogen bonds between the nitrogen-containing organic bases (**purine** and **pyrimidine**). The bases of one strand are bonded to complementary bases of the other strands (Fig. 6.19). When a purine or a pyrimidine base is joined with a pentose sugar and a phosphate, the resulting compound is a **nucleotide**. Although hundreds or even thousands of nucleotide units may polymerize to form the gigantic nucleic acid molecule (Fig. 6.18), only four different nucleotides are found in a nucleic acid molecule (Figs. 6.20, 6.21). The two pyrimidine bases occurring in DNA are **cytosine** and **thymine**; in RNA, **uracil** replaces thymine (Fig. 6.20). The two purine bases in both DNA and RNA are **adenine** and **guanine** (Fig. 6.21). Notice in Fig. 6.18 that the bases have a linear arrangement along the background strands of the helix. The sequence of purines or pyrimidines in one of the strands of the spiral determines the sequence of the bases in the other strand, since adenine of one strand always pairs with thymine of the other strand, and guanine of one strand always pairs with cytosine of the other. In order to illustrate this arrangement and the great

Figure 6.18. Schematic diagram of the double helix in a portion of a DNA molecule. Alternating sugar, deoxyribose (S), and phosphate (P) groups make up the backbones of the strands. Attached to each sugar unit is one of the purine or pyrimidine bases, adenine (A), thymine (T), guanine (G), cytosine (C). The strands of the helix are held together through hydrogen bonds between the base pairs, adenine to thymine and guanine to cytosine. A, One strand is displaced along the axis of the helix for convenience of diagraming; B, Diagram of the parallel strands in the helix.

Figure 6.19 (*right*). General formulas for purine and pyrimidine bases. Each particular base has characteristic groups attached at the unfilled valence bonds in the diagrams. These groups are shown in the complete formulas of Figs. 6.20 and 6.21.

Figure 6.20 (*above*). Pyrimidines found in nucleic acids.

Figure 6.21 (*right*). Purines found in nucleic acids.

variety of nucleic acids that may arise from this simple set of four compounds, let us use the letters A, G, T, and C. We may first write them down in a line using each as many times as we wish and in any order that we wish. Now below these write again A, G, T, and C but be careful that A always matches T and G always matches C.

Thus, if the first strand is represented by

A G C A A C A T C

The second strand would be

T C G T T G T A G

The two complementary bases (AT or GC) joined by hydrogen bonds are frequently called **base pairs.** A sequence of three base pairs, or a **triplet,** codes an amino acid. The sequence of triplets determines the order in which amino acids will be joined in a protein, thus determining the nature of the protein.

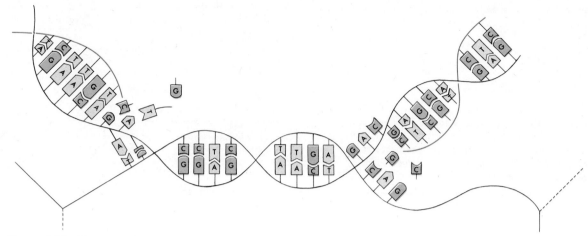

Figure 6.22. Diagrammatic representation of the replication of the DNA molecule.

We have seen that in the replication of the DNA helix, one strand is maintained intact and a new complementary strand is formed (Figs. 6.16, 6.17). It is thought that this happens in the following manner. The original double helix untwists (Fig. 6.22), and a new complementary strand is synthesized. In this synthesis, guanine is always joined to cytosine and adenine to thymine. We may suppose that a free base will act as a seeding mechanism for the formation of its complementary nucleotide. A new helix apparently starts to form immediately after the untwisting of the old helix.

RNA

RNA, or ribonucleic acid, differs from DNA in a number of ways. In RNA the sugar component, ribose, has one more atom of oxygen than deoxyribose (Fig. 5.1). Uracil replaces thymine and RNA forms only a single helix. One form of RNA, **messenger RNA,** transfers information from DNA to the site of protein synthesis. However, it is actually the sequence of the bases in the single helix strands of RNA in the ribosomes that directs the type of protein formed. In the transfer of information from DNA to RNA, the following change in nucleotides takes place: Adenine goes to uracil; thymine goes to adenine; cytosine goes to guanine; and guanine goes to cytosine. The RNA sequence of bases determined by the DNA sequence given above would be

DNA   A G C A A C A T C
RNA   U C G U U G U A G

This series of bases, in RNA, codes for the amino acids alanine (UCG), cysteine (UUG), and glutamic acid (UAG).

SUMMARY

1. Cell division always precedes differentiation and growth.

2. Cell division appears to be required because only relatively small amounts of cytoplasm can interact with a single nucleus.

3. The presence in chromosomes of linearly aligned "qualities" controlling inheritance was first postulated in 1887. The importance of chromosomes as the physical basis of inheritance was established between 1900 and 1904.

4. Nucleic acid was first isolated in 1868; its location in the chromosomes was shown by Feulgen in 1924, and the molecular structure of the DNA molecule was described in 1957.

5. Replication of DNA involves the uncoiling of the DNA double helix, the separation of the complimentary polynucleotide chains, and the formation of new polynucleotide chains adjacent to the old chains.

6. Both nucleus and chromosomes are characterized by fibers of undetermined length and about 175 Å in diameter. The diameter of the double helix of DNA is 25 Å.

7. The DNA molecule consists of two parallel helices formed by alternating pentose sugars (deoxyribose) and

phosphorus. The helices are joined together through hydrogen bonding between pyrimidine and purine bases. The unit, deoxyribose, phosphate, and pyrimidine or purine is a nucleotide.

8. The two associated pyrimidine and purine bases constitute a base pair. A pyrimidine base always joins with a purine base.

9. In DNA the purine bases are adenine and guanine, and the pyrimidine bases are cytosine and thymine. In RNA uracil replaces thymine.

10. The sequence of base pairs codes for amino acid synthesis.

11. The information coded in the DNA molecule is transferred to RNA (**messenger RNA**) which probably, in ribosomes, directs the synthesis and assemblage of amino acids into a protein.

12. Three base pairs or a triplet code for one amino acid.

7

The Stem—Primary Growth

We have seen that the mechanical strength of cellular structure was obviously a prerequisite for growth on land (page 85). Once on land, plants evolved and raised into the atmosphere the photosynthetic organs, leaves, thus facilitating the trapping of the radiant energy of sunlight. This increase in plant body required anchorage in the soil and a conducting system for the transportation of water, food, and nutrients throughout the plant body (Fig. 3.2). Also required was a means of protecting meristematic and reproductive structures from desiccation. Furthermore, some mechanism must have evolved providing for survival over the annually recurring period unfavorable for growth. With the successful development of specialized structures to meet these requirements there evolved the dominant present-day land flora.

FUNCTIONS OF STEMS

Stems provide mechanical support and raise leaves into the air, thus facilitating photosynthesis. Flowers and fruits are also produced in positions facilitating pollination and seed dispersal. **Stems support.**

Stems provide a pathway for the movement of water and mineral nutrients from roots to leaves, and for the transfer of foods, hormones, and other metabolites from one part of the stem to another. **Stems conduct.**

The normal life span of plant cells is from one to three years. Water and mineral salts in dilute solution move in dead cells, but this movement depends upon the activity of living cells in leaves and roots generally less than three years old. Stems in herbaceous perennials (Fig. 7.1*B*) and in 2000-year-old redwoods, or bristle-cone pines (Fig. 7.1*A*), annually provide new living tissue for the normal metabolism of the plant. **Stems produce new living tissue.**

Stems thus have three major functions; (1) support; (2) conduction; and (3) the production of new living tissue.

GROWTH IN STEMS

All stems grow in length; new living tissue is added at the shoot tips. Some stems grow in girth; new living tissue is added to the circumference of the

Figure 7.1. A. (left) bristle cone pine (*Pinus aristata*); gymnosperms and woody angiosperms achieve longevity through secondary growth, which annually results in a cylinder of new tissue around the trunk. The older tissues soon die; B, (above) *Iris*; most other seed plants attain longevity by the continued growth of new primary tissue at the apex of the shoot. Older parts of the stem soon die. New leaves form at the apex, roots arise in nodes behind the apex, and older portions of rhizome die.

stem. All stems that increase in girth, first grow in length; hence, growth in length is **primary growth.** Growth in girth, which follows primary growth, is **secondary growth.** Secondary growth not only produces additional mechanical support but also, in both roots and stems, produces anew each year young and active cells for conduction. Secondary growth makes possible the attainment of great age by individual plants. The connection of new primary growth at the widely separated points of root and shoot apices (Fig. 3.1) by new secondary growth provides an efficient mechanism enabling trees of considerable age to carry out all life processes in cells one to five years old (Fig. 7.1A).

In general, secondary growth is confined to the woody angiosperms and gymnosperms. With few exceptions, the stems of all other land plants (ferns, horsetails, club mosses, herbaceous angiosperms and monocotyledons) do not increase in girth year after year.

There are many annuals or biennials among the dicotyledons and monocotyledons. These plants, alive for one or two years, live out their life-span with tissues produced only by primary growth. They produce flowers and seeds and die. Longevity, with only primary growth, is attained in two general ways. First, the stem may continue to grow indefinitely in length. This generally results in a horizontal stem, either underground or creeping along the surface of the ground. Roots are formed at nodes close to the growing tip of the stem (Fig. 7.1B). Upright portions, leaves or shoots, may be produced only at the growing tip or also at special regions behind the tip, the nodes (Fig. 3.2). The old portions of these horizontal stems die and rot away, leaving separate plants. The example in Fig. 7.1B is an iris rhizome. A second method of continuing growth with only primary tissues is regular production of new short stems, from which grow upright flowering stems. Each year's crop of stems may die back during the winter. This occurs in bulbs, such as daffodil (Fig. 7.31D), or corms such as gladiolus (Fig. 7.31B). Here the older bulb or corm is used up to produce the current flowering stalk, and a new bulb, or corm, is formed by primary growth in anticipation of the next season. In long-lived monocotyledons, such as the palms, Joshua trees, *Dracena* and *Pandanus* (Fig. 7.33), other mechanisms are invoked (page 138).

The principles of primary growth are generally similar throughout the vascular plants. We will first discuss primary growth as it occurs in a woody angiosperm. This will be followed by a discussion of primary growth in a monocotyledon and then by a fuller account of the ways longevity may be attained without secondary growth. Secondary growth is considered in Chapter 8.

BUDS

The basic structure of the vascular plants is an axis with potentially immortal apical meristems at opposite ends of the axis. Specialized protection for the root apex involves only a root cap (Fig. 9.4), which serves to protect the root as it grows through the soil. More elaborate protection is needed for the stem apical meristems, particularly in perennial stems which overwinter. In such cases the shoot meristem is dormant during the unfavorable season. It may be protected from desiccation by small modified leaves known as **bud scales.** Such a dormant shoot apex with its protective scales is a **bud** (Figs. 7.2, 7.3, 7.4). In a few instances the protective scales are lacking and the unprotected dormant apex is a **naked bud.** Actively growing shoot apices are also supplied with a protective covering of young leaves. Such active shoot apices are not generally called buds and they occur most generally in the annual or biennial species.

Kinds of Buds

We shall see (page 292) that growth terminates in shoot apices that produce flowers. Yet the production of flowers early in the season is frequently advantageous for woody plants; therefore, at the time of resumption of growth in the spring, new primary growth must provide for the production of new shoots and flowers. Consequently, there are buds that produce only stems and leaves **(leaf buds),** buds that produce only flowers **(flower buds),** and buds that produce both leaves and flowers **(mixed buds).** *Camellia japonica* always produces, frequently at the same node, both flower and leaf buds (Fig. 7.2A). Apples and pears produce mixed buds on short spurs (Fig. 7.2B). These spurs grow less than an inch a year, but a severe pruning will induce the growth of a long vegetative shoot rather than the usual production of a short shoot with fruit! Almonds produce both flower and leaf buds on short spurs (Fig. 7.2C). Apricots often exhibit three buds at a node; the middle bud is a leaf bud and the two accessory buds are flower buds (Fig. 7.2D). In walnut, there are both leaf buds and staminate flower buds (catkin buds) (Fig. 7.2E). The large flower bud of lilac is flanked by two small leaf buds (Fig. 7.2F).

A terminal bud of buckeye is shown in Fig. 7.3A. Removing the bud scales reveals the largest of the rudimentary leaves (Fig. 7.3B). When all of the rudimentary leaves are removed, there remains a small floral apex and leaf bud primordium in an axil of a rudimentary leaf (Fig. 7.3C, 7.5B). This bud is a mixed bud. Figure 7.4 shows several stages, over a twenty-four hour period, in the expansion of a terminal bud similar to that of Figs. 7.3A and 7.5A. When the expanding leaves are removed, only several small unexpanded rudimentary leaves remain. This bud is a leaf bud.

What determines whether a shoot tip will produce leaves alone, flowers alone, or both leaves and flowers? The answer is still very elusive. But it is known that, in some cases, the induction of a floral apex is related to day length, to a hormone, and to two specific wave lengths of red light. The stimulus is transmitted from leaves to the shoot apex. The problem of floral initiation can be studied to better advantage in herbaceous species, such as pigweed (*Chenopodium album*) or pea (*Pisum sativa*), than it can in woody species.

Freehand, median, longitudinal sections of a terminal bud of buckeye are shown in Fig. 7.5. The bud is well protected by bud scales. The rudimentary leaves are attached at nodes (Fig. 3.2) to the conical shoot apex, which is protected, here, by several of the smallest leaves. The midribs and veins are already apparent in the rudimentary leaves. The internodes (Figs. 3.2, 7.5) are very short in the bud. The elongation of these internodal regions will account for the growth of the shoot in length. Bud primordia in the axils of the rudimentary leaves are not commonly seen in leaf buds (Fig. 7.5A). However, if shoot growth is to occur in a mixed bud,

Figure 7.2. Buds. *A,* flower and leaf buds of *Camellia japonica,* ×1; *B,* mixed buds of pear on short fruit spur, ×2; *C,* flower and leaf buds of almond (*Prunus amygdalus*), ×1; *D,* a single leaf bud of apricot (*Prunus armeniaca*) with two accessory flower buds at a node, ×2; *E,* the naked staminate bud of walnut (*Juglans regia*), ×2; *F,* a flower bud of lilac (*Syringa*) with two opposite leaf buds, ×2.

7 The Stem—Primary Growth

Figure 7.3. A, terminal mixed bud of buckeye (*Aesculus californica*), ×2; B, bud scales removed to show rudimentary leaves just prior to opening of bud, ×2; C, rudimentary leaves removed to show a floral apex, ×15.

Figure 7.4. A, terminal bud of buckeye (*Aesculus californica*) with bud scales opening, ×½; B, C, stages in expansion of terminal bud of buckeye over a 24-hour period, ×½; D, most of the expanding leaves removed, with a few very small remaining leaves protecting a vegetative shoot apex, ×15.

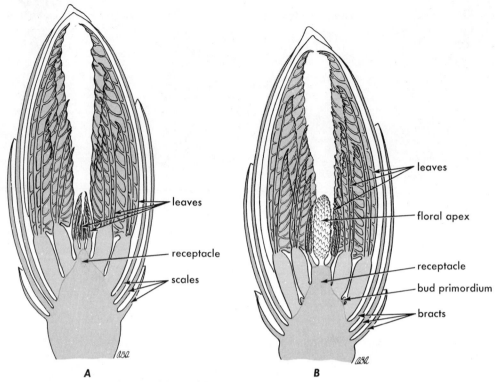

Figure 7.5. Diagrams of terminal buds of buckeye (*Aesculus californica*). *A*, leaf bud; *B*, mixed bud; note axillary leaf bud already formed.

provision must be made for vegetative growth. In the buckeye mixed bud with a terminal flower primordium, bud primordia do occur in the axils of rudimentary leaves (Fig. 7.5*B*).

Arrangement of Buds

A three-year-old twig of walnut in winter condition is shown in Fig. 7.6. Buds are conspicuous structures of stems at this time of the year. The tip of the twig generally bears a large **terminal leaf bud** (Fig. 7.6). At regular intervals along the stem, other buds may be seen; they are called **lateral buds.** Note that below the base of each lateral bud there is a scar that was made when a leaf fell from the twig; this is a **leaf scar. Vascular bundle scars** (Fig. 7.6) may be seen within each leaf scar; strands of food- and water-conducting tissues passing from the stem into the leaf stalk were broken when the leaf fell, leaving these scars. Buds and leaves are usually borne in this relationship to each other; the buds form in the angle made by the stem and the leaf stalk. This angle is termed the **leaf axil,** and conse-

quently these buds may also be called **axillary buds.**

The passage of strands of conducting tissue into the leaves and buds must mean that the internal structure of the stem is so arranged as to make possible a continuous connection of leaves and buds with all other parts of the stem. We have thus a specialized stem region—a region to which are attached buds and leaves and within which conducting strands unite with other conducting strands of the stem. This region is known as a **node** (Fig. 7.6). The region between any two adjacent nodes is an **internode.**

Protecting the young immature cells within the bud is a series of overlapping scales, the **bud scales.** They are usually shed when the bud develops into a new shoot and they also leave scars, **bud-scale scars.** The part of a stem or twig between sets of terminal-bud-scale scars is generally formed during one growing season. For instance, growth made by the twig this year is set off from the growth made last year by means of a ring or girdle or terminal-bud-scale scars (Fig. 7.6). When the scales of a terminal bud fall off in the spring, they leave a number of closely crowded scars that form a distinct ring. Examination of several-year-old twigs shows that growth in length may vary from year to year. This fact is shown by the spacing of the terminal-bud-scale scars. It is significant that there is no increase or decrease in the length of any portion of a stem after that portion is a year old.

The slightly raised areas on the bark are **lenticels.** They are composed of cells that fit loosely together, with air spaces between, which permit the passage of gases inward and outward.

Position of Buds on a Woody Twig

In the walnut twig there are just *one leaf bud* and *one leaf at each node.* This arrangement of buds and leaves on the stem is spoken of as **alternate** (Fig. 7.7). It is the most common type of bud and leaf arrangement. Ash, maple, lilac, and many other plants have two leaves opposite each other at each node, and a bud in the axil of each leaf (Fig. 7.7). This arrangement of leaves and buds is spoken of as **opposite.** When three or more leaves and buds occur at each node, as in *Catalpa,* the leaf arrangement and the bud arrangement are said to be **whorled** (Fig. 7.7).

Some plants have several buds in or near the leaf axil. For example, the apricot often has a group of three buds in the leaf axil: a central bud, which develops into a side branch, and two lateral ones, which are flower buds (Fig. 7.2). All but the central one are called **accessory buds.** The walnut may also have more than one bud in the leaf axil (Fig. 7.6).

Not infrequently, buds may arise on the plant at places other than leaf axils. They may appear on stems, roots, or even leaves and give rise to new shoots. Such buds are called **adventitious buds.** Their formation may be stimulated by injury, such as occurs in pruning.

Dormant or **latent buds** are ones that arise in a regular fashion in the leaf axil but whose development is usually inhibited by the dominance of the terminal bud.

In the identification of deciduous woody plants, it is usually considered desirable to have leaves, flowers, and even fruits. But, by careful study of the twigs of such plants in their winter condition, it is possible in most instances to determine the species (Fig. 7.8).

From the above discussion it is seen that buds may be classified by their **arrangement on the stem,** which may be (1) **alternate,** (2) **opposite,** or (3) **whorled;** by their **position on the stem,** which may be (1) **terminal,** (2) **lateral (axillary),** (3) **accessory,**

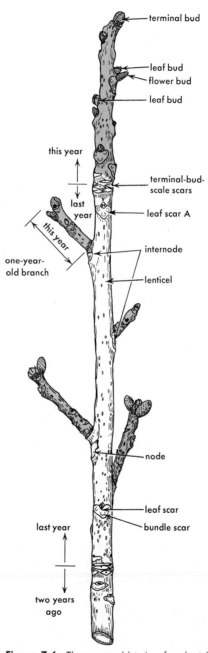

Figure 7.6. Three-year-old twig of walnut (*Juglans regia*), ×1.

Figure 7.7. Twigs showing three methods of bud and leaf arrangement. The position of leaves is shown by the leaf bases and scars. *A,* alternate, walnut (*Juglans regia*); *B,* opposite, lilac (*Syringa vulgaris*); *C,* whorled, *Catalpa.* × ½.

or (4) **adventitious;** and by the **nature of the organs into which they develop,** which may be (1) **leaf,** (2) **flower,** or (3) **mixed.**

As a rule, the terminal bud of a stem is the most active and grows more vigorously than any of the axillary buds. Usually, the lowest lateral buds on a year's growth of the shoot remain dormant and do not develop into branches. If the terminal bud is removed, however, as may be done in pruning, lateral buds, otherwise dormant, may become active.

Internal Characteristics of Buds

A longitudinal section of single leaf bud is shown in Fig. 7.9. The bud is essentially a **miniature shoot,** with nodes and very short internodes and with leaf primordia at each node. Primordial buds may occur in the axils of the more advanced leaf primordia.

Note (Fig. 7.9*A*) that the shoot tip is composed of cells and that the cells are not all alike. Groups of cells are known as **tissues.** A given tissue or cell group has definite functions to perform. For instance, some cells are constructed and grouped to protect, others to support or to conduct. If we have one cell type performing but a single function, we have a **simple tissue.** Frequently, however, several cell types will be organized into a tissue performing more than one function, such as support and conduction. In this case we may speak of a **complex tissue.**

Figure 7.8. Leaf scars of different species of woody plants. *A*, walnut (*Juglans regia*), ×2; *B*, catalpa (*Catalpa bignonioides*), ×2½; *C*, tree of heaven (*Ailanthus altissima*), ×2; *D*, box-elder (*Acer negundo*), ×2; *E*, European plane (*Platnus acerifolia*), ×3; *F*, buckeye (*Aesculus californica*) ×2.

Let us examine the cells and tissues of the bud. At the apex of the shoot within the terminal bud, and also at the apices of shoots of axillary bud primordia, there are dome-shaped masses of cells, the **apical meristem.** The cells of the apical meristem have large nuclei, a compact cytoplasm, and small vacuoles (Figs. 4.3, 6.4). While the bud is dormant these cells are relatively quiescent, but when conditions are favorable some divide rapidly and the daughter cells increase in size, becoming in some instances larger than the cell from which they originated. This increase in number and size of cells results in the elongation of the young shoot. But in conjunction with increase in number and size of cells, **differentiation** occurs; that is, the cells change morphologically and physiologically from the meristematic cells from which they arose. Thus, a short distance (usually a few millimeters) below the apical meristem we recognize three fairly distinct **primary meristematic tissues,** namely **protoderm, ground meristem,** and **procambium** (Fig. 7.9*A, B, C, D*). These three tissues are derived directly from apical meristem.

The term **shoot tip** is applied to the youngest part of the shoot and includes apical meristem cells associated with the leaf primordia. Some of these meristematic cells are elongating and others are in early stages of differentiation. The rapid elongation of stems results from the formation and growth of thousands of new cells in the shoot tip. It is in the shoot tip that we see the origin of **leaf primordia, bud primordia,** and the **primary meristematic tissues.**

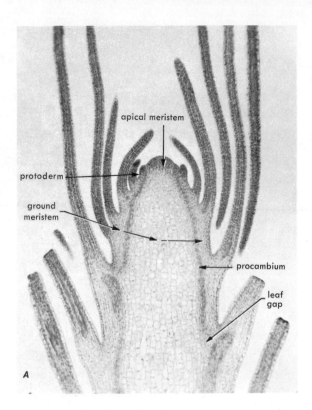

Figure 7.9. Sections of shoot of flax (*Linum*). *A*, (*above*) photomicrograph of longitudinal section of apex; ×300; *B*, (*right*) diagram of longitudinal section from apex to completion of primary growth, length of stem shortened. Apical meristem, dark green; procambium, light green; primary xylem and primary phloem, light gray; *C, D, E, F,* diagrams of cross sections of shoot taken from longitudinal section at positions indicated. (*A*, courtesy of E. M. Gifford.)

DEVELOPMENT OF TISSUES OF THE PRIMARY PLANT BODY OF A WOODY STEM

The Primary Meristems

The process of stem development is gradual. The youngest cells are those of the apical meristem, and they are similar morphologically. But in the process of differentiation, different cell types and tissues are formed. Thus, as seen in Fig. 7.9, and as mentioned previously, we recognize three different tissues, protoderm, ground meristem, and procambium. These three primary meristematic tissues differentiate into the **primary tissues.**

The **protoderm** (Fig. 7.9*A, D*) is the outermost layer of cells. It develops into the **epidermis**—that special primary tissue which covers and protects all underlying primary tissues. The epidermis prevents excessive water loss and yet allows for the exchange of the gases necessary for respiration and photosynthesis.

The **ground meristem** (Fig. 7.9*A, B, D*) comprises the greater portion of the meristematic tissue of the shoot tip. The ground meristem cells are relatively large, thin-walled, and isodiametric. The primary tissues forming from the ground meristem are (1) the **pith,** in the very center of the stem, and (2) the **cortex,** in a cylinder just beneath the epidermis and surrounding the vascular tissues. Sometimes the pith

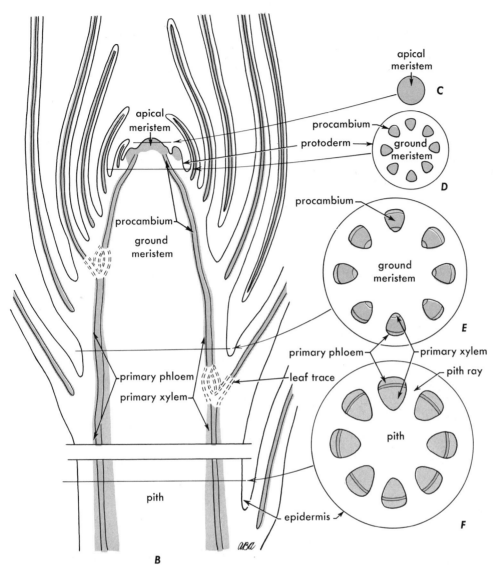

Figure 7.9. (continued)

and cortex are connected by (3) **pith rays,** also formed from the ground meristem (Fig. 7.9*D, E, F*).

Procambium cells usually appear first as strands among the ground meristem cells (Fig. 7.9*A*). In cross section (Fig. 7.9*D*), the strands appear as isolated groups of cells arranged in a circle. Sometimes a continuous **procambium cylinder** is formed. As seen in a transverse section of a single procambium strand, the procambium cells are smaller than those of the surrounding ground meristem; in lengthwise section they are much longer, and some of them may be pointed at the ends. The procambium cells give rise to the **primary vascular tissues** (Figs. 7.9, 7.10, 7.11). These primary tissues carry out several

functions and are divided rather rigidly into three groups according to these functions. The food is conducted in the outer group of primary vascular cells, which is the **primary phloem.** Water and mineral salts are conducted in the inner group of primary vascular cells, which, together with strengthening cells, constitute the **primary xylem.** In many stems, a meristematic region remains between the primary xylem and the primary phloem to become the **vascular cambium** (Fig. 7.9*F*).

It is well to repeat here that the **primary tissues** of the stem are those differentiated from the three primary meristematic tissues—protoderm, ground meristem, procambium—and that these three are

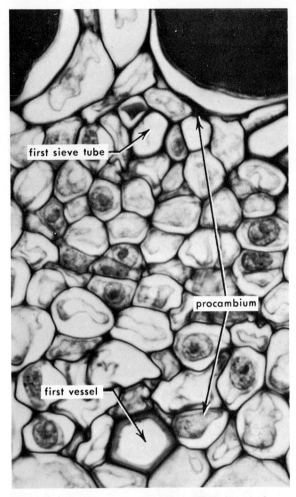

Figure 7.10. Cross section of a procambium strand of an elderberry (*Sambucus*) stem at the start of vascular differentiation. Compare with Fig. 7.9E. (Slide courtesy of Triarch Products.)

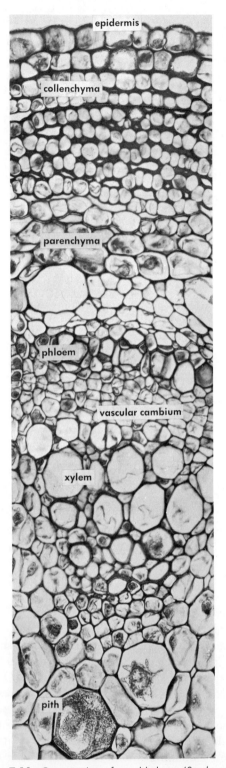

Figure 7.11. Cross section of an elderberry (*Sambucus*) stem at the completion of primary growth. Compare with Fig. 7.9F. (Slide courtesy of Triarch Products.)

derived from the apical meristem of the shoot tip. In woody plants, we must look for primary tissues of the stem a very short distance behind the stem tip. Even before the end of the first season's growth, the differentiation of these primary tissues from primary meristematic tissues is completed, and **secondary tissues** may be formed in abundance. Whereas primary tissues are derived from the primary meristematic tissues of the shoot, secondary tissues are the result of the production of new cells by **vascular cambium** and by **cork cambium.** The origins and nature of these two types of cambiums are discussed further on.

A Summary of Primary Development

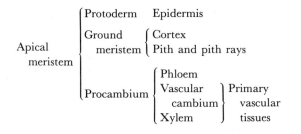

Apical meristem
- Protoderm — Epidermis
- Ground meristem
  - Cortex
  - Pith and pith rays
- Procambium
  - Phloem
  - Vascular cambium — Primary vascular tissues
  - Xylem

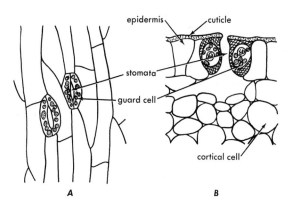

Figure 7.12. Epidermis of bean showing epidermal and guard cells. A, face view; B, cross section, ×80.

The term **stele** is applied to that part of the stem which includes the primary vascular tissues, the pith, and the pith rays. The so-called **primary plant body** is composed of the above primary tissues.

The main functions of these primary tissues may be summarized as shown below.

Epidermis: Protection of underlying tissues.
Vascular tissues:
  Phloem: Conducts food.
  Vascular cambium: Produces secondary phloem and secondary xylem.
  Xylem: Conducts water and mineral salts, and gives strength to stem.
Cortex: Stores food and, in young stems, manufactures food, strengthens, and protects.
Pith: Stores food.
Pith rays: Store food, and conduct water, mineral salts, and food radially.

Primary Tissues

The primary structural organization in the stems of most gymnosperms and dicotyledonous angiosperms consists of several tissues, possessing different cell types. The different types of cells described in the following few pages may be more or less isolated or they may be grouped to form a tissue.

The Epidermis

The epidermis is usually a single superficial layer of cells covering all other primary tissues and protecting them from drying out and, to some extent, from mechanical injury. It is the limiting layer of cells between the plant and its environment. In surface view (Fig. 7.12), epidermal cells are elongated in the direction of the stem's length; in transverse section they are usually isodiametric. The protoplasm forms a thin layer lining the cell cavities and normally retains its living properties for a long time.

The outer tangential wall of cells exposed to the air is usually thicker than other walls of the cells, and its surface layer is usually coated with a waxy substance called **cutin.** This superficial layer of cutin is termed the **cuticle.** By means of certain stains and chemical tests, the cutin can be readily distinguished from the cellulose of the rest of the wall. The cuticle is quite impermeable to water and gases. The inner walls, parallel to the stem surface, are the thinnest, and the radial walls, at right angles to the surface, often taper in thickness toward the inner wall.

Young stems usually possess specialized epidermal cells called **guard cells** (Fig. 7.12). Between each pair of guard cells is a small opening through which gases enter and leave the underlying stem tissues. The guard cells and the opening are known as a **stoma.** Guard cells, unlike ordinary epidermal cells, possess chloroplasts. They also differ from ordinary epidermal cells in their crescent shape, as seen in surface view. Guard cells are usually thought of as structures peculiar to foliage leaves. While they are much more common in the epidermis of leaves, they may occur also in the epidermis of young stems, of floral structures, and of fruits.

Epidermal appendages, such as hairs, may occur on young stems (Fig. 7.13).

The Cortex

This complex tissue, derived from ground meristem, forms a cylindrical zone beneath the epidermis extending inward to the primary phloem (Figs. 7.9F, 7.11). The following simple tissues or cell types may be found within it: **parenchyma, collenchyma, sclerenchyma,** and **secretory tissue.**

Parenchyma. The principal tissue of the cortex is parenchyma (Figs. 7.11, 7.14*G*). It usually consists of isodiametric cells with thin walls, mostly of cellulose, and with protoplasts that remain alive for a long time. We may speak either of a parenchyma tissue or of a parenchyma cell, one of the units composing the tissue.

This tissue is characterized by the presence of intercellular air spaces which vary greatly in size; in some parenchyma tissues they are difficult to find, while in others they are very apparent (Fig. 7.14*G*). Because parenchyma cells retain active protoplasts, they function in the storage of water and food, or in photosynthesis, and sometimes in secretion. The green color of many stems is due to the presence of chloroplasts in the parenchyma cells, **chlorenchyma,** of the cortex.

Parenchyma tissue is not confined to the cortex of the stem but occurs in practically all parts of the plant: flowers, fruits, seeds, leaves, and roots.

Collenchyma. The outermost cells of the cortex of young stems, lying just beneath the epidermis, often constitute a tissue known as **collenchyma.** This tissue may form a complete cylinder or it may occur in separate strands. The cells are elongated rather than isodiametric as are those of parenchyma, and have pointed, blunt, or oblique ends. In the most common type of collenchyma, the cell walls are thickened at the corners where the cells meet (Figs. 7.11, 7.14*D, E*). Because of these thickenings, collenchyma serves as a strengthening tissue. The walls are composed of pectin and cellulose. As in parenchyma, cells of collenchyma have long-lived protoplasts. Chloroplasts are often present.

In addition to the cortex of the stem, collenchyma cells may occur in other plant parts comprised of primary tissues. For instance, collenchyma is frequently associated with the veins of leaves.

Sclerenchyma. The main functions of sclerenchyma cells are support and, in many cases, protection. Their shape and the thickness and toughness of their walls contribute to the ability of these cells to support and protect the young stem. The thickness and toughness of the walls are increased by the deposition, within the original cellulose wall, of a substance known as **lignin.** The process of deposition is called **lignification.** The lignin is elaborated by the protoplast, and when deposition

Figure 7.13. Epidermal hairs. *A,* simple; *B,* star-shaped; *C,* glandular, ×50.

Figure 7.14 (*right*). Cell types and tissues. *A,* and *B,* fibers in lengthwise view; *C,* fibers in cross section; *D,* collenchyma in cross section; *E,* collenchyma in lengthwise view; *F,* sclereid; *G,* parenchyma; *H,* stone cells; *I,* woody parenchyma. (*A, B,* redrawn from Carl C. Forsaith, *The Technology of New York State Timbers,* New York State College of Forestry Publications 18; *F,* after A. S. Foster.)

middle lamella

collenchyma

pectin cellulose

D

pit

A   *B*

lumen

lignified wall

middle lamella

C

E

nucleus

lumen

F

Sclereid
after
Foster

pit membrane

vacuole

nucleus

pits

after
Livingston

lamella
middle

cytoplasm

cell wall

middle lamella

nucleus

cytoplasm

vacuole

intracellular
air space

G

lignified wall

lumen

canal

H

Figure 7.14. Development of Tissues of the Primary Plant Body of a Woody Stem

sieve plate

sieve elements

phloem parenchyma cells

phloem parenchyma cell

companion cell

sieve tube plastids

parenchyma plastid

Figure 7.15. Phloem tissue from the stem of tobacco (*Nicotiana*), ×400. (Courtesy of A. S. Crafts.)

ceases, the protoplast usually dies. There are two types of sclerenchyma cells, (1) **sclereids** and (2) **fibers** (Figs. 7.14*A, B, F*).

Of the various types of **sclereids,** the most common are the so-called stone cells (Fig. 7.14*H*), which are more or less isodiametric. Other types of sclereids are much branched, resembling very irregular stars (Fig. 7.14*F*). Some sclereids are derived from parenchyma cells by pronounced thickening of the cell walls; others arise from separate meristematic

cells. Minute canals, the so-called **pit canals,** extend outward through the thickened walls, being separated only by the primary walls from the ends of similar canals in the walls of adjoining cells (Fig. 7.14*H*).

Sclereids occur not only in the cortex of stems but also in the hard shells of fruits, seed coats, and bark, in the pith of stems, and in certain leaves.

Fibers are elongated, thick-walled cells, usually pointed at the ends (Fig. 7.14*A, B, C*). They give strength to the tissue in which they occur. Each fiber is **one cell.** The walls may or may not be lignified. When lignified, the walls may be so thick that the cavity, **lumen,** of the cell almost disappears. Fibers are usually very elastic and can be stretched to a great degree without losing their ability to return to their original length. Various types of fibers occur in vascular tissue of stems as well as in the cortex. The protoplasts of fibers often disappear as they attain maturity.

Secretory cells. Secretory cells are parenchyma-like, with dense protoplasmic contents that secrete various substances, such as resinous materials (Fig. 8.13) and nectar. Many epidermal hairs are secretory cells (Fig. 7.13*C*).

Summary of the Cortex

Thus, we may find in the cortex of stems several kinds of tissues and cell types. Usually it is not possible to find all kinds in the cortex of any one species of plant. Parenchyma and collenchyma occur in most species and usually predominate. Cortical fibers and secretory cells are rarer.

THE PRIMARY VASCULAR TISSUES

The term "vascular" pertains to tissues that conduct various substances in liquid form. For example, the veins and arteries that carry blood in higher animals are vascular tissues. In vascular plants, water and different water-soluble inorganic salts from the soil,

as well as food substances, are conducted throughout the plant in well-defined vascular tissues.

In a young dicotyledonous stem, very near the bud (Figs. 7.9*A*, *B*, *D*), the vascular tissues occur as separate bundles, the **primary vascular bundles.** Each primary vascular bundle is differentiated from a procambium strand. Each mature primary vascular bundle of all vascular plants consists of primary xylem and primary phloem. If secondary growth is to occur, a thin band of meristematic tissue destined to become vascular cambium remains between the primary xylem and primary phloem (Figs. 7.9*F*, 7.10, 7.11). Between the vascular bundles are parenchyma cells of the **pith rays,** and in the center of the stem is the **pith,** also of parenchyma (Figs. 7.9*F*, 7.26, 7.27).

The Primary Phloem

The phloem in angiosperms may possess four types of cells: **sieve-tube elements, companion cells, fibers,** and **parenchyma.** In gymnosperms, companion cells are lacking; there are **sieve cells** rather than sieve elements, and in the primary phloem of many species there may be no fibers. The important role of phloem is the conduction of organic solutes. It is probably through the sieve tubes, rather than through other types of phloem cells, that organic substances move rapidly.

A **sieve tube** is a vertical row of elongated cells, each cell known as a **sieve element** (Fig. 7.15). Among angiosperms, a sieve element (one cell) and

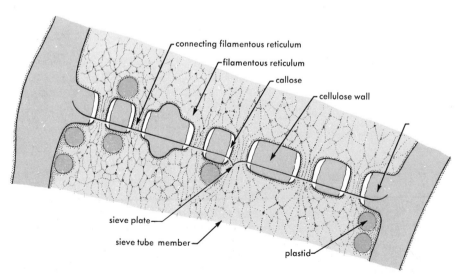

Figure 7.16. Diagrammatic representation of the ultrastructure of a sieve plate. (Courtesy of A. S. Crafts.)

a companion cell are sister cells; that is, they originate by division of a common procambial mother cell. The young sieve element is rich in cytoplasm and possesses a normal nucleus. As the element matures, its protoplast becomes greatly modified.

The nucleus disintegrates and the plastids lose almost all of their internal membranes but do usually retain a carbohydrate ganule. The mitochondria are small. Most important, the cytoplasm is limited to a thin peripheral layer, and the central

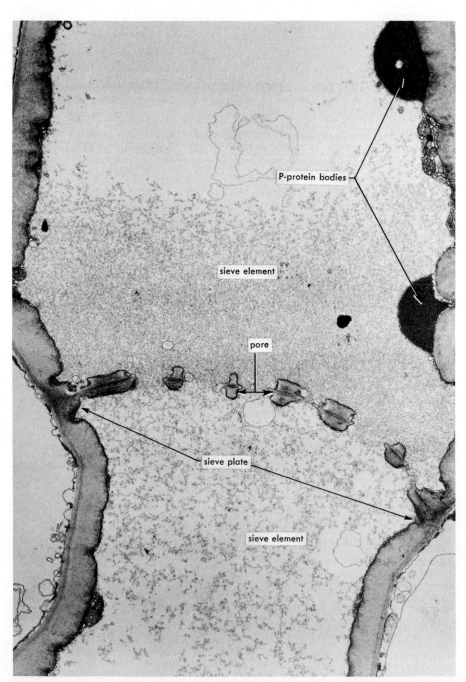

Figure 7.17. Electron micrograph of sieve plate between two sieve elements of *Cucurbita maxima.* ×7000. (Courtesy of J. Cronshaw and K. Esau, *J. Cell Biol.*, **38,** 292.)

part of the cell is occupied by a large mesh or tangle of fibers or tubules (Figs. 7.15, 7.16). This mass may be seen with the light microscope and has been called **slime.** Since it is now known to be a protein it is referred to as **P-protein** (Figs. 7.17, 7.18). At maturity the sister cell **(companion cell)** has a normal protoplast with a full complement of organelles (Fig. 7.18).

A characteristic structural feature of mature sieve tubes is the **sieve plate** (whence the term sieve tube).

Figure 7.18. Longitudinal section through portions of three sieve elements and one companion cell. A nucleus and all cytoplasmic organelles may be identified within the companion cell. Most of the organelles are absent from one sieve element, but may be recognized in the other two sieve elements, ×9000. (Courtesy of J. Cronshaw and K. Esau, *J. Cell Biol.* **38,** 25.)

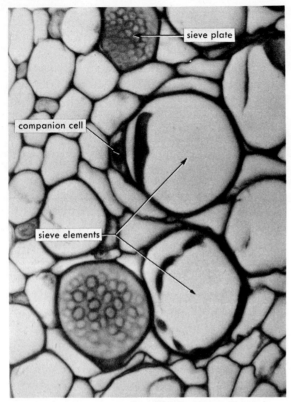

Figure 7.19. Cross section of phloem tissue of squash (*Cucurbita*). Sieve plates are visible in two sieve elements. The dark circles around the pores in the sieve plates are callose. Sieve plates do not appear in two other sieve elements because of sectioning. ×200. (Slide courtesy of Triarch Products.)

Figure 7.20. Cross section of a stem of flax (*Linum usitatissimum*). Note phloem fibers. ×100. (Courtesy of K. Esau.)

It may occur in end walls (Figs. 7.15, 7.16, 7.17, 7.19) or in side walls. In end walls, sieve plates are seen to the best advantage. The end wall between two adjacent sieve elements is thickened, and strands of cytoplasm pass through pores in it. Hence, the protoplasts of adjoining sieve elements are connected.

The structure of the mature sieve element is still in doubt, even though it must move the main food stream of the plant body. It is a living, functioning cell, but the absence of a nucleus and other modifications point to profound changes in its metabolism. With the exception of a few trees such as the palms, the sieve elements live and function for about three years. New ones are formed annually. In many studies on the structure of the mature sieve element, a carbohydrate known as **callose** is seen around the margins of the pores in the sieve plate (Fig. 7.19). In some instances protein may collect at the sieve plate. Obviously, such a development would block the pores of the sieve plate and obstruct movement of food materials. It has been demonstrated in other tissues that callose forms very rapidly in response to wounding. Phloem cannot be studied without cutting and wounding cells. Furthermore, considering the relatively large amount of material moved through sieve tubes, it is quite possible that slime could pile up against a sieve plate when an osmotic equilibrium is upset by cutting a sieve tube. Because of the experimenter's own interference with the process of this experiment, it becomes practically impossible to know which appearance of fixed and stained images corresponds to the living uninjured sieve element. The best guess is that in the living, functioning condition there is little or no callose on the margins of pores to obstruct movement of foods through sieve plates. The P-protein, or slime, is present in the form of a fine reticulum or threads (Figs. 7.15, 7.16). It extends through the pores, and in some sieve elements, the P-protein, or slime, may be present as small P-protein bodies (Fig. 7.17).

The fibers of phloem have the same general characteristics as those described as occurring in the cortex. In most plants the development of fibers is greater in the phloem than in the cortex (Fig. 7.20).

The Vascular Cambium

Xylem and phloem elements that have differentiated from procambium cells cease to be meristematic. Certain cells derived from procambium, however, do not lose their meristematic character. They form a narrow tissue between xylem and phloem known as **vascular cambium** (Fig. 7.9*F*). Procambium and cambium may be regarded as two developmental stages of the vascular meristem.

Figure 7.21. Bordered pits in tracheids of pine wood *(Pinus).* ×85. (Courtesy of Artschwager.)

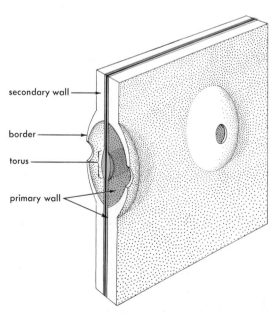

secondary wall

border

torus

primary wall

Figure 7.22. Diagram of a bordered pit from a pine tracheid.

primary wall

pit chamber

pit membrane

aperture

torus

middle lamella

border

secondary wall

primary wall

Figure 7.23. Bordered pit of black locust (*Robinia pseudo-acacia*). Note that the torus is part of the primary wall and that the pit membrane is lignified, ×4,300. (Courtesy of Myron C. Ledbetter.)

Figure 7.24. Photomicrographs of longitudinal section showing stages in development of vessels in squash (*Cucurbita*). A, transverse wall between vessel members intact; B, transverse wall beginning to dissolve, whereas thickening material is being laid down on longitudinal walls; C and D, transverse walls dissolved. ×150. (Courtesy of K. Esau.)

The Primary Xylem

The conducting cells that occur in the primary xylem of vascular plants are **tracheids** and **vessel elements.** These cells conduct water and mineral salts. Associated with them may be **fibers** (xylem fibers) and **parenchyma** (xylem parenchyma).

The **tracheid** is a single elongated cell more or less pointed at the ends (Fig. 8.6). Functioning tracheids are not alive. The tracheid wall may not be the same thickness throughout. All the wall may be thickened except for numerous small, circular or oval areas called **pits** (Figs. 7.21, 7.22). There are two types of pits in xylem cells, **simple pits** and **bordered pits. A simple pit** is shown in Fig. 7.14*I*. Here it will be observed that, where two cells lie side by side, the pit (depression or recess) in the wall of one cell is opposite the pit in the wall of the adjacent one. The term **pit-pair** is used to designate this condition. A pit-pair is *not* a hole in the wall; the two cells are separated by a membrane, the so-called **pit membrane,** comprised of the primary wall of each of the two cells and the intercellular cementing substance between these walls. Pits of this type commonly occur in the walls of parenchyma cells, and also in collenchyma cells and vessel elements. The type of pit known as **bordered pit** (Figs. 7.21, 7.22, 7.23) occurs in tracheids, vessel

Figure 7.25. Types of vessels that occur in primary xylem in an elongating branch. *A*, diagram. Annular vessels were formed first, and are therefore the oldest and most stretched. The pitted vessel formed last is the youngest. Elongation has stopped, so the pitted vessels will not be stretched; *B*, photomicrograph of vessels during primary growth in cleared node of *Alternanthera*, ×30. (*B*, courtesy of L. M. Srivastava.)

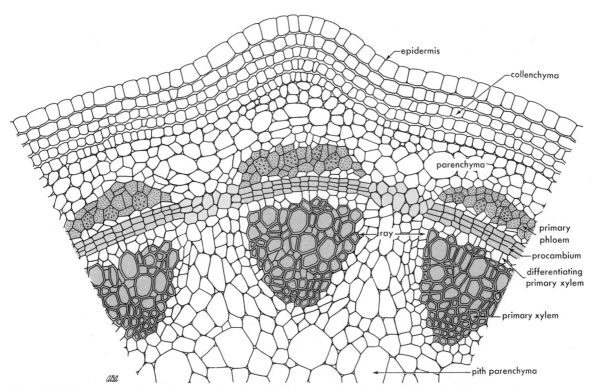

Figure 7.26. Diagram of a woody stem at the close of primary growth. Some procambium is still present and the last primary xylem and primary phloem cells are still undergoing differentiation. Ray parenchyma separates vascular bundles. Cortex consists of collenchyma and parenchyma. Epidermis is intact.

elements, and xylem fibers. Sometimes the thickening in tracheids is in rings, spiral bands, or bars, and sometimes it forms a network.

A **vessel element** is a *single cell* with oblique, pointed, or transverse ends. A **vessel** is a *series of vessel elements,* end to end, most of whose end walls are perforated or dissolved. Thus a vessel is a long tube. Vessels are often several centimeters long and in some vines and trees they may be many meters in length. A row of procambium cells becomes transformed into a vessel. Before the protoplasts disappear, the vessel walls become thickened, forming a secondary wall (Figs. 4.20, 7.24A), the thickening material being laid down on the primary walls in various patterns so that parts of the secondary walls are thick and others thin. The material deposited is cellulose; later, the layers of cellulose become lignified. The dissolution of end walls of vessel elements also takes place before the protoplasts disappear (Fig. 7.24B, C, D). Thus, the deposition of thickening material forming the secondary walls and the dissolution of end walls are functions of living cells. Following these processes, the protoplast dies. Therefore, functional vessels have no living contents.

The secondary walls of vessels in angiosperm stems are deposited in several different patterns (Fig. 7.25A, B). **Annular vessels** have the material in the form of separate rings. In **spiral vessels** the material is in the form of spiral bands. **Scalariform vessels** have the thickenings in the form of transverse, interconnecting bars. In **reticulate vessels,** the material forms a network on the wall. And in **pitted vessels,** the walls are pitted. The ends of the vessel elements (Fig. 8.6) are generally on a slant, and although open, they may have bars of wall material

Figure 7.27. Cross section of elderberry (*Sambucus*) stem at stage of completion of primary growth. ×100. (Slide courtesy of Triarch Products.)

across them (Fig. 7.25B). Vessels do not occur in small veins in leaves, and are lacking completely in most gymnosperms and lower vascular plants. In these forms, tracheids occur in elongating regions and they then may have annular and spiral types of secondary walls.

Xylem parenchyma cells outlive vessels, tracheids, and most xylem fibers. They function in the storage of water and foods, which, as we have learned, is one of the principal functions of parenchyma wherever it occurs in the plant. Parenchyma may also conduct materials for short distances.

Xylem fibers are similar to the fibers described elsewhere.

Pith and Pith Rays

Pith is composed of large-celled parenchyma with numerous intercellular spaces (Figs. 7.11, 7.14G,

7.26). The storage of food is its principal function. In some stems, primary vascular bundles are separated by wide strips of parenchyma that extend from the pith to the cortex. Such strips of parenchyma are called pith rays and may be considered radial extensions of the pith (Figs. 7.9F, 7.26). Like the pith, one of their functions is food storage; they also conduct materials short distances radially. They merge with the parenchyma of the cortex so that no line of separation is visible. In many woody species the primary vascular bundles are separated by very narrow rays of perhaps two to four cells in width (Figs. 7.27, 8.4, 8.9).

THE HERBACEOUS DICOTYLEDONOUS STEM

Many species of angiosperms lack secondary growth and therefore do not become woody. They are gen-

Figure 7.28. Cross section stem of alfalfa (*Medicago sativa*) at the close of primary growth, ×100. (Slide courtesy of Triarch Products.)

erally referred to as **herbs.** They are usually annuals or biennials and in the temperate zones they are generally small plants. Their external appearance is similar to that of the young woody twig, except that, in keeping with their short life, typical buds do not form, and they do not achieve a winter condition similar to that of the woody twig. They do have apical growth, nodes, and internodes. Branches, or modifications thereof, may be present in the leaf axils.

Their cellular anatomy has been described in detail on pages 114 to 127.

The tissues of both herbaceous and woody stems originate in the same fashion. The primary tissues are arranged in bundles in alfalfa (Fig. 7.28) as they are in *Sambucus.* Slightly different arrangements occur in other species.

These primary tissues are, from the epidermis inward: the epidermis; the cortex, composed of collenchyma, groups of fibers, and parenchyma; the phloem with sieve tubes, companion cells, fibers, and parenchyma; the xylem with vessels, tracheids, fibers, and parenchyma; the pith and pith rays composed of parenchyma cells. The outermost cells of the primary phloem in alfalfa develop into fibers.

THE MONOCOTYLEDONOUS STEM, AS EXEMPLIFIED BY THE GRASSES

With few exceptions the monocotyledons do not increase in girth after their initial period of rapid growth. They do not have secondary growth. Some, such as the palms and *Pandanus* that do achieve considerable size, do not produce a woody stem suitable for lumber. There are many perennial monocotyledons, so mechanisms have been devised to circumvent the generally short life span of primary tissues. Because of this fact, the external appearances and cellular anatomy of the monocotyledons shows considerable variation. At this point we will consider only monocotyledonous stems as they appear in the grasses, particularly corn (*Zea mays*). Corn is an annual. In external appearance its stem is not unlike that of an herbaceous dicotyledon. It has terminal growth, nodes and internodes. Leaves, branches, flowers (tassel), and fruits (ear) appear in leaf axils.

In stems of dicotyledonous plants, both herbaceous and woody types, and in those of gymno-sperms, the procambium strands, and hence vascular bundles, are usually arranged *in the form of a single ring*. With but few exceptions the procambium strands, and hence the vascular bundles, in stems of monocotyledonous plants are *scattered* throughout the ground meristem or at least through the outer region of it (Fig. 7.29). However, in some grasses like wheat and barley, the stems are hollow and the vascular bundles are in a definite ring. In most monocotyledonous stems, *all* the procambium cells differentiate into primary xylem and primary phloem elements; there is no vascular cambium and, as a consequence, no production of secondary tissues. Bundles of this sort which are "closed" to further growth are called **closed bundles** (Fig. 7.30). As has been observed, bundles of other seed-bearing plants that possess a vascular cambium are "open" to further growth; they are known as **open bundles.**

In transverse sections of the internodes of most grasses (barley, rye, wheat), the following tissues are evident (Fig. 7.29): (1) a single layer of epidermal cells, (2) strengthening tissue variously arranged beneath or near the epidermis, (3) ground parenchyma, and (4) vascular bundles. The limits of cortex and pith are indistinct. Stomata occur in the epidermis. The strengthening tissue consists of elongated fibers with thick lignified walls. In some grasses the vascular bundles are arranged in two rings near the periphery of the stem, and in such grasses there is usually a continuous ring of fibers some distance from the epidermis, with the small bundles of the outer row embedded in it. On the outer sides of the small bundles of the outer ring occur strands of fibers that reach to the epidermis. Bands of chlorophyll-bearing ground parenchyma are enclosed between these strands. Ground parenchyma may extend to the center of the stem, as in corn (Fig. 7.29), sorghum, and sugar cane, or its central portion may become destroyed during the growth of the stem, leaving a hollow pith cavity, as usually happens in wheat (Fig. 7.29), oats, barley, and rye.

The vascular bundle of the corn stem is an example of a common type of bundle occurring in monocotyledonous stems (Fig. 7.30). The large pitted vessels of the xylem are prominent features. There are usually two of these, and one or two smaller annular or spiral vessels may also be present. In addition, older bundles invariably contain a large air space or intercellular passage. Close

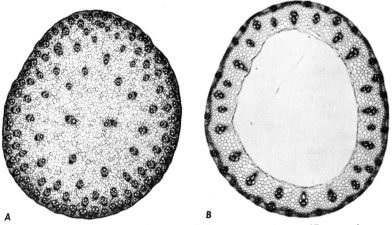

Figure 7.29. Cross sections of monocotyledonous stems. A, corn (*Zea mays*); B, wheat (*Triticum*), ×30. (Courtesy of L. Feldman.)

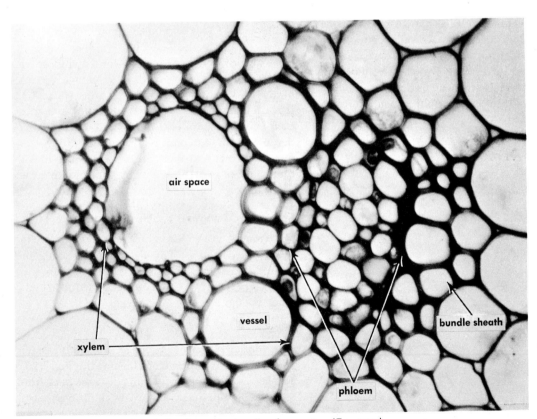

Figure 7.30. Cross section of closed vascular bundle of corn stem (*Zea mays*), ×300.

examination of this space is likely to reveal within it a lignified ring or portion of a ring. This is evidence that the air space was originally an annular vessel which, due to growth of the young stem, has been stretched and broken, thus giving rise to the relatively large air space. These vessels, together with the intervening fibers, tracheids, and parenchyma cells, comprise the xylem tissue, which is always located on the side of the bundle toward the center of the stem. The cells of the phloem form a regular pattern of thin-walled cells exterior to the xylem. Sieve plates may usually be seen. The companion cells are small, generally square or rectangular in cross section, and, because of their cell contents, stain more heavily than the sieve tubes. The bundle is generally surrounded by lignified fibers forming a tissue called the **bundle sheath.**

In grasses, a tissue at the base of each internode usually remains meristematic long after the tissues in the rest of the internode are fully differentiated. Thus, such plants contain a meristem at the apex of the shoot and also a meristem at the base of each internode. Each internode has its own growing zone. These internodal meristems are called **intercalary meristems,** and growth of the cells derived from them is termed **intercalary growth.** The flowering stems of such cereals as wheat, oats, barley, and rye shoot up very quickly. This rapid elongation results not only from the growth of cells derived from the apical meristem but also from the growth of cells derived from intercalary meristems at the base of each internode.

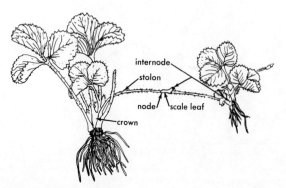

Figure 7.31. Runner of strawberry (*Frageria*), × ⅛; roots and shoots are produced at every other node.

PERENNIAL STEMS WITH ONLY PRIMARY GROWTH

Since primary growth occurs only at apices, longevity of a primary plant body can be attained only by continued elongation of the stem and root axis, resulting in a separation of root and shoot apices in space and time. Since these plants do not have secondary growth, there is no provision for connecting primary plant bodies at opposite ends of an elongating axis as the intervening primary tissues live out their life spans and die. Instead, adventitious roots arise at nodes a short distance behind the shoot apex. Nodes of perennial plants, lacking secondary growth, bear the usual leaves and buds, and also roots. Such stems are likely to have differentiated upper and lower surfaces, with leaves and buds occurring on the upper surface and roots on the lower surface. Since these stems are not strengthened or enlarged by secondary growth, they are generally prostrate, frequently underground, and often small. There are numerous modifications among these stems. Comparable modifications of a primary plant body, which permit the attainment of longevity, also occur in such widely diverse groups as the algae, horsetails, club mosses, ferns and dicotyledons. A long-lived gymnosperm having only primary growth is unknown. Because of the wide diversity of these groups (page 12), it is thought that similar stem modifications must have evolved separately from several different starting points in different plant groups during different epochs, and therefore several times during the course of evolution. This is known as **parallel evolution.**

Stolon (or Runner)

Bermuda grass has above-ground horizontal stems called **stolons.** These stems creep along the ground and at each node shoots and roots arise. In strawberry (Fig. 7.31), roots and leaves arise at every other node.

Rhizome

In most iris species, leaves and flowering stalks are produced at the growing **rhizome** tip (Fig. 7.1*B*). The leaves die a relatively short distance back from the growing tip, so that many iris plants bear senescent leaves. Roots are also formed at nodes and they may remain for the life of the rhizome. Other rhizomes, like canna (Fig. 7.32*A*), produce upright leafy

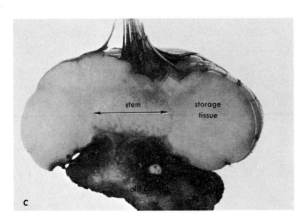

Figure 7.32. Plants illustrating continued development with only primary growth. *A,* roots and shoots are produced at every third node by the rhizome of canna, ×½; *B,* corm of gladiolus, ×1; *C,* corm sectioned to show short stem and storage tissue of current corm and disintegration of corm of preceding year, ×1; *D,* Longitudinal section of young daffodil plant, showing two bulbs, one with two shoots, united by the short stem; ×1; *E,* Potato tuber; note spiral arrangement of the eyes and the short sprout, ×1.

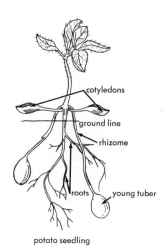

Figure 7.33. A potato seedling showing development of young tubers at the end of slender rhizomes.

potato seedling

stems with terminal flowers at every third node. The intervening nodes are marked by only small sheath leaves.

Corms

A shortened vertical, thickened underground stem is a **corm** (Fig. 7.32*B*). In gladiolus it consists of a short stem with much stored food. Nodes are, as usual, indicated by leaves; the bases of some are shown in Fig. 7.32*B*. Small buds occur in the axils of some of these leaves. A median section of the corm (Fig. 7.32*C*) distinguishes between stored food and the central portion containing the single bud that will produce a single leafy, flowering shoot. The food stored in a dormant corm is used in the production of the leafy shoot. New corms will develop from the axillary buds. In addition, short underground stems may form, each giving rise, at its tip, to a single small corm.

Bulbs

A **bulb** differs from a corm in that in a bulb, food is stored in leafy scales. The stem portion is small and has at least one central terminal bud that will produce a single upright leafy stem. In addition, there is at least one axillary bud that will produce a bulb for the subsequent year. In the longitudinal section of the sprouting daffodil bulb shown in Fig. 7.32*D,* the stem is producing three leafy stalks, one of which is forming a new bulb.

The food stored in the leafy scales of the bulb is used up by the initial growth of the leafy shoot. Food to be stored for the new bulb is supplied from the leafy shoot. The table onion is a good example of a commercially valuable bulb.

Tubers

Tubers are enlarged terminal portions of slender rhizomes (Figs. 7.32, 7.33). The potato, a dicotyledon (*Solanum tuberosum*), is a good example. The potato plant possesses three types of stems: (1) ordinary **aerial stems,** (2) slender rhizomes underground, which become enlarged at the tips and form (3) **tubers.** In the mature potato the scar left where the tuber was broken from the rhizome is clearly visible. On the potato tuber there are nodes and internodes, lateral buds, and a terminal bud. The buds develop into stems (Fig. 7.32*E*). The "eyes" of the tuber are groups of buds; each group along the sides represents a lateral branch with undeveloped internodes, and at the unattached "seed end" of the tuber, the "eye" is in reality a terminal branch on which only one bud is strictly terminal. In an elongated potato, it is possible to make out the spiral arrangement of the eyes, for there is only one eye at a node. Beginning at one end of a tuber and proceeding toward the other end, while at the same time turning the tuber, usually makes it possible to follow the spiral arrangement (Fig. 7.32*E*).

Figure 7.34. Arborescent monocotyledons. *A,* California fan palm (*Washington filifera*) in a native stand; *B,* palms, and presumably Pandanus, have long-lived phloem; *C,* the Joshua tree (*Yucca brevifolia*) has secondary growth but produces typical closed vascular bundles from meristematic tissue resembling procambium strands. (*A,* courtesy of Richard Parker–National Audubon Society; *B,* courtesy of Diamon T. Smithers–National Audubon Society.)

THE ARBORESCENT MONOCOTYLEDONS

Palmaceae

The arborescent monocotyledons, in achieving longevity with only primary growth, pose a perplexing problem. Palms (Fig. 7.34A) have only primary growth. They continue to elongate, reaching a height of as much as 100 feet and living upwards of 200 years. The bases of their trunks are slightly swollen because the first internodes are shorter and broader than those produced later. There is also some continued enlargement of the parenchyma cells in the trunk bases, with the deposition of woody material in the cell walls. There may also be an elaboration of a corky tissue (periderm). Otherwise, the diameter of the tall palm trunk is constant. The palms have a massive apical cone from which new leaves and stem structure differentiate. Primary growth is presumably normal, with procambium strands branching out into the differentiating leaves as well as growing upward in pace with the advancing apical meristem. New leaves are thus connected with the root system by way of living and functioning primary vascular tissue in palms over 200 years old. Even the parenchyma cells in the oldest parts of the stem are still able to enlarge and elaborate new wall material. Presumably, the sieve elements living and functioning during the 200-year life span of the palm tree lack nuclei. Though their ultrastructure is not known, the light microscope shows them to be similar to the one-to-two-year-old sieve elements of herbaceous plants. The only other living, functioning cell lacking a nucleus is the mammalian red blood cell, and its life span is measured in weeks. The century life span of the cells of a palm tree is without explanation.

Pandanaceae

Although many species of *Pandanus* develop a massive conical apex similar to that in palms, their continued arborescent growth is not anticipated (as it is in palms) by the initial production of an enlarged buttressed base (Fig. 7.34B). Instead, root primorida are produced in the axils of the leaves of the first five to ten nodes. At the summit of the stem, leaves are produced and die as in the palm. Buds, as usual, are present in the leaf axils. While most of these buds remain dormant, a specific proportion grow, producing a regularly branched tree. Since the branches also taper, many *Pandanus* trees resemble dicotyledonous trees. Each new branch has root primordia at the first formed nodes. Additional support and probably increased capacity for conduction for the enlarging tree is obtained by the development of roots from the root primordia at the lower nodes. Since secondary growth has not been observed in any of the Pandanaceae, it too must have long-lived primary vascular tissue in at least portions of its older stems. It is, however, not known whether after the formation of the prop roots, the primary tissue in the initial stem remains functional.

Agavaceae

In *Dracena, Yucca* (Fig. 7.34C), and other members of this family, a cylinder of cambium is present around the stem close to its surface. This cambium, however, produces secondary xylem and phloem and parenchyma cells mostly on the inner side of the cambium. At precisely separated points, small aggregations of parenchyma cells develop into a meristematic tissue resembling a procambium strand. From this xylem and phloem differentiate in a manner similar to the differentiation of primary vascular tissue in a procambium strand. Thus, secondary growth occurs in the Agavaceae. Stems increase in girth, and primary vascular tissues at shoot and root tips are connected by young vascular tissues in the stem. However, in these cases a strand resembling a procambium strand is secondarily produced and the elaboration of xylem and phloem follows the pattern of primary development.

Other Arborescent Monocotyledons

There are still other small assemblages of arborescent monocotyledons, such as bamboo (*Bambusa*), *Philodendron,* and others that occur as isolated groups in large herbaceous families. The methods of production of the vascular tissue in most of these forms is too unclear for consideration here.

SUMMARY OF PRIMARY GROWTH IN STEMS

1. Buds are characteristic of woody stems. Bud scales enclose and protect rudimentary leaves surrounding an apex. The apex may be either a vegetative shoot or a floral apex. A floral apex terminates growth. Woody plants require each year new vegetative growth which is normally accompanied by flowers. Various devices have evolved that insure the formation each year of buds enclosing floral and or vegetative shoot apices.

2. Primary growth brings about the elongation of the stem and lays down the basic pattern of cells and primary tissues characteristic of the particular stem and upon which the functioning and future growth of the stem depend.

3. The primary meristematic tissues are: apical meristem, protoderm, ground meristem, and procambium.

4. The primary plant body is composed of the following different kinds of cells and tissues: (a) the epidermis, with possibly three cell types (epidermal, guard cells, and epidermal hairs); (b) the cortex, composed of collenchyma, sclerenchyma (fibers and sclereids) and parenchyma; (c) the vascular tissue, composed of xylem (fibers, tracheids, vessel elements, and parenchyma) and phloem (fibers, sieve-tube elements, companion cells, and parenchyma); (d) pith, composed largely of parenchyma with occasional fibers; (e) pith rays, composed of parenchyma cells.

5. The functions of these cells and tissues are as follows. Parenchyma is used for the storage of water and food and the conduction of materials for short distances. Collenchyma, sclereids, fibers, and tracheids are strengthening or mechanical tissue elements. Tracheids and vessels (series of vessel elements) conduct water and mineral salts. Sieve tubes (series of sieve-tube elements) conduct foods. Vascular cambium is a meristematic tissue, the cells of which are capable of division.

6. Collenchyma and sclerenchyma may occur in patches or completely surround the stem just underneath the epidermis. Sclerenchyma is frequently associated with vascular bundles.

7. In woody and herbaceous stems the vascular tissues are arranged in bundles, generally forming a definite circle. In monocotyledonous plants the bundles are irregularly distributed throughout ground parenchyma.

8. In woody dicotyledonous plants some procambium, together with parenchyma cells of the pith rays, unite to form a continuous cylinder of vascular cambium which proceeds to lay down the secondary plant body.

9. With the exception of a few genera such as the palms, the life span of plant cells does not exceed five years. Longevity of individual woody species is attained by secondary growth. In species wih only primary growth, longevity is attained by continued primary growth at the shoot apex and the formation of roots at nodes. Older portions of the stem die.

8

The Stem—Secondary Growth

Whereas primary growth is responsible for increases in stem length and the laying down of the basic tissue pattern, secondary growth is responsible for the development of new secondary vascular tissue, increasing the girth of the stem and providing a continuous connection of living cells between the newly formed primary tissues of the developing shoots and roots. This device makes possible the attainment of great age by individual plants, although living tissue may be only three years old or less.

SECONDARY GROWTH

Formation of Vascular Cambium

This immediate discussion is concerned with perennial woody stems like those of common shrubs and trees, including pines and their allies; but even some herbs, such as alfalfa, may develop some secondary tissues before the end of their relatively short lives.

Indeed, the formation of a complete circle of cambium in alfalfa makes this plant a convenient one for an explanation of the manner in which cambium develops (Fig. 8.1). This occurs in the following manner. Not all of the procambium between the primary xylem and primary phloem is used up; some remains and proceeds to divide to form phloem toward the exterior of the stem and xylem toward the interior. It becomes a vascular cambium and forms **secondary xylem** and **secondary phloem,** thus causing the stem to increase in girth (Figs. 8.1*A*, 8.2). Since it forms within vascular bundles it is called a **fascicular cambium.** What happens between the vascular bundles? In stems such as alfalfa, in which the vascular bundles are separated by strips of parenchyma, the pith rays, certain of these cells may become meristematic (Fig. 8.1*A*). Thus, cambium within the bundles may be joined by cambium between the bundles, forming a complete cambium ring. (Figs. 8.1*B, C*). Cambium formed between bundles is called **interfascicular cambium.** The circle of vascular cambium thus formed in alfalfa and other herbaceous stems is not active for long. In woody stems it remains active as long as the tree lives, over a thousand years in redwoods. Interfascicular cambium thus formed joins the fascicular cambium to form a complete ring of vascular cambium (Fig. 8.1C). It should be noted that fascicular cambium cells are derived from procambium cells that have never ceased to behave as a meristem; interfascicular cambium originates from cells that have differentiated into a parenchyma tissue.

This development results in a circular sheet of vascular cambium completely surrounding the xylem and surrounded on the outside by the phloem (Fig. 8.3A). In cross section these cambial cells are uniformly bricklike with thin walls (Fig. 8.3C). However, all cells of the vascular cambium thus formed are not alike, as reference to the longitudinal view (Fig. 8.3B) will show. The small cells packed in regular groups will develop into vascular rays;

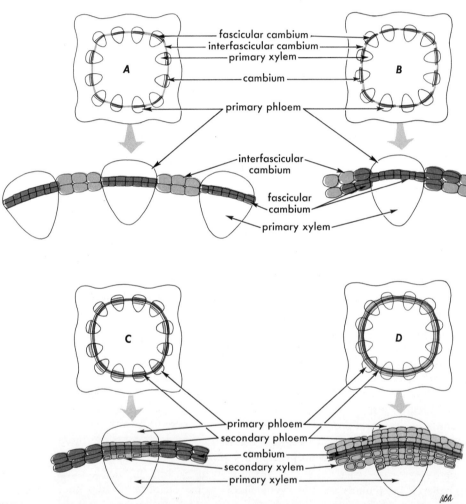

Figure 8.1. Formation of a complete cylinder of vascular cambium. A, at completion of primary growth, some meristematic cells remain between primary xylem and primary phloem (fascicular cambium, shown in dark green). Parenchyma cells appear in pith rays between vascular bundles (light green); B, some of the parenchyma cells of the pith ray become meristematic (interfascicular cambium; dark green); C, parenchyma cells of pith between the meristematic cells of the bundles have returned to a meristematic state, forming a cylinder of cambium (dark green); D, complete cylinder of secondary xylem and secondary phloem is formed by the vascular cambium (light gray). Parenchyma cells are shown in light green, vascular cambium in dark green, and secondary tissues in light gray.

Figure 8.2. Cross section of a stem of alfalfa (*Medicago sativa*), showing the fascicular cambium within the bundle and interfascicular cambium between the bundles. Some secondary vascular tissue has been produced. ×300.

they are **ray initials.** The elongated cambial cells will develop into fibers, tracheids, vessel elements, companion cells, or sieve tube elements; these cambium cells are called **fusiform initials.**

In most woody species, the interfascicular cambium differentiates typical secondary xylem internally and typical secondary phloem externally, so that a solid cylinder of xylem and phloem is formed (Fig. 8.1*D*). In many vine types of stems, the interfascicular cambium continues to produce parenchyma cells only; thus these stems are composed of a number of well-formed vascular bundles separated by distinct pith rays.

Sections of Woody Stems

A woody stem may be viewed in three different planes (Fig. 8.4). A cross-wise view is logically termed a **cross section.** There are two types of longitudinal sections. A section cut in a plane passing through both the center of the stem and the circumference is cut along a radius and is termed a **radial section;** the other longitudinal section is cut at right angles to a radius and is therefore called a **tangential section.**

Differentiation of Phloem and Xylem from Vascular Cambium

Figure 8.5 shows stages in the differentiation of vascular cambial cells to form secondary phloem and secondary xylem. Cambial cells divide most frequently by a tangential wall, forming two daughter cells. As a rule, the inner daughter cell, next to the xylem, develops into a secondary xylem cell, and the outer daughter cell, next to the phloem, remains meristematic and again divides. When new phloem elements are produced, the inner daughter cell retains the power of division, and the outer

Figure 8.3. Vascular cambium. *A,* diagram showing relationship of cambium and cambial initial to stem; *B,* tangential section ×140; *C,* cross section of quiescent cambium of locust (*Robinia*) ×160; *D,* active cambium of *Fraxinus.* ×200. (*B* and *C,* courtesy of R. Evert; *D,* courtesy of L. M. Srivastava.)

daughter cell gives rise to one or more secondary phloem cells. Generally, more secondary xylem than secondary phloem is produced during the season. The cambium cells continue thus to divide throughout the growing season, adding secondary xylem on the *outside* of the old xylem, and secondary phloem on the *inside* of the old phloem. Thus, secondary xylem is superimposed upon primary xylem, and secondary phloem tends to exert pressure on primary phloem and cortex and to push them outward. Although cambial cells, by repeated divisions, are differentiating into xylem and phloem elements,

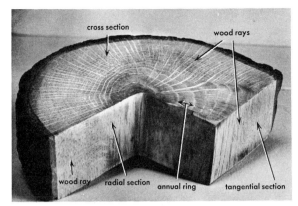

Figure 8.4. Portion of stem of oak (*Quercus*), showing cross, radial, and tangential sections and their gross characteristics. × ½.

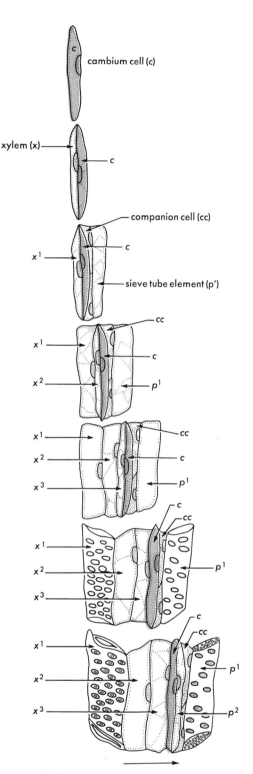

Figure 8.5. Diagram as seen in radial section, showing stages in differentiation of vascular cambium cells (c, cambium; cc, companion cell; p^1, p^2, phloem; x^1, x^2, x^3, xylem).

some daughter cells remain meristematic, and so a cambium is always between the xylem and the phloem. In addition, as the stem increases in circumference, the cambium keeps step with this increase by radial divisions of cambial cells, thus increasing the number of cells and the circumference of the cambium cylinder. During the winter season the cambium is inactive, only to begin divisions again in the spring.

The resulting cellular composition of the secondary xylem (Fig. 8.6) and phloem is, in general, quite similar to that of the primary vascular tissues. It differs in that annular and spiral vessels are usually lacking, and secondary phloem contains more fibers.

The Tissues of a Young Woody Stem

In a section of a young woody twig, such as that of the basswood shown in Fig. 8.7, the relationships and arrangements of the various secondary and primary elements may be seen distinctly. Note the layers of cork cells on the exterior of the stem. The collenchyma of the cortex (primary tissue) is intact, but there appears to be some crushing of the cortical parenchyma. The primary phloem is the outermost tissue of the phloem region; it is not distinguishable in this figure. There are two types of **phloem rays,** and the sieve tubes and companion cells are sandwiched between masses of fibers. Individual cambium cells are difficult to distinguish, but a cambial zone is apparent. The secondary xylem is well differentiated into **spring wood** and **summer wood,** which together form a year's growth or an **annual**

Figure 8.6. (*above*) Tracheids and vessel elements from secondary wood. A, tracheid from spring wood of white pine (*Pinus*); B, tip of tracheid from wood of oak (*Quercus*); C, tip of vessel element from wood of magnolia; D, tip of vessel element from wood of basswood (*Tilia*); E, vessel end walls in xylem of *Alseuosmia*. (A to D redrawn from Carl C. Forsaith, *The Technology of New York State Timbers*, New York State College of Forestry Publication 18; E, courtesy of L. M. Srivastava.)

Figure 8.7. (*right*) Cross section of 3-year-old stem of basswood (*Tilia*). ×185. (Slide courtesy of Triarch Products.)

Figure 8.8. Cross sections of A, oak (*Quercus borealis*); B, poplar (*Populus* deltoides). ×30. (Slides courtesy of Triarch Products.)

ring. Spring wood is frequently characterized by cells having larger lumens and thinner walls than those found in summer wood. Vessels are prominent, but in cross section the distinction between tracheids and fibers is difficult to see. **Xylem rays,** composed of parenchyma cells, are conspicuous, although none occur which are as large as the larger phloem ray in the illustration. Points of primary xylem may be distinguished adjacent to the parenchyma cells of the pith (primary tissue).

The Tissues of an Old Woody Stem

As the woody stem increases in girth, the primary phloem and tissues external to it are completely lost, as will be described below. The primary xylem, on the other hand, together with the pith, remains in the center of the stem, but both are reduced to insignificance by the great accumulation of secondary xylem. The possible variations in size, shape, and arrangement of the vessels, tracheids, fibers, and parenchyma cells composing secondary xylem are so great that individual species may generally be

identified by the structure of their wood. The secondary phloem, formed in lesser amounts and continuously sloughed off, constitutes a complex tissue.

Xylem. The cell types of the xylem have already been described in the discussion of primary xylem, and though detailed differences do occur between the cells of the primary and secondary xylem, we shall not consider these differences. Thus in the following description of secondary xylem, the fibers, tracheids, vessels, and parenchyma cells, unless otherwise specified, will be considered as having the same characteristic as those of primary xylem.

A comparison of the cross sections of oak and poplar (Fig. 8.8) shows how differences in the size and distribution of the vessels may be responsible for important characteristics of wood. Note that summer and spring wood are present in both cases. In oak the vessels are of two distinct sizes and less numerous than in poplar. Furthermore, the larger vessels of oak are grouped in the spring growth. The character of the rays and grouping of the fibers are also different in each species. Because of the distribu-

Figure 8.9. Sections of wood of oak (*Quercus borealis*). A, cross section; B, tangential section; C, radial section. ×40. (Slides courtesy of Triarch Products.)

Figure 8.10. Tyloses in vessels of walnut (*Juglans cinerea*). ×150. (Slide courtesy of Triarch Products.)

tion of the vessels in rings in the oak, it is said to be **ring porous,** while poplar is **diffuse porous.**

Under higher magnification the cellular detail of the oak wood becomes more apparent (Fig. 8.9*A*). Note the large vessels in the spring wood and the compact summer wood. The rays are of two sizes, one but a single cell in width and the other massive. The radial view (Fig. 8.9*C*) shows the rays, one cell in width, lying perpendicular to a group of fibers. Note the intertwining fibers and the parallel rows of wood parenchyma interspersed among the fibers. A single large vessel is present in this section and, as it was not perfectly parallel with the section, the cut passes through it. Note the pits here and the overlying parenchyma cells. One complete vessel element and portions of two others are visible. The locations of the original end walls are indicated by projections into the lumen of the vessel. The tangential section (Fig. 8.9*B*) shows again the intertwining fibers and the vertical rows of parenchyma cells. The two types of rays are strikingly evident. Note that the vessel present in this section is smaller in diameter than that of the vessel seen in the radial section, but it is composed of elements of comparable length, as indicated by the projections of the original end wall. The presence of the large rays in radial section is one of the characteristic features of quartersawed oak.

As a tree grows older the wood in the center of the stem changes in character; it becomes known as **heartwood** and it no longer serves to conduct. The outer unchanged conducting wood is called **sapwood** (Fig. 8.19). Parenchyma cells adjacent to the vessels may grow into the lumen of the vessel through the pits and completely plug the vessel. Such cells form **tyloses** (Fig. 8.10).

The wood of gymnosperms, with few exceptions, has no vessels. A three-dimensional reconstruction is diagrammed in Fig. 8.12. A cross section from a small piece of *Sequoia* is shown in Fig. 8.11*A*. It is extremely regular, almost like a wire netting. The tracheids appear in cross section as square hollow cells. Those formed in the spring are the largest in diameter; as the season progresses, newly formed tracheids are smaller in diameter with thicker walls. The heavy-walled cell is called a **fiber-tracheid.** Note the **xylem rays** or **wood rays** composed of parenchyma cells, and the occasional small square cells with dark-staining contents. These latter are also parenchyma cells. The radial view (Figs. 8.11, 8.12) shows mainly tracheids, with a small strand of fiber-tracheids. Pits are present in the radial walls of tracheids. An elongated parenchyma cell is made apparent by its dark-staining contents. Wood parenchyma cells are represented by dark green in Fig. 8.12. They are shown in the tangential and cross sections. A horizontal ray in the radial section lies perpendicular to the tracheid. Pits occur between ray cells and the tracheids. The ends of the rays are apparent in the tangential section (Fig. 8.11*B*). Here the fiber-tracheids have pits in their tangential walls. The pits in the radial walls of the tracheids are just visible. In this view several parenchyma cells with heavily stained contents are present.

Sequoia does not produce resin as do many other

Figure 8.11. Sections of wood of redwood (*Sequoia sempervirens*). A, cross section; B, tangential section; C, radial section. ×100. (Slides courtesy of D. Graham.)

Figure 8.12. Block diagram of secondary xylem of redwood (*Sequoia sempervirens*).

gymnosperms. This substance occurs in specialized cavities which are surrounded by groups of parenchyma cells. Together they are called **resin ducts** and occur in many gymnosperms, where they may be continuous over considerable distances. One is shown in Fig. 8.13.

We have learned that water and mineral salts are conducted upward in the xylem (in tracheids and vessels), and that foods are conducted upward or downward in the phloem (in the sieve tubes).

Radial conduction occurs in *vascular rays*. These are composed of relatively long-lived parenchyma

cells which are somewhat elongated radially.

Vascular rays in the xylem are spoken of as wood or xylem rays, and those in the phloem as phloem rays. Phloem rays are always contiguous with xylem rays. They both originate from the ray initials of the cambium.

The parenchyma cells in vascular rays may differ in detail from the parenchyma cells of the pith and the cortex. The parenchyma cells of the cortex, pith, xylem, and phloem form a living network extending throughout the stem, at least until the formation of heartwood.

Formation of Cork

Cork Cambium

Figure 8.14 shows the origin of cork cambium and the development of secondary tissues from it. Cork cambium in most plants arises from outer cortical cells. Cortex cells divide by tangential walls, producing daughter cells. The outer daughter cells generally differentiate into cork cells, thus forming a layer of **cork** beneath the epidermis. Should the outer daughter cells remain meristematic, the inner daughter cells may give rise to a tissue known as **phelloderm,** which is composed of parenchyma-like cells. Cork cambium may originate also from epidermal or phloem cells.

Cork

This tissue (Figs. 8.7, 8.14) is composed of flattened, thin-walled cells with no or small intercellular spaces. A fatty substance called **suberin** is deposited in the walls, rendering the cells almost impermeable to water and gases. Hence, cork tissue provides protection for the stem against excessive loss of water and also against mechanical injury. The protoplasts of cork cells are short-lived. When cork tissue with its impermeable cell walls is formed, the epidermis and other tissues outside it are cut off from water and food supplies and soon die. In

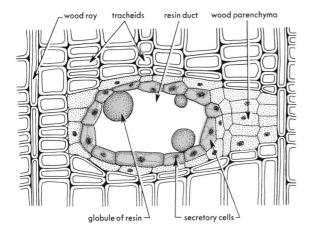

Figure 8.13. Resin duct in pine wood (*Pinus*) as seen in cross section.

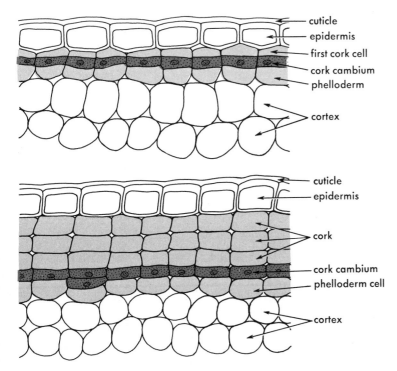

Figure 8.14. Diagram showing origin of the first cork cambium and the first layer of cork. New cork cambia and new layers of cork form in a similar manner each spring.

Figure 8.15. Cross section of a portion of a young stem of elderberry (*Sambucus*). From the outside to the inside of the stem, the following tissues may be seen: cork, cork cambium, cortex, secondary phloem and xylem. The vascular cambium is probably the two layers of thin-walled cells on the phloem side of the xylem. Note the lenticel in the cork. ×100.

woody stems, cork replaces epidermis as a protective tissue.

In a young stem the bark is made up of the following tissues, in order, from the *outside* to the *inside:* cork, cork cambium, phelloderm (if present), cortex, and phloem (Figs. 8.7, 8.15). Microscopic examination may reveal the presence of epidermal cells still clinging to the cork. In old stems, the epidermis, cortex, and primary phloem become separated from the adjacent inner tissues by successively deeper layers of cork formation. As this happens, the tissues outside of newly formed cork die for lack of water and nutrients. They dry up and eventually weather away.

Lenticels

An impervious layer of cork would effectively cut off the oxygen supply of the living tissues beneath, if it were not for the development, here and there, of groups of parenchyma cells called **lenticels** (Fig. 8.15), through the intercellular spaces of which, gases can move freely. Stomata occur in the epidermis of very young stems. Gases may enter and leave through these openings. A lenticel is usually formed just beneath a stoma and takes its place as an aerating organ in older stems. Beneath

a stoma, cork cambium cells differentiate into parenchyma rather than into cork. These masses of parenchyma tissue enlarge and finally break through the epidermis. The shape and other characters of lenticels are usually quite distinctive in different genera or species. They are especially conspicuous in cherries, plums, birches, maples, dogwoods, and hazelnuts.

Anatomy of the Stem at Nodes

Buds occur in the axils of leaves, and this whole region is called the node. Now, since both buds and leaves are joined by vascular tissue to the conducting tissue of the stem, anatomically the node must be characterized as a region where vascular bundles are connected to bring about a continuity of vascular tissue in leaves, buds, and stem. It is often a rather complex region. The bundles leading to the leaves and buds are referred to as **leaf traces** and **bud traces,** respectively. The xylem of the traces is continuous with the xylem in the leaf and stem and the phloem of the trace is continuous with the phloem in the leaf and stem.

Where vascular tissue leaves the vascular cylinder of the stem, a gap exists in the cylinder immediately above the point where the trace departs. This is called a **leaf gap** or a **bud gap.** They are shown diagrammatically in Fig. 8.16A. Branch gaps are already present in twigs in the winter condition (Fig. 7.7), and they may be seen by cutting a free-hand cross section through a bud (the future branch) and its accompanying node. Figure 8.16B shows a free-hand cross section through a bud and node of young branch of fig (*Ficus carica*). The section makes a nearly longitudinal cut through the bud (the future branch), and the continuity of the vascular tissue

of the bud with the vascular cylinder is apparent. The plane of the cross section of Fig. 8.16B is roughly indicated by the line A-A on Fig. 8.16A. The gaps are also prominent features of herbaceous stems from which pith, phloem, and cortex have rotted away. The more resistant woody tissues remain, and the gaps are holes through the xylem cylinder. This is shown in Fig. 8.16C by the vascular cylinder of a tomato stem.

Now, as secondary development of the branch

Figure 8.16. Branch and leaf gaps. A, diagram of branch and leaf gaps; B, freehand cross section of a twig of fig (*Ficus carica*) through bud and accompanying node. Gap in vascular cylinder is apparent ×15; C, gap in the vascular cylinder of a tomato (*Lycopersicon esculentum*) stem ×3. D, branches form a continuous pathway from the center of the trunk to the exterior, thus constituting a gap in the xylem cylinder of the trunk ×1/10 (*Pinus jeffreyi*).

and trunk proceed, both will increase in diameter; new vascular tissue of the stem must be laid down in such a manner as to bury the younger portions of the branch in the trunk. Furthermore, there must be a crowding of the new tissue in the upper acute angle formed by the branch and the stem. This is shown diagrammatically in Fig. 8.17. This entire process results in a knot. Can you explain why it is that in some gymnosperm lumber the knots are held firmly in place while in others they fall out?

Primary and Secondary Growth

In spite of the relatively limited life span of cells, some plants having only primary growth do reach considerable age. Primary growth, as usual, takes place at the shoot apex, but in such plants roots form at nodes a short distance behind the apex. The older portions of the stems with associated roots die after several years. No provision is made to connect the most recently formed primary growth of roots and shoots that are separated by large distances of space or age. Secondary growth does just this; it connects the active functional primary growth of shoots and roots separated by considerable distances and by many generations of living cells. This mechanism is illustrated by a longitudinal section of a three-year-old shoot (Fig. 8.18). The center of the stem is occupied by pith. The strands of primary xylem form the innermost ring of vascular tissue for all three years. The cylinder of secondary xylem produced the first year extends for only a short distance up the stem (Fig. 8.18*A*1). In cross section it is represented only in the oldest section (Fig. 8.18*D*). At the level shown in Fig. 8.18*C* only the most recently formed cylinder of secondary xylem is present. However, the primary xylem extends upward from the base of the stem to the shoot apex (Fig. 8.18*B*). The significant observation is that the active functional primary xylem is always in contact with the most recently formed cylinder of secondary xylem. Furthermore, older portions of the primary xylem are also in contact with all other annually produced cylinders of secondary xylem. The primary phloem on the periphery of the expanding stem is continually cut off from the inner portions of the stem by the stem's increase in girth and the successive formation of layers of cork (Fig. 8.18*D*). However, the young functional primary phloem, too, will always be in contact with the most recently formed layers of secondary phloem (Fig. 8.18*B*, *C*). This arrangement provides a continuous pathway

of living cells less than five years old from the shoots to the roots of trees which, as individuals, may be many hundreds of years old (Fig. 7.1*A*).

A SUMMARY OF SECONDARY GROWTH

1. Vascular cambium originates from the procambium and adjacent parenchyma cells in the rays between vascular bundles.

2. Vascular cambium lying between xylem and phloem produces new secondary xylem and secondary phloem each year; fusiform initials differentiate into elongated conducting and strengthening cells and ray initials differentiate into ray cells.

3. Primary xylem (next to the pith) becomes more and more widely separated from primary phloem (next to the cortex) as secondary xylem and phloem are produced.

4. Vascular cambium persists as a meristem between the last-formed xylem and the last-formed phloem.

5. Cork cambium is short-lived; new cork cambiums may arise each year, producing new layers of cork.

6. Cork cambium may originate in successively deeper tissues from epidermis, cortex, and phloem.

7. Secondary xylem provides a continuous pathway of young functioning cells between the active primary growth of shoot and root. This mechanism makes possible the attainment of both great age and great size by individual plants, in spite of the normal three-to- five-year life span of plant cells.

Figure 8.17. Relationship of wood and bark in main stem to wood and bark in a branch. (Redrawn from A. J. Eames and L. MacDaniels, *Introduction to Plant Anatomy.* Copyright 1947 by McGraw-Hill Book Company. Used with permission of McGraw-Hill Book Company.)

primary phloem

primary xylem

cambium

B

cortex

epidermis

cortex

secondary xylem

primary xylem

primary phloem

secondary phloem

cambium

cork

C

cambium

primary xylem

secondary xylem

cork

pith

secondary phloem

bark

D

1 2 3

A

Figure 8.18. Changes in a stem as it increases in age and in girth. *A*, longitudinal section through a 3-year-old stem; *B, C, D*, cross section at indicated levels.

Figure 8.19. Cross section of a branch of mulberry (*Morus*). × ½.

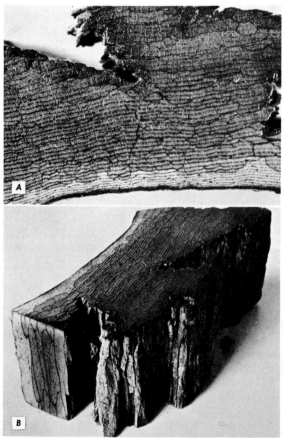

Figure 8.20. Bark of incense cedar (*Libocedrus decurrens*). A, section of bark; B, three dimensional view. × ½.

Macroscopic Appearance of an Old Woody Stem

If we look at the end of a several-years-old stem or at the end of a log, two prominent regions are distinguished—the bark and the wood (Fig. 8.19). Pith may be evident in certain species or in young stems. In old stems, it may be impossible to find even a trace of pith, and it may be also very difficult to find primary xylem, which is at the very center of the stem just outside the pith. The vascular cambium can be seen only by microscopic examination.

Bark

The bark (Fig. 8.19) includes all of the tissues outside of the vascular cambium and its exact cellular composition will depend upon the age and species of the twig or tree trunk being examined. In this discussion we shall consider only bark as it occurs on an old woody tree trunk in which all of the epidermal and cortical tissues have disappeared and the activity of the cork cambium has resulted in the formation of considerable cork tissue. Bark as such may easily be separated from the woody cylinder or trunk, especially in the spring or early summer, because the walls of the cells in the "cambial zone" are thin and easily ruptured. When the bark is peeled from the wood, some of the cells of the "cambial zone" adhere to the bark, and some adhere to the wood. Examination of a piece of bark of incense cedar (Fig. 8.20) shows that it is made up of a series of layers of two different-appearing tissues. Recalling that the vascular cylinder of a tree is increasing in diameter, let us examine the manner of formation of this bark.

At the close of primary growth the primary vascular tissues, in strands, formed a ring around the peripheral area of the stem (Figs. 7.27, 8.18*A*). The central portion of the stem was occupied by pith. The cortex and epidermis surrounded the ring of vascular strands. The fascicular cambium joined with the interfascicular cambium to produce a ring of secondary vascular tissue (Fig. 8.1). This increases the girth of the stem (Fig. 8.18). As this takes place, what will happen to the cortex and epidermis?

The cells composing the epidermis are fully differentiated, and except in rare cases (apple), do not divide. As the stem increases in diameter through the activity of the vascular cambium, the epidermis must be stretched and torn. When this

Figure 8.21. Bark of three species of trees. A, western yellow pine (*Pinus ponderosa*); B, cork oak (*Quercus suber*); C, plane tree (*Platanus orientalis*).

happens, the delicate underlying cells would be exposed to drying out and weathering were it not for the development of a cork cambium and a layer of cork as previously described (Figs. 8.14, 8.18*A, C*).

As the stem continues to grow in girth the cylinders of cork, in their turn, will be ruptured. Exposure of underlying tissue is prevented by generation each spring, or sometimes more frequently, of a new cork cambium. Finally, even the parenchyma cells of the secondary phloem may produce a new cork cambium (Figs. 8.18*A, D*). This is why bark from some old tree trunks is layered (Fig. 8.20). This layering is brought about by the formation of successive layers of cork, cutting off exterior phloem tissue. The tissue is thus deprived of water by the impervious nature of the cork cells. Now, with further increase in circumference, the dead exterior sheets of phloem and cork cells are stretched and torn, resulting in bark patterns characteristic of many trees (Figs. 8.21, 8.18*A, D*). Weathering causes a continual wearing away of the surface of the bark, and in regions where sand particles are carried by the wind, the trees have a beautiful polished appearance. Commercial cork is the bark of an oak (*Quercus suber*) (Fig. 8.21*B*). It is not layered, because it develops in a somewhat different fashion, the cork cambium never developing within phloem tissue. Thus, the outer bark of the cork oak may be comprised almost entirely of cork cells.

Wood

The commercial product, **wood,** is xylem. It occurs in all vascular plants, even the tenderest herbs, but only trees that have lived for a number of years and gradually built up a large trunk capable of being converted into lumber furnish us with commercial wood (secondary xylem).

Annually, the vascular cambium forms a new layer of wood which is usually many cells thick. As seen on the end of a log, these layers appear as concentric rings (Fig. 8.19), the so-called **annual rings.** It should be kept in mind, however, that the vascular cambium in a tree like the pine, for example, has the form of a hollow cone; consequently, the wood produced annually has the same form (Fig. 8.22). The wood produced in any one year is laid down over that of the preceding year.

Annual rings of wood are particularly conspicuous in woody plants of temperate climates, where a season of active growth alternates with a season of inactivity. In such plants, the wood elements formed in the spring and early summer are quite different from those produced in late summer and fall. That portion of the ring formed in the early part of the season is known as **spring wood.** That of angiosperms has larger and more numerous vessels than does the so-called **summer wood,** produced later in the season (Figs. 8.7, 8.9). Also, tracheids and fibers of spring wood are less abundant and thinner-walled than those of summer wood. As viewed with a hand lens, or even with the naked eye, summer wood appears darker and denser than spring wood. Of course, in any annual ring, summer wood lies outside of spring wood, and the summer wood of any year is bounded on the outside by the spring wood of the following year. Therefore, small, thick-walled elements of summer wood of one year stand out in contrast to the adjoining, relatively large, thin-walled wood elements of the spring growth of the following year (Fig. 8.7). Thus a sharp line of demarcation lies between them.

Annual rings or growth increments are just as conspicuous in gymnosperms, which possess no vessels, as they are in woody angiosperms. In gymnosperms, the tracheids of spring wood are radially larger and have thinner walls than those of summer wood.

In the moist tropics, where there are no marked seasonal changes, growth rings are almost completely absent or, at best, poorly defined. If they occur, they do not necessarily represent annual growth increments and hence could not be used in determining the age of a tree.

Examination of the cut end of a log shows that in the individual tree annual rings vary greatly in thickness. In other words, the amount of wood produced each year is not the same. As a rule, the first few rings produced in the early life of the tree are thicker than those formed later. Also, the thickness of rings, or, to state it differently, the amount of wood added during any season, is influenced by many factors, such as the age of the tree, light, temperature, moisture, leaf area, and competition both above and below ground. As would be expected, naturally fast-growing tree species have thicker annual rings than slow-growing species living under the same environmental conditions.

The **age of a woody stem** can be determined by counting the annual rings of wood. It is obvious that the true age of a tree can be ascertained only by counting the rings at the very base of the main trunk. As shown in Fig. 8.22, the number of annual rings decreases from the base of the trunk to its apex.

Infrequently, two or more rings of growth may be formed during one season. For example, in irrigated apricot orchards, an application of water may stimulate cambium activity, resulting in the differentiation of wood elements resembling those of spring growth. Then, as the soil dries out, the wood

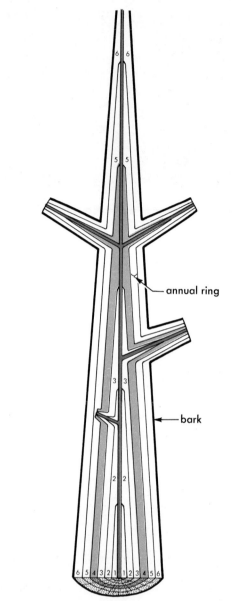

Figure 8.22. Diagram showing cross and longitudinal sections of a 6-year-old tree.

elements formed may be smaller and thicker-walled. A second irrigation again stimulates the vascular cambium to greater activity, and thus a line of demarcation is formed between two rings of growth, both made in one season. Sometimes a drought or an insect attack may destroy the foliage and thus interrupt the activity of the cambium. Later, a new crop of leaves may be produced, the cambial activity renewed, and another ring of wood produced.

Under these conditions two (or more) growth rings of wood will be formed in one season.

The forester is primarily interested in the amount and quality of wood laid down in the main trunks of commercial timber trees. It is of practical value to know in terms of board feet the volume of wood formed by a certain species, say over a 10-year period. The volume may be determined by counting the annual rings on cross cuts of tree trunks, or on stumps of a considerable number of trees, or by using an instrument known as the increment borer to determine the age of a number of standing trees. This borer has a hollow bit that bores out a "core" of wood, on which the rings may be counted. The hole that it makes does not seriously injure the tree.

Sapwood and heartwood. The cross sectional end of a log usually shows the presence of two zones of wood: (1) a light-colored outer zone, the **sapwood,** surrounding a generally darker-colored zone, the **heartwood** (Fig. 8.18). The sapwood functions in sap conduction and food storage, whereas the heartwood performs only the function of mechanical support. Sapwood is gradually transformed into heartwood. In this transformation, various substances infiltrate into the cell walls and even accumulate in cell cavities. The substances include tannins, gums, resins, pigments of various sorts, salts of organic acids, and other materials. These depositions may impart a characteristic color to the heartwood, and they also may increase its weight, hardness, and durability and decrease its permeability. In some species the vessels and even tracheids may be plugged by protrusions from adjacent living parenchyma cells. These plugs are called **tyloses** (Fig. 8.10). The penetration of wood preservatives is rendered difficult by the presence of tyloses.

Industrial Products from Trees

The most obvious use of wood is as lumber. Trees, however, are grown in many countries for other purposes, some of which do not involve the destruction of the trees. For instance, turpentine, natural rubber, and chewing gum are derived, respectively, from the sap exudates of southern pine, particularly the long leaf pine (*P. palustris*) and the slash pine (*P. cubensis*); the rubber tree (*Hevea brasiliensis*); and the sapodilla tree (*Achras sapota*). In all three cases, gashes are cut in the trunks of the trees and the sap is collected in buckets or on sticks, after which it is processed in factories into the well-known articles of commerce.

When the wood itself is utilized, the trunk may be sawed into lumber or completely disintegrated into the fibers and tracheids of which it is composed. In the manufacture of ordinary lumber the log moves against the saw so that tangential cuts rip each board from the log. The large rays of oak wood gives boards cut from these trees a characteristic pleasing appearance if they are cut in a radial plane (Fig. 8.24B).

Sheets for plywood manufacture are produced in a different manner. The logs are steamed and then placed on a large lathe and made to turn against a large and sharp knife which literally peels off thin sheets of wood from the slowly revolving log (Fig. 8.23A, B). Note (Fig. 8.24) the different appearances that these processing procedures give to wood. Can you account for the "grain" in A and B, in terms of spring and summer wood? Compare B with Fig. 8.4.

For other industrial uses wood is not sawed into boards but placed into machines which tear it apart, breaking it down in some instances into fibers and tracheids. The paper you have in your hands is the end product of such a process (Fig. 8.25), and if examined under a microscope will reveal the presence of the cellular elements of the trees from which it was manufactured. Dacron shirts may have started as trees, although other cellulose products may have been the original raw material. To make such fibers, the cellulose is dissolved in various solvents to produce one of several modern man-made fibers. In paper making the lignin is undesirable and must be removed from the fibers and tracheids. It is a waste product and huge piles of it have accumulated outside of paper mills. Finding a use for it has been a major research project of a number of the large paper companies. It is now being marketed and used among other things, as a soil conditioner, an aid in drilling oil wells, and an adhesive in the manufacture of plywood and linoleum.

SUMMARY OF STEM GROWTH

1. Primary growth provides the basic requirements for the production of a complete plant body from seed to seed.

2. The predominently short life span of three to five years for cells, in general, limits the size and longevity of individual plants having only primary growth. The mechanism of secondary growth overcomes this limitation.

3. If all of the meristematic cells of the procambium strand become differentiated into primary vascular tissue, the vascular strand is closed to further growth.

4. If there remains an active meristematic region between primary xylem and primary phloem, continued growth is possible. These meristematic cells become the fascicular cambium.

5. An interfascicular cambium forms from the ray parenchyma cells between the vascular strands.

6. Union of fascicular and interfascicular cambium produces a uniform cylinder of vascular cambium.

7. Divisions in the cambium are longitudinal divisions, so the stem now increases only in girth.

8. The smooth structural transition from primary to secondary growth means that the primary vascular tissue is always in physical and physiological contact with the most recently formed secondary tissues.

9. In a 1000-year-old redwood or bristle cone pine, the primary tissues in the young shoots and roots have been in continued physical and physiological contact with each and every layer of secondary vascular tissue as these layers have formed during the past 1000 years.

10. The production of secondary vascular tissue thus makes possible the attainment of great size and great age by individual trees, even though the functioning cells may be no older than those in a perennial herbaceous plant such as an *Iris*.

Figure 8.23. Manner of preparing plywood. A, sheet of veneer being peeled from a slowly revolving log; B, diagram showing relation of grain in plywood to annual growth rings of the log. (A, courtesy of American Forest Institute.)

Figure 8.24. Characteristics of finished woods. *A*, plywood, a perfect tangential section; *B*, quarter-sawed oak, a radial section; *C*, knotty pine. ✕ ¼.

Figure 8.25. Manufacture of paper. The "wet" end of a Fourdrinier paper machine. A thin suspension of pulp flows onto a moving endless wire screen from the head box in the background. The water drains through the screen leaving behind a mat of fibers in the form of a sheet of paper shown in the foreground. (Courtesy of Crown Zellerbach Corporation.)

Summary of Stem Growth | 161

STEM MODIFICATIONS

Stems may become adapted for functions other than support, conduction, and production of new growth. They may, for instance, become attachment organs for vines; they may carry on photosynthesis, store water in preparation for periods of water stress and develop protecting devices.

Stem Tendrils

Tendrils are slender, coiling structures that are sensitive to contact stimuli and attach the plant to a support. Tendrils of plants, morphologically, are of two sorts: leaf and stem. In the trumpet flower, for example, several of the uppermost pairs of leaf-lets have no blades, but instead form very slender tendrils (Fig. 10.15*E*). Obviously, these are leaf tendrils. In grapes and Virginia creepers, the tendrils are modified stems (Fig. 8.26*B*). This is evidenced by the fact that they arise at nodes in leaf axils and sometimes bear small leaves or flowers.

In the Virginia creeper, each tendril ends in a knob which flattens out when it comes in contact with a surface, and it adheres to that surface by a mucilaginous disk-shaped structure.

Cladophylls (Cladodes)

These are stems which are leaflike in form, are green, and perform the functions of leaves. They may bear flowers, fruit, and temporary leaves. Examples of plants with cladophylls are *Ruscus* (Fig. 8.26*D*), asparagus (Fig. 8.26*E*), smilax, and various species of cacti (*Opuntia*).

Stem Spines and Thorns

Most spines and thorns of plants are modified stems or outgrowths of the stems. Leaf spines, however, occur in certain plants, such as barberry and black locust, and in a few cases even roots become modified as spines. Good examples of stem thorns are those of *Pyracantha* (Fig. 8.26*F*) and honey locust (Fig. 8.26*G*). They are borne in the axils of leaves as ordinary branches are. Sometimes thorns bear leaves (Fig. 8.26*F*), which is further evidence that they are stems. In the honey locust the stem thorns may be branched (Fig. 8.26*G*).

Storage of Water in Stems

The so-called succulent plants are the best examples of those whose stems may store large quantities of water. In various cacti, for example, water storage

Figure 8.26. Types of stem modifications. *A* and *B*, stem tendrils; *C*, *F*, and *G*, thorns; *D*, leaflike stems.

tissue constitutes a large proportion of the stem, and the water stored therein may be as much as 90 percent of the weight of the entire plant (Fig. 18.32). This water reserve may be used during drought.

Production of Other Materials by Stems

A great range of substances are produced in stems. A few of these may be mentioned.

Mucilaginous substances (complex carbohydrates very similar to gums) accumulate in the stems of some plants, such as many ferns, cacti, and other succulents.

Tannins, substances that impart an astringent,

bitter taste to tissues in which they are present, occur in the bark and wood of many plants. They are derived commercially chiefly from the bark of hemlock, tan oaks, mangrove, and wattle (certain *Acacia* species), and from the wood of chestnut.

Latex is a milky secretion that occurs in many different families. It is a very complex mixture containing such materials as water, resin, oil, proteins, gums, tannins, sugars, alkaloids, and salts of calcium and magnesium. Latex may occur in special cells or in tubes (**latex tubes**), which are distinct from the vascular system. Common latex-producing plants are the fig, dandelion, spurge, and milkweeds. Of particular economic importance is the latex of those plants from which crude natural rubber is obtained. The most important rubber-yielding species are as follows:

(1) *Para rubber tree* (*Hevea brasiliensis*), a native of the tropical forests of the Amazon and Orinoco river valleys in South America, but since 1929 grown in large plantations, chiefly in Ceylon, Malaya, Java, and Sumatra; (2) *Panama rubber tree* (*Castilla elastica*), a native of Mexico and Central America; (3) *Manihot glaziovii*, a native tree of Brazil, yielding "ceara rubber"; (4) *guayule* (*Parthenium argentatum*), a native American shrub which, during World War II, was grown in large plantations in western United States; and (5) *kok-saghyz* or Russian dandelion (*Taraxacum koksaghyz*), a plant grown quite extensively in Russia, and to some extent, experimentally, in the United States during the period of rubber shortage.

Resins are complex substances, secreted by glands into special ducts known as **resin ducts** (Fig. 8.13). Crude resins contain, in addition to resinous materials, a considerable amount of essential oils. The principal product derived from resin is turpentine, obtained exclusively from coniferous trees.

9

The Root

The roots of a plant, considered collectively, form the **root system.** The two principal functions of roots are **anchorage** and **absorption.** In addition, the roots of all plants usually **store** a certain amount of food, at least for a short time; the roots of such plants as sugar beet, carrot, sweet potato, and others are specialized food storage organs. Besides anchorage, absorption, and storage, roots perform the function of **conduction.** Water and mineral salts absorbed from the soil, and foods that may be stored in roots, are conducted by the roots to the stems, and thence to the leaves and other organs aboveground. Foods manufactured in the leaves are conducted by the stems to the main roots, and then by the latter to branch roots, so that these foods are carried to the extremities (growing tissues) of all the smallest roots.

BALANCE BETWEEN SHOOT AND ROOT SYSTEMS

In a normal, healthy plant there is a balance between the shoot system and the root system. Of particular importance is the relation of total leaf surface to total root surface; or, to state it differently, total surface exposed to the sun, from which energy is absorbed and used in the manufacture of carbohydrates, in relation to the total root surface in contact with the soil solution, from which the plant absorbs water and mineral nutrients. The root system must be able to supply the shoot with sufficient water and mineral nutrients, and the shoot system to manufacture enough food for maintenance of the root system. Thus the balance is physiological.

Roots absorb water, stems conduct water, and leaves use water in photosynthesis and transpire to the air much of the water absorbed by the roots. We shall see that **water absorption** is a principle function of the young primary roots. Furthermore, the efficiency of the water-absorbing activity of a root system is of considerable importance in the successful competition of the plant. We have seen a great range in stem morphology. Root systems, too, show variation, which, in general, is more or less closely associated with stem morphology and the ecological niche the plants inhabit.

Small garden plants or weeds easily pulled up by the roots show a mass of soil-clad roots arising from the base of the stem. This constitutes a **fibrous root system** (Fig. 9.1*A*). Some fibrous root systems do not penetrate deeply into the soil, but extend for a distance outward from the base of the plant (Fig. 9.1*A*). Other fibrous systems are more deeply rooted, the cluster of roots forming a rather uniform mass as they penetrate the soil (Fig. 9.1*B*).

The *main roots* are approximately of the *same size;* each gives off numerous side roots, or roots of the second order; these bear branches of the third order, which, in turn, bear roots of the fourth order. All the roots are slender and fiberlike; no one root is more prominent than the others. Other young seedlings do not come up from the soil so easily. The stems of an oak or hackberry seedling 10 inches tall will break before the root system releases its hold on the soil. The oak root system must be dug up and it will then be seen to consist of a long slender root, with short lateral rootlets. It has grown directly downward deeply into the soil. This is a **tap root system** (Fig. 9.1*C*).

The fibrous root system is characteristic of many plants, including most of the monocotyledons. It is best adapted to shallow soils or to regions of light rainfall that may not penetrate the soil deeply. The tap root system is characteristic of most dicotyledons and the gymnosperms. It is well adapted to deep soils and to those in which the water table is relatively low. Most tap roots are woody but some, like carrots and sugar beets, are also storage organs.

DEVELOPMENT OF ROOT SYSTEMS

A seed contains a young plant, the **embryo.** The soaking of a seed, such as the bean, in water to remove the seed coat more easily, makes it possible to observe the entire embryo. It will be seen that certain rudimentary organs are already in evidence. One of these organs is the **radicle** or **rudimentary root** (Fig. 3.1). When the seed germinates, the radicle is usually the first structure to appear; it becomes the *first* root and is called the **primary root.** It has branches and subbranches, all of which are called **secondary roots.** In many plants, such as beet, radish, and carrot, the primary root remains the principal root throughout the life of the plant. The tap

root system is usually the primary root and its branches.

The development of a fibrous root system, like that of a cereal, is quite different from that of the tap root system. In the embryo of barley or wheat, for example, a primary root (Fig. 16.23) is evident. When the seed germinates, this root takes the lead. Soon it is followed by two pairs of threadlike roots, which, however, are not branches of the primary root; that is, they are not strictly secondary roots. The primordia of the two pairs of roots can be detected by microscopic examination of the ungerminated seed. Therefore, we may speak of the primary root and the two pairs of roots which follow it as **seminal (seed) roots,** because the primordia of all these roots are evident in the embryo of the seed. Thus, the seminal root system of barley or wheat usually consists of five roots (sometimes six). In most grasses the seminal root system is usually temporary, although under some conditions it may function throughout the life of the plant.

Within a short time after the seminal root system is formed, permanent roots arise from the lower nodes of the young stem. All the fibrous root system of a corn plant is formed from roots growing from the lowest nodes (Fig. 9.1*D*). These roots are not branches of the primary root. They are **adventitious roots.** It is customary to speak of all roots that arise from organs other than roots as adventititous roots.

Roots do not have nodes as do stems (Fig. 7.6). Indeed, the lack of nodes is a distinguishing external trait differentiating roots from stems. But roots can, and frequently do, grow at nodes on stems (Figs. 7.1*B*, 7.32). In fact, the nodes of most monocotyledons, like that of the grain just discussed and the corn shown in Fig. 9.1*D*, are provided with root primordia which may grow or remain dormant. Bermuda grass, a monocotyledon, and *Oxalis,* a dicotyledon (Fig. 9.1*E*), both regularly produce roots at nodes. This device enables them to be active and invasive competitors in lawns and gardens.

Root systems are generally thought to occur only at the lower end of the plant axis (Fig. 3.1). Such a restriction is true of all gymnosperms and of many dicotyledons, but only of some monocotyledons, and probably not of any lower vascular plants. Roots, forming fibrous root systems, regularly occur at nodes along the stem (Fig. 9.2). In many plants, like corn and wheat, there is also a fibrous root system at the lower end of the main axis of the plant complimenting the root system at designated nodes. The formation of roots at nodes is very frequently associated with some type of horizontal stem, rhizomes, or stolons (Figs. 7.1*B*, 7.32). The formation of roots at the nodes of such stems, whether monocotyledon or dicotyledon, makes possible the continued devel-

Figure 9.1. Different types of root systems. *A*, shallow, spreading, fibrous root system; *B*, fibrous root system penetrating the soil evenly from three to five feet; *C*, tap root system, in which main primary root penetrates soil eight feet or more; *D*, fibrous root system developed from adventitious roots growing at lower nodes of stem; *E*, small fibrous root systems developed from nodes of a horizontal stem.

Figure 9.2. Although roots do not ordinarily produce buds, adventitious stems do form on some roots as in *Rumex acetosella*.

opment of primary stem and root systems without resorting to secondary growth (page 132). Roots forming at nodes a short distance behind the growing shoot apex assure a continued supply of water and nutrients to the developing apex. Secondary tissues in either stem or root are thus not needed.

Adventitious roots may form on stems at places other than nodes. Canes of roses and blackberries readily initiate roots along their length where they contact moist soil. Roots will also form on the ends of stem cuttings and their formation may be stimulated by auxin.

While roots are not divided into nodes and internodes, shoot buds can form on roots. This adaptation seems to be species-specific, regularly occurring in some species and not at all in others. The common plant, sheep sorrel or dock (*Rumex acetosella*), produces long, somewhat horizontal, primary roots. These produce, at intervals, adventitious buds giving rise to shoots which grow upward to form the typical aboveground sheep sorrel shoot. A fibrous root system also grows downward from the primary horizontal roots (Fig. 9.1*E*, 9.2).

ROOT SYSTEMS AND PLANT COMPETITION

Where rainfall and temperatures permit, plants of the same and different species grow close together and compete for available soil water, mineral nutrients, and light energy. As discussed in chapter 18, plants reduce competition by utilizing different parts of the environment: some utilize full sunlight, while others grow normally in shade; some are ephemeral and complete their life-cycle only during periods of abundant moisture, while others are perennial and grow slowly even during periods of water shortage. In addition, as documented by the American ecologist J. E. Weaver, the root systems of different species occupy and utilize different parts of the soil profile.

Weaver and his students carefully excavated root systems of hundreds of prairie plants and concluded that there were three general categories of rooting depths. As shown in Fig. 9.1, some grassland species are like blue grama (*Boutelous gracilis*), which possesses a very shallow root system. Most of the roots are within the top $\frac{1}{2}$-foot of soil. Another group of species is like buffalo grass (*Buchloe dactyloides*), which has an even distribution of roots to a soil depth of about 5 feet. A third group is like loco weed (*Oxytropis lamberti*), which has an even distribution of roots to soil depths much below 5 feet. Loco weed happens to have a tap root system, but there are some grasses with fibrous root systems that also fall into the last category. These three categories of plants, then, reduce competition for moisture when growing together by utilizing different regions of the soil profile.

Competition for moisture may be the reason why one species is replacing another in the northern intermountain region of the United States. About 125 years ago, the dominant plant of this grassland area was bluebunch wheatgrass (*Agropyron spicatum*), a perennial. About the middle of the 19th century, an annual grass called cheatgrass (*Bromus tectorum*) was accidentally introduced to the area from Europe. From that time to the present, graziers have noticed an enormous increase in abundance of cheatgrass and an equally impressive decrease in the abundance of bluebunch wheatgrass. What caused the shift?

Both species have a similar life-cycle: they germinate (or break dormancy if perennial) in fall, grow slowly during the winter, grow rapidly in spring, form flowers in early summer, and die in June (or begin dormancy in mid-July if perennial). Dr. Grant Harris of Washington State University studied growth and survival of these two species during a year, starting from seed for both. He found that the presence of cheatgrass greatly reduced growth and survival of bluebunch wheatgrass. In one field trial, bluebunch wheatgrass and cheatgrass were sown together in two plots; in one plot the amount

Figure 9.3. Seasonal leaf and root growth of *Agropyron spicatum* and *Bromus tectorum* in glass tubes in the field. (Redrawn from Harris, *Ecological Monographs*, **37**, 102.)

of cheatgrass sown was much higher than in the other. After a normal growth period (October–June), the number and weight of wheatgrass plants in each plot were tabulated. Table 9.1 shows that survival and plant weight was much lower when cheatgrass was more abundant than when it was scarce.

Dr. Harris then planted seeds of each species in long glass tubes which were filled with soil. The tubes were inserted in the field, level with surrounding soil. Every month the tubes were lifted and depth of rooting was measured. The graph in Fig. 9.3 shows that root growth during the winter was much greater for cheatgrass than for wheatgrass. At the start of rapid spring growth in April, cheatgrass roots had penetrated ninety centimeters, in contrast to twenty centimeters for wheatgrass. This difference in depth allowed cheatgrass to absorb water from a much greater part of the soil profile than wheatgrass could. Consequently, when the upper soil became dry in early summer (with both species drawing water from it), only wheatgrass suffered, for water still remained available at deeper depths (where only cheatgrass roots penetrated). Harris measured soil water availability in the two soil regions at this time and found water potential was below minus fifteen atmospheres in the upper region (near the wilting point of many species), but was only minus one atmosphere in the lower region (essentially still saturated).

The recent success of cheatgrass, then, seems due to its winter root growth, which gives it a spring and early summer advantage over wheatgrass that started from seed at the same time. Cheatgrass each year increases in density and each year this results in greater competition for moisture in the upper soil and greater stress on wheatgrass.

Table 9.1	Survival and plant weight of bluebunch wheatgrass after a season's growth in two levels of cheatgrass competition

RELATIVE CHEATGRASS COMPETITION	SURVIVAL OF WHEATGRASS (%)	AVERAGE DRY WEIGHT OF WHEATGRASS (g)
low	86	10.4
high	39	1.8

SUMMARY OF ROOT SYSTEMS

1. The primary root develops from the radical in the embryo. It generally penetrates the soil to some depth and may either form a tap root system or develop an extensive series of lateral roots to form a fibrous root system.

2. More frequently, the fibrous root system is formed by adventitious roots arising from nodes of the stem. This is the general situation in monocotyledons and occurs in many dicotyledons; it is most generally associated with horizontal stems.

3. Adventitious roots may arise in the internodal regions of the stem.

4. Shoots may arise from adventitious buds developing on roots.

STRUCTURE OF ROOTS

The root, like the stem, grows, conducts water and nutrients, and stores food. Unlike the stem, it absorbs nutrients and water from the soil and anchors the plant. Accompanying similarities and dissimilarities in function are corresponding variations in structure. We shall find, in the root, meristematic and vascular tissues similar to those found in the stem. Both primary and secondary growth occur. There are three tissues in the root (root cap, endodermis and pericycle) that are not present in the stem, and the arrangement of vascular and meristematic tissues is slightly different. A cuticle does not develop on the epidermal cells of the young root. Instead, some root epidermal cells become hair cells and facilitate absorption (Fig. 9.9).

Primary Plant Body

External Features

The general structural features of roots can best be studied by means of seedlings that have been grown either in sand or on filter paper in a moist atmosphere. Figure 9.4 shows such a seedling. Observe here the portion of the root that is clothed with root hairs; this region is usually known as the **root-hair zone.** Note that root hairs do not extend to the very tip of the root. The root-hair zone and the hairless tip constitute a region of especial interest. It is in this part of the root where (1) growth in length, (2) most of the absorption of materials from the soil, and (3) development of the primary tissues

take place. Below the root hairs there occur, in descending order, the **region of elongation,** a **meristematic zone,** and the **root cap. Differentiation** starts in the upper cells of the meristematic zone and extends upward through the region of elongation into the lower zone of root hairs.

In stems, the apical meristem is either naked or protected by rudimentary leaves and bud scales. In contrast, the apical meristem of roots is protected by a thimble-shaped mass of cells, the **root cap.** Cells of the root cap are constantly being sloughed off at the very tip, but at the same time new cells are being added to it by apical meristem cells. In the **apical meristem** the cells are actively dividing, adding new cells to the root cap and others to the region of elongation. Rapid growth in root length, however, is largely the result of elongation of cells behind the apical meristem. This is the **region of elongation** (Figs. 9.4B, C, 9.6). Thus it is seen that a very short portion of the root at the tip, usually from two to five millimeters in length, is constantly being forced through the soil. The method of growth in length of the root just described explains the protective function of the root cap; the delicate cells of the apical meristem are protected from mechanical injury as they are pushed through the soil. The region of elongation grades into the region of differentiation of the primary vascular tissue internally and the formation of root hairs in the epidermis. Thus, the mechanism for absorption (root hairs) differentiates as the conduction tissue is forming from the procambium (Figs. 9.6, 9.7).

It is significant that elongation of cells does not occur in the root-hair zone. If elongation did occur in this zone, root hairs that wrap around and adhere to soil particles would be torn loose.

Internal Features

Apical meristems. The meristematic cells of root apices have precise arrangements that differ from those of the shoot apex (Figs. 7.9, 9.4A, B). In the monocotyledons, as illustrated by apices of onion and corn (Fig. 9.5) five cell groups are discernable. The lowest of these groups forms (1) the **root cap.** The meristematic cells which will give rise to the mature tissues of the root form long rows of cells. Rows of cells destined to become (2) epidermis **(protoderm),** (3) cortex **(ground meristem),** and (4) the central vascular cylinder **(procambium cylinder)** may be recognized. These rows of cells diverge

cotyledons

A

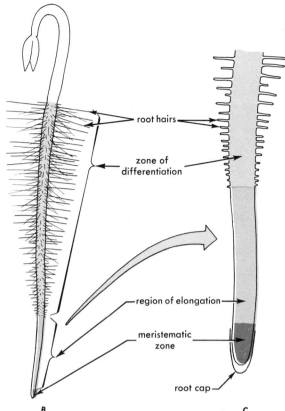

root hairs

zone of
differentiation

region of elongation

meristematic
zone

root cap

B   C

Figure 9.4. Radish seedling (*Raphanus sativus*). *A,* photograph showing root hair growth ×2; *B, C,* diagrams showing growth regions.

from a smaller fifth zone of less regularly arranged cells. There is good evidence that cell divisions rarely occur in this central cell mass or (5) **quiescent zone** (Figs. 9.5*B,* 9.6). Numerous divisions take place in the cells surrounding the quiescent zone. The function of the quiescent zone is not known: it may be of significance in determining the pattern of root development.

The root cap is composed of short-lived parenchyma-like cells. Since new cells are continually being added to it, the root cap persists throughout the growing life of the root. It is present in all common land plants but is lacking in many aquatic plants and certain other specialized types.

The apical meristem is composed of thin-walled cells, that are very much alike and are practically without intercellular spaces. Sections in this region

usually show many cells in which the nuclei are in some stage of mitosis (Fig. 9.5*A*).

In the region of elongation, there is less uniformity in the appearance of cells than there is in the apical meristem. Some differentiation has taken place. As in the corresponding region of the stem, three primary meristematic tissues are evident, although the exact sequence of their development may vary in different plants: (1) the **protoderm,** (2) the **procambium,** and (3) the **ground meristem.** These three meristematic tissues differentiate into the primary tissues of the roots (Figs. 9.5*B,* 9.6, 9.7).

Differentiation of primary tissues. Differentiation of the meristematic cells into the primary body proceeds in the root as it does in the stem, with slight modifications. The protoderm gives rise to an epidermis devoid of a cuticle and guard cells but

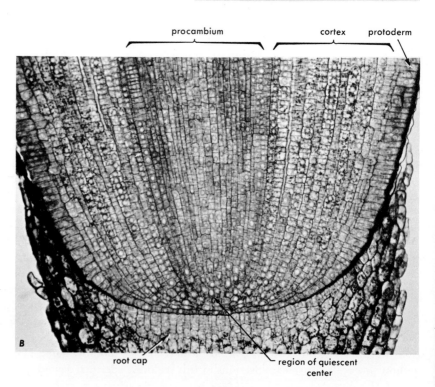

Figure 9.5. Longitudinal sections of root tips. *A*, section of an onion (*Allium cepa*) root tip showing growth regions. Processing emphasizes nuclei. ×85. *B*, region of apical meristems of corn (*Zea mays*) root tip, showing the very early differentiation of tiers of cells that eventually become specialized in the mature region of the root. ×300. *C*, diagram of apical region showing tiers of cells. The relative number of cell divisions in a given region is indicated as the Incorporation of tritiated thymidine is shown here by the relative depth of color. (*A* and *B*, courtesy of Lewis Feldman and Elizabeth Cutter.)

well-provided with root hair cells (Figs. 9.6, 9.9). A cortical tissue of parenchyma cells arises from the ground meristem. While strengthening cells, such as collenchyma, are not formed, the innermost layer of cortical cells does become specialized to perform an important role in the absorption of water and nutrients. This is the **endodermis** (Figs. 9.6, 9.7, 9.11). Note that it starts to form a very short distance behind the apex (Fig. 9.6). The procambium cylinder gives rise to a cylinder of vascular tissue (Fig. 9.6). There is no pith. The outermost layer of procambium cells differentiates early into a specialized cylinder of cells known as the **pericycle** (Fig. 9.6). The first cells of vascular tissue to form are two to four phloem cells (sieve elements) which arise singly at the outer periphery of the procambium cylinder (Fig. 9.6). The differentiation of sieve elements is soon followed by an equal number of vessel elements that arise at the periphery of the procambium cylinder in pockets between the sieve elements (Figs. 9.6, 9.7, 9.8). Further differentiation of the vascular tissue occurs toward the center of the root. When the primary plant is complete, it appears in transection as in Fig. 9.7. A cross section through the primary vascular tissue of a young root at the completion of primary growth is shown in Fig. 9.8; the following primary tissues can be seen.

(1) The epidermis. This region, with its root hairs is a single layer of cells derived from the protoderm. Figure 9.9 shows stages in the development of a root hair. Essentially, it is a lateral outgrowth of an epidermal cell. When grown in moist air, each root hair has the form of a slender tube, but in the soil it may be greatly contorted in its growth between and around soil particles. The root hair and the epidermal cell from which it grows constitute a single cell. The walls are thin, composed principally of cellulose and pectic substances. The protoplast has a high water content. The nucleus is usually near the end of the hair, and there is a large central vacuole.

In most plants, the life of any one root hair is short; it functions only for a few days or weeks. New

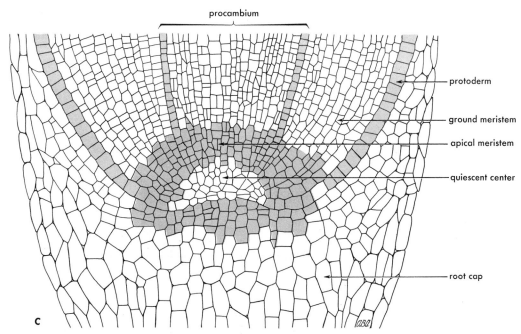

procambium

protoderm

ground meristem

apical meristem

quiescent center

root cap

C

Figure 9.5C. *(continued)*

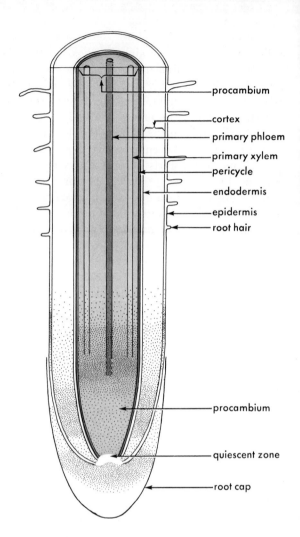

procambium

cortex

primary phloem

primary xylem

pericycle

endodermis

epidermis

root hair

procambium

quiescent zone

root cap

Figure 9.6. Diagrammatic representation of development of a root tip from the root cap to the initiation of the first cells of the primary phloem.

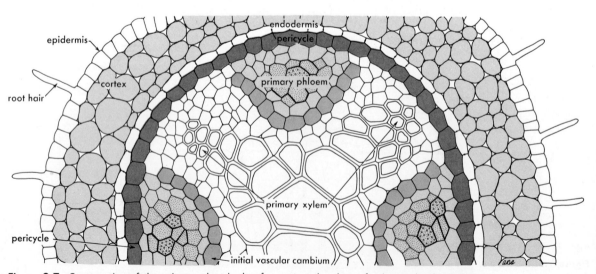

Figure 9.7. Cross section of the primary plant body of a root at the close of primary development.

9 *The Root*

Figure 9.8. A photograph of the vascular cylinder, including pericycle and endodermis, of young root of *Ranunculus* at close of primary growth × 200.

phloem

secondary xylem

cortex

endodermis

primary xylem

pericycle

Figure 9.9. Development of a root hair.

Figure 9.10. Three-dimensional view of two endodermal cells. The suberized Casparian strip is shown in gray. Water from solution outside of the root wets the cellulose walls until it reaches the Casparian strip. Since water cannot wet and cross the Casparian strip the cell walls inside of the Casparian strip are wet by water that has passed through the protoplast of the endodermal cell.

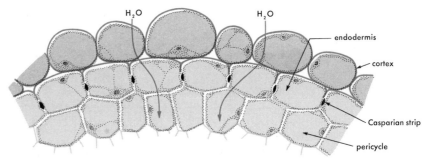

Figure 9.11. Diagram showing the relationship between cortex, endodermis, and pericycle. Water present in the walls of cortical cells cannot diffuse across the suberized Casparian strip. Water inside the endodermal cells has been filtered by the protoplast of the endodermal cells.

hairs are constantly forming at the anterior end of the root-hair zone, while those at the posterior end are dying. Thus, as the root advances through the soil, fresh, actively growing root hairs are constantly coming into contact with new soil particles. In the rye plant, it is estimated that new root hairs develop at an average rate in excess of 100 million per day. The root hairs of such plants as redbud, honey locust, and a number of others may persist for several years. *Root hairs do not become roots.*

Although nearly all ordinary land plants possess root hairs, a few, such as the firs, redwoods, and Scotch pine, are devoid of them. Also, many aquatic plants have no root hairs. Moreover, land plants (corn, for example) that normally develop root hairs when the root system grows in the soil or in moist air, develop no root hairs when the roots grow in water. In plants devoid of root hairs absorption is through the thin-walled epidermal cells. Root-hair development is often inhibited by a concentrated soil solution and by high or low soil temperatures. Root hairs develop in the light and dark about equally well if moisture and oxygen are adequate.

(2) The cortex. This region, relatively thicker than that in stems, is derived from the ground meristem. The cortex is composed chiefly of storage parenchyma with large intercellular spaces (Fig. 9.7). In many species secretory cells and resin ducts are present. The innermost layer of the cortex is a single row of cells, the **endodermis,** which is usually a conspicuous feature of roots. As a rule, in the primary state the endodermal cell walls are thin except for a bandlike thickening running around the cell on the radial and transverse walls.

This thickened strip, known as the **Casparian strip,** is suberized or cutinized (Figs. 9.10, 9.11).

Cellulose walls are permeable to water. The cell walls and intercellular spaces of the cortex may freely imbibe water and dissolved nutrients from the surrounding soil solution. The suberized strip that forms a complete ring around the endodermal cell in the radial and transverse walls prevents water from reaching the pericycle except by passing through the protoplast of the endodermal cell. This situation is shown diagrammatically in Figs. 9.10 and 9.11. The Casparian strip in Fig. 9.10 is represented by a gray band completely encircling the cell. The dark green portion of the cell represents the side toward the soil solution, and consequently these walls have imbibed water directly from the soil solution. In Fig. 9.11 the Casparian strip is represented by a black bulge in the center of the radial walls of the endodermal cells. The dark green walls of the cortical cells and the outer portion of the endodermal cells freely imbibe water from the soil solution (dark gray). But water is not imbibed by the suberized Casparian strip. Consequently, any water entering the pericyclic cells must pass through the protoplasts of the endodermal cells. Thus, the protoplasts of the endodermis may exercise a degree of control over the water and dissolved substances entering and leaving the root. The cell walls and intercellular spaces (light green) of the pericyclic cells have imbibed water that has passed through the protoplasts of the endodermal cells into the pericyclic cells (light gray). Frequently, in older root regions, the inner, radial, and transverse walls of endodermal cells are thickened.

(3) The pericycle. At an early stage of development, a special layer of parenchyma cells is differentiated from the outer region of the procambium cylinder. This is the pericycle (Figs. 9.6, 9.7, 9.8). It persists as a rather unspecialized type of meristematic tissue until secondary development interferes with its activity. It gives rise to lateral roots; certain of its cells develop into portions of the vascular cambium, and others give rise to cork cambium; and it perpetuates itself by cell division. In roots without secondary growth it may eventually form a sclerenchyma tissue.

(4) The vascular cylinder. (Figs. 9.6, 9.7, 9.8) This originates from the remaining internal portion of the procambium. Since there is generally no internal ground meristem as occurs in the tip of shoots, pith does not usually develop in dicotyledonous roots,

although it may occur in the roots of monocotyledons.

In some discussions of root structure, the pericycle and vascular cylinder, which both originate from the procambium, are considered a single general region called the **stele.**

It will be recalled that in stems, xylem and phloem of a vascular bundle occur in such a manner that a radial line cuts through both tissues (see Figs. 7.9, 7.11, 8.1). In roots, in contrast, the primary xylem and phloem occur in such a manner that a radius passing through the xylem arms does not pass through the phloem. The primary xylem usually consists of a central mass or "core" of xylem elements, with several radiating arms, between which are groups of phloem elements. Between these two tissues are one or more layers of procambial cells, represented by the dark green line on Fig. 9.12*A*. In roots with secondary growth, these cells give rise to vascular cambium (which is shown in Fig. 9.12*B* as a dark green line), while in roots without secondary growth they often mature into sclerenchyma. Many roots have no pith, but in most monocotyledons and some herbaceous dicotyledons, the central core of the stele is parenchyma that resembles the pith of stems. Whereas the pith of stems is derived from gound meristem, that of roots is derived from procambium.

The cells that occur in the primary xylem of stems may also be found in the primary xylem of roots, although spiral and annular vessels are relatively rare in roots. The primary phloem of roots does not differ essentially from that of stems. It consists of sieve tubes, companion cells, and parenchyma.

Origin of Lateral Roots

It will be recalled (page 113) that lateral or branch stems originate from cell layers at or near the shoot tip. In contrast, lateral or branch roots in gymnosperms and angiosperms originate from cells of the pericycle. Figure 9.13 shows the manner of origin of branch roots. Often the point of origin of a branch root is opposite a primary xylem strand. For example, in the beet root, there are two primary xylem strands and two vertical rows of branch roots. It is seen from Fig. 9.13 that the tip of the branch root must penetrate the cortex and the epidermis in order to reach the surface. These tissues are stretched and finally ruptured as a result of the mechanical pressure of the growing root. It has also been suggested that cells of the branch root secrete substances which digest the tissues ahead of it. This method of root formation leaves the xylem portion of the vascular cylinder intact, in contrast with the stem where leaf and bud gaps result in openings in the xylem cylinder. This difference is shown in Fig. 9.13E, which illustrates that portion of cabbage xylem where root and shoot merge.

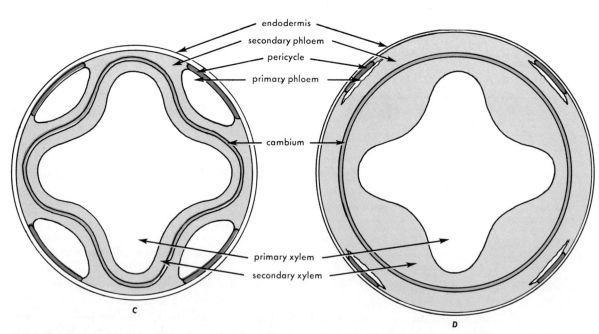

Figure 9.12. Diagrammatic representation of the development of the secondary plant body in a root. A, At the completion of primary growth a row of procambium cells remains (dark green) and a complete circle of pericycle is present (light green); B, The procambium (dark green) joins with those pericyclic cells outside of the xylem arms to form a continuous line of vascular cambium (dark green); C, The vascular cambium forms secondary xylem internally and secondary phloem externally (light gray). The primary tissues are white. The primary phloem is being pushed outward and a small amount of pericycle (light) green is still associated with it; D, A smooth circle of vascular cambium forms, producing secondary xylem and phloem. The primary xylem remains in the center of the stem, the primary phloem has been crushed and only a small amount of the pericycle remains.

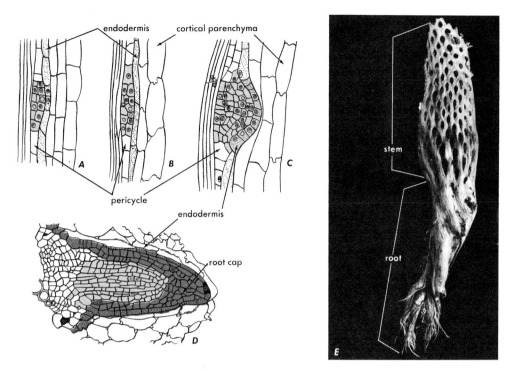

Figure 9.13. Branch roots. *A*, initiation of branch carrot (*Daucus carota*) root through formation of meristematic cells in pericycle; *B, C*, enlargement of meristematic region; *D*, young root pushing through cortex, ×50; *E*, roots forming in this manner do not leave a gap in the vascular cylinder, as is shown by the xylem framework of a rotted cabbage (*Brassica oleraceae*) at junction of stem and root × ½. (*A–D*, after K. Esau.)

Summary of Primary Tissues

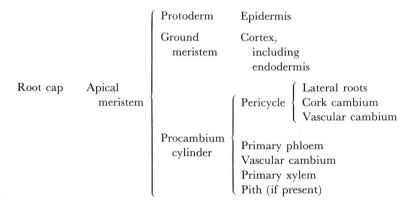

THE SECONDARY PLANT BODY

If we examine old roots of woody plants, some of which may attain a diameter of many inches, we observe annual rings of growth resembling those in stems (Fig. 9.14). At the completion of primary growth, procambium remains between the radiating arms of the primary xylem separating it from the primary phloem. This remaining procambium is represented by the dark green cells of Fig. 9.7 and

Figure 9.14. Cross section of an old root of cherry (*Prunus avium*). ×⅞.

production of secondary xylem being more rapid between the arms of primary xylem. This development soon results in a continuous smooth circle of vascular cambium (Fig. 9.12). With continued secondary growth, the epidermis and cortex are usually ruptured and finally sloughed off.

The cambium thus formed gives rise to secondary xylem and phloem after the manner of the vascular cambium of stems. Annual rings are generally formed, and macroscopically the wood of an old root cannot be distinguished from that of a stem. However, microscopic examination shows that, in general, the secondary xylem of the root contains fewer fibers, more storage parenchyma, and larger, more numerous thin-walled vessels than does that of the stem. The secondary phloem of the root as compared with that of the stem has more storage parenchyma and less mechanical tissue.

The formation of cork cells accompanies the development of the secondary xylem and phloem. They are derived from a cork cambium which in most cases originates first in the pericyclic cells. As the root increases in diameter, this first-formed cork is stretched and torn as in stems. Cork cambium forms anew from deeper lying cells, generally in the phloem. This results, as in stems, in the development of a thick protective layer (Fig. 8.18).

Summary of Secondary Growth in Roots

A vascular cambium originates from the procambium cells between the primary xylem and phloem and from the pericyclic cells exterior to the radiating points of the primary xylem. This vascular cambium forms secondary xylem internally and secondary phloem externally. The resulting increase in diameter stretches and tears the endodermis, cortex, and epidermis. A cork cambium develops from the pericycle and forms cork. Woody roots, consequently, are very similar in structure to woody stems (Fig. 9.14).

THE FUNCTIONS OF ROOTS

The functions of the root system are **absorption, anchorage, conduction,** and **storage.** From the soil, roots absorb water, mineral salts, and oxygen. They anchor the plant firmly in place. They conduct water, mineral salts, and sometimes previously

the dark green lines of Fig. 9.12*A*. No procambium remains at the points of the radiating primary xylem arms, so a continuous band of procambium is not present. In this early stage of development no vascular cambium is present, but it is soon formed in the following manner. (1) Vascular cambium forms from the procambium remaining between the primary xylem and primary phloem, and (2) cells of the pericycle on the outside of the radiating arms of primary xylem become meristematic. This situation is represented in Fig. 9.12*B* by the continuous dark line. Note that the light green, representing the pericycle (Fig. 9.7) becomes dark green just outside of the xylem arms (Fig. 9.12*B*). This change indicates its return to a meristematic condition.

Thus, a continuous layer of vascular cambium is formed from procambial cells that have not lost their meristematic character, and from pericyclic cells that have returned to a meristematic condition. As seen in cross section (Fig. 9.12*A*), this layer has the form of a wavy band; it passes inside each primary phloem group and outside each of the arms of the star-shaped primary xylem. This vascular cambium gives rise internally to secondary xylem and externally to secondary phloem, with initial stored foods, *to* the stems and leaves aboveground, and they conduct foods *from* the leaves to all parts of the system underground. Most roots usually store foods at least for a short period and in small quantities, and special storage roots accumulate rather large amounts of foods.

Absorption by Roots

Absorption of Water

In land plants, large amounts of water are absorbed by the roots and lost from the leaves. Most of the water that is absorbed moves upward to the leaves and passes out as *water vapor* through stomata in the epidermal layers of the leaves.

We have seen that, while essentially similar patterns occur in both roots and shoots, there are differences related to their different functions. The response to water is different in the root and in the shoot, and some of the difference in the pattern of root and shoot appear to be related to the water economy of the plant. About 95 percent of the plant body is water. Water enters into all of its metabolic processes. A part of the water absorbed by most land plants enters through the root hairs (Fig. 9.9); some enters through the thin walls of ordinary epidermal cells. The water entering through the root hairs is conducted throughout the plant by the vascular system. Much of the water leaves the plant from the leaves in the process of transpiration. The epidermis of the shoot is provided with a cuticle and guard cells to conserve water while allowing for exchange of gases. The epidermis of the root lacks a cuticle and is provided with root hairs for the absorption of water. In the shoot the primary xylem is internal to the phloem. In the root the phloem alternates with arms of xylem, with only a pericycle and the protoplasts of the endodermis intervening between the outermost vessel and the imbibed water of the cortex. Water need not pass through phloem tissue to reach the xylem. We have learned that the area of contact of the root system with the soil may be enormous; that the numerous rootlets are constantly elongating and exploring new soil areas; that root hairs are being formed anew just behind the growing tip of the root; and that these young, thin-walled root hairs are flattening out and surrounding soil particles, and thus coming into very close contact with the film of water that surrounds the soil particles.

Separating the vacuole of a root hair and the soil solution is the cell wall and a thin layer of cytoplasm. The wall itself is chiefly of cellulose and pectic substances. There is usually a pectic coating on the outside of the wall. Due to the gummy nature of this coating, the root hairs adhere closely to the soil particles.

If a molecule of water in the soil is to reach the cell vacuole, it must penetrate the cell wall of the root hair and a layer of cytoplasm (Figs. 9.7, 9.9).

As seen from Fig. 9.7, the pathway of water from the root hair is through several layers of thin-walled cortical cells, then through the endodermis and one or more layers of cells of the pericycle, to the vessels of the xylem. Once in the vessels, the water moves from the very small root branches to larger roots, then to the stem, and on upward to the leaves and other organs of the plant. From the leaves and young stems, the water passes off to the atmosphere as water vapor. Thus there is a continuous stream of water through the plant that has been in existence since the germination of the seed.

Absorption of Solutes

Land plants derive their inorganic salts from the soil. Usually, every kind of chemical element found in the soil can also be found in plants. Certain mineral elements, such as potassium, sulfur, phosphorus, calcium, and magnesium, constitute a relatively large portion of the inorganic components of plants. Other elements, such as iron, manganese, boron, zinc, copper, iodine, and selenium, occur in small quantities or as mere traces.

At present there is no reason for believing that every element found by chemical analysis in plants has an essential role in the plant's life. Although a number of elements are known to be essential, others are apparently carried into the plant but may be of no particular use. Certain elements, however, such as boron, zinc, copper, and manganese, formerly regarded as nonessential, now are known to be indispensable. Some elements are absorbed in considerable excess of the plant's actual needs. It is very likely that most salts of the soil enter chiefly in the form of ions.

The process of the absorption of solutes by roots will be discussed in Chapter 11.

Anchorage by Roots

We have seen (Fig. 9.1) how extensive are the root systems of ordinary plants and how completely they may occupy the soil. Thus, the roots function as very effective anchorage organs.

Storage by Roots

All roots, even slender ones whose primary function is absorption, may have a small amount of food stored temporarily in them. For example, when sugar moves into the roots more rapidly than it can be utilized by the growing cells, it may be converted to starch and as such be stored for a time, particularly in cortical cells. During the dormant season rather large quantities of starch are stored in the woody roots of orchard trees. This food constitutes a reserve that is called upon when active growth is resumed in the spring. The roots of the wild morning-glory (bindweed) and other perennials store large quantities of food. This stored supply enables the plant to send up new shoots when the "tops" of the plants are destroyed. The seasonal trend in the storage of foods in the roots of wild morning-glory is seen in Fig. 9.15. It will be noted that, in undisturbed plants, readily available carbohydrates rapidly build up during the summer, whereas, in plants cultivated at two-week intervals, food storage is greatly hindered.

Food storage in roots may occur in the cortex, phloem, and xylem. Among our native plants, the most striking examples of fleshy-rooted plants occur in arid regions. Such roots generally contain a large quantity of stored water that can be used by the leafy shoots during periods of drought.

Reproduction by Means of Roots

Ornamental crab apple, cherry, plum, quince, hawthorn, and a number of other plants are usually propagated by root cuttings. The shoots that arise from the cuttings are from adventitious buds. Injury to roots may induce the development of such buds. The tap root of a dandelion may be cut into many small pieces, and each section be capable of producing new shoots.

SUMMARY

1. The principal functions of roots are the absorbtion of water and nutrients, the conduction of absorbed materials and food, and the anchorage of the plant in the soil.

2. There are two general types of root systems, a fibrous root system and a tap root system.

Figure 9.15. Seasonal trend in the storage of reserve carbohydrates in the roots of wild morning-glory (*Convolvulus arvensis*). (From Barr.)

3. Roots differ from all stems in lacking nodes and internodes. In some species, however, leafy shoots and associated root systems do form from adventitious buds.

4. In the dicotyledons and gymnosperms, roots generally form at the lower axis of the stem; in many monocotyledons, roots, or root primordia, occur at nodes.

5. The root tip is divided into four zones of specialization: (1) the root cap, which protects (2) the meristematic region as it moves through the soil; (3) a region of elongation; and (4) a region of differentiation characterized externally by root hairs and internally by the formation of the primary vascular tissues.

6. Soil water and nutrients move easily through the epidermal and cortical tissues.

7. The endodermis forms the innermost cell layer of the cortex. A suberized band, the Casparian strip, in the radial and transverse walls, completely encircles the endodermal cells.

8. Water cannot move across the Casparian strip.

Therefore, all water with dissolved nutrients which is inside the endodermis, has passed through the protoplasts of the endodermal cells.

9. The pericycle, a row of cells internal to the endodermis, represents the first row of cells of the vascular cylinder; they have differentiated from procambium. Cells of the pericycle may eventually form part of the vascular cambium or give rise to branch roots.

10. In cross section, the primary xylem is star-shaped and generally has from three to five arms. The primary phloem arises between the arms of the primary xylem. There is no pith.

11. Water and inorganic nutrients enter the primary xylem from parenchyma cells without passing through the cells of the phloem.

12. Water and inorganic nutrients enter root hair cells, pass through the cortex, are filtered by the protoplasts of the endodermal cells, cross the pericycle, and move directly into the vessels or tracheids of the primary xylem.

13. In roots having secondary growth, a vascular cambium is formed by pericyclic cells over the xylem arms and procambial cells remaining between the primary xylem and phloem. This vascular cambium at first forms a wavy line. It eventually becomes circular, and there is little difference in the appearance of wood from root or stem.

10

The Leaf

FORM AND FUNCTION OF THE LEAF

Green plants are the only real producers; all other inhabitants of the earth consume the production of green plants. In the seed plants, leaves are the organs of production. Chloroplasts within leaf cells are the sites in which light energy is trapped and is converted into chemical energy. Chlorophyll, membrane-bound within the chloroplasts, performs the first step in the trapping of light energy. Leaves are green because wave lengths of red and blue light are absorbed by the chlorophyll and wave lengths of green light are transmitted or reflected. The absorbed light gives up energy in exciting an electron in the chlorophyll molecule to a higher energy level (see page 242). This is the first step in **photosynthesis,** a primary function of the leaf.

The water economy of plants calls for the absorption of much more water than can be metabolized. This excess water is returned to the atmosphere by leaves. A second function of leaves is **transpiration.**

The form and anatomy of the leaf are such that it is peculiarly adapted to carry on the two primary functions of photosynthesis and transpiration. The blade provides a large surface for the absorption of light energy and carbon dioxide needed in photosynthesis; it is thin, and hence none of the cells lie far from the surface. These characters facilitate the absorption of light, carbon dioxide, and oxygen by the cells inside the leaf, and also the outgo of gases, including water vapor, oxygen, and carbon dioxide, from the cells inside the leaf. Leaves are thin. The water they use and transpire comes from the roots. The food they manufacture must pass to all other parts of the plant. Veins support the thin parenchyma tissue of the leaf. Xylem and phloem within the veins perform their usual functions of conducting water and transporting food.

EXTERNAL MORPHOLOGY OF THE FOLIAGE LEAF

Leaves are varied as to their arrangement on the stem, their form, the distribution of their veins **(venation),** their structure, and many other characters. Leaf characters are used extensively in the description of species. It is often possible to distinguish certain species of the same genus by a glance at the leaves. For example, in Fig. 10.1, note the marked differences in the leaves of several common species of *Populus*. On the other hand, plants distantly related may have leaves that are quite similar. Thus, reliance cannot usually be placed on leaf characters alone in identifying plants. As a rule, a combination of plant characters needs to be taken into consideration.

Dicotyledonous Leaves

The leaves of dicotyledons are generally distinctly different from those of monocotyledons. A typical foliage leaf of a plant belonging to the dicotyledons is composed of two principal parts: (1) **blade** or lamina and (2) **petiole** or stalk (Fig. 10.2*A*). The blade is thin and expanded, the petiole slender. The thin blade is supported by the very distinct network of veins which are composed of vascular tissues and fibers or collenchyma. The veins collect at the base of the leaf into several strands and, passing through the petiole, join the leaf blade with all parts of the plant. In addition to forming a supporting framework for the softer tissues of the blade, the veins carry water, mineral salts, and food to and from the leaf.

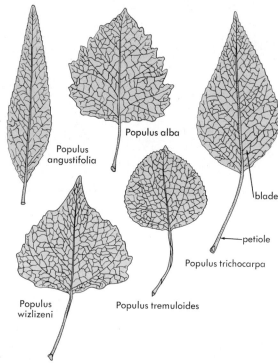

Figure 10.1. Leaves of different species of *Populus*. Leaf characters alone often enable one to identify a species.

Venation

The arrangement of the veins of a leaf is called **venation.** There are two principal types of venation: (1) **parallel venation,** characteristic of monocotyledonous leaves (Fig. 10.3*A, B*) and **netted venation** characteristic of dicotyledonous leaves (Fig. 10.3*C, D*). In the latter type there are one or more prominent veins from which the smaller veins branch off to join with other small veins, thus forming a conspicuous net. When there is a single midrib from which branches diverge like the pinnae of a feather, the leaf is **pinnately netted-veined** (Fig. 10.2*A*). When several principal veins diverge from the base of the blade, each of which branches, the leaf is **palmately netted-veined** (Fig. 10.2*B*).

Parallel venation, characteristic or monocotyledons, is discussed below.

The Blade

The chief variable features of the leaf blade are its overall shape, its **apex,** its **margin,** and its **base.** This range of variation is great and many terms are employed by taxonomists to describe leaf shape accurately. A few of these terms are given below:

Shape. The leaf may be long and slender (linear), or oval or heart-shaped (cordate), or triangular (deltoid).

Apex. The apex of a leaf may be pointed, rounded, or flattened.

Margin. The margin may be entire, that is, with no indentations whatsoever; the margin may be toothed or scalloped or wavy, or cut into a number of lobes. The marginal indentations may be shallow, or they may be deep clefts extending almost to the midrib. Lobed leaves that are pinnately veined are said to be pinnately lobed (Fig. 10.2*C*); those that are palmately veined are palmately lobed (Figs. 10.2*B*, 10.3*C*).

Base. The base of the leaf may be rounded, heart-shaped (cordate), or flattened (truncate).

Figure 10.2. Different kinds of leaves. *A,* poplar (*Populus deltoides*); *B,* castor bean (*Ricinus communis*); *C,* oak (*Quercus lobata*); *D,* rose (*Rosa odorata*); *E,* Virginia creeper (*Parthenocissus quinquefolia*). All × ½.

The Petiole

Some leaves, such as those of peas, beans, roses, tulip trees, and many others, have two small, leaflike outgrowths at the base of the petiole, known as **stipules** (Fig. 10.2*D*). The petiole itself may be long or short, rounded, or occasionally flat. It is usually attached to the base of the leaf, but in some plants, such as nasturtium and castor bean, it is attached, not to the edge of the leaf, but to its underside. This leaf is called a **peltate** leaf (Fig. 10.2*B*). The petiole is sometimes absent, as in *Zinnia,* the blade being mounted directly on the stem. Such a leaf is said to be **sessile** (Fig. 10.14).

Simple and Compound Leaves

As to configuration of the blade, there are two kinds of leaves: (1) **simple** and (2) **compound.** A simple leaf is one in which the blade is all in one piece (Figs. 10.1, 10.2*A, B, C*). A compound leaf is one in which the blade is composed of a number of separate leaflike parts, the **leaflets** (Fig. 10.2*D, E*). Inasmuch as leaflets have the characteristics of a simple leaf, one may be in doubt at times as to whether the structure is a simple leaf or a leaflet, especially if the leaflets are large. The distinctions are as follows: (1) buds occur in the axils of leaves, but not in the axils of leaflets; (2) leaves stand in different planes on the stem, whereas leaflets lie in a single plane; (3) in trees and shrubs that shed their leaves, the leaf as a whole usually falls, being cut off at the base of the petiole, whereas leaflets seldom fall separately. Since individual leaflets possess, in common with simple leaves, apices, margins, bases and shapes, the leaflets may be described in the same terms used to describe the simple leaves.

When the leaflets of a compound leaf arise from the **rachis** (continuation of the petiole), as do the pinnae of a feather, the leaf is said to be **pinnately compound,** like the sweet pea and rose (Fig. 10.2*D*); when the leaflets diverge from a common point at the tip of the petiole, the leaf is said to be **palmately compound,** like the horsechestnut and Virginia creeper (Fig. 10.2*E*). Compound leaves may be once, twice, or thrice compound.

Monocotyledonous Leaves

The leaves of members of this subclass generally have parallel veins, as in corn and *Canna* (Fig. 10.3*A, B*). As contrasted with netted venation, the veins in leaves with parallel venation are all of about equal size, usually very numerous, and all parallel with each other. They are sparsely joined by inconspicuous cross veinlets. In many monocotyledonous leaves the veins run from the base to the apex of the leaf; in other monocotyledonous leaves they extend from a prominent midrib outward to the margin and generally join at the leaf margin.

Leaves of the monocotyledons show as great a diversity of form as do those of the dicotyledons. In onions they are tubular, and in palms they are large and fanlike. The grasses have a very characteristic type of leaf. Crabgrass and corn (Fig. 10.4*A, B*) may be taken as examples. The grass leaf is divided into

Figure 10.3. Leaves, showing parallel and netted venation. *A,* Canna leaf, about one-fifth natural size × ¼; *B,* portion of similar leaf magnified about 10 times; *C,* maple (*Acer*) leaf about one-half natural size; *D,* portion of similar leaf magnified about 20 times.

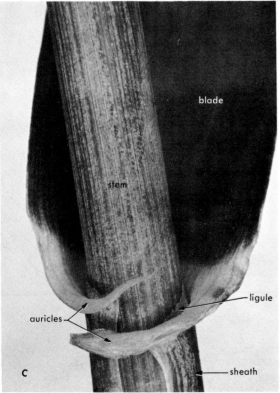

Figure 10.4. Leaves of various members of the grass family. *A*, crabgrass (*Digitaria sanguinatis*) showing sheath and blade; *B*, corn (*Zea mays*) showing ligules (sheath pulled away from stem); *C*, ligules and auricles of barley (*Hordeum vulgare*). All about 1.

two parts, the **sheath** and the **blade.** The blade is the typical thin, expanded portion. The sheath is green, perhaps nearly as large as the blade, but it is not a flat expanded structure. Instead, it completely sheaths the stem, extending, in many species, such as corn (Fig. 10.4B), over at least one complete internode. In crabgrass (Fig. 10.4A) the sheath covers only about one-half of an internode. If the region of union between the blade and the sheath is examined carefully, a small flap of delicate tissue extending upward from the sheath may be seen, closely investing the stem. This is called the **ligule** (Fig. 10.4B). It may in some cases serve to keep water and dirt from sifting down between the stem and the sheath. In many species, of which barley (Fig. 10.4C) is a good example, the base of the blade, at its union with the sheath, is carried around the stem in two earlike points. These points are called **auricles.** Ligule and auricles may both be present, or one or the other may be absent.

Gymnosperm Leaves

The leaves of all gymnosperms native to the United States and Europe are needle-like (Figs. 10.5A, 28.10) or scale-like (Fig. 28.16B). They are discussed in some detail in the chapter on the conifers (page 16). The needles of the dawn redwood (*Metasequoia glyptostroboides*) and the fan-shaped

Figure 10.5. Gymnosperm leaves. A, Dawn redwood (*Metasequoia glyptostroboides*); B, *Ginkgo biloba*; C, *Podocarpus*. All about ¾.

leaves of *Ginkgo biloba* are shown in Fig. 10.5*A*, *B*. Both of these trees are native to China and represent very primitive seed plants. *Ginkgo* is known only in cultivation, and the dawn redwood, like the California redwood, has a very restricted distribution. The leaves of many gymnosperms of the southern hemisphere are linear (Fig. 10.5*C*) or expanded.

ANATOMY OF THE FOLIAGE LEAF

The anatomy of a leaf blade is best shown in a cross section (Figs. 10.6, 10.7*A*, *B*). In these figures we observe three principal tissues: (1) **epidermis,** (2) **mesophyll** (middle of leaf), and (3) **veins** or **vascular bundles.** The **epidermis** usually consists of a single layer of cells that covers the entire leaf surface. It protects the tissues within from drying out and from mechanical injury. The **mesophyll** is composed of parenchyma cells, most or all of which contain chlorophyll and thus are able to carry on photosynthesis. The **veins** possess xylem and phloem elements and hence conduct water, inorganic salts, and foods. Fibers and collenchyma may be associated with the conducting elements of the midrib and larger lateral veins.

The **petiole** has its own specialized structures, enabling it to support the leaf blade, to conduct food, water, and inorganic salts, and to disconnect itself from the stem at the close of the growing season without exposing living stem tissue to drying out or to infection.

Let us now consider each of these regions in detail.

The Epidermis

The epidermis covers the entire surface of the leaf and is continuous with that of the stem to which the leaf is attached. In most leaves, the epidermis is a single layer of cells. It may consist of several kinds of cells: (1) **ordinary epidermal cells,** (2) **guard cells,** (3) **hair cells,** (4) **glandular cells.** The ordinary epidermal cells show a variety of shapes depending upon the species. One form is shown in Fig. 10.8*A*. They are similar in shape to an irregular pavement block, whose depth is usually less than its breadth or length. Epidermal cells are generally covered on their outer surfaces by a waxy cuticle secreted by their protoplasts. The deposition of the cuticle frequently forms a pattern, quite specific for the leaves on which it is found. The cuticles on the lower epidermis of a leaf of the Gambel oak and of a petal of the Shasta daisy, as seen with the Stereoscan electron microscope, are shown in Figs. 10.9*A* and *B* respectively.

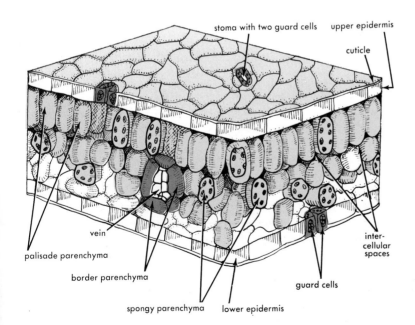

Figure 10.6. Three-dimensional diagram of a section of a foliage leaf. ×20.

upper epidermis

palisade parenchyma

stoma

lower epidermis

spongy parenchyma

A

Figure 10.7. Photomicrographs of sections of a lilac (*Syringa*) leaf. A, cross section of leaf ×75; B (*Right*), section cut parallel to surface of leaf. ×75. (Slides courtesy of Triarch Products.)

Guard Cells

Guard cells occur in pairs, and each is crescent-shaped or semicircular in form, as seen in surface view (Fig. 10.8A). Chloroplasts occur in guard cells but are lacking in ordinary epidermal cells. Both kinds of epidermal cells have long-lived protoplasts. The outer wall of epidermal cells, including guard cells, has a cuticle (Figs. 10.6, 10.7B) like that of the epidermis of stems. It is effective in limiting the movement, either inward or outward, of water vapor and other gases. The cuticle is usually thicker on the upperside of the leaf than on the underside.

A guard cell is a special type of epidermal cell. Guard cells occur in pairs, and between them is an opening or pore (Fig. 10.8C).

A cross sectional view of the guard cells shows that the walls are typically unevenly thickened with the thicker, less elastic walls, adjacent to the stomatal pore. It is convenient to apply the term **stoma** (**stomata,** plural) to the entire unit, the two guard cells and the pore they surround (Fig. 10.8). The stomata are the only openings in the leaf epidermis, and it is chiefly through them that gases pass into or out of the leaf. Although the cuticle is nearly

Table 10.1 *Average number of stomata per square centimeter*

NAME OF PLANT	UPPER EPIDERMIS	LOWER EPIDERMIS
Alfalfa	16,900	13,800
Apple	0	29,400
Bean	4,000	28,100
Cabbage	14,100	22,600
Corn	5,200	6,800
English oak	0	45,000
Nasturtium	0	13,000
Oat	2,500	2,300
Potato	5,100	16,100
Tomato	1,200	13,000

impermeable to gases, small amounts of gases pass directly through the outer wall and the cuticle of epidermal cells.

Stomata vary considerably in size in different species of plants, and even in any one plant. Some representative measurements in microns are as fol-

Figure 10.7. B *(continued)*

lows (length × breadth): bean, 7 × 3; geranium, 19 × 12; corn, 19 × 5; oat, 38 × 8; sunflower, 22 × 8. The number of stomata per unit area varies widely, depending upon the species of plant and the environmental conditions under which it is growing. Usually more stomata are on the lower surface than on the upper. Table 10.1 gives the average number of stomata per square centimeter on the upper and lower surfaces of some common plants.

Stomata are of widespread occurrence in the plant kingdom. With the exception of a few aquatic plants, stomata are present in all angiosperms and gymnosperms, and functional stomata have been found in liverworts, mosses, horsetails, ferns, and cycads. In the angiosperms they can occur on stems, petals, stamens, and pistils as well as on leaves. The two major groups of plants that lack stomata are the algae and fungi. Stomata are basically similar in the many different groups in which they are found.

Opening and Closing of the Stomata

The physical opening and closing of the stomata is a result of changes in guard cell turgor, particularly with reference to the turgor of the adjacent epidermal cells. Water movement from the adjacent epidermal cells into the guard cells results in an increased turgor in the guard cells and an elastic stretching of the guard cell walls. In two associated guard cells, the thin walls adjacent to the epidermal cells stretch more than the thicker inelastic wall bordering the pore. This results in the formation of an elliptical aperture. The change in shape of the guard cells can easily be visualized by imagining that you are blowing up a long balloon with one side slightly thinner than the other. As air pressure increases in the balloon, the thin side will stretch more rapidly than the thicker side and the balloon will assume a kidney shape. The central regions of two similar balloons, attached only at their ends and with their thicker walls parallel and touching, would separate when blown up. Thus, if the structural features of the guard cells are understood, it is easy to see how changes in the turgor can cause either an opening or closing of the stomata. But how are changes in turgor brought about? If one observes the influence of the external environment on stomatal opening, it soon becomes apparent that stomatal opening and closing can be triggered by a number of factors: light generally causes opening and darkness, closing; wilting, or a water deficit in the leaves, or an increase of CO_2 around the leaves may induce closure; an increase in temperature frequently results in stomatal opening.

Since such widely different environmental changes may result in opening or closing of the stomata, it is evident that the osmotic regulatory mechanism of this stomatal action is complex—in

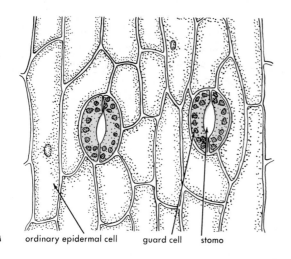

A   ordinary epidermal cell   guard cell   stomo

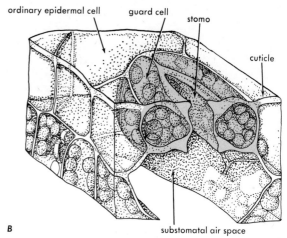

ordinary epidermal cell   guard cell   stomo   cuticle

B   substomatal air space

epidermal cells

chloroplasts

vacuoles

pore

nuclei

mitochondria

C

Figure 10.8. Epidermal cells of leaf. A, lower surface, ×60; B, diagram showing surface view and cross section, ×150; C, a paradermal section through the epidermis of *Opuntia* showing two guard cells with a small pore between them and several associated epidermal cells, ×1000. (B, redrawn from R. M. Holman and W. W. Robbins, *A Textbook of General Botany*, John Wiley and Sons; C, courtesy of W. W. Thomson.)

fact it is still not clearly understood. One of the earliest suggestions made to explain changes in turgor associated with light-induced stomatal opening was that light caused photosynthesis in the chloroplasts in the guard cells. Sugar, an osmotically active solute, would be produced by photosynthesis and would then cause a decrease in water potential in the guard cells. Thus, water would diffuse into the cells, increasing turgor and causing stomatal opening. It was soon discovered, however, that the effects of light were very rapid and, in some instances, the light intensity that induced opening was too low to produce significant photosynthesis.

An extension of this photosynthesis theory was based on the observations that in the light the sap of the guard cells became less acid and the total amount of starch decreased. It was thought that the decrease in acidity resulted from the use of carbon dioxide during photosynthesis (CO_2 combines with water to form an acid). In the presence of phosphate in the cell, the decrease in acidity induced an enzymatic breakdown of starch to a sugar phosphate which is turn formed sugar and liberated the phosphate. Such a series of reactions would lead to a net change from starch to sugar, with the associated lowering of the water potential in the guard cells. Such a series of enzyme reactions would not directly require energy from photosynthesis or respiration. Water would diffuse in and the stomata would open. Stomatal closure was envisioned as a reverse process, in which respirationally produced CO_2 would result in a more acid environment in the guard cells. Starch synthesis from sugar glucose would be induced, the osmotic pressure would decrease, water potential would increase, and water would diffuse out of the guard cells. The resulting decrease in turgor would cause closure.

Although this theory is based on a number of valid observations, it cannot be used to explain all of the reactions of the stomatal apparatus to environmental changes. It has been observed that both the opening and closing reactions may require cellular energy which may be supplied by the generation of ATP in the normal processes of photosynthesis and respiration (page 257).

Epidermal Hairs

Several different types of hair grow out from the epidermis of leaves, and resemble those from the epidermis of stems. They may be unicellular or multicellular, simple or branched, scalelike or glandular. The unicellular hair is the simplest kind. It may be an extension of an epidermal cell, or it may be provided with a basal cross wall. Unicellular hairs are sometimes branched. Multicellular hairs

Figure 10.9. Cuticle on the epidermis of leaves as seen with the Stereoscan electron microscope. A, lower epidermis of leaves of the Gambel oak (*Quercus gambelii*). ×3000. B, epidermis of a petal of the Shasta daisy (*Chrysanthemum maximum*). ×2000. (A, courtesy of J. Tucker; B, courtesy of R. Falk.)

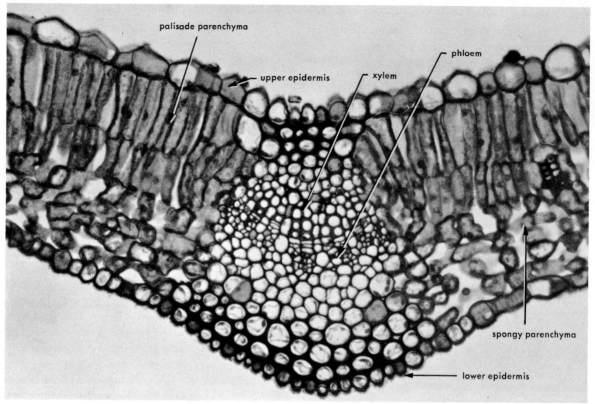

Figure 10.10. Photomicrograph of a cross section of a midrib of a lilac (*Syringa vulgaris*) leaf. ×150.

may consist of a single row of cells, or they may be branched. Glandular hairs bear, at the upper end, a single large cell or a group of cells; such cells excrete ethereal oils, which often impart a stickiness to leaves.

THE MESOPHYLL

The mesophyll is the photosynthetic tissue between the upper and lower epidermal layers. It is parenchyma tissue, traversed by veins. The chloroplasts are present in the mesophyll cells, which may be divided in two distinct layers: **palisade parenchyma** and **spongy parenchyma** (Figs. 10.6, 10.7*A*). The palisade parenchyma is just below the upper epidermis and usually consists of from one to several layers of narrow cells with their long axes at right angles to the leaf surface. The spongy parenchyma extends from the palisade parenchyma to the lower epidermis. Cells of the spongy parenchyma are irregular in shape and loosely arranged. Large air spaces, or stomatal chambers, are generally present above each stoma.

A section cut obliquely through the leaf shows, essentially, cross sections of the leaf tissues. Such a view is given in Fig. 10.7*B*. The cuticle occupies the upper portion of the figure, with sections of cells of the upper epidermis just beneath it. Below them the palisade parenchyma cells have a circular outline and fit together very loosely. One obtains the impression of a greater amount of air space in the palisade tissue than is apparent in the cross section of leaf. The anastomosing veins with their occasional endings are very apparent.

The palisade parenchyma is, in general, found above the veins and the spongy parenchyma below them. Note also that in this plane the cells of the spongy parenchyma fit together so as to form an open network. The lower epidermis is at the bottom of the section (Fig. 10.7*B*) and guard cells may be seen within it.

Intercellular spaces are prominently developed in the mesophyll but are much larger in the spongy parenchyma than in the palisade parenchyma. The air spaces between cells of the mesophyll are interconnecting, and many cells are in contact with an intercellular space. Thus, most food-making cells have free access to carbon dioxide and oxygen.

Veins of the Leaf

The veins or vascular bundles form a network extending throughout the leaf. The conducting elements are xylem and phloem, and fibers and collenchyma are sometimes associated with these. Hence, veins conduct water, mineral salts, and foods and also mechanically support the mesophyll tissue. In addition to the midrib and larger lateral veins that are visible to the naked eye, innumerable minute branch veins occur that can be seen only with the aid of a microscope. The large veins contain vessels, tracheids, sieve tubes, and companion cells and also some mechanical tissue (Fig. 10.10). Such veins may have both primary and secondary vascular elements. The smaller veins have few vascular elements and few or no mechanical elements. The very end of a vein is usually a single tracheid of the spiral type (Fig. 10.11). The free ends of veinlets are usually surrounded by one or more layers of parenchyma cells, **border parenchyma,** which may or may not possess chloroplasts. Through these cells, water and solutes must pass from the conducting elements of the veinlet to cells with chloroplasts. The smallest veinlet has an unbroken connection with the vascular elements of the midrib, the petiole, and the stem to which the leaf is connected.

In larger veins that have both xylem and phloem elements, the xylem is toward the upper surface of the leaf, the phloem toward the lower surface (Fig. 10.10).

The Petiole or Leaf Stalk

In the vascular tissues of the petiole, the phloem and xylem maintain their relative positions, as in the stem (Fig. 10.12*B*); thus, phloem is on the underside of the petiole (and leaf blade) and xylem on the upperside. One or more vascular bundles are embedded in parenchyma. Fibers may be associated with the vascular tissues of the bundles, and, not infrequently, groups of collenchyma cells occur beneath the epidermis (Fig. 10.12*A*).

LEAF SHAPE

Genetic Control

Leaf shape is under direct genetic control. Several single-gene mutations of tomato (*Lycopersicon esculentum*) result in striking changes in leaf shape (Fig. 10.13). Cell cytoplasm, however, may also influence the final form assumed by the leaf. In hybridizing experiments with the evening primrose (*Oenothera*), the leaf shape of the progeny always resembles that of the maternal parent. The leaves of *Oenothera berteriana* are broader and longer than those of *Oenothera odorata*. The pollen parent in breeding experiments between these two species has no influence on the shape of the leaves borne by the progeny. When *Oenothera berteriana* is the maternal parent, all the leaves of the progeny resemble *O. berteriana* leaves. Conversely, when *O. odorata* is the maternal parent, the leaves of the progeny resemble *O. odorata* leaves. The genes contributed by the paternal chromosomes do not influence the leaf form of the progeny. This type of **maternal inheri-**

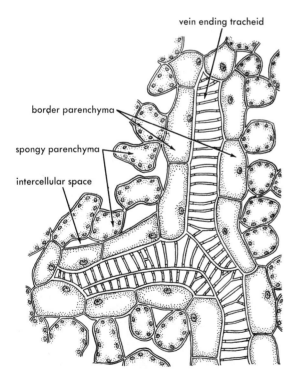

Figure 10.11. Diagram of vein ending in a leaf. Note the parenchyma tissue bordering the vein and the presence of only annular tracheids in the vein.

Figure 10.12. Photomicrographs of celery (*Apium graveolens*) petiole. *A*, outer tissues; *B*, vascular bundle (vein). ×50. (Courtesy of K. Esau.)

Figure 10.13. Different leaf shapes induced in leaves of tomato (*Lycopersicon esculentum*) by single gene mutations. (Courtesy of Tomato Genetics Cooperative.)

Figure 10.14. Adult (A) × ¼ and juvenile (B) leaves of *Eucalyptus*. × ½.

tance demonstrates the dominant influence of the egg cytoplasm in determining leaf form in these species.

Environmental Factors

Leaf shape is also influenced by environmental factors such as light, moisture, and temperature. The total intensity of light, the wave length of light, and the daily duration of light have separate but interrelated effects. Plants growing in intense sunlight usually have thick leaves with a thick palisade tissue and a dense, spongy parenchyma. Their intercellular spaces are small, and the epidermis is heavily cutinized and generally glossy, with the stomata confined to the lower epidermis. These plants may also have wooly epidermal hairs. Leaves of the same species growing in the shade have contrasting traits: they are thin, with a single palisade layer and a spongy parenchyma with many intercellular spaces. The epidermis has a thin cuticle which is usually dull, and stomata may be present on both upper and lower epidermal surfaces.

When bean or pea plants are grown in red light of 660 μ wave length, the internodes elongate and the leaves become small and slender. A short exposure to red light of wave length 730 μ will bring about a normal development of nodes and leaves. Plants grown in blue light will also have normal nodes and leaves even though they may not be able to carry on photosynthesis. Seedlings grown in darkness will have long internodes and small yellow leaves lacking chlorophyll; this condition is called **etiolation.**

The duration of light may also influence the shape of leaves. When *Kalanchoë* was grown on a regime of 8 hours of light and 16 hours of darkness, the leaves were small, succulent, and sessile, with entire margins. Under the reverse situation, 16 hours of light and 8 hours of darkness, the leaves were large, thin, and had petioles. Their margins were notched.

The submerged leaves of semi-aquatic plants may be vastly different in shape from the aerial leaves of the same plants. Similar changes may be induced by reducing the CO_2 content and lowering the temperature of the aerial portion of the shoot.

Age of Plant

Leaf shape may vary considerably with the physiological age of the plants. Thus, some plants have distinctive juvenile leaves for the first few years of growth. Subsequent adult leaves of a different form are characteristic of the older or adult plant (Fig. 10.14).

Leaf Modifications

Leaf shape may be considerably modified for functions other than photosynthesis (Fig. 10.15).

Leaf Shape   **199**

Figure 10.15. Leaf modifications. A, stipular spines of the black locust (*Robinia pseudoacacia*); B, spines from stipules and midrib in *Parkinsonia*; C, insect-capturing leaf of *Darlingtonia*; D, modified petioles of *Acacia*; E, tendrils from leaflets of *Bignonia*; F, plantlets on leaf of *Bryophyllum*, ×¼ to ×1.

Bud scales are short, thick, sessile, often covered with dense hairs on the outer surface, and sometimes waxy or resinous. When present, they protect the delicate meristematic tissue of the shoot tip, and the rudimentary leaves, from drying out.

The **spines** of various species of cacti and those of the *Fouquiera* represent entire transformed leaves. In the black locust (Fig. 10.15*A*) the stipules are spines and in *Parkinsonia* the stipules and midrib are spines.

In some species of *Lathyrus*, the **tendrils** are transformed leaflets; in others, the whole leaf is trans-

formed into a single tendril and leaflike stipules perform the normal functions of the leaves. In *Bignonia carpreolata* the third leaflet is transformed into a tendril (Fig. 10.15*E*). Both leaf tendrils or stem tendrils serve to attach the plant to a support. In certain species of *Acacia*, the petiole is leaflike (Fig. 10.15*D*).

Leaves are sometimes modified as food or water **storage organs.** The thick, fleshy bases of leaves that make up much of the daffodil bulb (Fig. 7.32*D*) accumulate large quantities of food. The so-called succulents of deserts and of saline soils have thick,

Figure 10.16. *A*, An insectivorous plant, sundew (*Drosera rotundifolia*). Glandular hairs secrete a protein-digesting enzyme. ×4. The endoplasmic reticulum (*B*) is well developed, and many dictysomes with large vesicles (*C*) indicate a high rate of synthesis in these cells. ×15,000. (*A*, courtesy of P. Jones; *B*, *C*, courtesy of D. Forbes.)

fleshy leaves with special water-storage tissue. This tissue consists of large parenchyma cells that usually lack chloroplasts. During the short period when water is available, it accumulates in the special storage tissue, and the plant draws upon it during periods of drought. Examples of plants bearing water-storing leaves are stonecrop (*Sedum*), ice plant, (*Mesembryanthemum*), Russian thistle (**Salsola**), *Sempervivum,* and *Bryophyllum.*

A striking adaptation of leaves to a special function occurs in the so-called **insectivorous plants.** In these plants the leaves have taken on forms and various structural features that enable them to capture insects and obtain food from their bodies. Well-known insectivorous plants are sundew (*Drosera*) (Fig. 10.16) and pitcher plants (*Nepenthes, Sarracenia,* and *Darlingtonia*) (Fig. 10.15*C*).

Leaves generally do not function effectively in vegetative reproduction. However, African violets are regularly reproduced vegetatively by the placing of leaves in moist sand. In certain species of *Bryophyllum* (Fig. 10.15*F*), patches of tissue in notches along the leaf margins remain meristematic. In time this tissue will develop small new plants while the parent leaf is still active. The little plants eventually drop from the leaf to the ground where under favorable conditions they may develop into new individuals.

LEAF DEVELOPMENT

Morphology of the Shoot Apex

Leaf development is intimately associated with the differentiation of the young shoot apex. The following section briefly reviews some of the problems of leaf and shoot development. The series of events is much more complex than in the root and is directly related to stem development. In spite of much work, many aspects of shoot development remain unclear.

The leaf is initiated in the promeristem of the shoot, close to the apex (Fig. 10.17). It first appears on the flanks of the apex, as a slight bulge resulting from the enlargement of several cells in the outer layer of the apical promeristem (Fig. 10.17). The promeristem cells, including the leaf intials, are true meristematic cells, small in size, having large nuclei, and with a densely-staining cytoplasm. The ground meristem rapidly differentiates from the promeristem and may be recognized by the more elongated, more vacuolated cells, below the promeristem (Fig. 10.17). Directly below the leaf initials, however, the cells remain truly meristematic; they are small, have large nuclei and have a dense, heavily-staining cytoplasm, with few, if any vacuoles. This is the procambium and it forms a continuous connection from the leaf initials to the primary vascular tissues below (Fig. 7.9). Differentiation of the procambium strand and the leaf initials is very closely correlated.

The Initiation of a Leaf

What causes the initiation of a leaf? Why, in lilac, do two leaf initials arise 180° apart and at the same time? The next two leaf initials will also be 180° apart, but oriented at 90° from the leaf initials below them. Leaves and axillary buds in lilac are opposite (Fig. 7.7B). Leaves in walnut are alternate (Figs. 7.7A). Taking bud A on a walnut twig (Fig. 7.6) as a starting point, it will be found that a line drawn through successively lower buds on the stem trace a spiral around it. The fifth bud in the series will be directly below bud A, and the spiral trace passes twice around the stem before arriving at the fifth bud. This relationship—the number of buds

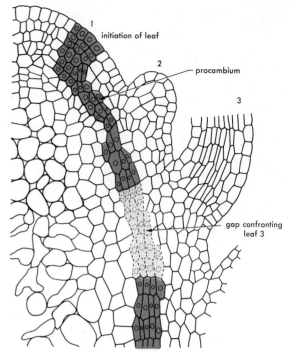

Figure 10.17. Diagram of a shoot apex showing the direct relationship between the initiation of a leaf and the procambium strand. The second leaf shows a definite primordium and the third leaf is well-formed. The procambium strands associated with these leaves are not shown.

between two aligned buds and the number of turns around the stem required to arrive at an aligned bud—may be expressed as a fraction, which for walnut is $\frac{2}{5}$. The relationship in lilac is given by the fraction $\frac{1}{2}$. Other common bud arrangements are $\frac{1}{3}$ and $\frac{3}{8}$. The relationship is constant for a given species and the arrangement makes it possible to precisely locate the point of leaf initiation on a shoot apex before it becomes visible to the eye. If that precise point of future leaf initiation is injured, that leaf primoridium will not develop. It is currently thought that new leaf initials arise in the next available space. When, in the growth of the shoot apex, such a space becomes available, new leaf initials will appear.

The degree of autonomy of the shoot apex is graphically demonstrated by the ability of cultured apical tissue of a fern, consisting only of promeristem, to continue to differentiate and to finally produce a whole plant. In the angiosperms, exised promeristem alone will not differentiate when placed in a complex growth medium. However, if the first leaf initials are included with the excised apex, differentiation follows with the eventual formation of a whole plant.

Development of a Young Leaf

The young leaf develops first as a slender pencil of tissue (Fig. 10.18), gradually flattening against opposing leaves, and a marginal meristem appears (Fig. 10.18). The lamina of the leaf is produced by this marginal meristem, which is also apparently responsible for the shape of the leaf. Uniform activity of the marginal meristem results in an entire leaf like the almond (Fig. 10.2*A*). If the activity is more closely related to the veins, lobed lamina arise (Fig. 10.2*C*). The cellular organization of the leaf is completed while it is very small. In horse chestnut, cell divisions are even completed in the fall while the leaves are still enclosed within the bud. Expansion of the leaves as the bud opens in the spring (Fig. 7.4) is due entirely to cell enlargement.

LEAF ABSCISSION

The separation of plant parts from a parent plant is a normal, continually occurring phenomenon. Leaves fall, fruits drop, flower parts wither and fall away, and even branch tips or whole branches may be separated normally from the parent plant. The autumn fall of leaves from woody dicotyledons is the most common example of this. In practically all cases separation or **abscission** is the result of the formation of definite, specialized cell layers known as the **abscission zone** at, or close to, the base of the petiole (Fig. 10.19*A*, *C*). In some instances the abscission zone is formed early in the development of the leaf; in other species it does not appear until the leaf is fully developed. The parenchyma cells comprising the abscission layer may be smaller and lack the lignin which may occur in significant amounts in the cells of adjacent tissues. Even the vascular elements may be shorter and fibers absent from the bundle in the abscission zone. These anatomical features definitely make this zone an area of weakness.

Previous to leaf fall, changes may normally occur in the zone. Cell divisions, though apparently not necessary, frequently take place. When they do occur, a layer of brick-shaped cells is formed across the petiole. Actual separation of the leaf may be brought about in a number of ways. In some species the middle lamella is dissolved away and the cells simply fall apart. In other plants, the cells themselves are dissolved and thus the separation is effected. In a third small group of plants a layer of cork forms across the petiole so that the leaf simply withers in place and is blown away. If a cork layer has not formed before leaf fall, it develops immediately afterward (Fig. 10.19*B*, *D*). Furthermore, the vessels are likely to become plugged with tyloses or gums. Thus the fall of a leaf does not leave an open wound or point of entrance for organisms that might cause disease.

The function of the abscission zone is two-fold: (1) to bring about the fall of the leaf or other plant part, and (2) to protect the region of the stem from which the leaf has fallen against insect damage or rot caused by lower plants.

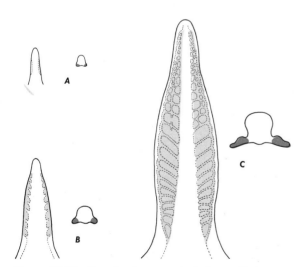

Figure 10.18. Growth of a young leaf. A, pencil stage, formation of marginal meristems; B, activity of marginal meristems initiates lateral expansion of a young leaf; C, continued lateral growth through activity of marginal meristems. (Redrawn after G. S. Avery, Amer. J. Bot. **20,** 565.)

Hormones and Leaf Abscission

It has been known since 1953 that auxin, particularly IAA, can inhibit abscission. It has become standard orchard practice to spray with IAA to prevent fruit drop. This is particularly important in apples, where a large drop of young fruit is a common occurrence. In the course of studying the influence of auxin on cotton, a substance was isolated from cotton buds that accelerated abscission of cotton leaves and was antagonistic to auxin. This

Figure 10.19. Leaf abscission. *A, C,* formation of the abscission zone; *B, D,* separation within the abscission zone and initiation of periderm; (*A* and *B,* courtesy of F. T. and A. B. Addicott; *C* and *D,* courtesy of Hall.)

substance was chemically characterized in 1963. It is now called **abscisic acid** or **ABA,** and it promotes abscission in many other plants.

At the same time other workers were isolating a material which was known to induce dormancy in buds. It was given the name **dormin.** It is now known that dormin and abscisic acid are the same substance.

Abscisic acid is widespread in the plant kingdom. In small amounts it stimulates abscission of both leaves and fruits, induces dormancy of apical buds, prolongs dormancy of potato buds, inhibits seed germination, and plays some part in flowering.

The gibberellins will also induce leaf abscission, but are only about one-half as active as abscisic acid. The action of the two hormones is somewhat different and may account for the variation in the morphological changes during abscission that were noted above. When abscission is induced by abscisic acid, cell division is restricted and no definite separation layer is formed. The cell walls of the parenchyma cells of the abscission zone break down. Separation occurs about 24 hours after application of ABA and may be initiated at any place in the abscission zone. In contrast, gibberellin application causes a great increase of cell division and the creation of a definite separation layer. Actual cell separation occurs by a breakdown of the middle lamella. The leaf falls about 48 hours after the application of the gibberellin.

In general, gibberellin, auxins, and phytokinins stimulate growth. Abscisic acid retards growth. Knowledge of the hormonal properties of abscisic acid is limited, because of its recent isolation. However, it is tempting to speculate upon the possible interaction of ABA with the auxins, gibberellins, and phytokinins, as a partial explanation of the yearly rhythm of plant growth and senescence. During the fall, an increase of abscisic acid could account for leaf and fruit abscission and dormancy of buds, seeds, and other vegetative parts. The dominance of the stimulating hormones in the spring could induce the germination of seeds, break the dormancy of buds, and in general account for the flush of spring growth. But little is yet known about the detailed interactions of these hormones.

SUMMARY

1. The functions of leaves are photosynthesis and transpiration.

2. Angiosperm leaves are generally supplied with flat, thin blades attached to the stem by petioles or sheaths. Veins strengthen the blades and transport food and water.

3. Gymnosperm leaves of the northern hemisphere are needle- or scale-like.

4. A cross section through the blade of a leaf shows the following tissues: upper epidermis, palisade parenchyma, spongy parenchyma and lower epidermis. Generally, a waxy cuticle coats the epidermis. Guard cells of the epidermis form stomata through which gases pass.

5. Increased turgor inside the guard cells enlarges the stomatal opening; decrease in turgor pressure decreases the size of the stomatal opening. Increased turgor may result from increased intensity of sunlight and increased temperature around the leaves. Wilting, or an increase of CO_2 around the leaves, brings about a reduction of turgor.

6. The mesophyll tissue is comprised of palisade and spongy parenchyma tissues and veins. Chloroplasts are present in both parenchyma tissues. The mesophyll is adapted for photosynthesis.

7. Leaf shape is under both nuclear and cytoplasmic control, and is also influenced by light, moisture, and the physiological age of the plant.

8. Leaves may become modified, serving as bud scales, food or water storage organs, and insect traps.

9. Leaf primordia develop in the next available space in a definite spatial sequence on the flanks of the shoot apex.

10. The excised promeristem of a fern apex will develop a new shoot, but leaf initials of an angiosperm shoot must be included with the excised apical meristems before shoot differentiation will occur.

11. The young leaf lamina is produced by marginal meristems which are also responsible for the shape of the leaf.

12. Cell divisions are completed while the leaves are enclosed in the bud; leaf expansion results mainly from cell englargement.

13. The formation of a definite abscission zone across a petiole or fruit stem is responsible for leaf fall or fruit drop.

14. Abscisic acid induces the formation of the abscission zone; it is also responsible for dormancy of apical buds. It is antagonistic to the growth hormones.

Soil and Mineral Nutrition

We have been studying the morphology of seed plants—the structure of their organs, tissues, and cells. With this knowledge to help us, we are now ready to consider in more detail the physiology of the plant—the ways in which the various parts function together in the processes of absorption, conduction, transpiration, photosynthesis, and respiration.

SOIL—THE ENVIRONMENT OF ROOTS

Soil may be defined as the weathered superficial layer of the earth's crust that typically is made up of decomposed and partly decomposed parent rock material with associated organic matter in various stages of decomposition.

Soil is the natural medium in which the roots of most plants grow. From the soil the plant absorbs the water and solutes necessary for its continued well-being. If a soil is fertile, it contains in a readily available form all the chemical elements essential for plant growth. It is through the soil that the agriculturist can effectively alter the environment of the roots and thus control plant growth, at least partially. The time and method of fertilizer used, as well as cultivation and irrigation practices, are all directed toward increasing the production of plant products through the effects that these practices have on the soil and root relationships and ultimately on the growth and development of the plant. There are many kinds of soils and many different soil conditions. The character of the natural plant covering and the behavior of crops depend upon soil conditions as well as upon climatic conditions. Environmental factors that operate through the soil are called **edaphic factors;** those that act upon the plant through the atmosphere are called **climatic factors.**

Soil Formation

Each environment creates a soil type unique to it. These soil types have their own history of development, morphology, and chemical attributes. Soil is a complex system which includes mineral (inorganic) matter, organic matter, water, air, and organisms.

Most soils consist largely of mineral particles formed by a slow continual process of weathering of the parent rock. The mineral matter is derived from fragmented rock. The kind of parent rock (granite, feldspar, sandstone, limestone, shale, etc.) and the degree of weathering determines the nature of the mineral or inorganic components of the soil.

Weathering may involve simply **mechanical breaking** of the parent rock. For instance, the action of strong winds or wave action may hurl sand against rock outcroppings and wear them down, or water collecting in pores of the rocks may expand at freezing temperatures to create fissures which ultimately fragment the rock. **Chemical weathering** results in more profound changes in the mineral matter itself. Atmospheric gases, such as CO_2 or SO_2, become dissolved in rainwater and produce acidic solutions which dissolve the parent rock material. Plant roots secrete weak acids. Certain algae, bacteria, and lichens hasten the decomposition of some of the least resistant of the mineral matter.

Weathering of the parent material under the influence of climate and organisms proceeds along a definite series of stages from a young soil in which the processes of soil development are continuing to a mature soil that has approached a state of equilibrium.

Soil formation may proceed relatively rapidly if the type of parent material and type of climate are favorable. Conditions that hasten the rate of soil development are a warm, humid climate, forest vegetation, flat topography, and a parent material which is easily broken down. Fort Kamenetz, built of limestone slabs in the Ukrainian part of Russia, was abandoned in 1699. Today, a mature soil 4–16″ thick has developed from the limestone. Conditions which retard soil development are a cold, dry climate, grass vegetation, steeply sloping topography, and parent rock material not easily broken down. It has taken from 1000 to 10,000 years for mature soils to develop in northern areas scoured by the Wisconsin glaciation.

In a wet temperate region, soil development is principally caused by percolating rainwater that continuously dissolves nutrient salts in the upper portion of the soil and carries them, along with particles of finely divided, chemically weathered mineral matter and bits of organic matter, downward in a process called **leaching.** In time, the soil consists of a series of superimposed layers or **horizons**

which differ in color, texture, and chemical attributes (Fig. 11.1).

The upper, or *A*, horizon is rich in organic matter only in the few uppermost inches, where freshly deposited organic litter is decomposing. The lower portion of the *A* horizon is sandy and light-colored, and may be deficient in nutrients, organic matter, and clay. Gradually, one to two feet below the surface, the *B* horizon begins. The *B* horizon is often reddish-brown from oxidized iron which has accumulated in it; it is relatively rich in mineral nutrients and clay and usually has a more neutral pH than the *A* horizon. The *B* horizon, then, is a layer of accumulation. Finally, three to four feet below the surface, the *C* horizon, made up of slightly weathered parent material, appears. The relative

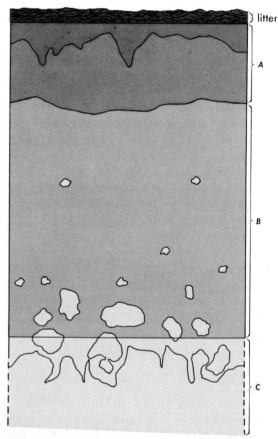

Figure 11.1. Profile of a typical soil of a wet temperate region. The A horizon, except for a thin region just beneath the litter, is leached of organic matter, clay, and salts. These leached substances accumulate in the B horizon. The C horizon, removed from weathering processes, represents parent material from which the A and B horizons have been formed.

Table 11.1	Classification of soil mineral matter according to size of particles

TYPE OF PARTICLES	RANGE IN DIAMETER OF SOIL PARTICLES IN MILLIMETERS
Coarse sand	2.0–0.2
Fine sand	0.2–0.02
Silt	0.02–0.002
Clay	0.002 and smaller

thickness of the *A* and *B* horizons depends upon the climate and amount of plant cover.

Mineral Matter of the Soil

The mineral fractions in most soils are primarily responsible for the texture of the soils.

In referring to the size of the mineral fractions of a soil, it is customary to speak of coarse sand, fine sand, silt, and clay. The size of soil particles decreases in the order given (Table 11.1). Clay is not only the smallest mineral soil particle but it is also the most changed from the parent rock by chemical weathering. The clay fraction, together with some of the organic material, is of colloidal dimensions and, as we will see, gives to soils some of their most important properties, such as the ability to hold water and nutrients.

Organic Matter of the Soil

The organic matter of the soil is derived from plants and animals. Throughout the centuries of soil formation, the plants and animals that lived in and on the soil have left their residues. Annual herbs die each year; the whole plant, including the roots and tops, contributes to the organic matter of the soil. Also, trees and shrubs shed their leaves, twigs, bark, and fruits, and their roots die. All this plant material decomposes, and thus, with unimportant exceptions, soils are a mixture of mineral matter and organic matter, the latter in various stages of decomposition.

We might be led to believe that during the centuries the organic material in the soil would gradually increase. This material, however, is being continually decomposed by bacteria and fungi. Cultivation hastens the loss of organic matter from a soil by increasing aeration.

Generally, for good crop growth, most soils should have considerable organic matter. Those soils naturally low in organic residues may be improved by adding barnyard manure or other fertilizers, or by plowing under green manure crops. Organic matter improves the physical condition of most soils, especially those with much clay. It makes them easier to cultivate, and it may increase slightly their ability to hold water, but added organic matter may not increase the amount of water the plant is able to extract from the soil. The gradual decomposition of organic matter in the soil continually liberates mineral elements that are essential for plant growth. Sulfur, phosphorus, and nitrogen are among the most important elements thus liberated in forms available for plant growth.

The amorphous, dark colored, partially decomposed organic matter in soils is called **humus.** The chemical nature of humus is very complex; in part, humus is a polymer of rings and straight chains of C, H, O, and N.

Mineral and Organic Soils

Although most soils contain both mineral and organic matter, in the majority of soils the organic matter constitutes only two to five percent of the soil by weight. These predominantly **mineral soils** may be classified according to the relative amounts of the sand, silt, and clay that they contain. Soils which contain roughly equal amounts of sand, silt, and clay are called **loams.** Depending upon the relative amounts of each component in the soil, we speak of clay, clay loam, silt loam, loam, sandy loam, loamy sand, and sand (Table 11.2). Soil texture and the suitability of a soil for plant growth are greatly influenced by its sand, silt, and clay content.

Table 11.2	Composition of three soils according to the size of the mineral particles		

SOIL TYPE	COARSE SAND %	FINE SAND %	SILT %	CLAY %
Sandy loam	67	18	6	9
Loam	27	32	21	20
Clay	1	9	22	68

Organic soils are defined as soils having a layer, 12 or more inches thick, that contains more than 30 percent organic matter. Such soils occur in wet, cool depressions that become acidic and anaerobic as the rate of litter accumulation becomes greater than the rate of decomposition.

Organic soils are divided into two categories, **peat** and **muck,** depending on the degree of decomposition. In peat, most of the original plant material can be identified as to species, while in muck the original plant material has been decomposed beyond recognition. Peat accumulates to such an extent that, in treeless areas, it is cut out in blocks and used for fuel.

Soil Water and Its Dissolved Substances

From the standpoint of plants, an important difference in various soils is their ability to hold water. It is greatest in clay and organic soils, least in coarse sand.

If water is applied to a soil in the field, the spaces between the soil particles **(pore spaces)** become filled only for a short time to the depth wetted. With drainage the water begins to move downward under the influence of gravity. After a while this movement downward stops; the soil particles hold a certain amount of water against the pull of gravity. The amount of water held by the soil after drainage is called the **field capacity** of that soil. When a soil is at field capacity, or even at a moisture content well below this amount, a film of water completely surrounds each soil particle, and water also exists in the form of wedges between the soil particles. Clay particles and organic matter in the soil constitute **colloidal systems,** and such particles hold water by **imbibition.** Water imbibed by soil particles is much more difficult to remove from the soil than that which exists as a film or as wedges.

The soil, then, is a reservoir of water. The water holds in solution various inorganic salts (nitrates, phosphates, sulfates, etc.) and other water-soluble substances. The composition and concentration of this solution is ever-changing, depending upon the higher plants and the microorganisms growing on and in the soil. All substances that enter the plant must be in solution. This makes possible their passage through the cell wall and the cytoplasmic membranes of root hairs. In most agricultural soils, the soil solution has a very low concentration—usually lower than that of the cell sap. The water potential in the soil is usually higher than that in the plant. As long as this relationship is maintained, water will diffuse inward.

In the soil, nutrient elements may exist in a relatively unavailable form. For example, most organic nitrogen compounds are not available as such to crop plants; that is, nitrogen in the form of either plant or animal proteins cannot be utilized directly by green plants. However, these organic compounds are acted upon by soil organisms, and some of the products of decomposition do become directly available. Organic sulfur compounds are also broken down by living organisms to inorganic compounds. In general, it can be said of soils that at all times chemical changes are taking place which set free inorganic nutrient substances that will dissolve in soil water and thus assume a form that can be taken up by the roots.

The soil solution is usually very dilute and a large number of the soluble inorganic compounds are present as ions. For example, sodium nitrate, $NaNO_3$, exists as Na^+ and NO_3^- ions.

The soil particles themselves, particularly the finely divided clay and organic matter, act as giant negatively charged ions which attract a cloud of positively charged ions such as Ca^{++} and K^+. Thus the colloidal soil particles, while not in solution themselves, serve as reservoirs to which many of the various ions essential for plant growth are loosely attached and from which these ions may be released into solution.

Negatively charged ions such as nitrate, NO_3^- and sulfate, SO_4^{--}, are not held by the soil particles but are in solution. Phosphate, PO_4^{---}, presents a different picture in that these ions enter into the composition of certain clays or are easily precipitated and do not usually exist in high concentrations in soil solution.

Acidity of the Soil

The acidity of a soil influences the physical properties of the soil, the availability of certain minerals to plants, and the biological activity in the soil. It consequently strongly influences plant growth. Certain plants such as azaleas, camellias, and cranberries grow best in acid soils; most plants do best in soils near neutrality, while a few plants grow satisfactorily in slightly alkaline soils (Table 11.3). Soil acidity is a measure of the concentration of hydrogen ions in solution and is usually expressed in pH units. The pH scale is a logarithmic scale in which each unit represents a hydrogen ion concentration ten times more, or less, than the next unit. Thus, a solution having a pH of 7, which is the neutral pH, has ten times the hydrogen ion concentration of a solution of pH 8, a solution that is slightly alkaline. The same solution of pH 7 has one tenth of the hydrogen ion concentration of a solution of pH 6, a slightly acid solution.

Acid soils, such as those beneath northern coniferous forest with acidic foliage, range from 3.5 to 5.0 pH; agricultural soils in the humid areas range from 5.0 to 7.0; and soils of arid or saline regions may range as high as 11.0, but pH 8.0–9.0 is more common. Acidity *per se*—that is, H^+ or OH^- ions—is not directly responsible for limiting plant growth. Rather, soil pH affects the availability of plant nutrients in two ways. In an acid soil, hydrogen ions (H^+) replace other **cations** (like K^+, Ca^{++}, and Mg^{++}) on the negatively charged clay particles, allowing the nutrient ions to be easily leached from the soil. Soil pH also affects the solubility of plant nutrients. As shown in Fig. 11.2, some nutrients, like iron, manganese, and aluminum, increase in solu-

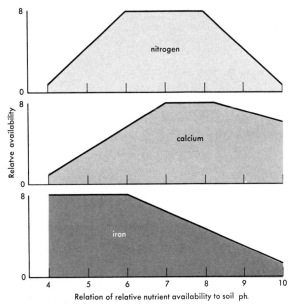

Figure 11.2. Relationship of relative nutrient availability to soil pH.

bility as pH decreases. Aluminum may reach toxic levels in some acidic soils. Other nutrients, like calcium and magnesium, increase in solubility as pH increases. Soil pH may also affect plant growth indirectly by suppressing bacterial growth at pH extremes.

Acid soils may be improved by the application of lime (calcium hydroxide or calcium carbonate), which tends to neutralize the acid. Alkaline soils may be improved by the addition of sulfur or ammonium sulfate. If elemental sulfur is applied, it may become oxidized to sulfate and make the soil more acid. The ammonium of ammonium sulfate is either converted into nitrate by bacterial action or is absorbed by the plants, resulting in a net increase in sulfate and acidity.

Air of the Soil

We have spoken of the pore space in soils. The air of the soil contains the same gases as the atmosphere above the soil. The relative proportions of these gases in the soil air, however, may be somewhat different from those in the atmosphere. The air of the soil is considerably richer in carbon dioxide and poorer in oxygen than that of the atmosphere. In the latter, the average percentage by volume of carbon dioxide is 0.03 percent; in soil air

Table 11.3 *Representative values for the acidity (pH) of soils in which certain plants grow well*

RELATIVE ACIDITY	pH RANGE	PLANTS GROWING WELL IN SOILS OF THIS ACIDITY
High acidity	3.2–4.6	*Sphagnum* moss in peat bogs
Moderate acidity	3.5–5.0	Blueberry
Slight to moderate acidity	4.5–6.0	Azalea, camellia
Nearly neutral	6.0–7.5	Many forest and field plants
Alkaline	7.5–9.1	Greasewood, rabbit brush, stonewort, various plants growing around alkali flats

Figure 11.3. Effect of aeration on root growth in tomato (Lycopersicon esculentum). Left, plants growing in complete nutrient solution through which air was bubbled; right, plants growing in same solution without aeration. (Courtesy of D. R. Hoagland.)

the percentage may go as high as 5 percent. This high percentage of carbon dioxide in soil air is due to the respiration of soil bacteria and fungi and, to a lesser degree, the respiration of roots themselves. The soil air usually has a relative humidity near 100 per cent, a condition very favorable to the growth of soil organisms.

The living cells of roots must have oxygen to support respiration. Normally, the oxygen necessary for respiration reaches the roots by absorption through the root hairs and other epidermal cells. This inward diffusion of oxygen from the soil air ordinarily goes on readily because of the thinness of the walls of these cells and the absence of a cuticle. Roots at all levels in the soil secure oxygen directly from the air in the pore spaces and, to a limited extent, from the oxygen dissolved in water; in ordinary land plants there is no system of air-conducting tubes that conveys oxygen from the atmosphere through the plants to the extremities of the roots.

The amount of air in the soil depends not only upon the pore space but also upon the water content of the soil. If water occupies the pore space, air is forced out. A water-soaked (saturated) soil contains practically no air save that which is dissolved in water. If the soil about the roots is continuously

water-soaked, plants die because of insufficient oxygen and possibly as a result of the accumulation of carbon dioxide. Most species of land plants will grow normally with their roots in a water solution, if it is well aerated by bubbling air through it (Fig. 11.3). Evidence indicates that inadequate soil aeration results in a reduction in the rate of water and mineral absorption.

Whereas most land plants, including agricultural plants, will not long survive with the root system submerged in unaerated water or surrounded by a soil that is water-soaked, some plants flourish under such conditions. Among them may be cited rice, various swamp and marsh plants, and the bald cypress (*Taxodium distichum*) (Fig. 11.4). Almost all such plants contain in the stem and roots large communicating air spaces. Thus, air absorbed into the leaves and stems may reach the living cells of the root in sufficient quantity. Bald cypress and certain species in tropical mangrove swamps develop special root branches that grow upward until their ends are above the water level. These special root branches have a central core of loose tissue through which air moves downward to the submerged organs.

In most agricultural soils, air of the soil is not a limiting factor in plant growth. However, stirring of the soil (tillage) may facilitate its aeration and is especially beneficial if the supply of air is inadequate to begin with. But, as stated, it is very improbable that in ordinary agricultural soils an actual deficiency of soil oxygen ever exists, unless the soil is water-soaked; the latter condition is one that cannot be corrected by tillage. If not to improve soil aeration, what are the benefits of tillage? The student may care to explore this subject further.

Organisms of the Soil

Ordinary soils teem with living organisms, both plants and animals. These organisms are an important part of the environment of roots. The soil flora includes bacteria, fungi, and algae. Living roots of higher plants also are an important part of the soil. When the roots die they leave a considerable quantity of organic matter. The soil fauna includes protozoa, nematodes, earthworms, various insects, and burrowing animals. Most of these are beneficial in some way or another; others may be harmful. Certain bacteria are absolutely essential in the maintenance of soil fertility. This point will be pursued further in Chapter 18, but it should be indicated here that certain bacteria and fungi that live in the soil are responsible for the decomposition of organic matter (plant residues, manure, and other organic fertilizers), and for nitrogen fixation (see Fig. 18.33). As a result of their activity, they keep up the supply of soil nitrogen, one of the chief factors in soil fertility. Earthworms affect soil structure by moving and mixing the soil; they pass large quantities of soil through their bodies. Charles Darwin, in 1885, published *The Formation of Vegetable Mold,* in which the importance of earthworms in soils is stressed. He points out that earthworms in an acre of soil may pass through their bodies annually as much as 15 tons of dry earth.

Temperature of the Soil

The rate of absorption of both water and solutes by roots may be reduced by extremes of temperature. Plants native to cool climates absorb these substances more freely at low temperatures than do plants of warm climates. A plant may wilt in a soil containing ample water if the soil temperature sinks below, or rises above, a certain degree. In cold, dry climates, winter killing may be the result of a cold soil that slows up absorption, accompanied by a high transpiration rate. It is believed that in winter-killing, the plant is as frequently killed by direct drying as by actual freezing.

All chemical and biological activities of a soil are influenced by soil temperature.

The soil temperature is not always the same as the temperature of the air above it. It may be lower or higher than the air temperature (see chapter 18).

MINERAL ELEMENTS ESSENTIAL FOR PLANT GROWTH

For over 100 years physiologists have been studying the mineral nutrition of plants. Extensive investigations have been carried out to determine which elements are essential, how the plant absorbs them, how they are utilized, why they are essential, and what effects are produced in the plant when a particular essential element is lacking (see color plate 3).

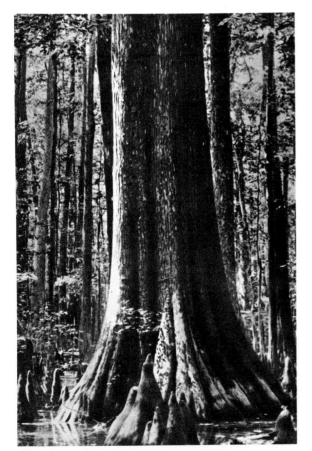

Figure 11.4. Aerial "stump roots" of bald cypress (*Taxodium distichum*). (Courtesy of U.S. Forest Service.)

Since 1860 it has been known that plants required at least seven elements: nitrogen, sulfur, phosphorus, potassium, calcium, magnesium, and iron. Since then, nine more required elements have been discovered. These elements are required in such minute amounts that their importance remained long undetected because the air about the plants or the containers holding the growth medium supplied enough of them to support plant growth. There are other elements that are required by some plants in very small amounts but which most plants do not require. Sodium, for example, is required in very small amounts by the desert shrub *Atriplex*, but is not required by most other species. It is not considered an essential element.

Plants utilize elements in three basic ways. (1) The elements may form part of structural units, such as carbon in cellulose or nitrogen in protein. (2) Elements may be incorporated into organic molecules important in metabolism, like magnesium in chlorophyll or phosphorous in ATP. (3) Elements may function as enzyme activators, necessary as catalysts in certain enzymatic reactions. Magnesium is used as an enzyme activator in several of the enzymatic steps of glucose degradation in the process of respiration.

What methods have been employed to determine the chemical elements that are essential to plant growth? Of course, there are the accumulated experiences of agriculturists which have shown that applications of various nutrients to the soil result in healthier plants, increased yields, and better quality. But these observations have thrown little light on the specific role played by chemical elements in the plant. Plant physiologists have used more critical methods—careful weighings and measurements of quantities of elements used and results obtained. The solution-culture method has long been employed to determine what elements are indispensable in plant growth. Most kinds of plants that normally grow in soil can be grown to maturity

(fruit and seed production) in water to which soluble essential nutrient salts have been added (Fig. 11.5; see color plate 3). By growing plants in a nutrient solution containing all elements believed to be essential, except the particular element being investigated, the response of plants when this element is absent can be ascertained. In this case, all plants grown in deficient solutions were first grown in a complete nutrient solution for two weeks before being transferred into the deficient solution.

From such experiments a list of elements that are known to be essential for the growth of plants has been compiled. Long recognized as essential to the continued growth and development of green plants are the following ten elements, with their appropriate chemical symbols: carbon (C), hydrogen (H), oxygen (O), phosphorus (P), potassium (K), nitrogen (N), sulfur (S), calcium (Ca), iron (Fe), magnesium (Mg).

If the symbols are arranged in a line they can be used to remember the elements thus:

C HOPK'NS CaFe Mg

"See Hopk'ns Cafe, mighty good." Of these ten elements, the last seven are in the nutrient medium, that is, their various salts are dissolved in the soil water. Relatively large quantities of nitrogen, sulfur, phosphorus, potassium, magnesium, and calcium are required; very small quantities of iron meet all plant needs for this element. These elements, with the exception of iron, have been called the **macronutrient elements,** because they are needed in relatively large amounts. More recently it has been discovered that at least six additional elements are needed by higher plants but in much smaller amounts. The **micronutrient elements** (trace elements or minor elements) include, in addition to iron (Fe), the elements boron (B), manganese (Mn), copper (Cu), zinc (Zn), chlorine (Cl), and molybdenum (Mo). Table 11.4 shows the relative amount of macro- and micronutrients re-

Table 11.4 *Amounts (in pounds per acre) of macro- and micronutrients removed from the soil in one growing season by a wheat crop*

	MACRONUTRIENTS						MICRONUTRIENTS				
Element	N	K	P	Ca	S	Mg	Fe	Mn	B	Zn	Cu
Amount	76	42	14	12	11	8	0.7	0.5	0.3	0.2	0.03

moved in pounds per acre by wheat during a growing season.

A plant growing in soil or in solution cannot distingusih between those elements that are essential to it and those that are not essential or that might be harmful. If the element or ion containing it is in solution, it will probably be absorbed by the plant. Thus we find in the plant almost all the elements present in the soil. Gold has been isolated from plants growing in goldbearing soils. Selenium, an element poisonous to livestock, when present in soils is absorbed by certain plants in sufficient amounts to be harmful to animals grazing on them.

Macronutrient Elements

Carbon, Hydrogen, and Oxygen

The absorption of water, carbon dioxide, and oxygen brings into the plant large quantities of the elements carbon, hydrogen, and oxygen, from which is formed the major part of each of the hundreds of organic compounds found in the plant. We have seen how in the process of photosynthesis these elements are combined into carbohydrates. Simple fats also are composed entirely of these three elements, whereas proteins, alone of the foods, contain appreciable amounts of other elements, particularly nitrogen and, to a lesser extent, sulfur and phosphorus.

Nitrogen

This element is a component of proteins, which form an essential part of protoplasm and also occur as stored foods in plant cells. Nitrogen is also a part of other organic compounds in plants, such as chlorophyll (chloroplyll *a* has the formula $C_{55} H_{72} O_5 N_4 Mg$), amino acids, alkaloids, and at least some plant hormones.

Ordinary green plants cannot utilize elemental nitrogen, which represents about 78 percent of the air. It has been estimated that above every acre of land surface there are about 145,000 to 150,000 tons of this gas. Chief sources of nitrogen for green plants are the nitrates, such as sodium nitrate (Chilean saltpeter), potassium nitrate, ammonium nitrate, and calcium nitrate. Green plants can also utilize nitrites and ammonium salts, and evidence indicates that some plants can derive nitrogen from certain organic nitrogenous compounds.

Nitrogenous fertilizers, both natural and commercial, are usually the most important fertilizers applied to growing plants. The chief commercial nitrogenous fertilizers are as follows: (1) those with nitrogen in the nitrate form; (2) those with nitrogen in ammonia or its compounds; (3) those with nitrogen in organic compounds, such as tankage and cottonseed meal; and (4) those with nitrogen in the amide form, such as urea and calcium cyanamide. Organic nitrogen in complex molecules cannot be used directly by plants but must first be converted into available forms through the action of organisms in the soil.

The rate of growth of plants is influenced to a large degree by the available nitrogen. An early symptom of nitrogen deficiency is a yellowing of leaves, particularly the older leaves, although other factors may cause this condition; then follows a stunting in the growth of all parts of the plant (Fig. 11.5*E*). On the other hand, an excess of available nitrogen results in vigorous vegetative growth and a suppression of food storage and of fruit and seed development. The nitrogen cycle in nature is discussed more fully on page 378.

Sulfur

Sulfur forms a part of the molecules of proteins. Plant proteins containing sulfur may have from 0.5 to 1.5 percent of this element.

Ordinary green plants cannot utilize elemental sulfur; it must first be oxidized to sulfates, in which form it is normally absorbed by the roots.

Certain agricultural soils may become deficient in sulfur. The most noticeable responses to applications of sulfur fertilizers are an increased root development and a deeper green color of the foliage. Sulfur deficiency symptoms in tomato are shown in Fig. 11.5*F*.

Phosphorus

This element is also a component of some plant proteins. The highest percentages of phosphorus occur in those parts of the plant that are growing rapidly, such as meristematic regions and maturing fruits and seeds. Some of the very important roles that phosphorous plays in plant metabolism are discussed in Chapter 20.

Methods have been developed by which elements are "tagged." For example, radioactive phosphorus in the form of phosphate has been introduced into the culture solution in which tomatoes were growing. By means of the Geiger-Müller counter, or by radiograms (Fig. 20.14), the path of radioactive phosphorus through the plant has been followed.

After a time, the greatest accumulation of phosphorus was found in the young developing fruit; very little occurred in the ripe fruit. Within the fruit itself the greatest accumulation of phosphorus was in the developing seed.

Applications of phosphorus to soils deficient in this element promote root growth and hasten maturity, particularly of cereals. Phosphates are the principal source of phosphorus for plants. A tomato plant grown in culture solution without phosphorus is shown in Fig. 11.5 *B, C*.

Potassium

This element, indispensable to plants , accumulates in those tissues that are growing rapidly. It will migrate from older tissues to meristematic regions; for example, during the maturing of a fruit crop there is a movement of potassium from the leaves into the fruit. In plants, potassium occurs chiefly in inorganic form.

The specific functions of potassium in the plant are not clearly understood. It is known, however, that plants do not grow normally in a soil deficient in this element. In fact, many plant diseases, abnormal responses, and low yields of crops are associated with a lack of potassium; they are cured or alleviated by applications of potassium. Processes in the plant that appear to require an adequate supply of potassium are: (1) normal cell division, (2) synthesis and translocation of carbohydrates, (3) synthesis of proteins in meristematic cells, (4) reduction of nitrates, (5) development of chlorophyll, and (6) stomatal opening and closing.

Any water-soluble inorganic compound of potassium, such as potassium sulfate, potassium phosphate, or potassium nitrate, can be utilized by plants as a source of potassium. Potassium deficiency symptoms in tomato are shown in Fig. 11.5D.

Calcium

All ordinary green plants require calcium. It is one of the constituents of the middle lamella of the cell wall, where it occurs in the form of calcium pectate. Calcium affects the permeability of cytoplasmic membranes and the hydration of colloids. Calcium may be found in combination with organic acids in the plant. Oxalic acid, for example, is a by-product of protein metabolism. It is a soluble substance and is toxic to the protoplasm if it reaches a high concentration in the cell. When united with calcium, however, the soluble oxalic acid is converted into the highly insoluble calcium oxalate, which does not injure the protoplasm. There is also evidence that calcium favors the translocation of carbohydrates and amino acids and encourages root development. Calcium deficiency is frequently characterized by a death of the growing points (Fig. 11.5G).

Magnesium

This element is a constituent of the chlorophyll molecule. A deficiency of magnesium results in the development of pale, sickly foliage, an unhealthy condition known as **chlorosis** (Fig. 11.5I). This disease is a common one of economic plants, and in many instances applications of magnesium have effected a cure. However, chlorosis can be caused by deficiencies in other essential elements such as iron. The optimum activity of a number of enzymes is dependent on the presence of the magnesium ion in the cell.

MICRONUTRIENT ELEMENTS

Until recently, it was not known that the elements boron, zinc, manganese, chlorine, molybdenum, and

Figure 11.6. Disease of olives (*Olea europaea*) known as "monkey face," caused by a deficiency of boron. (Courtesy of Scott and Thomas.)

Figure 11.7. Disease of peach (*Prunus persica*) known as "little leaf," caused by a deficiency of zinc. Branch at left untreated; branch at right cured by driving in zinc-coated nails. (Courtesy of E. L. Proebsting.)

Iron. Although iron is not a component of the chlorophyll molecule, it is essential for the synthesis of chlorophyll. Chlorosis may be caused by iron deficiency as well as by a deficiency of magnesium (Fig. 11.5*B, H*). The quantity of iron required is very small. As an example, chlorosis of pineapples in Hawaii, due to the unavailability of iron from the soil, is remedied by spraying the plants with iron salt solutions. Orchard trees suffering from iron chlorosis may be cured by injecting iron compounds into the trunk, or by applying various kinds of iron salts to the soil.

Boron. That plants require boron is well established. Symptoms of boron deficiency include darkening of tissues and various growth abnormalities and disturbances. Some of the physiological diseases of plants due to boron deficiency are internal cork of apples, top rot of tobacco, cracked stem of celery, browning of cauliflower, heart rot of sugar beets, and "monkey face" of olives (Fig. 11.6). Very small applications (20 to 50 pounds) of sodium tetraborate (borax) per acre may be sufficient to cure these diseases. In contrast, fertilizers containing the macronutrient elements are frequently applied to fields at rates of several hundred pounds per acre.

Although plants require boron for normal development, the quantity in the soil or culture solution must be very small or injury will result. In fact, in certain agricultural sections severe injury to crops occurs because of excessive amounts of boron in the soil or in the irrigation waters. Plants vary in their tolerance of boron. Borates are used as weed-killers.

Zinc. Zinc is essential to the normal development of a variety of plants; probably it is required by all plants. As with other micronutrients, proof has come from careful water-culture experiments; also, certain diseases of plants have been cured by zinc applications. Large quantities of zinc are toxic to plants. Two well-known diseases caused by zinc deficiency are "little leaf" of deciduous fruit trees (Fig. 11.7) and "mottle leaf" of citrus trees. These abnormal conditions are corrected by spraying the trees with zinc salts, by injecting dilute solutions of zinc salts into the trunks, or by driving zinc brads into the trunks. These corrective measures show how minute are the quantities of zinc required by plants for normal development. Zinc is a constituent of certain enzymes and plays a role in the synthesis of indoleacetic acid, an important growth hormone in plants.

copper were indispensable for normal plant growth. The reasons were that they are needed only in very minute quantities (mere "traces") and that the chemicals used in culture solutions contained unsuspected traces of these elements sufficient to supply the needs of the plant's growth in the water culture. For example, a bottle of ferric sulfate labeled "chemically pure," taken from the laboratory shelf, may show on the label traces of chlorine, phosphoric acid, manganese, zinc, copper, sodium nitrate, and ferrous salts. If a culture solution containing this "chemically pure" ferric sulfate were used, most likely it would possess enough manganese, zinc, chlorine, and copper to meet the plant's requirements. In a study of micronutrient elements needed, or suspected of being needed, by plants, it has been necessary to employ highly refined techniques, employing redistilled water and highly purified salts. It may even be necessary to prevent atmospheric dust from falling into the culture, as the dust may contain traces of critical elements. A tomato plant grown in a nutrient medium lacking all of the micronutrient elements except iron is shown in Fig. 11.5*J*.

Figure 11.8. Chlorosis of tomato (Lycopersicon esculentum) leaf caused by a deficiency of manganese in the nutrient solution. (Courtesy of D. R. Hoagland.)

Figure 11.9. Tomatoes (Lycopersicon esculentum) growing in solution with all essential chemical elements except copper. Leaves at left were sprayed with solution containing copper. (Courtesy of D. R. Hoagland.)

Manganese. The most striking symptom of plants with a deficiency of manganese is chlorosis (Fig. 11.8), but it is a somewhat different type from that caused by iron deficiency. Whereas in iron chlorosis the young leaves may become yellow or white with prominent green veins, manganese chlorosis results in the leaf taking on a mottled appearance. Spraying or dusting crops suffering from manganese deficiency with as little as 20 pounds of manganese sulfate per acre frequently effects a cure.

Copper. Abnormalities in the growth of many plants, especially those in marsh and peat soils, have been corrected by the application of copper compounds (Fig. 11.9). Copper is also a constituent of certain enzyme systems, but its exact function is not definitely known.

Molybdenum. Molybdenum is now recognized as an essential micronutrient element, as least to certain plants. For example, it was found that molybdenum is required for normal growth of tomato seedlings, that only .01 parts per million of the element in the nutrient solution is needed, and that concentrations exceeding 10 parts per million cause injury. One of the roles of molybdenum in plants appears to be associated with the reduction of nitrates. For some Australian soils, only 2 ounces of MoO_3 per acre, applied once every 10 years, produces optimum forage improvement.

Chlorine. Chlorine is present in the soil solution as the very soluble, negatively charged chloride ion, Cl^-. Because of the very small quantities required by plants and its almost universal occurrence in soils, chlorine is never added intentionally as a fertilizer although it occurs in small amounts in all fertilizers. Rainfall may contain enough chlorine to supply plant requirements and some plants may obtain enough from the air to prevent deficiency symptoms. It is very difficult to prepare a nutrient solution completely free of chlorine. Consequently it is difficult to demonstrate chlorine deficiency symptoms in plants. Although the functions that chlorine plays in plant metabolism are obscure, it is known that chlorine takes part in the reactions of photosynthesis leading to oxygen evolution.

SUMMARY

1. Soil is a complex system composed of rock particles and plant and animal remains in various stages of decay, and in which an abundance of soil microorganisms is growing. Throughout the soil structure there is an interconnecting system of large and small pores filled with air or with water and dissolved materials.

2. All the essential elements except carbon, hydrogen, and oxygen are absorbed through the roots, generally in the form of ions from the soil.

3. The macronutrient essential elements are carbon, hydrogen, oxygen, nitrogen, sulfur, phosphorus, potassium, calcium, and magnesium.

4. The micronutrient essential elements are iron, zinc, copper, boron, manganese, chlorine, and molybdenum.

5. A plant's growth can be influenced markedly by the chemical nature and physical structure of the soil.

12

Transpiration, Conduction, and Absorption

The most abundant compound in an active cell is water. It plays a varied role in the life of a plant, for it serves as (1) a raw material in the synthesis of organic compounds, (2) the solvent in which vital reactions take place, (3) the medium through which solutes move from cell to cell, and (4) the source of turgor in plant cells. Plant tissues vary in water content; those actively growing have a water content equal to as much as 85 to 95 percent of their fresh weight, while a dormant structure such as a seed may have a water content as low as five to 10 percent. The entire shoot system of an herbaceous plant such as barley may be 82 percent water (Fig. 12.1; see color plate 4).

A plant from which the water has been removed is made up chiefly of organic material synthesized by the plant primarily from carbon dioxide, water, and inorganic nitrogen such as is found in nitrates or ammonium. Actually, the mineral matter absorbed from the soil is relatively little in terms of total amount but is, nevertheless, essential for the life of the plant. The small quantity of mineral matter contained in a plant can be readily demonstrated by burning the dry tissue. The carbon, hydrogen, and oxygen, as well as the nitrogen and some of the phosphorus and sulfur, pass off in the combustion process, while most of the minerals remain in the ash, combined with a small amount of oxygen (Fig. 12.1). These mineral elements may have occurred in the plant as inorganic ions dissolved in the cell or may have been a part of complex organic molecules. For instance, the magnesium in the ash may have come from the destruction of chlorophyll of which it is a part, while the carbon, hydrogen, oxygen, and nitrogen of the chlorophyll would have been lost during burning.

While minerals that are once absorbed by a plant are not generally lost again in large quantities, except during leaf fall, water is constantly being lost from the aboveground portions of the plant. Because this water loss profoundly influences the rates of water absorption and movement, it will be studied first.

TRANSPIRATION

By far the greater part of the water lost from the plant passes into the atmosphere as invisible water vapor. This loss of water in vapor form from a living plant is called **transpiration.** Although a small amount of water may be transpired directly through the cuticle, most of the water lost during the day diffuses out through the stomata. This is called **stomatal transpiration,** in contrast with **cuticural transpiration.** Leaves are the principal transpiring organs.

A knowledge of leaf structure is essential to an understanding of the mechanism of transpiration. It will be recalled that the mesophyll of most leaves is a very loose tissue with large intercellular air spaces. Even within the palisade parenchyma, a large portion of the walls of most cells is exposed to intercellular air spaces (Figs. 10.6, 10.7A). These are connected with similar air spaces throughout the leaf and, through the stomata, with the outside air. The extensive intercellular surface, loose internal structure, and numerous stomata of the leaf allow for rapid gaseous exchange between the internal leaf cells and the outside air. An appreciation of this structure is gained when we consider a large leaf such as a squash leaf that may be 70 percent internal air space, and may have 20 times as much total internal cell surface bordering this air space as leaf surface exposed to the outside air. In addition, a squash leaf may have as many as 60 million stomata connecting this internal air with the outside environment and may have 6000 vein endings per square centimeter.

The Process of Transpiration

Water permeates the living turgid leaf cells, filling the vacuoles, making up a large part of the cytoplasm, and penetrating the walls. Thus the cell walls bordering on intercellular air spaces are moist, and because of this the air in these spaces is almost saturated with water vapor. The water potential or vapor pressure of water in these spaces is then generally higher than in the outside atmosphere. When the air around a leaf is not saturated but relatively dry, water vapor molecules diffuse from the saturated air in the leaf through the stomata into the less saturated outer air where the vapor pressure, and hence the water potential, is lower. This loss in water results in a slight drying of the air in the intercellular spaces (Fig. 10.6). Water molecules then evaporate from the wet walls and diffuse into this drier air in the leaf. As water leaves the walls, the walls in turn become drier and imbibe water from the enclosed cytoplasm and vacuole.

Thus a diffusion gradient for water exists from the vacuole through the cytoplasm, the wall, and the intercellular air spaces to the drier outer air surrounding the leaf. Water diffuses along this gradient, in liquid form within the cell, and in vapor form once it evaporates from the cell walls. Each cell, however, is in contact with several other cells and is only a few cells distant from the water-filled xylem elements of a vein.

As water is lost from a cell by outward diffusion, the cell loses some of its turgor and the concentration of solutes in it increases; hence the water potential within the cell sap decreases. The cell is then able to gain water by osmosis from adjacent more-saturated cells. These in turn gain water from adjacent cells and eventually from the water-filled tracheid of a vein ending. A water potential gradient soon becomes established, along which water diffuses from the vein through one or several cells to a cell wall bordering an intercellular space, into and through the intercellular spaces and through the stomata to the outside air.

The Quantity of Water Transpired

A number of quantitative methods of measuring the loss of water by transpiration have been developed. Figure 12.2 shows the type of apparatus used in many studies. This includes a can holding sufficient soil to support the root system of large plants, even small trees. The top of the can is covered, and so all water escaping from the container must be that absorbed from the soil by the roots. This water passes up through the stems to the leaves and is transpired to the atmosphere. Provision is made for measuring accurately the amount of water that is added to the container from time to time. Also, suitable arrangements are made for weighing the whole container, including soil, water, and plant. Thus, for any time interval, the transpiration loss (weight of water that passes through the plant from the soil) can be determined. Of course, some loss of weight results from respiration, and some gain in weight occurs owing to photosynthesis, but these amounts are relatively small.

The amount of water transpired by plants is very

B

A

Color Plate 3

Figure 11.5. Tomato plants grown in nutrient culture solutions to show visual symptoms of mineral deficiencies. *A*, complete solution; *B*, deficiency symptoms in leaves; *C*, minus phosphorus; *D*, minus potassium; *E*, minus nitrogen; *F*, minus sulfur; *G*, minus calcium; *H*, minus iron; *I*, minus magnesium; *J*, minus all micronutrients. All plants subjected to deficient solutions were first grown in a complete nutrient solution for two weeks.

C

D

E

F

G

H

I

J

A

B

D

C

Figure 12.1. Moisture content, dry weight, and ash content of fresh grass leaves. *A*, 100 grams of fresh leaves yield *B*, 85 grams of water and *C*, 18 grams of dry leaves; *D*, the dry leaves yield, on burning, 1.5 grams of ash.

Figure 13.7. Chlorophyll is required for photosynthesis. Variegated leaf from *Coleus*. *A*, living leaf; *B*, same leaf after extraction of chlorophyll and iodine staining to show starch. Compare the blue areas indicating the presence of starch with the distribution of chlorophyll.

Figure 13.16. Light is necessary for photosynthesis. *A*, leaf from a bean plant which was kept first in darkness for 24 hours; a portion of the leaf was then covered by foil as shown, and the plant was kept in the light for 8 hours. *B*, the iodine test shows that starch was formed only in the portion of the leaf receiving 8 hours of light.

A

B

A

B

great. A single corn plant in Kansas, between May 5 and September 8, transpired 54 gallons of water. An acre of such plants (6000 plants) would transpire, during that season, 324,000 gallons of water, which is equivalent to a sheet of water 11 inches deep over the entire acre. It has been estimated that an acre of red maple trees, growing in a soil with ample moisture, may lose in a growing season an amount of water sufficient to cover each acre with 28.3 inches of water. A soil clothed with plants is depleted of its moisture at a much more rapid rate than one that is bare. Nearly all the water loss from a soil below the first 6 to 8 inches results from absorption and transpiration by plants.

Of the total quantity of water absorbed by the roots of plants, as much as 98 percent of it escapes from the plant by transpiration. The small quantity of the water that is retained by the plant includes the water of vacuoles and protoplasm, that in cell walls and in the conducting elements, and that entering into chemical combination.

By means of such an apparatus as shown in Fig. 12.2, the total transpiration during the life of the plant (or any part of its life) may be determined. This quantity may then be compared with the dry matter produced during the same period of time. For example, it was found that an alfalfa plant transpired 900 units of water for each unit of dry matter produced and that a millet plant transpired 248 units of water for each unit of dry matter. Although these values vary with the environmental conditions under which the plants are grown, it is evident that plants differ in their water economy.

Evapotranspiration

In evaluating an area of land in terms of the amount of water needed to produce maximum plant growth, one must take into consideration both evaporation of water from the soil surface and transpiration of water. The total amount of water that a unit of land area loses to the atmosphere through evaporation and by transpiration from plants growing on the land is called **evapotranspiration.**

A number of rather complex and expensive instruments have been designed to determine evapotranspiration. One of these, the **lysimeter,** is shown in Fig. 12.3. The lysimeter, essentially, is a large metal tank containing a soil mass which may weigh (together with the soil water) as much as 100,000 pounds. It is mounted in a pit so that the soil surface in the tank is level with the surface of the soil in the field. The tank is so arranged that a sensitive and automatic balance will continuously record any changes in weight of the container and its contents. The sensitivity of the balance may be to the nearest two pounds, which is a sensitivity equal to 0.001

Figure 12.2. Apparatus used in measuring loss of water by transpiration. (Courtesy of F. J. Veihmeyer.)

of an inch of water applied to the soil surface. All water draining through the soil in the lysimeter is collected and measured. Hence, if plants are grown in the lysimeter, a continuous record of weight changes and of the volume of runoff or water percolating through the soil is recorded, and the evapotranspiration can be calculated. The gain in weight due to the growth of the plants must be taken into consideration, but this weight change is usually relatively small compared to the changes caused by moisture gain or loss.

Under conditions of adequate water supply to an area covered with vegetation, the plants will act as channels through which water passes from the soil to the atmosphere, and the amount of water lost by evaporation and transpiration will be primarily a function of the climate. Many attempts have been made to use weather data to calculate the maximum or potential evapotranspiration for a given area. These calculations have been only partly successful.

Factors Affecting Transpiration Rate

The factors influencing the rate of transpiration include two principal groups, **environmental factors** and **morphological factors.** The former include conditions external to the plant, whereas morphological factors include structural features and habits of growth of the plant, sometimes referred to as internal factors.

Environmental factors. The important environmental factors that influence transpiration rate are (1) relative humidity of the atmosphere, (2) air movements, (3) air temperature, (4) light intensity, and (5) soil conditions. These factors affect transpiration through their effect on the vapor pressure of water in the intercellular space or of water in the air.

Atmospheric humidity. The humidity of the air surrounding the plant is an important external factor determining the rate of transpiration. The drier the air above the plant, the greater the transpiration rate. For example, air with a relative humidity of 20 percent is considered dry because it has the capacity for holding a great deal more water; air with a relative humidity of 80 percent is regarded as moist because its capacity for holding more water is not great. A dry atmosphere has a low concentration of water vapor; a moist atmosphere, a high concentration.

The air in intercellular spaces of a turgid leaf is

Figure 12.3. A lysimeter is a device for measuring the evaporation and transpiration from a given area of turf. A shows the appearance of the turf; the area being measured is delineated by the circle and the size of the area may be judged by the individual present; B, the lysimeter laboratory beneath the turf is crowded with electronic instruments. (Courtesy of W. O. Pruitt.)

very moist—in fact, usually almost saturated. A diffusion of water vapor from intercellular spaces through stomata to the air surrounding the leaf takes place as long as the vapor pressure of water in the intercellular spaces exceeds that of the air outside. Diffusion becomes more rapid as the difference between these two pressures increases. When the difference is great, we say the **diffusion gradient** is steep. If the relative humidity of the air surrounding the plant is low, the diffusion gradient between the moist intercellular spaces and the outside air is steep, and hence transpiration is rapid.

Air movements. As transpiration occurs there is a tendency for a moist layer of air to form next to the leaf surface, particularly in still air. This will decrease the diffusion gradient between the leaf and the atmosphere and transpiration will consequently decrease. On the other hand, air movement carries away this layer of humid air, replacing it with drier air, resulting in an increase in transpiration. The more rapid the air movement, the faster the moist air will be carried away and the faster the rate of transpiration. If the wind is quite strong, stomata may close, probably as a result of excessive water loss, and transpiration is then reduced.

It is a common observation that on a warm, bright, windy day, plants may lose water to the point of wilting. On a cool, cloudy day with little wind movement, however, seldom is the water loss sufficient to result in wilting. Even though plants may wilt during the hottest part of the day, this may be but a temporary condition, as is shown by the fact that the next morning the leaves are again fresh and turgid. Return to the turgid condition indicates that available water in the soil is not exhausted.

Air temperature. In direct sunlight the temperature of a leaf is usually higher than that of the air about it; the difference may be as much as 10°C. About 80 percent of the radiant energy of sunlight that falls upon green leaves may be absorbed by them. Part of this absorbed energy is changed to heat and raises the temperature of the leaf; part is utilized in vaporizing water; and another small part is used in photosynthesis (Fig. 13.17). When the leaf temperature rises, the vapor pressure in intercellular spaces becomes greater than that in the air surrounding the leaf; as a result, the diffusion gradient becomes steeper, and the rate of transpiration increases.

Light intensity. As the light intensity increases, the internal temperature of the leaf is rasied, with the result that water loss is accelerated. Another effect of illumination upon transpiration is that it stimulates, at least in many species of plants, the opening of stomata (page 193). When the stomata are closed, transpiration virtually ceases. Thus light, through its effect upon leaf temperature and upon the opening of stomata, influences the rate of transpiration.

Soil conditions. Any soil condition that influences absorption of water by the roots affects the transpiration rate. When the soil becomes very cold, absorption of water is retarded, even though soil water may be available. An increase in the concentration of the soil solution, as often occurs in alkali soils, reduces the rate of water intake. Poor aeration of the soil may result in diminished absorption. The availability of soil water also determines (see page 210) the rate of intake. The rate at which water is absorbed by the roots greatly influences the rate of transpiration. If water loss exceeds water absorption, wilting will occur. Under these conditions the mesophyll cells do not give up water vapor as freely as when they are well supplied with water. Consequently, the rate of transpiration is reduced.

Morphological factors. On days when water loss is high, various kinds of plants growing side by side, exposed to the same environmental conditions, respond differently with respect to wilting. Some may be severely wilted; others may show no signs of distress. It is apparent that some plants have better provisions for transpiration regulation than do others. Thus plants have certain structural features and habits of growth that influence the rate of transpiration. They may be referred to as morphological factors, that is, conditions in the plant itself.

Certain plants are able to survive in extremely dry habitats, whereas certain other plants succumb. Such surviving plants undoubtedly maintain a balance between water outgo and water intake, whereas in those plants that fail to survive the dryness, the loss of water, at least for a period, exceeds the absorption of water. Plants of dry habitats often possess roots that penetrate deeply into the soil; or, like many desert annuals, they have an extensive surface root system that can rapidly absorb limited rainfall from spring or summer showers. Many desert plants possess special water storage tissue, as is well shown in so-called succulents, that is, plants with fleshy stems or leaves (Fig. 18.31).

The anatomical features that are advantageous from the point of view of prevention of water loss are (1) the cuticle of leaves, young stems, and fruits, (2) sunken stomata, (3) distribution of stomata, and (4) reduction of the transpiring surface. In addition, stomatal behavior is an important factor in controlling water loss.

Figure 12.4. Cross section of leaf of rubber plant (*Ficus elastica*). (Redrawn from Van Tieghem and Constantin.)

upper epidermis

palisade parenchyma

cystolith

spongy parenchyma

guard cells

lower epidermis

epidermal hairs

Cuticle. Most of the water lost from a plant passes out through the stomata. Some water, however, is lost through the cuticle. Various modifications in leaf structure reduce cuticular transpiration; for example, thickening of the outer wall of the epidermal cells and the presence of a waxlike material, cutin, in this wall. Most plants of arid and semiarid climates have a thicker cuticle than do those of humid climates.

Stomatal behavior. We have seen (page 193) that a stoma may open and close as a result of changes in the turgor of guard cells. When guard cells become turgid, the stomatal aperture widens; when they become flaccid, the aperture narrows or completely closes. Often after a period of rapid transpiration, there is a deficit of water in all leaf cells, and stomata close, thus reducing the water loss due to transpiration. This is an advantageous behavior. But, from the standpoint of photosynthesis, the stomata must serve also as entrances for carbon dioxide. If they close during the daytime because of excessive transpiration, the diffusion inward of carbon dioxide, as well as the diffusion outward of water vapor, is restricted. Stomatal closure, in this event, works to the disadvantage of the plant by decreasing photosynthesis, although it is a favorable behavior in that it limits water loss.

Although light is a factor that stimulates stomatal opening, in many plant species stomata may remain closed throughout the day if the water deficit in leaf cells is pronounced. On the other hand, there are species in which stomata are open all night, and in some instances stomata may remain open even though transpiration is proceeding at a high rate.

Sunken stomata. In some plants (Fig. 12.4) the stomata are below the general level of the leaf surface. When this condition occurs the water vapor must diffuse through a relatively long and sometimes tortuous passageway, with the result that the diffusion rate is lessened.

Distribution of stomata. The leaves of many plants have stomata only on the undersurface, or, if on both surfaces, mostly on the lower. This distribution is shown in Table 10.1, page 192. Hence, the loss of water from leaves that have stomata only on the lower surface, or fewer on the upper than the lower, is usually less than the loss from leaves that have an equal number on both the upper and lower sides.

Reduction of transpiring surface. In most plants the principal organs of transpiration are leaves. Therefore, any decrease in leaf surface will reduce transpiration and conserve water absorbed by roots. In corn and other grasses, and also in oleander, the leaves roll up during drought, thus exposing less surface to the air than do full expanded leaves. Cacti and some euphorbias have no foliage leaves; in these plants the transpiring surface is restricted to the stem surface (Fig. 18.31).

Epidermal hairs. The leaves of many plants are clothed with epidermal hairs (Fig. 7.13). In some plants, such as common mullein and dusty miller (*Centaurea cineraria*), the hairs may form a dense, cottony covering. There is no clear evidence that surface hairs reduce transpiration; in fact, experiments with some plants have shown that water loss from the leaves is greater when the hairs are present than when they are lacking.

Guttation

The loss of liquid, as contrasted with the loss of vapor, from leaves of *intact plants* is termed **guttation.** This will occur especially in herbaceous plants when conditions favor rapid absorption of water and low transpiration. For instance, when a well-watered, vigorously growing tomato plant is placed under a bell jar, transpiration ceases as the atmosphere in the jar becomes saturated. Continued water absorption then results in a slow exudation of water from the tips of the leaves (Fig. 12.5). Many plants have specialized openings called **hydathodes** at the tips of their leaves, through which the liquid passes outward. Experiments with barley plants show that, when the roots are immersed in distilled water, guttation is very slight or ceases altogether. This is true even though the water is aerated amply. In dilute salt solution, but without aeration, guttation is also slight. On the other hand, if the roots are immersed in a dilute salt solution, with good aeration and favorable temperature, guttation is ample and continues for a long period in a humid atmosphere. Guttation is associated with salt absorption and salt movement into the xylem. The liquid of guttation is not pure water but a dilute salt solution (Fig. 12.6).

SUMMARY OF TRANSPIRATION

1. Transpiration is the loss of water vapor from plant surfaces, largely those of leaves.

2. Evaporation removes the water vapor from the external surface of the leaf.

3. The amount of water transpired during the life of a plant is large, from 200 to 1000 times its dry weight.

4. The rate of transpiration is affected by the relative humidity of the air, air movements, air temperature, light intensity, and soil conditions.

5. Morphological details such as the cuticle, epidermal hairs, and location and distribution of the stomata also influence the rate of transpiration.

6. Guttation is the escape of water as liquid from leaves; it may occur when the relative humidity is high. Guttation fluid contains solutes.

Figure 12.5. Guttation from tips of barley (*Hordeum vulgare*) leaves.

Figure 12.6. Deposit on blades of grass following the evaporation of guttation fluid. Deposit composed largely of glutamine with a small amount of potassium chloride and undetermined organic matter. (Courtesy of Curtis.)

Water in the plant moves upward from the roots to the stems and hence into leaves, from which it escapes into the atmosphere. That the path of movement is through the xylem is shown by "ringing" or "girdling" the stem, that is, removing a complete ring of bark (thus all of the active phloem). This procedure has little immediate effect on the movement of water; the plant does not wilt. The path of movement of water may also be demonstrated by placing a leafy branch that has been cut underwater into an aqueous solution of a dye, such as eosin. If after a short period the stem is split, it can be observed that the walls of vessels and tracheids are stained by the dye, whereas other tissues are not stained.

TRANSPIRATION PULL

It is estimated that a force of 20 atmospheres (Fig. 5.13), almost 300 pounds per square inch, is necessary to lift water to a height of 350 feet through the xylem of a tree. This estimate takes into consideration the resistance to water movement offered by the xylem, as well as the weight of the water column. How does water get to such a height in a tree? What are the forces involved? Although plant physiologists are not agreed on the relative importance of the various processes, the explanation that is now most generally accepted as most plausible will be described here. It is spoken of as the **transpiration-pull and water-cohesion theory.**

Evaporation

When evaporation of water occurs from the cell walls in a leaf during transpiration, forces are brought into play that result eventually in the movement of water through the plant. The diffusion of water molecules from the wet cell wall surfaces into the intercellular spaces of the leaf mesophyll results in partial drying of the walls.

Imbibition

The walls of a leaf cell, losing water as a result of evaporation, imbibe water from the enclosed cell contents or from adjacent cell walls with a higher water content. The forces of imbibition that hold the water in the walls (see page 72 and Fig. 5.12) are very great. Although the first water that is lost

from a saturated mesophyll wall is held loosely, the forces holding the remaining water in the wall are stronger. Thus, during transpiration the water potential in the wall gradually decreases because some of the imbibitional forces are not satisfied.

Osmotic Forces

This decrease in the water potential in the wall results in the movement of water from the protoplast into the wall. As water moves from the protoplast of a mesophyll cell into the walls, the turgor of the cell decreases. This causes a decrease in the water potential of the cell and water will move into the cell from adjacent cells down a water potential gradient. Because of the relatively high osmotic concentration in mesophyll cells, these cells can have a lowered water potential and still retain considerable turgor pressure. The disturbance in water equilibrium is transmitted from cell to cell through both cellular contents and walls. Water moves into these cells from the tracheids in the veinlets of the leaves. Thus a force equivalent to the imbibitional force created in the leaf cell walls, as a result of the loss of water from them during transpiration, is pulling on the water in the tracheids. Because of the relatively low osmotic concentration in the tracheids, as the water potential in the tracheids gradually decreases the contents may be subjected to reduced pressures and to tensile pull.

Water Continuity and Cohesion

Water in the tracheids of leaf veinlets is continuous from the leaves to the roots. This continuity exists throughout the life of the plant. In a sense, the water columns "grow" with the plant. However, the living cells of the root, the xylem elements themselves, and the living leaf cells offer resistance to the free flow of water. This resistance, and the gravitational pull on the water itself, are overcome by the forces developed in the leaf when water is moving up in the plant. The question now arises: Do these unbroken columns of water have sufficient tensile strength to prevent them from being broken by the forces acting on them? Experiments using both water and plant sap indicate that this requirement is met. As a matter of fact, plant sap in vessels and tracheids may withstand tensions greater than 300 atmospheres without being broken. Thus, it is evident that the attractive forces of the water molecules for each other, **cohesion,** and of cell walls for the water molecules, **adhesion,** are very great. It is because of these attractive forces that the liquid columns possess tensile strength.

The columns of water are raised by the force of evaporation of water from the leaf cell walls and the imbibitional forces that develop here.

SUMMARY OF THE TRANSPIRATION-PULL THEORY

1. The evaporation of water from the walls of mesophyll cells of leaves result in a decrease in the water potential in these walls.

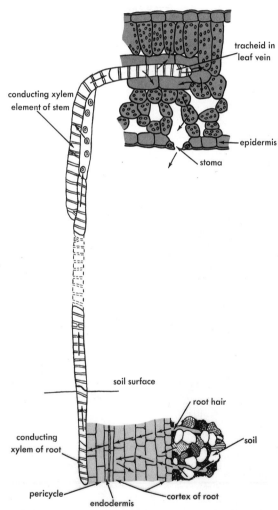

Figure 12.7. Diagram showing entry of water from the soil to the root hair and its path of movement through the plant to parenchyma leaf cells, from which it escapes as water vapor into intercellular spaces, then passes into the atmosphere, mainly through stomata.

2. This disturbance in the water balance of the cells causes a series of changes that set in motion the entire train of water through the plant.

3. The living leaf cells next to the tracheids in veinlets eventually lose water to neighboring cells.

4. Water moves from the veinlets into the adjacent cells. This results in the upward movement of water in the continuous liquid columns in the xylem.

5. Continuity of water is maintained even under conditions of considerable strain because of the strong cohesive forces between water molecules.

ABSORPTION

Absorption of Water

The large quantities of water moved through the plant body and lost by transpiration, then, must be replaced continually if the plant is to live.

In the absorbing region of the root (Figs. 9.5, 9.6), the phloem does not lie outside the xylem but occurs in groups of cells that alternate radially with xylem masses or with the radiating arms of a single central mass of xylem. Hence, it is possible for the water and mineral salts to be absorbed by epidermal cells or root hairs, pass through the cortex, endodermis, and pericycle into the primary xylem, and move upward without traversing the phloem (Fig. 12.7). This dilute solution of water and mineral salts that moves upward through the conducting elements of the xylem to the stems and leaves is called xylem sap.

Mechanisms of Absorption

The uptake of water by roots involves two mechanisms, which may or may not act simultaneously.

(1) When transpiration is slow and water is available in the soil, absorption of water may exceed transpiration. This condition results in a pressure being developed in the xylem, and may lead to guttation. If the shoot of a plant is cut off a few inches above the soil level, absorption of water by the roots will often force the sap out of the cut ends of vessels and tracheids, a phenomenon known as **bleeding.** It appears that this type of absorption depends upon the *activity of living root cells.* The pressure developed in the xylem is called **root pressure.**

This mechanism of absorption of water by roots is generally explained as resulting from the osmotic movement of water into the root. As a result of the activities of the living root cells, soil solutes are absorbed and accumulate in the cell sap in high concentrations (page 181). The total concentration of the cell sap of the root hairs, as well as of the other root cells, is thus normally greater than that of the soil solution. The total amount of solutes present in the xylem elements of the root is also generally higher than that of the soil solution. The greater amount of solutes in the cell sap reduces the potential of water in the cell sap below that of the water in the soil solution. Under this condition movement of water is inward. The rate of movement inward depends upon the difference in the water potential of the two solutions.

(2) When transpiration exceeds absorption, no root pressure can be demonstrated. In the xylem elements of plants in this condition, the water is under tension, and the pressure on the water in the vessels is lower than atmospheric pressure. Under these conditions a high rate of transpiration results in a water deficit in the plant tissues. Water then moves into the roots passively. In fact, under such conditions, water may even be absorbed by dead roots. Water absorption may then result from osmotic forces originating in roots or from forces that originate in the leaves—that is, forces set in motion by the loss of water in transpiration.

Factors Affecting Absorption

We have seen that roots grow in a very complex environment—the soil. The activities of roots are influenced to a marked degree by this environment. Many environmental factors determine the nature and extent of the root system, the rate of growth of roots, the rate at which they absorb water and mineral nutrients—in fact, all their activities.

Unavailable soil moisture. Plants cannot absorb all the water in a soil. There is always some water in the soil which is not available to the plant; it is held so tightly by the soil particles that the roots cannot absorb it rapidly enough to prevent wilting. To illustrate this relationship, we shall describe a simple, direct method of determining the **unavailable moisture** in a soil (Fig. 12.8). The container used here is a standard No. 2 tin can, with a tight-fitting lid in which there is a hole. The can is filled with soil which is wet to field capacity (see page 210). Several seeds are planted in the soil, and after they have germinated the best seedling is selected and led through the hole in the lid; the other seedlings are discarded. The space between the stem and the edges of the hole is filled with cotton to prevent evaporation from the soil. The roots penetrate the volume of soil in the can. It is apparent that all the water lost from the soil is that absorbed by the roots and transpired to the atmosphere. After a time, if no water is added, the plant begins to wilt, as is evidenced by drooping of the lower leaves. A degree of wilting known as **permanent wilting** is soon attained from which the plant will not recover if placed in a darkened moist chamber; the only way

Figure 12.8. Direct method of determining the available water in the soil. The narrow range of soil moisture contents within which wilting takes place is also shown. Percentage of moisture in cans is as follows: plant 97, 3.3%; 105, 3.6%; 101, 3.7%; 94, 3.9%; 102, 4.1%; 104, 4.1%. Plants 104 and 102 are turgid; 94 shows slight dropping of first leaves, 101 is in a more advanced stage of wilting, 105 is permanently wilted, and 97 has passed this stage. (Photograph and data courtesy of F. S. Veihmeyer.)

of bringing about its recovery is to add water to the soil. At the time that the plant has permanently wilted, the percentage of water left in the soil is determined. This percentage (based upon the dry weight of soil) is called the **permanent wilting percentage of the soil.** This water represents that which the plant cannot absorb readily from the soil. Only that part of the total soil moisture which is above the permanent wilting percentage is available for plant growth.

Different kinds of ordinary plants growing in the *same soil* reduce it to approximately the same permanent wilting percentage.

Influence of soil type. *Different kinds of soils,* however, differ in their permanent wilting percentage. Let us perform an experiment to illustrate this point. Fill one container with a clay soil, and another with a sandy soil. Then wet each soil to field capacity. We know the clay soil will hold more water than the same volume of sandy soil. In each grow the *same kind of plant.* Allow each to reach permanent wilting, and then determine the permanent wilting percentage of the two types of soil. It will be found that the permanent wilting percentage of clay soil is much higher than that of sandy soil.

Moisture content of soil. From the standpoint of plant growth, the readily available water of a soil is the important consideration. This is really of more significance than the water-holding capacity of a soil.

It has been determined experimentally for many crop plants, including orchard trees, that if a soil has a moisture content *above* the permanent wilting percentage, there is no optimum moisture content, that is, a content at which the plants grow best. For example, if the percentage of water in a soil at field capacity is 30 percent, and its permanent wilting percentage is 15 percent, the plant growing therein will do equally well with the moisture of 16, 17, 18, or any other percentage up to 30 percent.

Under field conditions, soon after a rain or irrigation, followed by drainage, the soil to a certain depth (depending upon the amount of rain or irrigation water applied) is up to its field capacity of moisture. The roots of plants begin to absorb the water, and soon the soil is below its field capacity. The soil moisture constantly decreases in the zones occupied by roots. Moreover, the movement of water by capillarity, from moist soil to that which is drier immediately surrounding the roots, is so slow

that it may be regarded as negligible in influencing water available to the roots.

Root growth and soil moisture. Since the capillary movement of water in soils is too slow to meet the requirements of growing plants, roots must continually penetrate into new soil if they are to utilize the water which it holds. The total root elongation of a plant may be very great. Root extensions and the formation of new root hairs are constantly bringing about new contacts with moist soil.

There seems to be no reliable evidence that the roots of ordinary crop plants are able to penetrate more than a fraction of an inch into dry soil, that is, a soil at or below the permanent wilting percentage. Some evidence, however, indicates that certain drought-resistant plants, such as occur in deserts, are able to push their roots considerable distances into a dry soil.

The erroneous impression is sometimes held that roots "go in search of water." A more accurate statement is that they grow only in moist soil (soil above the permanent wilting percentage). In practice this would mean that if the upper foot of soil is moist, and dry soil lies below it, the roots of plants will be confined to the upper, moist soil. In irrigation practice, this relationship of soil moisture and root growth is important. In irrigating a garden or lawn with a hose, one is easily misled into believing, because the soil surface is wet, that sufficient water has been applied. Examination may reveal that at depths of five or more inches the soil is very dry—too dry for the growth of roots. In growing plants it is nearly always desirable to stimulate maximum root growth and a root system that attains its normal depth and spread. This may be accomplished in part by keeping the soil moist to the proper depth.

It may be possible for a plant to obtain sufficient moisture to keep it alive, even though only a part of the root system is in moist soil. In early spring or after an irrigation, when the soil is moist from the surface to deep levels, the roots may develop in the surface foot of soil and continue to penetrate to greater depths. Later, the loss of water by direct evaporation from the superficial soil layers and the removal of water by transpiration may bring the moisture content of the top layers of soil below the permanent wilting percentage; but the plant may continue to survive from water absorbed by the deeply penetrating roots.

The loss of water from soils at levels below the first few inches is essentially that due to transpiration from plants. Soils are dried out because of the presence of absorbing roots. One of the principal reasons for removing weeds from a growing crop is that they deplete the soil of water.

Solute concentration of soil solution. As the concentration of the soil solution approaches that of the cell sap, the rate of water absorption declines. When the concentrations of the two solutions are the same, water intake ceases. And, if the soil solution becomes more highly concentrated than the cell sap, owing to excessive applications of fertilizers or of saline water, water will be withdrawn from the root cells. Moreover, root growth is inhibited at high salt concentrations, and the roots are not able to extend into new soil areas.

Transpiration and absorption. Within limits, if a plant is placed under conditions that cause it to transpire rapidly and if soil water is available, the rate of water intake will increase. At times when transpiration is very rapid, absorption lags behind transpiration to a point which causes wilting, even though the water in the soil is ample.

Absorption of Solutes

In the discussion of absorption by plant cells, the principle of diffusion of water and solutes was discussed. Diffusion of a solute was described as a movement of solute particles from a place where their activity is higher to one where their activity is lower. Also, it was pointed out that the different solutes diffuse independently of the movement of each other, and of water. Investigations indicate that the intake of solutes by the roots cannot always be explained as being the result of simple diffusion. For example, it has been shown that the root cells of higher plants absorb nutrients and build up within their vacuoles a concentration of certain ions many times that of the external solution. This process is called **accumulation** or **active solute absorption.** There is a movement of ions from a region of the lesser concentration of any particular ion to a region of its greatest concentration. This movement *against* a concentration gradient is accomplished by an expenditure of energy and may constitute much of the work the plant cells are required to perform. It appears that high absorption rates are usually attended by high respiration rates, and it seems logical to conclude that respiration provides the energy required for active absorption and accumulation of solutes.

SUMMARY OF ABSORPTION

1. Absorption of water by roots when transpiration is low or absent occurs only in living root systems.

2. This absorption of water results when the solute concentration of the xylem sap is higher than that of the external solution.

3. Root pressure, guttation, and bleeding are expressions of active absorption.

4. Passive absorption of water may result from the transmission to the roots of forces that originate in the leaves; it may even take place if the roots have been killed.

5. A plant is said to be permanently wilted when it will not recover in a saturated atmosphere but only when water is added to the soil.

6. The percent moisture remaining in soil in which a plant is permanently wilted is called the permanent wilting percentage of that soil.

7. Only that part of the total soil moisture which is above the permanent wilting percentage is available for plant growth.

8. The rate of water absorption is influenced by the available water in the soil, the air in the soil, the soil temperature, the concentration of the soil solution, and the rate of transpiration.

9. The absorption of ions by root cells may occur against a concentration gradient. This process is called accumulation and is dependent upon energy released by the living cells during respiration.

CONDUCTION

Within a living plant, substances are constantly moving from one place to another. Although they are interrelated, we can detect four types of movement based on the rates at which movement occurs. (1) Slow *diffusion* of molecules and ions. (2) Moderate movement of materials carried by *cytoplasmic streaming* in living cells. (3) More rapid *flow* of material *in the sieve tubes.* (4) Very *rapid conduction* of water and mineral solutes *in the xylem.*

Diffusion and Protoplasmic Streaming

The movement of water or of solute molecules and ions into a cell across the cell wall is a relatively slow process. Diffusion occurs along activity gradients and may take place through plasmodesmatal connections between cells as well as directly through the permeable cellulose walls of most cells. Once a solute molecule or ion diffuses into a cell, its rate of movement may be increased many-fold as it is picked up by the protoplasmic stream. Cytoplasm may stream at rates of a few to several centimeters per hour. The highest rate of protoplasmic streaming that has been measured was observed in the protoplasm of a slime mold in which the cytoplasm was streaming at a rate of 48.6 centimeters per hour.

In terms of the dimensions of a single cell, cytoplasmic streaming is frequently very rapid, since a substance may be transported across the length of a cell in a matter of seconds. In terms of the whole plant, however, this is a very slow process. It would take days, for instance, for a molecule in the leaf to be carried upward to the growing shoot tip or downward to the roots. Figure 12.9 diagrams the way in which one solute may be moving by diffusion and protoplasmic streaming in one direction through a series of cells and another solute may be moving in another direction. Diffusion and protoplasmic streaming are the chief methods by which solutes move through living plant tissues, except in the sieve tubes.

Conduction in the Phloem

The rapid translocation of foods throughout the plant takes place chiefly in the phloem and more specifically in sieve tubes. That the phloem is the major path of food transport is borne out by the results of ringing experiments followed by chemical analyses of tissues and cells above and below the ring. When a ring of bark is removed from a stem down to the cambium (thus removing the phloem), the carbohydrate content and especially the sugar concentration of the sap in the leaf, bark, and wood *above* the ring increases after a few hours. *Below* the ring, the total carbohydrate content and the sugar concentration in the sap of the bark and the wood decreases. These conditions would not prevail if foods were moving downward in the xylem tissue.

Both carbohydrates and nitrogenous substances move to tissues that are utilizing them. This means that foods may move upward or downward in the phloem. The direction of food movement is usually from a region of the shoot where excessive food is present to a region where a lesser amount is present. Food may move from the place of manufacture (leaves) to a storage place (roots, fruits, seeds) or to a region of growth (buds, cambium). It may also move from storage tissue to a region of growth.

The phloem stream in the sieve tubes is under a constant pressure that is built up by the diffusion of water into the sieve tubes in the regions of the plant where food is produced. According to one of the most generally accepted theories, a mass movement of water and solutes occurs in the sieve tubes from these regions to the sites of food storage or utilization where solutes are removed from the sieve tubes and water diffuses out into the adjacent cells

Figure 12.9. Diagram illustrating how two solutes may be moving in opposite directions by diffusion and protoplasmic streaming in the same tissue.

and into the xylem. Translocation of material in the phloem may reach a rate of 100 centimeters per hour. When this is compared to the much lower rates of protoplasmic streaming, it is evident that normal protoplasmic streaming alone is insufficient to cause this rapid transport. In fact, evidence indicates that the mature functional sieve tube elements have highly modified cellular contents (absence of nuclei and definite vacuoles, page 123) and that protoplasmic streaming does not occur in them.

Lateral Transport

There is lateral (radial) movement of foods in stems along the vascular rays, bringing the foods into close contact with conducting elements of both phloem and xylem. Thus the rays carry foods from sieve tubes radially into the cambium and xylem.

In some plants, such as the sugar maple, the xylem parenchyma cells serve as places where large quantities of carbohydrates are stored. In the spring, the presence in the xylem of soluble carbohydrates results in a high osmotic concentration. This, in association with cold nights and warm days, causes sap to bleed from cuts made in the trunk in the spring of the year.

Conduction in the Xylem

If a stem is girdled by removing a layer of bark around the circumference of the stem down to the cambium, the mineral (inorganic) salts that are absorbed from the soil and conducted in the stems and roots move upward past this girdle. Girdling may or may not decrease this upward movement, depending upon the species and other factors. This fact would serve to indicate that normally the greater amount of salt translocation in the stem is in xylem elements. An especially convenient method of studying the path of movement of salts in stems is now being used. Certain ions, such as potassium, phosphate, and bromide, are rendered radioactive and added to the solution in which the plants are growing. By appropriate means, the path of movement of these radioactive **tracers** has been followed accurately, even for short periods of time, both in girdled and ungirdled stems. The results of these experiments seem to show that inorganic solutes move upward chiefly in xylem tissues. As the salts move upward, however, they may transfer radially from xylem through vascular rays into phloem tissues; and they may also accumulate in living cells along their path of movement. When radioactive phosphorus was supplied to the nutrient solution in which tomato plants six feet high were growing, it could be detected throughout the plants after 40 minutes. There is also evidence that certain minerals carried upward in the xylem to the leaves may be moved from the leaves and conducted downward via the phloem.

Frequently not all of the xylem vessels or tracheids function in the conduction of water and mineral salts at any one time. In trees particularly, the heartwood (Figs. 8.10, 8.19) ceases to play a role in conduction (page 159). In some instances, only the outermost layers of sapwood contain xylem elements filled with water.

The rate of movement of water and solutes in the xylem depends in large measure upon the rate of transpiration. If plants are growing in a very humid atmosphere, the movement of water through the xylem is very slow; if they are in a dry, warm atmosphere, which heightens the rate of transpiration, the movement may be as rapid as 131 feet per hour. In the tobacco plant, water conduction was found to go on at a rate of about 4.7 feet per hour. In the larch tree, the rate of water movement was found to be approximately 5.5 feet per hour at a period of the day when water loss was highest; but in the early morning, when the rate of water loss was at the minimum, the rate of water movement was but 0.2 or 0.4 inch per hour.

Absorption of inorganic salts by roots is *not proportional* to the absorption of water. Some increase in salt absorption accompanies an increase in water absorption, but not in the same ratio. Moreover, the different ions are not absorbed at the same rate. Thus, the amount of water moving through the plant is not the only factor that determines the amount of salts or of any particular ion which is absorbed and conducted through the plants to the leaves. Other factors, such as the photosynthetic activity of leaves, respiration, and other chemical processes in the plant, determine to a degree the quantity and kind of ions absorbed and transported.

SUMMARY OF CONDUCTION OF SOLUTES

1. Normally the greater amount of mineral salt translocation in the stem is in the vessels and tracheids of the xylem. The movement is usually upward.
2. Salts may also be conducted in the phloem.

3. The major path of food movement is through the phloem.

4. Food movement is from a region of high food content or place of manufacture or storage to a place of food utilization, or from a place of manufacture to storage tissues.

5. Rapid lateral movement of solutes takes place along the vascular rays.

6. The mechanics of food movement are not well understood, and much information still needs to be obtained before a detailed explanation of food movement can be given.

CLASSIFICATION OF PLANTS ACCORDING TO THEIR WATER NEEDS

In many agricultural sections of the country, and particularly in semi-arid and arid regions, water is the principal limiting factor in crop production. Water is a most important factor in determining both the distribution of plants over the earth's surface and the character of the individual plant. Probably no single factor is so largely responsible for the diversity of plants in various habitats as is the difference in supply of water. So important are the water relationships of plants that various attempts have been made to classify plants on the basis of these relationships. One such classification divides plants into (1) **xerophytes** (Fig. 18.31), which are able to live in very dry places, (2) **hydrophytes,** which live in water or in very wet soil, and (3) **mesophytes,** which thrive best with a moderate water supply.

Plants with xerophytic characteristics, which limit transpiration or in other ways balance water outgo and water intake, occur in different climatic zones, but those of deserts are most typical. Xerophytic species do not necessarily have a lower transpiration rate than do mesophytes when water is ample, but they do possess one or more characteristics that enable them to survive periods of drought. The most effective of these are thick cuticle, stomatal closure, reduction of the transpiring surface, and water storage.

It should be kept in mind that all plants that fall into any one of the above three classes cease active growth when the soil moisture reaches the permanent wilting percentage. Xerophytes and mesophytes alike, growing side by side, reduce the soil to approximately the same permanent wilting percentage. Mesophytes may not be able to long survive such low soil moisture, whereas plants with xerophytic characteristics will.

Photosynthesis

The maintenance of life in a plant or animal cell requires the continual use of energy. This energy directly or indirectly comes from the sun. Two major processes that are carried out by terrestrial green plants and that use energy directly from the sun are transpiration and photosynthesis. Light influences other processes such as flowering, seed germination, certain growth curvatures, stomatal movements, and pigment production.

During transpiration, energy from the sun is used to evaporate the water from the cell walls. Although this results in a movement of the water in the xylem, this energy is neither stored nor used to bring about the vital reactions involved in the synthesis of foods, in assimilation, growth, and reproduction. On the other hand, photosynthesis is a process in which energy from the sun is transformed into chemical energy. In most photosynthesizing plants, the process of photosynthesis results in the utilization of carbon dioxide and water and the production of complex sugar molecules and oxygen gas. During this process, light energy from the sun is changed into the energy holding the atoms together in the sugar molecule. Neither carbon dioxide nor water will burn, nor would either make a very good meal. Carbohydrates (sugars, starches, cellulose), on the other hand, burn easily and release considerable energy in the process. Furthermore, many carbohydrates are excellent foods. The great importance of photosynthesis lies in the transformation of low-energy compounds, carbon dioxide and water, into high-energy compounds (sugars). Energy stored as chemical energy in foods (carbohydrates, fats and proteins) is continually released in living cells during the process of respiration and is lost as heat or converted into energy-rich bonds (for example, ATP, adenosine triphosphate). Many of the energy-requiring reactions carried out by the living cell occur at the expense of the energy temporarily stored in ATP. *Photosynthesis stores the energy. Respiration releases it,* enabling the cells to perform the work of living.

Perhaps some day man will use atomic energy to drive the energy-requiring steps in the production of food and fiber, but today photosynthesis is man's only source of food. Will the capacity of plants to carry out photosynthesis finally determine the number of people who can live on the planet earth?

There are now (1969) about three billion people living on earth; in 1900 there were about two billion, and in 1800 a little over one billion (Fig. 13.1). If the population increases at its present rate, it will double by the year 2000, reaching between six and seven billion.

All organisms, including man, increase in numbers by well-established growth curves. While many factors may be involved, rate of growth is really a function of two factors, the food supply and contamination of the environment. The familiar sigmoid curve of growth is shown in Fig. 21.16. It indicates a slow start when pioneers are paving their way, a rapid growth at a uniform rate when all conditions are favorable, and then a slowing down to a constant plateau when food and contamination become limiting factors. There is no evidence that 20th century man's population increase will not follow this sigmoid growth curve.

If the potential productivity of the earth and the population it could support are calculated, one arrives at the startling conclusion that if all of the land surface of the earth could support plants, it would potentially be able to produce enough food for 1,000 billion people. This figure is of course unrealistic since only a small part of the land can be used for agricultural purposes. Land is needed for urban and recreational use and some land is unsuitable for any of these uses. Even if only 7 percent of the earth's land surface is considered agriculturally productive, it should still be able to produce enough food to support 79 billion people.

Such calculations lead to two far-reaching conclusions: (1) the number of people who can live on the planet earth ultimately will be limited by the amount of space that a man needs to live in and to work in with reasonable comfort and not by the production of food, and (2) starvation of thousands of people today is occurring in a world in which technical means of preventing this are known and available.

Our present understanding of photosynthesis is closely linked with the development of modern scientific thought and the development of modern chemistry. Prior to the early 17th century the general belief was that plants derived the bulk of their substance from soil humus. This idea was overthrown by a simple experiment performed by Flemish physician and chemist Jan van Helmont, who planted a 5-pound willow branch in 200 pounds of carefully dried soil. He supplied rain water to the

Figure 13.1. Graph showing the world population projected to the year 2000.

plant as needed and in five years it grew to a weight of 169 pounds. The soil had lost only two ounces according to his measurements; consequently, van Helmont reasoned that the plant substance must have come from water. This was a logical deduction, though we now know it was incorrect.

Our knowledge of photosynthesis begins with the observations of a religious reformer, philosopher, and spare-time naturalist, Joseph Priestley. In 1772, Priestley reported that he had found almost accidentally that a sprig of mint could restore confined air that had been made impure by the burning of a candle. The plant was able to change the air so that a mouse was able to live in it. The experiment was not always successful, probably because Priestley, who did not know the role of light, did not always have adequate light. Seven years later Jan Ingen-Housz noticed that air was revitalized only when green parts of plants were in the light.

It was three years later, in 1782, that a Geneva pastor, Jean Senebier, discovered another important part of the process—that "fixed air", carbon dioxide, was required. Thus in the new chemical language of Antoine Lavoisier, it could be said at this time that green plants in the light use carbon dioxide and produce oxygen.

But what was the fate of the carbon in the carbon dioxide? Ingen-Housz answered this question in 1796 when he said that the carbon went into the

　　　13 *Photosynthesis*

nutrition of the plant. Now the experiment that van Helmont made almost 200 years earlier could be explained. Finally in 1804, 32 years after Priestley's early observations, the final part of the overall reaction of photosynthesis was explained by the Swiss botanist and physicist, Nicolas Th. de-Saussure, when he observed that water was involved in the process.

$$\text{carbon dioxide} + \text{water} \xrightarrow[\text{green plants}]{\text{light}}$$
$$\text{carbon dioxide} + \text{organic matter}$$

Almost 50 years elapsed before the carbohydrate nature of the organic matter formed during photosynthesis was recognized.

In a remarkably simple experiment, Engelmann demonstrated the quality (color) of light that is effective in photosynthesis. He placed together on a microscope slide an algal filament and some bacteria that would migrate toward high concentrations of dissolved oxygen. He then viewed the alga in different wave lengths of light. The bacteria were most active near the algal filament in the red and in the blue light. So red and blue light were trapped

and supplied the energy to drive photosynthesis and liberate oxygen.

By the early 20th century it was known that some of the steps of photosynthesis involved enzyme reactions. Thus, the reactions of photosynthesis may be divided into two general types, (1) photochemical reactions, and (2) enzymatic reactions. The sequence of events carried out by the enzymatic reactions was worked out in detail by Melvin Calvin and his co-workers. The enzymatic reactions of the carbon cycle of photosynthesis are discussed on page 245.

Of importance in the study of the photochemical reactions was the demonstration by Hill, in 1932, that chloroplasts isolated from the cell could still trap light energy and liberate oxygen, and later the demonstration by Arnon and others that isolated chloroplasts could convert light energy to chemical energy and use this energy to reduce CO_2.

With the advent of electron microscopy, it has been generally recognized that the photochemical reactions are bound to membranes found within the chloroplasts, while the enzyme reactions of the carbon cycle occur in the stroma of chloroplasts.

CHLOROPLASTS—THE SITE OF PHOTOSYNTHESIS

Chloroplasts seen with the light microscope in cells in a perfectly natural habit appear a homogeneous green. Such chloroplasts, in a filamentous stage in the life history of a moss, are shown in Fig. 13.2. When chloroplasts are isolated, two different kinds may generally be seen on a microscope slide; they may again be perfectly homogeneous or may show the presence of darker spots or grana. But electron micrographs of these plastids show some to be complete, with internal membranes embedded in stroma and surrounded by the outer envelope. The others have lost the envelope and stroma and consist only of a system of membranes (Fig. 13.3). Under certain conditions these latter membranes will carry out part of the reactions of photosynthesis and will liberate oxygen. But only the complete chloroplasts will both liberate oxygen and carry out the enzymatic steps that convert CO_2 into a carbohydrate.

A section of a plastid in a mesophyll cell of a corn leaf is shown in Fig. 13.4. The double nature of the envelope is apparent. The stroma is granular; some of the darker, larger granules are chloroplastic ribosomes. There appear to be two types of mem-

Figure 13.2. Protonema of a moss, showing numerous chloroplasts × 30.

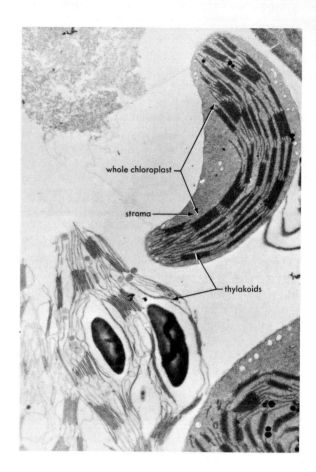

Figure 13.3. Electron micrograph of isolated chloroplasts of *Vicia faba*, showing one plastid with stroma and grana and one plastid that has lost its outer envelope and stroma ×2,500.

Figure 13.4. Electron micrograph of chloroplast from corn (*Zea mays*) mesophyll.

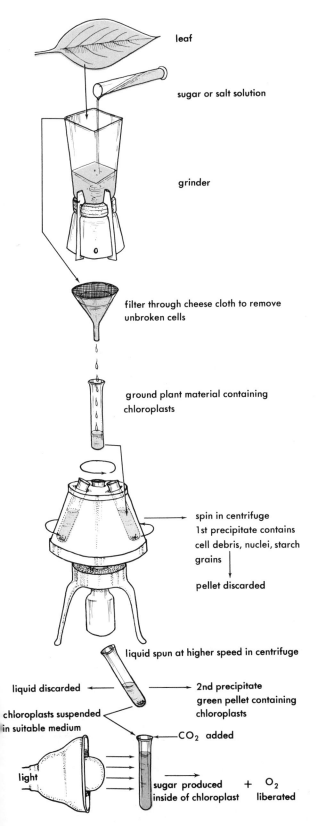

leaf

sugar or salt solution

grinder

filter through cheese cloth to remove
unbroken cells

ground plant material containing
chloroplasts

spin in centrifuge
1st precipitate contains
cell debris, nuclei, starch
grains

pellet discarded

liquid spun at higher speed in centrifuge

liquid discarded

2nd precipitate
green pellet containing
chloroplasts

chloroplasts suspended
in suitable medium

CO_2 added

light

sugar produced + O_2
inside of chloroplast   liberated

branes, those forming the grana, and those interconnecting the grana (the fret membranes). The two
end membranes of the grana are only about one-
half as thick as the internal membranes (the partitions), and it is thought that the partitions are
formed by the appression of two fret membranes.
The location of the chlorophyll is not definitely
known, except that it is membrane-bound. Some
workers consider the internal membranes to be
simply a series of alternating large and small flattened sacs known as **thylakoides.**

It is believed that the chlorophylls and carotenoid
accessory pigments are in close contact with lipid
material and with some of the photosynthetic enzyme systems in the grana compartments in a definite protein, lipid, pigment pattern. Such a structural
arrangement would facilitate the trapping of light
energy and its transfer into chemical energy during
photosynthesis. As we will discuss later, light energy,
when trapped by chlorophyll, initiates a series of
reactions in which electrons from chlorophyll may
flow to other compounds **(electron acceptors).** It is
known that such electron acceptors do occur in
chloroplasts. These include a lipid soluble material
called **plastoquinone,** a water soluble iron-contain-
ing protein called **ferridoxin,** and a relatively insoluble compound called **cytochrome** *f.*

The chemical reactions going on in photosynthesis are directly associated with the structure of
the chloroplast. This can be very clearly shown by
isolating chloroplasts from living leaves in such a
way that they are still able to carry out many of
the reactions that they normally do in the leaf.
When cells are cut or ground carefully in a dilute,
buffered sugar or salt solution, many chloroplasts
remain intact and can be separated from the rest
of the cell contents by placing the ground material
in a centrifuge and spinning it at high speed. If the
correct speed is chosen, large and heavy particles,
nuclei, starch grains, and cell wall fragments will
collect at the bottom of the tube (Fig. 13.5). If the
liquid containing the chloroplasts is now poured
into another tube and centrifuged again at a slightly
higher speed, the chloroplasts will separate out at
the bottom of the tube.

These isolated chloroplasts (Fig. 13.3) may then
be resuspended in a suitable solution, and in the

Figure 13.5. Diagram of the method of isolating
chloroplasts.

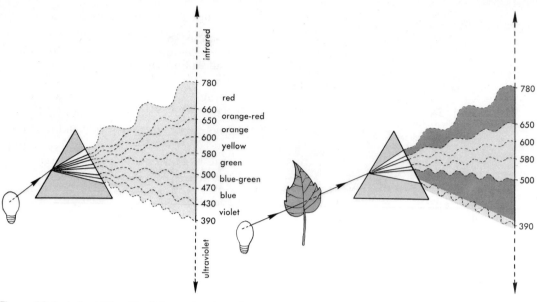

Figure 13.6. Left, visible white light passing through a prism is broken up into a range of wavelengths from the 390 millimicrons of violet to the 780 millimicrons of red; right, the sunlight passing through a leaf placed in front of the prism reveals dark bands in the spectrum between 390 and 500 mu and between 650 and 740 mu, indicating that chlorophyll has absorbed light of these wavelengths.

presence of carbon dioxide and light, they will synthesize carbohydrate and liberate oxygen. Although carbon dioxide fixation by isolated chloroplasts occurs at a slow rate compared to the rate in an intact cell, it demonstrates that all of the reactions of photosynthesis go on in the chloroplast. Chloroplasts isolated in this way have been used to study many of the reaction steps in the complex process of photosynthesis.

LIGHT—THE ENERGY USED IN PHOTOSYNTHESIS

The Nature of Light

It can be demonstrated that photosynthesis does not go on in the absence of light. If leaves are kept in the dark (or both surfaces covered with black paper or tinfoil), they do not increase in dry weight. But during the exposure of green leaves to light, they increase in dry weight, owing chiefly to the production of sugar and starch, using the carbon dioxide of the surrounding atmosphere. Before we can discuss the way in which light is used in the process of photosynthesis, we should consider the nature and properties of light.

White light, as it comes to us from the sun, is composed of waves of different lengths, ranging from the relatively long waves of red light, through successively shorter waves to violet light. When passed through a glass prism, white light is resolved into these colors. The band of colors is the **visible spectrum.** The complete visible spectrum is composed of the following colors (starting with the longest rays): red, orange, yellow, green, blue, indigo, and violet. Wave lengths exist that we are unable to perceive with our eyes. Beyond the red are still *longer,* invisible rays, the infrared; and beyond the violet are *shorter,* invisible rays, the ultraviolet. Thus, the visible spectrum represents only a part of the radiant energy that comes to the earth from the sun. But only a part of the visible spectrum is effective in photosynthesis (Fig. 13.6).

Numerous experiments have shown that light not only acts as if it were traveling in waves but also resembles particles or packets of energy called **photons** or **light quanta.** The energy of the photons determines the color of the light.

Absorption of Light Energy

The fact that chlorophyll is green to the eye is evidence that some of the blue and red wave lengths of white light are absorbed, leaving proportionally more green to be transmitted or reflected and seen. It is the absorbed light that is used in photosynthesis. The part of the white light that is absorbed by chlorophyll can be determined if the light is passed through a solution of chlorophyll in alcohol and then caused to fall upon a glass prism so that it is broken up into its component colors. The spectrum thus formed is different from the spectrum of white light that has not passed through a chlorophyll solution. In the chlorophyll spectrum, the wave lengths that are absorbed appear as dark bands, the **absorption bands** (Fig. 13.6). Thus, the positions of dark bands in the chlorophyll spectrum indicate which wave lengths are absorbed. They show that much of the red, blue, indigo, and violet are absorbed; these are the wave lengths that are utilized the most in photosynthesis. Part of the red and most of the yellow, orange, and green are scarcely absorbed at all unless the chlorophyll solution is very concentrated.

Only absorbed light can be utilized in photosynthetic reactions. If a green leaf, instead of a chlorophyll solution, is placed between a light source and a prism, the absorption bands are quite similar to, though not identical with, those from the chlorophyll solution (Fig. 13.6). The leaf contains the yellow carotenoid pigments that absorb blue light; also, the chlorophyll in the leaf is probably in association with protein and fat-like materials, while the chlorophyll in the solution is not. These factors contribute to the difference between the absorption spectra of a green leaf and of a chlorophyll solution.

Some plants, such as certain varieties of *Geranium* and of *Coleus,* have variegated leaves, that is, leaves with white bands or blotches. Microscopic examination of the mesophyll cells of the white areas shows them to be devoid of chloroplasts. These cells, however, are living and functioning. If a variegated leaf is given the starch test after several hours exposure to light, starch will be found only in those cells that contain chloroplasts (Fig. 13.7; see color plate 4), demonstrating that photosynthesis and starch storage have occurred in association with the chloroplasts.

What happens when light is absorbed by a chlorophyll molecule? How is it used in the process of photosynthesis? We know that molecules are composed of atoms having positively charged nuclei and one or more electrons spinning in certain definite positions, **orbitals,** around the nuclei. Energy is required to move an electron away from the positive nucleus. When a unit of light energy, a **quantum,** is absorbed by a molecule, the energy causes an electron of an atom in the molecule to be moved into an orbital farther from the atomic nucleus.

The molecule now is said to be in an **excited state.** This is an unstable condition and the electron tends to move rapidly (within a fraction of a second) back into its original orbital. As it does this, the trapped energy is released. It may be given off as heat, or as light, or as fluorescence (Fig. 13.8). It is evident that in the process of photosynthesis the energy that has excited the chlorophyll molecule is in some way used to bring about a chemical reaction rather than being lost simply as light or heat; exactly how this is done is not known.

The Conversion of Light Energy into Chemical Energy

Although we do not yet know exactly how light energy absorbed by the chlorophyll in a chloroplast is trapped as chemical energy, a great deal has been learned in recent years about this process. Some of this energy we know is stored in the two compounds **ATP** (adenosine triphosphate) and **NADPH** (reduced nicotinamide adenine dinucleotide phosphate) that are formed in photosynthesis. It is believed that the absorption of light energy by the chlorophyll starts a chain of events, the first step of which is the *excitation* of an electron in the chlorophyll molecule itself. Instead of falling back to its normal position in the chlorophyll, the excited electron may escape and be captured by certain chemicals, **electron acceptors,** in the chloroplast that will combine readily with electrons. Once the electron acceptor has gained an electron, it is able to transfer an electron to another suitable electron acceptor. In this way electrons may flow from one compound to another in much the same way as electricity flows in a wire.

Figure 13.8. Diagram showing the fate of the energy in a quantum of light absorbed by a pigment molecule.

NADP (nicotinamide adenine dinucleotide) is an electron acceptor, and if it is in the electron flow series, it may capture an electron and combine with a hydrogen ion in the medium to become reduced to NADPH.

At the same time, the chlorophyll that lost an electron (to become ionized) gains another electron from its environment and indirectly from water. The net result of this process is that hydrogen from the water is used to reduce NADP to NADPH in the light and oxygen gas is formed.

During the flow of electrons initiated by the absorption of light energy by chlorophyll, some of the energy is trapped in the formation of ATP (adenosine triphosphate) from ADP (adenosine diphosphate) and Pi (inorganic phosphate).

energy from electron flow

$$ADP + Pi \rightarrow ATP$$

The two compounds NADPH and ATP are the chemicals in which the energy absorbed as light by chloroplasts is first stored as chemical energy. In final steps in the process of photosynthesis, the energy stored in these two chemicals drives the energy-consuming reactions that are involved in the reduction of CO_2 during the formation of sugar. ATP and NADPH are not accumulated in the cell in any appreciable amount, but are continually being formed and broken down as energy is transferred through them from chlorophyll to sugar molecules.

The production of ATP as a result of photochemical reactions like those just described is called **photophosphorylation.** Two types of photophosphorylation are recognized. In **cyclic photophosphorylation,** the flow of electrons from the light-excited chlorophyll molecules to electron acceptors proceeds in a cyclic fashion through the acceptors and back to the chlorophyll. In this case, no NADP

13 Photosynthesis

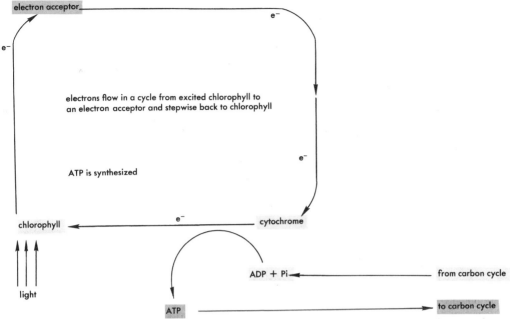

Figure 13.9. Diagram showing some steps in the process of photosynthesis: the production of ATP, cyclic photophosphorylation.

is reduced, since it does not receive electrons. ATP is formed during the electron flow and light energy is converted into chemical energy in the ATP molecules (Fig. 13.9). In the second type of photophosphorylation, both ATP and NADPH are produced. In this series of reactions, the electrons from the excited chlorophyll are trapped by the NADP in the formation of NADPH and do not cycle back to the chlorophyll. One postulated scheme for this **noncyclic photophosphorylation** is diagrammed in Fig. 13.10.

Although some scientists believe that only one light reaction is involved in photosynthesis, most workers in the field believe that the experimental evidence can best be explained by assuming that there are two light reactions, as shown in Fig. 13.10.

THE NECESSITY OF ENZYMES FOR PHOTOSYNTHESIS

Photosynthesis will not take place in the dark, yet many of the steps in the process are controlled by enzymes and are therefore not sensitive to light.

These reactions are often called **dark reactions,** since light is not directly involved. It should be pointed out, however, that under favorable conditions they proceed both in the light and in the dark. This can readily be shown by comparing the rates of photosynthesis of a plant exposed to a bright light at 20°C and then exposed to the same light at 5°C. It is known that photo-reactions (reactions brought about by light energy) are relatively unaffected by temperature in contrast with enzyme reactions that are temperature-sensitive. When one takes pictures on a hot beach or in the snow, one makes the same exposure for the same light conditions irrespective of the temperature. This is because the reaction of light on the film is a photo-reaction and not particularly sensitive to temperature. Likewise, if photosynthesis were only a light reaction or series of light reactions, it should not be influenced by a change in temperature. In our example above, however, it would be found that the rate of photosynthesis would be several times faster at 20°C than at 5°C, as long as the plant was in bright light.

This means that some of the intermediate reac-

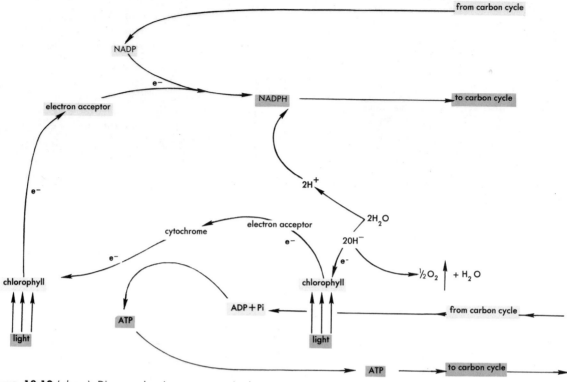

Figure 13.10 (*above*). Diagram showing some steps in the process of photosynthesis: the production of ATP and NADPH, noncyclic photophosphorylation.

Figure 13.11 (*right*). The carbon from carbon dioxide passes into several intermediate compounds (A, B, C, etc.) before sugar is formed in photosynthesis.

tions in the process of photosynthesis are temperature-sensitive and controlled by enzymes rather than by light. As we have seen, the production of ATP and NADPH involve light-sensitive photo-reactions that go on in the chloroplast. Other steps in photosynthesis, however, are controlled by enzymes, and only indirectly by light.

CARBON DIOXIDE REDUCTION AND THE FORMATION OF SUGAR

Physiologists have long been interested in determining the sequence of carbon compounds that are

formed in photosynthesis. Although early physiologists analyzed the carbohydrate content of photosynthesizing leaves and could easily recognize that starch as well as sugars usually accumulate during photosynthesis, it was not until two sensitive analytical techniques were developed that the actual sequence of reactions was understood. These new methods were the use of carbon dioxide in which the carbon was radioactive and the use of a special process, **chromatography,** with which the investigator could easily and accurately separate minute amounts of different organic compounds from one another. These techniques permitted the investigator to follow carbon in the series of reactions taking place in the plant.

The problem confronting physiologists was to determine the sequence of compounds (A, B, C, etc., Fig. 13.11) through which carbon atoms taken into the leaf pass during the synthesis of sugar. When the radioactive isotope of carbon (C^{14}) became available, scientists treated leaves with carbon dioxide containing radioactive carbon ($C^{14}O_2$) and determined the carbon compounds formed. This may be easily done by the method outlined in Fig. 13.12. It is first necessary to prepare "maps" of the rates of movement of known compounds across the large sheets of filter paper. Once the positions (on the filter paper) of the known compounds, under controlled, precise experiments, have been located, a drop of an unknown extract may be placed on the corner of the filter paper. After the usual washing and rotating, the presence of a spot at a definite position on the filter paper is taken to mean that a certain specific compound was present in the unknown sample. A group of scientists at the University of California under the direction of Dr. Melvin Calvin were particularly successful in using this technique for working out the early carbon pathways of photosynthesis. Dr. Calvin eventually received the Nobel Prize for this work.

The method finally adopted by this group as most successful was to grow algae in an inorganic nutrient medium, place such an algal culture in a glass flask in the light, expose the cells to radioactive $C^{14}O_2$, and kill the cells immediately in boiling alcohol. A similar technique of treating leaves with $C^{14}O_2$ could also be used (Fig. 13.13). The alcohol extract contained essentially all of the radioactivity incorporated into the cells after short time intervals. By varying the time to which the plant was exposed to $C^{14}O_2$, the first compounds synthesized during exposure to C^{14} could be determined. Not only is photosynthesis very rapid but also the first products are quickly changed into other substances. Thus, very short exposures to $C^{14}O_2$, in the range of five to 10 seconds, were used. Figure 13.14 shows a radioautogram of the products formed when a tobacco leaf is treated wtih $C^{14}O_2$ for 10 seconds.

If photosynthesis proceeds for an hour or so in an atmosphere containing radioactive carbon dioxide, most of the labeled carbon will be found in carbohydrates (sugar or starch). If photosynthesis is stopped after only a few seconds, most of the labeled carbon is found in a three-carbon-atom acid

containing phosphorus, **phosphoglyceric acid** (PGA).

Phosphoglyceric acid, formed during photosynthesis, is thus an intermediate between carbon dioxide and sugar. By stopping photosynthesis at increasingly longer intervals (from a few seconds to several minutes), investigators have found that many carbon compounds become labeled with radioactive carbon.

It has been possible by means of such experiments to draw up an outline representing the various reactions that occur. Some of the reactions are shown in Fig. 13.15. This figure primarily shows the carbon cycle and its association at two points with photoreactions. Carbon dioxide diffusing into the mesophyll from the atmosphere finally reaches the cytoplasm of the mesophyll cells. The initial contact of CO_2 with the carbon cycle is to combine with ribulose diphosphate, a five-carbon-atom sugar phosphate continually being produced in the cell. A six-carbon sugar is not produced from this reaction; instead two molecules of phosphoglyceric acid (PGA), a three-carbon compound, are formed. PGA is now reduced to form two three-carbon-atom sugar phosphates. The energy that drives this reaction comes from NADPH and ATP photo-reactions.

Once the PGA is reduced, the two three-carbon-atom sugar phosphates (triose phosphates formed from PGA) may be combined to form a six-carbon-atom sugar phosphate (fructose diphosphate). The plant by this process has essentially added one CO_2 molecule to a five-carbon-atom sugar to produce one molecule of a six-carbon-atom sugar. As this process continues, some of the fructose phosphate may be transformed through other reactions into other carbohydrates, including sucrose and starch. However, some fructose phosphate molecules are used to form more five-carbon-atom sugar phosphate, which in turn accepts more CO_2. Thus a cycle of carbon compounds exists, with CO_2 from the air and hydrogen from water entering the cycle and various sugars being produced.

THE PRODUCTS OF THE REACTION

Oxygen

The evolution of oxygen during photosynthesis in higher plants can be readily demonstrated. Some plants, such as photosynthetic bacteria, can carry out photosynthesis without the production of oxylight

$C^{14}O_2$

end of paper is in H_2O

algae

concentrated and placed on filter paper

boiling alcohol

wash with butanol propionic acid

partial separation of compounds

rotate paper

paper rotated

wash with phenol water

final separation of compounds

Place on x-ray film in dark.   Develop film after two weeks.
Radioactive compounds on the paper are indicated by
darkened areas on the film.

final radiograph on the x-ray film

spots on film matched with paper
and compounds eluted and identified.

Figure 13.12. Diagram showing steps in development of an
autoradiogram as used in the study of photosynthesis.
Radioactive carbon is supplied in the leaf, and metabolites are
separated by paper chromatography, C^{14} produces a latent
image on x-ray film which, when developed, shows the
presence of the initial C^{14} in the metabolites formed by
photosynthesis in the chloroplasts after exposure to $C^{14}O_2$.

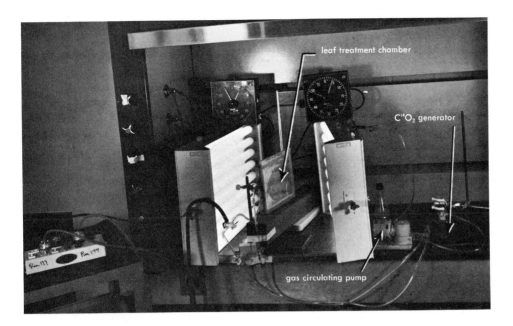

leaf treatment chamber

C¹⁴O₂ generator

gas circulating pump

← PHENOL/WATER

↑ BUTANOL/PROPIONIC ACID/WATER

Figure 13.13 (*above*). The apparatus used to treat a leaf with $C^{14}O_2$ in order to study the path of carbon in photosynthesis. The $C^{14}O_2$ is fed to the leaf in a closed system.

← PHENOL/WATER

↑ BUTANOL/PROPIONIC ACID/WATER

Figure 13.14 (*left*). Radioautogram of the products of photosynthesis after treating a tobacco leaf with $C^{14}O_2$ for 10 seconds.

The Products of the Reaction

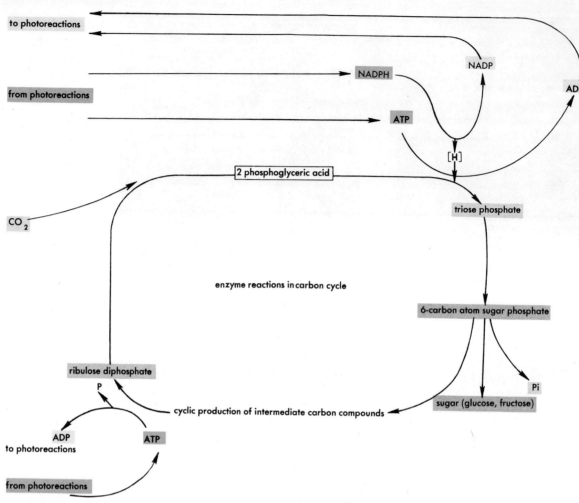

Figure 13.15. Diagram showing some steps in the process of photosynthesis: the carbon cycle.

gen. In these cases the hydrogen used in the process does not come from water but from some other electron donor such as H_2S. If an inverted glass funnel is placed in water over a mass of green water plants and the funnel is completely filled with water and then closed, the gas bubbles given off by the plants may be collected in the funnel tube. If a glowing splinter is inserted into the tube, it will glow more brightly, showing that the gas contains oxygen.

The volume of oxygen liberated in photosynthesis is approximately equal to the volume of carbon dioxide absorbed. This ratio, $O_2/CO_2 = 1$, is known as the **photosynthetic quotient.** Under like conditions of temperature and pressure, equal volumes of different gases have the same number of

molecules; hence for every molecule of carbon dioxide absorbed during photosynthesis one molecule of oxygen is liberated. Thus the equation

$$6CO_2 + 6H_2O \xrightarrow[\text{green plant cell}]{\text{light energy}} C_6H_{12}O_6 + 6O_2$$

is supported by experimental data. This equation, however, does not indicate whether the oxygen evolved comes from the water or from the carbon dioxide. In order to answer this question, plants were exposed to carbon dioxide in which the oxygen was heavier than ordinary oxygen, having an atomic weight of 18 rather than 16. When this was done, the oxygen evolved during photosynthesis was found to be composed of oxygen atoms with an atomic weight of 16. However, when a photosynthesizing plant was placed in water made with O^{18}, the oxygen evolved contained the O^{18}. This indicated that the process of photosynthesis in higher plants involves a splitting of the water molecules and a liberation of oxygen from them.

A more complete equation for photosynthesis that indicates this fact is

$$6CO_2^{16} + 12H_2O^{18} \xrightarrow[\text{green plant cell}]{\text{light energy}}$$

$$C_6H_{12}O_6^{16} + 6O_2^{18} + 6H_2O^{16}$$

Sugar

The demonstrations that carbon dioxide and water are the raw materials and that oxygen is one of the end products of photosynthesis are easy to make and may be carried out in any botany laboratory. Proof that sugar is the other end product of photosynthesis was more difficult to obtain before $C^{14}O_2$ was available.

It has been pointed out that the photosynthetic quotient, O_2/CO_2, is equal to 1; in other words, the amount of carbon dioxide taken up by the leaf is equal to the amount of oxygen given off. This condition is met by the elaboration of a carbohydrate as is seen from the overall equation.

J. H. C. Smith, in 1943, successfully accounted for all the carbon dioxide absorbed by a leaf during a period of several hours. The analyses were accomplished by carefully measuring the amount of carbon dioxide the plants absorbed during a definite period of illumination. Then he located the carbon assimilated within the leaves. His results showed that about 70 percent of the carbon absorbed by the leaves was in the form of sugar. The remainder was present in starch and in a compound allied to cellulose. All of these compounds are carbohydrates.

Starch grains occur only in the chloroplasts of the higher plants, and if leaves containing starch are kept in darkness for some time, the starch grains will disappear (Fig. 13.16; see color plate 4, page 222); then if these leaves are exposed to light, starch reappears in the chloroplasts. Starch is the *first visible product* of photosynthesis.

EFFICIENCY OF PHOTOSYNTHESIS

The green leaf is not particularly efficient in utilizing the sun's energy. We speak of the efficiency of a machine, such as a Diesel engine, a gasoline motor or an electric motor. We calculate the energy value of the fuel used and compare it with the energy output of the machine. We know that much of the energy is lost. The ratio of energy outgo to the energy intake represents the efficiency. Of the total radiant energy that falls upon green leaves, about 80 percent is absorbed (Fig. 13.17). Of the remaining 20 percent, a part is reflected from the leaf surface and a part passes through the leaf. Part of the radiant energy absorbed is changed to heat and raises the temperature of the leaf; a large part of what is absorbed is used up in transpiration; the remainder is utilized in photosynthesis and stored in the carbohydrate molecules. Thus only about 0.5 to 3.5 percent of all the light energy that falls on a leaf is used in the process of photosynthesis. By standards of machine efficiency, this is a very low percentage. But the supply of solar energy is continuous and abundant.

However, there is reason to believe that the leaves of higher plants, under the most favorable conditions, should be about ten times as efficient as this in converting solar energy into chemical energy by photosynthesis. This is a reasonably efficient chemical energy. Intensive research is needed to raise the low efficiency of photosynthesis of plants in the field to the efficiency that is theoretically possible.

We have emphasized that photosynthesis involves the transformation of two low-energy inorganic compounds to a high-energy organic compound. We know that **oxidation** usually involves the union of oxygen with some other element or some compound and that energy is released during oxidation. Note that in photosynthesis oxygen is given off and that hydrogen combines with carbon dioxide to form a carbohydrate. This type of reaction is **reduction,**

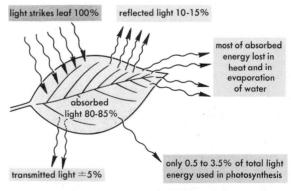

Figure 13.17. Diagram showing what happens to light that strikes a leaf.

and requires energy. This energy requirement can be measured. We know, for instance, how much energy is needed to separate the oxygen from the hydrogen in the water molecule. The oxygen-hydrogen bonds in water are among the strongest known. The bonds that hold together carbon and oxygen in carbon dioxide are also strong, but in sugar or starch they may be broken easily. Carbon dioxide and water do not burn, but sugar does. In other words, the heat of combustion of carbon dioxide and water is zero, but the heat of combustion of glucose is 673 kilocalories for each gram molecule of glucose.

It is interesting to note that the plant is able to rearrange the elements carbon, hydrogen, and oxygen, which compose sugar, to form compounds of even higher energy content. For instance, sugar may be transformed into fats—compounds also containing carbon, hydrogen, and oxygen but with relatively less oxygen.

CONDITIONS AFFECTING THE RATE OF PHOTOSYNTHESIS

The rate of photosynthesis is of great importance in food and fiber production. The rate at which sugar or other carbohydrates is formed will affect the yield of any given crop.

Internal Factors

The rate of photosynthesis is influenced by a number of **internal factors,** that is, conditions inherent in the plant itself. Chief of these are (1) the structure of the leaf and its chlorophyll content; (2) accumulation within the chlorophyll-bearing cells of the products of photosynthesis; and (3) protoplasmic influences, including enzymes. Considering the interplay of these various internal factors, it is evident that different species of plants vary considerably in their photosynthetic efficiency, even when growing under the same environmental conditions.

Leaf Structure

The structural features of the leaf influence the amount of carbon dioxide that reaches the chloroplasts. These features include size, position, and behavior of the stomata, and the amount of intercellular space. Also, the intensity and quality of the light that reaches the chloroplasts are influenced by thickness of the cuticle and epidermis, the presence of epidermal hairs, the arrangement of mesophyll cells, the position of chloroplasts in the cells, and so forth.

Products of Photosynthesis

With an increase in the concentration of the products of photosynthesis in mesophyll cells, there is a decrease in the photosynthetic rate. Starch may accumulate in chloroplasts during the day, when sugar is manufactured at a more rapid rate than it is transferred from the cell; a result of this accumulation may be a retardation in the rate of photosynthesis.

Protoplasm

The rate of photosynthesis is affected by conditions associated with the protoplasm itself. If the cells lack water and the protoplasm is dehydrated, photosynthesis slows down. Moreover, a disturbance of certain enzyme activities influences the photosynthetic rate.

External Factors

These factors as they influence the rate of photosynthesis are, to a degree, under the control of the plant grower. Crop yields can be influenced by modifying these factors. The principal external conditions that affect the rate of photosynthesis are (1) temperature, (2) light intensity, quality, and duration, (3) carbon dioxide content of the air, (4) water supply, and (5) mineral elements in the soil.

Temperature

Plants of cold climates carry on photosynthesis at much lower temperatures than do those of warm climates. The process is known to occur in certain evergreen species of cold regions, even at temperatures below 0°C. Algae in the water of hot springs may carry on photosynthesis at a temperature as high as 75°C. Most ordinary temperate-climate plants, however, function best between temperatures of 10° and 35°C. If there is adequate light intensity and a normal supply of carbon dioxide, the rate of photosynthesis of most ordinary land plants increases with an increase in temperature up to about 25°C; above this range there is a continuous fall in the rate as the temperature is raised. At these higher temperatures the time of exposure is of importance. At a given constant high temperature (for example, 40°C), the rate of photosynthesis decreases with time.

Under conditions of low light intensity, an increase in temperature beyond a certain minimum will not produce an increase in photosynthesis (Fig. 13.18). These conditions may occur in the winter in greenhouses. If the temperature is raised too high the plants will suffer because the rate of photosynthesis has not been changed but respiration has been increased by the higher temperature.

Light

In discussing the effect of light upon the rate of photosynthesis, three elements must be considered: (1) intensity, (2) quality (wave lengths), and (3) duration. With proper temperature and sufficient carbon dioxide carbohydrates produced by a given area of leaf surface increase with increasing light intensity up to a certain point (optimum light intensity), after which they decrease. It is not the intensity of light that falls upon the leaf surface that is of importance as much as it is the intensity to which the chloroplasts are exposed. The light intensity diminishes from the leaf surface to the chloroplasts, owing to surface hairs, thick cuticle, thick epidermis and other structural features.

Intense light appears to retard the rate of photosynthesis. Many plants that live in deserts and other places where the light is very bright often have structural adaptations which tend to diminish the intensity of light that reaches the chloroplasts. The usual light intensity in arid and semiarid regions is well above the optimum for photosynthesis in many plants, especially introduced crop plants. In these regions on days when the sky is overcast, the light intensity is probably nearer the optimum for photosynthesis than on clear, sunny days. Leaves on the surface of plants receive light of greater intensity than those beneath that are shaded. Therefore, some of the leaves receive light of optimum intensity, whereas others may receive light either above or below the optimum.

Carbon Dioxide

The carbon dioxide utilized by land plants is absorbed by the leaves from the atmosphere. About 78 percent of the atmosphere is nitrogen and about 21 percent oxygen, the remaining small percentage being composed of carbon dioxide, argon, and traces of hydrogen, neon, helium, and other gases. Surprisingly, only about 0.03 percent of the atmosphere is carbon dioxide.

Carbon dioxide in the air surrounding the leaves ultimately reaches the chloroplast. Its inward diffusion path is through the stomata to the intercellular spaces, through walls of palisade and spongy parenchyma cells, to the cytoplasm, and thence into the chloroplasts. The walls of palisade and spongy parenchyma cells contain water. Carbon dioxide is readily soluble in water; hence, carbon dioxide passes through the cell walls in aqueous solution.

In the process of photosynthesis, the cells remove carbon dioxide from solution in the cell sap. As a result, there is diffusion of that gas inward from the wet cell walls. This loss of carbon dioxide from the cell wall allows the water in the wall to dissolve more carbon dioxide from the air in the intercellular spaces. The carbon dioxide content of the intercellular spaces is lowered below that of the outside atmosphere. Diffusion of carbon dioxide inward through stomata tends to make up this deficiency. Thus, during active photosynthesis in the chloroplasts, a diffusion gradient for carbon dioxide is set up between the outside atmosphere and the chloroplasts. At the same time, oxygen liberated in photosynthesis is used in respiration, or it diffuses outward in aqueous solution through the cell walls to intercellular spaces and thence through stomata to the atmosphere surrounding the leaf (Fig. 10.6).

We may rightly consider the problem: If all the

Figure 13.18. Graph showing interaction of light and temperature on the rate of photosynthesis. When light is limiting, the rate of photosynthesis is independent of temperature. When temperature is limiting, the rate of photosynthesis is independent of light intensity.

chlorophyll-bearing cells of the plants of the world are constantly taking carbon dioxide from the atmosphere during daylight, the quantities of this gas used must be enormous, and there must necessarily be processes in nature that are continually replenishing this supply. It is known that the amount of carbon dioxide in the air is low (0.03 per cent) and that it remains fairly constant. It has been estimated that an acre of corn (10,000 plants) during a growing season of 100 days will accumulate 5585 pounds of carbon; all this carbon is derived from the carbon dioxide of the atmosphere. It would require 20,480 pounds of carbon dioxide to furnish this quantity of carbon. These estimates serve to emphasize the fact that enormous quantities of carbon dioxide are used in the photosynthetic process of green plants.

Obviously, the amount of carbon dioxide is limited. The present atmospheric supply would be used up in about 22 years were it not constantly being renewed. Several natural processes are continually releasing carbon dioxide to the atmosphere. They are as follows. (1) The living cells of all plants (both green and nongreen) and of all animals release carbon dioxide in the respiratory process (Chapter 14). (2) The dead bodies of plants and animals, and the excretions of animals, contain large quantities of carbon and other elements in the form of organic compounds; in the decay of these compounds, resulting from the activities of bacteria and fungi, large quantities of carbon dioxide are released to the atmosphere. (3) Carbon dioxide is also added to the atmosphere when wood, coal, oil, gas, or any other carbon compound burns. (4) Carbon dioxide is released to the atmosphere from mineral springs and volcanoes. (5) The oceans are important reservoirs of carbon dioxide, and carbon dioxide probably escapes from the oceans whenever its concentration in the atmosphere decreases.

If light intensity and temperature are favorable, the carbon dioxide of the atmosphere frequently limits the rate of photosynthesis. This may be particularly true in greenhouses kept closed in the winter. Under these conditions the carbon dioxide in the air may be reduced much below the 0.03 percent average.

It has been determined experimentally that at usual temperatures and light intensities an artificial increase of carbon dioxide up to a concentration of 0.5 percent may give an increased rate of photosynthesis, but only for a limited period. It appears that this high level of carbon dioxide is injurious to plants; after ten to fifteen days' exposure the plants show injury.

Water

In land plants, the soil is the source of water. The water used by the green cells enters the roots from the soil and is conducted upward through the xylem of the roots and stems to the leaves. It moves through the xylem of the petiole and veins of the leaf blade to the finest veinlets. Then it passes from the tracheids of the veinlets to the mesophyll. Although the water content of an actively photosynthesizing cell is high and large amounts of water are lost from the cell by transpiration, only about one percent or less of the water absorbed by the roots is actually used in photosynthesis. However, the rate of this process may be changed by small differences in water content of the chlorophyll-bearing cells. In some instances the rate of photosynthesis is increased by mild dehydration (15 percent water loss) and retarded by vigorous drying (45 percent water loss). Since stomata tend to close when the plant is deprived of water, conditions of drought tend to reduce the rate of photosynthesis. Thus, though water is one of the raw materials in the process, it rarely, if ever, is directly a limiting factor in photosynthesis.

Minerals

The chemical formula of chlorophyll a is $C_{55}H_{72}O_5N_4Mg$ and, of chlorophyll b, $C_{55}H_{70}O_6N_4Mg$. Thus it is seen that the synthesis of chlorophyll depends upon a supply of nitrogen and magnesium, both derived from salts in the soil. Moreover, chlorophyll is not formed unless iron is available, although this element is not a component of the chlorophyll molecule. Leaves of plants deficient in nitrogen, magnesium, or iron are pale and yellow, a condition termed **chlorosis.** This abnormal condition may also be caused by other factors, but when it occurs the rate of photosynthesis is lowered.

SUMMARY OF PHOTOSYNTHESIS

Some of the important points that have been mentioned concerning photosynthesis are summarized in the schematic diagrams shown (Figs. 13.10, 13.15).

1. Photosynthesis is the major energy-storing process of life in which light energy is stored as chemical energy in organic compounds.

2. Carbon dioxide and water are the raw materials.

3. The products of photosynthesis are sugar and oxygen.

4. Light energy is absorbed by chlorophyll in the chloroplasts and drives the reaction of photosynthesis.

5. Adenosine triphosphate (ATP) and reduced nicotenamide adenine dinucleotide phosphate (NADPH) are formed by photo-reactions.

6. The hydrogen of the water is transferred to organic compounds and eventually reduces the carbon dioxide.

7. The oxygen of water is liberated as a gas.

8. Phosphorus is essential to the formation of sugars in plants and is found in some of the intermediate products, one of which is phosphoglyceric acid.

9. Many steps in the process of photosynthesis are controlled by enzymes.

10. Both external and internal factors affect the rate of photosynthesis.

Respiration

The term respiration was first used to indicate the exchange of gases between an organism and its environment. Even today respiration and breathing of animals are often popularly considered synonymous. Nevertheless, breathing is only an outward indication of the fundamental chemical reactions going on in the animal cells and characteristic of all life. Every living cell, in order to stay alive, must break down complex organic molecules (foods) and obtain energy from them. This breakdown process, as we have seen (page 80), is brought about through oxidation reactions and often, though not always, yields carbon dioxide and water as end products. Today the term respiration has come to have a wider and more fundamental meaning than mere gaseous exchange or breathing. In its broadest sense, *respiration is defined as the oxidation of organic substances within cells and is accompanied by the release of energy.* Some biochemists reserve the term respiration for the final sequence of reactions which terminate in the use of molecular oxygen. However, we will use respiration in its broader chemical sense.

In this process the chemical energy that holds the atoms together in the food molecules is transformed into the chemical energy in ATP that can be readily used to do work in the cell. Not all of the energy is released as usable energy; much may be lost as heat. But the synthesis of new cellular material, such as amino acids, proteins, and fats and the complex processes of growth, protoplasmic movement, accumulation, and solute transport, to mention only a few, all require energy—energy released by the oxidation of food.

Respiration plays another very essential role in biology, for it is during the many steps in the respiratory breakdown of food that essential intermediate compounds are formed. Many of these intermediate compounds are used by the living cells as sources of carbon building blocks from which new compounds such as proteins, fats, nucleic acids, and vitamins are synthesized. Respiration thus supplies (1) energy in an available form and (2) usable intermediate carbon compounds essential to the continued growth and metabolism of cells.

Figure 14.1. Diagram showing the coupling of energy-yielding reactions to energy-requiring reactions through ATP production and utilization.

THE USE OF THE STORED ENERGY IN FOODS TO DO WORK IN THE CELL

Since the significance of respiration for the living organism is the release of the potential energy in the food to do work, some mechanism must exist in the cell whereby the energy liberated in these breakdown and oxidation reactions is captured in the chemical bonds of ATP and then delivered to the energy-consuming reactions of the cell. In other words, energy-yielding reactions are *coupled* or linked to energy-requiring reactions. The coupling of a diesel motor, driven by the burning of a petroleum fuel, to a dynamo, producing electrical energy, and the use of this electrical energy to run the motors of a locomotive is a common example of the coupling of machines for energy conversion to perform work.

Living cells are unable to utilize heat in the performance of their functions. Instead, the cell utilizes dozens of oxidative enzymes, each specialized to catalyze one reaction in the series of reactions involved in the oxidation of food to CO_2 and H_2O. During many of these oxidation reactions, much of the energy liberated is not lost as heat but is temporarily stored in the universal intracellular carrier of chemical energy, ATP. Fig. 14.1 is a diagram of this energy-coupling sequence. Once ATP is formed it may move in the cell and be used to supply energy where needed to drive synthetic reactions along, to maintain an osmotic system, and to aid other energy-requiring processes. Large amounts of ATP are never formed in the cell at any one time since the total amount of ADP from which it is formed is small. The energy of the cell is stored in large amounts in food.

If a cell is at rest and is not using energy rapidly, most of its ADP may be converted to ATP. Respiration may be very slow under these conditions. If such a cell is stimulated to do work (for example, to synthesize new materials, or to increase the rate of protoplasmic streaming, or the amount of salt uptake), then ATP is utilized in these reactions and ADP and inorganic phosphate become available to the respiratory machinery, where they may be synthesized again into ATP, trapping more energy. Under these conditions, the rate of respiration may be increased. Figure 14.1 shows some of these interrelationships.

THE CHEMICAL REACTION

A knowledge of the process and effects of respiration in plant cells is of paramount importance to growers and handlers of plants and plant products. For example, let us consider the storage of a ton of apples. Every apple contains thousands of living cells, each of which is respiring. If sufficient oxygen is available to the cells, each cell will be using oxygen to break down some of its sugar and will be producing carbon dioxide, water, energy, and, in addition, various other products in low concentration. Much of the energy is not used by the cells but is lost as heat.

If the fruit has been packed tightly together in a closed storage room, the environment around the apples will change. The air will be depleted of oxygen and become richer in carbon dioxide. The apples will begin to warm up. Such changes will affect the storage life of the fruit and must be controlled if the fruit is to be held in storage with a minimum of loss.

Respiration involves a whole series of chemical reactions, each of which is controlled by a particular enzyme and each of which proceeds more slowly at low temperatures than at high temperatures. Cold storage slows down respiration and reduces the changes in fruit during storage. Refrigeration is needed not only to cool the fruit but also to prevent the release and accumulation of heat by the fruit itself as a result of respiration. How much ice would be required to prevent the heat released during respiration from heating the fruit? For example, if we desired to keep 40,000 boxes of apples in storage at 5°C, it would require between two and three tons of ice every day to prevent the heat released by respiration from raising the temperature of the apples above 5°C. It is apparent that shippers of fruits and vegetables must know how much refrigeration should be supplied to keep the produce at the proper temperature. Likewise, they must know how the products of respiration affect the stored plant material. Carbon dioxide at high concentrations can be toxic. If it is allowed to accumulate, injurious effects may result. Various gases, such as ethylene, are given off by some cells during respiration and affect the ripening processes of certain fruits. These and many other problems make a knowledge of plant respiration essential to agriculturists.

THE OVERALL PROCESS OF RESPIRATION

The process of respiration is summarized by the following equation, which represents the oxidation of a simple hexose sugar in the presence of oxygen from the atmosphere.

sugar + oxygen → carbon dioxide + water + energy

$$C_6H_{12}O_6 + 6O_2 \rightarrow 6CO_2 + 6H_2O + 673 \text{ kcal}$$

The equation for respiration indicates the two reacting substances, sugar and oxygen, and the two usual end products, water and carbon dioxide; it tells nothing of the intermediate steps or of other possible end products.

In our consideration of this process, all of the detailed steps whereby sugar is completely oxidized will not be discussed. Rather, the process of respiration will be considered as consisting of several gen-

eral phases, each of which represents a whole series of individual reactions. Furthermore, although these reactions have been studied in some organisms, it is not necessarily true that the course of respiration follows the same pathway in all plant cells.

When a cell uses sugar in the process of respiration, the sugar first undergoes a series of changes and is broken apart. No atmospheric or molecular oxygen enters into these first reactions, but small amounts of energy are released and intermediate compounds are formed. This series of reactions is usually referred to as **glycolysis** and consists of the following three steps.

1. The preparation of the sugar for reaction by the addition of phosphorus—**phosphorylation.**
2. The splitting of the sugar into two fragments—**sugar cleavage.**
3. The oxidation of the fragments to form an intermediate product of respiration—**pyruvic acid formation.**

The final reactions of respiration are influenced by the presence or absence of oxygen. If oxygen is present, as is usual, the intermediate product formed by glycolysis is oxidized to carbon dioxide and water. This process that requires molecular oxygen is called **aerobic respiration.** If oxygen is absent, a state that prevails under certain conditions, the intermediate compounds of glycolysis are usually broken down to form carbon dioxide and alcohol. This process, which does not involve molecular oxygen, is termed **anaerobic respiration.**

We may summarize respiration as follows:

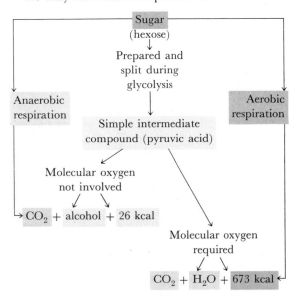

Preparation of Sugar for Reaction

Sugar itself does not combine readily with oxygen at temperatures at which life reactions proceed, nor is it easily broken down into intermediate products. You will recall that phosphorus compounds serve as intermediate steps in photosynthesis and that sugar phosphates are formed. So also in respiration, before sugar can be broken down and its stored energy released, it must be combined with phosphorus to form sugar phosphate. This preparation of sugar in the process of respiration is called **phosphorylation** and, like the other steps in respiration, is controlled by enzymes. Phosphorylation is an energy-requiring reaction and during the process, one of the three phosphate groups of ATP is transferred to the sugar molecule that becomes phosphorylated. Actually, two phosphorylation steps are involved before a hexose sugar is ready for sugar cleavage.

Thus two ATP molecules are involved in the complete phosphorylation of the sugar molecule. Adenosine triphosphate and similar organic phosphorus compounds are formed during the later stages of respiration and are very important as carriers of energy in cell metabolism. They might be likened to cogs in the respiration wheel that mesh with other wheels of cell metabolism and through which the energy of respiration can be directed to turn the other reaction wheels.

Sugar Cleavage

After sugar has been prepared by phosphorylation, enzymes can now act on it to break the molecule into two parts. The original six-carbon-atom sugar becomes changed into two three-carbon-atom fragments that are each still combined with phosphorus. You will recall that photosynthesis also involves a three-carbon-atom intermediate compound containing phosphorus. In some of its stages, respiration appears to be a reversal of photosynthesis.

Pyruvic Acid Formation

The sugar fragments containing phosphorus lose their phosphorus, some of their hydrogen atoms, and a small amount of energy in a series of reactions that result in the formation of the three-carbon-atom intermediate compound called **pyruvic acid.** During the changes that have so far been described, molecular oxygen has played no part. Oxidation of the sugar has occurred through a rearrangement of the atoms within the molecule and the removal of some of its hydrogen.

The molecule receiving the hydrogen in respiratory reactions, the **hydrogen acceptor,** is frequently NAD or NADP. In this way NADH or NADPH are formed by some oxidation reactions in respiration. They may be oxidized in later stages of respiration and the energy that they contained used to form ATP from ADP and inorganic phosphate. In this process there is a flow of energy from the food to the hydrogen acceptor to ATP in a manner not unlike that described for the flow of electrons during the formation of ATP and NADPH in photosynthesis (page 244).

The fate of the final product of glycolysis, the final intermediate compound common to both aerobic and anaerobic respiration, now will depend upon the presence or absence of molecular oxygen.

Anaerobic Respiration

Normally, higher plants are not able to live long in the absence of oxygen. Under this condition, insufficient energy is available for life processes and certain products formed may be poisonous. In contrast, some fruits, notably apples, may be held for long periods in an atmosphere containing very small amounts of oxygen and continue to give off carbon dioxide. Other plants, such as yeast, may live actively in an atmosphere with very small amounts of oxygen and produce relatively large amounts of carbon dioxide and alcohol. Still other microorganisms can live in an atmosphere completely devoid of oxygen.

Under completely anaerobic conditions the products of glycolysis are most frequently converted into carbon dioxide and ethyl alcohol. This process is often called **alcoholic fermentation.** In this process the original sugar is only partially oxidized and only a small amount of its energy is released; the rest remains unavailable to the organism. This energy is still locked in the molecular structure of the alcohol. Alcohol and carbon dioxide are not the only possible products of anaerobic respiration. There is a great variety of microorganisms that respire anaerobically and many produce other products beside alcohol (see page 423).

Aerobic Respiration

Although some cells of higher plants are capable of living under anaerobic conditions, such conditions are harmful to the majority of the tissues and to the entire plant. Aerobic respiration is the common way in which these cells obtain energy and the necessary intermediate compounds. When oxygen is dissolved in the cell fluid and glycolysis has occurred, the oxygen itself is still unable to combine directly with this intermediate compound (pyruvic acid). Rather, a whole series of reactions must occur before the final products, carbon dioxide and water, are formed. These reactions can be summarized in two main steps.

1. **The organic acid cycle of respiration.** This involves the gradual breakdown of pyruvic acid. Other organic acids in the plant combine with pyruvic acid, carbon dioxide is lost, and hydrogen is removed in several stages (Fig. 14.2).

2. **Terminal oxidation, the final stage of aerobic respiration.** The hydrogen that has been removed from the substance being respired is transferred by a series of enzymatic reactions to molecular oxygen, and water is formed (Fig. 14.2).

The Organic Acid Cycle of Respiration

Sometimes this is called the **Krebs cycle** after a physiologist whose research contributed a great deal to our knowledge of respiration. It has been found that, after glycolysis has broken down sugar into pyruvic acid, the hydrogen in this compound is

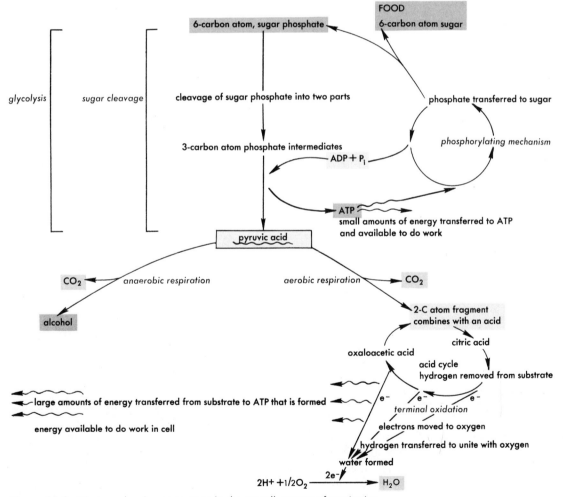

Figure 14.2. Diagram showing some steps in the overall process of respiration.

removed by a series of reactions. One of the ways in which this may occur in the cell is through combination of pyruvic acid with another organic acid (oxaloacetic acid) that is present in the cell. In this process a larger molecule (also an acid) is produced and a molecule of carbon dioxide is lost. Now the new acid is gradually broken down a small piece at a time. Hydrogen is removed and transferred to other substances, carbon dioxide is evolved, and finally a new molecule of oxaloacetic acid is formed which in turn can combine with more pyruvic acid to keep the cycle going, as shown in Fig. 14.2.

Because this is a repeating process and because organic acids are involved in the process, it has been called the **organic acid cycle** of respiration. It is believed that in many plants, animals, and microorganisms, such a cycle is a part of aerobic respiration. In this stage, some of the energy contained in the pyruvic acid being oxidized is lost as heat, but much of it is stored in energy-rich compounds, such as adenosine triphosphate (page 80), that are produced during this oxidation.

The Final Stages of Oxidation

We have seen how in the presence of oxygen, sugar may be broken down, the hydrogen of its molecule removed, and carbon dioxide produced. The final stages of aerobic respiration involve a transfer of this hydrogen by means of a series of enzyme reactions to molecular oxygen; thus, water is produced and respiration is complete. It is during these final stages of hydrogen transfer that most of the **ATP** molecules that are synthesized from ADP and inorganic phosphate during respiration are formed. The process of the formation of ATP as a result of the oxidative release of energy is known as **oxidative phosphorylation.**

Complete aerobic respiration of sugar thus involves a stepwise breakdown of the sugar molecules. CO_2 and H_2O are produced, some of the energy is lost as heat and some is transferred to the ATP that is formed as O_2 is used.

RESPIRATION AND THE SYNTHESIS OF ORGANIC MOLECULES

We have seen that all living cells respire and that respiration provides chemical energy and certain intermediate compounds, both needed for many cell functions. This intimate relationship between respiration and other cellular reactions can readily be seen when one considers the formation of carbohydrates, fats, amino acids, and proteins in the cell. All of these compounds are ultimately derived from the sugars produced during photosynthesis.

During development of a fat-storing seed such as squash, the seed may store as much as 20 percent of its dry weight as fat (Fig. 16.27). However, it is sugar and not fat that is translocated through the phloem into the seed. Obviously a mechanism exists in the cells of the seed to convert sugar into fat. How is this done?

The fat molecule is made up of two parts, (1) a three-carbon-atom alcohol (glycerol) combined with (2) one or more organic (fatty) acids. These two parts of a fat molecule are synthesized from respiratory intermediates; glycerol may be synthesized from the cleavage products of glycolysis, and fatty acids are formed by a complicated series of enzymatic reactions acting on the two-carbon-atom

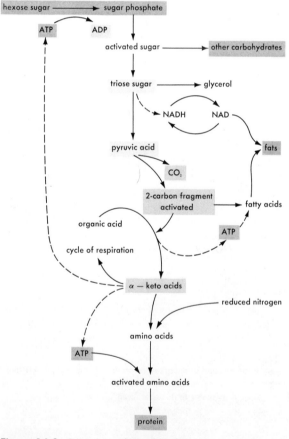

Figure 14.3. Some interrelationships between respiration and the synthesis of carbohydrates, fats, amino acids, and proteins.

Figure 14.4. *A*, electronmicrograph of a thin section of a mitochondrion, the cellular site of aerobic respiration. Mitochondrion is from companion cell in phloem of squash (*Cucurbita*) ×8,000. *B*, portion of mitochondrion from *Carteria* showing the double mitochondrial membrane and the inward extension of the inner component of the membrane to form the cristae. ×43,000. (*A* courtesy of K. Esau; *B* courtesy of Carole Lembi.)

product formed when pyruvic acid loses CO_2 (Fig. 14.3). The production of fat is further tied to respiration, since the energy from ATP as well as from reduced NAD is needed for fat synthesis.

Similarly, amino acid synthesis is linked to respiration. However, some of the same organic acids that are formed as intermediates during the organic acid cycle of respiration, are used in amino acid synthesis. Again, the ATP produced by respiration is used for the necessary activation of the amino acids used in protein synthesis.

Only a few of the many sugars and other carbohydrates found in a plant cell are directly products of photosynthesis. The interconversion of one sugar into another or the polymerization of a sugar in the synthesis of a high-molecular-weight carbohydrate such as starch or cellulose involves the activation of the sugar. Energy for these sugar interconversions is again derived from respiration (Fig. 14.3).

RESPIRATION AND CELL MORPHOLOGY

We know that photosynthesis occurs within the chloroplasts. We may ask ourselves. "Where in the cell does respiration take place?" In a very active cell, such as a meristematic cell, the rate of respiration is very high. If the cell is in an inactive tissue, such as in a dried seed, the rate of respiration is very slow indeed. Just as investigators of photosynthesis have literally taken the cell apart in order

to study its photosynthetic apparatus, the cell has been broken apart and each part (nucleus, chloroplasts, dictyosomes, and mitochondria) has been studied to determine its role in cellular metabolism.

We have seen that the complete oxidation of a sugar molecule involves a whole series of reactions. Some of these, particularly those of glycolysis, take place in the cytoplasm of the cell. Others, particularly those involving the aerobic oxidation of pyruvic acid (the acid cycle and terminal oxidation steps of respiration), are carried out by mitochondria.

Mitochondria have been isolated from plant as well as from animal cells and when properly prepared will break down pyruvic acid in the presence of oxygen. Carbon dioxide is produced and ATP may be synthesized during the oxidation reactions. Thus respiration can be studied outside the cell in the test tube with preparations from living cells.

Mitochondria, though so small that little structural detail can be seen with the light microscope, nevertheless have a distinctive structure that can be easily recognized by use of the electron microscope. Figure 14.4*A* shows a mitochondrion from a companion cell of squash (*Cucurbita*), and Fig. 14.4*B* from the alga *Carteria*. Here the characteristic double boundary membrane that separates the mitochondrion from the rest of the cytoplasm is clearly seen. The circular and elongated dark bodies inside the mitochondrion are sections through an internal membrane system that is made up of finger-like protuberances extending inward from the inner mitochondrial membrane. It is believed that many of the enzymes necessary for respiration are located on this membrane system in a definite relationship to each other. We see that the respiratory activity of mitochondria is closely associated with its structure in a manner similar to that in which the photosynthetic activity of chloroplasts is related to their structure. We see also that an integration of activities exists between the cytoplasm, where the early stages of food breakdown occur and the mitochondria, where the final stages of aerobic respiration take place.

During the life of the cell, there is a continuous traffic of molecules across the mitochondrial membranes. Pyruvic acid, inorganic phosphate ions, and ADP molecules are three important substances moving into the mitochondria from the cytoplasm. Here certain of the energy-releasing reactions during the final stages of respiration are coupled to the synthesis of ATP from the ADP and inorganic phosphate. The newly formed ATP, carrying with it some of this energy, may move from the mitochondria to some other place in the cell. As work is done there, the ATP is hydrolyzed to ADP and inorganic phosphate, which again may enter a mitochondrion. In this way, chemical energy in the substrate (pyruvic acid) molecules is continuously moving into the mitochondria, where it is transformed into a more readily usable form of energy (ATP), which is constantly moving from the mitochondria to all parts of the cell.

FACTORS AFFECTING THE RATE OF RESPIRATION

Since respiration is the process by which some chemically-stored energy in foods is liberated in a form immediately available to do work in the cell, any conditions either inside or outside the cell that affect the rate of respiration will influence cellular activities. A few of the factors affecting respiration are discussed below.

Cell Hydration

Many seeds remain viable for long periods of time when stored in relatively dry air. The water content of the cells at this time is very low and is only a small percentage of the seed weight. In contrast, the water content of active protoplasm may be as high as 90 percent or more. Under low moisture condi-

tions, some cellular activities slow to a very low rate; others stop completely. Respiration goes on at a very slow rate, so little oxygen is consumed, a little carbon dioxide is given off, and only very minute amounts of heat are released. Energy is needed only to maintain a certain little-understood steady state in the quiescent protoplasm. Growth has ceased, mineral salts are not being absorbed, and cell divisions have stopped, as have many, if not all, reactions involving synthesis of new materials. Because of this dormant state, seeds may be kept in large storage bins such as grain elevators for long periods of time.

If even a small amount of water is added to the seeds, imbibition occurs, the seeds swell, respiration increases rapidly and, if the seeds are closely confined, as in a grain elevator, the temperature inside the mass of seeds that are insulated from the outside may increase to a point where the seeds will be killed. In this example, the rate of respiration is greatly influenced by the water content of the cell. A number of other factors, both external and internal, influence the rate at which a cell respires and consequently affect all of the cell's activities that depend upon the use of energy released in respiration. In addition to the degree of hydration of the cell, factors such as temperature, oxygen supply, food availability, carbon dioxide concentration, the presence of certain chemicals, and the age of the cell commonly influence cell respiration.

Temperature

Temperature has a particularly marked effect upon most biological reactions, especially those that are controlled by enzymes. In the temperature range from near freezing (0°C) to about 30°C, an increase of 10°C approximately doubles the rate of respiration. Above 30°C the harmful effects of high temperature on the cell may become marked. At these higher temperatures, cellular enzymes progressively become inactivated and respiration decreases.

Food

Since respiration involves the utilization of food, the presence of easily available food is necessary if a cell is to continue to respire and stay alive. Photosynthesizing leaf cells, of course, are able to produce their own food in the light and under normal conditions must also form enough food to supply the needs of all of the nonphotosynthetic cells in the

Comparison of Photosynthesis and Respiration

Photosynthesis	Respiration

Photosynthesis

1. CO_2 and H_2O are used.
2. Food (carbohydrate) and O_2 are produced.
3. Energy from light is trapped in chlorophyll and in food.
4. ATP is produced by use of light energy (photosynthetic phosphorylation).
5. Hydrogen is transferred from water to NADP in the formation of NADPH.
6. ATP and NADPH are used primarily to drive reactions involving sugar synthesis.
7. Only chlorophyll-containing cells carry out photosynthesis.
8. Occurs only in the light.
9. Occurs in chloroplasts in eukaryotic plants.
10. Total photosynthesis must exceed total respiration for growth to occur.

Respiration

1. O_2 and food are used.
2. CO_2 and H_2O are produced.
3. Energy in food is released and may be temporarily stored in ATP or lost as heat.
4. ATP is produced by oxidation of food (oxidative phosphorylation).
5. Hydrogen is transferred from food to NAD or NADP to form NADH or NADPH.
6. ATP and NADH or NADPH are available to do many types of work in the cell.
7. Every living cell carries out respiration.
8. Occurs both in light and in darkness.
9. Glycolysis occurs in cytoplasm, while aerobic respiration occurs in mitochondria.

plant. If a plant is kept in the dark, continued food usage without food synthesis will rapidly deplete the available food. Stored food reserves, particularly starch and sugars, may be drawn upon by all living cells. If the plant is prevented from carrying out photosynthesis and if food is not supplied, the plant will eventually starve to death. In fact, any factor which limits photosynthesis and thus food availability must indirectly influence respiration.

Oxygen and Carbon Dioxide

Although most plant cells can continue for a time to oxidize foods even in the complete absence of gaseous oxygen (anaerobic respiration), molecular oxygen is normally necessary for the health of higher plant cells. Rarely does the concentration of oxygen in the atmosphere deviate enough from the normal 21 percent to appreciably affect the rate of respiration. However, underground stems, seeds, and roots may be in an oxygen-poor environment, since the microorganisms in the soil as well as the plant parts themselves may use the oxygen in the soil atmosphere faster than it is replaced from the air. Under these conditions, respiration in the roots may be decreased. Similar conditions of low oxygen level and high carbon dioxide concentration may occur in the internal cells in bulky plant organs such as large fleshy fruit. The effect of aeration on root growth is graphically shown in Fig. 11.3.

SUMMARY OF RESPIRATION

We may summarize respiration as shown in Fig. 14.2.

1. Sugar is prepared for oxidation through the process of phosphorylation, during which phosphorus is transferred from an organic phosphorus donor to the sugar molecule. Sugar cleavage then occurs, resulting in the production of two three-carbon-atom phosphate intermediates. These in turn are oxidized to the common intermediate, pyruvic acid.

2. If molecular oxygen is absent, pyruvic acid is changed to ethyl alcohol and carbon dioxide in the final stages of anaerobic respiration. Only small amounts of energy are released in this process.

3. If molecular oxygen is present, the pyruvic acid becomes further oxidized to carbon dioxide and water through the organic acid cycle. In this process (aerobic respiration), large amounts of energy are stored in energy-rich compounds that are used when work is done in the cell.

4. The two major functions of respiration are (1) the transformation of potential energy stored in food to an energy form readily available to do work (ATP) and (2) the production of intermediate products that are used in the synthetic reactions of the cell.

5. Energy-yielding oxidative reactions are coupled to energy-requiring reactions.

6. Respiration is carried out through an integration of reactions going on in the cytoplasm (glycolytic reactions) and the mitochondria (acid cycle and terminal oxidation reactions).

15

The Flower

The flower initiates the sexual reproductive cycle in all Anthophyta, and in so doing it terminates the growth of the shoot bearing the flower. Flowering poses some very interesting questions. (1) If flowering terminates the growth of a shoot, what provision is made for continued vegetative growth of a stem during and after flowering? (2) What determines the change from vegetative growth to flowering? (3) How are the flower parts adapted for sexual reproduction? (4) Are the flower parts modified vegetative leaves or is the flower a totally new structure unrelated to vegetative leaves? (5) What is the evolutionary history of the flower?

We have seen that in woody plants, provision is always made, through mixed buds or associated shoot buds, for continued vegetative growth at the same time that flowers are produced.

Flower induction of a vegetative apex is not yet understood. It seems as though the apical bud loses its ability to elongate so that the young floral apex has many crowded nodes with very short internodes (Fig. 15.25*A*). Furthermore, and more important, the potentiality for continued vegetative growth is lost. It is only regained after the fusion of sperm and egg cells (**gametes**), each with one-half the chromosome number of the parent plant. **Meiosis,** two cell divisions in which the chromosome number is halved, and **fertilization,** in which the normal vegetative chromosome number is restored, occur before vegetative growth is resumed. The function of the flower is to facilitate the formation and fusion of gametes.

The nature of the flower parts themselves, their relation to vegetative leaves, and their adaption for sexual reproduction form the subject matter of this and the succeeding chapter.

The evolutionary history of the flower is followed in some detail in chapters 27 to 30, which deal with the comparative morphology of the vascular plants. It is possible to trace a satisfactory evolutionary development of the floral parts of the Anthophyta back to leaf-like structures in very primitive vascular plants.

Of all the characteristics of flowering plants, the flower is the least affected by changes in the environment. For instance, we have seen that leaf shape

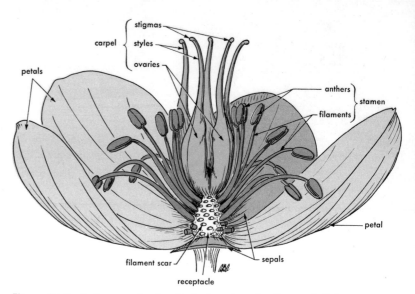

Figure 15.2. A diagram of a longitudinal section of a flower of Christmas rose (*Helleborus*). The perianth consists of two similar whorls; there are numerous stamens arranged in a spiral on a cone-shaped receptacle. Five separate carpels form the central whorl of floral parts.

is influenced by age, by light, by water, and by nutrition. Flowers and fruits are not so affected. This constancy of floral structures makes flowers and fruits the most important plant parts for classification.

In addition, flowers have a high esthetic value and the fruits and seeds which accompany them are of major importance in food production.

The essential steps of sexual reproduction, meiosis and fertilization take place in the flower. The complete sexual cycle involves: (1) the production of special reproductive cells, accompanied by meiosis, (2) pollination, (3) fertilization, (4) fruit and seed development, (5) seed and fruit dissemination, and (6) seed germination. The seed completes the process of sexual reproduction in the angiosperms and the embryo in the seed is the first stage in the life cycle of new individuals.

THE MORPHOLOGY OF THE FLOWER

Flowers are of many different forms. They also vary greatly in size, color, number of parts, and arrangement of parts. There are flowers so small that their organs are scarcely visible to the unaided eye; for example, the flowers of the duck-weeds, free-floating

plants common in ponds throughout the world. Then there are the flowers of a plant (*Rafflesia*), growing on the floor of dark tropical forests of the Malay archipelago, that are three to four feet in diameter.

A typical flower is composed of four whorls of modified leaves: (1) **sepals,** (2) **petals,** (3) **stamens,** and (4) a **carpel** or **carpels,** all attached to the **receptacle,** the modified stem end that supports these structures (Figs. 15.1, see color plate 5, 15.2).

The **sepals** enclose the other flower parts in the bud. Generally, they are green. All the sepals taken collectively constitute the **calyx;** that is, the calyx of a flower is composed of more or less distinct parts, the sepals.

The **petals** are usually the conspicuous, colored, attractive flower parts. Taken together, the petals constitute the **corolla.**

The **stamens** form a whorl, lying inside the corolla. Each stamen has a slender stalk or **filament** at the top of which is an **anther,** the pollen-bearing organ. The whorl or grouping of stamens is called the **androecium.**

The **carpel** or **carpels** comprise the central whorl of modified floral leaves (Fig. 15.1, see color plate 5; Figs. 15.2, 15.3). Collectively the carpels are

A

B

Figure 15.1. Flowers showing three different arrangements of flower parts. *A, Nymphaea odorata*, four sepals, many petals, numerous stamens, many carpels compressed in inner whorl; *B, Sedum* sp., five sepals, five petals, ten stamens, five carpels; *C, Lilium tigunum*, perianth consists of two whorls, each with three similar sepals and petals, six stamens, and three carpels coalesced in one pistil.

C

A

B

C

Figure 15.16. Elevation of floral parts. *A*, hypogyny in tulip flower (*Tulipa*); *B*, perigyny in almond flower (*Prunus amygdalus*); *C*, epigyny in daffodil flower (*Narcissus pseudo-narcissus*). *A* and *C* about × ⅓; *B*, ×1.

Color Plate 6

A

Figure 15.28. Floral guides for insects collecting nectar or pollen. *A*, viola, splashes of color and guide lines leading to throat of corolla; *B*, section of *Nasturtium* flower showing tubular spur for nectar and guide hairs leading to entrance to tube; *C*, mimicry—the flower of the orchid *Orphys* resembles a female wasp.

C

B

spoken of as the **gynoecium.** Each *individual structure* in the gynoecium is commonly referred to as a **pistil.** A pistil may be composed of a single carpel, or of several united carpels in the center of the flower. Thus a pistil may be formed from *one* or *more* carpels. The pistil may be dissected from a flower as a unit. When one carpel forms one pistil, a carpel may be physically separated from a flower as a whole unit (Figs. 15.2, 15.3, 15.5D, F). When several carpels form a single pistil, the pistil must be cut apart in order to separate out the carpels (Figs. 15.5H, 15.6B). The name "pistil" is ancient and is derived from the resemblance of numerous commonly occurring pistils to the pestle of the mortar-and-pestle set always present in the medieval pharmacy, where it was frequently used to compound medicines from floral parts. The term is a convenient one, but modern concepts of flower structure have rendered its meaning somewhat inexact.

There are generally three distinct parts to each pistil: (1) an expanded basal portion, the **ovary,** in which are borne the **ovules;** and (2) the **style,** a slender stalk supporting (3) the **stigma** (Figs. 15.2, 15.3B, 15.5). The pollen is deposited on the stigma.

The **receptacle** is the enlarged end of the flower stem or stalk to which the sepals, petals, stamens, and pistils are attached.

The term **perianth** is applied to the calyx and corolla collectively. It is frequently used to describe those flowers, such as the tulip, in which the two outer whorls, though present, are morphologically indistinguishable.

Sometimes individual flowers or compact clusters of flowers will have a whorl of small leaves or **bracts**

Figure 15.3. The essential floral organs. A, magnolia, showing many separate stamens and separate carpels arranged in a spiral on the receptacle; B, regal lily (*Lillium regale*), showing six stamens with anters and filaments, and three carpels in a single pistil with stigma, style, and ovary.

standing close below them. Such a collection of bracts subtending flowers is called an **involucre.**

The parts of the flower in outline are as follows:

Receptacle

Calyx, consisting of sepals ⎫
Corolla, consisting of petals ⎭ Perianth

Androecium, consisting of stamens
Gynoecium, consisting of carpels

Essential Organs Of The Flower

The essential floral organs are the androecium (stamens) and gynoecium (carpels) (Figs. 15.2, 15.3A, 15.5). They are essential because they produce reproductive bodies. The perianth, composed of calyx and corolla, is a protective covering of the stamens and pistil.

The Androecium

Each stamen consists of an anther supported on a stalk or filament (Figs. 15.2, 15.3B, 15.5). The anther usually has two longitudinal lobes united by a band of tissue. Each lobe has two longitudinal pollen sacs, within which the **pollen grains** are produced. Before the distribution of pollen, the tissue separating the two pollen sacs in each lobe of the anther breaks down. In most species a longitudinal slit then develops in the wall of each anther lobe. The edges of the slit separate and the pollen escapes through the opening thus formed.

The anther consists at first of a small mass of meristematic cells. Differentiated early in this mass are four separate groups of cells, called **pollen mother cells** (Fig. 15.4A, B). Each pollen mother cell undergoes meiosis by two successive divisions and forms four haploid cells, each with a single nucleus. These cells are commonly called **spores** (Fig. 15.4D). Shortly after they are formed the single nucleus divides by mitosis to form a two-cell haploid plant. The surrounding wall thickens and may become elaborately sculptured. This is the **pollen grain** (Figs. 15.4E, 15.27). The two cells are called the **generative** and **tube cells** (or sometimes nuclei), respectively (Fig. 15.4E). At this stage the pollen is generally shed and in some manner transferred to the stigma of adjacent flowers. This process is **pollination.** On the stigma the pollen grain germinates to form a **pollen tube.** Within the pollen tube the tube nucleus degenerates and the generative cell divides to form two **sperm** cells (Fig. 15.4F).

The Gynoecium

The structure of the gynoecium depends upon the number and arrangement of carpels comprising it (Fig. 15.5). In the pea flower (Figs. 15.5A, B, 15.6A, 15.7A), there is a single carpel forming the gynoecium (Fig. 15.6A). It consists of three parts, (1) the **ovary,** an expanded basal portion, and (2) the **style,** a slender stalk that ends in (3) a hairy irregular portion, the **stigma.** In the flower of the Christmas rose (Fig. 15.5C, D) there are five separate and distinct carpels comprising the gynoecium, and each carpel has its own ovary, style, and stigma (Figs. 15.2, 15.5D). In the flower of *Cotyledon* (Fig. 15.5F) there are again five carpels which, while not actually fused together, grow so compactly as to appear on casual observation to form a single central pistil. In the tulip (Figs. 15.5G, H) there are three carpels so completely fused that only a single structure is present to represent the central whorl of floral leaves. A single pistil composed of one carpel is called a **simple pistil** (Fig. 15.2). When two or more carpels fuse to form a pistil, it is known as a **compound pistil** (Figs. 15.5H, 15.6B).

The ovary is a hollow structure having from one to several chambers, or **locules** (Fig. 15.6). The number of carpels in a compound pistil is generally related to the number of stigmas (Fig. 15.5G, H), the number of locules (Fig. 15.6), and, sometimes, the number of faces of the ovary (Fig. 15.6B). The pea has a single stigma and the ovary has one locule (Figs. 15.5A, B, 15.6A). A similar situation holds for each pistil in the gynoecia of Christmas rose (*Helleborus*) and *Cotyledon* (Figs. 15.2, 15.5). There are three stigmas in the tulip gynoecium (Fig. 15.5H), the ovary is three-sided (Fig. 15.5H) and contains three locules (Fig. 15.6B); it is composed of three carpels.

Cross sections of the gynoecia of the garden pea and of the tulip are shown in Fig. 15.6. Assuming the carpel to be a leaf, the structure of these two gynoecia may be interpreted as follows. The single modified leaf of the pea gynoecium has been folded along its midrib (Fig. 15.6A). The margins of the leaf are thus brought together. They infold slightly, and regions just back from the margins are fused together. There is a single vascular bundle to represent the midrib and a vascular bundle for each of the fused margins. The ovules are attached to the margins. The mature pea pod will open by splitting

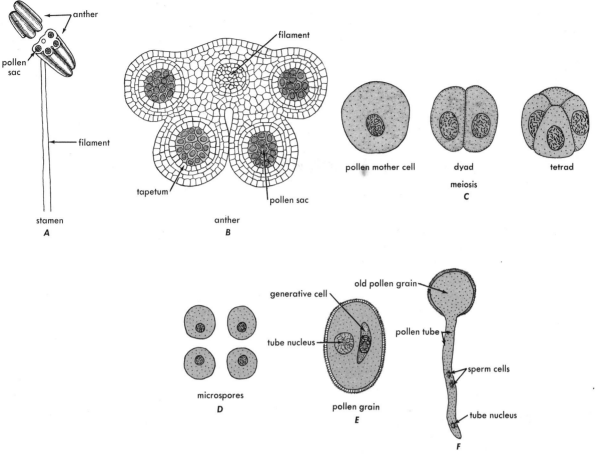

Figure 15.4. Development of pollen from a pollen mother cell to the pollen grain. *A,* stamen; *B,* cross section of anther; *C,* development of tetrad of cells from the pollen mother cell by meiosis; *D,* four microspores; *E,* pollen grain; *F,* germination of pollen grain into pollen tube.

both along the midrib and between the margins. These lines of splitting represent respectively, the **dorsal** and **ventral sutures.** The region where the ovules are attached is the **placenta** (Fig. 15.6*B,* 15.8). In the tulip there are three infolded carpels (Fig. 15.6*B*); the margins of all three carpels fold inward into the locules and fuse to form a central placental region. Thus, here too, the ovules are borne on the margins of infolded modified leaves. The midrib of each carpel is marked by vascular tissue, as are the margins. When the tulip ovary is mature it will split along the midrib or dorsal suture.

The following chart shows the relationship between the parts of the gynoecia in the flowers shown in Fig. 15.5.

Plant	Carpels	Pistil
Pea	1 carpel	1 simple pistil
Christmas rose	5 single and separate carpels	5 simple pistils
Cotyledon	5 distinct but closely appressed carpels	5 simple pistils
Tulip	3 fused carpels	1 compound pistil

In all of these cases the term gynoecium refers to the carpels taken collectively, without reference to their number or their manner of association with each other. The term simple pistil refers to a single carpel, while compound pistil denotes the union of several carpels to form a single structure.

The carpel as a modified leaf. The question arises as to why a carpel is considered a modified leaf.

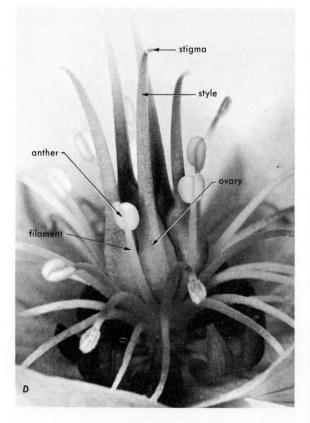

Figure 15.5. Flowers and essential organs. *A*, flower of garden pea (*Pisum sativum*); *B*, gynoecium of garden pea consisting of a single carpel; *C*, flower of Christmas rose (*Helleborus*) with stamens and carpels; *D*, maturing gynoecium of Christmas rose consisting of several loosely associated carpels; *E*, flower of *Cotyledon*; *F*, opened flower of *cotyledon*, showing stamens adnate to corolla tube and five separate compact carpels; *G*, flower of *Tulipa*; *H*, enlarged view of stamens and compound pistil of *Tulipa*. *A, C, G* ×½; *B*, ×2; *E, F, D, H* ×1.

Figure 15.5 (continued)

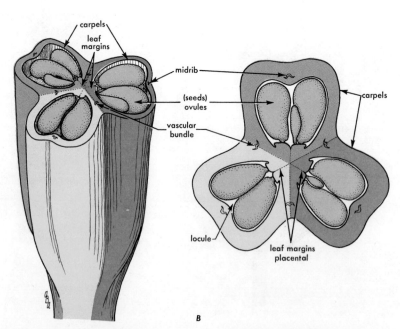

Figure 15.6. Diagram comparing gynoecia consisting of a single carpel and of three coalesced carpels. *A*, section of ovary of pea consisting of a single carpel; *B*, section of ovary of *Tulipa* showing three coalesced carpels.

There are two main reasons for so considering it. One of these reasons is its manner of development, which in early stages resembles that of a foliage leaf.

The second reason for considering the carpel a modified leaf concerns the evolutionary development of the flower. Fern leaves frequently bear spores or specialized reproductive cells resulting from meiosis (Chapter 17) on their margins or along veins (Fig. 27.28). If a young pea pod, which is a single carpel, is opened carefully along its ventral suture, the margins may be folded back, showing the seeds, which have developed from ovules, along the margin of a leaflike carpel (Fig. 15.7A). The situation is even more striking in *Sterculia platanifolia*, for in this plant there are five simple pistils united only by their stigmas which, when mature, open to show five very leaflike carpels bearing seeds on their margins (Fig. 15.7B, C). Carpels may thus be compared with the leaves of ferns, which are not only photosynthetic organs, but are reproductive structures as well. The evolutionary steps in the transition of leaves like those of the fern to the carpel of the flowering plants will be discussed in a later chapter.

Placentation. The tissues within the ovary to which the ovules are attached are called **placentae** (singular, **placenta**) (Fig. 15.6). The manner in which the placentae are distributed in the ovary is termed **placentation** and its determination is helpful in classification. When the placentae are on the ovary wall, as in the violet, currant, gooseberry, and bleeding heart (*Dicentra*) (Fig. 15.8D), the placentation is **parietal.** When they arise on the axis of the ovary which has several locules, as in lilies and fuchsia, the placentation is **axile** (Fig. 15.8C). Less frequently the ovules are on the axis of a one-loculed ovary, in which event the placentation is **central,** as in the primrose family (Fig. 15.8A, B).

Style and Stigma. The style is a slender stalk that terminates in the stigma. It is through stylar tissue

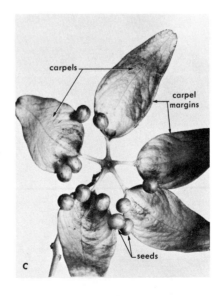

Figure 15.7. (*right*), Carpels suggestive of foliage leaves. A, pea pod (*Pisum sativum*), ×½; B and C, *Sterculia platanifolia*. B shows the young gynoecium with distinct ovaries and coalesced styles. ×1. In C the ovaries have opened, showing seeds attached to the margins of leaf-like carpels. ×½.

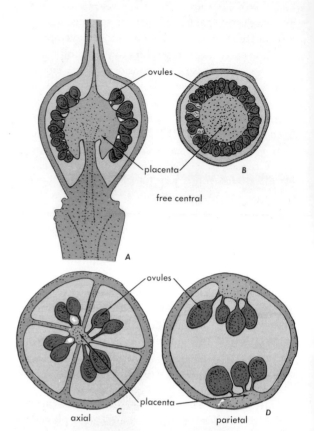

Figure 15.8. Three types of placentation. *A* and *B*, free central (primrose); *C*, axile (*Fuchsia*); *D*, parietal (*Dicentra*). (*A* and *B* redrawn from Priestley and Scott, *An Introduction to Botany*, fourth edition. Used with permission of Longmans Group Ltd.)

Figure 15.9. Inflorescences of corn (*Zea mays*). Left, long silks (styles) arising from the kernals (ovaries), many of which make up the ear (inflorescence). Right, stamen-bearing flowers of the tassel. × ⅙.

that the pollen tube grows. In some flowers the style is very short or entirely lacking; in others it is long. In corn (*Zea Mays*), the corn silks are the styles (Figs. 15.9, 15.20C, D). As a rule, the style withers after pollination, but in some plants (for example, *Clematis*) it persists and becomes a structure that aids in the dispersal of the fruit (Fig. 15.11B). The stigmatic surface often has short cellular outgrowths that aid in holding the pollen grains; and sometimes it secretes a sugary and sticky solution, the **stigmatic fluid.** In many wind-pollinated plants, such as the grasses, the stigma is much branched, or plume-like.

Structure of the ovule. The ovule, the structure that becomes the seed, first appears as a slight dome-shaped mass of cells upon the surface of the placenta (Fig. 15.25A). This group of cells develops into a mass of tissue called the **nucellus.** All of the cells at this stage are meristematic in nature. From the sides of the nucellus two ring-shaped layers of tissue usually arise; they grow upward and enclose the apex of the nucellus, except for a narrow opening at the end, which is the **micropyle.** These two layers are called the **inner integument** and **outer integument.** The integuments become the seed coats.

Embryo sac development. In a very young ovule all the cells that compose the nucellus are alike, as far as can be discerned by microscopic examination. Soon, however, one of the nucellar cells, at the micropylar end of the sac, becomes differentiated from the surrounding cells by its larger size and denser protoplasmic contents. This is the **megaspore mother cell** (Fig. 15.10B). The nucleus of this mother cell undergoes two successive divisions (**meiosis**), and as a result a row of four cells (the **megaspores**) is produced in the nucellus (Fig. 15.10C, D). As a rule, three of the cells (the ones nearest the micropyle) disintegrate and disappear, whereas the one farthest from the micropyle enlarges greatly. This megaspore, with its one nucleus and cytoplasm, now develops into the mature embryo sac. The usual stages are as follows: (1) *a series of mitotic divisions;* the megaspore nucleus divides, forming a two-nucleate embryo sac, each of these two nuclei divides forming a four-nucleate embryo sac, and then each of the four nuclei divides forming an eight-nucleate embryo sac; (2) *migration of nuclei;* after the first division the daughter nuclei migrate to opposite poles of the embryo sac, and after the last division one nucleus from the sets of four at the opposite poles of the embryo sac migrates toward the center; (3) *cell formation about nuclei;* the nuclei, with their surrounding cytoplasm, form cells. Six of these nuclei, each with its associated cytoplasm, become complete individual cells. The two polar nuclei form a single, binucleate cell. This results in a seven-celled embryo sac. These cells have names as follows: At the micropylar end of the embryo sac there is one **egg cell** associated with two **synergid cells.** Since it is frequently difficult to differentiate the egg cell from the other two, these three cells are sometimes referred to as the **egg apparatus.** The two nuclei that migrated centerwards approach each other; they are called **polar nuclei** (Fig. 15.10H), and form a binucleate cell called the **endosperm mother cell.** The three nuclei remaining at the end of the embryo sac opposite the micropyle form the **antipodal cells.** The embryo sac, which may be considered to be a seven-celled haploid plant, is now mature and ready for fertilization (Fig. 15.10H).

Fertilization

Fertilization is a most critical stage of development in the life cycle of all sexually reproducing plants and animals (Fig. 15.10I). It involves the union of two cells called **gametes.** In the flowering plants the gametes differ in size and form and are known as **sperms** and **eggs.** Fertilization is not completed until a sperm nucleus fuses with an egg nucleus to form a single nucleus. The cell resulting from the union of a sperm and egg is the **zygote** and its nucleus the **zygote nucleus.** This nucleus must have two sets of chromosomes, one contributed by the sperm and another contributed by the egg. In the higher plants, in addition to the union of a sperm and an egg, the nucleus of the second sperm cell present in the pollen tube unites with the two polar nuclei in the endosperm mother cell to form the **primary endosperm nucleus.** The cell containing the *primary endosperm nucleus* is the **primary endosperm cell.** The zygote arising from the union of egg and sperm will give rise to the embryo, which, having two sets of chromosomes, is *diploid.* The primary endosperm cell arising from the union of endosperm mother cell and sperm will develop into the **endosperm,** a food tissue of varying degrees of importance in different species. Note that the primary endosperm nucleus has three sets of chromosomes and is *triploid.* These relationships may be outlined as follows on page 280.

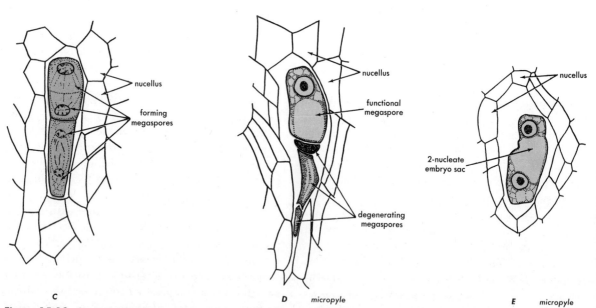

Figure 15.10. Stages in the development of the embryo sac. *A,* cross section of a flower bud of lily; *B,* an ovule, megaspore mother cell in prophase of meiosis; *C,* late telophase of second meiotic division; *D,* four megaspores, three degenerating; *E,* two-nucleate embryo sac; *F,* four-nucleate embryo sac; *G,* eight-celled embryo sac; *H,* mature embryo sac; *I,* fertilization, synergids degenerating.

nucellus

4-nucleate embryo sac

F

nucellus

8-celled embryo-sac

G   micropyle

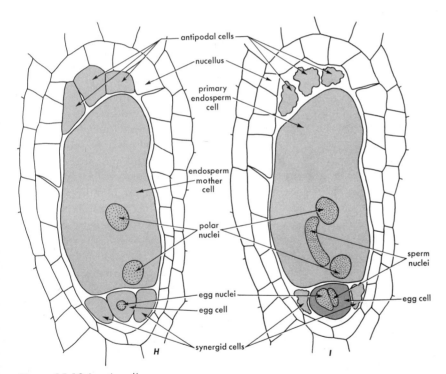

antipodal cells

nucellus

primary endosperm cell

endosperm mother cell

polar nuclei

sperm nuclei

egg nuclei

egg cell

egg cell

synergid cells

H   I

Figure 15.10 (*continued*)

| sperm | + egg | → zygote |
| one set of chromosomes (n) + | one set of chromosomes (n) | → two sets of chromosomes ($2n$) |

| sperm | + endosperm mother cell | → primary endosperm cell |
| one set of chromosomes (n) + | two sets of chromosomes ($2n$) | → three sets of chromosomes ($3n$) |

Note that two sperm cells are discharged from the pollen tube and that the nucleus of one of them fuses with the egg nucleus and the nucleus of the other fuses with the two polar nuclei. This constitutes *double fertilization,* as both fusions are necessary for the development of fruit and seeds. Double fertilization occurs only in the Anthophyta.

After fertilization, the synergids and antipodal cells usually degenerate, leaving the zygote and primary endosperm nucleus within the ovule (Fig. 15.10*H,* 16.21*A*). The other tissues constituting the ovule, the nucellus and the integuments, now enter into the further development of the ovule, eventually resulting in the seed (see Chapter 16).

VARIATIONS IN FLORAL STRUCTURE

General Description

Complete and Incomplete Flowers

The parts of a typical flower are sepals, petals, stamens, and carpels, all attached to the receptacle. A flower with all four sets of floral leaves is said to be a **complete flower.** An **incomplete flower** is one in which one or more of the four sets are lacking. For example, there are (1) flowers that have no perianth, (2) flowers that have a calyx but no corolla, (3) flowers with carpels but no stamens, and (4) flowers with stamens but no carpels. The flowers of the calla lily are examples of flowers without a perianth (Fig. 15.11*A*). In the goosefoot family, which includes spinach, and beet, the flowers have greenish sepals but no petals; but in *Clematis* and a number of other plants, although there are no petals, the sepals are colored and petal-like (Fig. 15.11*B*). Unisexual flowers (with either carpels or stamens but not both) are fairly common in the plant kingdom. Examples are corn (*Zea Mays*) (Figs. 15.9, 15.20), oak, walnut (Fig. 15.12*A, B*), willow, poplar, squash (Fig. 15.12*C, D*), hemp, asparagus, hop, and date palm.

Perfect and Imperfect Flowers

Unisexual flowers are either **staminate** (stamen-bearing) or pistillate (pistil-bearing). Unisexual flowers are said to be **imperfect,** whereas bisexual flowers are **perfect** or **hermaphroditic.** When staminate and pistillate flowers occur *on the same individual plant,* as they do in corn, squash, walnut (Fig. 15.12*A, B*), and many other species, the species, or the plant, is said to be **monoecious.** In corn (Figs. 15.9, 15.20), for example, the tassel (borne at the top of the stalk) consists of a group of staminate flowers, and the young ear is a group of pistillate flowers. When staminate and pistillate flowers are borne *on separate individual plants,* as in asparagus, willow, and many other species, the species, or the plant, is said to be **dioecious.** For example, in a commercial asparagus field, approximately half of the individual plants bear only staminate flowers, and half bear only pistillate flowers. In such circumstances, we speak of staminate or "male" plants and of pistillate or "female" plants. Only pistillate (female) plants bear fruit (berries). Another example of dioecism is the date palm (Fig. 15.13). Some individual palms are staminate, others pistillate. The edible fruit (date) is produced only by pistillate palms. In such instances there will be no fruit production unless staminate plants and pistillate plants are growing near enough together for the pollen to be transferred. In commercial plantings of dates most of the individual palms are pistillate, that is, fruit-bearing. Dates are propagated by offshoots, which arise chiefly near the base of the stem in the early years of the palm's life. Offshoots from a staminate palm grow into staminate palms, and offshoots from a pistillate palm into pistillate palms. In a commercial date garden it is economically desirable to have as many pistillate or fruit-bearing individuals as possible, and this is secured by vegetative propagation. A relatively few staminate palms are scattered throughout the garden—only enough to supply pollen. Pollen may be carried naturally by wind or it may be collected and dusted by hand on the pistils (Fig. 15.13*B*).

Some species produce three kinds of flowers: staminate, pistillate, and perfect (hermaphroditic). For example, in the red maple, one may find both unisexual and bisexual flowers in the same flower cluster.

Figure 15.11. Flowers with perianth parts lacking. *A,* spathe of calla lily (*Zantedeschia althiopica*), sepals and calyx absent; *B,* flower of *Clematis,* petals absent, sepals prominent.

Often staminate flowers have rudiments of a pistil, and pistillate flowers the rudiments of stamens.

Whorled and Spiral Arrangements of Flower Parts

In most flowers the sepals, petals, stamens, and carpels are in **whorls** or circles on the axis of the flower. For example, in *Sedum* (Fig. 15.1*B*) and *Cotyledon* the sepals arise in a circle at the outer lowest level on the receptacle; slightly above this circle are the petals; above these is a circle of stamens attached to the petals; and in the center, two united carpels. In contrast with this **cyclic** or **whorled** arrangement of flower parts is the arrangement in which one or more sets of flower parts are in **spirals.** Examples of flowers with spiral arrangement of certain flower parts are Christmas rose (Figs. 15.2, 15.5*D*) and magnolia (Fig. 15.3*A*).

Floral Symmetry

In many flowers such as the water lily (Fig. 15.1*A*), lily (Fig. 15.3*B*) and *Sedum* (Fig. 15.1*B*), the corolla is made up of petals of similar shape that radiate from the center of the flower and are equidistant from each other. Other flower parts have a similar arrangement. Such flowers are said to be **regular.** In these cases, even though there may be an uneven number of parts in the perianth, any line (Figs. 15.1*B,* 15.14*A*) drawn through the center of the flower will divide the flower into two similar halves. They may be exact duplicates or mirror images of each other.

Irregular flowers, such as the sweet pea (Fig. 15.5*A*) and the mints (Fig. 15.14*B*), have whorls either (1) with dissimilar flower parts, (2) with parts that do not radiate from the center, or (3) not equidistant from one another. In most of these flowers only one line will divide the flower in equal halves; the halves are usually mirror images of each other. Some flowers (bleeding heart, Dutchman's breeches), though irregular, may be bisected by any number of lines into similar mirror images.

The irregular bean or pea flower has a corolla composed of the following (Fig. 15.5*A*): one broad

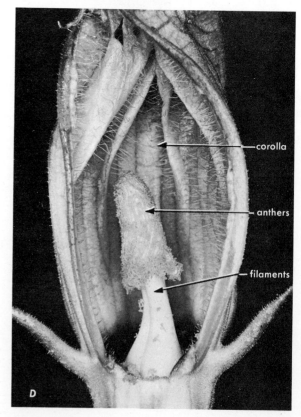

Figure 15.12. Imperfect flowers. *A*, Pistillate flowers; ×15; *B*, staminate flowers in catkins in walnut (*Juglans regia*), ×½; *C*, pistillate and *D*, staminate flowers of squash (*Cucurbita*), ×½.

Figure 15.13. Date palms (*Phoenix dactylifera*); A, commercial grove; B, artificial pollination of date flowers; a, strands of male flowers being placed in the center of the female cluster; b, freshly opened flower cluster ready for pollination; c, flower cluster after pollination, with the strands tied to hold male flowers in place. (A, courtesy of California Date Administrative Committee; B, courtesy of Nixon.)

conspicuous petal (the **banner** or **standard**); two narrower petals (**wings**), one on each side; and, opposite the banner, two smaller petals that are united along their edges to form the **keel**. Other examples of irregular flowers are mints, violets, orchids, and snapdragons.

Union of Flower Parts

Coalescence. In the flower of *Sedum*, illustrated in Fig. 15.1*B*, all parts of the flower are separate and distinct; that is, each sepal, petal, stamen, and car-

pel is attached at its base to the receptacle. In many flowers, however, the members of one or more whorls are to some degree united with one another, or are attached to members of other whorls. The union with each other of members of a given whorl is termed **coalescence.** Partial or complete union of the sepals along their edges occurs in the flowers of mints (Fig. 15.14*B*), violets, evening primroses, peas (Fig. 15.5*A*), and many other plants. This condition is referred to as **synsepaly.** Petals may also be attached to one another (Figs. 15.5*F,* 15.12*C,* 15.15*B, C,* 15.16*C*) and they are said to be **sympetalous** (the noun would be **sympetaly**). When the corolla is sympetalous it may have the form of a bell (canterbury bells, *Campanula*) or a tube (red hot poker, *Tritoma*) or a funnel (*Petunia*).

Stamens may coalesce, as is seen in the legume and cotton families (Figs. 2.4, 15.15*B*). This coalescence is **synandry.** In the cotton flower the filaments of many stamens are united to form a sheath, whereas the anthers are separate. In the flower of orange there are 20 to 60 stamens, united at their bases to form groups. And in the thistle family the anthers are united into a tube, whereas the filaments are distinct.

When the carpels of the gynoecium coalesce, the situation is called **syncarpy** and one may say that the pistil is compound. This condition occurs in many families of plants and may be seen to advantage in such flowers as those of lily (Fig. 15.3*B*), tulip (Figs. 15.5*H,* 15.6*B*) and squash (Fig. 15.12*C*).

When the parts of a flower have not joined, the prefix **apo-** (separate) may be used to describe the flower; for example, **apopetalous.** These situations may be simply outlined:

Figure 15.14. Floral symmetry. A, regular flower of columbine (*Aquilegia*); B, irregular flower of a mint (*Salvia*). × ½.

Whorl	No Coalescence	Coalescence
sepals	aposepalous	synsepalous, synsepaly
petals	apopetalous	sympetalous, sympetaly
stamens	apoandrous	synandrous, synandry
carpels	apocarpous	syncarpous, syncarpy

Adnation. We have just discussed the union of different members of the same whorl. Union of members of two different whorls also occurs, **adnation.** For example, in such flowers as *Cotyledon* (Fig. 15.5*F*), almond (Figs. 15.16*B,* 15.17*B*), snapdragons, and honeysuckle, the stamens are attached to the corolla rather than to the receptacle. In these cases the stamens are said to be adnate to the corolla.

Elevation of flower parts. In flowers of magnolia (Fig. 15.3*A*), lily (Fig. 15.3*B*), and tulip (Fig. 15.16*A,*

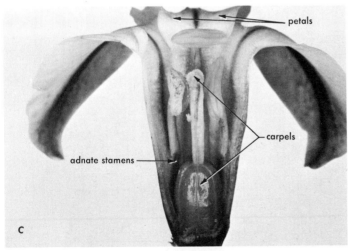

Figure 15.15. Coalescence, adnation, and reduction in number of floral parts. *A, Anemone,* many floral parts, no coalescence, no adnation; *B, Hibiscus,* reduction in number of perianth parts and coalescence of petals, stamens, and carpels; *C, Allium,* coalescence of perianth parts and of carpels; adnation of stamens to corolla tube. ×¼.

see color plate 6; Fig. 15.17A), the receptacle is convex or conical and the different flower parts are arranged one above another. They occur in the following order (beginning with the lowest): sepals, petals, stamens, and carpels. A flower of this type is said to be **hypogynous** (Figs. 15.2, 15.17A), even if the stamens are adnate to the corolla as in *Allium* (Fig. 15.15C). If the sepals, petals, and stamens appear to come from the top of the ovary, as in the

daffodil flower (Figs. 15.16C, 15.17C), and sunflower (Figs. 15.23, 15.24), the flower is said to be **epigynous.** Intermediate between the hypogynous and epigynous types are those flowers such as cherry and almond (Figs. 15.16B, 15.17B), in which the sepals, petals, and stamens have their origin, not above or below the pistil, but *around* the ovary. This type of flower is said to be **perigynous.**

The terms **inferior** and **superior** are frequently

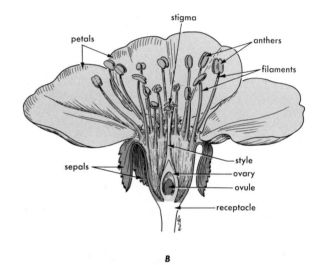

Figure 15.17. Diagram of elevation of floral parts in tulip flower. A, hypgyny (*Tulipa*); B, perigyny in almond flower (*Prunus amygydalus*); C, epigyny in daffodil flower (*Narcissus pseudonarcissus*).

Figure 15.18. Head, or spike, of barley (*Hordeum vulgare*). ×1.

the groups of flowering plants. We shall now proceed to describe the specialized flowers of two very important families, the Gramineae (grasses) and the Compositae (the sunflower family). Several members of the Gramineae are of great economic importance, supplying our single largest source of food directly in the form of flour and indirectly as feed for cattle, sheep, and hogs. The Compositae family, while not a major source of food products, is made up of a large number of species with a wide distribution, and colorful flowers; zinnias, marigolds, dahlias (Fig. 15.21), dandelions, and many similar flower types belong to this family.

The Grass Flower

Grass flowers generally grow in a **head** or **spike** (Fig. 15.18) as in wheat or barley, or are loosely grouped in a **panicle** (Fig. 15.19) typical of oats. In all instances the individual small flowers or **florets** are associated in groups known as **spikelets** (Fig. 15.19). Each spikelet is separated from its neighbors by two small modified leaves or bracts, known as **glumes.** They are found at the base of the spikelet and in some cases (wheat, oats, barley), they may fairly well enclose it. A separation of the glumes reveals the individual florets of the spikelet attached to a slender stalk, the **rachilla.**

Each floret is in turn protected by two additional bracts, the **lemma** and the **palea** (Fig. 15.19C). In wheat, barley, oats, and other grains, the lemma is large and may have a long slender **awn** attached to it. The palea is small and frequently enclosed by the lemma. The essential parts consist of the androecium composed of three stamens and the gynoecium of two fused carpels (Fig. 15.19D). When the flowers are mature the palea and lemma separate slightly, exposing two feathery stigmas and allowing the anthers to hang free from the spikelet on greatly lengthened filaments (Fig. 15.18). At the base of the floral parts, two small protuberances, the **lodicules,** may be distinguished. These are thought to be the greatly modified perianth parts and are supposed to function in the separation of the lemma and palea at pollination time.

In corn (Fig. 15.20), the flowers are imperfect, with the staminate flowers grouped in the tassel and the pistillate flowers eventually developing into the ear. The peculiar and very important difference between the corn flowers and those of other grasses is the absence of bracts of any sort around the pistillate flowers.

used to describe the relationship between the points of origin of the various floral parts. An inferior ovary is one that is completely or partially attached to the calyx tube, while a superior ovary is completely free from the calyx tube. Taxonomists frequently use the terms superior and inferior to describe the elevation of flower parts with particular reference to the position of the ovary. The words hypogyny, perigyny, and epigyny are employed more generally to designate the position of stamens, and less frequently of petals and sepals. For instance, the ovary is described as being superior, while the stamens may be described as being hypogynous or perigynous.

We have discussed in some detail the various aspects of flower structure as it occurs throughout

Figure 15.19. Diagram of inflorescence and flower of oats (*Avena sativa*). *A*, inflorescence; *B*, spikelet; *C*, floret; *D*, opened flower. *A*, ×1/10; *B,C,D*, about ×1.

The Composite Flower

As the name implies, the composite flower is a group of small flowers arranged to give the appearance of a single typical flower. Several types of composite flowers are shown in Fig. 15.21. This characteristic grouping of flowers is called a **head.**

In the dahlia (Fig. 15.21*B*), sunflower (Figs. 15.22, 15.23), and many other composite flowers, there are two distinct types of individual flowers, **ray flowers** and **disk flowers.** Some species may, however, have all ray flowers (cactus, dahlia, Fig. 15.21*A*), others all disk flowers (globe thistle, Fig. 15.21*C*). Ray flowers are frequently sterile or pistillate.

Figure 15.20. Flowers of corn (*Zea mays*). *A*, the tassel ×⅕; *B*, exserted anthers of staminate flowers, ×1; *C*, pistillate flowers forming a very young ear, ×½; *D*, several pistilate flowers with attached styles. ×1.

The disk flower is generally complete. It is epigynous, the small tubular corolla arising from the top of the ovary (Figs. 15.22, 15.23). There are two carpels, as evidenced by the forked stigma (Figs. 15.22, 15.23*C*, 15.24*B*) which, when receptive to pollen, protrudes from the corolla tube. The stamens, five in number, have separate filaments, but their anthers are coalesced to form a tube around the style. In sunflower, the stamens usually mature before the stigma and protrude from the corolla tube. These details are shown in Figs. 15.22, 15.23*B*, 15.24*A*. Referring to the figures, note that the flowers on the periphery of the flower head have been pollinated, and both stigmas and stamens have

Figure 15.22A

Figure 15.21. Three types of composite flowers. A, (*top left*) cactus dahlia (*Dahlia juarezii*), all ray flowers; B, (*center left*) dahlia (*Dahlia pinnata*), both ray and disk flowers; C, (*left*) globe thistle (*Echinops exaltatus*), all disk flowers. All ×½.

B

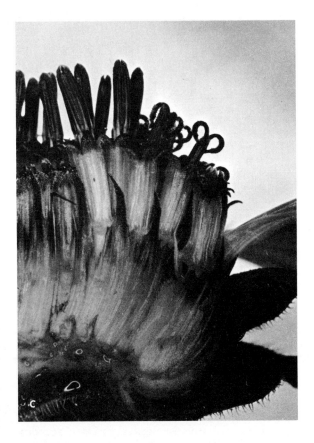

C

Figure 15.22. The composite inflorescence. A, head of golden glow (*Rudbeckia laciniata* ×½); B, median section of head of sunflower (*Helianthus annuus* ×1); C, enlarged view of flowers of sunflower head ×2. In all figures, fertilized flowers are outermost, the stigmas reflexed; the next inner row shows exserted recurved stigmas; next lies a circle of flowers with stamens ready to shed pollen; and in the center are the young unopened flowers of the head.

Figure 15.23. Individual flowers of a sunflower head (*Helianthus annus*). *A*, a ray flower; *B* to *E*, disk flowers. *B*, flower after fertilization, both stigmas and stamens have withered; *C*, flower with stigmas exserted and recurved; *D*, flower with staminal column exserted; *E*, unopened flower. ×2.

withered. Adjacent flowers show protruding stigmas dusted with pollen. Next come several rows of dark anthers capped by white masses of pollen. The flowers in the center have not yet opened. A calyx is not present as such but many composite flowers are surrounded by a conspicuous tuft of hairs or otherwise modified calyx parts. This is known as the **pappus.** In sunflower these are simply two small sepal-like structures (Fig. 15.24). In dandelion the pappus is feathery and aids in the dispersal of the fruit.

Two kinds of **floral bracts** sometimes occur in composite flowers. The whole head may be surrounded by green **involucral bracts.** Each individual flower may have its own **receptacular bract** arising from the receptacle near the base of the flower (Fig. 15.22*C*).

THE DEVELOPMENT OF THE FLOWER

As noted previously, the floral apex loses its ability to elongate and all of it eventually differentiates into

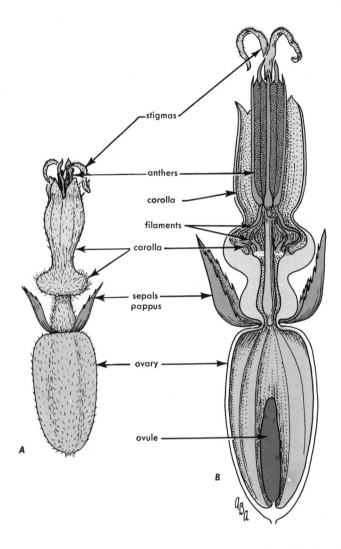

labels from top to bottom, left side:
stigmas
anthers
corolla
filaments
corolla
sepals
pappus
ovary
ovule

Figure 15.24. Disk flowers from a sunflower head. The pollen has been shed, and stigmas are raised and exserted. *A*, external view ×2½; *B*, enlarged longitudinal cross section ×4.

floral leaves. Moreover, the floral leaves are very much crowded together, not being separated by long internodes (as for foliage leaves), and they lack axillary bud primordia. Compare Fig. 7.9, which shows the shoot tip of flax, with Fig. 15.25*A*, which is a floral apex of *Ranunculus* (buttercup). As a rule the sepals are in a whorl; that is, they arise on the stem axis at approximately the same level. Petals and stamens also are usually in one or more whorls; and, if there are several carpels, they too may be in one or more whorls. In some of the more primi-

tive types of flowers, such as *Ranunculus*. (Fig. 15.25*A*), the stamens and pistils are in a close spiral on the receptacle. In these the distance between two successive individual stamens or individual pistils is very short; in other words, the internodes are very short.

The development of lettuce flowers (*Lactuca sativa*) illustrates, in a general way, some of the steps of floral development (Fig. 15.25*B*, *C*, *D*, *E*). The floral apex of lettuce, a composite flower, supports many flower primordia (Fig. 15.25*B*). Examination

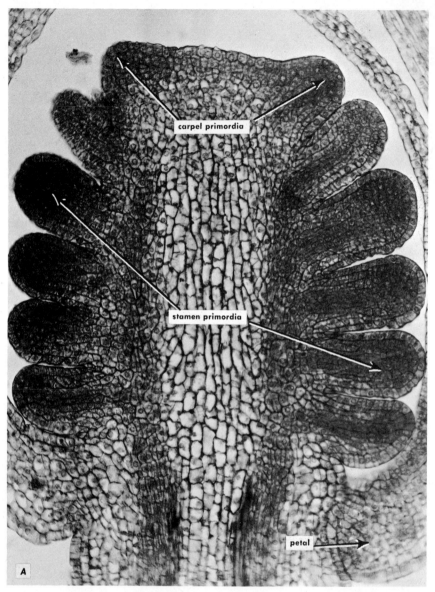

Figure 15.25A

of these primordia shows that petals appear first and rise above the developing stamens and carpels (Fig. 15.25C). Lettuce is epigynous and the two carpels form a depression in the center of the flower primordium (Fig. 15.25C, D). However, regardless of flower type, the carpels always originate as hollow flasks in the center of the flower primordia. As development continues the petals and stamens elongate. The carpels also elongate and their margins grow toward each other, finally fusing. The result is the hollow ovary which, at this early stage,

may already contain the primordium of an ovule (Fig. 25.25E). Upward growth of the carpel margins continues and results in the style and stigma.

INFLORESCENCES

In most flowering plants, the flowers are borne in clusters or groups. Morphologically, an inflorescence is a flower-bearing branch or system of branches. In manuals of flowering plants, there are many different terms descriptive of the various kinds of

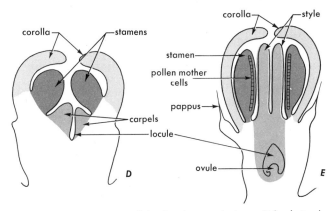

Figure 15.25. Development of the floral apex. A, (*opposite*), photomicrograph of floral apex of *Ranunculus* showing numerous stamen primordia and the earliest carpel primordia. ×300. B through E, diagram of the developing floral apex of lettuce (*Latica sativa*); B, young head showing several flower primordia; C, young flower showing carpel, stamen, and corolla primordia; D, slightly more advanced stage showing carpel just prior to fusion of the margins; E, carpel margins have fused, forming the hollow ovary in which occurs a primordium of an ovule; corolla tube stamens, and stigma are in early stages of differentiation. (A, courtesy of S. Tepfer; B to E redrawn after H. A. Jones, *Hilgardia*, **2,** 454.)

inflorescence in a flower-bearing branch or system of branches. Only the most common ones are discussed and illustrated here (Fig. 15.26).

Raceme Type of Inflorescence

A very simple type of inflorescence may be found in such plants as currant, hyacinth, wild mustard, and radish. It is called a **raceme.** In this type, note that the main axis has short branches, each of which terminates in a flower. Thus, each flower is on a short branch stem, the **pedicel.** The main axis of a raceme continues to grow in length more or less indefinitely; the apical meristem persists. The primordia of leaves arise in the usual manner along the margin of this apical meristem, and in the axil of each leaf a flower is borne. The oldest flowers are at the base of the inflorescence and the youngest at the apex. In such an inflorescence (mustard, for example), mature fruit may be near the base, while at the upper end are minute buds containing very rudimentary flower parts. In plants like cabbage, mustard, and others with inflorescences of this type, seeds may be shattering from the fruits at the base, while flowers are still forming near the tip.

In the simple raceme, the flowers are on pedicels which are about equal in length. In the **spike** (Fig. 15.26), the main axis of the inflorescence is elongated, but the flowers, each in the axil of a bract, are sessile (without a pedicel). The **catkin** is a spike that usually bears only pistillate or only staminate flowers. The inflorescence as a whole is shed later. Examples are willow, cottonwood, hazel, and walnut.

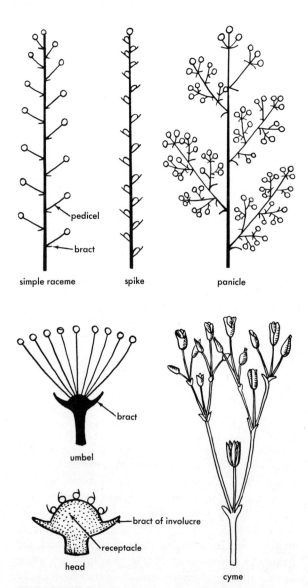

Figure 15.26. Diagram showing types of inflorescence.

In all the foregoing kinds of inflorescences, the flowering axis is elongated; that is, the internodes are quite long. If the internodes are very short, the flowers appear to be arising umbrella-like from approximately the same level. An inflorescence of this kind, in which pedicels are of nearly equal length, is called an **umbel** (Fig. 15.26). The onion is a good example. The **head** is an inflorescence in which the flowers are sessile and crowded together on a very short axis. Members of the family Compositae, including thistle, sunflower, cosmos, dahlia, etc., have this type of inflorescence (Fig. 15.26).

Inflorescences of the raceme type may be compound, that is, branched. A branched raceme is called a **panicle** (Fig. 15.26), as in oats and rice. Compound spikes occur in wheat, rye, and certain other grasses. Umbels, also, may be compound as in carrot, parsnip, and other members of the family Umbelliferae (which gets its name from its characteristic umbel inflorescence).

Cyme Type of Inflorescence

In contrast to the raceme types of inflorescences, just described, is the *cyme* (Fig. 15.26). In the cyme, the apex of the main axis produces a flower which involves the entire apical meristem; hence, that particular axis ceases to elongate. Other flowers arise on lateral branches farther down the axis of the inflorescence, and thus usually the youngest of the flowers in any cluster occurs farthest from the tip of the main stalk. The flower cluster of chickweed is an example of the cyme.

THE POLLEN GRAIN

The function of the external floral parts is the formation and transfer of pollen from anthers to stigmas of the same species of flower. The mature pollen grain (Fig. 15.4*E*), at the time of its discharge from the anther, consists of a two-celled male haploid plant enclosed in a thickened and elaborately sculptured cell wall (Fig. 15.27). The role of the male haploid plant is to produce the two sperm nuclei required for double fertilization. The function of the pollen grain wall is to protect the two-celled haploid plant as it is being transferred from the stamen to the stigma.

The elaborate cell walls of pollen grains present many intriguing problems (Fig. 15.27). Their intricate details are species-specific and their chemical constitution (sporopollenin) is so resistant to decay that pollen may still be present in fossil and sub-fossil plant deposits when all other traces of biolog-ical material have been lost. How do genes ex-ercise control over the deposition of such inert materials outside of the protoplast? The deposi-tion of the wall begins during meiosis and passes through a number of steps, some of which occur after the developing pollen grains lie free in the pollen sac in the anther. There is evidence of some change in ribosomes in developing pollen and of ribosome association with the plasmalemma. There is also evidence that the anther cells surrounding the pollen sacs may be influential in the wall pattern formation. The sculpturing of the pollen wall is so precise and consistent in its major fea-tures that it forms the basis for a well-developed field of **pollen taxonomy.**

POLLINATION

Pollination may be brought about by either **wind** or **insects,** and occasionally by water or by birds. Flower structure and the type of the pollen is gen-erally adapted to one or the other of these pollinat-ing vectors. In the case of insect pollination the adaptation may be very complex, indicating a long association between insect vector and plant.

Pollinating Vectors

Wind

Pollen is carried chiefly by wind and insects. Rarely, water and birds are the agents of pollina-tion. **Wind pollination** is the common type in those plants with inconspicuous flowers, as in the grasses (Figs. 15.19, 15.20), poplars, walnuts (Figs. 15.12A, B), alders, birches, oaks, ragweeds, and sage. Such plants usually produce pollen in enormous quan-tities. The flowers of grasses are inconspicuous, and most of them lack odor and nectar and hence are unattractive to insects. Furthermore, the pollen is light and dry and easily blown, and the stigmas, in many grasses, are feathery and expose a large surface to flying pollen. In cottonwoods, alders, birches, oaks, walnuts, and hickories, the flowers are in **catkins** (Fig. 15.12B). The staminate catkins are pendulous and move easily in the wind, and the light pollen is shaken from the anthers and readily

carried away by breezes. In many catkin-bearing trees, the flowers open before the leaves unfold so that pollen movement is unhampered. In pines, the pollen grains themselves have two winglike struc-tures (Fig. 28.22C), which assist in their dispersal by wind. Pine pollen is produced in such tremen-dous quantities that showers of pollen may be wit-nessed in a pine forest.

Living Vectors

The living vectors include a variety of insects, a few species of birds, and even some bats. Plants which are pollinated by living vectors usually share certain characteristics. They possess a colorful peri-anth which, together with the other floral parts, may be arranged into a complex architecture (Fig. 15.28; see color plate 6); and they produce both a sweet-tasting fluid, called nectar, and volatile com-pounds having unique odors. The work of many investigators has shown that these characteristics are highly adapted for the attraction of pollinators.

Bees, for example, are able to detect and distin-guish between many odors and colors and degrees of sweetness. Bees can distinguish at least three colors in addition to ultraviolet, and these colors include the most common perianth shades (yellow and blue). Bees are directed to nectar-producing glands by splashes or lines of contrasting color (Fig. 15.28A). When the insect attempts to reach the nectar or collect pollen, the floral architecture en-sures transfer by the insect of pollen to the stigma. In *Nasturtium* (Fig. 15.28B), for example, the nectar is held in a long narrow tube or spur. A foraging bee directs his proboscis down the tube and usually rubs it against the stigma along the way. In this way, pollen, which may have been picked up in earlier visits to other flowers, is deposited on the stigma.

Orchid flowers have many very specialized mechanisms to ensure pollination. The petal shape and color of one species (*Ophrys*) closely resemble the appearance of a female wasp (Fig. 15.28C). Male wasps, emerging from the pupal stage before the females, mistake the *Ophrys* flower for females. The male lands on the flower and attempts copulation; repeated pseudo-copulation results in pollination.

There is a very close interdedpendence between the life cycle of the commercial fig (*Ficus carica*) and that of its pollinator, the fig wasp (*Blastophaga psenes*). There are two types of fig trees, one which

Figure 15.27. Pollen grains as seen with the Stereoscan scanning electron microscope. A, whole pollen grain of *Pedilanthus tithymaloides*, ×4000; B to D (opposite), portions of surfaces of pollen grains: B, *Jatropha curcas*, ×2500; C, *Euphorbia bubalina*, ×2300; D, *Corton xalepensis*, ×2000; E, *Phyllanthus grandifolius*, ×6000. (Courtesy of M. J. York and G. L. Webster.)

produces edible fig fruit and one which produces only small, hard, inedible fruit which is nothing more than an incubator of fig wasp larvae.

An immature, inedible fig is an inflorescence. Small, unisexual flowers line the interior of a hollow, flask-shaped structure which has a pore at the enlarged end (Fig. 16.8). Male flowers lie near the pore and female flowers line the rest of the cavity (Fig. 15.29*A*). The female flowers have pistils with short styles (Fig. 15.29*B*), and female wasps are able to enter the inflorescence through the pore and deposit eggs in each ovary. The eggs hatch into larvae and later the larvae change to wasps, eat their way out of the ovaries, and mate inside the immature fig. The males usually remain inside and die, but the females escape through the pore, becoming dusted with pollen from the male flowers (Fig. 15.29*C*) as they do so.

Searching for ovaries in which to deposit eggs, the female may enter the inflorescence of a edible fig. This inflorescence differs from that of an inedi-ble fig by having only female flowers, and the pistils having much longer styles (Fig. 15.29*D*). The wasp is unable to oviposit because of the elongated styles, but she succeeds in pollinating the flowers. Seeds and an edible fig fruit will later result. Eventually, she will find an inedible fig inflorescence, lay her eggs, and start the cycle over again. A single branch of an inedible fig can be grafted to an edible fig tree, and it will provide sufficient numbers of pollinators for the rest of the tree.

Types of Pollination

There are essentially two different kinds of pollination, determined by the genetic similarity of the plants involved. If the anthers and stigmas, the essential organs in pollination, have the same genetic constitution, (that is, have chromosomes bearing identical genes, whether or not they are produced on the same or different individual plants), the transfer of pollen from anther to stigma

Figure 15.27 (continued)

is **self-pollination.** If two parent plants with different genetic constitutions are involved, the transfer of pollen from the anther of one to the stigma of the other is **cross-pollination.**

Forms of Self-pollination

1. Transfer of pollen from an anther to the stigma of the same flower. This is the normal method in cereals (except rye and corn), in garden peas, and in some other plants.

2. Transfer of pollen from anthers to stigmas of other flowers on the same plant. This occurs in many plants with perfect flowers and is usual in monoecious plants.

3. Transfer of pollen from anthers to stigmas of flowers on other plants.

(*a*) For centuries many horticultural plants have been reproduced vegetatively. For example, potatoes are usually propagated by cutting the tubers into a number of sections. Obviously, all the plants which grow from the tubers or sections of tubers from one plant have the same genetic composition.

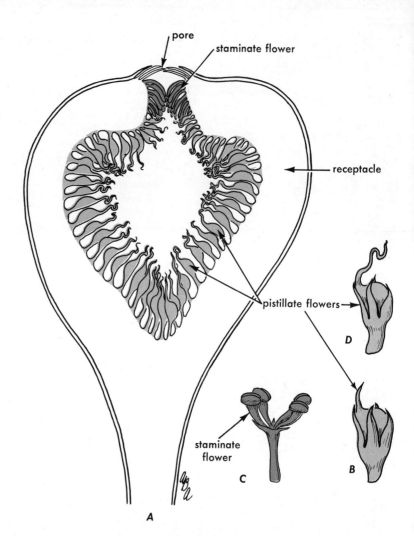

Figure 15.29. Diagram of a fig; A, B, and C, inedible caprifig; A, section of receptacle showing location of staminate and pistillate flowers ×1; B, short-styled abortive pistillate flower, ×3; C, staminate flower; ×3; D, fertile, long-styled pistillate flower of the edible Calimyrna fig, ×3. See also Fig. 16.8.

Such a group of asexually reproducing plants is called a **clone.** Another example: all the individual pears trees in an orchard may be parts of a single parent tree; that is, they may have been propagated by cuttings or buds from one parent tree. In this event, the individual pear trees have the same genetic composition. Pollination within clones is essentially self-pollination in that, although pollen goes from one plant to another plant, both plants have the same genetic constitution because they reproduced by vegetative means.

(*b*) The two individual plants may belong to the same pure line. In wheat, for example, the stigma of a flower usually receives pollen from its own anthers. Following fertilization, the ovary of the self-pollinated flower develops into a grain. The

plant that develops from this grain in turn produces many flowers. If these flowers are self-pollinated, the many individual plants that develop from the grains constitute a **pure line.** If pollen from one individual of a pure line is placed on the stigma of a flower of another individual of the same pure line, the result, genetically, is the same as when pollen is placed on the stigma of the flower that furnished the pollen.

Cross-pollination

Cross-pollination always involves two plants with different genetic constitutions. The differences may involve one or many characters. The two parents may be of the same variety or of the same species; they may be of different species or even of different genera.

GERMINATION OF THE POLLEN GRAIN

The pollen grain adheres to the stigmatic surface. In many flowers this surface has short outgrowths to which the grains adhere; and in some species the stigma produces a sticky secretion, the **stigmatic fluid.** The presence of this fluid, which may give the stigmatic surface a shining or glistening appearance, is evidence that it is receptive to pollen. The hybridizer soon learns to recognize a receptive stigma and thus the proper time to apply pollen artificially. The pollen grain germinates on the surface of the stigma. The protoplast of the grain absorbs water and swells, breaking the outer wall and forming a protoplasm-lined tube. The pollen tube is a slender threadlike growth; it penetrates the tissue of the stigma, grows down through the style (Fig. 15.30), and enters the ovary. The styles of some flowers are hollow, but most of them are not, and so the passage of the tube apparently involves the secretion of tissue-dissolving enzymes by the advancing tip of the tube.

In short-styled flowers the pollen tube needs to grow only a short distance. For example, in the pea and tulip, the style has a length of only two or three millimeters (Fig. 15.5B, H). On the other hand, in common corn, the distance from the stigmatic surface at the end of the corn silk to the young corn grain (ovary) may be as much as 50 centimeters (Fig. 15.20C). In a long style, the pollen tube is not a continuous, unbroken tube extending from the grain to the ovary. It consists of a very short apical portion containing the two sperm cells, and the upper part of the tube gradually disintegrates.

The rate of growth of pollen tubes varies widely. In wheat, for example, fertilization normally occurs between one and two days after pollination. Under favorable conditions the pollen tube of corn may grow six inches in 24 hours. In some oaks the tube is almost a year in growing two or three millimeters.

As a rule, when the pollen grain is shed from the anther it is a two-celled haploid plant, containing a generative cell and a tube cell (Fig. 15.4E). Before or during growth of the pollen tube, the generative cell undergoes division, forming two sperm cells (Fig. 15.4F). The two sperms, and in some species the tube nucleus, move to the tip of the pollen tube, retaining that position as the tube grows. In other species the tube nucleus degenerates either before the tube starts to grow or shortly thereafter. The two sperm nuclei bear the hereditary characters of the male parent.

The pollen of many plants will germinate in water. That of other plants requires a nutrient medium, such as a solution of cane sugar. For example, it was found that sugar beet pollen germinates abundantly on culture media containing $1\frac{1}{2}$ percent agar and 40 percent cane sugar. But the pollen of some species will not germinate in any artificial medium thus far tried.

Pollen varies considerably in the length of time it will remain viable, depending particularly upon the moisture and temperature conditions surrounding the grains. Plant breeders may desire to keep pollen viable for a considerable length of time in order to make use of it in cross pollination. It is sometimes shipped long distances for special hybridizing purposes. Pollen of the date palm will retain its viability for several months, if kept dry. Experiments show that sugar beet pollen will

Figure 15.30. Crushed stigma and style stained to show pollen grains and innumerable pollen tubes in the style, ×100. (Courtesy of Triarch Products.)

germinate fairly well after 50 days if kept at low temperature and low humidity. Apple pollen has been successfully stored for $4\frac{1}{2}$ years at 50 percent relative humidity and at a temperature of 2° to 8°C. Dry pollen will withstand greater temperature extremes than moist pollen. For example, pollen of apple, pear, and plum will withstand a temperature of 0.5°C, whereas as much as 50 percent of peach and apricot pollen grains is killed by exposure to this temperature.

Fertilization

Reaching the ovary, the pollen tube grows toward one of the ovules, usually enters the micropyle, penetrates one or more layers of nucellar cells, and enters the embryo sac (Fig. 15.10*I*), approaching, in one manner or another, the egg cell and the endosperm mother cell. These stages are not easy to study, but it appears that the tip of the tube ruptures, one sperm cell enters the egg, and the other sperm enters the endosperm mother cell. Within the egg, the sperm and egg nuclei fuse; within the endosperm mother cell, the other sperm and two polar nuclei fuse (Fig. 15.10*I*). This double fusion of egg with sperm and polar nuclei with sperm is sometimes called **double fertilization.** The zygote and primary endosperm cell have now been formed, the antipodal cells and synergid cells will degenerate, and conditions are set for the further development of seed and fruit.

Only one pollen tube functions in the fertilization of each embryo sac. If an ovary contains many ovules, each of course with one embryo sac, one pollen tube normally enters each ovule. Sometimes, however, fertilization of the egg is not effected, in which event it does not develop into an embryo. The presence of a hundred or more mature seeds in a watermelon fruit is evidence that at least as many individual pollen tubes grew down the style of the flower and discharged their contents into separate embryo sacs. As a matter of fact, however, usually many pollen tubes disintegrate at some point along their path of growth from the stigma to the embryo sac.

The immediate external evidence of fertilization is the withering of the stigma and style and, in many flowers, the dropping of the petals. If flowers are bagged and pollination prevented, the petals usually remain fresh for a much longer time than they do in pollinated flowers.

The fusions that occur in the embryo sac initiate a number of changes in the entire ovary, which are as follows.

1. Development of zygote to form the embryo plant.

2. Development of the primary endosperm cell to form the endosperm (reserve food supply of the seed).

3. Development of the integuments to form the seed coat.

4. Absorption or disintegration of nucellar tissue. In some plants, however, a portion of the nucellus may become the storage tissue of the seed rather than the endosperm. Such tissue of nucellar origin is called **perisperm.**

5. Development of ovary tissue to which ovule or ovules are attached to form the fruit.

6. Possible stimulation of accessory flower parts, such as the receptacle, sepals, or petals, to increased growth and incorporation into the fruit.

Thus, the importance of fertilization is to stimulate growth of certain floral parts and to bring about a withering of others. The net result is the development of the ovary wall into the fruit and the ovule into the seed.

Apomixis

This term refers to a type of reproduction in which no sexual fusion occurs but in which structures usually concerned in sexual reproduction are involved. Normally the egg nucleus will not start on the series of changes that result in the embryo plant unless a sperm nucleus fuses with it. In other words, in most plants, the embryo is the result of a sexual process (the union of male and female cells, gametes). Each gamete contributes n chromosomes, and hence the zygote and all the cells derived from it have $2n$ chromosomes. Rarely, however, an embryo develops from an *unfertilized egg*. This occurrence is called **parthenogenesis.** Two types of parthenogenesis are as follows: that in which the embryo develops from an n or haploid egg cell (the individuals are usually sterile) and that in which the embryo develops from a $2n$ or diploid egg cell. In the latter type meiosis is omitted in the development of the embryo sac. Another form of apomixis is that in which the embryo plant arises from tissue surrounding the embryo sac. These so-called "adventitious" embryos (diploid) occur in *Citrus, Rubus,* and other plants.

Apomixis has variations other than those given

above, but all forms of this phenomenon involve the origin of new individuals without nuclear or cellular fusion.

In some plants deposition of pollen on the stigma is a prerequisite to apomictic embryo development, even though no pollen tube grows down the style and no nuclear fusion takes place. In such instances, there is evidence that hormones formed in the stigma, or furnished by the pollen, are transferred to the unfertilized egg cell, initiating there those changes which result in embryo development.

SUMMARY

1. The flower, the distinguishing structure of the Anthophyta, is formed of four whorls of leaves specialized to carry out sexual reproduction, including pollination, fertilization, and seed production.

2. A floral apex has lost its ability to elongate and its potentiality for vegetative growth. The ability for vegative growth is not regained until the completion of the two nuclear phenomena, meiosis and fertilization.

3. The four whorls of floral leaves are (a), the calyx, composed of sepals; (b), the corolla, composed of petals; (c), the androecium, composed of stamens; and (d), the gynoecium, composed of carpels.

4. There are two parts to a stamen, the pollen-producing anther and the filament.

5. A single unit of the gynoecium is frequently called a pistil. If it consists of one carpel it is a simple pistil; if it consists of two or more fused carpels it is a compound pistil.

6. A pistil consists of three parts, the stigma, style, and ovary. The stigma is receptive to pollen, and the ovary encloses the ovules. At the time of fertilization, the ovules consist of integuments, nucellus and embryo sac.

7. Meiosis takes place both in the anthers and in the ovule.

8. Pollen may be considered a two-celled haploid plant protected by a cell wall which is frequently elaborately sculptured. Two sperms are produced by one of the two cells.

9. The embryo sac may be considered an eight-celled haploid plant. Two of its cells, the egg cell and the primary endosperm cell, are stimulated to development by fusion with sperm cells.

10. Pollination is the transfer of pollen from the anthers to stigmas. A pollen tube growing down the style to the embryo sac serves as a pathway by which the two sperms reach the embryo sac.

11. Floral architecture facilitates pollination, which is most frequently carried out by wind or by insect vectors.

12. Double fertilization, the union of sperm with the egg cell and of the second sperm with the primary endosperm cell, occurs only in the Anthophyta.

13. The embryo plant develops from the zygote. The endosperm, developing from the fusion of sperm and primary endosperm nucleus, supplies some nourishment for embryo and sometimes for seed development.

14. Variation in floral architecture arises (a), from variation in number of parts in a given whorl; (b), from regularity in floral parts; (c), by coalescence of floral parts; (d), by adnation of floral parts; (e), by elevation of floral parts; and (f), by the presence or absence of certain whorls.

15. Flowers are either solitary on flower stalks or grouped in various inflorescences such as heads, spikes, catkins, umbels, panicles, racemes, or cymes.

16. The primordia of the floral leaves are crowded on the floral apex. The carpels, the central whorl, originate as depressions or cups; their margins fuse during floral development, enclosing the ovules.

17. The potentiality for vegetative growth has been transferred from the apex to the ovule.

18. There are essentially two types of pollination, cross-pollination, which results in much variation in the progeny, and self-pollination, which results in progeny having great similarities.

19. Occasionally seeds may be formed without fertilization. This is a form of vegetative reproduction and is known as apomixis. Navel orange plants reproduce by apomixis.

The Fruit, Seed, and Seedling

THE FRUIT

A fruit is a ripened ovary of a flower. Sometimes other floral parts may be closely associated with the mature ovary and thus be part of the fruit. In many cases fruit development depends upon pollination. Sometimes a large amount of inactive, or dead, pollen will stimulate fruit development, and indoleacetic acid, or other growth substances, may substitute for pollen. In the pineapple, banana, and navel orange, fruits normally develop without pollination. Without pollination there can be no fertilization, so these fruits must generally be seedless, and frequently they differ from normal seeded fruits. Fruits are auxiliary structures in the sexual life cycle; they occur only in the Anthophyta. Their development has unquestionably been an important factor in the successful evolution of land plants. Fruits protect seeds, aid in their dissemination, and may be a factor in timing their germination. Fruits are highly constant in structure and are thus very important plant parts in the classification of the Anthophyta. In everyday usage the term "fruit" usually refers to a juicy and edible structure, such as an apple, plum, peach, cherry, orange, or grape. Such structures as string beans, eggplant, okra, squash, and cucumbers, which are commonly called "vegetables," and the "grains" of corn, oats, wheat, and other cereals are not popularly thought of as fruits. However, all the above are **fruits** in the botanical sense.

Figure 16.1 shows changes in the appearance of an almond as it matures from the flower to the fruit. Note that the petals fall, leaving the calyx tube with attached stamens intact (Fig. 16.1*B*). Soon the developing ovary bursts the calyx tube, after which the ovary rapidly enlarges and differentiates into three distinct layers. If this ovary is split longitudinally while still green (Fig. 16.1*C*), several regions become clearly visible. An unripened seed is present in the center. The ovary wall, which is called the **pericarp,** is divided into a firm **endocarp,** a spongy **mesocarp,** and a skin or **exocarp.** As the maturing almond dries, the mesocarp and exocarp split (Fig. 16.1*D*). These parts, commonly called the shucks, fall away, leaving the mature almond of commerce which is comprised of the endocarp or shell and the enclosed

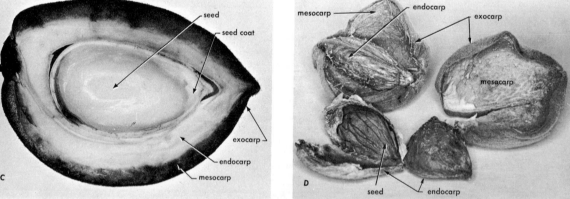

Figure 16.1. A fruit is a ripened ovary. The ovary of an almond flower (*Prunus amygdalus*) develops into the almond. *A*, almond blossom; *B*, developing ovary bursting calyx tube; *C*, section of a maturing fruit showing divisions of pericarp and seed (*A, B, C* ×1). *D*, mature almonds showing split outer portion of ovary wall (shucks) and intact inner portion (shell) ×½.

seed. (Fig. 16.1*D*) Strictly speaking, no fruits are produced in any of the plant groups except the flowering plants (Anthophyta).

Development of the Fruit

The development of the fruit (pod) of pea and bean is here described as an example of the various changes that may occur during the transformation of the ovary into fruit. The pistils of bean and pea flowers are each composed of one carpel (Fig. 16.2), that is, one ovule-bearing leaf. The ovules are attached along the ventral suture, along which the edges of the carpel are fused; the dorsal suture corresponds to the mid-rib of the carpel. At the time of fertilization, the ovary and the ovules within are small. A cross section of the bean ovary (Fig. 16.3) shows the ovary wall, the carpellary bundles, the ovule, and the locule. The ovary wall may be divided into three distinct layers, (1) an outer epidermis, (2) an inner epidermis, and (3) a middle zone consisting of several layers of cells. There are three carpellary bundles, one on each side of the ventral suture and one on the dorsal side opposite them.

Fertilization initiates a series of changes in the embryo sac and other tissues of the ovule which lead to the development of the seed. The stimulus of fertilization extends not only to the ovule; other parts of the flower are also influenced. The tissues of the ovary wall undergo marked changes. The three layers of the ovary wall, however are evident in the mature fruit. The fruit wall (developed from the ovary wall) is called the **pericarp,** and the three more or less distinct parts, in order beginning with the outermost, are the exocarp, mesocarp, and endocarp. When the pod is mature, floral structures, such as the pedicel, the calyx, withered stamens, and even remnants of the corolla may also be present. Quite generally, withered remains of the style and stigma persist at the tip of the ovary (Fig. 16.2). Seeds have developed with the matured ovary.

In all kinds of mature fruits the various structures and tissues can be traced back to those of the ovary or ovaries, and to any other floral organs that may constitute a part of the fruit.

Mature pods usually show the presence of small, underdeveloped (abortive) ovules. It is probable that these ovules were not fertilized. If none within the bean ovary is fertilized, the ovary does not enlarge. In most plants, normal fruit development takes place only if pollination is followed by fertilization

and if fertilization is followed by seed development. The fruits of tomato are sometimes distorted and irregular in form; this may be due to the lack of seeds in one or more locules and to the resultant poor development of such locules.

Parthenocarpy

The stimulus of fertilization usually exerts an influence on all parts of the flower, particularly on the ovary and its contained ovules. In Fig. 16.4 normal and seedless fruits of tomato are compared. Note the greatly increased development of the fruit that accompanies normal seed development. In some plants, *normal fruit development may take place without fertilization.* Such unfertilized fruits may or may not be seedless depending upon the occurrence of parthenogenesis and the subsequent development of adventitious (page 302) embryos. On the other hand, fertilization may occur, but the ovules may fail to develop into mature seeds even though fruit may form in a normal fashion. In a strict sense only those fruits that develop without fertilization are called **parthenocarpic fruits.** Practically, it is not always possible to know without extensive study, whether a given fruit is parthenocarpic. For instance, Thompson seedless grapes were thought to be parthenocarpic until it was shown that fertilization takes place and the ovules fail to mature into seeds. In many citrus fruits the seed formed by fertilization is accompanied by seeds with adventitious embryos. To complicate definitions further, the sexually formed seed is not as robust as the adventitious seed and frequently does not develop. This situation has led to a loose use of the word parthenocarpy to mean simply seedless fruits.

Parthenocarpic (seedless) fruits are quite regularly produced in such cultivated plants as the English forcing cucumber, certain varieties of eggplant, the navel orange, the banana, the pineapple, and some varieties of apple and pear.

In certain plants seedless fruits may be induced by pollen that is incapable of fertilizing the ovules. For example, in some orchids, the placing of dead pollen or a water extract of pollen upon the stigma may start fruit development.

Recently, parthenocarpy has been induced in some plants by spraying the blossoms with dilute aqueous solutions of certain growth substances. Commercial application of this technique is being made.

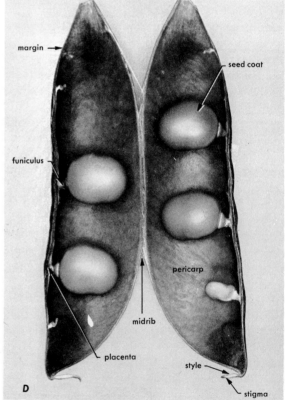

16 *The Fruit, Seed, and Seedling*

ventral carpellary bundles

ovule

locule

ovary wall

dorsal carpellary bundle

Figure 16.3. Cross section of bean ovary after fertilization ×20.

Figure 16.2 (*left*). Development of legume fruits. *A*, bean (*Phaseolus vulgaris*) from flowers to young pods, ×1; *B*, pistil from a pea (*Pisum sativum flower*), ×5; *C*, pea pod, unopened; *D*, opened pea pod showing developing seeds attached to carpel margins, ×1.

Kinds of Fruits

The flower of garden beans has a **single ovary** that is composed of one carpel. After fertilization, the seeds develop, the ovary enlarges, and the ovary wall becomes a dry, parchmentlike hull. When the fruit is mature, it splits open along both edges (sutures). The fruits of peas and beans are called **legumes** or **pods** (Fig. 16.2). Thus, the characteristics of the legume or pod are as follows: (1) it is derived from a single ovary, (2) the ovary is composed of one carpel, (3) the pericarp is dry, and (4) the pericarp usually splits along both sutures at maturity.

The lily type of flower has a single ovary, but, unlike that of the bean, it is composed of three united carpels (Fig. 15.1*C*). As in the bean, the ovary wall develops into a dry structure, each locule of which splits open at maturity, allowing the seeds to escape. This type of fruit is called a **capsule** (Fig. 16.5).

These two types of fruit are discussed to point out some of the characters used in describing and classifying fruits. In classifying the different kinds of fruits, taken into account are: (1) the structure of the flower from which the fruit develops, (2) the number of ovaries involved in fruit formation, (3) the number of carpels in each ovary, (4) the nature of the mature pericarp (dry or fleshy), (5) whether or not the pericarp splits (**dehisces**) at maturity, (6) if dehiscent, the manner of splitting, and (7) the part that sepals or receptacle may play in the formation of the mature fruit.

The so-called **simple fruits** are those derived from a *single ovary*. They may be dry or fleshy; the ovary may be composed of one carpel or of two or more carpels; and the mature fruit may be dehiscent or indehiscent. On the other hand, **aggregate fruits** and **multiple fruits** are formed by *clusters* of simple fruits. The difference between these types of fruits depends upon the number of flowers involved in their formation. In the strawberry (Fig. 16.6) and blackberry (Fig. 16.7) there are many simple fruits, each derived from the individual ovaries of a *single flower*. These matured ovaries or fruits are all attached to a common receptacle. Such groupings constitute *aggregate* fruits, and the simple fruits comprising them may be classified according to the scheme of classification of simple fruits given below. The mulberry, fig, and pineapple are *multiple* fruits. The multiple fruit consists of the enlarged ovaries of *several flowers*, more or less grown together into one mass. In some plants,—mulberry, for example—associated floral structures form a part of the fruit, and in the fig the receptacle enlarges to become the sweet edible portion (Fig. 16.8).

Simple Fruits

Legume or pod. This type of fruit, characteristic of nearly all members of the pea family (Leguminosae), arises from a single carpel, which at maturity generally dehisces along both edges (Figs. 16.2 and 16.9). In the bean or pea pod, the "shell" is the pericarp, and the "beans" or "peas" are the seeds. Pods may be spirally twisted or curved as in alfalfa. A number of legumes such as honey locust, alfalfa,

and bur clover have pods that do not dehisce.

Follicle. Examples of the follicle are the fruits of Christmas rose and magnolia (Fig. 16.10*A* and *B*). The follicle develops from a *single carpel,* and opens along *one suture,* thus differing from the pod, which opens along both sutures.

Capsule. The capsule is derived from a compound ovary, that is, an ovary composed of two or more *united carpels.* Each carpel produces a few to many seeds. Capsules dehisce in various ways: (1) *lengthwise* (*Amaryllis, Iris,* and *Datura,* Fig. 16.11*A,* *B* and *C*); (2) by *pores* toward the top of each carpel (poppy, Fig. 16.11*E*); (3) by a *transverse lid* (plantain, Fig. 16.11*H*).

Figure 16.4. Tomato (*Lycopersicon esculentum*) fruits. A and B, normal development after fertilization; B, cross section; C and D, seedless or parthenocarpic fruits; D, cross section. × ¾.

Figure 16.5. Capsule of lily (*Lilium regale*). *A,* nearly mature capsule; *B,* view of three stigmas; *C,* section; *D,* section separated into portions of three carpels; *E,* longitudinal view showing ovules; *F,* dehisced capsule. ×1.

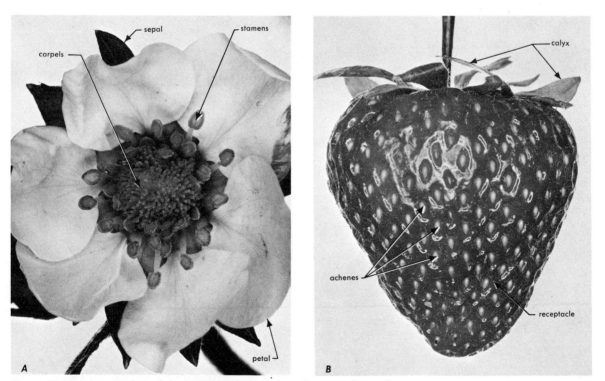

Figure 16.6. Aggregate fruit of the strawberry (*Fragaria*). *A,* flower; *B,* fruit, achenes with enlarged receptacle. ×2. (Courtesy of R. S. Bringhurst.)

Figure 16.7. Aggregate fruit of the blackberry (*Rubus ursinus*). *A*, flower; *B*, flower shortly after petals have fallen; *C*, section of fruit. ×2.

16 *The Fruit, Seed, and Seedling*

Figure 16.8. The fruit of the fig (*Ficus carica*); *A*, end of fleshy receptacle, showing pore, which is lined with staminate flowers; *B*, receptacle turned inside out showing pistillate flowers that have matured into small drupes. ×1.

Figure 16.9. Pod of Scotch broom (*Cytisus scoparius*). *A*, unopened; *B*, partially split; *C*, dehisced down both sutures.

Figure 16.10. Follicles. *A, Magnolia,* each carpel shown in Figure 15.5D has now developed into a follicle, × ½; *B,* Christmas rose *(Helleborus),* five follicles present in one flower, × ¾; *C,* dehiscing follicle of a South American milkweed *(Oxypetalum caeruleum).*

Silique. This is the type of fruit (Fig. 16.12) characteristic of the members of the mustard family (Cruciferae). One interpretation suggests that the silique is a dry fruit derived from a superior ovary consisting of two carpels. At maturity the dry pericarp separates into three portions; the central persistent portion has the seeds attached to it.

Achene. Examples of this type of fruit are buckwheat, strawberry (Fig. 16.6) and members of the Compositae family, such as sunflower. These fruits are commonly called "seeds," but as in sunflower (Fig. 16.13), a carefully broken pericarp reveals the seed within, attached only by its funiculus to the placenta. This pericarp may be separated easily from the seed coat, that is, from the layer of cells just beneath it. Achenes are indehiscent.

Grain or caryopsis. This is the fruit of the grass family (Gramineae), which includes such important plants as wheat, barley, oats, rye, corn, and rice. Like the achene, the grain is a dry, one-seeded, indehiscent fruit (Fig. 16.14). It differs from the achene, however, in that the pericarp and the seed coat are firmly united all the way around and it is difficult to separate the two except by special milling processes.

Figure 16.11 (*right*). Capsules. *A, Amaryllis; B, Iris; C, Datura; D,* poppy capsule, side view before dehiscence; *E,* mature poppy capsule dehiscing by pores at the top; *F,* cross section of poppy capsule, showing its many carpels; *G,* mature capsule of plantain; *H,* capsule of plantain opening by transverse lid; *I,* capsule of tulip dehiscing lengthwise; *J,* cross section of tulip ovary. All approximately × 1. (*A, B,* and *C,* courtesy of W. Russell; *D, E,* and *F,* redrawn after Korsmo.)

wall of capsule

wall of capsule

Figure 16.12. Siliques. *A, B,* of stock (*Matthiola incana*): *A,* unopened, ×¾; *B,* partially split to show orientation of valves and septum, ×¾; *C, D,* of moonwort (*Lunaria annua*), *C,* unopened, ×¾; *D,* opened to show valves and septum, ×1. (Courtesy of W. Russell.)

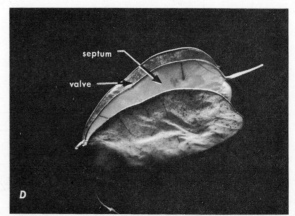

Samara. This is a dry, indehiscent fruit, which may be one-seeded, as in the elm, ash, and "tree of heaven" (*Ailanthus*), or two-seeded, as in maple (Fig. 16.15) and box elder. These fruits are typified by an outgrowth of the ovary wall, which forms a winglike structure.

Schizocarp. This is the fruit characteristic of the carrot family (Umbelliferae), which includes such common plants as carrot, parsnip, celery, and parsley. The schizocarp is a dry fruit that consists of two carpels which split, when mature, along the midline into two one-seeded indehiscent halves. The carrot "seed" of commerce, for example, is the one-seeded half of the schizocarp. The one true seed in each carpel completely fills the whole cavity and is usually grown fast to the pericarp.

Nut. The term nut is popularly applied to a number of hard-shelled fruits and seeds. A typical nut, botanically speaking, is a one-seeded, indehiscent dry fruit with a hard or stony pericarp, the shell. Examples are chestnut, walnut, hickory nut, acorn (Fig. 16.16), hazelnut, and beechnut. The chestnut develops within a bur or prickly involucre. An acorn, the fruit of the oak, is partially enclosed by a hardened involucral cup. The husk or shuck of the walnut, which has been removed before the product reaches the market, is composed of involucral bracts, the perianth, and the outer layer of the pericarp; the hard shell is the remainder of the pericarp. The unshelled almond, sometimes called a nut, is, in reality, a drupe, from which the hull (exocarp and mesocarp) has been removed. Brazil nuts are seeds, not fruits. And the unshelled peanut is a pod, corresponding to the fruit of bean or pea; the edible portions are the true seeds within the pod.

Drupe. Examples of this type of fleshy fruit are plum, cherry, almond, peach, and apricot, all of which are members of a subfamily in the Rosaceae (rose family). The olive fruit is also a drupe. The drupe is derived from a *single carpel*, and is usually one-seeded. However, if one examines the young ovary of a flower of the almond (Fig. 16.1), cherry, plum, or other members of the group to which they belong, two ovules will be found, one of which usually aborts (fails to develop into a seed) (Fig. 16.17). The drupe has a hard endocarp consisting of thick-walled stone cells. The exocarp is thin, forming the skin, and the mesocarp forms the edible flesh. The pit or stone of a cherry, for example, is one seed, with thin seed coats, surrounded by the stony endocarp. In other words, the pit is composed

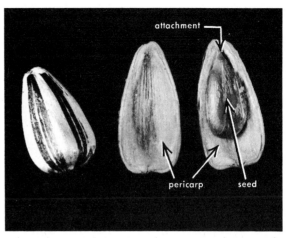

Figure 16.13. Achene of sunflower (*Helianthus annuus*), unopened, and opened to show attachment of seed. ×1.

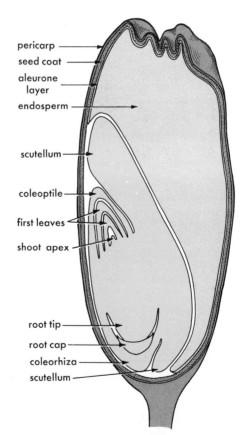

Figure 16.14. Median lengthwise section of mature corn (*Zea mays*) grain.

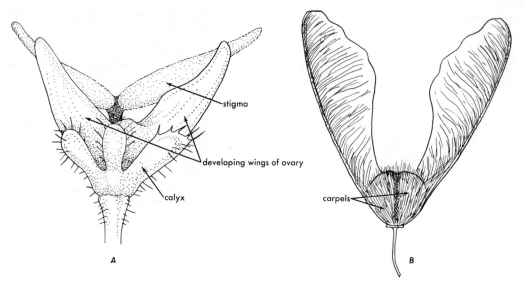

Figure 16.15. Samara or key fruit of maple (*Acer*). *A,* flower; *B,* mature fruit. (Redrawn from A. M. Johnson, *Taxonomy of Flowering Plants.* Copyright 1931, 1959 by Meredith Corporation. Used with permission of Appleton-Century-Crofts.)

Figure 16.16. Acorns of oak (*Quercus*). The fruits are a type of nut and the cups are fused involucral bracts. ✕ ¾.

16 The Fruit, Seed, and Seedling

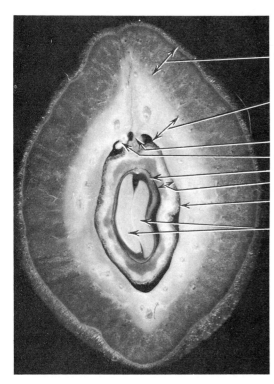

mesocarp

endocarp

funiculus
abortive ovule
nucellus
integuments
ovary cavity

cotyledons

Figure 16.17. Median transverse section of young almond fruit (*Prunus amygdalus*) (drupe). (From R. M. Holman and W. W. Robbins, *A Textbook of General Botany*, John Wiley and Sons.)

of the seed plus the stony inner layer (endocarp) of the ovary wall.

In the almond fruit (Figs. 16.1 and 16.17) the mesocarp is fleshy like a typical drupe when the fruit is young, but it becomes hard and dry as it develops, and forms the hull. The shell of the almond is endocarp. In the coconut, also a drupe, the pericarp tissues become dry at maturity.

Berry. This fleshy type of fruit is derived from a *compound ovary*. Usually many seeds are embedded in a flesh, which is both endocarp and mesocarp, although the line of demarcation may be difficult to discern (Fig. 16.18). The tomato is a common example of a berry. The wild form of the garden tomato has a two-celled fruit; cultivated forms have several locules in the fruit, and the placentae are fleshy (Fig. 16.4).

The date is a one-seeded berry, derived from a three-carpelled ovary. After pollination, one of the carpels rapidly enlarges and suppresses the other two, which soon dry up. The stone of the date is the seed, and it does not include a stony endocarp as do the drupes of peaches, plums, cherries, and the like.

The citrus fruit (lemon, orange, lime, and grapefruit) is a type of berry called a **hesperidium.** It has a thick, leathery rind (peel), with numerous oil glands, and a thick juicy portion composed of several wedge-shaped locules (Fig. 16.19). The peel of a citrus fruit is exocarp and mesocarp; the pulp segments are endocarp. The juice is in pulp sacs or vesicles; they are outgrowths from the endocarp walls (Fig. 16.19*C*), and each mature vesicle is composed of many living cells filled with juice. The common sweet orange usually has ten two-seeded locules, each representing a carpel. The navel orange is a seedless variety, an example of a parthenocarpic fruit.

Another berrylike fruit is that of members of the family Cucurbitaceae, which includes the watermelon, squash, pumpkin, cantaloupe, cucumber, and others. It is called a **pepo** (Fig. 30.11). The outer wall (rind) of the fruit consists of receptacle tissue that surrounds and is fused with the exocarp. The flesh of the fruit is principally mesocarp and endocarp.

Pome. This type of fruit is characteristic of a subfamily of Rosaceae, to which belong the apple (Fig.

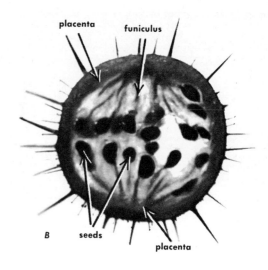

placenta   funiculus

B   seeds

placenta

Figure 16.18. Gooseberries (*Ribes*). True berries, A, typical appearance, ×¾; B, cross section, showing parietal placentation, ×1¼.

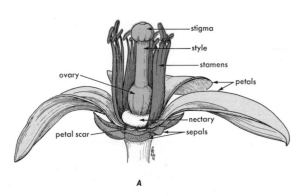

stigma

style

stamens

ovary

petals

nectary

petal scar

sepals

A

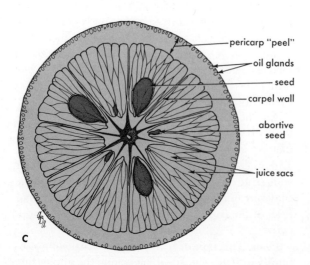

pericarp "peel"

oil glands

seed

carpel wall

abortive seed

juice sacs

C

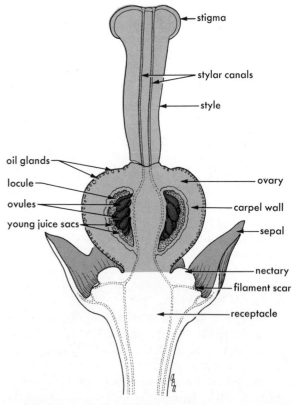

stigma

stylar canals

style

oil glands

locule

ovules

young juice sacs

ovary

carpel wall

sepal

nectary

filament scar

receptacle

B

Figure 16.19. Flower and fruit (hesperidium) of orange (*Citrus sinensis*). A, flower, dissected to show pistil and stamens; B, lengthwise section of maturing ovary; C, cross section of mature fruit.

16.20), pear, and quince. Cross and lengthwise sections of the mature apple fruit are shown in Fig. 16.20. Various interpretations have been given as to the morphological nature of the pome. The one accepted here is that most of the flesh is derived from the **floral tube** (fused bases of sepals, petals, and stamens) and that the five-carpelled ovary is fused with this tube; that is, the fleshy exocarp is united with the tissues of the floral tube. The so-called core line is the outer limit of the carpels. The endocarp is parchment-like, and both mesocarp and exocarp are fleshy and not easily distinguished one from the other.

Aggregate Fruits

An aggregate fruit is one formed from the numerous carpels of one individual flower. These fruits, considered individually, may be classified as types of simple fruits. Thus the fruits of the strawberry are achenes and the fruits of the raspberry and blackberry are small drupes (Fig. 16.7). The strawberry flower and fruit are shown in Fig. 16.6. The flower has numerous separate carpels on a single receptacle. The ovary of each carpel has one ovule, and the ovary develops into a one-seeded dry fruit (achene). The receptacle to which these fruits are attached becomes fleshy; the whole structure, which we call a strawberry, is an aggregate of simple fruits, each an achene. The receptacle is stem tissue and consists of a fleshy pith and cortex with vascular bundles between them. The so-called hull of the strawberry fruit is composed of the persistent calyx and withered stamens. The achenes are usually spoken of as seeds.

The flowers of dewberry, raspberry, blackberry (Fig. 16.7), and other species of *Rubus* have essentially the same structure as those of strawberry. In these flowers the many separate carpels attached to the one receptacle develop into small drupes (Fig. 16.7C), instead of into achenes. An aggregate fruit is derived from a *single flower* having many simple pistils (carpels). The various kinds of aggregate fruits differ in the structure of their individual fruitlets, which may be classified as simple fruits.

Multiple Fruits

A multiple fruit is one formed from the individual ovaries of several flowers. These fruits, considered individually, may be classified as types of simple fruits. The fruits of the mulberry, fig, and pineapple are examples of multiple fruits, and the individual fruits composing them are nutlets in the mulberry and fig and parthenocarpic berries in the pineapple.

The flowers of mulberry are of two kinds: staminate and pistillate. The pistillate flowers have a deeply four-lobed calyx surrounding a single pistil, the ovary of which is one-seeded. The ovary develops into a nutlet and is enclosed by the persistent juicy calyx lobes. The separate fruits become very much crowded together, making up a collection of fruits that commonly go by the name mulberry.

The fig fruit we eat is an enlarged fleshy receptacle (Fig. 16.8). The flowers are very small and are attached to the inner wall of this receptacle. Both staminate and pistillate flowers occur and may be borne in the same or in different receptacles. The pistillate flower has a single one-celled (and one-seeded) ovary, surrounded by a two- to six-parted calyx (the corolla is absent). Each ovary develops into a nutlet that is embedded in the wall of the receptacle. Thus, the fig is derived from many flowers, all attached to the same receptacle. Common edible figs (Mission and Kadota) are parthenocarpic. Smyrna figs and several other varieties do not mature their fruits unless fertilization and seed formation have taken place (Fig. 15.29).

In the pineapple of commerce, the edible portion is largely the greatly thickened pulpy central stem in which the berries are embedded. Tissues, other than the ovary wall, which form part of a fruit are referred to as **accessory.** Thus, much of the fleshy fruit of pineapple, apple, and strawberry is accessory.

Summary of Fruits

1. A fruit is a ripened ovary, plus any other closely associated floral parts.

2. The initiation of fruit development generally depends upon pollination and fertilization.

3. Fruit development is accompanied by the development of the ovule into the seed.

4. There are *three different kinds of fruits,* classified on the basis of the number of ovaries and flowers involved in their formation:

(a) *simple fruits,* derived from a single ovary;

(b) *aggregate fruits,* derived from a number of ovaries belonging to a single flower and on a single receptacle;

(c) *multiple fruits,* derived from a number of ovaries of several flowers more or less grown together into one mass.

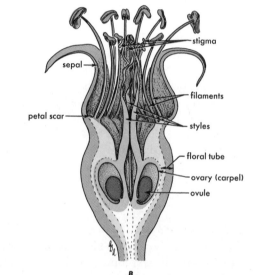

Figure 16.20. Flower and fruit (pome) of apple (*Malus sylvestris*). A, median longitudinal section of flower; B, median longitudinal section of a maturing fruit; C, median section of fruit; D, cross section of fruit. (C, D, redrawn from W. W. Robbins, *Botany of Crop Plants.* Copyright 1931 by Blakiston. Used with permission of McGraw-Hill Book Company.)

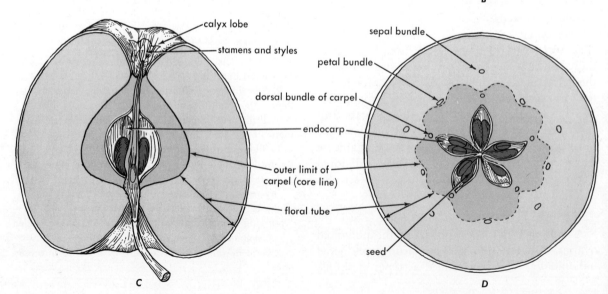

5. Simple fruits may have a dry pericarp or a fleshy pericarp. If the pericarp is dry it may be dehiscent (splitting at maturity, allowing seeds to escape) or indehiscent (not splitting).

THE SEED

Seed Development

The seed completes the process of reproduction initiated in the flower, and it always consists of an embryo surrounded by seed coats. The embryo developed from the zygote, and the seed coats developed from the integuments of the ovule. Recall (page 277) that at the time of fertilization the egg cell, the primary endosperm cell, the synergids, and the antipodals constituted the embryo sac. The embryo sac is surrounded by the nucellus and all are enveloped by one or two integuments. This complete structure is the ovule (Fig. 15.10B, 15.10H, 16.21A). Double fertilization involves (1) the fusion of egg and sperm nuclei to form the zygote nucleus, and (2) the fusion of polar nuclei with a second sperm nucleus to form the primary endosperm nucleus (Fig. 16.21B).

After fertilization, the zygote nucleus remains quiescent for a time. The primary endosperm nucleus, however, divides rapidly and soon builds up an **endosperm tissue** which, at first, may lack cell walls (Fig. 16.21C). Division of the zygote nucleus awaits the development of the endosperm. The first cell divisions of the new generation result in a filament of from four to eight cells (Fig. 16.21D). Now the cell closest to the micropyle elongates, pushing the other cells of the filament into the endosperm. At about the same time, the cell furtherest from the micropyle divides at right angles to the axis of the filament (Fig. 16.21E). This is the first in the series of divisions that will finally result in the embryo of the seed. Further divisions result in a spherical or globose stage (Fig. 16.21F). The major distinction between the embryos of monocotyledons and of dicotyledons is the presence of one or of two cotyledons. There is always a root apex and a shoot apex. In the dicotyledons, two cotyledons overshadow the shoot apex (Fig. 16.21H), while in the monocotyledons only a single cotyledon rises above the shoot apex (Fig. 16.21G). The course of events just described is very general, for there are many variations in details of embryo development.

While the development of the embryo is taking place, the nucellus, endosperm, and integuments are also undergoing changes which are characteristic of the group of plants to which the seeds belong. In the great majority of plants, the nucellus and endosperm are required only for the initial stages of development. This is particularly true of the nucellus, which is generally used up in the early stages. It persists as a food storage tissue, the **perisperm,** in the seeds of the sugar beet and a few other species. The persistence of the endosperm as a food reserve occurs in many monocotyledons (Fig. 16.21G), is highly developed in the grasses, and is of economic importance (Fig. 16.14). The endosperm persists as a food storage tissue in relatively few dicotyledonous seeds. The seeds of the castor bean (*Ricinus communis,* Fig. 16.23) and of coffee (*Coffea arabica*) are examples of a persistent endosperm in seeds of dicotyledons.

In most seeds the integuments of the ovule become the hard and horny seed coats of the mature seed.

Stimuli in the Course of Seed Development

The seed thus completes the process of sexual reproduction initiated in the flower. This development has been marked by a series of crises in which further development is blocked and cannot proceed without precise stimulation. Meiosis in the anthers results in pollen—dormant cells which are stimulated to produce pollen tubes only when brought into contact with a specific stigmatic surface (Fig. 15.4, 15.28). The polar nuclei do not undergo further development until fusion with a sperm nucleus takes place. The egg nucleus does not divide until it fuses with a sperm nucleus, and even then, the first zygote division may still await the stimulus provided by an endosperm tissue (Fig. 16.21). The seed, too, blocked from further development, is a dormant structure, awaiting definite stimuli before (Fig. 16.21) it again becomes active in germination.

The nature of the zygote, coupled with the blocks to growth which occur at various stages of development, poses definite questions and has stimulated much experimentation. What keeps the zygote from dividing? What stimulates the zygote to divide? What induces the changes in cell polarity to produce changes in the form of the developing embryo? What is responsible for dormancy in seeds? How

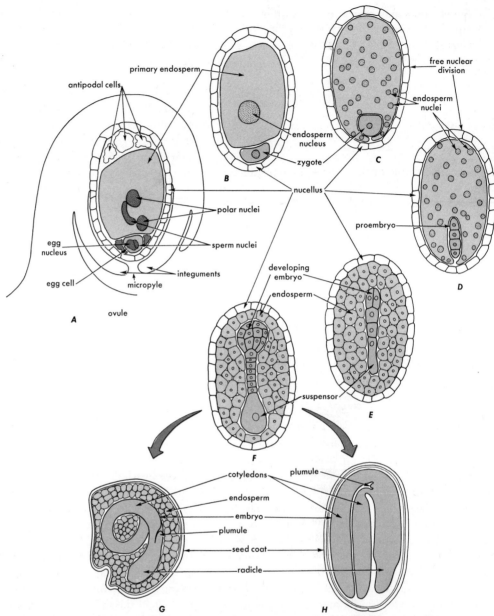

Figure 16.21. Diagram of the development of an angiosperm seed. *A,* ovule after fertilization; *B,* after fusion of gamete nuclei to form zygote and endosperm; *C,* free nuclear divisions in the endosperm; *D,* filamentous stage of the proembryo; *E,* elongation of suspensor cell and division of proembryo cell; *F,* globular stage of the embryo; *G,* embryo of a monocotyledon in seed; *H,* embryo of a dicotyledon in seed.

long can a seed remain dormant? What breaks seed dormancy?

Kinds of Seeds

The seed provides a highly efficient mechanism to: (1) carry plants over periods unfavorable for growth; (2) maintain a reserve supply of potential plants in case of successive unfavorable seasons; and (3) disseminate plants. Plants are unusually profligate in the production of reproductive cells. The tremendous production of seeds insures the renewal or continuance of plant populations. Many animals, including man, cooperate in this population renewal and dissemination.

Table 16.1	*Production of selected fruits in the United States, 1960–1968*				
CROP	1960	1962–1966°	1967	1968	1000 UNITS
apples		5,930	5,425	5,400	tons
grapes		3,719	3,069	3,579	tons
oranges		115,619	183,610	124,820	boxes
corn (grain)	3,906,949			4,374,840	bushels
4 feed grains	149,527			168,082	tons
wheat	1,354,709			1,570,433	bushels
4 food grains	44,318			53,029	tons
59 crops		289,008	300,446	293,900	acres
corn, grain		57,517	60,557	55,707	acres
all wheat		47,677	58,771	55,309	acres

*average

Food storage. The large amount of food stored in fruits and seeds may be simply illustrated by considering the production of fruits and seeds in the United States. Table 16.1 shows the production of apples, grapes, oranges, grain for animal consumption, and grain for human consumption in selected years since 1960. Note that the unit of measurement varies. The first seven items are units of production, the last three are acres.

The seeds of the angiosperms differ in two ways: (1) they have one or two cotyledons and (2) they store food either in the embryo, in the endosperm, or more rarely, in nucellar tissue, the **perisperm,** which lies between the embryo and seed coat. In most dicotyledons, the food is stored in the cotyledons (Fig. 16.22). The castor bean and coffee bean are examples of dicotyledonous seeds in which food is stored in an endosperm (Fig. 16.23). The endosperm is the principal food storage tissue in the seeds of monocotyledons (Figs. 16.14, 16.25).

When food storage occurs within the embryo, the normal vascular tissues of the embryo convey the solubilized food to meristematic regions, where it is required for growth. Food stored in the endosperm, outside the embryo, is in many cases absorbed through normal protodermal or epidermal cells. In other cases the cotyledon becomes specialized to act as an absorbing organ. In lilies, onions, and others, the tip of the cotyledon is modified for food absorption. The seed coats of those seedlings with an enclosed endosperm remain attached to the tip of the cotyledon for some time after it has emerged (Fig. 16.31). In most grasses the entire cotyledon has become a highly specialized absorbing organ known as the **scutellum** (Fig. 16.24).

Common bean. The fruit of bean is a pod. Within it are the seeds (beans). The external characters of the bean seed are more clearly seen after soaking it in water. The points of interest are the **hilum, micropyle,** and **raphe** (Fig. 16.22). The hilum is a large oval scar, near the middle of one edge, left where the seed broke from the stalk or funiculus when the beans were harvested. The micropyle is a small opening in the seed coat (integument) at one side of the hilum and was observed in the ovule, as the opening through which the pollen tube entered the ovule. The raphe is a ridge at the side of the hilum opposite the micropyle. It represents the base of the funiculus, which is fused with the integuments. Conducting tissue present in the funiculus will be continued in the raphe. At the end of the raphe, this conducting tissue may fan out over the ovule or seed and lose its identity as conducting tissue. This region is known as the **chalaza.** It is always present at the end of the ovule or seed opposite the micropyle. An elevation or bulge of the seed coat adjacent to the micropyle marks the position of the radicle (embryo root) within the seed.

When the seed coat of a soaked bean is removed, the entire structure remaining is the embryo; no endosperm is present. The following parts of the embryo can be observed: (1) the shoot consisting of (*a*) two fleshy cotyledons, (*b*) a short axis, the hypocotyl, below the cotyledons, and (*c*) a short axis, the epicotyl, above the cotyledons, bearing several minute foliage leaves and terminating in a shoot tip; and (2) the root or radicle. In Fig. 16.22, the bean seed is opened to show the relationship of some of the parts just mentioned.

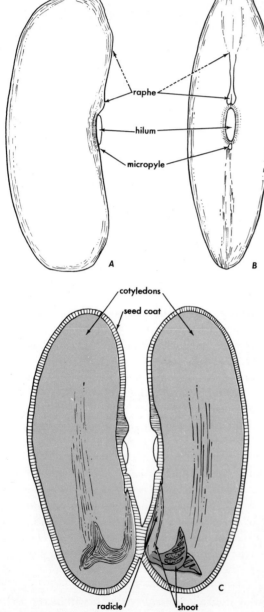

Figure 16.22. Bean seed. *A*, external side view; *B*, external face or edge view; *C*, embryo opened.

Castor beans. The external points of interest of this seed are: (1) the **caruncle,** a spongy structure, an outgrowth of the outer seed coat; (2) the hilum and (3) the micropyle, which are beneath the caruncle; and (4) the raphe, which runs the full length of the seed (Fig. 16.23).

The chalaza region is marked, in the castor bean, by a protuberance at the end of the seed opposite the caruncle. In contrast to the bean seed, the castor bean seed has a massive endosperm in which the embryo is embedded. The shoot of the embryo consists of (*a*) two thin cotyledons with conspicuous veins, (*b*) a very short **hypocotyl** (shoot axis below the cotyledons), and (*c*) a minute **epicotyl** (shoot axis above the cotyledons). The root portion of the embryo axis consists of a small radicle.

Grasses. Wheat is a member of the grass family (Gramineae). The so-called seed of corn, or of any other grass, is in reality a fruit (grain), as explained on page 314. It is a one-seeded, dry, indehiscent fruit with the pericarp (ovary wall) firmly attached to the seed; the pericarp and seed coats are so firmly attached to each other and to other tissues of the grain that it is impossible to peel them off as one can the seed coats of the common bean and castor bean. Therefore, in order to study the internal structure of the grain, it is necessary to employ sections.

A longitudinal section of the grain of corn is shown in Fig. 16.14. Note here that the endosperm constitutes the bulk of the grain. The endosperm is composed of (1) an outermost layer (single row of cells) known as the **aleurone layer** and (2) a **starchy endosperm.** The cells of the aleurone layer contain proteins and fats but little or no starch. The starchy endosperm cells are filled with starch grains.

The embryo of grasses is quite unlike that of most other plant families. Like all embryos of seed plants, it has an axis with a **shoot apex** and a **root apex** (Fig. 16.24). The shoot apex of the corn embryo, and the several rudimentary leaves, are surrounded by a sheath, the **coleoptile.** The rudimentary root (radicle) is also surrounded by a sheath, the **coleorhiza.** At the juncture of the shoot and root is a very short stem structure. A relatively large part of the grass embryo is the single **cotyledon,** which has for a long time been called the **scutellum.** It is a shield-shaped structure that lies in contact with the endosperm. The outer cells of the scutellum secrete enzymes that digest the stored foods of the endosperm. These digested foods move from endosperm cells

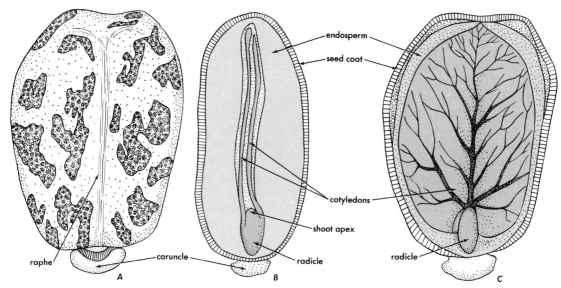

Figure 16.23. Castor bean seed. A, external appearance; B, section showing edge view of embryo; C, section showing flat view of embryo.

through the scutellum to the growing parts of the embryo. Unlike the bean and castor bean seed, the cotyledon of grasses remains within the seed during germination and never develops into a green leaflike structure aboveground.

Lily or onion. The onion or lily seed consists of seed coats enclosing a small amount of endosperm. The embryo is a simple axis, the radicle and single coytyledon being quite prominent. The shoot apex is located at about the midpoint of the axis and appears as a simple notch. The embryo is coiled within the seed coats, the radicle usually pointing toward the micropyle in the seed coats (Fig. 16.25).

Dissemination of Seeds and Fruits

Agents in Seed and Fruit Dispersal

The chief agents in seed and fruit dispersal are **wind, water,** and **animals,** including **man.**

Wind. The structural modifications of seeds and fruits that aid in dissemination by wind are of several types. The most common of these are winged types (Fig. 16.15*B,* 16.26*C*) and plumed types (Figs. 16.10*C,* 16.26*A* and *B*).

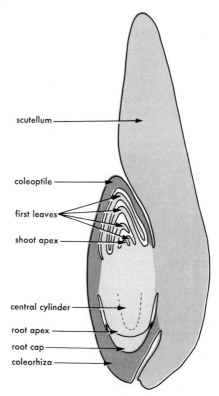

Figure 16.24. Longitudinal section of an embryo of corn (*Zea mays*).

Water. Fruits with a membranous envelope containing air like those of sedges, or with a coarse, loose, fibrous outer coat, as in the coconut, are well adapted for dispersal by water. A great variety of fruits and seeds float in water, even though lacking special adaptations to insure buoyancy, and are readily transported long distances by moving water in the form of surface runoff, natural streams, irrigation and drainage channels, and floods. In the irrigated districts, irrigation water is a very important means by which weed seeds are distributed.

Animals. Many seeds and fruits are carried by animals, both wild and domesticated. Those seeds with beards, spines, hooks, or barbs adhere to the hair of animals (Fig. 16.26D and E). Examples of seeds carried in this manner are wild barley, puncture vine, star thistle, sandbur, and cocklebur.

The seeds of many plants pass through the digestive tract of animals without having their viability impaired. Fleshy, edible fruits may be eaten by birds (Fig. 16.26G) and carried by them long distances, and then the seeds regurgitated or discharged with the excrement. Squirrels carry nuts, such as those of walnut and hickory, and the seeds of pines. The seeds of aquatic and marsh plants and of mistletoe, which are covered with a sticky material, are carried on the feet of birds.

Summary of Seeds

1. Four types of seeds have been described in some detail; other variations occur. Seeds may have one or two cotyledons and they may store food within the embryo or outside the embryo, generally in an endosperm. In most dicotyledons such as the cucumber and bean, the food is stored in the two cotyledons. The seeds of coffee have two cotyledons that serve as absorbing and, later, as photosynthesizing organs, and the food is stored in an endosperm. In monocotyledonous seeds, the food is stored in an endosperm.

2. In most grasses the single cotyledon has become a highly specialized absorption organ, while in seeds like those of lilies and onion the cotyledon emerges from the seedcoats and becomes green. The tip of the cotyledon serves to absorb food from the endosperm. In many monocotyledonous seeds the food is absorbed through epidermal cells.

Seed Dormancy

If the seed is the culmination of the process of sexual reproduction initiated in the flower, it is also the start of a new generation! How long will seeds remain viable? In the fall of 1879 Dr. Beal, at the Michigan Agricultural College, buried 1000 freshly gathered seeds representing 20 different plants. Fifty seeds of each species were mixed with sand and placed in a pint bottle, so that the 1000 seeds were distributed equally among twenty bottles. These bottles, each containing fifty seeds of each kind, were buried at a depth of about eighteen inches. Every 5 or 10 years up to 1931, one bottle was dug up and the percentage of germination of each species was determined. Then the time interval was changed and bottles were dug up each 10 years. After 50 years, the seeds of five of the 20 original species were still viable and after 60 years seeds from four species were found to be living.

One of the most authentic cases of longevity of seeds under natural conditions is that reported for seeds of lotus (*Nelumbo nucifera*). Seeds of this plant are enclosed in a hard shell that is almost impervious to water. This hard seed coat must be ruptured before the seeds can germinate. Viable seeds of lotus buried for over 200 years have been found

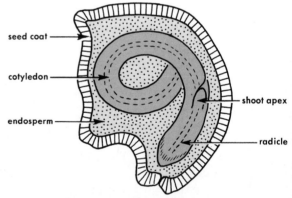

Figure 16.25. A longitudinal section through an onion seed (*Allium cepa*) showing the embryo coiled within the endosperm. (Redrawn after Wilhelm Troll.)

Figure 16.26 (*right*). Fruits and seeds showing various devices aiding in dissemination. **Wind:** A, *Clematis*, ×4; B, dandelion (*Taraxacum vulgare*), ×10; C, seed of Coulter's big-cone pine (*Pinus coulteri*), ×2. **Attachment:** D, cranesbill (*Geranium*), ×3; E, foxtail (*Hordeum hispida*), ×3; F, bur clover (*Medicago denticulata*), ×5; G, fleshy **edible** fruit of Cotoneaster, ×4. Violent **dehiscence** of pericarp; H, vetch (*Vicia sativa*), ×2; I, California poppy (*Eschscholtzia californica*), ×2.

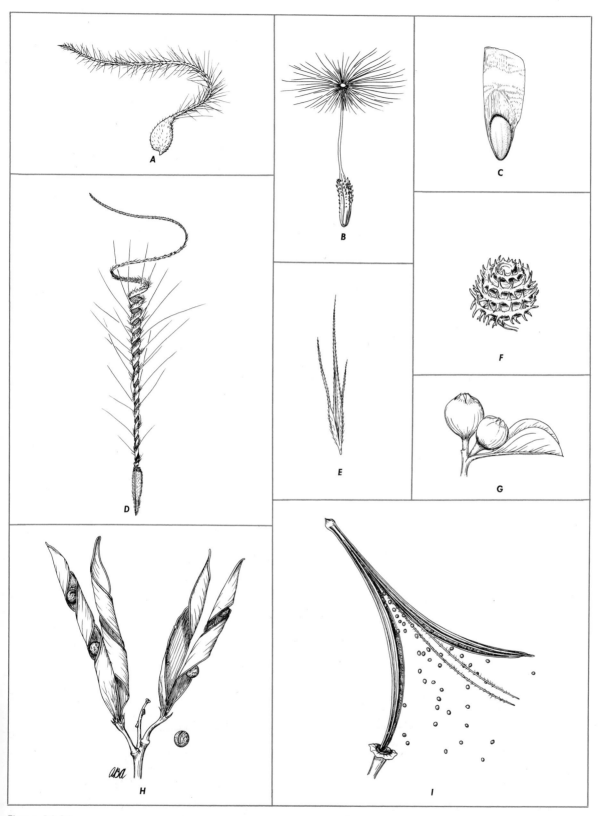

Figure 16.26

in old peat beds in South Manchuria; some scientists estimate their age at over 1000 years.

What induces germination? Many seeds will germinate when provided with moisture, oxygen, and a favorable temperature. The water content of seeds is low, between five and ten percent. The cytoplasm with contained organelles is scarcely recognizable crowded between the large amount of reserved food materials (Fig. 16.27A and B). Water is, at first, imbibed very rapidly and results in the swelling of

Figure 16.27. Electron micrographs of cotyledons of a squash (*Cucurbita maxima*) seed. *A,* food stored in the mature dry cotyledon crowds the dehydrated protoplasm into narrow strips between the lipid and protein granules, ×3000; *B,* after 24 hours hydration, the protoplasm has imbibed water; lipid and protein bodies still crowd the cytoplasm into narrow strands, and organelles are not apparent, ×41,000; *C,* after 72 hours hydration, organelles become apparent, ×41,000. (Courtesy of John Lott.)

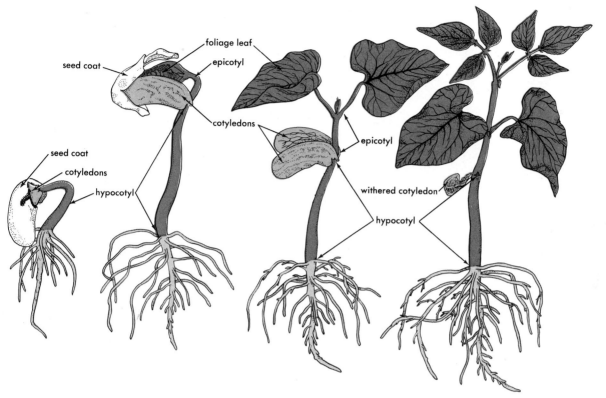

Figure 16.28. Stages in the germination of a bean seed (*Phaseolus vulgaris*).

the protoplasm and the reappearance of the organelles (Fig. 16.27C). Water uptake now slows down. The increase in water content makes possible an increased metabolic activity. From this time on the seed must have an adequate water supply for survival.

Many seeds will not germinate even when supplied with water, oxygen, and a favorable temperature. There are a number of different factors in different plants that are associated with the breaking of dormancy. For instance, light is necessary for the germination of some grass seed and for some strains of lettuce. Water and gases must pass through the seed coats. If seed coats restrict the movement of water and gases or are impermeable to one or the other or to both, the seed coats must decay, be broken, or be scratched, allowing water and oxygen to reach the embryo before germination can start. In some seeds the embryo itself fails to germinate when provided with favorable conditions. In orchids, the seeds are disseminated while the embryos are rudimentary and the embryos must develop further before germination.

Many kinds of seeds are known that will not

germinate unless the seeds, while on a moist substrate, are subjected for a time to temperatures close to freezing. At the other extreme some seeds will not germinate unless they have been subjected to the rather high heat of a flash fire. Another type of dormancy is brought about by the presence of natural chemical inhibitors. These occur in many fruits and serve to keep the seeds dormant while they are enclosed by the fruit. In other cases, inhibitors are present in the seeds themselves or may be produced by decaying leaves, or forest litter. The influence of the plant growth substances on dormancy and seed germination is still unclear.

Patterns controlling dormancy and germination are complex. They have evolved over a long period and confer certain competitive advantages that assure survival.

Seed Germination

That the processes of germination have started, is generally first made evident by the swelling of the radicle. In all cases the radicle imbibes water rapidly and, bursting the seed coats and other cov-

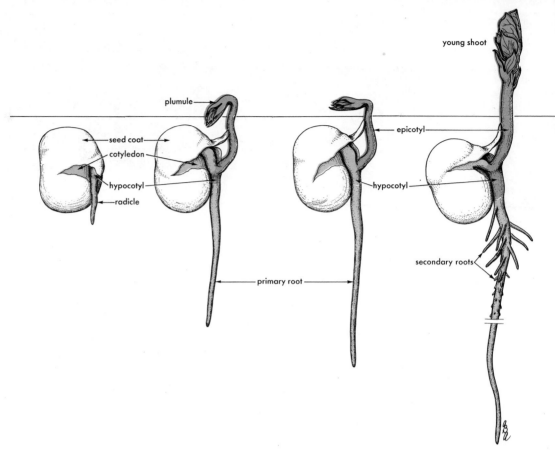

Figure 16.29. Stages in the germination of a pea seed (*Pisum sativum*).

erings that may be present, starts downward growth into the soil. This helps to assure the young seedling an adequate supply of water and nutrients when the shoot breaks through the surface of the soil (Figs. 16.28, 16.29, 16.30, 16.31). While the succeeding steps of germination are essentially similar, there are variations. For instance, in the germination of beans, peas, and onions, a structure with a sharp bend or hook is first forced through the soil (Figs. 16.28, 16.29, 16.31), but the structure forming the hook is different in each case. In the bean (*Phaseolus vulgaris*) the hypocotyl (that part of the shoot below the cotyledons) elongates (Fig. 16.28). In the pea (*Pisum sativum*) the epicotyl (the shoot above the cotyledons) elongates (Fig. 16.29). In both cases cotyledons and shoot apex remain below the ground for a short time. Above the ground, the hook now straightens. In the case of the bean, a straightening of the hypocotyl raises the cotyledons and shoot apex above the ground (Fig. 16.28); a lengthening and straightening of the epicotyl will now pull the shoot apex away from the cotyledons (Fig. 16.28).

Upon straightening out of the pea epicotyl, however, the cotyledons will remain in the ground and only the apex and first leaves will be raised upward (Fig. 16.29).

The cotyledons, whether raised above ground or remaining below, do not carry on effective photo-

Figure 16.30. Stages in germination of a castor bean seed (*Riccinus communis*).

synthesis for a long time in most cases. In the onion or lily the primary root first penetrates the soil. A sharply-bent cotyledon first breaks the soil surface; this is the single cotyledon emerging through the soil. The cotyledon slowly straightens out—in many instances, before all of the nourishment has been absorbed from the endosperm. In such a case, the endosperm and surrounding seed coats may be lifted upward (Fig. 16.31*A, B*). The cotyledon of the onion is tubular and its base encloses the shoot apex (Fig. 16.31*A, B*). There is a small opening at the base of the cotyledon through which the first leaf finally emerges. In grasses the situation is more complex. The shoot and root apices are enveloped

by tubular sheaths known as the **coleoptile** and **coleorhiza,** respectively (Fig. 16.24, 16.32). The primary root rapidly pushes through the coleorhiza. The root continues to grow, but is eventually replaced by adventitious roots that surround the nodes of the stem (Fig. 9.1*D*). The coleoptile elongates and emerges above ground, becoming from one to one and a half inches long. At this time the uppermost leaf pushes its way through the coleoptile and, growing rapidly, becomes part of the photosynthesizing shoot.

The life cycle of a flowering plant is now complete. We started with a seedling in Chapter 3 and we have now returned to a seedling.

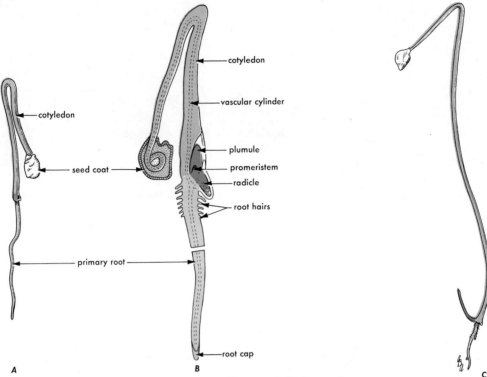

Figure 16.31. Germination of an onion seed (*Allium cepa*). (Redrawn after Wilhelm Troll.)

Figure 16.32. (*right*) Germination of grain of wheat (*Jriticum*).

16 *The Fruit, Seed, and Seedling*

Summary of Dormancy and Germination

1. Mature seeds are dormant and, depending upon the species and the immediate environment of the seeds, they may remain viable and dormant from a few months to many years.

2. Dormancy is usually broken by the provision of moisture, oxygen, and a favorable temperature. Other factors, such as light, removal of chemical inhibitors, or destruction of the seed coats, may be required in some instances.

3. In germination the cotyledons may be elevated above the ground, sometimes to become photosynthetically active, or in the case of some monocotyledons, to continue absorption of the endosperm. In other cases the cotyledons remain below the ground.

4. In the grasses the single cotyledon functions as a specialized food-absorbing organ.

17

Inheritance

The individuals of a single-celled species living in a pond have been multiplying by dividing in two since life started, and they will continue to divide and grow as long as they, or other species to which they may give rise, continue to exist on this earth. The simplest type of life cycle is this division of a single-celled plant to form two new plants. Most individuals, plant or animal, are not so simply constructed; specialized cells are set aside for reproductive processes. For example, in the flowering plants the flowers and seeds possess such specialized cells. Even in these plants the special reproductive cells are in a sense immortal, for one can trace their history back to the beginning of life through a series constituted mainly of meristematic cells and reproductive cells. Every plant and animal living today must have a chain of ancestors stretching back to the beginning of life. During this long period the characteristics of present-day plant and animal life have been derived. How has this happened? Are species of animals and plants constant? Do they change? How similar to each other are the individuals of a given species? If individuals of a given species vary, what are the causes of variation? Can this variation be controlled or predicted? How do the specialized reproductive cells function? What is the importance of sex? These and many other questions arise when one considers the reproductive cycle of plants and animals. Partial answers may be given, but to no question stated above is there, as yet, a final and complete answer.

AN ANGIOSPERM LIFE CYCLE

Briefly in the life cycle of a flowering plant—corn, for example—seed is planted and gives rise to the vegetative plant, which in time produces staminate flowers in the tassels and pistillate flowers in the ears. Pollen is formed by the staminate flowers and carried to the stigmas of pistillate flowers in the young ears, either on the same individual plant or on adjacent individuals. The pollen tube grows down the long style and eventually penetrates the embryo sac, where a sperm and an egg fuse to form the zygote, which develops into the embryo. A second sperm from the pollen tube fuses with the two polar nuclei of the embryo sac. The body resulting from this triple fusion develops into the endosperm. The embryo and endosperm constitute the

337

major portion of the corn "seed," or, more properly, the fruit (Fig. 16.14). Of very great importance in this cycle is the behavior of the **chromosomes,** which are the carriers of the **genes,** the units responsible for the morphological and physiological characteristics of every individual. During the formation of the pollen and of the embryo sac, the chromosomes go through characteristic division, the **meiotic division** or **meiosis.**

At fertilization, nuclei with the gene-carrying chromosomes from two parents are generally combined to form the zygote nucleus. These two phenomena, **meiosis** and **fertilization,** involving, as they do, the assortment and combination of chromosomes, form the basis of the sexual reproductive cycle. Indeed, the great diversity of plant and animal life is due largely to the changes that occur in a chromosome set during meiosis and fertilization. In many plants (asparagus and date palm, for example), two separate parents are involved in sexual reproduction; in others, such as pea and wheat, the reproductive cells are produced on the same parent. Thus chromosome behavior at these two critical stages constitutes a mechanical basis for inheritance, and inheritance itself may be placed on a firm mathematical foundation.

REVIEW OF MITOSIS

In the prophase of mitosis, extended chromosomes divide and contract to form two chromatids per chromosome. The kinetochores then become aligned at the equator of the cell; during anaphase, sister chromatids move to opposite poles of the cell; during telophase, daughter nuclei are reconstituted; and the development of a cell wall completes the formation of two identical daughter cells. In this process there has been no change in the chromosome number, and no assortment of chromosomes; each daughter nucleus has the same complement of genes. It should be emphasized, however, that all cells containing 2n chromosomes have two sets of chromosomes. In corn, for instance, there are twenty chromosomes, two sets of ten each. The chromosomes in corn have been numbered from one to ten. Each nucleus of the corn plant therefore contains the following chromosomes:

$$
\begin{array}{cccccccccc}
1 & 2 & 3 & 4 & 5 & 6 & 7 & 8 & 9 & 10 \\
1 & 2 & 3 & 4 & 5 & 6 & 7 & 8 & 9 & 10
\end{array}
$$

The two number 1 chromosomes are said to be **homologous** chromosomes, as are the two number 2 chromosomes, and so on. Each pair of chromosomes (3 + 3, 4 + 4, 5 + 5, and so forth) constitutes a pair of homologous chromosomes. Each member of the homologous pair carries, at identical positions, or **loci,** a gene affecting the expression of the same trait. Such a pair of genes are known as **allelomorphs,** or **alleles.**

CHARACTERISTICS OF MEIOSIS

The division stages of chromosomes during meiosis are identical to the division stages of chromosomes undergoing mitosis. Meiosis opens with the chromosomes present as long slender uncoiled threads (Figs. 17.1*A,* 17.2*A*). They proceed to shorten and thicken and split into chromatids as in mitosis (Fig. 17.2*C*). The chromatids separate at anaphase (Fig. 17.2*F, I*). Replication of DNA occurs in the interphase preceeding meiosis and again in the interphase after the completion of meiosis when separate chromatids are present. The single basic difference between mitosis and meiosis lies in the *pairing of homologous chromosomes* so that four chromatids, two derived from each of the pairing homologous chromosomes, are associated at mid-prophase (Fig. 17.2*B, C*). In mitosis, only the two sister chromatids, derived from a single chromosome, are associated at this point (Fig. 6.6*D*). This introduces a complication, for the chromatids of paired homologous chromosomes are able to exchange pieces with each other (Fig. 17.2*F*). Such exchanges between chromatids derived from opposite members of a homologous pair of chromosomes (nonsister chromatids) (Fig. 17.2*E, F*) may be expressed as visible changes in the offspring. In order for this to take place, there appear to be two divisions in meiosis, which result in four cells. However, DNA replication does not take place during meiosis, so each of these four cells has only one complete set of chromosomes. Thus,

Figure 17.1. (*right*) Meiosis in lily anther. Photomicrographs showing A, early prophase I. Note paired threads; B, late prophase I. Each body represents two paired chromosomes; note chiasmata; C, late prophase I. Paired chromosomes; two chiasmata present. Compare with chromosomes of Figure 16.2E; D, metaphase I. Note tractile fibers extending from chromosomes. E, anaphase I; F, telophase I; G, prophase II; H, metaphase II; I, anaphase II; J, telophase II. (B, courtesy of Gankin.)

Figure 17.2. Diagram of meiosis, two pairs of homologous chromosomes are shown. *A*, early prophase with unpaired chromosomes; *B*, the homologous chromosomes have paired; *C*, the chromatids have formed. Those of a single chromosome are sister chromatids. Here and in the following diagrams they are represented as paired parallel lines (this is not the normal situation); *D*, markers on the chromosome, future location of crossover indicated by arrows; *E*, metaphase I. Nonsister chromatids of homologous chromosomes have broken and, at the crossover region, have exchanged pieces with each other;

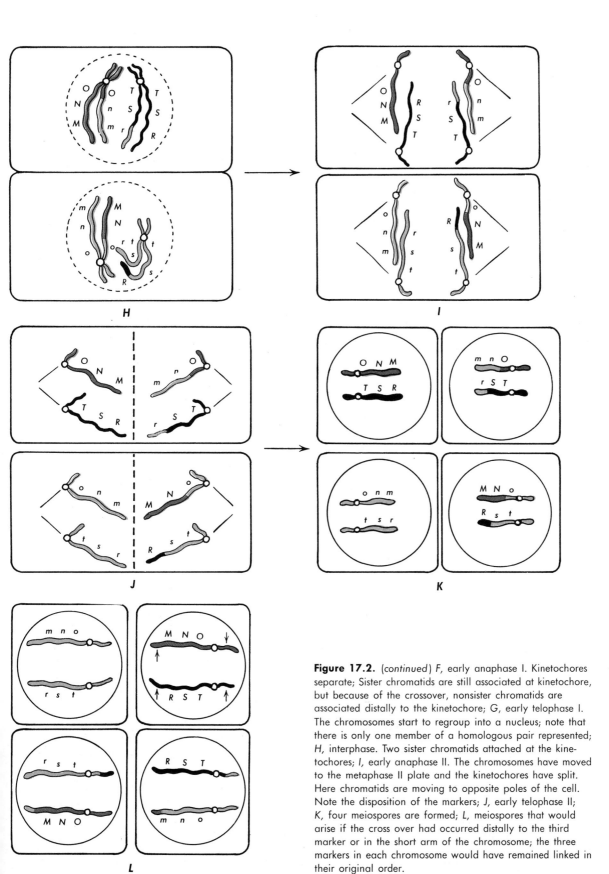

Figure 17.2. (continued) F, early anaphase I. Kinetochores separate; Sister chromatids are still associated at kinetochore, but because of the crossover, nonsister chromatids are associated distally to the kinetochore; G, early telophase I. The chromosomes start to regroup into a nucleus; note that there is only one member of a homologous pair represented; H, interphase. Two sister chromatids attached at the kinetochores; I, early anaphase II. The chromosomes have moved to the metaphase II plate and the kinetochores have split. Here chromatids are moving to opposite poles of the cell. Note the disposition of the markers; J, early telophase II; K, four meiospores are formed; L, meiospores that would arise if the cross over had occurred distally to the third marker or in the short arm of the chromosome; the three markers in each chromosome would have remained linked in their original order.

during meiosis the two sets of homologous chromosomes found in the **diploid** parent are reduced to a single set of chromosomes, and the cell or plant bearing this single set is said to be **haploid.** The cells in which meiosis occurs are **meiocytes.** If they produce **spores**—cells able to produce new plants directly—as generally happens in plants, the meiocytes may be called **spore mother cells.** The spores resulting from meiosis are **meiospores.** Since meiospores give rise to a new generation of plants directly, it follows that plants arising from meiospores must be haploid. In order to regain the diploid state, two haploid cells must fuse. Cells capable of fusing are generally specialized cells and are called **gametes.** The diploid cell resulting from the gamete fusion is a **zygote** and will give rise by mitosis to a diploid, or 2n, plant body.

Sometimes conditions arise that result in three, four, or more sets of homologous chromosomes per cell. This condition is known as **polyploidy** and, depending upon the number of homologous chromosomes, the plants are **triploid** (3n), **tetraploid** (4n), or **hexaploid** (6n).

Since, apparently, two divisions are required for the completion of meiosis, it is common to designate the two divisions by the numerals I and II and the phases as prophase I, metaphase I, anaphase I, telophase I, prophase II, metaphase II, anaphase II, and telophase II.

Meiosis—First Division

Prophase I

During prophase of meiosis (Figs. 17.1*A*, 17.2*A, B, C*) the nuclear threads of the vegetative nucleus, present in the diploid number, contract to form chromosomes, each with two chromatids; in this respect the process resembles mitosis. The process is complicated, however, because before the chromatids become apparent, the homologous chromosomes pair (the two number 1 chromosomes, for example). In so doing, they approach and coil about each other (Figs. 17.1*B*, 17.2*B*). The two chromatids of a chromosome appear only after the pairing of the homologous chromosomes is well advanced (Fig. 17.2*C*). The resulting figure is composed of two paired homologous chromosomes and four chromatids (Figs. 17.1*B*, 17.2*C*). This stage of prophase involves two phenomena: (1) the pairing of homologous chromosomes and (2) the formation of

chromosomes with two chromatids each. The chromatids thus formed do not remain unchanged; they normally break at from one to ten places and rejoin in such a way that a given reconstituted chromatid may be composed of parts of four chromatids (Fig. 17.2*C*). This breaking and rejoining of the chromatids is called **crossing over,** and the cross formed by the chromatids involved in the interchange is known as a **chiasma** (**chiasmata,** plural). (Fig. 17.2*E*). Chiasmata may occur at any point along the paired chromosomes. Figure 17.2*D* represents two pairs of homologous chromosomes, each with three pairs of allomorphic genes to serve as markers; Mm, Nn, and Oo on the green homologs and Rr, Ss, and Tt on the gray homologs. If the chiasmata occurred between Oo and the kinetochore, or to the left of Mm, the relationship of the allomorphic pairs would remain unchanged: M, N, and O would remain **linked** together as would m, n, and o on the chromatids of the original chromosomes. However, in Fig. 17.2*E*, chiasmata are shown between Nn and Oo on the green chromosome and between Rr and Ss on the gray chromosome.

Thus a third step in the prophase I of meiosis is (3) the formation of chiasmata due to the breaking and rejoining of chromatids from homologous chromosomes. As in mitosis the nucleolus disappears, material from it probably passing to the developing chromosomes. The nuclear membrane also disappears during the later stages of prophase.

Metaphase I

At metaphase I (Fig. 17.2*E*) the kinetochores of the paired homologous chromosomes pass to the equator of the cell. Tractile fibers grow toward the poles (Fig. 17.1*D*).

Anaphase I

In anaphase I the kinetochores of *whole chromosomes* separate from each other and, with their associated chromosomes, move to opposite poles of the cell. However, because of chiasmata formation, whole, complete chromosomes are not separated from each other, for on the side of the chiasma away from the kinetochore, sister chromatids will separate just as in mitosis (Fig. 17.2*F*). Anaphase I of meiosis is thus the separation of *two chromatids.* Since the chromatids have become variously modified, the separation involves both a separation of whole chromosomes (at the kinetochores) and a separation of sister chromatids as in mitosis (across the chiasma) (Fig. 17.2F).

Telophase I

Following anaphase I the chromatids group together at opposite poles of the cell (Figs. 17.1F, 17.2G) and immediately prepare for the second meiotic division. While each telophase chromosome consists of two chromatids, and in this sense the n or haploid number of chromosomes is present, it must be remembered that structurally the chromosome is composed of both sister chromatids and chromatids derived from homologous chromosomes.

Meiosis—Second Division

A second meiotic division now follows, which separates these rearranged chromatids. During prophase II (Figs. 17.1G, 17.2H) the chromosomes again form, each with two composite chromatids. The kinetochores approach the equatorial plate, forming metaphase II (Figs. 17.1H, 17.2I). As in meiosis the kinetochores now split and separate. In anaphase II single chromatids move to opposite poles of the cell (Figs. 17.1I, 17.2I, J) and are reconstituted in telophase II into nuclei (Fig. 17.2K). Each nucleus thus formed contains one of the four chromatids derived from the pairing of homologous chromosomes in prophase (Fig. 17.2C).

Walls develop about each new nucleus and associated cytoplasm, thus forming cells with the n or haploid number of chromosomes. Since the chromatids within each of these four cells have been variously modified by crossing over, each of the four cells may be genetically different. In just what way do they differ and of what importance is this variance in the life cycle of plants?

DISTRIBUTION OF GENES DURING MEIOSIS

The distribution of genes during meiosis depends upon the separation of the chromatids, and the location and amount of interchange taking place between the chromatids. If, for instance, as in Fig. 17.2D, the interchange took place to the left of Mm, then M N and O would move with the kinetochore at anaphase I, and the two telophase I nuclei would contain the markers M, N, O and m, n, o. At metaphase II and anaphase II the chromatids,

bearing M, N, O and m, n, o, would separate and the three markers, M, N, O, m, n, o, would remain linked together so only two types of meiospores M, N, O and m, n, o (Fig. 17.2L) would result. The same thing would happen if the exchange occurred to the left of Rr. T, S, and R would remain linked, as would t, s, and r. The number of different kinds of meiospores to be derived from one to several linked allelic pairs is two or four, depending upon whether crossing over has occurred between the alleles. Without crossing over, two of the meiospores would be similar; with crossing over, all four meiospores would be different. The proportion of meiospores having crossover chromosomes in a large number of meiotic divisions will depend upon the number and location of interchanges.

For example, if exchanges occur between Oo and Nn in four out of one hundred meiocytes, eight of the resulting 400 meiospores would show interchanges. This is because one meiocyte gives rise to four meiospores, and because a single interchange results in two crossover and two noncrossover chromatids. In this case, two percent of the resulting meiospores would show interchanges. The allelomorphs Mm are further removed from Oo than are Nn (Fig. 17.2D) It turns out that the greater the distance separating genes, the greater the number of chromatid exchanges. We might expect that exchanges would occur between Mm and Oo in eight of 100 meiocytes. Sixteen of the 400 meiospores would show exchanges resulting from crossing over between Oo and Mm. The percentage of exchanges between genes linked together on the same chromosome is taken to indicate the distances separating them and is expressed as **crossover units.**

Now, consider the distribution of the genes, Oo and Tt, which are on nonhomologous chromosomes. Each of the four meiospores are different in the example shown; they are OT, Ot, oT and ot.

Combining segregation of chromatids during meiosis with chromatid interchanges, or crossing over, brings about variation of the genetic constitution of the resulting meiospores.

Meiospores give rise to haploid plants of varying genetic constitutions; they will all produce gametes that fuse to give rise, once again, to a new diploid generation. The gametes carry the genetic constitutions endowed them by the meiospores, and if meiospores differ, gametes must also differ. The fusion of gametes having different genetic constitutions further greatly enhances the possibility for variation.

FERTILIZATION

While crossing over is probably always present, the combined phenomena of meiosis and fertilization are greatly simplified by ignoring it for the present.

Let us assume that the Tt genes occur in a cross-pollinating plant. Recall that meiosis occurs in the formation of pollen and also in the development of the embryo sac. One-half of the pollen grains and eggs formed will carry the gene T; and the other half will carry the gene t. Since large numbers of both pollen and egg cells are formed, and since the fusion of any two may occur, we should expect the following combination.

The many hundreds of sperm cells produced in pollen tubes will carry either T or t genes in approximately equal numbers. These sperms will unite with eggs carrying either the T or t gene.

Notice that there is one chance for the formation of zygote TT, one chance for the formation of zygote tt, and two chances for the formation of zygote Tt. If the genes Tt governed height, we should expect to find plants of different heights occurring in certain very definite ratios in a breeding plot. That this situation actually does occur has been well established by a long series of experiments.

BREEDING SYSTEMS IN PLANTS

The art of plant and animal improvement is very ancient. It has been practiced by many different races of men for thousands of years. Wheat, corn, and rice are examples of plants that have been improved by men of three different continents. The domestication of these grains was accomplished before the beginning of recorded history and ranks as an accomplishment of major importance. The task was so well done that the tribes that grew these grains were assured of a food supply and could turn their energies to activities other than food getting. Civilization resulted. Attempts to increase crop yields and the nutritive values of plants, or to in-troduce new and better food plants, have been continued down through the ages to the present time. The last 50 years has seen a great increase in plant and animal improvement and in a knowledge of the mechanism of inheritance in plants and animals.

Mutation

Most genes have been discovered by examining the individual plants in large uniform populations to find the rare individual that differs from its neighbors. It has been found that such differences occur in rather characteristic ratios. When these new plant types are bred to the standard wild type, they are generally recessive to the wild type and the new traits seems to be deleterious to the plants that bear them. These new genes are called **mutations.** They may be due to a variety of causes, but a common explanation is that an error has occurred in the duplication of the DNA helix. How might this occur?

In our discussion of the DNA molecule and its replication, it was pointed out that gene specificity resided in the arrangement of the base pairs, adenine-thymine and cytosine-guanine, in the DNA helix. It was further pointed out that the helix uncoiled in replication and, as the two strands separated, a new strand formed exactly like the disjoining member. The perfect replication of thymine opposite adenine and cytosine opposite guanine results in the new identical DNA helix. Sometimes there may arise a **copy error** caused perhaps by a modification of the sugar-phosphate bonds, perhaps by a small deletion of a portion of a chromosome. Such a copy error would give the replicating DNA helix a slightly different base sequence than that found in the original helix. Now if our hypothesis is correct that the DNA molecule is involved in the production of specific cellular protein through messenger RNA, we would expect the change of sequence in the base pairs to be reflected in all events leading to the final form of the trait related to the gene in question. Such a change in the replication of the DNA molecule may be one causes of mutations.

Variation

Variation in a population of plants or animals may be due to a number of conditions. Variation is probably always related to genes. The number of genes involved in the expression of a trait, and the relationships between genes, play an important

Figure 17.3. Gregor Mendel. (Courtesy of The Bettmann Archive, Inc.)

part in the expression of the trait and in the relative importance of the environment upon the type and amount of variation. Mendel's revolutionary approach to plant breeding lay in his plan to follow the numerical presence or absence of single traits through a number of generations of carefully controlled pollinations and seed collections. The differences between the progeny were large and easily separated the progeny into distinct categories. This is **discontinuous variation.** It may be contrasted with **continuous variation,** in which the variation is evenly and widely distributed about a mean (Fig. 17.7). Continuous variation is a very important aspect of modern genetics. However, the principles established through studies on discontinuous variation form the foundation for explaining the mechanism underlying continuous variation as well as other breeding systems.

Discontinuous Variation

Modern **plant breeding** and **genetics** date from 1900, when Gregor Mendel's (Fig. 17.3) experiments on hybridizing garden peas were first appreciated.

The importance of his experiments lies in the introduction to plant breeding of several procedures which, in 1900, were new and revolutionary. They are as follows:

1. The study of the inheritance of single or **unit** characters.

2. The keeping of accurate records of the number of times a given unit character appears in the offspring of selected parents.

3. The maintenance of the pollination of the experimental plants under the complete control of the investigator at all times.

While these procedures are accepted today as self-evident, they were so revolutionary in 1865 when Mendel's work was done that the standard scientific magazines of the time refused to publish Mendel's results. Instead, his experiments were presented before a small society of men in his home town. This society was not greatly unlike some of the present-day luncheon clubs and, like some of them, it published its proceedings. Thus Mendel's work was placed on record, where it lay unnoticed for 35 years. In 1900 it was discovered by three prominent

European plant breeders and has become the cornerstone of all modern work in plant and animal genetics.

The monohybrid cross. Mendel selected garden peas for his experimental plants. He carefully took pollen from the anthers of a dwarf-growing variety and dusted it on the stigma of a tall-growing variety. The seeds resulting from this cross pollination were collected and planted the following season. All the plants that grew from these seeds were tall. The same results were obtained when the tall variety supplied the pollen.

Tall plant × Dwarf plant (pollen parent)
↓
All tall plants

or

Dwarf plant × Tall plant (pollen parent)
↓
All tall plants

Then flowers of the tall progeny were self-pollinated. Peas are normally self-pollinated, and so in order to obtain this seed Mendel merely had to keep stray pollen from reaching the stigmas of his experimental plants. The seeds resulting from these self-pollinated flowers were collected and planted the following spring. Upon counting the plants that grew from these seeds, he found that 787 were tall plants and 277 were dwarf plants. In other words, about three-fourths of these pea plants resembled one of the original pair of parents and about one-fourth resembled the other parent (Fig. 17.4).

It has been found convenient to give designations to these generations of plants. The two original parents are designated by **P.** The first generation, comprising only tall pea plants, is the **first filial generation,** or the F_1; the second generation of three-quarters tall and one-quarter dwarf pea plants is the **second filial generation,** or the F_2. Subsequent generations of self-pollinated plants would be consecutively the F_3, F_4, F_5, and the like.

Note in the cross that, even though we are dealing with two different plants, we are concerned with only **one character**—height of growth. A cross dealing with a *single unit character* is a **monohybrid cross.** Mendel studied, in all, seven monohybrid crosses. The results were similar to those just described for height. They were as follows:

1. The F_1 always resembled one of the parents: tall × dwarf gave tall; wrinkled × round seeds gave round seeds; and the like. From this result, Mendel concluded that one expression of a given character was **dominant** (tall, or round seeds), whereas the other aspect of the same character was **recessive** (dwarf, or wrinkled seeds).

2. Selfing of the F_1 resulted in an F_2 generation in which the original forms of a given character **segregated** to give three times as many dominant plants as recessive plants.

Mendel further postulated a possible mechanism for this genetic behavior. He knew that the F_1 plants must contain factors, now called **genes,** responsible for both tallness and dwarfness, even though all the plants were tall, because (1) tall and dwarf plants were crossed to produce the tall F_1, and (2) upon selfing, the tall F_1 gave rise to both tall and dwarf individuals. If we then represent the gene for tallness by T and the gene for dwarfness by t, we may diagram the genes determining height that are present in the tall F_1 by Tt.

Tt = tall-growing F_1 pea plant

Since the F_1 hybrid contains two genes, T and t, determining height, each true breeding parent plant should contain two similar genes, thus:

TT = tall-growing parent
tt = dwarf-growing parent

Mendel assumed that the pollen would contain but one gene, and that the eggs similarly would contain but one gene. Thus, the tall parent, TT, would produce T pollen and T eggs, and the dwarf parent, tt, would produce t pollen and t eggs.

The parent plants whose nuclei contain the homologous chromosomes with the identical genes TT or tt are said to be **homozygous** for either the dominant or recessive genes. The F_1 plant with the homologous chromosomes carrying the genes Tt is said to be **heterozygous.**

Pollen produced by the heterozygous tall F_1 plant, Tt, will contain either gene, T or t. Likewise, eggs produced by this F_1 plant will contain either gene, T or t, and the sperm nuclei formed from the pollen will contain either gene, T or t. Many thousands of pollen grains, sperms, and eggs will be formed by the tall heterozygous F_1 plant. A sperm containing the gene T may unite with either a T or a t egg, as follows:

T sperm + T egg = TT zygote
T sperm + t egg = Tt zygote

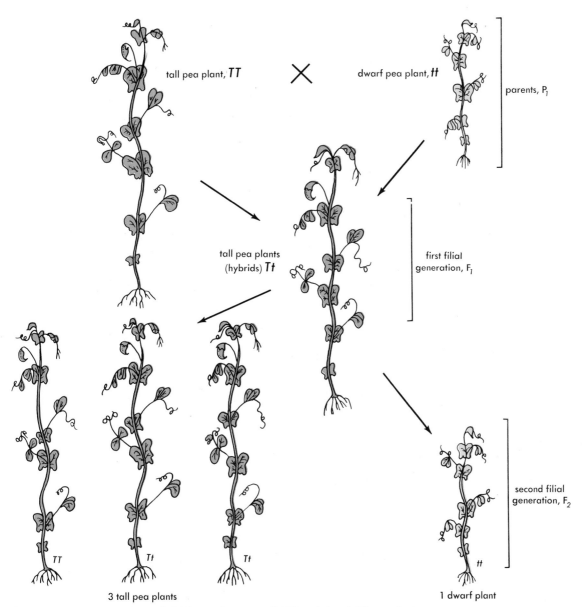

Figure 17.4. Diagram of monohybrid cross between tall (*TT*) and dwarf (*tt*) pea plants.

A sperm containing the gene *t* may unite with a *T* egg or a *t* egg, as follows:

t sperm $+$ T egg $=$ Tt zygote
t sperm $+$ t egg $=$ tt zygote

These four possible combinations of genes will occur at fertilization and may be diagrammed as follows at right:

		Possible sperms	
		T	t
	T	Zygote 1	Zygote 2
		TT	Tt
Possible eggs			
	t	Zygote 3	Zygote 4
		Tt	tt

The possible types of eggs are arranged along the left of the checkerboard, and the possible types of sperms at the top. The squares of the checkerboard represent the possible zygotes, and they are derived by combining the gametes opposite the squares. For instance, zygote 1 is derived from sperm T and egg T; zygote 2 is derived from sperm t and egg T, and so on. The genes, Tt, or TT, or tt, for example, present in the plant, constitute the **genotype** of the plant. Reference to the diagram will show that there are three different genotypes: TT, Tt, and tt. The appearance of the plant is the **phenotype**. For instance, the appearance or the phenotype of the plant TT is tall; of the plant Tt is tall; and of the plant tt is dwarf. Note that there are three times as many tall plants represented on the checkerboard as there are dwarf plants. Furthermore, if Mendel's assumption regarding these factors or genes is correct, the homozygous tall plant TT, when self-pollinated, should give only tall offspring; the homozygous dwarf plant tt, when self-pollinated, should give only dwarf offspring; the tall heterozygous Tt plants, when self-pollinated, should give tall and dwarf plants in the ratio of 3 tall plants to 1 dwarf plant. This is what actually happens. A complete monohybrid cross is shown in Fig. 17.4.

This cross gives us additional information about these genes. It might be thought that they could interact with each other to bring about some modification in their genotypes, either in the zygote nucleus, or in the line of vegetative nuclei leading to the next generation of spore or pollen mother cells, or that the cytoplasm could act upon them as they are replicated at each somatic division in the interval between fertilization and meiosis. But none of these things happens. T and t retain their identity and separate from each other inviolate at each succeeding meiosis, generation after generation. Another important fact about them is that they occupy equivalent spots or loci on homologous chromosomes. Such pairs of genes maintaining their integrity and having the same loci on homologous chromosomes are known as **alleles.**

The dihybrid cross. Let us consider a cross concerning two characters, height of the plant and form of the seeds. This is a **dihybrid cross,** that is, a cross involving *two unit characters* (Fig. 17.5). One parent plant is tall and has round seeds; the other parent plant is dwarf and has wrinkled seeds. We know that both these parent plants are homozygous for these particular characters because upon self-pollination they breed true. When cross-pollinated, all the resulting F_1 plants are **tall** and have **round** seeds. We know from this outcome that tallness is dominant over dwarfness and that roundness is dominant over wrinkledness. We may therefore write the genotype of the parent plant as follows:

The tall plant with round seeds, $TTRR$
The dwarf plant with wrinkled seeds, $ttrr$

Since all plants possess the characters of height and seed form, factors determining these characters must always be present in every plant.

After meiosis the pollen and eggs produced by the tall plant with round seeds will contain the genes TR. The pollen and eggs produced by the dwarf plant with wrinkled seeds will be tr. Union of sperms from a pollen grain TR with an egg tr (or a pollen grain tr with an egg TR) will result in a zygote having the genotype $TtRr$ (Fig. 17.5). This plant will be tall and have round seeds. It is heterozygous for both height and seed shape.

The four possible types of eggs and sperms that may be formed during meiosis from a plant with the genotype $TtRr$ are shown in Fig. 17.6. The sperms and eggs that result are:

Sperms TR, Tr, tR, tr
Eggs TR, Tr, tR, tr

Any one of the four types of sperms may unite with any one of the four types of eggs. The number of possible combinations is best shown on the checkerboard of Fig. 17.5. The four types of sperms are listed along the top of the checkerboard; the eggs are placed at the left. Each square represents a possible zygote. Examination of these squares will show that four phenotypes are possible:

Nine tall plants with round seeds, squares numbered 1, 2, 3, 4, 5, 7, 9, 10, 13.

Three tall plants with wrinkled seeds, squares numbered 6, 8, 14.

Three dwarf plants with round seeds, squares numbered 11, 12, 15.

One dwarf plant with wrinkled seeds, square number 16.

A dihybrid cross carried through the F_2 generation may be simply diagrammed as follows:

Figure 17.5. Diagram of a dihybrid cross between a tall pea plant with round seeds (*TTRR*) and a dwarf pea plant with wrinkled seeds (*ttrr*).

Parent Genotypes	$TTRR$	$\times$	$ttrr$
Parent Phenotypes	Tall plants with round seeds		Dwarf plants with wrinkled seeds

Gametes TR tr

F_1 Genotype $TtRr$

F_1 Phenotype   All plants heterozygous, tall with round seeds.

Self-pollination of F_1 $TtRr$ $\times$ $TtRr$

Gametes TR, Tr, tR, tr TR, Tr, tR, tr

F_2 Phenotypes

     9 tall plants with round seeds
     3 tall plants with wrinkled seeds
     3 dwarf plants with round seeds
     1 dwarf plant with wrinkled seeds

In this cross it is important to note that the genes for height and seed shape separate or *segregate independently of each other* during the formation of meiospores from the spore mother cell.

Soon after Mendel's work was discovered it was noticed that his results could be explained by assuming that the genes were located in the chromosomes. It has now been well demonstrated that the genes are carried in the chromosomes, and maps have been prepared showing their relative positions in the chromosomes of many plants.

The backcross. In a backcross, offspring are crossed to one of the parents. Usually the parent having homozygous recessive alleles is selected, because such a cross yields much information regarding the genotype of an offspring. For instance, there are four different phenotypes in a dihybrid cross and 9 different genotypes. There are nine tall F_2 plants with round seeds, and they may have any one of four genotypes: $TTRR$, $TTRr$, $TtRR$, $TtRr$ (Fig. 17.5). If the progeny of a backcross between one of these F_2 plants contains equal numbers of tall plants with round seeds, tall plants with wrinkled seeds, short plants with round seeds, and short plants with wrinkled seeds, the genotype of the F_2 plant must have been $TtRr$ (Table 17.1).

The backcross also provides information as to whether genes are linked. The crossover units separating them may be determined. For instance, there are genes in corn that are expressed as white sheaths on the leaves, and liguleless blades. The backcross here consists of crossing a plant homozygous for the recessive white sheaths (*ws ws*) and liguleless blades (*lg lg*) with a plant heterozygous for these traits: *Lg lg* and *Ws ws*. If these alleles are on different chromosomes, the four types should appear in the progeny in approximately equal numbers. If the alleles are linked on the same chromosome, the numbers of wild type plants should approximately equal the numbers of plants showing white sheaths and blades without ligules. A small number of plants with the dominant ligule leaves, but with the recessive white sheath, accompanied by an equal number of plants with the complementary traits, would indicate some crossing over between linked genes. It turned out that about forty-six percent of the progeny of such a cross resembled the heterozygous parent and another forty-six percent had white sheaths and ligule blades. Eight percent were

Table 17.1   *Progeny of backcross: TtRr $\times$ ttrr*

GAMETES OF $TtRr$	GAMETES OF tr	GENOTYPE IN PROGENY OF BACKCROSS	PHENOTYPE OF PROGENY OF BACKCROSS
TR	tr	TtRr	Tall, round seeds
Tr	tr	Ttrr	tall, wrinkled seeds
tR	tr	ttRr	dwarf, round seeds
tr	tr	ttrr	dwarf, wrinkled seeds

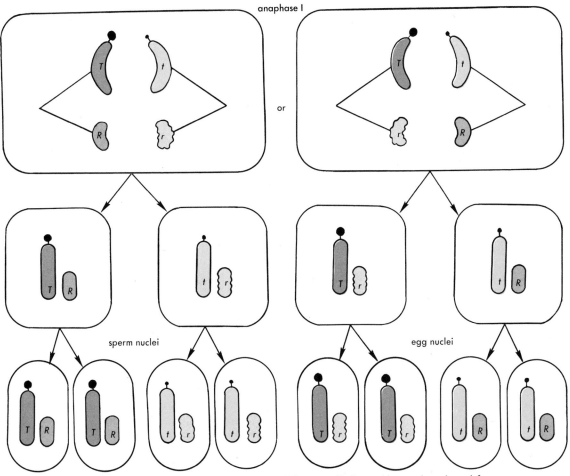

Figure 17.6. Types of sperms and eggs that may be formed from a tall plant with round seeds and from a dwarf plant with wrinkled seeds.

crossover types. Therefore these genes, *lg* and *ws*, are linked and separated by eight crossover units.

Other Breeding Systems

The breeding experiments so far discussed have involved one or two genes directly related to one or two traits. Height was controlled by one gene, leaf color or flower color, by another, and so on. These genes were independent of each other in activity, in segregation at meiosis, and in combining at fertilization. This situation probably occurs in only a relatively few cases. A variety of other more complicated breeding systems are known. (1) Numerous genes may play a part in determining the

expression of a trait. This results in **continuous variation,** rather than in the **discontinuous variation** just discussed. (2) We have considered the presence of a single pair of alleles, at one chromosome locus, which showed expression in the phenotype as the simple presence or absence of a trait. Sometimes there are several alleles at one locus and their action may show a series of slight differences in the expression of the trait. Several alleles at one locus are known as **multiple alleles.** (3) The number of heterozygous alleles in a genotype seems to be directly related to the vigor of an individual or population. This is **heterosis.** (4) More than two sets of chromosomes occur in many plant groups. This is

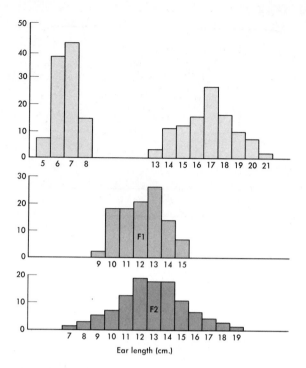

Figure 17.7. Continuous variation. Two varieties of corn are crossed. Black Mexican has long ears and Tom Thumb has short ears. The lengths of the ears of the F_1 show continuous variation between the ear lengths of the parents; there is continuous variation over a greater range in the F_2. (Information from E. M. East, graph adapted from S. L. Brewbaker, *Agricultural Genetics,* Prentice-Hall.)

known as **polyploidy** and it influences the vigor of a plant, as well as the genetic ratios to be expected in breeding experiments. Polyploidy rarely occurs in animals. (5) Genes interact with each other. (6) Portions of chromosomes may become **inverted** and, in the heterozygous condition, show interesting and unexpected expressions. (7) Parts of nonhomologous chromosomes may be exchanged, and such **translocated** portions of chromosomes greatly change the genetic expression of genes. We shall briefly consider selected aspects of continuous variation, multiple alleles, heterosis, and polyploidy.

Continuous variation. When E. M. East (in 1905) crossed a variety of corn (Tom Thumb) with ears five to eight centimeters long with a variety (Black Mexican) with ears thirteen to twenty-one centimeters long, the ears of the F_1 generation were from nine to fifteen centimeters long. The ears were from seven to nineteen centimeters long in the F_2 generation. A classification of the ears of the F_1 and F_2 generations by length showed a normal distribution:

the largest number of ears were close to the average length, with the number of ears gradually decreasing as extremes of ear length were reached (Fig. 17.7). Most traits of plants and animals involving growth rates, vigor, coloring, quality and amount of oil or starch, or quality of fiber or flour, show continuous variation. Traits showing continuous variation must be studied by statistical methods. It is necessary to count many individuals and to take many measurements, weights, and evaluations of color on a colorimeter. The collected figures are then processed by computers. These are **metrical traits.** They are influenced by numerous genes in different locations in the genotype. As the number of genes contributing to the expression of a trait increases, the importance of the contribution of an individual gene to the trait decreases. The many genes thus contributing to the expressing of a trait are **polygenes.** Polygenes are no different from other genes; some may show dominance, and they may interact with each or mutate. Traits controlled by polygenes more easily respond to the environment than do traits controlled by a single or a few genes. These factors make continuous variation difficult to study, but in the age of the computer, investigations on continuous variation are providing significant information on the mechanism of inheritance.

Multiple alleles and pseudoalleles. On page 348 we defined alleles as a pair of genes that retained their identity and separated inviolate at each succeeding meiosis. While it is true that genes do not appear to react with each other or to be greatly influenced by their environment in their normal process of replication, we are no longer sure that a gene can be strictly defined as a unit of crossing over. This is because all alleles do not neatly determine the expression of a character.

Sometimes a given locus may not consist of a single allele, but may apparently contain an allelic series in which one allele is dominant over a series of recessive alleles. Such alleles are frequently called **multiple alleles.** A type of recombination may rarely occur between the alleles of such a series, or in situations resembling such a series. In this case we may speak of **pseudoalleles.**

Alleles are paired genes and, by definition, at a given locus only two, one on each chromosome, can be present at a given time. Generally one of these is dominant, the other recessive, and in an allelic series this same relationship holds. One gene is generally dominant over all others, but the remainder show varying degrees of dominance over each other.

In corn at least eight alleles forming an allelic series control the expression of pigment in the leaves, stem, aleurone, and pericarp. They are all located on a chromosome at a point known as the *A locus.* At least three different types of pigments are influenced by the activity at this locus; purple anthocyanins, yellow-brown anthocyanins, and a red-brown pericarp pigment. The allele A^{st} is dominant over all other alleles with respect to plant and aleurone pigmentation, giving purple pigmentation in the leaves and aleurone and red in the pericarp. However, several in the series tend to dilute these colors when present in a heterozygous condition with A^{st}.

It is important to remember that only two alleles are present at a time, one on each homologous chromosome, and one is generally dominant over the other. This generalization holds in *Drosophila, Neurospora,* and many other organisms, when only a few thousand individuals are examined, and when multiple allelism is not involved. However, about once in 500,000 crosses, crossing over between multiple alleles occurs in Drosophila. This means that these multiple alleles are not strictly alleles, for they behave like separate genes in that crossing over occurs between them at rare intervals. They are pseudoalleles, and they prove that the classical gene may itself be subdivided into physical units. Indeed, in very rapidly multiplying organisms such as bacteria, it is possible to study chromosome interchange within the confines of a gene. In corn, pseudoalleles are known that are separated by 0.0015 of a map unit, a distance that matches the space occupied by about 1,500,000 of the nucleotide pairs of the DNA molecule. In bacteria pseudoalleles separated by only 0.000001 of a map unit are known, and this figures out to be about 10 nucleotide pairs in the DNA molecule.

Heterosis. Heterosis refers to the presence of a large number of heterozygous alleles in a genotype. The recessive alleles are generally less desirable. Therefore, an increase in the vigor of a plant should occur when one member of a pair of homozygous recessive alleles is replaced by a dominant allele. Thus, an increase in heterozygosity should increase vigor. An increase in heterozygosity is easily obtained. Inbreeding rapidly increases the homozygosity of a genotype. By crossing carefully selected inbred lines, it is possible to attain a very high degree of heterozygous alleles, and the progeny of such a cross exhibit great vigor, being larger and more robust than the initial parents of the pure lines. If the increased vigor were due simply to the dominant genes, genotypes having large numbers of homozygous dominant alleles should be as vigorous as plants with an equal number of heterozygous alleles. This is not the case. Vigor seems to depend upon the heterozygous condition and not upon the presence of many dominant alleles. A number of hypotheses have been developed to account for the relationship between heterozygosity and vigor, but the reason for the relationship remains unclear.

How then may vigor be fixed in a species? Given lines of hybrid corn certainly produce year after

| Original corn plant inbred for 7 generations Inbred strain *A* (detasseled) | Original corn plant inbred for 7 generations Inbred strain *B* (furnishes pollen) | Original corn plant inbred for 7 generations Inbred strain *C* (detasseled) | Original corn plant inbred for 7 generations Inbred strain *D* (furnishes pollen) |

Single-cross hybrid ($A \times B$) (detasseled)

Single-cross hybrid ($C \times D$) (furnishes pollen)

Double-cross hybrid ($A \times B$) $\times$ ($C \times D$)

year in a given predictable way. Asexual reproduction will, of course, fix the number of heterozygous alleles in a population. This does seem to be of importance because a great many plants do reproduce extensively by asexual means.

Strains of hybrid corn are obtained first by establishing pure lines of known constitution. This step requires about seven years of self-pollination, resulting in the development of many homozygous strains. Any two homozygous strains are then crossed to form F_1 hybrids, known to commercial corn breeders as *single-cross hybrids*. Since there are many homozygous strains differing in some respects, it is possible to obtain a large number of different single-cross hybrids, and, therefore, the selection of the best homozygous strains requires a thorough knowledge of the traits desired in the commercial strain. The seed produced in a field of such single-cross hybrids, $A \times B$, Fig. 17.8 for instance, would give an F_2 generation with its resulting segregation of characters. This is impractical commercially and necessitates a further step. Two carefully selected single-cross hybrids, $A \times B$ and $C \times D$, Fig. 17.8, are crossed, given what is known as *double-cross* seed. This is sold commercially and produces *double-cross* hybrid plants of remarkable uniformity, capable of yielding large amounts of grain. The steps in the production of double-cross hybrid corn seed are shown in Fig. 17.8 and are outlined above.

Polyploidy. Polyploidy is the increase in the number of chromosomes, or of chromosome sets,

above two. Polyploidy may arise in two ways: by the doubling of an homologous set of chromosomes (**autopolyploidy**); or by combining two complete sets of chromosomes from genetically different parent plants (**allopolyploidy**).

Autopolyploidy. The drug colchicine, derived from a crocus growing in the Mid-East, inhibits spindle formation when applied to cells during mitosis. The separation of chromatids occurs at anaphase, but they remain within the nuclear envelope. When the nucleus is reconstituted it contains four sets of homologous chromosomes. This is a **tetraploid,** and since its four sets of chromosomes are homologous, it is an **autotetraploid.** Such a doubling of the chromosome number in a zygote would result in an autotetraploid plant, and such plants are known. This plant would have four chromosomes with the *A* gene. If it is homozygous there would be four *A* genes or four *a* genes. There would be varying degrees of heterozygosity, *Aaaa, AAaa,* or *AAAa.* Pairing between four homologous chromosomes is irregular, and since gametic ratios depend upon the amount of pairing and crossing over, these ratios vary considerably. If a particular autotetraploid has the constitution *AAAa,* 8 chromatids would arise during meiosis and they could show random segregation. Thus the gametic ratios of *AA, Aa,* and *aa* could range from the extremes of $3AA:3Aa:0aa$ to $15AA:12Aa:1aa$. In addition, one gamete might receive three *AAA* chromosomes and its counterpart one *a* chromosome. Such irregularity renders autotetraploids unstable genetically, and unless they

Figure 17.8 (right). Diagram showing the manner in which commercial hybrid corn is produced. (Redrawn from *Farmers' Bulletin* 1744, U.S. Dept. of Agriculture, Bureau of Plant Industry, Soils and Agricultural Engineering.)

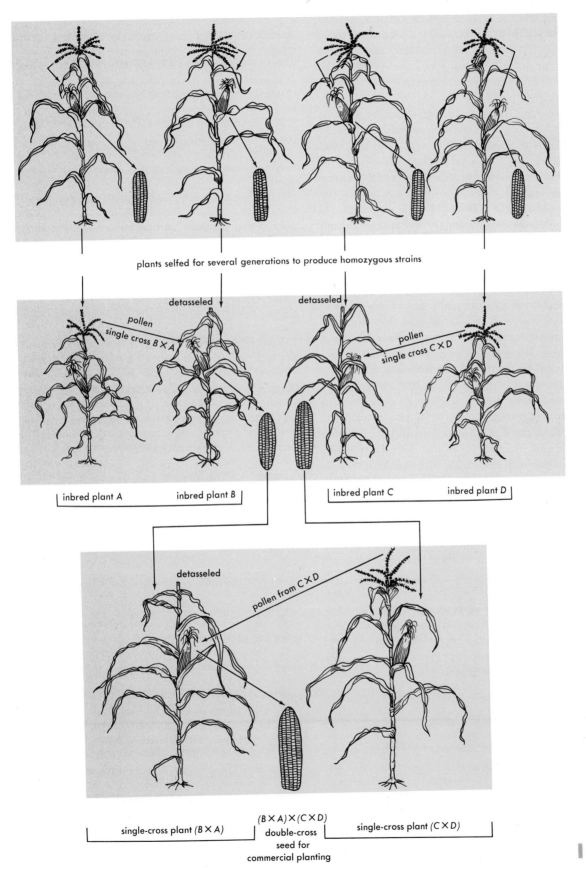

plants selfed for several generations to produce homozygous strains

detasseled

detasseled

pollen
single cross B × A

pollen
single cross C × D

inbred plant A   inbred plant B   inbred plant C   inbred plant D

detasseled

pollen from C × D

single-cross plant (B × A)   (B × A) × (C × D)   single-cross plant (C × D)
                             double-cross
                             seed for
                             commercial planting

reproduce asexually, they seldom become permanently established. Sugar cane is an autopolyploid that is reproduced asexually. Breeding experiments require the annual screening of some million seedlings derived from crosses between plants having different genotypes.

Allopolyploids. Hybrids between species are usually sterile. They do not have sets of homologous chromosomes; thus, pairing is very irregular, and viable seed is seldom set. However, should a doubling of chromosomes occur in a zygote having one complete set of chromosomes from each parent, the resulting tetraploid cell would have only two sets of homologous chromosomes. The plant arising from this kind of doubling of chromosome numbers is an **allotetraploid.** Pairing and crossing over are now normal; the parental chromosomes pair with each other as they would in the parent plants. This allotetraploid is a fertile plant. It will not successfully backcross with its parents and does not look like them. It is essentially a new species or even a new genus. Natural alloploids are known and several have been created in the breeding fields.

Aneuploids. It is not necessary that whole sets of chromosomes be added to those already present. One, two, three, or four chromosomes may be added to a plant with a diploid number. If one chromosome is added, a pair of homologous chromosomes will become a triplet of three homologous chromosomes. Instead of two *A* chromosomes (*AA*) there will be three *A* chromosomes (*AAA*). This plant is a **trisomic** plant. A diploid chromosome number of thirty indicates fifteen pairs of homologous chromosomes. Since a single chromosome may be added to each of these homologous pairs, a plant having fifteen pairs of chromosomes will have fifteen possible different trisomic plant types. Numerous cases are known in which all of the possible trisomics have been obtained. Two or more extra chromosomes may be present and different arrangements are possible. Plants having extra chromosomes are **aneuploids.**

SUMMARY

Summarizing, the following important facts have been established regarding the mechanism of inheritance:

1. The factors responsible for inheritance are located in small, definite regions of chromosomes. In other words, inheritance is due to small particles of chromosomes. These factors or chromosome particles are called **genes.**

2. Most higher plants have the diploid number of chromosomes, the vegetative cells containing two sets of homologous chromosomes.

3. Two genes thus interact in the development of many characters in the diploid plant.

4. The gametes are haploid and consequently carry only one member of each pair of homologous chromosomes.

5. Members of a gene pair, *TT* and *tt,* while influencing a single character (height), determine different aspects of that character, such as tallness and dwarfness.

6. When different forms of a character are involved, such as tallness and dwarfness, one form is generally dominant, the other recessive.

7. When a plant, heterozygous for a given character, such as *Tt,* is self-pollinated, the resulting progeny will number three times as many dominant plants as recessive plants.

8. When a plant, heterozygous for two given characters, such as *TtRr,* is self-pollinated, the resulting progeny will be of four different types, in the following ratio: nine plants will show only the dominant characters (tall and round), three will show one dominant and one recessive (tall and wrinkled), three will have the second dominant and second recessive (dwarf and round), and one plant will have both recessives (dwarf and wrinkled).

9. The chromosomes carrying the genes segregate independently of each other; thus, *T* may go with *R* or with *r.*

10. Genes are linked together on chromosomes. While such genes normally segregate as a unit, crossing over involving a recombination of characters occurs regularly between them.

11. Mutations may arise by an error in the replication of the nucleotide pairs in the DNA molecule.

12. The sum total of genes (particularly those under discussion) within a nucleus constitute the genotype, while the appearance of the plant due to these genes is called the phenotype.

13. The segregation of a progeny into sharply defined classes is discontinuous variation and is the result of the expression of one or two genes in the development of a precisely definable trait.

14. The expression of a trait may result from the interaction of many genes. The progeny in this case will show a continuous variation about a mean.

15. A series of modifications of a character occurring at a single locus is known as a multiple allele.

16. At rare intervals recombination takes place be-

tween what appear to be genes in an allelic series; these alleles are known as pseudoalleles.

17. A high degree of heterosis or a large number of heterozygous alleles in a genotype greatly increases vigor.

18. The diploid number of chromosomes is the optimum condition for the normal steps of meiosis. Polyploid individuals of groups of plants have numbers of chromosomes other than the diploid numbers. An increase in the number of sets of homologous chromosomes is autopolyploidy. Allopolyploidy occurs when the chromosome number is changed by the addition of sets of nonhomologous chromosomes. If two full sets of chromosomes are not added, the plants are said to be aneuploid.

19. Changes in chromosome number by polyploidy may result in increased vigor and great disturbance in meiosis and fertilization.

Plant Ecology

Existing plant species have achieved—by evolution—a balance with their environment which has resulted in stability for both the plants and the environment. In each type of habitat certain species group together. Fossil records indicate that some of these groups of species have lived together for millions of years. The species of these groups share incoming solar radiation, soil water, and nutrients to produce a constant biomass. They recycle nutrients from the soil to living tissue and back again, and they alternate with each other in time and space. Plant ecology attempts to discover what this balance between plants and their environment involves. What stresses does the environment put on plants, and how do plants respond to them? Plant ecology also attempts to determine how the natural balance may be artifically imitated or manipulated to achieve ends desired by man.

Plant ecology deals with levels of biological organization beyond the organism level: the population, the community, the ecosystem. A **population** is a group of closely related, interbreeding organisms. A **community** is composed of all the populations in a given habitat. An **ecosystem** consists of the living community plus the non-living factors of the environment. We will discuss ecology at each of these levels in this chapter. But first let us examine the factors of the environment which affect populations, communities, and ecosystems.

COMPONENTS OF THE ENVIRONMENT

The environment of an organism includes all the living and non-living things about it. Some important components of the environment are moisture, temperature, light, soil, and living organisms. All the components work separately and jointly to influence plant distribution and behavior. Species differ in their range of tolerance to these environmental factors. For example (Fig. 18.1), coast redwood (*Sequoia sempervirens*) is restricted to a narrow strip of California, some 5,000 square miles, which experiences heavy summer fog. Coast redwood is not tolerant of variations in moisture or temperature. Red maple (*Acer rubrum*), however, occurs over some 1,500,000 square miles of

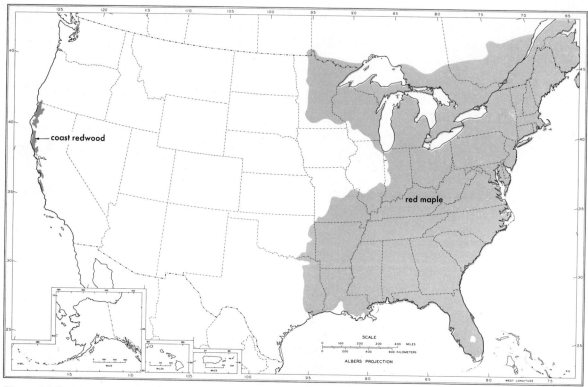

Figure 18.1. Distribution of coast redwood (*Sequoia sempervirens*) and red maple (*Acer rubrum*) in the United States.

eastern North America in wet, dry, cold, or warm environments. It is tolerant of wide variations in moisture and temperature.

There are two types of environment. The **macroenvironment** is the environment influenced by the general climate, elevation, and latitude of the region. Weather Bureau data on rainfall, wind speed, and temperature are measures of the macroenvironment. The measurements are taken at a standard height of 4.5 feet above the ground in a clear area away from buildings or trees.

The **microenvironment** is the environment close enough to the surface of an organism or object to be influenced by it. For example, bare soil tends to absorb heat, and consequently temperature just above or below the soil surface is much higher than air temperature during the day (Fig. 18.2); light quality and quantity is much different for herbs beneath a forest canopy than for leaves of the canopy (Fig. 18.3); air as far as ten millimeters from the surface of a leaf on a still day is less turbulent and higher in humidity than free air further away; in marshy areas, where the water table is close to the surface, minor dips or rises in the topography greatly influence the root microenvironment; and plants growing in shallow depressions are often subject to more frequent frost than those growing above them, because cold air will settle in the de-

Figure 18.2. Air and soil temperatures at several positions just above and just below the soil surface. Temperatures were measured during the warmest and coolest parts of the day. (Redrawn from W. D. Billings, *Plants and the Ecosystem*, Wadsworth Publishing Company.)

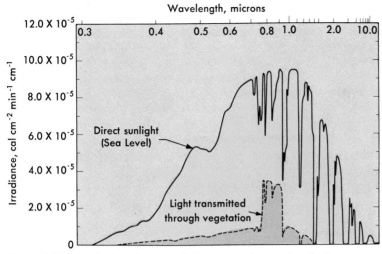

Figure 18.3. Distribution of solar radiation energy by wavelength. Energy was measured in direct sunlight and beneath a forest canopy. (Redrawn from D. M. Gates, *Ecology*, **46**, 1–13, 1965.)

pressions. Ecologists are convinced that the microenvironment is as important for plant growth as the macroenvironment.

Moisture

The distribution of precipitation in time is as important to plant growth as total annual amount.

Figure 18.4. Monthly rainfall of a rain forest near Georgetown, British Guiana and of a savannah near Bomako, Mali.

For example, equatorial and tropical areas[1] may both receive over 70 inches of rainfall a year, but the equatorial area receives an equal share of the total each month while the tropical area to the north or south experiences pronounced dry and wet seasons (Fig. 18.4). The result is that plants of the two areas are very different; equatorial regions support tall evergreen trees and vines in profusion (Fig. 18.29), but tropical areas support grasses, shrubs, or short trees which are often deciduous (Fig. 18.30).

Plants adapt to dry climates in several ways. Cactus plants (Fig. 18.33) develop water storage tissue in the stem and reduce transpiration by the loss of leaves; all of their photosynthetic activity is performed in stem tissue. Mesquite shrubs (*Prosopis*) and salt-cedar trees (*Tamarix*) possess long roots which tap ground water, sometimes at depths below 175 feet. **Epiphytes,** growing on tree trunks above the soil (Fig. 18.5), trap falling rain in leaf axils or in dead cells on the surface of roots. These epiphytes are not parasites; they depend on trapped water and debris for all growth requirements. Annual herbs

[1] For this chapter, equatorial means within the equatorial zone of latitude, and tropical means within the tropic zones of latitude. The "equatorial rain forest" of this chapter is often called the "tropical rain forest" by ecologists.

adapt to drought by completing a brief life cycle only during periods of sufficient soil moisture; *Boerrhavia repens* of the Sahara Desert is a small annual which can go from seed to seed in 10 to 14 days!

There is some evidence to show that plants can absorb dew through their leaves, but the amount of water obtained this way is not great. Ponderosa pine (*Pinus ponderosa*) (Fig. 2.16), a common tree of the western mountains, is one species which may take advantage of fog, mist, dew, or high humidity. This species sometimes grows far downslope in semi-arid climates, where soil moisture can be low for several months. In one experiment, one-year-old seedlings, growing in pots of soil, were exposed to fine mist every night. A control group was not exposed to mist. Both groups of pots were surrounded with waterproof material so that spray could only

fall on the foliage and not on the soil. Seedlings lived and transpired until soil moisture was depleted and death resulted. The group exposed to mist lived, on the average, a month longer than the control group, indicating that the former used atmospheric moisture.

Plants may also utilize fog by condensing the mist into large drops which trickle down the stem or drip from the branches, increasing soil moisture considerably. The amount of water added to soil this way by trees on the San Francisco Peninsula was measured as 2 to 59 inches, depending on the type of exposure. Normal precipitation (rain) for the area is only 25 inches!

Temperature

Temperature influences moisture availability. At freezing temperatures, water changes to ice and is unavailable for plant growth; consequently, in cold climates one of the principal factors limiting growth may be physiological drought! At high temperatures, evaporation may remove much of the soil moisture before roots can reach it. Transpiration will also be stimulated, resulting in further rapid depletion of soil water.

Brief extremes in temperature may be more important in determining plant distribution than are long-term moderate temperatures. Palms and many cactus plants seem excluded from areas which experience a certain frequency of frost. California lilac (*Ceanothus megacarpus*) seeds germinate poorly unless they have been exposed briefly to high temperatures. Many temperate zone plants are not frost-resistant until they have been exposed to moderately low temperatures for a short time.

Topography greatly influences soil and air temperature. Latitude and elevation combine, in mountains, to restrict species to certain belts (as shown for mountain hemlock in Fig. 18.6). Even at the same latitude and elevation, minor differences in the direction of slope (**aspect**) create different microenvironments. A gorge running east–west through an Indiana forest was studied to illustrate the effect of aspect on plant growth. The gorge was 200 feet wide at the top and 150 feet deep; its sides supported scattered trees, shrubs, and herbs. Meteorological instruments were placed six inches above and below the soil surface midway down each side. During spring, the south-facing slope was found to exhibit a greater daily range of topsoil

Figure 18.5. *Tillandsia*, an epiphyte native to Central America and Ecuador. (Courtesy of the New York Botanical Garden.)

Figure 18.6. Distribution of mountain hemlock (*Tsuga mertensiana*), showing the relationship between altitude and latitude.

temperature, a higher mean air temperature, a higher rate of soil water evaporation, and lower relative air humidity than the north-facing slope. Of nine spring-flowering species present on both banks, the average flowering time was six days earlier on the south-facing bank. A similar difference in flowering time could be achieved on level ground only over a distance in latitude of 110 miles!

Temperature is only an estimate of the *heat energy* available from solar radiation. Solar radiation at the limits of our atmosphere is equivalent to about two calories per square centimeter per minute. Much of this radiation is absorbed, scattered, or reflected within the atmosphere, and only half may reach the ground and heat it. In turn, the warm earth re-radiates energy back to space—*terrestrial radiation*. The difference between solar

(incoming) radiation and terrestrial (outgoing) radiation is termed **net radiation.** During a clear day solar radiation is greater than is terrestrial, and thus net radiation is positive; during the night the reverse is true and net radiation is negative. Over a 24-hour period, a month, or a year, net radiation may be positive or negative depending on the location. For example (Fig. 18.7), at Aswan, Egypt—a warm desert climate—net radiation is positive each month with a peak in June. Net radiation is positive only five months of the year at Turukhansk, Siberia—a cold subarctic climate—and the annual net radiation is very close to zero.

Light

Figure 18.3 shows how solar radiation is changed in quality and quantity by passage through green

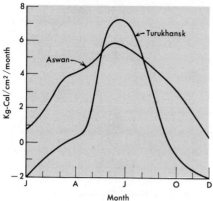

Figure 18.7. Monthly net radiation at Aswan, Egypt and Turukhansk, Siberia. (Redrawn from D. M. Gates, *Energy Exchange in the Biosphere*, Harper and Row.)

Figure 18.8. Photosynthetic response to different light intensities for sun and shade leaves of beech (*Fagus sylvatica*). (Redrawn from P. Boysen-Jensen and D. Muller, *Jahrb. Wiss. Bot.*, **70**, 493–502, 1929.)

Figure 18.9. Location of great soil groups in relation to gradients of temperature and moisture. (Redrawn from H. J. Oosting, *Study of Plant Communities*, second edition, W. H. Freeman and Company.)

leaves. Light intensity on the forest floor may be only 1 to 5 percent that of full sunlight. Some herbs of deciduous forests complete their life-cycle in early spring, before the trees above them completely leaf out and reduce light intensity. Leaves of plants which develop in shade exhibit different morphology, anatomy, and physiology than those which develop in sunlight, even if they are attached to the same plant. Shade leaves are larger, thinner, contain less chlorophyll per gram of tissue, have less well-defined palisade and spongy mesophyll layers, and achieve maximum rates of photosynthesis at much lower light intensities than do sun leaves (Fig. 18.8). Recent biochemical evidence has shown that the differences in photosynthetic rates is due to differences in the activity of the enzyme responsible for CO_2 fixation.

When a large tree dies in a mature forest, the canopy is opened and increased light intensity reaches the ground beneath, inducing seedlings there to begin rapid growth. Seedlings must be able to survive many years of slow growth in shade before such an opening "releases" them. Eastern hemlock (*Tsuga canadensis*) is remarkably shade-tolerant. Mature trees of this species live for 1,000 years. They can retain the capacity for rapid growth when released from shade to an age of 400 years! Saplings only six feet tall and one inch in diameter may have a ring count indicating an age of 60 years.

Trees which are not shade-tolerant may become established and grow to maturity if some disturbance such as fire or logging first removes the shade. Seedlings of shade-tolerant species grow beneath them, however, and replace them as they die, while their own seedlings do not remain alive. For this reason it is difficult to maintain pure stands of valuable shade-intolerant species such as teak, mahogany, Douglas fir, or the southeastern yellow pines. These species only invade disturbed sites and do not long maintain themselves.

Light is also reduced in intensity and quality by passage through water. Algae are not commonly found below a 200-foot depth except in very clear water. A record depth for algae may be found in Lake Tahoe, California, a very clear lake. A sample of water and bottom mud from a 450-foot depth contained *Chara*. Light intensity at that depth, despite the water clarity, is only 0.1 percent of full sunlight.

Brief exposure to light may be as important for some plants as long-term exposure. Seeds of some weedy species can be stimulated to germinate in the dark if first exposed briefly to full sunlight by simply shaking them in soil in a glass container for a few seconds. This treatment could be duplicated in nature by a plow turning the soil and seeds in it, briefly exposing the seeds to sunlight, then reburying them. Plowing aerates the soil, makes it more permeable for root growth, and removes plants at the surface—all conditions of advantage to an emerging weed seedling.

Soil

Soil is that part of the earth's crust which has been changed by contact with the living and non-living parts of the environment. Plants influence soil development in a number of ways: they move ions through the soil by absorbing them far below the surface through their roots, accumulating the ions in the leaves, and adding them to top soil after leaf abscission; plant roots and shoots decay and add nitrogen to the soil; and plant canopies shelter the soil surface, reducing erosion and water loss. But soil development depends on the physical environment as well. From unconsolidated deposits such as glacial debris, volcanic ash, and wind-deposited loess, young soils can develop in less than 100 years, and mature ones in 1,000 years. On the Kamenetz fortress in the Ukraine, a soil 4 to 16 inches thick has developed from limestone slabs since the fortress was abandoned in 1699. Soil development on harder rock, however, takes many thousands of years. Factors which affect the rate of soil development are discussed in Chapter 11.

The soil profile discussed in Chapter 11 is by no means typical of the earth's surface. Temperature and moisture factors of the environment combine to mold many variations of that basic profile. Figure 18.9 summarizes the relationship of major **soil groups** to moisture and temperature. Let us examine three of the major soil groups: podzol, chernozem, and laterite.

Podzol soils develop in cold, wet areas which support dense coniferous forests (Fig. 18.26). Acidic foliage accumulates as a thick layer of litter, and its slow decomposition helps create an acid soil of pH 3 to 5. Earthworm and bacterial activity is much less than in warmer soils. Below the litter is a sandy A horizon, the uppermost part of which is stained gray from organic matter but the rest of which is bleached white. Moisture is abundant enough to leach this horizon of all organic matter,

clay, aluminum, and iron. A brown B horizon begins abruptly 1 to 2 feet below the surface and may exhibit accumulations of iron, aluminum, and organic matter in alternating bands. The C horizon, of partially weathered rock, gradually merges with the B horizon.

Chernozem soils develop beneath grasslands which experience moderate rainfall and high summer temperatures (Fig. 18.31). Moisture is no longer sufficient to severely leach the soil, and the pH is 7 to 8. Fibrous root systems of grasses thoroughly permeate the soil and their decay evenly enriches it with organic matter. An A horizon, containing about 10 percent organic matter by weight, occupies the top 2 feet. It may be almost black in the upper part, but becomes dark brown below. Just beneath it is a layer of calcium carbonate accumulation, leached from above. This layer is sometimes called the A_c horizon. Below it is the parent material; there is no B horizon.

Laterite soils form in regions of high rainfall with continuously warm temperatures, such as those that support equatorial rain forests (Fig. 18.29). Mineral cycling is rapid because of fast plant growth and fast litter decomposition. Soil is nearly neutral in pH. Removal of the forest canopy quickly leads to soil deterioration because litter decomposition no longer compensates for leaching. Excessive amounts of fertilizer must be added if the land is farmed. Silicates (sand) are leached, but clay and iron are not; the iron in clay is oxidized to reddish Fe_2O_3, giving topsoil a brilliant color.

Depending on the type of parent material, soils may be abnormally low or high in particular salts. For example, **serpentine** soils, among other characteristics, are very low in calcium and many plants will not grow on them unless calcium is added. Some of the plants which do grow on serpentine are **endemic** to that soil (*restricted* to it and not normally found growing on other soils). Transplant tests, however, have shown that the serpentine endemics will grow perfectly well—in fact better—on normal soil *if* the surrounding plants are first removed. It appears, then, that the endemics are restricted to serpentine because they are poor competitors with other plants in non-serpentine areas, but they can tolerate serpentine soil and here find fewer competitors.

In arid regions, short and violent rainstorms create streams of excess water that run off and collect in low basins. In time, the ponds evaporate and leave the topsoil permeated with salts which had been in solution. The soil in the center of the basin (which lay under the most water) is highest in salt content; the soil at the edge is lowest. Plant distribution parallels salt distribution (Fig. 18.10): there may be no plants at all in the very saline center, but salt-tolerant plants such as salt grass (*Distichlis*) grow in a zone further out; less tolerant shrubs such as winterfat (*Eurotia*) grow still further out; and the usual shrubs of the area, such as sagebrush (*Artemisia*), surround the sink. Transplant experiments indicate that the seeds of salt-tolerant plants will germinate and grow very well in non-saline soil, and they appear to be restricted to saline soils only because they are poor competitors with other plants growing on non-saline soil.

Biological Factors

Typically, several plant and animal species occupy a given habitat. Very rarely does a single species, to the exclusion of all others, occupy a habitat. If the species in a group are individually examined, it will be seen that they utilize different parts of the environment or alternate with each other in time. Some plants (green plants) are pro-

ducers of carbohydrates, but others (parasitic and saprophytic fungi) are consumers or decomposers of it; some animals feed on plants, others feed on insects, others on small mammals; some plants are trees which utilize full sunlight, others are ferns which utilize weak light; some animals are nocturnal, others diurnal; some plants are evergreen, others deciduous; and so on. The portion of the environment utilized by each species is referred to as its **niche**. The niche has often been called the "occupational address" of a species.

In a particular area at a given time, one and only one species can occupy or fill each available niche. A laboratory experiment with two species of duckweed illustrates this point. Both species are small monocotyledons which float on the surface of quiet bodies of fresh water; their niches are identical. When *Spirodela polyrhiza* was grown in nutrient culture alone, its growth rate was spectacular (Fig. 18.11). But when it was grown with *Lemna gibba* it grew very poorly and eventually only *L. gibba* survived. The successful competitior in this case possessed a special type of parenchyma tissue which stored air, enabling it to stay on top of the water and not be submerged by the crowding population.

Figure 18.10. Zonation of plants around a saline basin, Grand Coulee, Washington. The saltiest soil in the center of the basin is devoid of plants. Surrounding this is a broad belt of salt grass (*Distichlis stricta*), then a narrow belt of dark shrubs (greasewood, *Sarcobatus vermiculatus*), and finally sagebrush (*Artemisia tridentata*). From R. F. Daubenmire, *A Textbook of Plant Autecology*, second edition, John Wiley and Sons.)

Figure 18.11. Growth of two species of duckweed in pure and mixed cultures. The species are *Spirodela polyrhiza* and *Lemna gibba*. (From John L. Harper, *Symposia of the Society for Experimental Biology*, **15**, 1–39, 1961.)

L. gibba was the better competitor, and it filled the available niche.

Competition is one form of biological interaction. It may be defined as decreased growth of two species or plants because of insufficient supply of some necessary factor(s). It is diagrammatically defined in Table 18.1. Sometimes only a single factor, such as a mineral element, is lacking, but usually two or more factors are limiting; it is difficult to separate them and determine which creates the greater competition stress. Competition is very important in determining plant distribution. It has already been mentioned that serpentine endemics and salt-tolerant species are restricted to their niches because they cannot compete with the numerous plants which populate more moderate niches. These poor competitors will die because the other plants have faster root and shoot growth rates, and remove water and sunlight. Sometimes introduced plants become wide-spread pests and actually replace native species because they are *better* competitors than the native species. This seems to be the reason for the enormous increase of cheatgrass (*Bromus tectorum*) in the intermountain region of the United States (see Chapter 9).

Amensalism, another form of biological interaction, may be defined as the inhibition of one species by another. In the coastal hills of southern California, sage shrubs (*Salvia*) cover the slopes, and grasses carpet the valleys. Occasional pockets of shrubs occur in the grassland. Figure 18.12 is an aerial view of these pockets of shrubs. The ground beneath and about the shrubs is bare of grass, and the grass is stunted as far as 30 feet from the shrubs. The zone of stunting is well beyond the limits of shrub root growth, so competition between shrubs and grass for soil moisture is not the cause of stunting. If clumps of grass are transplanted to the bare zone, their growth is severely retarded. Analysis of the soil beneath the shrubs, the bare zone, or the grasses, showed no major differences in physical or chemical attributes. Many volatile oils are emitted from sage leaves, and two of the oils in particular—cineole and camphor—are very toxic to grass seedlings. When fresh sage leaves were placed in a closed chamber with seeds of the grasses (Fig. 18.13), germination was often reduced to zero and radicle growth was considerably reduced. These same tox-

Table 18.1 *Categories of biological interaction in terms of the effect the relationship has on each partner. "On" indicates that the two organisms (#1 and #2) are in contact, while "Off" indicates that the two are not in contact. No effect = 0, stimulation = +, depression = −*

NAME OF INTERACTION	EFFECT			
	ON		OFF	
	#1	#2	#1	#2
Neutralism	0	0	0	0
Competition	−	−	0	0
Mutualism	+	+	−	−
Proto-cooperation	+	+	0	0
Commensalism	+	0	−	0
Amensalism	0	−	0	0
Parasitism/predation	+	−	−	0

Figure 18.12. Aerial photo of *Salvia* shrubs adjoining grassland. The light bands between and around the shrubs are bare of grass. (Courtesy of C. H. Muller.)

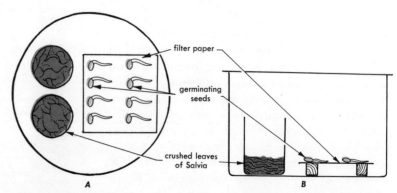

Figure 18.13. Chambers for the bio-assay of the effect of volatile substances evolved from *Salvia* leaves on germination of various grass species. *A* shows a top view and *B* shows a side view. (Redrawn from C. H. Muller, *Bulletin of the Torrey Botanical Club,* **93,** 332–351, 1966.)

ins were isolated from the air about sage shrubs, from the natural soil, and from seed coats of grasses in the soil. The conclusion was that the bare areas were caused by toxins emitted by *Salvia* which inhibited germination and growth of grasses.

Amensalism by chemical means may be very common in nature, for plants are leaky systems, passively contributing all sorts of substances to their environment. One investigator grew seedlings of 150 species in nutrient culture which included several isotopes as markers. The plants took up the isotopes through their roots and transported them to all parts of the plant, including the leaves. The plants were then exposed to a mist, and the water which condensed and ran off the leaves was collected for analysis. He found that 14 elements, 7 sugars, 23 amino acids, and 15 organic acids had been leached from the plants. In nature these substances would have been leached by rain from the leaves and would have accumulated in the soil.

Predation, another form of interaction, plays an obvious part in plant distribution. A glance along pasture fences reveals that normally abundant but palatable plants are absent from the pasture, yet are common outside it; weed species (which are not palatable) are common in the pasture, yet are rare outside it. Grazing animals are responsible.

Animal pollinators form a **mutualistic** relationship with plants. The bee, moth, and beetle which cross-pollinate flowers while feeding on nectar or pollen are familiar examples of mutualism. A more complex mutualism, exhibited by the fig tree and its wasp pollinator, is described in Chapter 15.

ECOLOGY AND PLANT POPULATIONS

The ecologist would like to use species as indicators of particular environments. He would like to know, for example, that a certain species only grows in a certain range of temperature or moisture. Then the temperature and moisture patterns of an area could be accurately predicted by just noting whether that plant species grows there. Unfortunately, most species are tolerant of a wide range of environments because they are not genetically uniform. Individuals of species *A* which grow in one environment are slightly different from others of species *A* which grow in another environment. Their genetic variation makes it possible for them to compete successfully in several different environments.

In the early 20th century, the Swedish botanist Göte Turesson collected seeds of species which had a wide range of habitats—from lowland, southern and central Europe to northeastern Russia in the Ural Mountains. When he germinated the seed and grew the plants in a uniform garden in Akarp, Sweden, he found that members of widespread species were not uniform. Those plants which grew from seeds collected in warm lowland sites often were taller and flowered later in the year than those which grew from seeds collected in cold northern sites. Yet these variants had very similar flower morphology and were inter-fertile, traits which indicated that the variants were all part of one species. Turesson called these variants within species **ecotypes.** He hypothesized that they were ecologically and genetically adapted to particular environments.

Evidence accumulated since that time indicates that most wide-ranging species are composed of several ecotypes, each differing slightly in morphology and/or physiology. Alpine sorrel (*Oxyria digyna*) is a small perennial herb which grows in rocky places above timberline in mountains and north of timberline in the artic. Both the alpine and arctic environments experience long periods of freezing weather, but alpine areas receive much higher light intensities during the growing season than do arctic areas. Physiological ecotypes have resulted. For example, the optimum light intensity for photosynthesis in alpine plants is much higher than that for arctic plants (Fig. 18.14). The two ecotypes also

Figure 18.14. Photosynthetic response to different light intensities for alpine and arctic ecotypes of alpine sorrel (*Oxyria digyna*). (Redrawn from H. A. Mooney and W. D. Billings, *Ecological Monographs,* **31,** 1–29, 1961.)

Figure 18.15. A community composed of essentially a single species: spinifex or porcupine grass (*Triodia basedowii*) in central Australia.

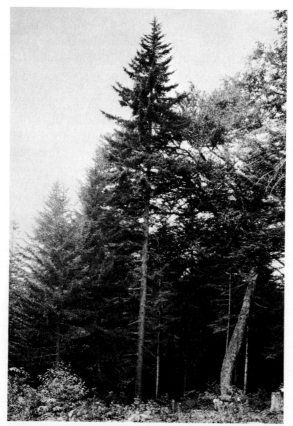

Figure 18.16. Subalpine forest in the mountains of North Carolina. The trees are red spruce (*Picea rubens*) and fraser fir (*Abies fraseri*); ferns carpet the floor of a local clearing caused by wind-throw of the trees. (Courtesy of U.S. Forest Service.)

differ morphologically in leaf color, leaf shape, and the frequency of rhizome production.

Applications of the ecotype concept have been especially useful in forestry. The United States Forest Service conducts tests to determine the best ecotype of a species to be used for reforestation in a given area. Seed is collected throughout a species' range and is germinated in a uniform garden; seedling growth is carefully watched, and the ecotype best suited for the garden climate is selected for reforestation programs.

ECOLOGY AND PLANT COMMUNITIES

Some environments support only one species. Dense clusters of cat-tail (*Typha*) grow in water-filled ditches, and all other species are excluded; widely-spaced spinifex (*Triodia*) covers hundreds of square miles of arid Australia, and all other perennial species are absent (Fig. 18.15). Most environments, however, support groups of species called **communities.**

Provided that distances are not too great, a given community repeats itself wherever the environment is suitable. For example, high peaks in the Smokey Mountains of North Carolina all exhibit red spruce (*Picea rubens*), mountain ash (*Sorbus americana*), Fraser fir (*Abies fraseri*) and a ground carpeted with several moss and fern species (Fig. 18.16). The spruce, fir, moss, and fern species are important members (but not the only ones) of this high elevation forest. High elevations at different latitudes or on different continents support other species, so there are other types of high elevation forest communities. Along slopes of the California Coast Range at moderate elevation, a shrubby community of scrub oak (*Quercus dumosa*), scattered pine and cypress trees, and some annual herbs consistently repeats itself. Black spruce (*Picea mariana*), *Sphagnum* moss, and the shrub Labrador tea (*Ledum palustre*) are repeatedly associated together in virtually every bog across Canada.

Communities are named after their largest, most common species—the plants which receive the full force of the macroenvironment. The spruce and fir of the high-mountain community in the Smokey Mountains occur in greater numbers than does the mountain ash, and their leaf canopies influence the microenvironment in which the carpet of ferns and mosses grow. The spruce and fir are said to **dominate** the community, and the community

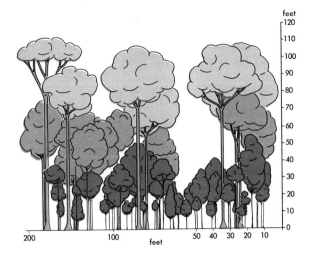

Figure 18.17. Profile diagram of the architecture of an equatorial rain forest in British Guiana. (Redrawn from P. W. Richards, *Tropical Rain Forest,* Cambridge University Press.) There are three tree strata, at approximately 105 feet, 60 feet, and 30 feet; shrub and herb species are uncommon. Small saplings are included in the 30-foot stratum.

name is **spruce–fir.** For similar reasons we have **black spruce** communities and **scrub oak** communities.

Communities exhibit a unique architecture or structure, produced by the leaf canopies of their dominant and subordinate species. In the equatorial rain forest of British Guiana, one group of species forms a canopy at 105 feet, a second group at 60 feet, and a third at 30 feet (Fig. 18.17). The ground floor is almost bare of shrubs or herbs. In deciduous forest communities of the United States, there is one tree canopy layer at about 60 feet, a layer of short trees (like dogwood) at 10 feet, and a rich shrub layer at 3 feet. Some desert communities exhibit only a single (shrub) layer.

Fossil impressions and preserved pollen in soil, rock, or bog deposits millions of years old sometimes reveal plant communities virtually unchanged from those which still exist today, but the deposits are usually geographically displaced from the living community. These fossil communities are so similar to present communities that paleo-ecologists are able to reconstruct past climates from them. A fossil community near the town of Goshen in central Oregon, dated to the Eocene epoch (50 million years ago), included the following genera: *Drimys, Magnolia,* holly (*Ilex*), oak (*Quercus*), *Ocotea,* and fig (*Ficus*). The closest living approximation of this fossil community today occurs 1,500 miles south of it in the temperate uplands of Central America. Apparently the climate in North America is now much colder than it was in the Eocene. The reconstruction of

past vegetation types and climates by use of fossil evidence is called **paleo-ecology.**

Another form of fossil evidence is pollen that has been trapped in lake sediments. As pollen is shed near a lake, some sinks to the bottom and is incorporated with silt and organic matter that forms sediment. A chronological sequence of pollen is thus preserved; the deeper the pollen occurs in the sediment, the older the pollen. Pollen of many species is resistant to decay under such anaerobic conditions, and may remain intact for thousands of years. Cores of sediments can be examined under the microscope and the pollen identified as to genus or species. Within limits, paleo-ecologists assume that the abundance of pollen in the core is proportional to the abundance of that species in the past.

Figure 18.18 summarizes the pollen profile of sediment from a lake in Nova Scotia. The present vegetation consists of a deciduous forest on hilltops with sugar maple (*Acer saccharum*), beech (*Fagus grandifolia*), and yellow birch (*Betula lutea*); and a coniferous forest in the valleys, with balsam fir (*Abies balsamea*) and white spruce (*Picea glauca*). But Fig. 18.18 shows that this pattern of vegetation is relatively recent. Sediments at a depth of five to six meters, corresponding to an age of about 9000 years, reveal pollen mostly from herbs and shrubs, which are species characteristic of meadow vegetation near permanent ice. The nearest such vegetation today occurs several hundred miles to the north of the lake. Shallower depths, corresponding to an age of about 6000 years, show peaks in spruce and fir,

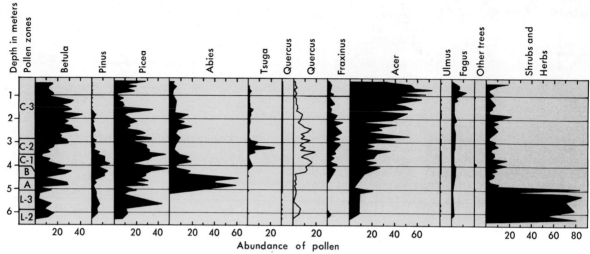

Figure 18.18. Pollen profile of sediment from a lake in Nova Scotia. Numbers along horizontal axis refer to relative abundance of each type of pollen. (From D. A. Livingstone, *Ecological Monographs,* **38,** 87–125, 1968.)

indicating dominance of typical northern coniferous forest. In yet shallower depths, deciduous species of birch and maple increase in abundance and coniferous species decline, indicating a continual warming trend in the climate.

Aerial photographs, using infra-red sensitive film, can reveal the species composition of communities. Different species stand out vividly because they reflect different amounts of infra-red (Fig. 18.19, plate 7). (Courtesy of S. Rae.) Such photographs can be used to map the distribution of species or communities over large areas.

How May Communities Be Measured?

If communities are to be useful as tools with which to estimate the nature of environment, they must be described more completely than just by their name, because minor variations in species composition indicate different environments. For example, where spruce and fir are both dominant and very abundant, the environment is probably different from where they are still dominant but less abundant. How can the abundance of a species or the composition of a community be measured?

1. One way to characterize a community is to measure the amount of ground covered by each species. The amount of ground cover is important because it affects the microenvironment of the community. To sample the ground cover, one could stretch a tape measure across the ground; the length

of tape covered by each plant canopy is proportional to the total ground cover of each species. This method of determining ground cover is called a **line transect.** Numerical information from line transects often contradicts common sense. In desert shrub communities, for example, a person looking across the landscape would think that the shrubs covered much of the ground (Fig. 18.32), but this estimate is in error because of the low perspective. If seen from above, when the canopy edges can be projected perpendicularly to the ground, it is apparent that actually only 10 to 20 percent of the ground is covered. Forest communities, on the other hand, exhibit more than 100 percent cover because of overlap in canopy layers at different levels.

2. Another method of estimating cover, especially useful when plants are low to the ground, is by use of quadrats. **Quadrats** are variously shaped frames which can be placed on the ground right over the plants. The amount of ground covered by each species can be estimated as a percentage of the area enclosed by the quadrat frame. A community can be sampled by placing quadrats at random or at regularly spaced locations throughout it. Usually, 10 percent or less of the total community area is included in the quadrat samples.

3. Other methods of sampling, based on mathematical assumptions, can be used to determine whether individuals are distributed at random, regularly (like trees in an orchard), or in clumps (Fig. 18.20). Non-random distribution (regular or

clumped) may result from propagation by asexual means, the method of seed dispersal, competition, or some pattern (like terracing) in the microenvironment. Most species are distributed in clumps.

Succession

As we have already mentioned, the microenvironment beneath a plant community is very different from that in the open; temperature, humidity, soil moisture, and light are all affected by the canopy. A stable community consists of species whose seedlings can survive in its unique microenvironment; seedlings of other species do not survive. If the stable community is removed by some catastrophe, leaving the soil exposed to full sunlight, the first species which colonize the site are not those of the old community. They are seedlings of species adapted to growth in full light intensity. In parts of the southeastern United States where oak–hickory communities are stable, the first species to invade following such disturbance, are annual and perennial herbs growing best in high light intensity, such as horseweed (*Conyza canadensis*), *Aster,* and broomsedge (*Andropogon virginicus*). Pine seed blows in from neighboring areas during the first several years after the disturbance, and within five years there are many pine seedlings. As the seedlings grow they compete for moisture with the herbs, and their expanding canopies begin to change the microenvironment. Within 30 years of the disturbance, a tall stand of pine has resulted. Examination of the forest floor, however, shows many oak and hickory seedlings and very few pine seedlings. Pine seedlings grow poorly in shade and under conditions of root competition, but those of oak and hickory do well. Within 50 years of the disturbance, the hardwood seedlings have formed a well-defined understory beneath the pines (Fig. 18.21). Whenever a pine tree dies, it is replaced in the upper canopy by an oak or a hick-ory. Within 200 years of the disturbance, the forest again consists only of mature oak and hickory trees, and the forest floor is covered with many seedlings of the same trees. The shrubs and herbs associated with an oak–hickory community are also present. Such a sequence of change in plant communities is called a **succession,** and the stable community which is capable of indefinitely maintaining itself is called a **climax community.** In this case, oak–hickory is the climax community.

Sometimes, climax communities are prevented from becoming established because of a recurring disturbance. There is some evidence that some of the prairie of the central United States which greeted pioneers could have been a forest were it not for recurring prairie fires. The fires destroyed slow-growing tree seedlings, but grasses quickly regenerated or germinated. When fire is controlled, prairie near the forest edge soon supports many tree seedlings. The Mediterranean area was once covered with a closed-canopy oak forest. The ground was densely shaded and soil moisture conserved. Beginning several thousand years ago, the forests were cut for lumber and fuel, and the cut-over land at first cultivated, then grazed by domestic animals. The exposed soil dried and eroded for many years until today it is so poor that the original forest species cannot grow on it, and will not return even if the area is left completely undisturbed. In place of the forest is a scrub community dominated by holly, heather, and gorse (Fig. 18.22; see color plate 7). Here again, the potential climax community is prevented from establishing itself.

The first ecologists to discover successional pathways—men like Josef Paczoski in Europe and Frederick Clements in the United States—had no experimental evidence to go on. They could not record changes as time passed, for most successions take hundreds of years to complete. The determination of successional pathways was arrived at by a

A   B   C

Figure 18.20. Diagrammatic representation of three types of plant distribution: A, clumped; B, regular; C, random. Courtesy of U.S. Forest Service.)

comparison of a climax community with adjacent small areas that had been disturbed at known times in the past. Examination of freshly disturbed or denuded areas (due to ice, landslides, fire, logging, wind-throw, etc.) revealed the likely **pioneer community,** or the first stage of succession. Sites which had been disturbed at some time earlier (verified from records of fires, grazing, or logging) exhibited later successional communities because the passage of time was greater. Complete sequences of succession were sometimes represented as bands of different communities about a gradually filling bog or behind a retreating glacier (Fig. 18.23; see color plate 7). The community closest to the lake or glacier was the most recently established and represented the pioneer community. The next closest community occupied soil which had been exposed for a longer time; at some earlier time it had supported the pioneer community, but enough time had passed for the next successional stage to appear. Beyond the second community was a third, representing the third successional stage, and so on until the climax community was reached. Ring counts of trees closest to the pond or glacier, coupled with estimates of rate of silting or of glacial movement, gave clues as to the time necessary for each successional stage to be reached.

The conclusions of these early workers have yet to be experimentally or empirically verified, and only recently have some of the causes of succession (such as competition, amensalism, and growth requirements of seedlings) been experimentally dealt with.

VEGETATION TYPES OF THE WORLD

The term **vegetation** refers to the life form of plants dominating a community: trees, shrubs, or herbs; deciduous or evergreen; coniferous or broad-leaved. A community dominated by evergreen angiosperms belongs to a different vegetation type than one dominated by shrubs or one dominated by conifers. On the other hand, many different communities dominated by different conifers belong to the same vegetation type. We will briefly examine six major vegetation types: tundra, taiga, deciduous forest, equatorial rain forest, savannah-prairie, and scrub. The distribution of these vegetation types in North America is shown in Fig. 18.24 (see color plate 9).

Figure 18.21. A stand of longleaf pine (*Pinus palustris*), about 50 years old, showing a well-developed understory of broad-leaved species of oak and hickory. In time, the pines will die and be replaced by maturing members of this understory. (Courtesy of U.S. Forest Service.)

Tundra

As one travels north or south towards the poles, trees gradually become stunted and less common, and finally disappear altogether. At the same time, low shrubs, perennial herbs, and grasses become dominant. The resulting meadow-like vegetation is called **tundra** (Fig. 18.25). Annual plants are very rare, and the reproduction of perennials is chiefly by vegetative means such as rhizomes. Shrubs are dwarfed, gaining normal height only in the protection of the lee of boulders or small hills. Winter wind, carrying ice, acts like a sand blast and prunes back stems wherever they project beyond the boulder or hill.

The warmest month has an average daily maximum temperature of 50°F or less and the growing season is short. Only two to four months of the year have average daily temperatures above freezing. The top foot or so of soil thaws during the growing season and roots may freely penetrate it, but below this level soil water may be permanetly frozen and the soil is called **permafrost.** Annual precipitation is about 10 inches, very little falling as snow. The

A

B

Figure 18.19. *A*, normal and *B*, infrared aerial photographs of the same area, showing mountain bog (red), sagebrush (green), and pine forest (dark brown) communities. Notice how different species show up in contrasting colors with infrared, and how rivulets of water in the bog become more obvious. (Courtesy of S. Rae.)

Figure 18.23. Succession of communities about a pond near Juneau, Alaska. Standing water in the pond supports leaves of cow lily (*Nuphar*); mats of *Sphagnum* moss and sedge (*Carex*) encroach on the pond along its edge; beyond is a shrub-dominated community (*Empetrum, Cassiope, Kalmia*) with some mountain hemlock (*Tsuga mertensiana*); finally, sitka spruce (*Picea sitchensis*), the tall, dark green trees, dominate the rest of the area. (Courtesy of J. Major.)

Figure 18.26. Taiga, northern Canada. (Courtesy of J. Major.)

Figure 18.27. Coniferous forest, Olympic Peninsula, Washington. Notice the moss and lichen-covered trunks and branches and the dense carpet of ground vegetation.

Figure 18.28. Change of leaf color in the deciduous forest during autumn.

Figure 18.29. Equatorial rain forest, New Guinea. The dense lower vegetation is absent farther into the forest. Notice the hanging lianas. (Courtesy of G. Webster.)

Figure 18.30. African savanna in Kenya. The flat-topped trees are *Commiphora*, source of the biblical myrrh. (Courtesy of O. A. Leonard.)

Figure 18.32. Desert scrub in southern New Mexico, dominated by creosote bush (*Larrea divaricata*).

Figure 18.36. Air pollution of a Tennessee foothill valley.

Figure 18.31. The original steppe, or prairie vegetation of central California, dominated by the bunch grass *Stipa pulchra*. (Courtesy of J. Major.)

Figure 18.22. Present-day shrub-dominated vegetation of the Mediterranean area. (Courtesy of H. A. Mooney.)

Figure 18.25. Alpine tundra, Beartooth Plateau, Wyoming, 10,690 feet elevation. (Courtesy of J. Major.)

terrain is generally flat, but small depressions are common and water tends to stand in them during the growing season. Day length fluctuates from 24 hours in June (northern hemisphere) to zero hours in December.

A similar tundra vegetation occurs in mountains at elevations above tree line. The climate, however, is slightly different. Solar radiation is more intense, day length does not fluctuate so extremely, and frosts are more common during the growing season. Many plants are tinged with red from excess anthocyanin, which may serve to protect the photosynthetic apparatus from damaging high light in-tensities by absorbing some of the light. The lower limit of these **alpine** areas depends upon latitude. At 60°N in Alaska, tree-line is at 3,000 feet; at 45°N in the Rocky Mountains, it is at 10,000 feet; at 20°N in central Mexico, it is at 13,000 feet; at 5°S in the Andes of South America, it is back to 12,000 feet; and at 50°S in Chile, it is back to 3,000 feet.

Taiga

Just south of the tundra is the **taiga,** a broad belt of communities dominated by conifers. A similar belt occurs in mountains, just below the alpine zone (Fig. 18.26; see color plate 7). Trees are slender, short (40–60 feet tall), and relatively short-lived (less than 300 years), but the forests are dense. The taiga is a patchwork of forests on upland sites alternating with bogs in depressions. Soils are acidic and relatively infertile, and in northern-most or upper-most parts of the taiga the soil is permafrost below a surface layer.

The growing season is as short as that of the tundra, but mean daily maximum temperature in the warmest month is much higher than 50°F (temperatures as high as 90°F are occasionally reached). Winter temperatures are very severe; the mean daily temperature is below freezing for six months of the year, and at Yakutsk, Siberia, the average January daily temperature is −46°F! Annual precipitation is less than 20 inches, and the air is exceptionally dry. The taiga is monotonous and silent; animal life is not abundant.

At lower elevations in some mountains, and along the coast (where temperature extremes are not so severe), coniferous forests are much more luxurious. Trees are taller, undergrowth denser. Many favorite recreation areas of the western United States are situated in these richer forests. The Olympic Peninsula of Washington receives over 80 inches of annual precipitation and supports the most spectacular coniferous forest in the world (Fig. 18.27; see color plate 7). Douglas fir, the most heavily cut lumber species of the United States, reaches its best growth in this area.

Deciduous Forest

In contrast to the taiga, the **deciduous forest** supports a diversity of plant life and is vibrant with animal activity. Deciduous angiosperm trees dominate communities of the deciduous forest. When day length becomes shorter and nights cooler in late summer, hormone concentrations in such trees change, and destruction of chlorophyll follows, allowing other pigments to show through in brilliant fall colors of red, yellow, and orange (Fig. 18.28; see color plate 8). In spring, before new leaves have fully expanded from buds, beautiful and distinctive herbaceous perennials and annuals come into flower. Shrubs are common, but not dense.

These forests occur in areas with a cold (though not severe) winter and a warm, humid summer. Snowfall may be heavy, but most of the annual precipitation falls as summer rain. Annual precipitation ranges from 30 to 50 inches. The growing season may be as long as 200 days, so it is no surprise that much of the deciduous forest in Europe, Asia and North America has been cut and replaced with cropland.

Equatorial Rain Forest

No other vegetation can quite equal the **rain forest** in sheer diversity of species and complexity of community structure (Figs. 18.17, 18.29; see color plate 8). The deciduous forest may exhibit 5 to 25 different tree species per acre, but the equatorial rain forest supports 20 to 50. The tree canopies typically form three layers, and beneath this dense overstory there are very few shrubs and herbs. Light intensity on the ground is less than 1 percent of full sun. Most of the herbaceous plants grow as epiphytes on trunks and branches of trees (Fig. 18.5). Other plants cling like vines to tree trunks and wind their way up to the canopies. Seeds of a few species, such as the well-named strangler fig, germinate in the canopy and grow down to the soil. Many of the tree trunks flare out at the base in peculiar buttresses.

The species are evergreen, that is, a given leaf may be shed or a given bud may break at any time of the year. The leaves tend to be very large, with long, tapering tips that may serve to drain the surface of excess moisture. Rainfall is high, often over 100 inches a year, but it is evenly distributed throughout the year (Fig. 18.4). Temperature is also uniform throughout the year, averaging 80°F. Warm temperatures and high humidity couple to produce an oppressive climate. Afternoon showers, which reduce the temperature, offer only short relief.

Farming peoples of the equatorial rain forest have for centuries practiced agriculture on a cut-burn-cultivate-abandon plan. The vegetation is removed in a small area by cutting and burning, cultivation follows for a few years, and the plot is abandoned. Abandonment follows because crop growth declines each year. The rain leaches nutrients from the exposed topsoil and the nutrients are not replaced by litter as they are beneath the normal forest canopy. Much of the equatorial rain forest of Asia and Africa has been disturbed in this way; this is less so in South America. Vegetation which invades abandoned plots is shrubby, dense, and often includes bamboo. It is this successional stage, which eventually will give way to rain forest, that serves as the "jungle" of television and films.

Savannah and Prairie

Forests give way to grassland as rainfall decreases and becomes more seasonally distributed. At first, trees remain close enough to retain a closed canopy, but gradually they become more widely spaced, and eventually they drop out altogether except along river banks and canyons. **Savannah** is the term applied to grassland with widely spaced trees (Fig. 18.30; see color plate 8); **steppe** or **prairie** is the term applied to grassland without trees (Fig. 18.31; see color plate 8).

Grasslands of the United States are heavily used for agriculture and grazing purposes, and little of the original vegetation or animal life remains today. Only from the records of explorers and early settlers are ecologists able to reconstruct in their imagination what these thousands of square miles once looked like.

The eastern part, with relatively high rainfall, supported a dense "sea" of grass some 3 to 6 feet tall. Here is a description from a journal dated 1837:

> "The view from this mound . . . beggars all description. An ocean of prairie surrounds the spectator whose vision is not limited to less than 30 or 40 miles. This great sea of verdure is interspersed with delightfully varying undulations, like the vast swells of ocean, and every here and there, sinking in the hollows or cresting the swells, appears spots of trees, as if planted by the hand of Art for the purpose of ornamenting this naturally splendid scene."

The drier, western part supported shorter and sparser stands of grass, but growth was still sufficient to allow graziers to cut and bale it as hay. Today the western prairie supports such a poor growth of grass that ranchers often plant seed to ensure enough fodder, and cutting and baling native grass is unheard of. Reasons for such a dramatic change include too much grazing pressure by herds of cattle and a slight drying trend in climate of the area over the past 75 years.

Grasslands are the home of the largest herbivores and carnivores. Vast herds of bison which roamed the North American prairie are gone, but animal preserves in African savannahs still excite visitors by a spectacular animal cast that includes eland, wildabeest, giraffe, elephant, and lion. Umbrella-shaped trees are characteristic of African savannah (see Fig. 18.30).

Scrub

Scrub is vegetation dominated by shrubs, and it occurs in areas with hot, dry summers and moderate winters. In areas with high, but seasonal, precipitation, the shrubs are tall and thorny, but if rainfall is 15 to 20 inches a year, scrub can be a nearly impenetrable thicket of shrubs with evergreen, hard, spiny, small leaves (see Fig. 18.31). Scrub of this sort is found around the Mediterranean Sea, on coastal hills of California and Chile, in southwestern Australia, and at the tip of South Africa. Despite the fact that the species of plants in these widely separate areas are entirely different, the vegetation has come to look strikingly similar. This type of scrub is called **chaparral.**

Desert scrub (Fig. 18.32; see color plate 8) grows in areas of rainfall below 10 inches a year. Shrubs are widely spaced and have far-reaching root systems (much of which runs just beneath the soil surface). The shrubs may be associated with **succulents,** plants which store water in stems or leaves, like the common cactus of the southwestern United States (Fig. 18.33). Some shrubs have deep roots which tap permanent supplies of ground water, while others reduce transpiration by dropping their leaves every dry period and developing new ones every rainy period. Seeds of desert annuals may remain viable for long periods in the soil until a favorable year when rainfall is sufficient to trigger germination and a colorful show of flowers a few weeks later. Although desert scrub seems to be made up mainly of perennial species, an actual count of all species appearing through the year reveals nearly as many annuals as perennials.

ECOLOGY AND THE ECOSYSTEM

Energy Flow

We have examined plant ecology at the population level, the community level, and the vegetation level. A final level of organization to consider is that of the ecosystem. The ecosystem of a particular habitat includes the plant and animal communities, the physical (nonliving) environment, and all the interactions between them.

The organisms of an ecosystem can be divided

into three categories: producers, consumers, and decomposers. **Producers** are green plants and certain bacteria which perform photosynthesis. **Consumers** are parasitic, herbivorous, or carnivorous plants and animals which feed on other plants or animals. **Decomposers** are saprophytic bacteria, fungi, and certain animals which obtain nutrition by breaking down dead organic matter. The three groups form a complex web of interdependence: consumers feed on producers, and decomposers feed on organic remains of both and convert them to smaller molecules that are taken up again by plants. The path of nitrogen through the ecosystem illustrates this interdependence (Fig. 18.34). Similar diagrams could be drawn for other nutrients like calcium, phosphorous, or sulphur.

Caloric energy is also passed through the ecosystem, and in this transfer organisms are linked to each other by a **food chain.** Food chains are short in arctic tundra: meadow plants trap light energy and convert it to caloric energy (protein, carbohydrate, etc.); caribou transfer some of this energy to themselves, retaining some of it, excreting some, and respiring some; and men transfer some of the caribou caloric energy by eating a fraction of the total caribou. Aquatic ecosystems exhibit longer food chains: algae (plankton especially) produce the caloric energy; some is passed to small animals which feed on plankton, then to small fish which eat the plankton feeders, then to larger fish, and then to shore birds or land animals like bear or man.

There is tremendous variation in net productivity from one ecosystem to another. **Net productivity** equals calories produced in photosynthesis minus calories lost in respiration. Desert ecosystems, for example, exhibit less than 0.5 grams dry matter produced per square meter per day; grasslands, 0.5 to 3.0; deciduous and coniferous forests, 3 to 10; agricultural areas utilized all year (as in sugar cane production), 10 to 25.

Energy transfer through the ecosystem seems very inefficient. Figure 18.35 summarizes a very simple food chain consisting of a pasture on which steers are grazed. Sample areas within the pasture were clipped of vegetation at the start of the experiment and were clipped again after a year's growth. The clippings were dried and weight converted to caloric content using a conversion factor of 4.3 kilocalories per gram of tissue (kcal/g). Root growth was estimated and also converted to calories. Total root and

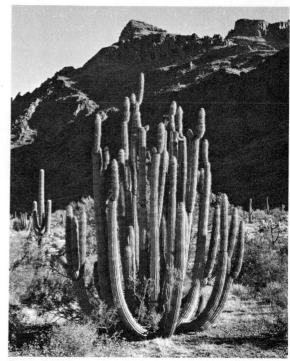

Figure 18.33. Desert scrub in southern Arizona with succulents (cactus species). (Courtesy of Ansel Adams.)

shoot growth represented net productivity. Weight increase of the steers during the year was noted and also converted to calories. Loss of plant material to insects, rodents, and rabbits was estimated as minor and ignored.

As shown in Fig. 18.35, annual solar radiation was 1,600,000 kcal/m^2 (of ground area). Of this, about 44 percent, or 700,000 kcal/m^2, was both within the spectrum of wave lengths which plants can use for photosynthesis, and actually reached the ground. Annual plant growth (net productivity) was 1,410 kcal/m^2, or only 0.2 percent of available light energy. Steer growth (net consumption) was 69 kcal/m^2, or only 5 percent of available plant energy (0.004 percent of annual solar radiation). From the energy standpoint, then, it is easy to see why large carnivores (like lions) require considerable territorial ranges; it takes a lot of savannah vegetation to support the herbivores, which in turn support large carnivores.

Pollution

Man is adding to the environment many gases, liquids, and solids whose effect on plants and ani-

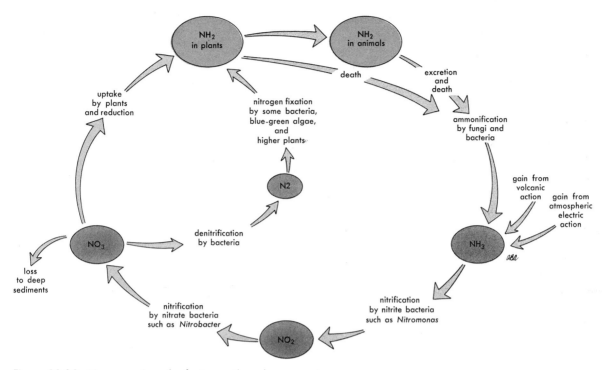

Figure 18.34. Diagrammatic path of nitrogen through an ecosystem.

mals is poorly known (Fig. 18.36; see color plate 8). If organisms in food chains or ecosystems shared by man are damaged by these substances, then man may be indirectly, but nevertheless severely damaged. The results of a spray program carried on at Clear Lake, California, illustrate this concept.

Clear Lake was an excellent recreation area except for intolerable summer gnats. The insecticide DDD (a close relative of DDT) was mixed into lake water in 1949, 1954, and 1957 to kill gnat larvae on the lake bottom. The amount of insecticide used was quite small—only enough to achieve a concentration of 0.02 parts per million (ppm)—but larval kills were 99 percent. In 1954, a few months after the second application, over 100 dead grebes (birds which feed on fish) were found around the lake. There was no external evidence of disease. Another large number of dead grebes was found after the 1957 application and this time fatty tissue was removed from some of them and analyzed for DDD. The tissue was found to contain 1,600 ppm DDD—a concentration some 80,000 times that of lake water! Components of the food chain leading to grebes

were then analyzed: water = 0.02 ppm DDD; plankton = 5 ppm; plankton feeders = 15 ppm; fish = 1000 ppm; grebes = 1,600 ppm. Each component of the chain had ingested large masses of the component below it, but had not metabolized or excreted the exotic new chemical, so a chain of accumulation resulted. In 1949 there were 1,000 pairs of nesting grebes around the lake with many young; in 1961 there were 32 pairs and no young. Had man been the ultimate carnivore in this food chain, he undoubtedly would have found the high level in DDD in fish or birds toxic to himself. There is some recent evidence, however, to show that man is able to metabolize DDD instead of accumulating it.

Large amounts of home and industrial sewage which are added to lakes and streams also affect food chains and, ultimately, man. Lake Erie covers some 10,000 square miles and has a shore lined with high population areas that introduce organically rich waste to it. Sewage settles on the bottom and bacterial activity is greatly increased. As bacteria respire they use up oxygen. In 1928 the amount of dissolved oxygen in bottom water was 78 percent

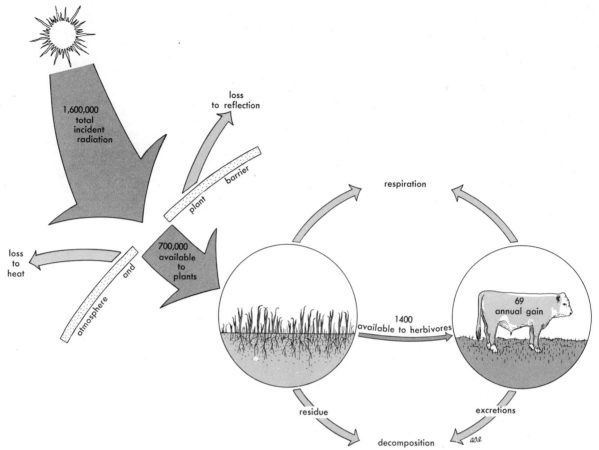

Figure 18.35. Transfer of caloric energy in a pasture-steer ecosystem. Numbers are calories per square meter. (Redrawn from W. A. Williams, *Journal of Range Management*, **19,** 29–34, 1966.)

of saturation, but in 1960 it was only 25 percent. It is on the lake bottom that commercially valuable fish such as blue pike lay their eggs, and their eggs require a certain minimum oxygen concentration in the water to survive. Recently, that minimum concentration may have been reached and passed. In 1956 the blue pike catch was 6.8 million pounds, worth $1.3 million; in 1963 the catch was 200 pounds, worth $120! Other commercial fish (trout, whitefish, herring, sauger, walleye) have suffered a similar decline, resulting in an overall loss of $3 million a year. Fish species now prevalent include commercially undesirable carp, shad, alewives, smelt, and goldfish. Algal species have changed or increased in number to the detriment of fishing and

boating. Bacterial counts occasionally reach levels which cause the closing of beaches. The time may not be distant when Lake Erie water is placed off limits for drinking. Such an end came to the lower part of the Hudson River in 1950 as a result of industrial pollution from Albany. Twenty billion gallons a day flow out the mouth of the Hudson, all of it unusable by New York City.

It is hoped that ecologists will eventually show us how ecosystems can be rationally manipulated to simultaneously satisfy all our needs: industry, agriculture, conservation, and recreation. Knowledge must be gained and compromises made between our needs and what the environment can support.

SUMMARY

1. Plant ecology is the study of how the environment influences plant distribution and behavior. It is also the study of the structure and function of nature. Ecology deals with the population, community, and ecosystem levels of organization.

2. Components of the environment include moisture, temperature, light, soil, and living organisms. Each species utilizes a different part of the environment, its niche.

3. A taxonomic species includes genetically similar plants which are inter-fertile. Within any wide-ranging taxonomic species, however, are many ecotypes which have become adapted to different environments.

4. A plant community is comprised of a group of associated species or populations which occupy a particular type of environment. One or more of the species dominate the community, but all the species together combine to form a unique architecture of canopy layers. The structure and composition of communities can be objectively determined with the use of sampling methods such as line transects or quadrants.

5. Climax communities perpetuate themselves, but successional communities are transitory.

6. Vegetation types of the world include: tundra, taiga, deciduous forest, equatorial rain forest, grassland, and scrub.

7. An ecosystem consists of the plant and animal communities of a habitat, their abiotic environment, and all the interactions between these components. Organisms of an ecosystem are tied together by food chains. Damage done to one component of a food chain (as through the action of pollution) may cause indirect but serious damage to the entire ecosystem.

19

Principles of Plant Taxonomy

Plant taxonomy deals with the **identification** and **classification** of plants. The term identification implies that many plant groups exist, and each group possessing unique characteristics can be separated from all other groups. These characteristics, as we shall see, involve morphology, anatomy, physiology, cytology, biochemistry, and geographic distribution. The term classification implies that plant groups can be ranked in some hierarchical relationship, on the basis of similarities in their characteristics. This ranking attempts to reflect real genetic relationships.

As mentioned earlier (see Chapter 2), taxonomy is building a library of the world's plant resources. Taxonomists are making a census of the unique characteristics of each of the half-million species that make up the world's flora. This census is far from complete for two reasons. First, many areas of the world have not been botanically explored and their floras can only be estimated; second, the process of evolution regularly produces new species even in well-known areas.

The important, exciting goal of the census is not identification, but rather classification. Taxonomists hope that, as the census becomes complete, patterns of similarity will become apparent within this maze of diversity; and that these patterns will enable them to reconstruct the evolutionary history of the present flora. They hope it will be possible to determine which characteristics—out of the hundreds available—are most reliable in showing pathways of evolution. Are they characteristics of flower morphology, wood anatomy, chemical composition, geographic distribution, breeding behavior, or chromosome number? Which traits in each category are **primitive** (imitating those of early, now-extinct plants)? Which traits are **advanced** (derived in time from primitive traits)? When the answers to such questions are known, it will be possible to construct a **phylogenetic classification**—one based on genetic, evolutionary relationships. Such a classification, in addition to summarizing the past, might prove useful in predicting future pathways of evolution.

Pre-Darwinian Period

The history of taxonomy can be very roughly divided into a pre-Darwinian (300 B.C.–1753 A.D.) period and a post-Darwinian (1753–present) period.

Before the time of written history, of course, man had learned to identify many plants at what we now consider the genus or species level. The plants named were those particularly useful as spices, medicines, food, drugs, or as religious symbols. The written history of botany begins with the Greek Theophrastus (ca. 300 B.C.), a student of Aristotle. He described and classified almost 500 kinds of plants in the book *De Historia Plantarum*. He utilized morphological characters like habit (tree, shrub, herb), length of life (annual, biennial, perennial), corolla form (petals fused together or free), and ovary position (superior, inferior) to distinguish among his plants.

Many of the taxonomists who followed the lead of Theophrastus through the dark ages and Renaissance are typified by Otto Brunfels (ca. 1500 A.D.) of Germany. He too classified plants on the basis of gross morphology, but he contributed to the record by expanding the list of known plants and accompanying many of the descriptions with drawings. Hans Weiditz was Brunfels' illustrator, and he did a magnificent job of carving accurate and intricate woodcuts from which prints were made (Fig. 19.1). The descriptions and drawings of many species were compiled in books referred to as **herbals.** Herbaceous plants used for medicinal purposes were often emphasized in the herbals. It was common to assign medicinal properties to a plant on the basis of its form: if the plant or parts of it imitated the shape of an organ then it was used for correcting ailments of that organ. For example, the thallus of liverworts, which resembles the shape of a liver, was used to treat liver ailments.

Carolus Linnaeus (ca. 1750) represents the culmination of the pre-Darwinian period. Born Carl von Linné in southern Sweden, he attended the University of Uppsala and received an M.D. degree from the University of Haderwijk in Holland. During his early days at Uppsala, however, he had become interested in the classification of plants and had developed his own system, called the sexual system. It was based almost entirely on the morphology and number of stamens and pistils, and proved very efficient in classifying the many new plants then being brought to Europe from Africa, America, and Asia and propagated in botanic gardens. After working on the collections of many large botanic gardens, Linneaus published the book *Species Plantarum* in 1753. The work included about 6000 species and 1000 genera.

Each species was described in Latin by a sentence

Figure 19.1. Part of one page from Brunfels' 16th century herbal. The plant illustrated is called stork's bill (probably *Erodium cicutarium*).

Figure 19.2. *Tradescantia virginiana* (spider wort). Note the jointed staminal hairs, from which the common name derives ×2.

Table 19.1 *Assumptions for the Besseyan phylogenetic scheme*

PRIMITIVE CHARACTERS	ADVANCED CHARACTERS
Plants woody	Plants herbaceous
Flowers bisexual	Flowers unisexual
Floral axis elongated	Floral axis short
Floral parts spirally arranged	Floral parts whorled
Floral parts numerous	Floral parts few
Sepals or petals free	Sepals or petals fused
Floral symmetry radial	Floral symmetry bilateral
Fruit single	Fruit aggregated

limited to twelve words which began with the genus name. Linneaus considered this sentence (a **polynomial**) to be the official, scientific species name, but he also coined a shortened form consisting of the genus name and one additional word from the polynomial. This shortened form (a **binomial**) served as his common name for the plant. He described spiderwort by this polynomial: *Tradescantia ephemerum phalangoides tripetalum non repens virginianum gramineum*, common name *Tradescantia virginiana*. Loosely translated, the polynomial means: The annual, upright *Tradescantia* from Virginia which has a grass-like habit, three petals, and stamens with hairs like spider legs, common name *Tradescantia* of Virginia. The genus name was in honor of John Tradescant, a gardener to King Charles I.

Not surprisingly, the binomial became favored over the polynomial by later taxonomists, and was accepted as the official, scientific name. Common names were left to individual choice and not restricted to Latin. The usual English common name for *Tradescantia virginiana* is spiderwort (spider plant), which comes from the jointed hairs on stamen filaments that resemble the jointed legs of a spider (Fig. 19.2).

Post Darwinian Period

After Darwin published *On the Origin of Species* in 1859, emphasis in taxonomy shifted to (1) a search for characters which reflected genetic (evolutionary) relationships, and (2) the construction of a phylogenetic classification scheme. Construction of phylogenetic schemes demanded that decisions be made as to which traits are primitive and which advanced. As there was little or no fossil evidence to support any of several hypotheses, early schemes were both

numerous and often in conflict. For example, the American C. E. Bessey started with the assumptions listed in Table 19.1.

Although Bessey's assumptions are still widely accepted today, some 60 years after he proposed them, disagreements continue to arise over specific interpretations. For example, some angiosperms such as willow (Fig. 19.3A) have flowers which lack petals or sepals. Are these plants primitive, never having evolved a complete flower, or did they once possess flowers with petals and sepals but have since lost them, hence really being advanced? Bessey thought they were advanced and consequently regarded plants like willow (*Salix*), birch (*Betula*), and maple (*Acer*) as closely related to advanced families. In contrast, equally competent taxonomists like Adolph Engler and Karl Prantl, living at about the same time, placed these plants among the most primitive angiosperm families.

Whether the absence of parts represents a primitive or derived condition can sometimes be determined by examination of the flower's vascular system. Vascular bundles which branch in the receptacle and extend to petals or sepals may still show the original branching even though the petals or sepals have been lost through time. Apparently the vascular system is very **conservative** (does not readily change). In other cases, the missing parts have not been lost altogether and are represented by small projections of glandular or apparently functionless tissue. Evidence of this sort, gathered since the time of Bessey, has confirmed his hypothesis that plants like *Salix* are indeed advanced. (Fig. 19.3B, C)

Another source of disagreement arises from answers to this question: do species which now resemble each other represent progeny from a common ancestor, or could it be possible they started from quite different sources and simply evolved to look alike? These two possibilities are diagrammed in Fig. 19.4. It is seen that species *A* and *B,* which are very similar today, actually arose from quite different stock. Similar environments, however, have shaped them along convergent paths. This pattern is termed **convergent evolution.** A striking example of convergent evolution is the present similarity between cactus plants of southwestern United States and certain *Euphorbia* species of Africa. Both groups grow in hot, arid climates, and as seen in Fig. 19.5 both have a similar morphology: they have no leaves, their stems are ribbed and swollen and contain water-storage tissue, and they are armed with spines. Yet, other traits, like flower morphology and geographic distribution, are so different that it is clear they must have arisen from different stock.

On the other hand, *B* and *C,* which are today less similar than *A* and *B,* actually did evolve from a common ancestor. This pattern is called **divergent evolution.** As time passes, *B* and *C* will become more and more different. The same pattern will be repeated by *C* and *D,* but at this point in time they are still quite similar.

If a certain group of plants arose from a single ancestral type, the group is referred to as **monophyletic** (page 13). If a group arose from several different ancestral types, the group is referred to as **polyphyletic.** Bessey considered each separate group of seed plants (the Anthrophyta, Coniferophyta, Ginkgophyta, etc.) to be monophyletic; some taxonomists considered all the seed plants together to be monophyletic; still others did not think that even one group, the Anthyophyta, was monophyletic. Unless fossil evidence is at hand, it is very difficult to separate the effects of convergent and divergent evolution. At present, most taxonomists agree that the largest group of plants which is *definitely* monophyletic is the family.

Figure 19.3. Willow (*Salix*) inflorescence and individual flowers: *A,* male catkin ✕1; *B,* single male flowers ✕15; *C,* single female flowers ✕15.

Considering the rich variety of hypotheses and plant characters upon which one could base a classification scheme, it is not surprising that the names of many plants have changed with time, being taken from one genus to another or being split into varieties or subspecies. *Tradescantia virginiana* L. is still an accepted name, more than 200 years after it was coined, but others that Linneaus selected have been changed. For example, in 1753 Linneaus described a plant from America as Canadian pine, *Pinus canadensis*. Later, the taxonomist Carriere realized it was a hemlock, not a pine, and transferred the species to the genus *Tsuga.* The new name was *Tsuga canadensis*, Canadian or eastern hemlock. How does this effect the old, valuable herbarium collections of the same plant which had been labeled *Pinus canadensis?* How would later taxonomists realize the connection between the two? There would be references, in older books, to something called *Pinus canadensis* which no longer is a name found in any taxonomic reference book. Who would know what plant was meant? There was a need, then, to allow for changes in name but at the same time preserve continuity and prevent a haphazard proliferation of new names.

The answer to the need was a set of international rules on botanical nomenclature, adopted by the International Botanical Congress of 1930, held in Cambridge, England. These rules were explicit guidelines for the determination and coining of valid plant names. Some rules dealt with the categories of classification available, others with the procedure of identifying new species, and others with the procedure for changing names.

One rule provided that, to be valid, a plant name must be published in Latin, accompanied by a short, descriptive paragraph which sets the new species off from related species, and that this description must be published in some recognized journal or flora. A book that summarizes all the plant species of a region is termed a **flora.**

Another rule provided that "each **taxon** (a general word for any plant group; pl. = taxa) . . . can bear only one valid name, the earliest that is in accordance with the Rules of Nomenclature." If two or more people independently proposed names for the same plant, then the earliest one proposed would be the valid one. Common sense indicated that there should be some limitation to this principle of priority, and for vascular plants the year 1753 (date of

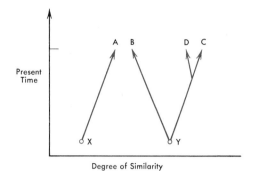

Figure 19.4. Hypothetical pathways of species evolution. X and Y represent ancestral forms from which A, B, C, and D have since evolved. The paths leading to A and B represent convergent evolution; the paths leading to B and C represent divergent evolution.

Figure 19.5. Convergent evolution of members of the Cactaceae (left) of the southwestern United States and members of the Euphorbiaceae (right) of Africa × 1/10.

publication of *Species Plantarum*) was selected. Names coined before that year were not valid, and the principle of priority only applied to names coined after that year.

Other rules dealt with changes of name. One of these is quite important, for it helps solve the problem of *Pinus (Tsuga) canadensis*. It provides that when a genus or species is changed, the name of the original author be placed in parentheses and followed by the name of the author who has now changed it. Thus, *Pinus canadensis* L. becomes *Tsuga canadensis* (L). *Carr.* It is now immediately apparent that *Tsuga canadensis* is not the original name for a plant and Linneaus had given it another name. The original name can be found in references like the Kew or Gray Herbarium Indexes which list all the previous names (sometimes there are several) for presently-named species. Portions of pages of *Index Kewensis* are shown in Fig. 19.6.

METHODS OF TAXONOMIC RESEARCH

Traditional (Morphological)

About ten years ago, the geneticist C. H. Waddington said it is ". . . an empirical fact that living organisms do not vary continuously over the whole range, but they fall into more or less well-defined groups, which are commonly called species." This philosophy is quite appropriate for a traditional taxonomist. He realizes that man's love of classifying nature often puts boundary lines between things where boundaries simply do not exist; but when it comes to species, he is convinced from extensive field observations that discrete species do exist and that classification at this level is not arbitrary or artificial.

A traditional taxonomist primarily examines plant morphology, searching for a few traits which consistently enable him to separate plants into "well-defined groups." In pre-Darwin days, a variety of traits were utilized, but now the choice is limited to those which presumably are (1) genetically controlled (rather than environmentally controlled) and (2) conservative in the evolutionary sense. Flower and fruit morphology, for example, meet these criteria, but leaf size does not.

It should not be supposed that the traditional taxonomist limits himself at the start of a study to the examination of only a few traits. He examines many characters, measures many plants, and keeps

PINUS :—

Boursieri, Carr. Conif. ed. I. 398 = contorta.

brachyphylla, Parl. in DC. Prod. xvi. II. 424 = Abies brachyphylla.

brachyptera, Engelm. in Wisliz. Tour N. Mexico, 89 = ponderosa.

bracteata, D. Don, in Trans. Linn. Soc. xvii. (1836) 443 = Abies bracteata.

Brunoniana, Wall. Pl. As. Rar. iii. 24. t. 247 = Tsuga Brunoniana.

bruttia, Tenore, Prod. Fl. Nap. p. lix = pyrenaica.

bullata, Roezl, Cat. Sem. Conif. Mexic. (1857) 16 = Montezumae.

Bungeana, *Zucc. in Endl. Syn. Conif.* 166.—China.

Buonapartea, Roezl, ex Gord. Pinet. 218 = P. Ayacahuite.

cairica, D. Don, ex Gord. l. c. 166 = halepensis.

calabrica, Hort. ex Gord. l. c. 168 = P. Laricio.

californica, Loisel. in Duham. Arb. ed. Nov. v. 243; Hartw. in Journ. Hort. Soc. ii. (1847) 189 = insignis.

Calocote, Roezl, ex Gord. Pinet. Suppl. 73 = P. Teocote.

canadensis, Bong. in Mém. Acad. Pétersb. Sér. VI. ii. (1833) 163 = Tsuga Mertensiana.

canadensis, Duroi, Obs. Bot. 38 = Picea alba.

canadensis, Linn. Sp. Pl. ed. II. 1421 = Tsuga canadensis.

canaliculata, *Miq. in Journ. Bot. Néerl.* i. (1861) 86.—China.

canariensis, *C. Sm. in Buch, Beschr. Canar. Ins.* 159.—Ins. Canar.

TSUBAKI, Adans. Fam. ii. 399 (1763) = **Camellia,** Linn. (Ternstr.).

TSUGA, Carr. Conif. ed. I. 185 (1855). *CONIFERAE,* Benth. & Hook. f. iii. 440.

MICROPEUCE, Gord. Pinet. Suppl. 13 (1862).

Balfouriana, *McNab, in Journ. Linn. Soc.* xix. (1882) 211.—Am. bor. occ.

Brunoniana, *Carr. Conif.* ed. I. 188.—Reg. Himal.

canadensis, *Carr. l. c.* 189.—Am. bor.

caroliniana, *Engelm. in Coult. Bot. Gaz.* vi. (1881) 223.—Am. bor.

diversifolia, *Mast. in Journ. Linn. Soc.* xviii. (1881) 514.—Japon.

Douglasii, Carr. Conif. ed. I. 192 = Pseudotsuga Douglasii.

Hookeriana, *Carr. l. c.* ed. II. 252.—Calif.

Lindleyana, Roezl, Cat. Grain. Conif. Mex. 8 = Pseudotsuga Douglasii.

Mertensiana, *Carr. Conif.* ed. II. 250.—Am. bor. occ.

Pattoniana, *Engelm. in Gard. Chron.* (1879) II. 756.—Am. bor. occ.

Roezlii, *Carr. in Rev. Hortic.* (1870) 217.—Calif.

Sieboldii, *Carr. Conif.* ed. I. 186.—Japon.

Tsuja, A. Murr. in Proc. Hort. Soc. Lond. ii. (1862) 508 = Sieboldii.

TSUSIOPHYLLUM, Maxim. in Mém. Acad. Pétersb. Sér. VII. xvi. (1871) n. IX. 12. *ERICACEAE,* Benth. & Hook. f. ii. 602.

Tanakae, *Maxim. l. c.*—Japon.

Figure 19.6. Part of two pages of *Index Kewensis*. Note that *Pinus canadensis* L. is shown to be a defunct synonym (by the = sign) for *Tsuga canadensis* Carr. The abbreviations refer to the publications where the original species description occurs. For example, *Pinus canadensis* is described in *Species Plantarum*, page 1421.

Table 19.2 *Characteristics of primitive and advanced vessel elements*

PRIMITIVE	ADVANCED
Similar in appearance to a tracheid	Not similar to a tracheid
Slender in cross section (100μ)	Broad in cross section (250μ)
Long (about 800μ)	Short (about 300μ)
Tapered ends	Truncated ends
Angular in cross section	Round in cross section
End wall multiperforated	End wall with single perforation

records of all of these. He eventually, however, subjectively selects only a few characters to serve in determining the number of taxa. The characters he selects are those which show discontinuities and are most helpful in separating related taxa.

Anatomical

We have already seen, with *Salix,* that the vascular system of a plant is very conservative and may be used to show genetic relationship between taxa. Anatomical studies of particular cell types within the vascular system reveal evolutionary patterns. Vessel elements, for example, show the changes summarized in Table 19.2. (See also Fig. 19.7.)

Evidence for this set of hypotheses comes from examination of fossils of lower plants that have vessels (such as *Selaginella*), and of *correlations* between changes of vessel characters with changes in other characters such as flower morphology (that is, plants with primitive flower morphology also show primitive vessel characters). Similar hypotheses of evolution have been applied to other cell types, such as fibers.

Differences in anatomy are usually minor at the level of genus or species; nevertheless these small differences can show evolutionary relationships. Creosote bush (*Larrea divaricata*) is a common desert plant in Argentina and the southwestern United States (Fig. 19.8). Although members of the species are very similar in morphology across its range from California to Texas, the species is not genetically homogeneous. Populations of creosote bush occur in three separate desert regions. The plants in each region have different numbers of chromosomes. Each desert, then, has its own subspecies of *L. divar-*

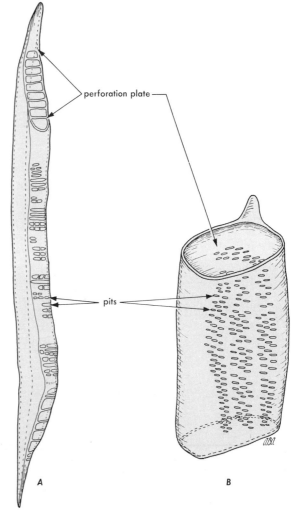

Figure 19.7. Advanced and primitive vessel elements: *A*, primitive type, from tulip tree (*Liriodendron tulipifera*); *B*, advanced type, from oak (*Quercus* sp.) ×100.

Table 19.3	*Length of fibers in creosote bush seedlings*
AREA WHERE SEED COLLECTED (DESERT)	AVERAGE FIBER LENGTH (μ)
California, Nevada (Mojave)	873
Arizona (Sonoran)	799
New Mexico, Texas (Chihuahuan)	628
Argentina (Argentinian)	533

icata. The eastern subspecies, in Texas and New Mexico, has the same number of chromosomes as the Argentina plants, indicating it is the most closely related of U.S. subspecies to Argentina stock. Fiber length shows the same pattern; Table 19.3 summarizes average fiber length of *Larrea* plants grown from seed in a uniform garden, all tested at the same age. The standard hypothesis for direction of fiber evolution is that advanced fibers are longer than primitive ones. Hence, the data in Table 19.3 indicate that western subspecies of *L. divaricata* have evolved from eastern ones.

Embryological

The pattern of embryo development is thought by some to be extremely conservative and to actually reflect in some measure the evolutionary history of the organism. For example, the embryo of man passes through early stages which are almost identical to those of lower groups of chordates (such as fish), and this is taken by some people as evidence of the evolutionary relationship between man and these other chordates.

Taxonomists have found that the pattern of embryo development in the Anthophyta is not uniform, and sometimes differences in the pattern correlate with differences in plant morphology. The position of the ovule within the ovary can be very distinctive and characteristic for entire families or orders; the ovule can be shaped so that the micropyle is directly opposite the point of ovule attachment to the placenta (**orthotropous**), or the micropyle can be facing the placenta (**anatropous**), or be somewhere between the two extremes (**campylotropous**) (Fig. 19.9). The number and position of cells in the embryo sac (prior to fertilization) can vary tremendously; there are eight cells in the embryo sac of lily (*Lilium*), four in evening primrose (*Oenothera*), and sixteen in *Peperomia* (Fig. 19.10). The number of cotyledons of course, is one of the criteria for separating members of the Anthophyta into two

Figure 19.8. Creosote bush (*Larrea divaricata*), an abundant warm desert plant of North and South America. (Courtesy of Allan B. Lang–National Audubon Society.)

Figure 19.9. Three types of ovule position within the ovary: A, orthotropous; B, campylotropous; C, anatropous. (Redrawn from G. H. M. Lawrence, *Taxonomy of Vascular Plants*, The MacMillian Company.)

classes, Monocotyledoneae and Dicotyledoneae. The degree of embryo differentiation at the time of seed dispersal may vary (members of the orchid family exhibit very rudimentary and incompletely formed embryos). The amount of endosperm may also be characteristic of large groups. For example, the endosperm of the grains is copious.

Biochemical

We have already seen (Chapter 2) that algal phyla are separated by biochemical characteristics, such as the type of chlorophyll or accessory pigments, the major component of the cell wall, or the principle form of food stored. Only within the past 20 years, however, has biochemistry been regularly applied as a taxonomic tool for higher groups of plants. Members of closely related taxa are chemically treated to extract particular compounds and the amounts (or presence vs. absence) of the compounds extracted are compared. The more similar the taxa are chemically, the closer their genetic relationship is presumed to be.

Almost every category of biochemical compound has been used at one time or another in these studies—sugars, amino acids, fats, oils, alkaloids, alcohols, terpenes, phenols—and they have been isolated by paper chromatography, electrophoresis, fractionation, or, more recently, gas chromatography. Many of these compounds are high in molecular weight and are structurally complex. It is reasoned by biochemical taxonomists that these complicated molecules probably evolved at only one time and in only one original group of plants; therefore, species or genera or families which possess the same complicated molecules are undoubtedly related to this original group of plants.

The existence of hybrids can sometimes be dramatically shown from biochemical evidence. Natural hybrids of two *Eucalyptus* species (Fig. 19.11) were suspected to occur in parts of Australia. The hybrids were hypothesized because unidentifiable *Eucalyptus*

Figure 19.11. Branch and foliage of a typical Eucalyptus tree ×½.

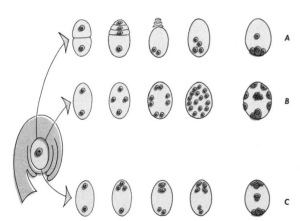

Figure 19.10. Number and position of cells within the mature embryo sac of three species: A, evening primrose (*Oenothera*); B, *Peperomia*; C, lily (*Lilium*). (Redrawn from A. S. Foster and E. M. Gifford, *Comparative Morphology of Vascular Plants*, W. H. Freeman and Company.)

Table 19.4 *Oil content (in relative units) of two* Eucalyptus *species and suspected hybrids* (from Pryor and Bruant, *Proc. Linn. Soc. N.S. Wales,* **83**:55, 1966)

CATEGORY	GERANYL ACETATE	CINEOLE	EUDESMOL
E. macarthuri	4.0	0.0	3.0
E. cinerea	0.0	4.0	0.5
Suspected hybrids	0.8	1.6	1.4

plants, somewhat intermediate in morphology, existed near the parent species. Leaf samples were collected from random members of each category and particular oils were extracted. Table 19.4 presents the results. The biochemical evidence of hybridization supported morphological evidence, because the suspected hybrids were intermediate in oil composition.

Biological

Biological taxonomists are interested in determining what the **natural biotic units** are. By natural biotic units, they mean populations of plants which maintain their distinctiveness because of **biological barriers** that genetically isolate them from other populations. These isolating barriers may be due to breeding behavior (time of flowering, type of pollinator), habitat or geographic isolation, or inability to form fertile hybrids with closely related groups (sometimes because of differences in chromosome number or chromosome morphology).

Biological taxonomists often deal with taxa at, and below, the rank of species because they are vitally concerned with detecting divergent evolution at an early stage. They are interested in discovering how species become distinct and how they are able to maintain their distinctiveness even though living in close proximity or in similar habitats. Biological taxonomists are often referred to as **biosystematists,** and their field as **biosystematics.**

Conflicts continue to arise between biosystematics and traditional taxonomy because their conclusions are not always the same. The "natural biotic units" do not always correspond to "well-defined groups" of morphological distinctiveness. The conflicts have, in part, a historical cause. As a result of world explorations in the 18th and 19th centuries, it became clear that some species within a genus differed only

slightly in morphology and habitat from region to region. One species might be in the mountains, another in the lowlands; one species might be along the coast of Europe, another along the coast of eastern North America. Were these really distinct species or should they be considered only as variants of a single species? The early answer was yes, they were distinct species. Much later, biosystematic research sometimes showed that fertile hybrids can result (or at least potentially can, once distances are overcome) from crosses between these species. The biosystematists then concluded that these "species" should be classed as subspecies of a single species because the morphological and geographic differences were less important indicators of relationship than breeding behavior.

On the other hand, biosystematists have sometimes discovered that breeding barriers exist between nearby populations of what traditional taxonomists had described as a single species (that is, the populations are morphologically identical). Yet, the genetic isolation undoubtedly is the first step of divergent evolution, and these populations represent distinct species at a very early stage. They are referred to as **sibling species.**

Many species which are widespread and occupy a variety of habitats may be composed of several gentically distinct subspecies. The subspecies are all interfertile, but hybrids are not common because hybrids are less suited to the habitats than are parental types. Drs. Clausen, Keck, and Hiesey of Stanford University have shown the existence of subspecies in plants such as five-finger, (*Potentilla glandulosa*), which grows in California from the coast near Stanford to high elevations near timberline in the Sierra Nevada mountains. Clausen, Keck, and Hiesey dug up plants of five-finger that were growing all along this range and transplanted them to-

| 98 ft. at Stanford | 4592 ft. at Mather | 10,000 ft. at Timberline |

Figure 19.12. Growth of *Potentialla glandulosa* ssp. *typica* after one growing season at each transplant garden. This coastal subspecies, or ecotype, grew best in its normal environment at Stanford; it grew very poorly at Timberline and was killed by the following winter temperatures. (Courtesy of W. M. Hiesey.)

gether in a uniform garden at Stanford. In the uniform garden, any differences among the plants from different locations would be due to genetic control. They found morphological differences; timberline plants were much shorter than lowland plants. They also found physiological differences; the timberline plants flowered earlier in the summer than lowland plants. When five-finger plants from different locations were grown together in a garden near timberline, these differences proved critical to survival: lowland plants which did not become dormant were killed during the winter, and lowland plants which flowered late in summer were nipped by frost before seeds could be produced. Clausen, Keck, and Hiesey named the lowland group *P. glandulosa* ssp. *typica* and the timberline group *P. glandulosa* ssp. *nevadensis* (Fig. 19.12).

Numerical

All the approaches to taxonomy just discussed share one common feature: each subjectively selects one form of similarity on which to base conclusions, and each weights its own choice of evidence as being more important towards understanding relationships than any other form of evidence. Traditionalists weight morphological evidence, biosystematists weight breeding behavior, and so on.

Another approach to taxonomy is to consider all forms of evidence with equal emphasis; all evidence has equal weight. This is a statistical approach and is called **numerical taxonomy.** A basic tenet of numerical taxonomists is that a great deal of evidence is required to objectively separate taxa from each other. They utilize 50 to 300 characteristics for each study.

As an example, let us say that the number and relationship of species within one genus is under study. To begin, the numerical taxonomist arbitrarily divides the genus into a number of species; this number may have no similarity to the number arrived at by traditional methods, and as we shall see it has no effect on the outcome of the study. These arbitrary or **operational species** are examined for each of the 50 to 300 traits and the data are

Figure 19.13. Dendrogram, summarizing percentage similarity between pairs and groups of operational species. If members of the same species were defined as showing at least 85 percent similarity (dotted line), then the 10 operational species would be lumped into six species as indicated by the intersections of the lines, 1 to 6, with the dotted 85 percent line.

recorded in relative terms (+ or − for present or absent and 1, 2, 3, or 4 for amount). Then all possible pairings between operational species are made and their similarity to each other is computed by a standard formula. Similarity is calculated as a percentage, 100 percent meaning complete similarity (identical) and 0 percent meaning no similarity (opposites).

The pairings are then ranked in order of increasing similarity and this ranking is diagrammed in a **dendrogram** (Fig. 19.13). The number of species can now be objectively determined by selecting an acceptable level of similarity. In Fig. 19.13 one taxonomist might select the range of 100–85 percent similarity as appropriate for members of the same species; if a line is drawn across the graph at the 85 percent level, we see that there would be six species. Another taxonomist might choose 65 per-

cent similarity, and there would be three species. And so on.

The numerical taxonomist, then, does not admit that discrete species exist in nature. He insists that when one considers many traits instead of a few, one will see a continuum of variation. Boundaries between species can be erected by taxonomists, and these boundaries may be objectively selected; but the boundaries are arbitrary and artificial in the final analysis. The viewpoints of traditional and numerical taxonomists, then, are quite different. Interestingly, classification schemes arrived at by numerical means often show good agreement with those arrived at by traditional means. Can we conclude that experienced, traditional taxonomists are just as efficient as the computers used in numerical studies?

SUMMARY

1. As a floral survey of the world becomes more complete, the goal of taxonomy becomes the construction of a phylogenetic classification scheme which summarizes the evolutionary history of our flora.

2. The history of taxonomy can be divided into a pre-Darwinian period and a post-Darwinian period. Linneaus culminated the pre-Darwinian period (ca. 1753) and organized plant nomenclature; nearly 200 years later, the rules of plant nomenclature were further elaborated.

3. Taxonomic research may take any of several directions, utilizing patterns of morphology, anatomy, embryology, biochemistry, and breeding behavior, any of which may be used to show genetic relationships between taxa. The numerical taxonomist utilizes information from all fields and does not give extra weight to any particular type of information. This is in contrast to other taxonomists who do choose to subjectively weight particular types of information.

20

Plant Growth and Development

The form any individual finally assumes is ultimately under the control of the genes present within its nuclei. However, the genes are only the starting point. Experiments with *Acetabularia* (Fig. 4.17) give evidence of the diffusion of a formative substance from the nucleus that plays an important part in the formation of the developing disc, and we already know much about the role of RNA in the transfer of genetic information during the synthesis of specific proteins. We have followed the development of shoots (Figs. 7.9, 7.27, 8.1, 10.17) and roots (Figs. 9.5, 9.6, 9.12) and noted that, in angiosperms, the apical cells of the shoot have considerable influence on the process of differentiation in the stem below them (page 202). The transformation of a vegetative shoot into a floral shoot changes the form and function of the apex (Figs. 7.2, 7.3, 15.25) and results in cessation of growth of the floral shoot (Fig. 15.25). Loss of a terminal bud induces the development of dormant lateral buds (page 69); fertilization stimulates the development of an embryo, seed, and fruit (Fig. 16.21). Stimulation to growth by one plant part may be brought about by the activity of distant parts. Growth in shoots, roots, and in development of the embryo is directional and polarized (Fig. 16.21). Expression of polarity in shoot or root growth is directly related to the polarity of cell division (Fig. 6.10). Development of root, shoot, flower, and probably even senescence have two things in common: (1) They are directed by hormones, and (2) direction of growth and form are related to the polarity of cell division. Growth is influenced by auxins, gibberellins, and kinins (page 94). Leaf abscission and dormancy are stimulated by abscisic acid (Fig. 10.19). Interaction between hormones is an important factor controlling growth and form.

A main problem of morphogenesis is to learn as much as possible about all of the factors influencing and regulating growth. Plant growth substances are one important group of factors. Where are they synthesized, what factors stimulate their synthesis, and what changes in growth rate or form do they induce? Work on these problems forms one of the most significant aspects of current botanical research.

TECHNIQUES USED IN STUDYING MORPHOGENESIS

An approach to problems of morphogenesis may be made by recognizing that most plant cells have a wide potential for development that is controlled and checked by factors in their normal environment *within* the organism. If such cells are not separated from their natural environment in the plant, all that they need from the external environment for normal development are relatively simple substances: water, mineral salts, and oxygen. However, from adjacent cells, they need and obtain foods and various regulatory compounds or hormones.

The control, or regulation, of development may be considered at three levels: (1) within the individual cell, **intracellular**, (2) between cells, **intercellular**, and (3) from outside of the organism, **en-**

Figure 20.1. Diagram showing different levels of regulation of plant development, such as intracellular, intercellular, and environmental.

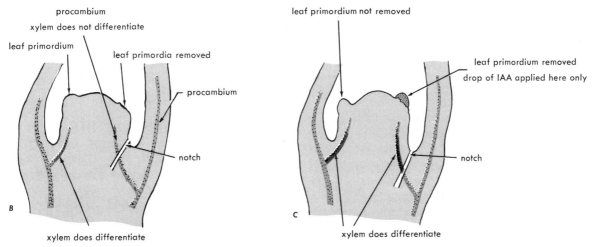

Figure 20.2. Diagram showing the effect on xylem differentiation of a notch cut in the axil of a young leaf directly below a leaf primordium, which is removed after the notch is made. A, notch is made, leaf primordium present; B, leaf primordium removed; C, drop of auxin in place of the leaf primordium.

vironmental. This is represented schematically in Figure 20.1.

By breaking down the organism into organs, tissues, and cells, by culturing these parts, and by studying the factors that induce normal growth, scientists are gradually answering some of the questions concerning growth and differentiation.

Perhaps one of the best ways to learn about the field of morphogenesis is to study some examples of the experimental methods used. Two of the most successful techniques used in this area of plant science are (1) **micurgical** or surgical operations and (2) **sterile culture** of plant parts.

Micurgical Experiments

It is known that the shoot apical meristem influences in some fashion the development of leaf primordia. Leaf primordia may influence the development of underlying tissues. Is this influence electrical, chemical, mechanical, or a result of geometrical relationships? Part of the answer has been obtained by careful microsurgical (micurgical) experiments on shoot apices. In order to understand these procedures, it will be well to refer to the diagrams of the shoot apex in Chapter 7 (Fig. 7.9). It will be recalled that xylem normally differentiates upward into the young leaf and toward a leaf primordium. If a cut is made between a leaf primordium and the next smallest leaf, differentiation toward the leaf primordium may be inhibited (Fig. 20.2A). If the leaf primordium is removed at the same time, no xylem is formed between the incision and the position of the excised leaf primordium (Fig. 20.2B). However, if a drop of a growth substance (**auxin**), indole-acetic acid (IAA) is placed on the wound caused by the removal of the leaf primordium, normal differentiation of the xylem strand takes place (Fig. 20.2*C*). In higher plants, such experiments have demonstrated that the differentiation of vascular tissue is dependent on a stimulus, probably a chemical growth substance, that passes from the leaf primordia in a **basipetal**

(downward) direction. In our example above, if leaf primordia are present above the cut, new xylem cells may be differentiated from the pith parenchyma and vascular continuity will be established around the cut (Fig. 20.2*A*). If no leaf primordia are present above the cut, or if they have been removed, no new xylem will form (Fig. 20.2*B*).

A logical conclusion from this experiment is that

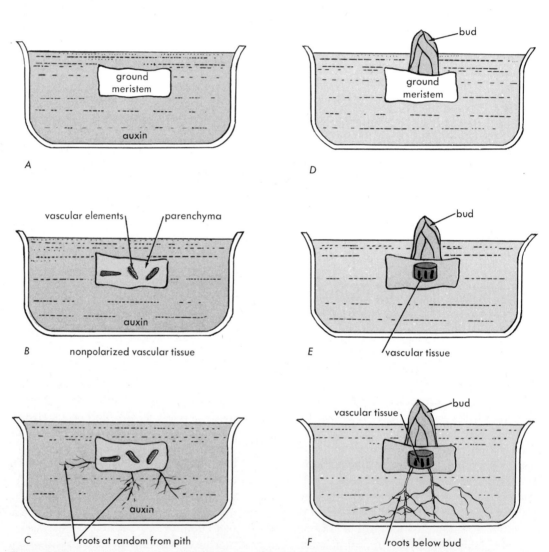

Figure 20.3. Diagram showing the effects of auxin and of a bud on the differentiation of ground meristem growing in sterile culture. *A, B, C,* auxin in sterile culture solution. *A,* ground meristem in culture solution; *B,* ground meristem differentiates into parenchyma and random xylem elements; *C,* roots grow at random from the block of ground meristem. *D, E, F,* no auxin in sterile culture medium. *D,* bud grafted to top of ground meristem callus; *E,* differentiation of parenchyma tissue and a cylinder of vascular tissue with polarized xylem elements; *F,* roots grow beneath the vascular cylinder.

a young developing leaf primordium is producing a compound (hormone) that stimulates meristematic cells below to differentiate into xylem. No information is given by this experiment on how the hormone regulates cell differentiation. Does it act on DNA, on certain specific enzymes, or in other ways?

Sterile Culture of Cells, Tissues, and Organs

Another very useful technique in studying morphogenesis is the isolation of cells, tissues, and organs and their culture under sterile conditions. For example, as we have already seen in chapter 6, it has been possible to isolate individual parenchyma cells from the mature, secondary phloem of wild carrot. If such a cell is carefully grown in sterile culture and supplied with a complex growth medium, it will give rise to a mass of relatively undifferentiated cells that is capable, under certain conditions, of producing a complete new plant (Fig. 6.2). Studies such as this demonstrate that the nucleus of a mature parenchyma cell contains all of the genetic information necessary for the development of the entire plant. During the normal life of this cell (in the phloem of the plant in this example), only certain genetic information stored in the genes is used. There must be a sequential activation of some genes and a repression of others. This is a delicately regulated process that is drastically altered when the cell is removed from its normal environment and cultured on an artificial medium. How is the regulation that results in normal ordered development accomplished?

Let us consider an experiment with meristematic tissue from the shoot apex. A small block of ground meristem is removed from the region just below the apical meristem of a lilac stem (*Syringa vulgaris*) (Fig. 7.9) and placed in a nutrient solution containing auxin (Fig. 20.3*A*). The meristematic cells will now differentiate into a disorganized mass of parenchyma cells with scattered areas of unorganized vascular elements (Fig. 20.3*B*). Randomly placed roots may arise from the differentiating parenchyma (Fig. 20.3*C*). Further organization of the cells in the tissue block will not occur even though the growing tissue is divided and subcultured for years.

If, instead of placing the auxin in the nutrient medium, a drop of auxin is placed in a localized area on top of a block of ground meristem, a cylinder of vascular tissue will differentiate just beneath the drop of auxin. Roots, however, do not arise. Grafting a small bud on the top of the block of ground meristem (Fig. 20.3*D*) results in the differentiation of a cylinder of vascular tissue from the ground meristem (Fig. 20.3*E*), followed by the appearance of roots below the vascular cylinder (Fig. 20.3*F*). The roots will eventually connect with the cylinder of vascular tissue.

CORRELATION OF GROWTH

The growth of the cells, tissues, and organs of a plant is not haphazard, but the various parts grow in relationship with one another. That is, the growth of the terminal bud influences the course of development of lateral buds, the activities of leaves influence the production of flowers, and so on. How is the growth of thousands of cells so correlated that the form and structure of a normal plant results? We have seen how the food is produced by leaves and translocated into other parts of the plant; how minerals and water are absorbed by roots and move throughout the organism; and how external factors influence these activities. Plant hormones also play an important part in growth correlations (page 70).

Plant Hormones and the Regulation of Growth

A plant **hormone** is defined as *an organic substance produced by the plant which, acting in very small amounts, regulates plant physiological processes.* Hormones in the plant usually move from the tissue where they are produced to a tissue where they act to produce special growth effects. Whereas, in animals, hormones are produced in special glands, in plants there are no special hormone glands; in plants, these substances may be produced in rapidly growing embryonic tissues, wherever they occur, or in older cells as well. One of the most active fields of plant physiological research today is the investigation into the nature and action of plant growth regulators. From this extensive research new knowledge is continually being gained—knowledge that increases our ability to control plant growth (pages 4–5). Less than forty years ago the existence of plant hormones was unknown. Today, many synthetic organic compounds have been produced that exert profound regulatory effects on growth. The general term **growth regulator** includes both naturally produced and synthetic compounds that delay fruit drop, induce the formation of roots on cuttings, cause flowers to set fruit, prevent sprouting, and hasten fruit ripening.

Four general types of hormones that together strongly influence cell division, enlargement, and differentiation are the **auxins,** the **gibberellins, kinins,** and **abscisic acid.** The history of the discovery of each of these classes of hormones is in itself an intriguing aspect of biology. Auxin was discovered as the culmination of many years of study of the growth response of seedlings, particularly the seedling's curvature toward a unilateral exposure to light. Kinetin was found almost accidentally after an intensive search for an unknown chemical, present in yeast extract and coconut milk, that stimulated the growth of tobacco callus cells. The discovery of gibberellin stemmed from observations on a disease of rice known in the Orient as the "foolish seedling disease." Seedlings of rice grew abnormally rapidly when infected with a fungus, *Gibberella fujikuroi.* The identification of the cause of this rapid growth and the isolation of the growth-promoting chemical, gibberellic acid, that was produced by the fungus and caused the abnormal growth in the seedling, was made by Japanese scientists in 1941, seventeen years before Western biologists became aware of the discovery in 1958.

Hormones long were postulated as agents inducing dormancy and leaf abscission. Indeed, substances able to induce dormancy and to bring about leaf abscission were found in different laboratories. Further study showed that one and the same substance was responsible for both phenomena. It was called abscisic acid.

It has long been postulated that a hormone must be involved in inducing the change from a vegetative apex to a flowering apex. No such hormone has as yet been isolated.

Much of our knowledge of the regulation of plant growth has come from studies of the growth of the coleoptile of oat (*Avena*) seedlings (Figs. 16.25, 16.32). When the seedling is grown in the dark and the coleoptile has attained a length of 25 to 40 millimeters, and before the first foliage leaf protrudes, it is ready for use in various experiments of the type to be discussed.

Auxins

It seems worthwhile to describe a number of early experiments with the grass coleoptile, chiefly that

of *Avena,* because they are basic to an understanding of growth regulators; furthermore, they demonstrate scientific methods and techniques of investigation.

As early as 1880, Charles Darwin observed that grass coleoptiles curve toward the light, a phenomenon known as **phototropism** (Fig. 20.4*A*). If the tip of the coleoptile is covered with a small cap of tinfoil that does not permit illumination of the cells of the tip (Fig. 20.4*B*), there is no curvature, even though the lower part of the coleoptile is still illuminated from one side. When the upper part of the coleoptile is illuminated from one side and the lower part is covered with tinfoil to exclude light, curvature of the coleoptile occurs in the normal way (Fig. 20.4*C*). If 2.5 to 4 millimeters of the tip of the coleoptile is removed, there is no response to one-sided illumination; the coleoptile remains straight and cells in the lower part do not elongate (Fig. 20.4*D*).

From simple but significant experiments like these, Darwin concluded "that when seedlings are freely exposed to a lateral light, *some influence is transmitted from the upper to the lower part,* causing the latter to bend."

Before describing experiments that demonstrate what the transmitted influence is and the manner of its transmission, let us consider the elongation of the cells of the coleoptile that result in the curvature. The actual bending of the coleoptile is the result of more rapid elongation of the cells on the shaded side than of those on the illuminated side. In the experiments described above, it is apparent that the coleoptile tip, rather than the lower part, must be illuminated if it is to produce the substance that we assume must move downward to the cells in the lower part of the coleoptile and stimulate those on the shaded side to more rapid elongation than those on the illuminated side.

Boysen-Jensen (1910–1911) showed that, if a transverse incision was made on the shaded side of the coleoptile and a thin plate of mica inserted (Fig. 20.4*E*), coleoptile curvature did not occur, but that, when the incision was on the illuminated side, curvature did occur (Fig. 20.4*F*). This response led to the conclusion that the stimulus moves down the shaded side of the coleoptile, from the tip to the cells in the region of elongation, which are stimulated in their growth. Boysen-Jensen also found that the stimulus could be transmitted through a thin layer of gelatine. He cut off the tip of the coleoptile, placed a drop of gelatine over the stump, and then,

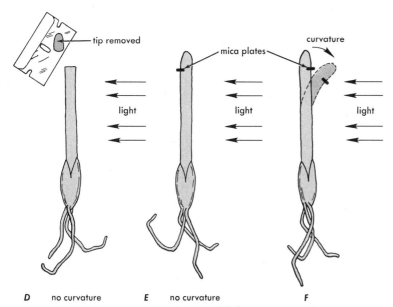

Figure 20.4. The influence of unidirectional light on curvature in an oat (*avena sativa*) coleoptile. *A*, coleoptile curves toward light; *B*, tip only, shaded, no curvature; *C*, tip only, exposed to light, curvature; *D*, tip removed, no curvature; *E*, mica plate inserted in tip away from light, no curvature; *F*, mica plate inserted in tip on side of light, curvature.

replaced the tip (Fig. 20.5*A*). With one-sided illumination there was curvature of the coleoptile.

These early experiments led to the definite discovery by Went (1928) of the hormonal nature of this growth stimulus. Went placed the tips of a number of coleoptiles on thin agar disks (Fig. 20.5*B*), where they remained for approximately two

hours. Then the disks were placed on one side of the decapitated coleoptile. Curvature of the coleoptile resulted; the direction of curvature depended upon which side the block was placed on (Fig. 20.5). This experiment demonstrated that a definite substance would diffuse out of the decapitated coleoptile tips into the agar and would in turn migrate

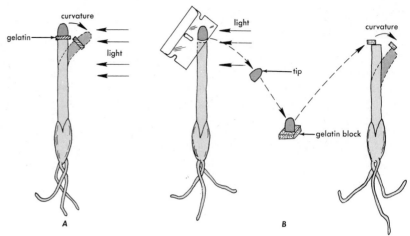

Figure 20.5. Demonstration of the production of a growth hormone in an oat coleoptile tip. *A*, tip of coleoptile is cut away, a piece of gelatin is placed on the stump, and the severed tip is placed on top of the gelatin; *B*, the tip of the coleoptile is severed from the coleoptile and placed on an agar block which, after a time, is placed on the stump.

from the agar to the cells in the stump. Moreover, it was found that the amount of curvature was proportional, within certain limits, to the concentration of the growth regulator in the agar blocks.

Other experiments showed that the growth regulator moved in greater amounts from the coleoptile tip to the shaded side of the coleoptile than to the illuminated side. When a coleoptile tip was placed on two agar blocks (separated by a thin metal plate) and the tip lighted from one side, there was an unequal distribution of the growth substance; it tended to move toward the shaded side. Thus, more rapid growth was induced on the shaded side of the coleoptile, resulting in curvature of the organ.

Plant growth regulators (of the type found in oat coleoptiles) that promote cell elongation are called auxins. Their effectiveness in promoting growth is frequently tested by using the oat coleoptile in a manner similar to that used by Went and briefly described above. Because the generic name for oat is *Avena,* this curvature test is known as the **Avena test.**

Auxins are widely distributed in the plant kingdom, in both higher and lower plants. In higher plants they may occur in foliage leaves, in the hypocotyl, stems, flower stalks, roots, fruits, seeds and even the pollen.

Although plants do not possess glands that secrete hormones, auxin is synthesized only in a few localized centers in the individual plant and is transported throughout the plant body from these centers. Thus, auxin helps to coordinate the various growth activities of the entire plant. In a normal green plant the apical buds and young expanding leaves are active centers of auxin synthesis and export. Older leaves in some plants may synthesize auxin, but generally, little auxin is exported from them.

A common growth regulator of the auxin type that has been identified in plants is 3-indoleacetic acid (IAA). This substance and several others of similar nature and action are now made synthetically.

Kinins

When young healthy tobacco stems are cut or wounded, a proliferation of parenchyma forms a **callus** around the wound. This callus may be excised and grown on an artificial nutrient medium much as was the ground meristem of the lilac stem (Fig. 20.3). It was found that the tobacco callus would not grow even when supplied with a normal complement of vitamins, nutrients, and auxin. The addition of coconut milk or yeast extract induced growth. When isolated and analyzed, this growth substance in coconut milk and yeast extract turned out to be an organic base related to the purines found in the nucleic acids (page 103). It was called kinetin. Alterations in growth patterns have been achieved by applying kinetin to plants.

Kinetin reacts with auxin in different ways. It may stimulate the continuous division and growth of cells in one instance, and in another instance may act in opposition to auxin. Thus, budding may be inhibited by auxin but be stimulated by kinetin.

The importance of the hormonal balance in regulating plant growth and differentiation may be illustrated by the growth of tobacco callus tissue in culture. Growth of undifferentiated tobacco callus tissue can be continued through an indefinite number of subcultures as long as certain major nutrients

such as amino acids and phosphate are supplied and as long as a proper balance between auxin and kinetin is maintained. If the ratio of kinetin to auxin is increased, by increasing the amount of kinetin or decreasing the amount of auxin in the culture, buds and leafy shoots are produced (Fig. 20.6).

Gibberellins

The most striking effect of gibberellins is their ability to increase elongation in intact plants. Several gibberellins have been isolated from higher plants. All of these complex chemicals are chemically related to the gibberellin first isolated from the fungus *Gibberella fujikurio*.

There is a dwarf corn mutant which may be

Figure 20.6. Effect of balance between kinetin, auxin, and vitamins on the growth and development of callus tissue in tissue culture from cotyledon of a *Streptanthus* seed, (Courtesy of R. D. Sjolund.)

| 0.0 | 0.0001 | 0.0005 | 0.001 | 0.0025 | 0.005 | 0.01 | 0.025 | 0.05 | 0.1 |

μg GA$_3$ / PLANT

Figure 20.7. Dwarf peas (*Pisum sativum*) showing the effect of added gibberellic acid. Gibberellic acid added to each plant seven days before the picture was taken. (Courtesy of Rappaport.)

stimulated to grow quite normally by the application of gibberellin. This mutation prevents, or at least limits, the synthesis of a natural gibberellin. The deficiency may be overcome by artificially supplying the required gibberellin. Dwarf plants of other genera may also be stimulated to additional growth by the application of gibberellin.

Since the rate of growth of such dwarf plants is frequently directly proportional to the concentration of gibberellin applied and is not influenced by low concentrations of other growth regulators, these dwarf plants can be used as a biological test material for determining the concentration of gibberellin in samples of plant extracts that contain unknown amounts of natural gibberellin. Figure 20.7 shows a tray of dwarf peas treated with varying amounts of gibberellic acid. The relationship between the amount of gibberellin applied and the total growth can be readily seen. Gibberellins not only affect vegative growth but in some instances affect flowering as well.

Abscisic Acid

Abscisic acid has been isolated from over thirty species of plants, including ferns, potato, apple, rose, and willow. It has been detected in many plant parts: leaves, stems, rhizomes, buds, seeds, endosperm, embryo, and fruits of all ages. It appears to be very common in mature and senescing tissues. Its physiological action seems to be involved largely in growth retardation, inhibition of physiological responses, abscission, and dormancy. When sprayed on leaves it frequently induces abscission and is known to lengthen periods of seed dormancy. When applied to plants, or plant parts, with growth substances, it frequently counteracts the effect of the accompanying growth stimulant. For instance, when it is applied with gibberellin, the expected response of increased growth does not occur. The action of abscisic acid on leaf abscission has been discussed in some detail on page 203.

Vitamins and the Regulation of Growth

Extensive research in the field of animal nutrition has shown that animals, including man, may suffer from certain so-called vitamin deficiencies. The diet may contain water, mineral salts, carbohydrates, fats, and proteins in sufficient quantities to supply their energy requirements, but something may be lacking which is necessary for normal development and good health. If this something, which we call a **vitamin,** is added, even in the minutest quantities, the individual recovers from the deficiency symptoms.

Today, a number of different vitamins have been discovered, each with specific physiological roles. Not only do vitamins function in animal metabolism, but also, many of these same chemicals have been found to regulate physiological processes in plants as well. One of the chief difficulties in studying the functions of vitamins in plants is that, in contrast to animals that are unable to synthesize vitamins, green plants in general produce their own supply. Thus it is difficult for a plant physiologist to produce a vitamin deficiency in a plant. If the plant manufactures the vitamin the physiologists cannot easily deprive it of the vitamin.

Certain organs in plants, however, are unable to produce all of the vitamins that they need. The culture of these organs isolated from the plant has led to a better understanding of the role that vitamins have in plant growth. As an example, let us look at the influence of vitamins on root growth.

Being devoid of chlorophyll, roots are unable to synthesize sugar, and hence are dependent upon the green organs of the plant. We may ask: do the leaves produce substances, other than sugar, essential for the normal growth of roots? If growing root tips are cut off and placed in a medium containing water, essential mineral salts, and sugar, the isolated roots will not continue to grow normally. They will grow for a prolonged period of time, however, if an extract of yeast cells is added to this medium. The growth substances in yeast extract are shown to be **thiamin, nicotinic acid,** and **pyridoxine,** all part of the vitamin B complex. The living cells of the roots of some species of plants cannot synthesize thiamin, nicotinic acid, or pyridoxine. These substances are manufactured in the leaves under the influence of light and move from them to the roots. Roots of other kinds of plants can produce pyridoxine but not nicotinic acid or thiamin.

Thiamin is widely distributed in the plant kingdom. It has been found in many kinds of bacteria, in yeasts and other fungi, in both fresh water and marine algae, and in mosses, liverworts, ferns, and the higher plants. Some bacteria, yeasts, and other fungi are able to synthesize thiamin, whereas others are not. Almost all the thiamin in a wheat grain is in the bran. It is now common practice to enrich white flour with thiamin.

Young plant tissues of higher plants, principally the leaves, are particularly active in synthesizing thiamin. From these tissues it moves to other parts of the plant. That thiamin is produced in leaves and translocated through the phloem to roots may be shown by removing a girdle of bark from the stem of a plant. It will be found that thiamin accumulates above the girdle. If the girdle is made on the petiole between the leaf and stem, thiamin accumulates in the leaf. As stated, thiamin is not produced in roots (at least in most plants that have been studied), but it is essential for root growth. Thus, thiamin may be classed as a root-growth hormone; it plays a role in the normal formation of roots in many seed plants.

Carefully controlled experiments lead us to believe that most ordinary plants growing with their roots in a soil containing the required mineral nutrients, and with their leaves in light, synthesize sufficient quantities of thiamin in the leaves to meet all requirements of the plant. Seeds usually are relatively rich in thiamin. Thus the seedling probably has an adequate supply of the vitamin to support normal growth until such time as newly formed leaves are synthesizing the substance. Cuttings, however, may be deficient in thiamin. A few leaves on stem cuttings promote root growth, not only because they contribute carbohydrates to the roots, but also because they may contribute hormones as well.

Mechanism of Hormone Action

The role of hormone balance in regulating the pattern of tissue development has been further clarified by the culture of excised roots. When roots of seedlings are cut off and grown in sterile nutrient culture, they fail to form any cambium or secondary vascular tissues. It was reasoned that since these roots normally would have formed secondary xylem if left attached to the plant, some factor, or factors, necessary for vascular cambium development is normally transmitted from the shoot to the root. Evidence for this was obtained when the cut basal ends of the roots growing in nutrient culture were treated with the hormone indoleacetic acid. Such an experiment again demonstrates the important regulatory role that hormones play in plant development.

How do substances that occur in such low concentrations as the hormones regulate cell growth? We have seen that certain of the micronutrient elements are necessary for the action of specific enzymes. Likewise it has been found that at least some of the hormones exert their action through their influence on the functioning of enzymes. Many of the vitamins, among them thiamin, are known to be actually a necessary part of specific enzyme systems. If they are lacking, the enzyme will not function. The reaction that it controls will not take place and metabolism and growth will be influenced. This is true in both plants and animals. For

Figure 20.8. Effect of day length on behavior of henbane (*Hyoscyamus*). Plants grown with 8-hour photoperiod until they were about a month old and then subjected to 24 photoperiods of 10, 11, 12, 13, 14, and 16 hours. Development of flower stalks occurs only when plant is exposed to the longer photoperiods. (Courtesy of Plant Industry Station, Crops Research Division, Agricultural Research Service, U.S. Dept. of Agriculture.)

example, thiamin is a part of an enzyme involved in the evolution of carbon dioxide during respiration; nicotinic acid is a component of enzymes that remove hydrogen during respiration; and pyridoxine is necessary for amino acid synthesis.

All of the major plant hormones—auxins, kinins, and gibberellins—cause multiple effects involving cell division, cell enlargement, and cell differentiation. It has been demonstrated that it is the interaction of all three of these hormones in the internal environment of the cell that results in various morphological expressions. In a remarkable series of experiments Folke Skoog showed that unorganized callus tissue derived from tobacco stem pith could be induced to differentiate buds or roots or entire plants simply by altering the relative amounts of hormones in the medium about the tissue (page 87, Fig. 20.6).

Since hormones induce such a variety of profound effects on cell development, it might be assumed that at least part of their regulatory activity is being exerted at some key point near the site of the genetic information, that is, DNA. In one instance, at least, this has been shown to be true. One of the first responses of germinating barley grain to the application of gibberellin is the formation of a specific enzyme (α-amylase) that will digest starch normally stored in the grain endosperm. The evidence is strong that this enzyme is formed as a result of the hormone, gibberellin, activating DNA. This results

in a stimulation of a specific RNA that controls the synthesis of the enzyme protein. Thus in this instance, at least, the hormone is acting at the level of the gene.

PHYSIOLOGY OF REPRODUCTION

The transformation of the vegetative apex that produces leaf primordia to a reproductive apex that produces flower primordia (page 292) is a dramatic event. What induces this change? Two factors are day length and temperature.

Photoperiodism and Vernalization

A knowledge of **photoperiodism,** the effects of day length on plant development, is of practical importance in agriculture. For example, it is possible for a floriculturist to delay the flowering of chrysanthemums and cause them to flower for the late fall market simply by extending the daylight period with several hours of artificial light. When flowering is desired the supplemental light is simply discontinued. More fascinating still is the fact that the same thing can be accomplished, and a considerable amount of electricity saved, simply by exposing the plants to a few minutes of light in the middle of the night. As long as these plants receive this small amount of light in the night they will not flower. If they are exposed to long periods of uninterrupted

Figure 20.9. Effect of day length on behavior of *Chrysanthemum*. A, plant that received light of natural short days of autumn and blossomed at the usual time. B, plant that received an hour of light near the middle of each night for several weeks, beginning just before flower buds would normally have been initiated; thus, each long dark period was divided into two short ones. This interruption was sufficient to delay flowering. (Courtesy of Plant Industry Station, Crops Research Division, Agricultural Research Service, U.S. Dept. of Agriculture.)

darkness, however, processes are initiated that lead to flowering.

Observations of plants growing under natural conditions and greenhouse experiments, have shown that some plants will flower only when the days are longer than a certain critical length that is characteristic for each species, or, more correctly, only when the nights are shorter than a certain critical length. Spinach, for instance, starts to flower when exposed for two weeks to days that are 13 to 14 hours or more long but will not flower if the days are less than 13 hours long. Plants that respond to day length in this way are called **long-day plants** and include beet, henbane (Fig. 20.8), spinach, and winter barley. Many long-day plants will not normally flower and reproduce by seeds in the tropics where even the days of midsummer are relatively short.

Other species of plants, such as *Chrysanthemum*, grow vegetatively in the summer, then normally bloom and set fruit in the autumn when the days are short and the night long. These species that flower only when the daily period of illumination is shorter than a certain critical length are called **short-day plants;** examples are *Coleus, Aster, Dahlia, Cosmos,* and *Chrysanthemum* (Fig. 20.9). There are also species that bloom throughout the growing season, apparently indifferent to the relative length of day and night (examples: tomato, buckwheat, snapdragon). These are **day-neutral** plants.

Long-day plants can be made to bloom in the short days of midwinter if the length of daily exposure to light is increased by artificial lighting. For example, if a long-day plant, like lettuce, is grown in the greenhouse during the winter, and the normal length of exposure to daylight is increased by artificial illumination, the plants will come into bloom. Or, a short-day plant, like the *Chrysanthemum,* can

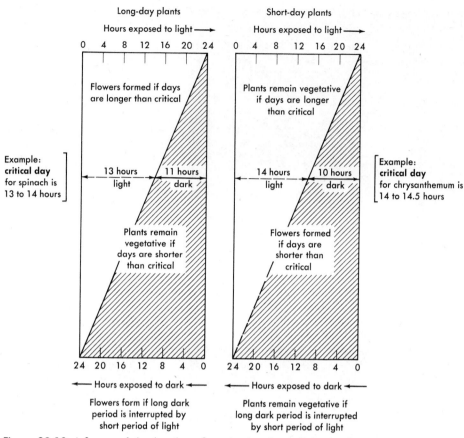

Long-day plants

Hours exposed to light ⟶

Short-day plants

Hours exposed to light ⟶

Flowers formed if days
are longer than critical

Plants remain vegetative
if days are longer
than critical

Example:
critical day
for spinach is
13 to 14 hours

13 hours
light

11 hours
dark

14 hours
light

10 hours
dark

Example:
critical day
for chrysanthemum is
14 to 14.5 hours

Plants remain
vegetative if
days are shorter
than critical

Flowers formed
if days are
shorter than
critical

⟵ Hours exposed to dark ⟵

⟵ Hours exposed to dark ⟵

Flowers form if long dark
period is interrupted by
short period of light

Plants remain vegetative if
long dark period is interrupted
by short period of light

Figure 20.10. Influence of day length on flowering in spinach (*Spinacia oleracea*) and *Chrysanthemum*.

be brought into bloom earlier in the fall by placing it in a dark room for a part of each day.

Intensity of light is not the critical factor in determining the behavior of long-day and short-day plants; rather, it is the relative length of the daily alternating periods of light and dark that determines flowering behavior. This relationship is shown in Fig. 20.10. In some instances a difference of an hour or even less may determine whether a plant will remain vegetative or will become reproductive. Apparently, fixed length periods of light and darkness focus their influence on floral initiation; that is, the photoperiod (light period) determines whether the shoot tip will remain in a state of vegetative activity (Fig. 7.9) or begin the initiation of flower primordia (Fig. 15.25). How can the change of less than an hour in length of day to which a plant is exposed produce such a drastic shift in development? Although our knowledge of the proc-

ess is still far from complete, several facts may be listed.

1. The photoperiodic stimulus is received by leaves. If a short-day plant has its growing point exposed to a short day, but its leaves to a cycle of long days, flowers will not be produced. If its leaves, or, in some plants, only a portion of one leaf, is subjected to an **inductive cycle** (a photoperiod of a length to induce flowering), the plant will flower even though the rest of the plant is under a long-day cycle.

2. A hormone or hormones must be involved in the correlation between the receptive organ (the leaf) and the reaction site (the growing point) that is induced to change in development. The stimulus that induces flowering may be transmitted through a graft union between a leaf that has been submitted to an inductive cycle and a plant that has not.

3. The photoreceptor pigment that absorbs the light energy important in photoperiodism is not chlorophyll or carotenoids but a pigment called **phytochrome.** It is a blue or bluish-green protein and is present in leaves in extremely small amounts.

Two forms of phytochrome are in equilibrium in the plant. One, an inactive form called P_{red} or P_{660}, strongly absorbs red light of a wave length of 660 mμ, and in the light it is rapidly changed to the other form, an enzymatically active form called $P_{far-red}$ or P_{730}. This form of phytochrome (P_{730}) absorbs the long wave lengths of red light, the far-red, and in doing so is converted into P_{660}.

absorbs red light → becomes

P_{660} P_{730}

becomes ← absorbs far-red light

P_{730} is also converted into P_{660}, the red absorbing, inactive form, in darkness.

Photoperiodism, as well as several other plant processes such as seedling growth, anthocyanin formation, and germination of some seeds, are sensitive to red and far-red light. In these cases, if red light activates a particular reaction, far-red light reverses the effect of red light. When a plant receives sunlight, the red wave lengths dominate the far-red so that phytochome is converted into the active P_{730} form.

4. The effects of temperature on reproduction are strongly interrelated with the photoperiodic response of plants. Certain plants have a low temperature requirement for flowering that is often a genetically controlled characteristic. Thus, winter wheat varieties, when planted in the fall, are subjected in their seedling stage to low temperatures. The following spring, vegetative growth takes place, and flowering and seed formation occur during the long days of summer. If the winter wheat is planted in the spring, the seedlings would not naturally be subjected to low temperatures. Vegetative growth would occur but flowering and grain formation would not.

Cold preconditioning of soaked, partially germinated winter wheat can be carried out artificially. The soaked grain is kept a few degrees above freezing temperatures for several days or longer. Plants which grow from this treated grain will complete their life cycles in one season even when planted in the spring. Such a cold temperature treatment which stimulates flowering is called **vernalization.** Spring varieties of wheat do not need a cold treatment and will flower in the same year if planted in the spring.

Many biennial plants such as beets and cabbage have a cold temperature requirement for flowering. They grow vegetatively the first year and, following a cold winter period, flower and produce seed the following year.

In contrast to photoperiodism, in which the leaf is the receptive organ, the shoot tip is the organ that directly receives the vernalization stimulus.

In some instances gibberellin may substitute for the cold requirement of plants and in other instances, it may substitute for the long-day requirement. Figure 20.11 shows a biennial carrot plant in which the control plant grew vegetatively under long days. A similar plant that received a gibberellin treatment, but no cold treatment, flowered. A

Figure 20.11. Substitution of gibberellic acid for cold treatment in the flowering of carrot (*Daucus carota*). A, control plant under long days only; B, long days plus gibberellic acid, no cold treatment; C, long days plus cold treatment; no gibberellic acid. (Courtesy of Lang.)

third plant that received a cold treatment as well as long days also flowered.

FACTORS INFLUENCING GROWTH

The growth behavior of a plant depends upon its genotype—its genetic make-up. For instance, some genotypes of wheat grow rapidly, maturing in 100 to 110 days. Other varieties, under the same environmental conditions, grow more slowly and come to maturity in 145 to 155 days.

Growth is also determined by environmental factors, such as temperature, water, light, and available supply of mineral nutrients in the soil. Any change in these external conditions is reflected in growth.

Various aspects of the influence of environmental factors upon the growth and distribution of plants are discussed in previous chapters on physiology and ecology. Here we will simply summarize certain of the general effects of the two environmental factors, light and temperature, on plant growth.

This will serve to illustrate how all of the various environmental factors such as humidity, carbon dioxide, soil water, and minerals interact on the plant.

Effects of Temperature on Growth

Every process in the plant is influenced by temperature: photosynthesis; assimilation; digestion; respiration; absorption of water, mineral salts, and gases; transpiration; formation of enzymes; permeability of membranes; rate of movement of materials within the plant; and reproduction. These processes determine the character and rate of growth. Each one may be critical and limiting. Modification or disturbance of any one may affect growth in some part of the plant.

We customarily speak of the *optimum temperatures* that influence any particular process. By this term we mean the temperature at which the process goes on *most actively*. The optimum temperatures for one process may differ considerably from those for another taking place in the plant at the same time. For example, the roots of many plants can absorb from the soil at temperatures much lower than those required for photosynthesis or for reproduction.

Temperature influences the structural development of plants. For example, at a temperature of 70° to 80° F, the white, Boston variety of lettuce fails to form heads; rather, an open rosette of leaves develops. At temperatures ranging from 60° to 70° F, however, the lettuce plants develop satisfactory heads. Also, it appears that either high temperatures or low temperatures may stimulate a premature formation of seed stalks, a phenomenon known as bolting.

Effects of Light on Growth

In discussing the effect of light upon growth, we must consider (1) light *intensity,* (2) light *quality,* and (3) light *duration.* The term light includes all the energy (radiant energy) that reaches the plant from the sun or from artificial sources. Light may affect the plant and its various activities (hence its growth), including the following:

1. The synthesis of chlorophyll, and other processes associated with photosynthesis. (Chapter 13)

2. The temperature of the plant body above ground. Part of the light energy that falls upon the leaves is changed to heat and raises their temperature. The important functions of the leaf—photosynthesis, transpiration, and synthesis of protein—are directly influenced by leaf temperature (Chapter 13).

3. The opening and closing of stomata. The area of the stomatal aperture affects the exchange of oxygen, carbon dioxide, and water vapor between the leaf cells and the atmosphere outside the leaf. Thus, the processes of respiration, photosynthesis, and transpiration are directly influenced by stomatal opening and closing (Chapter 10).

4. The direction of growth of certain plant organs (Fig. 20.12).

5. The form and anatomy of leaves. Plants that grow in the sun (so-called sun plants) usually have (a) thick leaves, (b) a thick palisade parenchyma, (c) a poorly developed spongy parenchyma, (d) small intercellular spaces in the leaves, (e) a thick and heavily cutinized epidermis, (f) stomata confined to or more abundant on the lower side of the leaf, (g) (frequently) glossy leaf surfaces, and (h) (often) densely hairy leaves. Plants that grow in the shade (so-called shade plants) usually have (a) thin leaves, (b) a thin layer of palisade tissue or none at all, (c) a well-developed spongy parenchyma, (d) large intercellular spaces in the leaves, (e) a thin and

Figure 20.12. Two tomato plants exposed for seven days; left, to diffuse light; right, to directional light.

slightly cutinized leaf epidermis, (*f*) stomata on both surfaces of the leaves, or the number on the two surfaces quite the same, (*g*) (usually) dull leaf surfaces, and (*h*) (generally) leaves devoid of hairs.

Light Intensity and Plant Growth

Some plants normally grow in dense shade, that is, in light of *low intensity;* and other plants can endure the noon sun of a bright, cloudless summer day, in a light of *high intensity.* Moreover, as long as there is a supply of reserve food, plants will grow in total darkness. As compared with plants grown in light, those grown in darkness are white or yellow in color, the internodes are long, the leaves are very much reduced in size, and the root systems are poorly developed. This condition is known as **etiolation.** Of course, such plants cease growth when the reserve food is exhausted. In general, absence of light increases, and presence of light decreases, the rate at which the stems elongate. This relationship may be observed in the growth of seedlings. Seeds buried in the soil germinate in darkness, and the stem grows upward very rapidly until it reaches the light; then the rate of stem elongation is reduced. The exposure of a seedling to light for only a few minutes per day is sufficient to prevent etiolation, even though this is not enough light to cause a significant amount of photosynthesis to occur. Light may thus control growth in ways other than through photosynthesis.

In many plants, there is a rhythmic night-and-day growth rate of the shoot—greater at night than during the day, provided that the temperature at night does not fall too low. Plants grown in full light have shorter and sturdier stems and somewhat thicker leaves than those grown in the shade. One result of crowding plants is a reduction in the light intensity to which they are exposed. Such plants have longer and more spindly stems than those grown under less crowded conditions.

At the other extreme is the behavior of plants under continuous illumination. Many common plants may be grown from seed to seed under continuous exposure to light, either continuous artificial illumination or natural daylight supplemented with artificial lighting. Examples are cereals, flax, buckwheat, peas, and beans. Squash, red clover, tomatoes, and potatoes, however, are among the plants that bloom but do not set seed when grown with continuous lighting.

Light Quality and Plant Growth

In the chapter on the leaf, the effect of different wave lengths of light upon photosynthesis was discussed. The red and blue portions of the light spectrum are the most effective in photosynthesis.

By filtering sunlight with different kinds of colored glass, it is possible to test the growth of plants when exposed to wave lengths that represent narrow bands of the different parts of the visible spectrum, and also to wave lengths not represented in the visible spectrum. The general conclusions of these experiments are as follows. Light of short wave lengths (blue light) tends to retard growth in length. Under the longer wave lengths (red light), plants show somewhat the characteristics of those grown

Figure 20.13. The effect of growth-regulating chemical on the rooting of American holly (*Ilex opaca*) cuttings. Row *A*, cuttings that stood in an aqueous solution of .01 percent indolebutyric acid for 17 hours before being placed in sandy rooting medium. Photograph taken after 21 days in rooting medium. Row *B*, untreated cuttings. (Courtesy of U.S. Dept. of Agriculture.)

in darkness. When exposed to infrared rays, plants have long internodes and a scarcity of chlorophyll. However, such plants may produce large leaves, in contrast to the scalelike leaves of plants grown in the dark.

Light Duration and Plant Growth

Although green plants are able to grow in total darkness if there is a supply of reserve food upon which they may draw, the length of day and night periods under which a plant is growing often profoundly affects its vegetative and reproductive activities, as has been discussed under the subject of photoperiodism (page 408).

THE USE OF GROWTH REGULATORS IN THE CONTROL OF GROWTH

Control of Sprouting

We have discussed the specific hormone of the auxin type, particularly its influence upon cell elongation and curvature in the coleoptile of oat seedlings. Another hormonal influence on growth is inhibition of the development of lateral buds. For example, if the terminal bud is cut off, the lateral buds grow; but, if indoleacetic acid is applied to the surface of the decapitated stem, the lateral buds fail to grow. Bud development may be delayed or inhibited by application of certain auxins. For example, potato tubers exposed to the vapors of a synthetic growth regulator, do not sprout in storage.

Rooting of Cuttings

Cuttings of certain species normally root slowly or produce a very small number of roots. It is now possible in some of these species to secure vigorous root production (Fig. 20.13) by treating the cuttings with various auxins, such as indoleacetic acid, indolebutyric acid, betanaphthoxyacetic acid, and numerous substituted phenoxy and benzoic acids. These growth regulators in proper concentration initiate cell divisions and stimulate the production of root primordia in the cuttings. The bases of the cuttings are immersed in the growth regulator solution for a period of from 12 to 24 hours, the strength of the solution ranging from 1 part of the substance in 5000 to 100,000 parts of water. The length of treatment and strength of solution that give optimum rooting vary with the species.

Root formation in cuttings of *Hibiscus* was found to depend upon a combination of indoleacetic acid and sugars and nitrogenous substances. The sugars and nitrogenous substances are contributed by the leaves, which explains the fact that rooting is very slight, even when cuttings are treated with a growth regulator, unless leaves are present.

Of significance is the fact that certain growth regulators affect the course of differentiation. For example, short segments of the tap root of dandelion normally develop buds from the upper cut surface, and callus and roots from the lower cut surface. If the auxin, indolebutyric acid, for example, is applied to these segments, roots differentiate from both cut surfaces; or, if the growth regulator in tissues is decreased by proper treatments, leaves can be caused to differentiate from both ends of the segments.

Control of Fruit Development

The development of fruits, particularly tomatoes, without pollination has been induced by application of certain growth regulators. A high percentage of seedless fruit results when the flowers are treated before pollination; but flowers treated after pollination set fruit, most of which have seed. Substances employed for this purpose include the auxins indoleacetic acid, betanaphthoxyacetic acid, and 4-chlorophenoxyacetic acid. They are synthetic compounds readily available. Flower production in the pineapple has been induced by the application of growth substances, particularly the auxin naphthaleneacetic acid.

Control of Abscission

Certain substances are effective in retarding the preharvest drop of fruit such as apples, pears, and citrus. After treatment with these chemicals, the fruits cling to the trees for a number of days longer than normal and may attain more satisfactory color and maturity. Growth regulators effective for this purpose are the auxins, naphthaleneacetic acid and 2,4-dichlorophenoxyacetic acid ("2,4-D"). The growth regulator apparently retards the processes that result in the formation of the abscission layer

at the base of the petiole (Fig. 000). Under natural conditions a special hormone called abscisic acid is closely associated with abscission.

Growth Regulators as Herbicides

Certain synthetic growth regulators, although stimulative in extremely small quantities, seriously disturb physiological processes in plants when added in larger amounts. Some of these compounds have turned out to be very potent weed-killers. Although not all growth regulators can be used to kill weeds, the so-called "phenoxy" compounds are particularly effective. Because these compounds are selective, killing broad-leaved plants and leaving the grasses relatively unharmed, they are particularly useful for destroying many kinds of weeds growing in combination with various grasses or crops of grain. Thus it is possible to spray a lawn with 2,4-D (2,4-dichlorophenoxyacetic acid) and, without injuring the grass, kill dandelions growing in it. The use of this and a related compound, 2,4,5-T (2,4,5-trichlorophenoxyacetic acid) in defoliation is being studied because of possible harmful side effects to humans possibly caused by impurities.

The fact that growth regulators are translocated throughout the plant in the phloem makes these herbicides particularly effective in killing deep-rooted perennial plants. Studies of the factors affecting their movement in plants have been greatly facilitated through the use of radioactive isotopes. Figure 20.14 shows the movement of 2,4-D containing radioactive carbon through a plant.

SUMMARY OF PLANT GROWTH AND DEVELOPMENT

1. Two experimental methods of studying morphogenesis are micurgical operations and the sterile culture of plant parts.

2. The influence of one part of a plant on the growth and development of another is an example of correlation of growth. Growth correlations may be a result of influences of one plant part on food, mineral, or water supplies of another plant part or on hormone production and distribution.

3. Plant hormones are natural compounds which are usually translocated and act in small amounts to regulate plant processes and development.

4. Four general types of hormones are auxins, gibberellins, kinins, and abscisic acid.

5. Auxins, the first group of plant hormones discovered, are translocated polarly in parenchyma tissue and stimulate elongation of stem cells, induce rooting,

Figure 20.14. A, silhouette of the entire *Zebrina* plant; B, autoradiogram of the *Zebrina* plant showing the path of movement of ATA (amitrol) containing radioactive carbon. ATA was applied to the leaf at the upper right. (Courtesy of A. S. Crafts.)

suppress lateral bud growth, and reduce the abscission of flowers and fruits.

6. Kinins, chemically related to the purines found in nucleic acids, may also affect growth and development in many ways such as the stimulation of bud development in callus culture.

7. Gibberellins, chemically different from auxins, may stimulate cell elongation, reduce abscission of some plant parts, and overcome dwarfism in certain dwarf plants.

8. Abscisic acid may act as a growth retardant. It may be involved in abscission and dormancy.

9. A number of vitamins such as thiamine, nicotinic acid, and pyridoxine have been shown to be essential for the growth of roots and are essential for the activity of certain enzymes.

10. All major plant hormones cause multiple effects

involving cell division, enlargement, and differentiation. Part of their regulatory effect must be exerted at some key point near the site of the genetic information, that is at or near DNA.

11. Photoperiodism is the response of plants to day length. On the basis of their flowering response to day length, plants may be generally classified into short-day plants, long-day plants, and day-neutral plants.

12. The pigment, phytochrome, in the leaves, absorbs red or far-red light during the photoperiod. If the photoperiod is of the correct length, a transmission of the stimulus, probably by means of a hormone, to the vegetative shoot apex results in an induction of flower primordia.

13. Vernalization is a cold temperature preconditioning of the plant that affects the subsequent time of flowering.

14. Growth is determined by many environmental factors such as temperature, water, light, minerals in the soil, and atmospheric gases.

15. Growth regulators are widely used to control growth and development of plants. Some examples are the induction of rootings, the control of fruit development, the control of abscission, and the removal of weeds.

21

Phylum Schizomycophyta

The Schizomycophyta comprise the bacteria, a group of prokaryotes, which obtain energy and nitrogen by a variety of metabolic pathways. The great majority of Schizomycophyta are heterotrophic forms obtaining energy and nitrogen from organic matter. Those utilizing dead organic matter are *saprophytes*, while others are *parasites* obtaining food from living organisms. Some may obtain food from both sources. Some bacteria are aerobic; others live only in the absence of oxygen and are strictly anaerobic. A number of species may live in either environment. For example, *Clostridium tetani* is a parasite in the absence of oxygen and a saprophyte in its presence. The autotrophic species are largely anaerobic forms obtaining energy in a variety of ways. Some oxidize NO_2 to NO_3, gaining energy from the process; others oxidize sulphur or iron compounds. Three species of bacteria are photosynthetic and anaerobic. These process a pigment similar to chlorophyll *a* and utilize the energy of sunlight. In fact, in past ages, these photosynthetic forms, as suppliers of oxygen, may have played a highly significant role in the evolution of our present atmosphere.

See Table 21.1 for a tabulation of the various ways in which bacteria obtain energy and nitrogen.

GENERAL CHARACTERISTICS AND DISTRIBUTION

This phylum of worldwide distribution comprises a large number of species, mostly single-celled or simple filamentous forms between 0.5 and 5 microns in diameter and up to 15 to 20 microns long. Multiplication results from the simple division of single cells, a process known as **fission.** Many members of the Schizomycophyta form spores, but the spores are more a means of carrying the species over periods unfavorable to growth than a means of multiplication. While the bacteria are not to be classified as fungi, certain types of bacteria have characteristics that resemble those of fungi. Some bacteria resemble slime molds, while others are filamentous.

Most bacteria are one-celled plants and among the smallest known living organisms. They are often referred to as "microbes" or "germs." They play an important role in the decomposition of organic matter, are indispensable in maintaining soil fertility, affect the quality of the water and milk we drink and of the food we eat, and cause diseases of animals and plants. We are likely to gain the impression that all bacteria are harmful, because we associate them with such dread human diseases as typhoid fever, tuberculosis, and many other types of infections. But many kinds of bacteria are beneficial, in fact, indispensable in the life of the world.

Discovery of Bacteria

Our knowledge of the shapes and morphology of bacteria depends largely on the type of microscopes with which they are studied. The Hollander, Anthony van Leeuwenhoek, first saw them in the summer of 1676. He was curious as to what made pepper hot, so he set some pepper corns aside in a cup of water and several days later examined this water under a simple single-lens microscope at a magnification of about 400 times. To his amazement he saw many actively moving, very small organisms which he called **animalcules.** He did not prepare any drawings of these small organisms until several years later, and these drawings, when finally prepared, showed organisms taken from his own mouth. These drawings, which are shown in Fig. 21.1, clearly indicate that he was describing bacteria for the first time.

While van Leeuwenhoek suggested that these small organisms might be related to fermentation, decay and disease, his suggestion was not taken seriously. Their relation to these processes was finally discovered quite accidentally. Pasteur, a young French chemist, was retained in 1860 by the French government to improve the quality of French beers. A chemist was considered the proper person for the task because it was then thought that fermentation was a purely chemical process. Pasteur, however, firmly established the part played by living organisms in fermentation. He definitely showed for the first time that yeasts were necessary for the production of wine from grapes. Pasturization, the flash heating of liquids to kill undesirable organisms, was first used to preserve French wines destined for export. During the 1890s the French Academy of Sciences, of which Pasteur was a member, was hotly debating whether or not life could arise spontaneously. The outcome of this debate was the definitive demonstration by Pasteur that mi-croorganisms always arise from preexisting organisms. Living organisms, whether bacteria or insect larvae, never arose *de novo* from decaying matter. A long series of experiments in Pasteur's laboratory succeeded in establishing the very definite relationship between bacteria (and yeasts) and the processes of fermentation, decay and disease.

Forms of Bacteria

Bacteria are of three general shapes: **spherical, rod-shaped,** and **spiral.** Spherical bacteria are called **cocci** (**coccus**, singular) (Figs. 21.2*A*, 21.2*B*, 21.2*C*), rod-shaped ones are called **bacilli** (**bacillus,** singular) (Fig. 21.2*D*), and spiral forms are known as **spirilla** (**spirillum,** singular). There is some inter-grading between these forms. It is sometimes difficult, for instance, to distinguish between a very short rod and a coccus that has elongated in preparation for cell division. Furthermore, shape depends to a certain extent upon the age of the bacteria and upon the environment.

Bacteria of the coccus form, in chains resembling strings of pearls, are called **streptococci** (Fig. 21.2*B*). Several species of the genus *Streptococcus* are very important. Some of the diseases they cause are scarlet fever, erysipelas, mastitis of cows, and sinus infections. *Streptococcus lactis* is an agent in souring

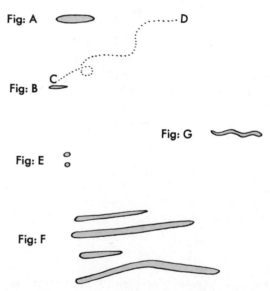

Figure 21.1. A copy of van Leeuwenhoek's drawings of the bacteria he obtained from his mouth. He described them thus: "I then most always saw, with great wonder, that in said matter there were many very little animalcules very prettily a-moving." (From *Arcana Natura Detecta Delphis Batavorum*, 1695.)

Figure 21.2. First electron micrographs of bacteria. A, Pneumococcus; B, Streptococcus; C, Staphococcus aureus; D, Bacillus anthrax. All ×10,000. (A, courtesy of Charles Pfizer and Company; B, C, D, courtesy of J. Hillier.)

milk. Spherical bacteria may sometimes remain associated in irregular flattened masses; these forms are frequently referred to as *Staphylococci* (Fig. 21.2C). One species of *Staphylococcus* is the usual causal agent of boils and abscesses. Some species of spherical bacteria remain associated in cubical packets of eight, 64 or even more cells; these species have been given the name *Sarcina*.

Size of Bacteria

Bacteria are among the smallest living organisms. Some of them are close to the limit of visibility with the most powerful light microscopes. Several million bacteria may be present in a cubic centimeter of soil, or milk, and not be crowded in the least bit. If milk containing 10 million bacteria in each cubic centimeter is examined under a lens magnifying 440

Figure 21.3. (*right*) Spatial relationship of bacteria and milk particles in a cubic centimeter of milk containing approximately 10,000,000 bacteria. (Redrawn from F. Lohnis and E. B. Fred, *Textbook of Agricultural Bacteriology.* Copyright 1923 by McGraw-Hill Book Company. Used with permission of McGraw-Hill Book Company.)

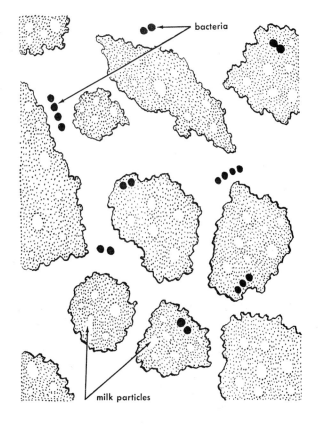

times, the relationship between bacteria and milk particles would approximate that shown in Fig. 21.3.

Most bacteria average from 1 to 2 microns in diameter. If the cube *A* in Fig. 21.4 is 10 millimeters along one edge, there are 100 square millimeters on each face and 1000 cubic millimeters in the cube. Ten thousand cocci, each 1 micron in diameter, could be placed along one edge of this cube. There would be room for 100 million of them on one face; 1000 billion would fit nicely into the cube. Thus, 1 million bacteria per cubic centimeter of soil, milk, or blood is really a small amount.

The very smallness of bacteria is a factor of major importance in their life. All the nourishment, all gases and inorganic salts that a bacterium requires must be taken in through its cell wall. All its waste products, all toxins or poisons, certain enzymes and other special substances that are associated with bacterial activity must pass outward through the surface of the bacterial cell. Therefore, the more surface a bacterium possesses in relation to its volume, the more readily will these substances diffuse.

Distribution of Bacteria

These minute organisms are ever-present about us. They occur in the air, water, and soil. They exist on the surfaces of all animal and plant bodies and on the surfaces of almost everything we touch. They are present in milk and in all foods that have not been sterilized. Even foods that have undergone the usual process of sterilization will sometimes contain the spores of bacteria. In 1939 viable spores of certain heat-resistant bacteria were found in canned roasted veal that had traveled with the explorer Parry to the Arctic Circle in 1824. Bacteria occur naturally in the digestive tracts of animals. Nitrogen-fixing bacteria are closely associated with the living cells in the roots of legumes.

After a rain or snow, the dust is washed from the air and carried to the ground, and then the air may be practically free of bacteria. Generally, the number of bacteria in the air decreases with altitude. The air at altitudes at which commercial airliners fly is bacteria-free.

Water from deep, cold wells or springs is usually devoid of bacteria. Whereas the surface few inches of most soils teem with bacteria, the number decreases as the depth increases.

Figure 21.4. Diagram showing relationship between volume and surface.

Although bacteria are almost universally distributed, it is possible, by taking great precautions, to keep rooms relatively free of these organisms. Such precautions are observed in hospital operating rooms and in certain types of biological and bacteriological laboratories. Surfaces are washed in disinfecting fluid, sterilized gowns are worn, and only instruments that have been placed in boiling water are used. Even the air may be purified by steam or ultraviolet light. This condition of freedom from bacteria is known as **asepsis. Sterilization** is the destruction of all living forms, including bacteria, that may be within or upon a particular object.

Movement

Many bacteria move. Some do so by simply flexing their bodies, others rotate and have a corkscrew type of motion. Motility may be obtained by a flexing of the protoplast or by the action of two sorts of exterior motor organelles. The bacterial surface may be provided with many short **cilia** (*Caryophanon latum*, Fig. 21.5) or with one, two, or several long whip-like filaments or **flagella** (Fig. 21.6 and *Escherichia coli*). However, neither of the bacterial motor organelles, cilia or flagella, have an ultrastructure similar to that of the flagella of higher organisms.

Figure 21.5. *Caryophanon latum* showing many flagella. ×5,000. (Courtesy of C. F. Robinow.)

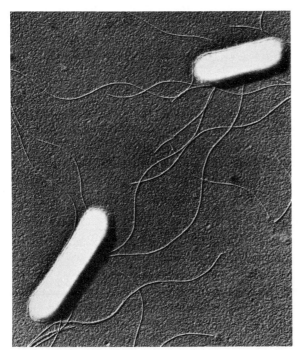

Figure 21.6. Diphtheroid bacteria with associated flagella. ×8,000. (Courtesy of R. C. Williams and M. P. Starr.)

In the latter case, the flagella are provided with a central complex of 9 + 2 strands (Fig. 22.24), while only a single central filament, about 120 A in diameter, is present in the bacterial cilia or flagella (Fig. 21.7). Ciliated bacteria may travel at the rate of some 50 microns per second, whereas those lacking motor organelles move at a much slower rate, about 5 microns per second. In proportion to its body size, a speed of $50\mu/\text{sec}$ is equal to a speed of 90 mph by a 6-ft long antelope.

Nutrition and Energy

Bacteria, like all other living organisms, need food for growth and multiplication. Food is any organic substance that supplies energy and basic organic materials for use in metabolism. Most bacteria utilize nonliving organic matter as a source of food. In other words, most bacteria are saprophytes. They may use relatively simple organic compounds, such as alcohol, sugars, and fats, or very complex organic compounds, such as proteins. For example, the bacteria that form acetic acid utilize ethyl alcohol as a source of food and secure from its oxidation energy for their own purposes. These bacteria are common in cider, beer, and wine. The reaction may be represented as follows:

Alcohol + oxygen → acetic acid + water + energy

Many bacteria depend on sugars as a source of food. The lactic acid bacteria use simple sugars in milk, and the souring of milk is caused by these bacteria. Many species of bacteria rely on proteins as foods, and in the process of decomposition proteins are broken down to simple products, many of which have offensive odors. Ammonia is one of the simplest products formed from proteins by bacteria.

In addition to the saprophytic bacteria, there are those that obtain their food from other living organisms. Such bacteria are parasites. Among them are the disease-causing bacteria of plants and animals. Sometimes the distinction between parasite and saprophyte is not clear-cut. For example, certain types of bacteria thrive as saprophytes in the soil. When they become embedded in wounded flesh, however, they may become parasites. Such are the bacteria causing tetanus and gaseous gangrene in man, and blackleg in cattle.

Some bacteria are able to obtain energy from simple inorganic salts, such as sodium nitrate, ammonia, and hydrogen sulfide. They bring about the

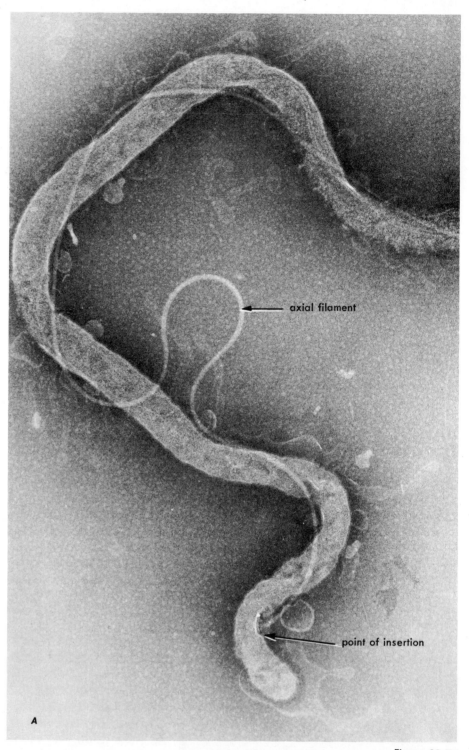

axial filament

point of insertion

A

Figure 21.7A

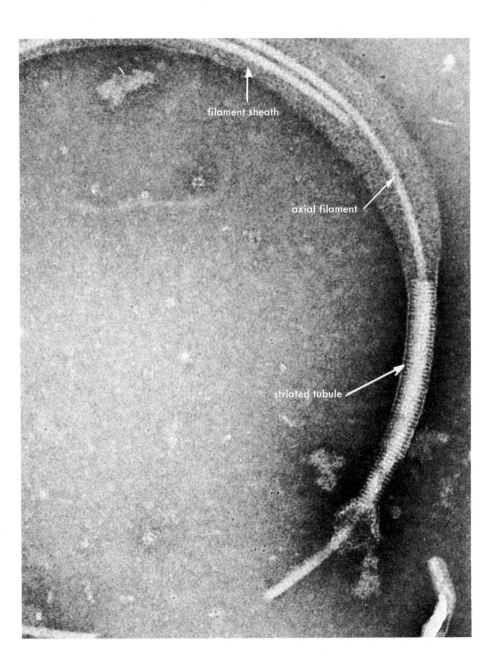

Figure 21.7. Axial filaments of *Spirochaeta stenostrepta. A, (left)* the axial filament is
attached to an end of the spirochete at an identation of the protoplasmic layer; ×43,000.
B, (above) segments of the axial filament are enclosed in a cross-striated tubule which in
turn is surrounded by an outer filament sheath. ×120,000. (Courtesy of S. C. Holt and
E. Canale-Parola, *J. Bacteriol*, **96,** 822, 1968.)

Table 21.2 *Nutritional and energy requirements of the Schizomycophyta*

BACTERIA	SOURCE OF NITROGEN	SOURCE OF ENERGY	PHOTOSYNTHETIC PIGMENTS PRESENT	HETEROTROPHIC OR AUTOTROPHIC
1. Many bacteria	Living organisms (parasitic)	Organic compounds	None	Heterotrophic
2. Many bacteria	Nonliving organic matter (saprophytic)	Organic compounds	None	Heterotrophic
3. Some bacteria	Free nitrogen of atmosphere	Organic compounds	None	Heterotrophic
4. Soil bacteria	Ammonia and nitrates	Inorganic salts	None	Autotrophic
5. Soil bacteria	Ammonia and nitrates	Organic compounds	None	Heterotrophic
6. Purple bacteria	Organic or inorganic nitrogen	Radiant energy	A purple pigment, and chlorophyll	Autotrophic

oxidation of these simple substances and utilize the energy so released to synthesize organic materials from which enzymes, protoplasm, cell walls, and other cellular materials are formed. This synthesis is carried out in a manner comparable to that by which green plants manufacture foods.

We may classify bacteria on the basis of their method of obtaining energy, as follows. (1) Most bacteria secure energy from the oxidation or breakdown of either nonliving or living organic matter. (2) A few obtain energy from the oxidation of inorganic materials and use this energy to synthesize their own foods. (3) One group of bacteria can utilize the energy of sunlight for the manufacture of food.

STRUCTURE OF BACTERIA

The Light Microscope

Most bacteria are single-celled organisms with a very primitive or simple organization. With the light microscope one sees little internal structure other than a central region which responds to nuclear stains and is known as the **chromatin body** (Fig. 21.8). We may consider the bacteria to possess a **cell wall**, which surrounds a **protoplast.** Many secrete an external gelatinous matrix. Though characteristic of the bacteria they are associated

with, these materials are not necessarily part of the living organism, for they may frequently be removed without in any way damaging the living cell. Some bacteria are surrounded by a thick capsular material; others possess a diffuse layer of loose slime or are entangled in a mesh or loose cellulose fibrils. For those bacteria lacking the capsules, slime, or

Figure 21.8. Dividing chromatin bodies in *Bacillus cereus* (Courtesy of C. F. Robinow.)

other dispensable surface components, the cell wall constitutes the outermost structure. While this cell wall may be completely separated from a bacterium and studied chemically in a test tube, its composition is not yet completely understood. Indeed, it apparently varies greatly from one organism to another. In most cases it seems to consist of a fat-like substance, a lipid, a polysaccharide, and an amino acid associated with a sugar. Experimental evidence indicates that penicillin owes its antibacterial effects, at least in part, to its interference in the synthesis of new cell wall materials. Cell walls of bacteria grown in the presence of penicillin become thin and eventually burst open, allowing the cell to separate into two hemispherical fragments. In some experiments, when *Escherichia coli* is incubated in a penicillium medium, the organism is converted in about two hours to spherical protoplasts. This appears to arise because the penicillin interferes with the normal stiffening component of the cell wall.

The protoplast is surrounded by a plasma membrane which, as in other cells, is differentially permeable. That such a membrane is present is borne out by the fact that the bacterial cell may be plasmolyzed. For example, when foods are preserved in sugar, sugar syrup, dry salt, or brine, the bacteria are unable to grow because of the withdrawal of most of the water in them by osmosis.

The protoplast can be isolated and studied intact by enzymic removal of the cell wall. Its cytoplasm, like that of the higher plants, is probably fluid. The enveloping plasma membrane, upon puncture, allows a rapid outpouring of the protoplasm. It is even thought that there actually may be some protoplasmic streaming in these extremely minute organisms. These isolated protoplasts frequently take the shape of spheres, a fact which furnishes additional evidence of the fluid nature of the protoplasm and indicates that it is the cell wall which determines the shape of the individual bacterial cell.

The Ultrastructure of Bacteria

Heterotrophs

Bacteria are prokaryotes, and the structure of the prokaryotic cell was discussed in Chapter 4. All bacteria possess a cell wall, of which the ultrastructure and composition is complex and varies with the genera. While a definite plasmalemma surrounds the protoplast, it has not always been easy to distinguish it with the electron microscope (Fig. 21.9). Internally the chromatin body consists of a tangle of filaments about 125 A in diameter (Fig. 21.9). Since digestion with the enzyme DNAase removes them, they are considered to be DNA fibrils. The fibrils may be isolated and spread on the surface of clean water (a Langmuir trough). They are then picked up on a grid and shadowed

Figure 21.9. A longitudinal section through a cell of *Bacillus subtilis.* ×80,000. (Courtesy of A. M. Glauert, E. M. Brieger, and J. M. Allen, *Exptl. Cell Res.,* **22,** 73, 1961. Copyright Academic Press.)

with platinum while the grid is rotated. DNA fibrils prepared by this technique are shown in Fig. 21.10. Such preparations demonstrate the presence of a single circular strand of DNA, or bacterial chromosome, embedded in the matrix of the chromatin body. The chromatin body is not separated from the cytoplasm by a membrane of any sort. Several types of granules may occur in the cytoplasm, some of which are ribosomes. It is interesting that the sedimentation constant of the bacterial ribosomes is similar to that of chloroplastic ribosomes, rather than to that of cytoplasmic ribosomes. The only internal membrane known in bacteria is an array of concentric membranes, the **mesosome,** which appears to arise from an invagination of the plasmalemma, and which seems to have an association with the chromatin body (Fig. 21.11). Some evidence indicates that the respiratory enzymes may be bound to the mesosome or to the plasmalemma.

Autotrophs

There are three families of photosynthetic bacteria, the purple sulfur bacteria, or the Thiorhodaceae; the purple non-sulfur bacteria, the Anthiorhodaceae; and the green sulfur bacteria, the Chlorobacteriaceae. They are all provided with membrane-bound pigments related to chlorophyll. The membranes of the purple bacteria which bind bacteriochlorophyll bear a resemblance to the grana of chloroplasts (Fig. 21.12), while the photosynthetic membranes of the green bacteria carrying chlorobium chlorophyll seem to be simply spherical bodies called **chromatophores** (Fig. 21.13). Bacteriochlorophyll resembles chlorophyll *a*; however the maximum absorption spectrum of bacteriochlorophyll is in the infrared. Photosynthesis may thus proceed in the absence of visible light.

The purple bacteria reduce sulfur compounds,

Figure 21.10. Fibrils of bacterial DNA produced by bursting a bacterial protoplast with distilled water and allowing the DNA fibrils to expand on the surface. ×100,000. (Courtesy of A. K. Kleinschmidt and D. Lang, *Fifth International Congress of Electron Microscopy*, 1962, Academic Press.)

Figure 21.11. Mesosomes. A, the array of coiled membranes in the cytoplasm of *Bacillus subtilis* constitutes a mesosome. B, (*right*) thin section of a vegetative cell of *Sarcina ureae*. Note the distinct association of the mesosomes with the areas of cross wall formation. Nuclear material is present in the less dense areas throughout the cytoplasm. A, B ×80,000. (A, courtesy of A. Ryter, *Bact. Rev.*, **32,** 45, 1969. B, courtesy of S. C. Holt and E. R. Leadbetter, *Bact. Rev.*, **32,** 346, 1969.)

frequently H_2S, in the photosynthetic process. In the Thiorhodaceae the sulfur is deposited within the bacterial cell, while in the green sulfur bacteria it is deposited outside the cell. The purple sulfur bacteria and the green sulfur bacteria are anaerobic. The purple non-sulfur bacteria are photosynthetic and anaerobic in the light but hetrotrophic and aerobic in the dark.

It is also of interest that the anaerobic chemotroph, *Nitrosocystis,* obtains its energy through the

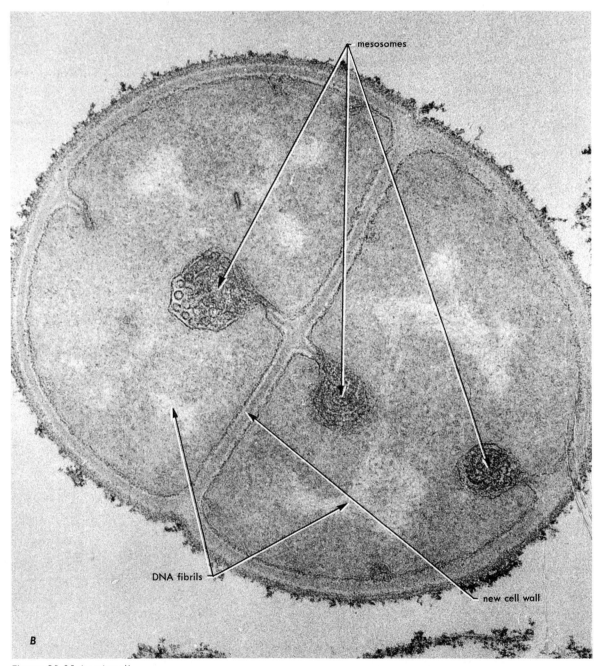

Figure 21.11 (*continued*)

oxidation of NO_2 to NO_3 and has an array of membranes resembling the grana of chloroplasts (Fig. 21.14).

REPRODUCTION OF BACTERIA

Asexual Reproduction

Most bacteria divide by simple fission, in which the cell is "pinched" in two much as a sausage might be cut by the tightening of a slip knot. Photographs taken with the electron microscope show the process (Figs. 21.8, 21.15). The slime layer surrounding

bacteria is distinctly demonstrated (Fig. 21.15). Near the middle of the central cell (Fig. 21.15) a band of some denser material may be noted. Adjacent to this band, a notch in the cell is apparent. A deeper notch occurs in Fig. 21.15. As the process nears completion, the two daughter cells pull apart, stretching the cell wall and plasma membrane (Fig. 21.15). This thin, stretched portion finally breaks and the ends of the two daughter cells round up.

The chromatin body apparently divides before the division of the cell (Fig. 21.8), although it appears from this figure that chromatin merely pinches in two.

Figure 21.12. Section of *Rhodomicrobium vannielii* showing array of peripheral photosynthetic membranes. ×60,000. (Courtesy of E. S. Boatman and H. C. Douglas, *J. Cell Biol.*, **11**, 469, 1961.)

Figure 21.13. Section of *Rhodospirillum rubrum* showing the chromatophores (CH). The cell wall (CW) and plasma membrane (CM) are also apparent. ×60,000. (Courtesy of E. S. Boatman and H. C. Douglas, *J. Cell Biol.*, **11**, 469, 1961.)

Figure 21.14. The chemotroph, *Nitrosocystis* oxidizing NO_2 to NO_3 anaerobically, has an array of parallel membranes. $\times 90,000$. (Courtesy of R. G. E. Murray.)

Under optimum conditions, a bacterium may divide and the daughters grow to full size, ready for another division, in 30 minutes. If such conditions were to prevail for $1\frac{1}{2}$ days, 200 trucks of 5-ton capacity would be required to haul away the progeny of a single cell. Obviously, these conditions have never prevailed for so long a period. Nevertheless, the reproductive capacity of bacteria is enormous. Bacterial growth is everywhere held in check. Probably the most important checks are (1) lack of food; (2) the production of substances unfavorable to growth by the bacteria themselves, by other bacteria, by fungi, by higher plants, or by the host in which the bacteria may be living; and (3) competition for oxygen. The rate of increase of a bacterial population is, in general, very similar to that of other organisms, including yeasts, fruit flies, and man. There is a short initial period during which increase in numbers is very slow; then the bacteria increase very rapidly and the population reaches a peak at which it remains constant for a time; and after the peak the number of bacteria decreases rather slowly (Fig. 21.16).

Spores

Several kinds of bacteria are able to form resistant cells, known as **spores,** at the end of their period

Figure 21.15. Changes in the surface of *Bacillus cereus* accompanying cell division. $\times 60,000$. (Courtesy of F. H. Johnson, *J. Bacteriol.,* **47,** 551, 1944.)

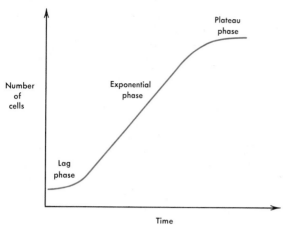

Figure 21.16. A growth curve showing an initial slow rate of growth followed by a rapid logarithmic rate of growth, and finally by the establishment of a colony with little or no growth.

Figure 21.17. Vegetative forms, sporing cells, and free spores of *Bacillus cereus*. (Courtesy of C. F. Robinow, *The Bacteria*, Vol. 1, eds. I. C. Gunsalus and R. Y. Stanier, Academic Press.)

of active growth (Figs. 21.17, 21.18, 21.19). Each spore is protected by several sets of membranes and probably contains a somewhat dehydrated cytoplasm. Spores may remain alive for many years under conditions quite unfavorable to growth. They resist desiccation, are not killed by long exposures to temperatures as high as 80°C, and may be suspended in boiling water for 20 minutes or longer. Their resistance to heat, coupled with their ability to become active immediately upon being placed in a suitable environment, makes them of great interest to food technologists, to doctors, and also to students of living cells. Most of the spore-forming bacteria are rod-shaped forms belonging to the genera *Bacillus* (Fig. 21.17) and *Clostridium* (Fig. 21.18). Only a few cocci and spirilla are able to form spores. While most bacteria form but one spore per cell, a few are known which may form two or more spores per cell. Such spores occur very commonly in soil bacteria. Fortunately, few disease-producing bacteria form spores. The form taken by the chromatin body in spores is shown in Fig. 21.19.

Sexual Reproduction

Sexual reproduction is definitively indicated by crossing over or gene recombination. It was through gene recombination that sexual reproduction was first demonstrated in the Schizomycophyta. The first suggestion of gene recombination occurred quite unexpectedly in studies on antibody formation and virulent and non-virulent colonies of *Diplococcus pneumoniae*. Here genetic information was passed from one living bacterium to another through the medium of dead bacteria. The process was called **transformation.** Its discovery stimulated a search for gene recombination in bacteria through a more normal sexual process.

Transformation

Pneumonia is caused by *Diplococcus pneumoniae*, and the virulent bacteria in the sputum of a patient are surrounded by large capsules formed by a carbohydrate. When grown on agar these virulent bacteria form large smooth colonies. It has been shown that the carbohydrates of the capsules are able to stimulate the formation of antibodies in the host, and that four strains of bacteria occur as indicated by the type of carbohydrate present in the capsule and its antibody response. These four types are known simply as Type I, II, III, and IV. Avirulent bacteria producing rough colonies on agar occur as frequent mutations of the virulent smooth form. Once obtained, these mutants do not revert to the virulent smooth form but may be carried unchanged through hundreds of culture transfers and tens of thousands of generations.

We have thus established for study three traits of *Diplococcus pneumoniae:* (1) virulence, (2) type of colony formed, and (3) type of carbohydrate in the capsule. Each of these traits may mutate, virulent to nonvirulent, rough to smooth colony form, and one type of capsule carbohydrate to another. We may study these characters just as we did the various characters of peas (Chapter 17). Thus avirulent bacteria producing rough colonies on agar frequently mutate to a virulent, smooth form.

In some studies on the activity of these bacteria, *living,* avirulent, rough colony bacteria with Type I capsules were injected into a mouse with a large number of *heat-killed,* virulent, smooth colony, Type II bacteria.

These injections should have been harmless because only *living* virulent bacteria can induce the disease. However, the mouse died, and *living,* virulent Type II cells were recovered from its dead body, although *no living* virulent cells of *D. pneumoniae* were injected into the mouse! We know that the dead bacteria cannot cause disease, and we do not believe

Figure 21.18. Sporulating cells and free spores of *Clostridium pectinovorum.* (Courtesy of C. F. Robinow, *The Bacteria,* Vol. 1, eds. I. C. Gunsalus and R. Y. Stanier, Academic Press.)

Figure 21.19. Arrangement of the chromatin at different stages in the development of spores of *Bacillus cereus.* (Courtesy of C. F. Robinow, *The Bacteria,* Vol. 1, eds. I. C. Gunsalus and R. Y. Stanier, Academic Press.)

that dead bacteria can become alive. The only other explanation is that somehow or other genetic information must have passed from the dead, virulent, smooth, Type II bacteria into the living, avirulent, rough, Type I bacteria. This hypothesis was confirmed when it was shown that an extract taken from the dead, virulent form could indeed transform the living, avirulent bacteria into a lethal culture. When this extract was purified, DNA turned out to be the material which brought about the change or the **transforming principle.** This change of the genetic information in a bacterial cell by the addition of DNA from another culture of bacteria is known as **transformation.** This experiment may be outlined as follows:

living bacteria		dead bacteria		living bacteria
avirulent		virulent		virulent
rough colony	+	smooth colony	→	smooth colony
Type I capsule		Type II capsule		Type II capsule

living bacteria		DNA from		living bacteria
avirulent		virulent		virulent
rough colony	+	smooth colony	→	smooth colony
Type I capsule		Type II capsule		Type II capsule

Gene Recombination in Bacteria

Transformation indicated that gene recombination (Chapter 17) occurred in bacterial cells in spite of the absence of the usual nuclear apparatus. It further stimulated a search for the presence of naturally occurring gene recombination in bacterial cells. Evidence for gene recombination was first obtained by Lederberg in a brilliant, yet simply designed, experiment. *Escherichia coli,* the common bacillus of the large intestine needs no specific growth factors. It is able to synthesis all of its protoplasm from a simple broth of meat extract. Lederberg obtained two mutant strains that were unable to synthesize from the meat extract certain compounds that were required for their metabolism. In order to obtain growth, biotin (B) and methionine (M) had to be added to the broth of one strain and threonine (T) and lucine (L) had to be added to the broth of the second strain. Since each strain could synthesize the metabolite that the other could not, *gene recombination, if it occurred,* should result in progeny that would grow on the minimal broth medium, thus:

$B^-M^-T^+L^+$
cannot grow because it cannot synthesize biotin (B^-) or methionine (M^-)

$B^+M^+T^-L^-$
cannot grow because it cannot synthesize threonine (T^-) or leucine (L^-).

after gene recombination
$B^+M^+T^+L^+$
can grow because it can synthesize all four compounds

A bacterium that now has all of the plus genes should give rise to a colony in the minimal broth culture to which these compounds have not been added, because it will be able to synthesize them from the meat extract in the broth. When this experiment was set up, a strain that was able to synthesize all four compounds was actually found. This demonstrated that recombination of genes did actually occur in bacteria. As it took place only once in about a million cells, its occurrence was too infrequent to allow for detailed study.

Eventually a strain of *E. coli* was found in which recombinations were sufficiently frequent to make a genetic analysis of inheritance in bacteria possible. This strain is known as the Hfr strain, and Fig. 21.20 shows a mating involving the Hfr strain. Note that

Figure 21.20. Mating between mutant strains of *Escherichia coli* (Hfr and F^-). Note the attachment of phage particles to the F^- mutant (upper bacterium) and their absence from the lower, Hfr mutant. The DNA from the phage heads has been injected into the bacterial protoplast. ×95,000. (Courtesy of T. F. Anderson, E. L. Wollman, and F. Jacob, *Ann. Inst. Pasteur,* **93,** 450, 1957.)

Figure 21.21. Diagram of an experiment to test whether a bacterial mating has taken place. If genetic information is not transferred across the sintered glass disc, a mating is necessary to bring about gene recombination.

the contact is rather tenuous, occurring at a small spot on the cell surface. This attachment is of a relatively short duration, and when the cells separate, each is able to form a colony in a normal fashion. With the knowledge of this Hfr strain and the 17 distinctive characters it possesses, it has been possible to prepare a genetic map for the *E. coli* chromosome just as for corn or *Drosophila,* although the methods are of necessity somewhat different than those used for higher forms.

It must first be demonstrated that a true mating has occurred and that a transforming principle is not involved. This is easily accomplished by placing a fine sintered glass disc in the base of a U-tube and growing the different strains of the bacteria in the opposite arms (Fig. 21.21). A transforming principle can diffuse through the sintered glass disc but bacteria cannot pass it. If a recombination occurs when strains are mixed in the same culture medium but does not occur when they are kept separated by a filter, the recombination must be the result of a mating. This experiment proves that a transforming principle was not present and that a mating occurred.

As with higher plants, it is possible to observe certain phenotypic changes. Hfr bacteria genetically marked for motility and shape may easily be distinguished in a mixed culture from a strain (F⁻) possessing the opposite characters (Fig. 21.22). The attachment between Hfr and F⁻ cells, which lasts only an hour, may be observed. When the cells separate, each produces a clone (page 300), but only the F⁻ progeny show any gene recombination. The progeny of the Hfr bacterium resemble the parent cell. Thus, transfer of genetic material is only from the donor Hfr cell to the receptor F⁻ cell (Fig. 21.22). Such an experiment tells us two things. (1) The transfer of genetic material goes only in one direction from Hfr to F⁻. (2) Since the Hfr bacterium still survives and reproduces in its original form it cannot have donated all of its genes to the receptor cell. Two factors may help to explain this survival. (1) There are probably at least two sets of genes in *E. coli* in separate DNA-containing bodies (Fig. 21.10) and (2) the Hfr cell does not donate a complete genetic complement.

Genetic evidence for these conclusions will not be discussed here, but separating the conjugations at successively longer times during the 60-minute period supplies interesting information. If the mating

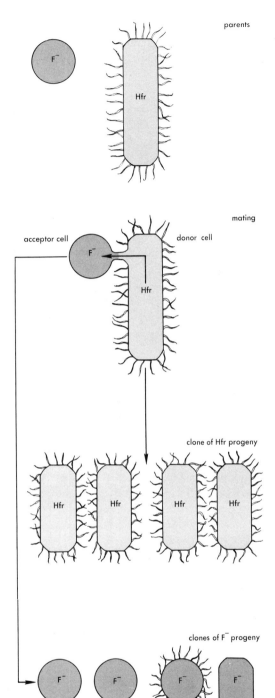

Figure 21.22. Diagram to show how gene recombination may be traced in bacterial matings. Experiment starts with mating of Hfr and F⁻ mutant strains of *E. coli*. After mating, these two bacteria are allowed to produce clones; the appearance of the individuals of the clones is shown.

period is interrupted after 10 minutes, one marker gene may have entered the F⁻ cell; after 20 minutes, two more marker genes have entered; after 30 minutes, another two genes have entered, and so on. Even after the full 60 minutes, when the mating is naturally ended, all genes have not been transferred. As explained in Chapter 17, genes are linked in a linear arrangement on a chromosome. The passage of genes takes time—60 minutes for only partial passage—and the bridge between the mating bacteria is narrow (Fig. 21.20). The suggestion is that the DNA or gene-bearing body is long and slender.

This method of breaking the mating period allows easy mapping of the genes, and the order of the genes so determined corresponds with that found by the conventional system of the percentage of recombinations occurring in a population (page 343). We can only analyze the Hfr chromosome; we know nothing about the chromosome in the F⁻ cell. Special conditions are necessary for the analysis of the F⁻ chromosome, but when they are obtained, it is possible to study them by the method of interrupted mating. This F⁻ chromosome turns out to have the gene order identical to the Hfr chromosome. Different clones obtained after periods of interrupted mating did not have the same genes, although those that were present were in the correct order. Thus one clone would have ABC, another CDE and a third, EAB. This may best be explained by assuming that the F⁻ chromosome is circular and that it may break at any point for injection by the donor cell into the receptor cell. Thus gene recombination may be demonstrated to occur in certain bacterial cells. The mechanism by which it occurs remains to be elucidated.

RESPIRATION OF BACTERIA

Bacteria, like all living organisms, require energy to carry on living processes. Bacteria that generally use atmospheric oxygen in respiration are called aerobes. Many bacteria, however, are able to respire without atmospheric oxygen and without the subsequent oxidation of food materials to carbon dioxide and water (page 259); they are able to obtain sufficient energy from glycolysis, the oxidation of food to organic acids. They do not absolutely require the greater amount of energy that is released when oxygen combines with alcohol, pyruvic acid, or some other simple organic compound to form carbon dioxide and water. Indeed, a few species of bacteria cannot grow in the presence of oxygen. Bacteria that grow in the absence of oxygen are called anaerobes. If small amounts of oxygen hinder their growth, they are known as obligate anaerobes.

Figure 21.23. Nodules produced by *Rhizobium leguminosarum* on the roots of a legume. (Courtesy of Francis Smith.)

Figure 21.24. Crown gall on almond. (Courtesy of Ark.)

INFLUENCE OF ENVIRONMENTAL FACTORS ON BACTERIA

The ability of bacteria to adapt to changing environments varies greatly from species to species. Some bacteria can grow only in certain very specific and constant surroundings. Others are able to adjust themselves to a wide variety of conditions. We have already noted that some bacteria require oxygen, some do not grow in its presence, and some may grow either with or without it. Factors besides food supply and oxygen that influence the rate of bacterial growth are water, temperature, and sunlight.

Water

All bacteria require water to grow. They will not grow in foods containing much less than 15 or 20 percent water. Strong salt and sugar solutions inhibit their growth; hence, salt and sugar may be used as preservatives. Most disease bacteria are unable to live if they become dried out in the air. When quick-frozen and dried in a vacuum, bacteria may remain viable for years.

Temperature

Bacteria may grow at temperatures ranging from about 0°F to around 160°F, but individual species usually have a narrow range. Spores of certain bacteria in a moist atmosphere may resist temperatures of 240°F for some minutes, or withstand boiling water (212°F) for hours. Most disease-producing bacteria do not form spores and are easily killed at 140°F after about 10 minutes' exposure. Disease-producing bacteria in milk may be killed by heating the milk to 160°F and holding it at that temperature for 15 seconds. This process is known as **pasteurization.** On the other hand, many vegetables must be heated to high temperatures for 20 to 40 minutes to kill the spores of soil bacteria that might cause spoilage.

Sunlight

Bacteria are rapidly killed by direct exposure to sunlight, ultraviolet rays being particularly effective.

CLASSIFICATION OF BACTERIA

Because of their small size and similar structure, bacteria are very difficult to classify. For instance, two species of bacteria are known which are almost indistinguishable from each other. Both grow in association with the roots of plants. One, *Rhizobium leguminosarum,* does not harm the host but instead makes nitrogen of the air available to it. The other species, *Agrobacterium tumefaciens,* rapidly kills the host plant. Thus the two species, although very similar both morphologically and physiologically, differ in one very important physiological process that results, in one instance, in a beneficial relationship and, in the other, in a very harmful relationship (Fig. 21.23, 21.24).

In order to name bacteria correctly, one must have complete data as to their size, shape, motility, and spore formation. In addition, many physiological characters must be studied. For example, it may be necessary to know how the bacterium grows in milk, on agar, and on potato, what sugars it is able to ferment, what pigments it produces, and a number of other physiological reactions.

The Schizomycophyta are separated into seven orders on the basis of obvious morphological and physiological differences. Some have characters that suggest molds or algae or one-celled animals. The largest order—the Eubacteriales, or true bacteria—have characteristics described in the previous pages. It is the only order that will be discussed here.

Order Eubacteriales

The Eubacteriales is divided into two suborders: (1) spore-forming bacteria, with a single large family, the Bacillaceae; (2) non-spore-forming bacteria, with eleven families.

The Spore-forming Eubacteriales

The family Bacillaceae contains several species of great economic importance. Most of them are nor-mal inhabitants of the soil or of animal wastes. Many of them will grow well in the absence of oxygen, and some are able to secrete powerful poisons or toxins. *Clostridium botulinum* is common in garden and farm soils. It not infrequently occurs on vegetables, such as corn and beans. If *Clostridium botulinum* is not killed in the process of canning, it grows and multiplies in the canned product, producing one of the most powerful poisons known. Food poisoning from **botulinus toxin,** however, has become quite rare. Commercial canning processes are adequate to kill all spores, and home canners have been so well educated that home-canned food is as unlikely to contain viable *Clostridium botulinum* as is commercially canned food.

Clostridium tetani, like *C. botulinum,* is also a normal inhabitant of well-manured soils. When introduced into deep wounds it may produce a very potent toxin. Tetanus infections were common during wartime. The organism is almost invariably present in dirty wounds, where it multiplies. The toxin secreted by it causes lockjaw. Fortunately, both soldiers and horses may be injected with a substance that gives them almost complete protection against the toxin of *C. tetani.*

The Non-spore-forming Eubacteriales

Although this suborder contains eleven distinct families, it is divided into three natural groups on the basis of food requirements.

1. One family (Nitrobacteriaceae) includes those bacteria that make their own food. These bacteria, like green plants, are autotrophic, although they do not possess chlorophyll.

2. Four families are heterotrophic but obtain the nitrogen they require from inorganic salts, such as sodium nitrate, or from atmospheric nitrogen.

3. The remaining six families are also heterotrophic but require organic substances as a source of energy and nitrogen. They may utilize dead plant or animal wastes; that is, they are saprophytes. Or they may require living cells for their source of nitrogen, in which event they are parasites.

SUMMARY OF BACTERIA

1. Bacteria are prokaryotes; they lack membrane-bound organelles and a nucleus. The light microscope shows them to possess a cell wall enclosing a protoplast. They contain a central chromatin body, so-called because it stains with certain nuclear stains and pinches in two before cell division. The electron microscope shows the presence of DNA fibrils in the chromatin body, and a special technique indicates only a single, circular strand of DNA in each chromatin body.

2. There are heterotrophic and autotrophic bacteria. The heterotrophic forms may be anaerobic or aerobic, but the autotrophic bacteria are anaerobic. The autotrophic bacteria may be chemotrophs oxidizing such substances as NO_2 to NO_3, or phototrophic, utilizing the energy of sunlight.

3. Respiratory enzymes may be bound to the mesosome or to the plasmalemma, or to both membranes.

4. The photosynthetic pigment of the photosynthetic bacteria resembles chlorophyll a and is membrane-bound. The photosynthetic membranes of the green sulfur bacteria are spherical chromatophores, while those of the purple bacteria resemble grana of chloroplasts. In both of these bacteria, photosynthesis is carried on anaerobically.

5. Transformation is the transfer of genetic information from cell to cell without cellular contact.

6. Gene recombination in bacteria results from the sexual process.

7. The chromosome of a bacterium consists of a single, circular DNA fibril.

22

The Algae

DISTRIBUTION OF THE ALGAE

With few exceptions the algae are autotrophic, aquatic, or semiaquatic plants. They are common in most semipermanent pools, in ponds, lakes and streams, along ocean shores, and in the surface waters of the oceans. The terrestrial forms grow on many moist surfaces, including the surface of, or in the top few inches of, moist soil. Although they occur all about us, only the common seaweeds (Figs. 2.7*C*; 22.1*H*, see color plates 2, 10; 22.7) along rocky coasts are large enough to impress themselves upon us. They range from microscopic forms (Figs. 22.1*A*, *B*, 22.3*A*, 22.4*A*) to plants many meters in length (Figs. 2.7*C*, 22.7*C*). They all contain chlorophyll *a*, yet they are so distinctly colored by other pigments, that pigmentation plays an important role in their classification.

Probably the most commonly noticed natural urban habitats for algae are the sides of glass fish tanks. They may also generally be found around leaking faucets and in garden or park pools that are not kept "pure" with chemicals. The "bloom" occurring during the summer on many lakes or the scum found on ponds are the result of algal activity. Microscopic forms occur in most natural waters, including the top 75 meters of the ocean. Here they constitute a primary food source for marine animals of all types and account for about 70 percent of the oxygen released into the atmosphere through photosynthesis.

ECONOMIC IMPORTANCE OF THE ALGAE

The direct economic importance of the algae is minimal, although they are of minor importance in the economy of many countries. The middle lamella and primary walls of many brown algae contain derivatives of **alginic acid.** These compounds, like the pectins found in the cell walls of higher plants, are non-toxic, highly viscous, and readily form gels. One tablespoon of alginic acid will give a quart of water the viscosity of honey. The alginates are used in many ways: they may find their way into marshmallows, ice cream, processed cheese, cosmetics, printing inks (for textiles), pharmaceuticals, and paints. As a food for man, only a few species of blue-green, brown, and red algae are useful to the populations of the islands and coasts of the eastern Pacific Ocean. The green alga *Chlorella,* because of its easy cultivation, has been proposed as a source of food, particularly for space travel. However, it has an unpleasant taste and seems to be quite unsatisfactory as a human food. The algae find some use in sewage disposal, where they may greatly facilitate oxygenation. They may become a nuisance in drinking waters, where they may impart an unpleasant taste or odor as well as clog filters. They are also an indication of pollution. For instance, the water of Lake Tahoe high in the Sierra Nevada has been crystal clear because its mineral deficiency prevented any extensive algal growth. With the increase in human population in the Tahoe basin, algae are appearing in Lake Tahoe waters. This indicates seepage of waste products into the lake.

ALGAL CLASSIFICATION

Grouping the aquatic, autotrophic, nonvascular plants under the general classification of algae is highly artificial but thoroughly practical (although, as we shall see, they have much in common). There does exist a far greater diversity of form and metabolism among them than among the mosses and the vascular plants. The algae have chlorophyll *a* as do all other autotrophic plants. However, some algae have, in addition to chlorophyll *a*, accessory pigments not occurring in the higher plants. Algae, including the blue-greens, require oxygen for respiration and evolve oxygen in photosynthesis, thus differing from the anaerobic photosynthetic bacteria and resembling higher plants.

Although large algae grow in the oceans (Fig. 22.7), there are no large terrestrial forms. Consequently, supporting sclerenchymatous tissue (page 120) is lacking in the algae. Their aquatic environment lacked survival pressures for the evolution of sclerenchyma and specialized absorptive organs such as roots. With the probable exception of the giant kelps, all of the cells of an alga can carry on photosynthesis or are close to cells that can. So a food-conducting, phloemlike, tissue has evolved only in the long stipes of the kelps (Figs. 22.1*G*, 22.7, 22.8). Algae lack roots, stems and leaves, and their plant body is therefore referred to as a **thallus.** All plants with such a plant body are sometimes called **thallophytes.**

Although it is generally easy to visually distinguish an alga from a moss, it does turn out to be difficult to find definitive traits that are common to all algae yet absent from all other phyla. Possibly the only structure possessed by all higher plants and rarely occurring in the algae is a protective jacket of sterile cells surrounding the gametes.

Separation of the algae into phyla is based largely upon differences in pigments, food reserves, the ultrastructure of the chloroplast, type of flagellation, and the chemical constitution of the cell walls. Thus the algal phyla are largely set up on biochemical and structural differences associated with photosynthesis. Detailed differences in reproductive cycles and vegetative structure, although associated with the biochemical differences, are of secondary importance in determining algal phyla, because parallel reproductive and vegetative evolution has occurred in several phyla.

The electron microscope, coupled with a more precise knowledge of algal metabolism, is providing information leading to a revision of the classification of the algal groups. For instance, the Cyanophyta are prokaryotic cells (Figs. 2.5, 22.19), and as such have recently been thought to form a natural group with the bacteria, particularly with the anaerobic photosynthetic bacteria (page 428). However, the thylakoids (page 241) differ in the two groups, and the metabolic processes occurring in the prokaryotic Cyanophyta more closely resemble those occurring in the eukaryotic algae and in the higher plants than they do the metabolic processes occurring in the prokaryotic bacteria. Photosynthesis and aerobic respiration definitely establish the blue-green algae as a separate algal phylum, the Cyanophyta. It stands as an intermediate group between the prokaryotic bacteria and the eukaryotic algae. The algae thus consist of one prokaryotic phylum, the Cyanophyta, and eight eukaryotic phyla, the Rhodophyta, Pyrrophyta, Chrysophyta, Bacillariophyta, Phaeophyta, Xanthophyta, Euglenophyta, and Chlorophyta. Of these, our discussion will mainly concern the Cyanophyta, the Rhodophyta, the Bacillariophyta, the Phaeophyta and the Chlorophyta.

Brief Description of Phyla to be Considered

Cyanophyta

These are the blue-green algae. They are ubiquitous. Characteristically they may be found on moist barren soil, footpaths, tidal flats, cliffs, and

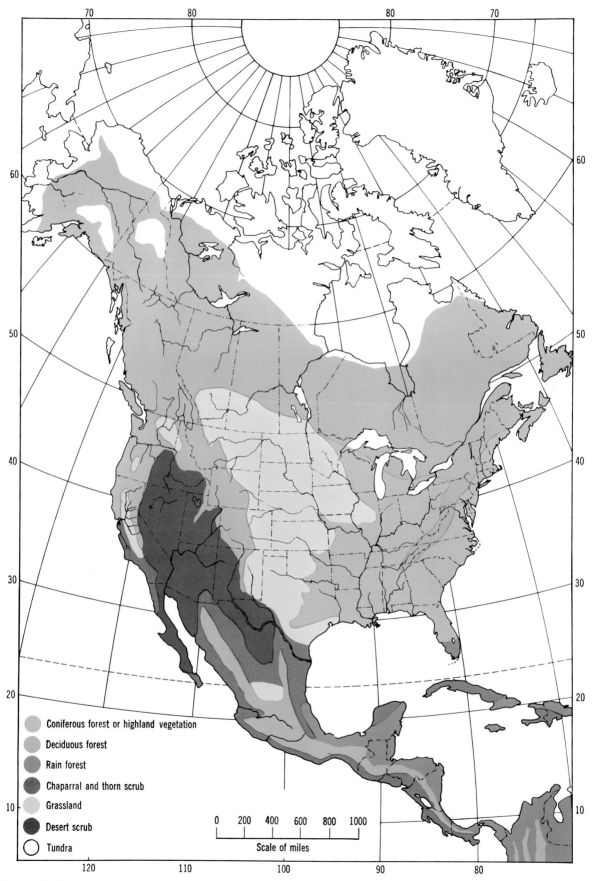

Coniferous forest or highland vegetation

Deciduous forest

Rain forest

Chaparral and thorn scrub

Grassland

Desert scrub

Tundra

0   200   400   600   800   1000

Scale of miles

Figure 18.24. Map of the vegetation types of North America.

Figure 22.1. Various algae. *A, Nostoc* colony, ×20; *B,* hormogonium of *Nostoc,* heterocysts at both ends, ×400; *C,* mat of *Cladophora* in a shallow pool, ×$\frac{1}{10}$; *D,* several cells of the red alga *Acrochaetium,* ×200; *E,* several cells of the brown alga *Pylaiella,* ×200; *F,* several cells of the green alga *Microspora,* ×200; *G,* sieve plate in sieve cell in the stipe of *Macrocystis,* ×1000; *H,* the brown alga *Fucus,* ×$\frac{1}{5}$; *I,* the red alga *Polysiphonia,* ×$\frac{1}{2}$. (*A, B, D, E, F,* courtesy of David Brown; *G,* courtesy of Knute Fisher; *I,* courtesy of N. J. Lang.)

Color Plate 10

sidewalks kept wet by rains, spray from falling water or leaking faucets. They occur in hot springs and in intimate association with other plants or animals. Their plant bodies consist of single cells, short filaments (**trichomes**), and simple colonies of various forms. The cells are almost always encased in mucilaginous envelope (Figs. 22.1*A*, see color plate 10, Fig. 22.2). They contain chlorophyll *a*, *β* carotene,

Figure 22.2. Blue-green algae. *A*, living cells of *Synechococcus*; *B*, killed and stained cells of *Synechococcus*; *C*, living cells of *Chrococcus*; *D*, living cells of *Nostoc*; *E*, killed and stained cells of *Nostoc*; *F*, killed and stained cells of *Micrococcus*; *G*, killed and stained cells of *Oscillatoria*. ×800. (*A* to *F*, courtesy of W. A. Cassel and W. G. Hutchinson, *Exptl. Cell Res.*, **6**, 134, Academic Press; *G*, courtesy of R. Norris.)

and two water-soluble pigments (phycobiliproteins), **c-phycocyanin,** and **c-phycoerythrin.** As a reserve they accumulate **cyanophyte starch** and certain characteristic proteins. No flagellated cells occur and sexual reproduction has not been established. Examples are *Nostoc* (Fig. 22.1*A, B*), *Oscillatoria. Synechococcus, Chroococcus,* and *Micrococcus* (Fig. 22.2).

Rhodophyta

These are the red algae (Figs. 2.7*D*, 22.1*D*). They occur mostly as feathery plants growing most abundantly on the rocky coasts of warm oceans. There are a number of freshwater Rhodophyta (Fig. 22.3*D*). They are frequently small and must be

Figure 22.3. Examples of freshwater algae. *A, Scenedesmus quadricauda,* green alga; *B,* young coenobium of *Scenedesmus* leaving the parent cell; *C,* the diatom, *Arachnoidiscus ehrenbergii; D,* a fresh water red alga, *Batrachospermum moniliforme.* (*A, B,* courtesy of T. Bisalputra; *C,* courtesy of CCM: General Biological, Inc.; *D,* courtesy of David Brown.)

searched for. The red algal plant bodies range from single unicellular (*Porphyridium*) forms through simple filaments to leaflike thalli (Fig. 2.7*D*) or to delicate (*Polysiphonia*) feathery plants (Fig. 22.1*I*). They contain only chlorophyll *a* and carotene and the water soluble phycobilins, b-, c-, r-phycocyanin and b-, c-, r-phycoerythrin. As a reserve they accumulate, outside the chloroplast, a carbohydrate **floridean starch.** It turns red, not blue, upon staining with iodine. No flagellated cells are present. *Batrachospermum* is a freshwater species. *Porphyridium* is a unicell of considerable laboratory interest because the phycobiliproteins can be seen with the electron microscope; they show a close relationship to the photosynthetic membranes. The genus *Polysiphonia* shows considerable evolutionary advancement and has an interesting reproductive cycle (Figs. 22.10, 22.18).

Bacillariophyta

The algae belonging to this phylum are the **diatoms.** They are widespread, especially in colder waters. They occur in the surface waters of temperate and northern oceans, in lakes, ponds, and semipermanent pools, on moist barren earth and in the top layers of soil. They are either single-celled forms (**unicells**), or a simple aggregation of cells forming filaments. They may be free floating, or attached to other algae, submerged roots or other material. Their cell walls consist of an intricate layer of silica sandwiched between two delicate organic membranes. They contain chlorophylls *a* and *c*, plus δ and ε carotene. While many diatoms show movement, flagellated vegetative cells are not present. Their ability to metabolize silica and to deposit it in their cell walls results in a delicate wall sculpturing, sometimes of great beauty (Fig. 22.3*C*).

Phaeophyta

These are the brown algae, or seaweeds, that grow abundantly along the rocky coasts of all oceans, but particularly those of the temperate zones (Fig. 22.7). In certain tropical areas the free floating *Sargassum* (Fig. 22.7*D*) forms dense beds. The brown algae range from small, almost microscopic forms to the giant kelps that may exceed 30 meters in length. No unicellular forms are known. They contain chlorophylls *a* and *c*, (Figs. 2.7*C*, 22.1*E*, *H*), carotene and a characteristic brown pigment, **fucoxanthin.** They store a carbohydrate known as **laminarin.** The reproductive cells of a number of species are flagellated.

Chlorophyta

These are the green algae (Figs. 2.7*B*, 22.1*C*, *F*, 22.3*A*, 22.4), which are the common freshwater algae found, frequently abundantly, in all freshwater streams, lakes, ponds and semipermanent pools. Three orders inhabit tropical oceans, and there are a few marine forms scattered through other groups. They are grass-green because of chlorophylls *a* and *b* (Fig. 22.1*F*). They also contain β carotene and store starch. They show great morphological diversity, both in vegetative structure and in reproduction. Flagella are lacking altogether in some genera and only gametes are flagellated. In others the vegetative forms are flagellated.

Green algal plant bodies consist of single cells, filaments or delicate flattened blades. There are many common green algae; almost any microscope slide of native water will contain some examples. *Chlamydomonas* and associated genera are common motile, unicellular, or colonial forms. *Spirogyra*, *Ulothrix*, *Oedogonium*, and *Cladophora* are common filamentous forms (Figs. 22.1*C*, *F*, 22.4 *C*, *D*, *E*, 22.5). Sea lettuce, or *Ulva*, is an example of a species whose thallus is flat and like a blade (Fig. 2.7*B*).

PHYLOGENY OF THE ALGAE

Since the photosynthetic pigments are important in algal classification, a comparison of pigments should reveal relationships between the phyla. Although such a comparison does supply some indication of evolutionary tendencies, the situation is complicated. Consider the relationship between the prokaryotic Cyanophyta (page 11) and the eukaryotic Rhodophyta. In spite of the great differences between them, they are thought by some to be related (1) because they both contain only chlorophyll *a* and the two phycobiliprotein pigments, phycocyanin and phycoerythrin, and (2) because flagellated cells are absent from both phyla. Otherwise, the great differences in cellular detail, vegetative form, and reproductive structures indicate a widely divergent evolution.

A different situation exists when pigments and form are compared among the eukaryotic phyla, including the Rhodophyta. Here, while each phylum is distinguished by characteristic pigments, they

nuclei

Pediastrum boryanum

A

nuclei

holdfast

Valonia
utricularia

B

chloroplast
nucleus
cytoplasm

C

Spirogyra

sporangium

oogonium

antheridium

nuclei

Vaucheria

G

chloroplast

branches

Draparnaldia

D

Hydrodictyon reticulatum

E

nuclei

F

all contain species with similar plant bodies and similar reproductive stages. Only the Euglenophyta consists solely of unicellular forms, and only the Phaeophyta lacks unicellular species. Otherwise, each phylum contains species ranging from unicells to filaments, branching filaments, aggregations of cells forming parenchymatous thalli, and intertwining filaments forming more complicated plant bodies. Reproductive structures have not been of great assistance in indicating relationships between the phyla. A recent account of algal phylogeny postulates three lines of development from a supposed common ancestor. The Cyanophyta and Rhodophyta, for reasons already discussed, form one line of ascent. A second line is made up of the Chrysophyta, the Pyrrophyta, the Bacillariophyta and the Phaeophyta, while the third line is comprised of the Xanthophyta, the Euglenophyta and the Chlorophyta. There is general agreement among botanists that ancient Chlorophyta were the ancestors of the higher nonvascular plants and the vascular plants.

DIVERSITY OF FORM WITHIN THE ALGAE

Within the phyla, evolutionary relationships are demonstrated by thallus morphology. It is widely accepted that unicells are the most primitive and that evolutionary advance proceeded, in general, along similar lines in all phyla. There were three tendencies in this evolutionary expression: (1) single cells cooperated to form rather complex colonies (the volvocine tendency) (Fig. 22.5A); (2) single cells became locally highly differentiated (the siphonous tendency) (Fig. 22.5B); and (3) single cells remain attached together after division to form filaments or simple sheets of parenchymalike tissue (the tetrasporine tendency) (Fig. 22.5C). While parallel changes have apparenly occurred in all phyla, they have not occurred to the same extent in all phyla. For instance, there are only unicells in the Euglenophyta, and unicellular forms are not known in the Phaeophyta. Further, the thalli of the Phaeophyta and the Rhodophyta show a much higher degree of specialization than is found in the other phyla. We shall discuss the morphology of

algal thalli, proceeding from less complex to more complex arrangements. The discussion will be illustrated by examples of Chlorophyta, Phaeophyta, and Rhodophyta.

The Volvocine Line

The volvocine tendency is well illustrated by the order Volvocales in the Chlorophyta (Fig. 22.5A). Here a unicell, *Chlamydomonas*, serves not only as a starting point for the volvocine line, but also as the starting point for the siphonous and tetrasporine lines. Most individuals in the *Chlamydomonas* genus are oval or oblong. They have two flagella at their anterior end, a cup-shaped chloroplast. Most species have a single red body of carotene with a characteristic ultrastructure; this is the **eye spot** (Fig. 22.25). The nucleus is centrally located within the cup formed by the chloroplast. Sometimes some species go through a **palmella** stage, during which time the individuals lose their flagella and become embedded in a common mulcilaginous matrix.

The permanent association of chlamydomonad-like cells into curved plates (*Gonium*, Fig. 22.5A) of four linear cells (*Scenedesmus*, Fig. 22.3A) may be considered the first step in advance in the volvocine line. Each new colony is formed within a parent cell and is known as a **coenobium.** No cell division takes place in the coenobium after its release from the parent cell wall. Further advance is occasioned by the formation of a hollow sphere of 32 similar individual cells. They are connected to each other by delicate strands. Increase in the number of cells forming a colony and specialization of some cells for reproduction culminates in the genus *Volvox* (Fig. 22.5A). The colonies contain from 512 to 20,000 cells, only a few of which are reproductive cells confined to a small area. The vegetative cells of *Volvox* still resemble *Chlamydomonas*.

The Siphonous Line

The outstanding characteristic of the individuals forming the siphonous line is the complete lack, or rare production, of septa. The cells of most siphonous species are thus multinucleate and coenocytic (Fig. 22.5B). In those forms completely lacking septa, the peripheral cytoplasm, with embedded nuclei,

Figure 22.4. (*left*) Freshwater algae. A to F, Chlorophyta; G, Xanthophyta. A, *Pediastrum*; B, *Valonia*; C, *Spirogyra*; D, *Draparnaldia*; E, portion of net of *Hydrodictyon*; F, several mutinucleate cells forming net of *Hydrodictyon*; G, *Vaucheria*.

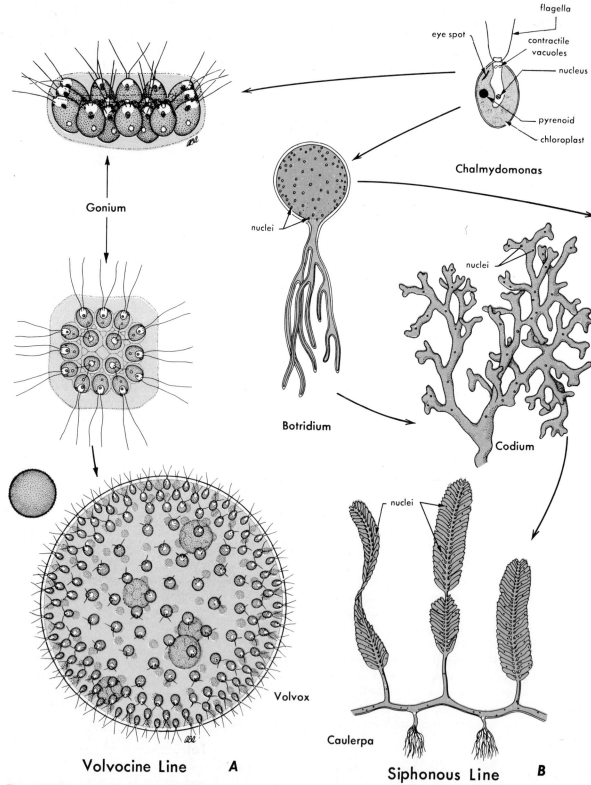

flagella

eye spot

contractile vacuoles

nucleus

pyrenoid

chloroplast

Chalmydomonas

nuclei

Gonium

nuclei

Botridium

Codium

nuclei

Volvox

nuclei

Caulerpa

Volvocine Line   **A**   **Siphonous Line**   **B**

Figure 22.5. Diagrammatic representation of the three lines of algal evolution. *Botrydium* belongs to the phylum Xanthophyta; all others are Chlorophyta. All three lines are thought to take their origin from a type similar to

Chlamydomonas. **A,** the volvocine line (light green); **B,** the siphonous line (medium green); **C,** the tetrasporine line (dark green). In the volvocine line the cells retain their identity; in the siphonous line mitosis is not followed by

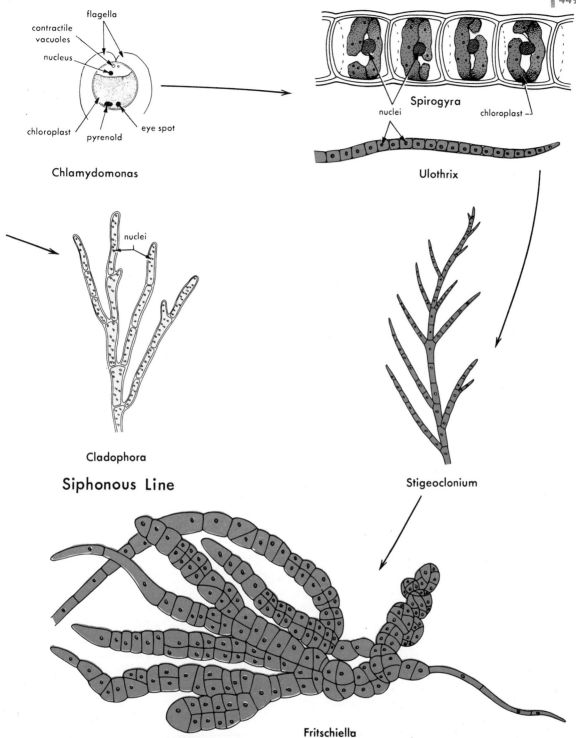

Chlamydomonas

Spirogyra

Ulothrix

Cladophora

Siphonous Line

Stigeoclonium

Fritschiella

C

Tetrasporine Line

Figure 22.5 (*continued*). cytokinesis; in the tetrasporine line mitosis and cytokinesis comprise a true cell division. Only plants with cell division successfully invaded the land, and Chlorophyta similar to *Fritschiella* may have been the progenitors of land plants. (*Codium* and *Caulerpa* redrawn after W. R. Taylor, *Marine Algae of the Eastern Tropical and Subtropical Coasts of the Americas,* University of Michigan Press. *Fritschiella* after M. O. P. Lyengar, *New Phytologist,* **31,** 329. *Botrydium* redrawn after G. M. Smith, *Freshwater Algae of the United States.* Copyright 1933 by McGraw-Hill Book Company. Used with permission of McGraw-Hill Book Company.)

plastids and all other organelles, is continuous throughout the plant body. This also means that there exists a single central vacuole extending through all filaments. Such an evolution of form may be envisioned with a chlamydomonad-like cell as the primitive starting point. Evolution of the siphonous plant body occurred several distinct times in the Chlorophyta and has been paralleled at least once (*Vaucheria*) in the Xanthophyta. The siphonous habit did not develop in other phyla.

The most primitive forms like *Protosiphon* (Chlorophyta) or *Botrydium* (Fig. 22.5B) (Xanthophyta) consist only of an aerial green, multinucleate vesicle anchored to the soil by colorless branching rhizoids (*Botrydium*) or by a single rhizoid (*Protosiphon*). The irregular branching or budding of this vesicle would result in the irregular thallus of *Valonia* (Fig. 22.4B) (Chlorophyta). Simple filamentous forms such as *Vaucheria* (Fig. 22.4G) occur in the Xanthophyta. Two types of multinucleate forms occur in the Chlorophyta. Some, like *Cladophora* (Fig. 22.5B) and *Hydrodictyon* (Fig. 22.4E), have cross walls separating multinucleate cells; others are strictly coenocytic. These coenocytic forms, belonging to the order Codiales, branch extensively and with a high degree of regularity, conferring considerable diversity of form on the adult thalli. The genus *Caulerpa*, which forms extensive mats on shallow tropical sea floors, is a good example of the degree of specialization attained by intertwining coencytic filaments (Fig. 22.5B). Large tracts of shallow areas of the floor of the Mediterranean Sea are covered by *Caulerpa*. The plants of this genus consist of a cylindrical rhizome up to three meters in length. It is formed by a characteristic intertwining of filaments. This rhizome bears branching rhizoids growing downward into the ocean floor. It also sends upward a number of photosynthetic shoots that take a great variety of shapes. In some species the shoots are featherlike, while in other species they are flat and leaflike. They, like the rhizome, are formed of branching coenocytic filaments arranged in a characteristic pattern. Growth in these forms is at the apex of each filament, where concentration of protoplasm also occurs. The filaments of some species may become appressed to form a compact parenchymatous tissue (Fig. 22.5B). Note that two types of filaments are formed, a horizontal rhizome and upright photosynthetic branches.

In contrast to this extensive growth and differentiation of coenocytic filaments, stands the uninucleate single-celled *Acetabularia* (Fig. 4.17). Here a high degree of specialization occurs within the limits of a single nucleated cell about 25 mm long and divided into three regions of specialization, (1) a branched rhizoid system, one branch of which contains the single nucleus; (2) an upright stalk; and (3) an expanded upper portion, the cap.

The Tetrasporine Line

Although a *Volvox* colony, or coenobium, in the volvocine line may consist of 20,000 cells with some degree of specialization, all of these cells were formed within a single parent cell. The daughter cells provided their own cell wall at a later stage in development. Cell divisions, as such, do not occur in a mature coenobium. Mitoses are confined to a definite period of the life cycle and the protoplasts so formed provide themselves with their own cell wall. In the siphonous line mitoses occur, but cytokinesis does not, and the resulting nuclei are not set apart in individual protoplasts. They remain within the parent cell wall, which enlarges and may take on a complex shape. In the tetrasporine line, cell divisions occur; each mitosis is accompanied by cytokinesis. Now the parent cell wall is retained as an essential part of the two daughter cells. Protoplasts and enclosing cell walls become cooperating units in a tissue or plant body. The need to confine cell division to a single specific period of the life cycle has been eliminated. Cell divisions, mitosis and cytokinesis, may occur through all, or much, of the life of a plant. Size and complexity of form may now be attained by continued cell division and by controlling (1) the polarity of cell divisions, (2) the relative rates of cell divisions, and (3) the relative rates of cell enlargement. As an example of the influence of polarity of cell divisions, if all mitotic spindles were aligned on a straight line, a simple unbranched filament would result (Fig. 22.6A).

The gametes of the green alga *Ulothrix* resemble mature chlamydomonad cells, as do the gametes of many other algae. When the zoospores of *Ulothrix* germinate, the new cells adhere to each other and the cells divide with spindles aligned in a straight line, so that repeated division of all cells results in a filament formed by a single row of cells. This is a **uniseriate** filament. Two very important things have happened: (1) cytokinesis results in adherent, cooperating daughter cells and (2) a definite polarity for all cells has been established for this simple multicellular plant. Adhering cells and genetic con-

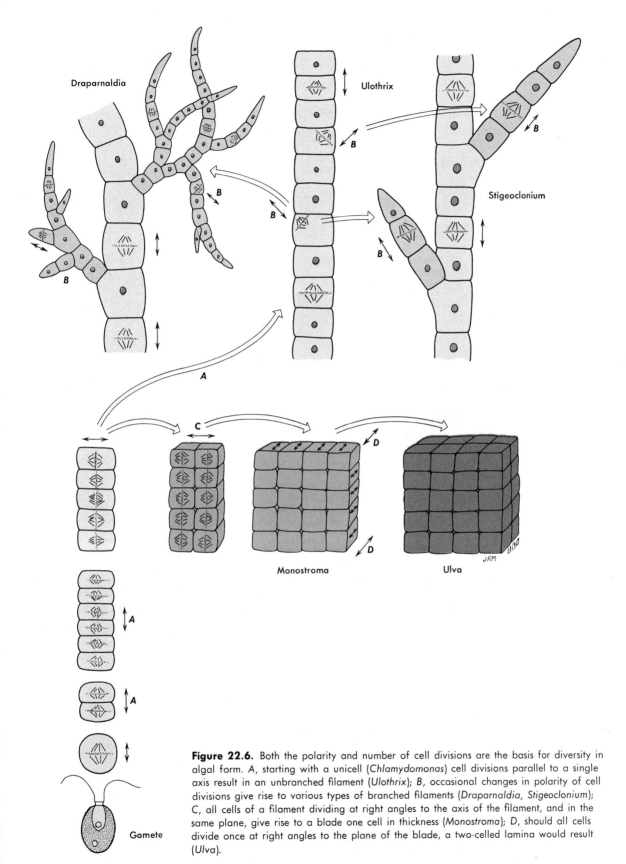

Figure 22.6. Both the polarity and number of cell divisions are the basis for diversity in algal form. *A*, starting with a unicell (*Chlamydomonas*) cell divisions parallel to a single axis result in an unbranched filament (*Ulothrix*); *B*, occasional changes in polarity of cell divisions give rise to various types of branched filaments (*Draparnaldia, Stigeoclonium*); *C*, all cells of a filament dividing at right angles to the axis of the filament, and in the same plane, give rise to a blade one cell in thickness (*Monostroma*); *D*, should all cells divide once at right angles to the plane of the blade, a two-celled lamina would result (*Ulva*).

trol of the polarity of cell division are two basic requirements for the attainment of great diversity of plant form (Figs. 22.6, 22.10).

In *Ulothrix* and other simple unbranched forms, all spindles are oriented in the same direction and all cells in the filament may divide. Upon establishing a second pair of poles at right angles to the axis of the first pair, a diversity of thalli will arise, depending upon the timing of cell divisions in the two planes. If all cells divided regularly at right angles to each other, but in the same plane, a thin tissue, one cell in thickness, would form a **monostromatic thallus** (Fig. 22.6*C*). This occurs in the green alga *Monostroma*. If, instead, every tenth cell divided, and continued to divide, at right angles to the line of divisions forming the filament, a flat filament with regular branches would result (Fig. 22.6*B*). If every twentieth cell could now divide in a plane at right angles to the plane of the flat branching filament, a branching filament with its branches at right angles to each other would result. If the rates and locations of cell divisions are confined to specific cells of the main filament, lateral branches will arise at definite points as in *Draparnaldia* (Fig. 22.6*B*). Note that the cells of lateral branches are much smaller than those of the central branch. If all cells early in the development of a monostromatic thallus divided but once in a plane at right angles to the plane of the monostromatic thallus, a thallus two cells in thickness (**distromatic thallus**) would result. Sea lettuce, or the marine green alga, *Ulva*, has a distromatic thallus (Figs. 2.7*B*, 22.6*D*). With division possible in three planes, form is determined by the relationship between the rates of division, the location of the divisions, the rates of cell expansion and the kinds of differentiation that occur in the three planes. Cell divisions under these conditions are generally confined to specific locations or meristems. These generally occur at the apex of the thallus, or at definite intercalary points. In many brown algae the epidermal cells may also be meristematic.

While polarity of cell division is of importance in determining plant forms, environmental factors may also be involved. Nevertheless, it would be exciting to discover the genetic mechanism that determined that given cells were to divide at right angles to their neighbors. Just what type of change in base pairs brings about the change from a filament to the monostromatic thallus of *Monostroma* or the distromatic thallus of *Ulva*?

Heterotrichy

Caulerpa, of the Codiales, produces a stout horizontal rhizome provided with rhizoids for anchorage and complex arrangements of upright filaments for photosynthesis. Two different types of filaments are formed. This type of growth habit is called **heterotrichy.** It is a type of specialization that has occurred numerous times in the Chlorophyta, Rhodophyta, and Phaeophyta. It is thought to represent a definite evolutionary advance.

Phaeophyta and Rhodophyta

With the exception of a lack of unicells in the Phaeophyta and of motile cells in the Rhodophyta, a series of thalli paralleling the tetrasporine line, as exemplified by the Chlorophyta, is to be found in both the Phaeophyta and Rhodophyta. However, both red and brown algae have developed much greater specialization than occurs in the thalli of the other algal phyla. This development has proceeded along separate pathways in these two phyla, and it will be necessary to consider the vegetative structure of the Phaeophyta and Rhodophyta separately.

Phaeophyta

The most primitive Phaeophyta, the Ectocarpales, are, as a group, small branched filamentous forms (Figs. 22.1*E*, 22.7*A*). The heterotrichous habit is developed in most, if not all, of them. The prostrate system consists of a few filaments and serves to anchor the upright system of branching photosynthetic and reproductive filaments to other algae or to rocks. This prostrate system of filaments exhibits apical growth. The upright system appears as feathers or compact brownish tufts (Fig. 22.7*A*).

Figure 22.7. (*right*) Habit sketches of brown algae. (*Macrocystis integrifolia* and *Laminaria* redrawn after G. M. Smith, *Marine Algae of the Monterey Peninsula*, second edition, incorporating supplement by G. J. Hollenberg and I. A. Abbott. Copyright 1969 by Stanford University Press. *Ectocarpus* redrawn from W. R. Taylor, *Marine Algae of the Northeastern Coast of North America*, University of Michigan Press. *Sargassum* redrawn after W. R. Taylor, *Marine Algae of the Eastern Tropical and Subtropical Coasts of the Americas*, University of Michigan Press.)

A

Ectocarpus

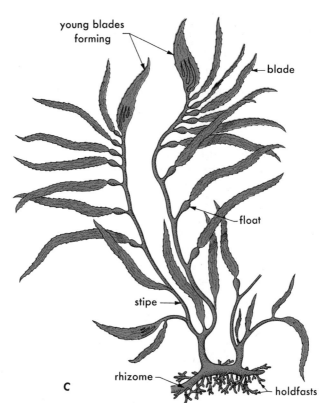

young blades
forming

blade

float

stipe

rhizome

holdfasts

C

Macrocystis integrifolia

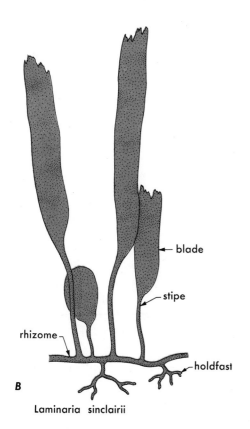

blade

stipe

rhizome

holdfast

B

Laminaria sinclairii

float

D

Sargassum hystrix

Diversity of Form Within the Algae

Cell division is diffusely distributed along the branching filaments. It is rarely apical or confined to special regions.

The most advanced Phaeophyta are large plants with regions specialized for anchorage, conduction, support, and photosynthesis. A holdfast attaches the plant to the rocky ocean floor; photosynthesis occurs in a leaflike blade or lamina, and a stipe connects the lamina with the holdfast. Bladders are found in many species and add to the buoyancy of the laminae. These features are well exhibited by *Laminaria* (Fig. 22.7*B*), *Macrocystis* (Fig. 22.7*C*). *Postelsia* (Fig. 22.8) and *Fucus* (Fig. 22.1*H*). This increase in size and complexity of brown algae over green algae is accompanied by an increase in the degree of cell specialization. Aggregates of cells form tissue systems such as those shown for the stipe of *Postelsia* (Fig. 22.7*C*). The outermost layer not only serves as a protective barrier around the stipe but is also meristematic, and cuts off new cells adding to the internal cortex. This outer layer of meristematic cells is appropriately called the **meristoderm.** The cells forming the cortex are parenchyma-like. Mucilage-secreting cells form definite canals throughout the cortex. The **medulla** or midrib contains mechanical or supporting cells as well as sieve cells. These latter have sieve plates, form callose (Fig. 22.1*G*), and strongly resemble sieve cells and sieve tube elements of the vascular plants. Work with C^{14} indicates that the carbohydrate mannitol is translocated and that the mechanism of transport resembles that found in the vascular plants.

The stipe of many Laminariales is perennial, but the meristems are not active during the winter months. Because of this, annual rings may sometimes be observed in cross sections of stipes of the large kelps. Meristematic activity is also found in a transition region between the stipe and lamina. Here cell formation adds new tissue to both stipe and lamina (Fig. 22.7*B*, *C*). Growth in the larger Phaeophyta may be apical. It arises from one or several apical cells.

Problems of polarity of cell division in the Phaeophyta have been studied experimentally in the germination of the zygote of *Fucus*. The first division of the zygote is preceded by the formation of a small protuberance in the zygote cell wall. The plane of the first division is at right angles to this protuberance (Fig. 22.9*A*). This division is unequal; the larger of the two cells formed receives most of the chloroplasts and develops into the vegetative and reproductive thallus. The smaller cell, with fewer chloroplasts, first forms a slender rhizoid that later develops into the stipe and holdfast (Fig. 22.9*A*).

The protuberance in the zygote cell wall develops in response to blue light, lower pH, or warmth. The plane of the first cell division of the zygote appears to be determined by purely physical factors. It does not seem to be under genetic or nuclear control. Polarity does not seem to play a prominent part in the next few cell divisions, for the successors of the larger cell simply form a spherical cell mass. However, this sphere soon elongates into a pear-shaped mass of cells, thus indicating that a degree of polarity in cell divisions has been established (Fig. 22.9*A*). Soon a small depression appears in the top of the young pear-shaped thallus. A group of from one to eight apical cells forms at the bottom of this depression (Fig. 22.9*B*). The young thallus now becomes flat, and cells in the apical region follow each other in an orderly sequence to produce the flat strap-shaped dichotomously branching *Fucus* plant body (Fig. 22.1*H*). In spite of a fairly good knowledge of the sequence of events from zygote to the mature thallus, the basic questions of why and how cells come to divide in such precise planes still require explanation.

Rhodophyta

Early evolution of diversity of form in the Rhodophyta proceeds along lines similar to that already described for the Chlorophyta. The more primitive subclass (Bangiophycidae) contains unicellular species (*Porphyridium*), unbranched filamentous forms (*Erythrotrichia*), and branching filamentous forms (*Goniotrichum*). There are also species whose thalli consist of single layer of cells bearing several upright threads (*Erythrotrichia*) while in still other forms the thalli have two cell layers (*Porphyra variegata*) forming blades similar to *Ulva* (Fig. 2.7*B*).

Morphological diversity in the much larger and

Figure 22.8. (right) Cellular organization of blade and stipe of *Postelsia*. A, cross section of stipe showing different tissue regions, ×25; B, enlarged portion of cross section of stipe showing cellular detail, ×2500; C, cross section of blade showing different tissue regions, ×50; D, enlarged portion of blade showing cellular detail, ×450.

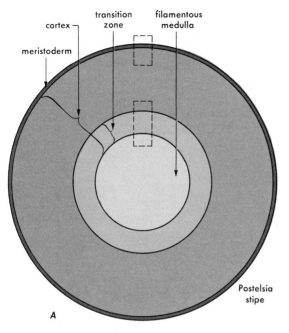

A

cortex — meristoderm — transition zone — filamentous medulla

Postelsia stipe

C

meristoderm — medulla — cortex

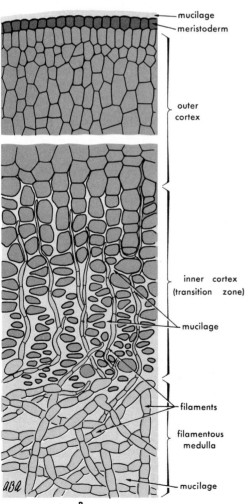

mucilage
meristoderm

outer cortex

inner cortex (transition zone)

mucilage

filaments

filamentous medulla

mucilage

B

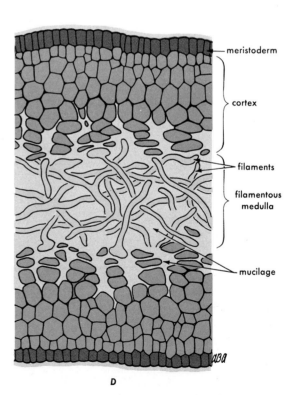

meristoderm

cortex

filaments

filamentous medulla

mucilage

D

more complex subclass Florideophycidae is attained by an aggregation of filaments. The simplest algae in the subclass are much branched, small, rather delicate forms. They have developed the heterotrichous habit, and, as in the brown algae, this type of growth disappears with increasing morphological complexity. In the more complex forms, the filaments are compacted into a parenchymalike mass of cells, and the plant body consists of a dissected blade, stipe, and holdfast such as exhibited by *Gigartina*.

Growth is almost always apical and, as in all other species, morphological form is related, among other things, to cellular polarity. This is well demon-

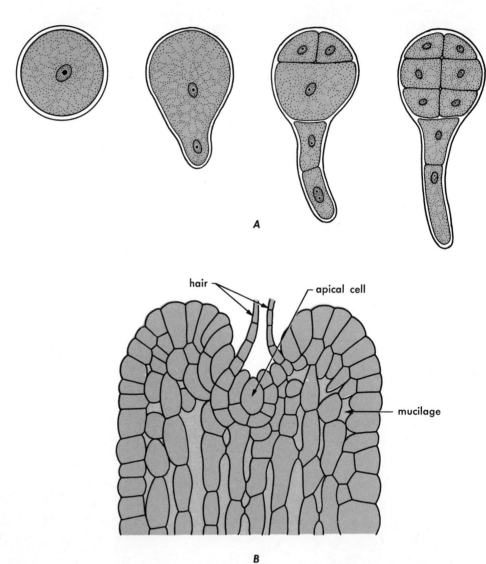

A

B

Figure 22.9. (*above*). Growth in *Fucus*. A, germination of zygote. The polarity of the first division is determined by environmental factors such as light; B, apical region of blade showing notch with meristematic apical cell.

Figure 22.10 (*right*). The early cell divisions in *Polysiphonia* are very precisely ordered and easily followed. They are responsible for the organization of the mature thallus. A, cell divisions only parallel to the axis of the filament would result in a uniseriate filament. Regularly spaced divisions at an angle to the axis produce regular branches; B, divisions at right angles to the axis of the filament, always in the order indicated, result in a thallus formed of a central cell surrounded by five pericentral cells. Arrows indicate polarity of cell divisions.

B

A

strated in *Polysiphonia* (Fig. 22.10). *Polysiphonia* has a single apical cell. Cell division with the mitotic spindle parallel to the long axis of the filament results in an increase in length of the filament (Fig. 22.10*A*). Cell division at right angles to the axis of the filament produces an array of five pericentral cells, precisely arranged around the central axial cell (Fig. 22.10*B*). This means that the planes of division giving rise to the pericentral cells are also precisely oriented to each other. Furthermore, the order of production of the new pericentral cells is the same in all plants: cells two and three are formed on opposite sides of cell one. Cell four forms beside two, and cell five forms between three and four (Fig. 22.10*B*).

Polarity at an angle to the main axis of the filament results in lateral branches (Fig. 22.10*A*). In this case, the daughter nuclei in the dividing apical cell end up in opposite corners of the cell, and the cell plate cuts the dividing cell at an angle to the axis of the filament. Normal polarity is restored in the apical cell. Polarity in the second cell is now established at an angle to the main filament, and a protuberance forms from this cell in line with this newly established polarity (Fig. 22.10*A*). Subsequent cell divisions are at right angles to this polarity, cutting off successive cells to form a lateral branch.

Thus the mature thallus of *Polysiphonia* is a branching thallus formed by a line of central or axial cells surrounded by a regular number of pericentral cells (Figs. 22.1*I*, 22.18). Axial and pericentral cells are the same length. A comparable structure occurs in many of the Florideophycidae. Morphological variation, which is considerable in the subclass, results from the number of cells forming the axis, from variation in the size of cells, and the degrees of branching.

SUMMARY OF MORPHOLOGICAL DIVERSITY IN THE ALGAE

Morphological diversity, which is very considerable in the algae, is attained with thalli consisting of (1) single cells (sometimes motile), (2) simple colonies of single cells, (3) unbranched filaments, (4) branched filaments, (5) heterotrichous filaments, (6) intertwining filaments, and (7) a simple parenchymatous tissue. Most phyla exhibit all or significant portions of this range of diversity, and it may be repeated in several families. Examples of phyla not showing the full range of diversity include the Euglenophyta, which consists only of unicells, many

of which are motile, the unicellular Rhodophyta which are not motile, and the Phaeophyta, which contains no unicellular forms. The simplest filaments of the Phaeophyta have developed the heterotrichous habit, and the most advanced Rhodophyta show extensive intertwining of filaments but appear to lack a true parenchyma. This latter tissue is, however, present in some of the primitive red algae.

Two conclusions can be drawn from these observations: (1) a high degree of morphological diversity can be obtained by different arrangements of single cells, of filaments of cells, and of sheets of cells; (2) the evolutionary pathway—single cells to parenchyma tissue with localized growth regions, or parts of this sequence—occurred numerous times in most of the algal phyla.

It is also important to note that phyla thought to have, at best, a very remote relationship with higher plants, followed evolutionary pathways leading to parenchyma tissue, apical growth, intercalary meristems and meristems giving rise to increase in girth—all things found in present-day vascular plants. Organs developed to anchor plants in place; specialized flattened laminae involved in photosynthesis, and specialized transport cells similar to sieve cells, develop in the long stipes of the kelps, whose holdfast is submerged and some distance from the photosynthetic laminae.

REPRODUCTION—DEFINITION

Details of reproduction as it occurs in higher plants have been discussed in the chapters on the cell (Chapter 4), the flower (Chapter 14), the fruit (Chapter 15), and inheritance (Chapter 17). In general the higher plants reproduce (1) sexually through seeds and (2) vegetatively through cuttings, grafts, and other asexual processes. In the fungi and algae we may distinguish between three general types of reproductive processes, asexual reproduction (1) by vegetative means, and (2) by specialized cells formed after mitosis (**mitospores**) and (3) sexual reproduction. In strictly vegetative reproduction, any portion, or specialized portions, of a vegetative plant may separate from the parent plant to form new individuals. We have seen examples of this in rooted cuttings (page 414) and the nucellar buds in oranges (page 302). As chromosomes pass unchanged from one individual to the next, succeeding generations have identical genotypes (page 348). Asexual reproduction, in the fungi and algae, also occurs through the formation of spores or single cells specialized for the replication of new individuals, while preserving unchanged, the genetic make up. Thus, succeeding generations have identical genotypes. Both types of asexual reproduction commonly occur under conditions favorable for growth and result in a rapid expansion of a given population. Sexual reproduction involves the union of two gametes, thus doubling the chromosome number. Meiosis must thus occur at least once in the life cycle of every sexually reproducing organism. The random distribution of chromosomes at anaphase I of meiosis and the recombination of genes in chromosomes during meiosis results in each individual of the progeny of a single mating having its own unique genotype. Sexual reproduction in the fungi and algae is commonly associated with conditions less favorable to growth, and in many instances the cells resulting from the union of gametes develop resistant walls that carry the species over periods unfavorable to growth. It will be recalled that, in the same way, seeds of most gymnosperms and angiosperms are adapted to remain viable for long periods of dryness, and that they too result from processes of sexual reproduction.

Knowledge of reproductive processes in the algae is incomplete and largely confined to a relatively few species in the Rhodophyta, Phaeophyta and Chlorophyta. In some other phyla, detailed knowledge is known only for a very few species, genera, or families.

Asexual Reproduction by Vegetative Means

Vegetative reproduction is widespread in many forms and probably occurs in all phyla. The longitudinal splitting of a biflagellate motile *Euglena* into two, still motile, but uniflagellate daughter cells is simple vegetative reproduction. The diatoms may divide vegetatively for upwards of five years. In this case the cell walls contain much silica. The cells are flat and their walls have two sections that fit into each other like the top and bottom of a petri dish. After division a new wall forms within the old wall (Fig. 22.11*A*). This means that the new cell receiving the lower and smaller wall will be smaller than the parent cell. Eventually a minimum size is reached and division stops. Full size can only be regained by sexual reproduction (Fig. 22.11*B*, *C*).

In still another type of vegetative reproduction exhibited by many Chlorophyta, a new, fully constituted young thallus is formed entirely within a parent cell. In *Scenedesmus* the parent plants consist of four flattened, linearly arranged cells (Fig. 22.3*A*,

B). Two mitoses occur simultaneously in each of the four cells. Within each of these cells, a new four-celled coenobium is now formed. These fully formed young colonies, coenobia, replete with an intricate surrounding net, are released simultaneously from each cell of the parent coenobium.

Vegetative reproduction in filamentous forms occurs by a simple fragmentation of the filament. In some species specialized cells do seem to be associated with breaking of the thallus. In species of the Cyanophyta, notably *Oscillatoria*, (Fig. 22.2*G*) the point at which fragmentation will occur is marked by a dead intercalary cell. Separation at these dead cells results in short filaments that have a gliding motion (**hormogonia**). This enables them to change their position in the gelatinous mass in which the filaments grow, and to gradually enlarge the mass. Many filamentous species of Cyanophyta are marked by occasional enlarged living cells (**heterocysts**). When located within a filament (*Nostoc*) (Figs. 22.1*A*, *B*, Fig. 22.2*D*) they appear to be associated with the fragmentation of the filaments.

Vegetative reproduction is probably common in the heterotrichous forms in which the prostrate system may behave in the manner of the stolons of higher plants (Fig. 7.31). In some species of Chlorophyta and Phaeophyta, specialized **propagules** consisting of small masses of meristematic cells serve as a means of vegetative reproduction. One of the most remarkable cases of fragmentation is exhibited by members of the family Sargassaceae, particularly the genus *Sargassum* (Fig. 22.7*D*). This genus is a floating kelp covering thousands of acres in the tropical Sargasso Sea. Fragmentation of its large thallus is apparently its only means of reproduction.

Asexual Reproduction Involving Mitospores

Asexual spores are always preceded by mitosis and may be called **mitospores** (Figs. 22.12, 22.13, 22.14) to distinguish them from spores formed after meiosis, or **meiospores** (Figs. 22.13, 22.15). The latter constitute a stage in sexual reproduction and will not be considered here. Since only mitotic divisions are involved the genotypes of all asexual spores arising from one and the same parent are identical with each other and with the parent plant. They will produce a population of genetically identical individuals. When a population has the genotype best fitted to a given growing condition, or locality, it will multiply by asexual reproduction

A
mitosis

B
meiosis

zygote

C
conjugation

isogametes

Figure 22.11 (*above*). A gametic life cycle as represented by a diatom. *A*, the daughter cells continually become smaller if, after cell division, the new cell forms within the silica shell of the old cell; *B*, meiosis occurs; three of the nuclei with the *n* number of chromosomes degenerate, and the single remaining haploid cell becomes a gamete; *C*, fusion of the two gametes to form the zygote. (*A* and *C*, redrawn after G. M. Smith, *Freshwater Algae of the United States.* Copyright 1933 by McGraw-Hill Book Company. Used with permission of McGraw-Hill Book Company.)

Figure 22.12 (*right*). Zygotic life cycle as represented by the green alga *Chlamydomonas*. Meiosis occurs during the germination of the zygote; it is the only diploid, or sporophyte, cell. (Redrawn after F. Moewus.)

and, frequently explosively, populate that locality.

Mitospores may be either motile or nonmotile. Motile spores are called **zoospores** (Figs. 22.13*B*, 22.14), and nonmotile mitospores are frequently called **aplanospores** or, in the case of *Polysiphonia*, **carpospores** (Fig. 22.18). They are usually borne in specialized cells called **sporangia.** In many algae the sporangia show little, if any, difference in appearance from ordinary vegetative cells (Figs. 22.13*B*, 22.14). The number of zoospores produced depends upon the species, being one for *Vaucheria* (Fig. 22.4*G*) and several hundred for *Cladophora* (Fig. 22.5). Most sporangia produce from 16 to 64 zoospores. In *Ectocarpus* (Fig. 22.14) the mitospores are produced in characteristic sporangia. These sporangia are divided into many cells and each cell produces a single spore (Fig. 22.14). Most of the

Rhodophyta produce nonmotile spores. Here the process is complex and variable. In general terms, short filaments arise from the zygote and, together with short filaments from nearby cells, form a structure known as a **cystocarp** (Fig. 22.18*D*). Mitospores or **carpospores** are cut off from the ends of those filaments that take their origin from the zygote.

Most zoospores are pear-shaped or spherical (Figs. 22.12, 22.13*B*, 22.14). Depending upon the species, they have two, four, or whorls of many flagella. After liberation from the sporangium the zoospores are actively motile for as short a time as four minutes (*Pediastrum*, Fig. 22.4*A*) or for as long as three days (*Ulothrix*, Fig. 22.13*B*). Their movement and periods of activity are frequently affected by light. After their period of activity they settle to the bot-

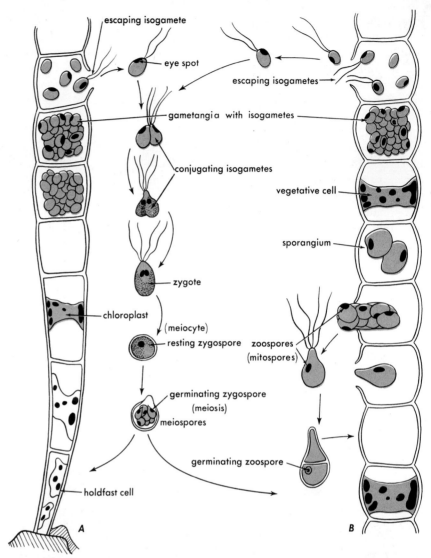

Figure 22.13 (*above*). Zygotic life cycle as represented by the green alga *Ulothrix*. Meiosis occurs in the germination of the zygote; it is the only diploid, or sporophyte, cell.

Figure 22.14 (*right*). Sporic life cycle, alternation of isomorphic generations, differing only in having diploid and haploid chromosome numbers, as shown by *Ectocarpus*.

tom of a pond or culture tank, lose their flagella, and secrete a cell wall (Figs. 22.13B, 22.14). Development of the new thallus is now initiated (Figs. 22.13B, 22.14).

Sexual Reproduction

Sexual reproduction, involving, as it does the segretation of genetic traits, complements asexual reproduction, in that sexual reproduction is responsible for variation within a population. Since sexual reproduction is frequently associated with resistant cells that carry the plants over seasons unfavorable for growth, the new plants formed in a given population by sexual reproduction will, upon the resumption of growth, all have different genotypes, and those best adapted for the new

zoospores (meiospores) n

meiosis

meiotic sporangia

asexual sporangia

haploid filament

diploid filament

$2n$

gametangia

n

gametophytes

A A

isogametes

n n

zygote

$2n$

zoospores (mitospores) $2n$

$2n$

$2n$

asexual sporangia

sporophyte

C

B

growing conditions will become established.

The cells that fuse in sexual reproduction are known as **gametes** (Figs. 22.11*B*, 22.12, 22.13, 22.14*A*, 22.16), and the single cell resulting from the fusion is a **zygote** (Figs. 22.11*C*, 22.12, 22.13*A*, 22.14, 22.16). Gametes have the **haploid** or *n* number of chromosomes, while the zygote has the **diploid** or 2*n* number. The number of chromosomes is returned to the haploid, or *n*, number through meiosis. The cells resulting from meiosis, with the *n* number of chromosomes, may be called **meiospores.** Gametes are produced in cells known as **gametangia** (Figs. 22.13*A*, 22.14*A*). Meiosis occurs at different places in the sexual cycle, and the cells in which it occurs have been called **meiocytes.** Plants that produce gametes are known as **gametophytes,** and those that

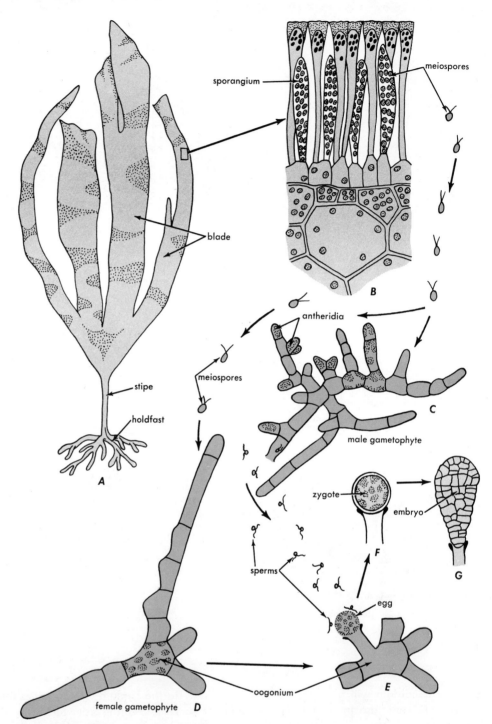

Figure 22.15 (*above*). Sporic life cycle, alternation of heteromorphic generations, as shown by the brown alga, *Laminaria*.

Figure 22.16 (*right*). Differences in gametes. *A*, isogamy, the gametes are identical; *B*, anisogamy, the gametes are slightly different in size or activity; *C*, oogamy, the gametes are very different in size and activity—they are eggs and sperms.

produce meiospores are called **sporophytes.** It follows that gametophytic plants are haploid and sporophytic are diploid (Figs. 22.11 to 22.18).

A gametangium, in the most unspecialized case, may be undistinguishable from an ordinary vegetative cell, except that several additional mitoses occur within it, giving rise to sixteen to thirty-two flagellated cells (Figs. 22.12, 22.13A, 22.14A). In the unicellular, motile *Chlamydomonas,* these cells strongly resemble the vegetative cells, except that they are smaller (Fig. 22.12). In the simple filamentous species *Ulothrix,* (Fig. 22.13) an ordinary vegetative cell gives rise to eight to sixty-four flagellated cells, again resembling the motile vegetative

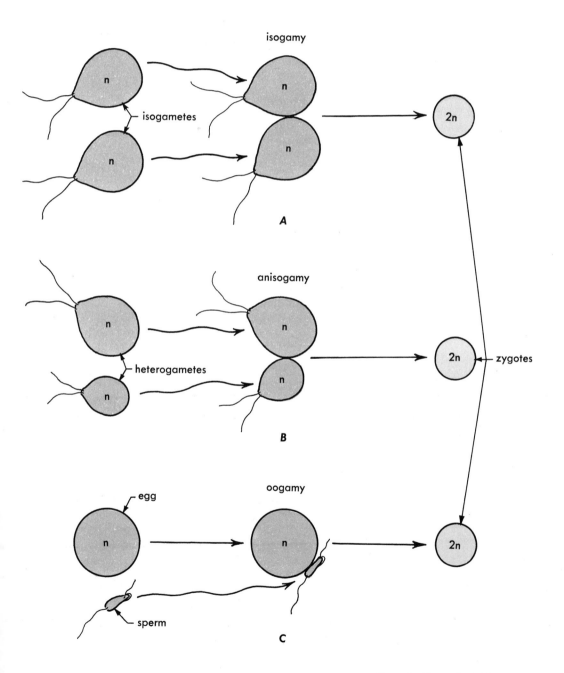

Chlamydomonas cells. When only eight motile cells are formed, they germinate directly, acting simply as zoospores; and the cell in which they are formed is called a **zoosporangium.** However, when a larger number of cells are formed, the cells are smaller. These smaller cells behave as gametes and fuse to form a zygote. When gametes are alike in appearance they are called **isogametes,** and the species producing them are said to be **isogamous** (Fig. 22.16*A*). Thus *Chlamydomonas, Ulothrix,* and *Ectocarpus* exhibit isogamy. The union of isogametes is known as **conjugation.** In some species the gametes differ in size, one being slightly smaller than the other. This situation occurs in two close relatives of *Chlamydomonas.* In *Gonium* (Fig. 22.5), large gametes fuse with small gametes. Occasionally in *Gonium,* similar-sized gametes have been observed to fuse. Gametes having only a slight size difference are said to be **anisogametes** (heterogametes), and species expressing this state are **anisogamous species** or exhibit **anisogamy** (heterogamy) (Fig. 22.16*B*). In *Volvox,* chlamydomonad cells are joined in a large hollow sphere (Fig. 22.5**A**). A few of these cells undergo rapid divisions, producing up to 512 slender biflagellate sperms. A sperm-producing gametangium is known as an **antheridium.** Another vegetative cell may become quite enlarged, sometimes becoming flask-shaped. The protoplast retracts slightly from the cell wall. It has become an **egg cell,** and the cell in which it is produced is known as an **oogonium** (Figs. 22.15*E,* 22.17*P,* 22.18*D*). The sperms are liberated from the antheridium and swim toward the oogonium. One sperm enters the oogonium, either through the funnel end or by a breaking down of the oogonial cell wall. Union of egg and sperm now ensues and the process is known as **fertilization** (Figs. 22.15*E,* 22.17*Q,* 22.18). When gametes differ in size and activity as they do in *Laminaria* (Fig. 22.15), *Fucus* (Figs. 22.1*H,* 22.17), and *Polysiphonia* (Figs. 22.1*I,* 22.18), the plants exhibit **oogamy** (Fig. 22.16*C*).

How does a sperm cell find an egg cell? It could happen by chance, as wind-blown pollen lands on, among other things, stigmas of flowers of its own species. And how are gametes able to recognize each other, perceiving not simply that they are members of the same species, but also that they have been produced by different individuals or by different strains of the same species? There is considerable evidence that the female gametangia or gametes, in plants, produce chemical agents that seduce the sperms to the close proximity of egg cells. For instance, a drop of a homogenate prepared from the mature female filaments of the alga *Oedogonium* may be drawn up into a capillary tube. When the tube is placed in a suspension of sperms the sperms will collect at the tip of the capillary tube and some will eventually enter the tube. This attraction seems to be species-specific for *Oedogonium.* This is important, because numerous general attractants for sperm cells are known. Most, if not all cells, probably have the ability to recognize each other and to accept or reject each other. Pollen tubes will grow only in styles of their own species (page 297). Species, and even different strains, must be taken into account when budding or grafting. And as has been recently demonstrated, human tissues frequently reject transplanted organs from other humans.

Alternation of Generations

Fertilization constitutes one of the two critical steps of a complete sexual life cycle; meiosis is the second critical step. We may distinguish three types of life cycles depending upon the relation of meiosis to fertilization. (1) In most animals, meiosis results in gamete formation, so the gametes are haploid. They are the equivalent of meiospores. In this case the mature individuals are diploid gametophytes and the haploid stage is limited to the gametes. This is a **gametic life cycle.** It occurs in the diatoms (Bacilliarophyta) (Fig. 22.11) and in the Codiales of the Chlorophyta. (2) In many of the more primitive algae, meiosis accompanies the germination of the zygote. In this case the mature plants are haploid and the diploid stage is limited to the zygote. This is a **zygotic life cycle.** It occurs in many Chlorophyta (Figs. 22.12, 22.13) and few primitive Rhodophyta. (3) In most plants, meiosis and fertilization are separated by distinct generations; one is haploid and produces gametes; the other is diploid and produces meiospores. This is a **sporic life cycle** (Figs. 22.14, 22.15, 22.17, 22.18).

In each of these cases haploid and diploid phases have alternated with each other. In a gametic life cycle the dominant generation is diploid; meiosis precedes the formation of gametes, which are the only haploid cells in the life cycle. In a zygotic life cycle, the dominant generation is haploid; the zygote constitutes the diploid generation and meiosis accompanies its germination. In a sporic life cycle mitoses occur in both haploid and diploid generations, giving rise to vegetative plants in both generations. The plants of the sporophyte and gametophyte generations may be identical in appearance. This is **alternation of isomorphic generations.** It occurs in some Chlorophyta, primitive Phaeophyta, and all higher Rhodophyta (Figs. 22.14, 22.18). In **alternation of heteromorphic generations** the haploid plants of the gametophyte generation have different forms from the diploid plants of the sporophyte generation. This occurs in a few Chlorophyta and in all of more advanced Phaeophyta (Figs. 22.15, 22.17).

Gametic Life Cycle

The diatoms (Figs. 22.3C, 22.11) are unicellular forms that may divide by cell division for periods as long as five years. With each division one of the daughter cells is smaller than the parent cell (Fig. 22.11A). At the end of a period of mitotic divisions, meiosis occurs and four gametes are produced, not all of which may function (Fig. 22.11B). Depending upon the species, these most generally are isogametes, but they may be anisogametes or even eggs and sperm. The zygote (Fig. 22.11C) increases greatly in size and, again depending on the species, may enter upon a rest period. The germination of the zygote is by mitotic cell division. The gametes are thus the only haploid cells and the diploid generation is a gametophyte.

Zygotic Life Cycle

Chlamydomonas and *Ulothrix* are examples of a zygotic life cycle (Figs. 22.12, 22.13). In each of these cases two haploid gametes fuse to produce a diploid zygote. Meiosis occurs in the germination of the zygote, which is the only diploid cell.

Alternation of Isomorphic Generations

Two genera, *Ulva* (Chlorophyta) and *Ectocarpus* (Phaeophyta) will serve to illustrate alternation of isomorphic generations. *Ulva*, allied to *Ulothrix,* has a distromatic thallus (Figs. 2.7B, 22.6D), while *Ectocarpus* (Fig. 22.7A) is a branched filamentous form. Both are marine, growing in the intertidal zone. In *Ulva* any cell of the haploid gametophyte thallus, except those near the base, may produce gametes. After conjugation of the gametes the zygote grows into a sporophyte thallus identical to the gametophyte except for the diploid chromosome number. Again, meiosis may take place in any cell of the sporophyte thallus. In some *Ulva* species the gametes, zygotes, and meiospores are similar in appearance, except that the gametes have two flagella, while the zygotes and meiospores have four flagella.

In *Ectocarpus* (Fig. 22.14), specialized reproductive cells are formed. The gametangia are grouped together, forming elongated, slightly curved structures. Each cell in this structure is a gametangium that has resulted from mitotic divisions and will produce one motile gamete with the haploid chromosome number. They are unable to germinate, but conjugate to form the usual diploid zygote (Fig. 22.14A). The zygote, upon germination, will grow into a diploid sporophyte (Fig. 22.14C) whose thallus is identical in appearance to that of the gametophyte thallus. Two different kinds of sporangia are formed on this sporophyte plant. One of these resembles exactly the group of gametangia found on the gametophyte plant. Like the gametangia, all sporangium cells have been formed through mitotic divisions, and each cell will produce a motile diploid zoospore that can germinate directly into a new diploid plant (Fig. 22.14B). The other type of sporangium found on the sporophyte plant is spherical; it will produce, after meiosis, numerous motile zoospores or meiospores with the haploid chromosome number (Fig. 22.14B). These meiospores grow into new gametophyte plants (Fig. 22.14A). Haploid gametes, haploid meiospores, and diploid zoospores are identical in appearance.

In the Rhodophyta an alternation of isomorphic generations is complicated by a second, distinctive sporophyte phase. The delicate feathery thalli of male and female gametophytes and tetrasporophyte of *Polysiphonia* are identical in appearance and growth (Fig. 22.1I). No motile cells are produced. The non-motile sperms (**spermatia**) are produced in great profusion near the tips of male gametophyte filaments (Fig. 22.18C). The oogonia (**carpogonia**) also form near the tips of the female gametophyte. The oogonia are flask-shaped cells with long slender necks (**trichogynes**) that protrude from a protecting envelope formed from the cells at the bases of the oogonia (Fig. 22.18D). Spermatia are carried by water currents to make contact with the necks of the oogonia; they enter it and passing downward unite with the female nuclei.

Fertilization stimulates the development of an

conceptacles

B

antheridium

sperms

D

mitosis

64 sperms

H

fertilization

Q

I

eggs

8 eggs

P

O

Figure 22.17

22 *The Algae*

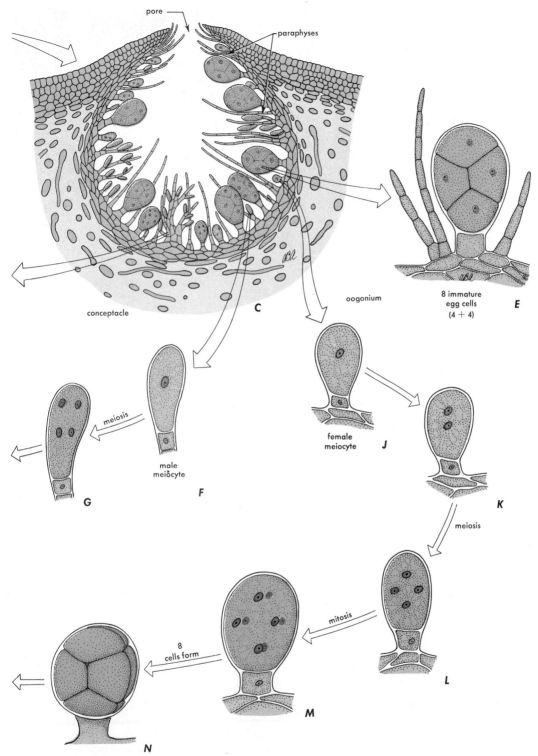

Figure 22.17. Sporic life cycle, as shown by *Fucus. A,* dichotymously branching blade with swollen tips, ×1; *B,* cross section of tip showing cavities, or conceptacles, ×8; *C,* the conceptacles contain paraphyses, short branching filaments bearing many small (antheridial) meiocytes and large (oogonial) meiocytes on short stalks, ×120; *D,* enlarged view of filaments bearing sperm; *E,* enlarged view of the oogonial meiocyte after meiosis; *F, G,* meiosis; *H,* mitotic division resulting in 64 sperms; *I,* sperms leaving antheridium; *J, K, L,* meiosis in the oogonial meiocyte; *M,* a single mitosis following meiosis; *N, O,* 8 egg cells form; *P,* the egg cells leave the oogonium; *Q,* eggs and sperms pass out into the sea water where fertilization takes place, ×2000. There is no free-living gametophyte generation. Magnification, *D-P,* inclusive, about ×850. (*F to Q,* adapted from Robert F. Scagel et al., *An Evolutionary Survey of the Plant Kingdom.* Copyright 1965 by Wadsworth Publishing Company. Reprinted with permission of the publisher.)

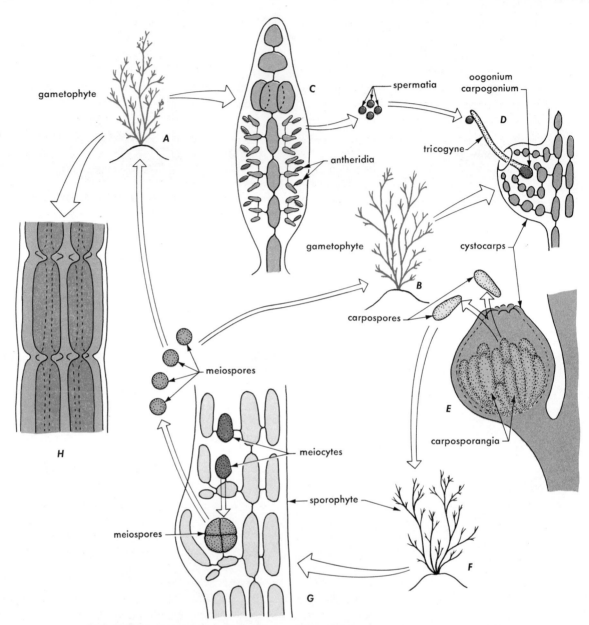

Figure 22.18. Sporic life cycle, alternation of isomorphic generations, as represented by the red alga *Polysiphonia*. *A*, male gametophyte plant; *B*, female gametophyte plant; *C*, production of spermatia at tip of male gametophyte; *D*, the carpogonium (oogonium) is protected by the cystocarp, and spermatia become attached to the protruding tricogyne; *E*, the zygote produces a body of 2n filaments (the carposporophyte) which terminate in carposporangia and give rise to 2n carpospores. The gametophytic cystocarp enlarges and surrounds the carposporophyte; *F*, a sporophyte, identical in appearance to the gametophyte, arises from the carpospores; *G*, some pericentral cells give rise to meiocytes which, through meiosis, produce four meiospores, or tetraspores; *H*, the vegetative structure of both gametophytes and the tetraspores are identical. (Adapted from Robert F. Scagel et al., *An Evolutionary Survey of the Plant Kingdom.* Copyright 1965 by Wadsworth Publishing Company. Reprinted with permission of the publisher.)

array of short diploid filaments (the **carposporo-phyte**), which will eventually produce diploid mitospores (**carpospores**). The original protective envelope of haploid filaments is also stimulated to grow. It enlarges and gives rise to a surrounding protective case formed of haploid filaments (**cysto-carp**) (Fig. 22.18E). Thus the first sporophyte generation (carposporophyte) in *Polysiphonia* and many other Rhodophyta is composed of short diploid filaments that produce diploid spores. This sporophyte is protected by an envelope of gametophyte cells. When discharged, the diploid carpospores give rise to sporophyte thalli (tetrasporophyte) identical in appearance to the two gametophyte plants. Meiosis occurs in sporangia in the sporophyte plant in cells formed between rows of axial and pericentral cells (Figs. 22.10B, 22.18G). Each sporangium produces four meiospores (**tetraspores**) which upon germination give rise to male or female gametophyte plants similar in appearance to the sporophyte plants.

Alternation of Heteromorphic Generations

Alternation of heteromorphic generations occurs in both the Chlorophyta and Phaeophyta but is most highly developed in the Phaeophyta, particularly in the kelps or Laminariales. The sporophyte generation of the kelps is the larger and better known. Its vegetative characteristics have been discussed in some detail on page 452 (Fig. 22.7B). The sporangia arise from meristematic cells, or meristoderm, that form the outermost layer of cells on the lamina (Figs. 22.8C, 22.15B). The sporangia usually arise in groups or **sori** and are accompanied by an elongation of adjacent cells into **paraphyses** that serve to protect the developing sporangia (Fig. 22.15B). Depending on the species each sporangium produces from eight to sixty-four motile meiospores.

The meiospores germinate, generally, into male or female gametophyte plants, both of which consist of a small branched filament (Fig. 22.15C, D). The antheridia are produced in large numbers, either singly or in groups, at the ends of short branches. There appear to be no specialized oogonia, as eggs may be produced in any of the cells of the female gametophyte (Fig. 22.15E). The eggs are usually discharged from the oogonium before fertilization but remain attached to it for some time. The egg is thus fertilized outside of the oogonium in open sea water.

Alternation of heteromorphic generations has reached its highest stage of advancement in *Fucus* (Fig. 22.17). The diploid generation is the conspicuous one (Fig. 22.1H), and the haploid phase has been reduced to single haploid cells that do not develop into free-living vegetative gametophytes (Fig. 22.17). Instead they become gametes and fuse to produce a new diploid generation. Meiosis and the gametes are so closely related that most algologists consider the meiocytes as gametangia rather than sporangia.

The ends of the diploid strap-shaped thallus of *Fucus* are swollen and notched. They bear small raised areas (Fig. 22.17A). A section through these areas shows them to be small cavities (**conceptacles**) that open by small conical pores to the sea water (Fig. 22.17B). The gametangia arise in these cavities. In Atlantic coast species of *Fucus* the male and female gametangia occur on separate plants, but they arise in the same cavities in the Pacific coast *Fucus*. The antheridia are formed profusely at the ends of short branching filaments and the oogonia are large spherical cells separated from the wall of the cavity by a single cell (Fig. 22.17C). Sterile hairs, or **paraphyses,** surround the gametangia, and may protrude outward through the pore.

The initial antheridial cell contains a single large nucleus (Fig. 22.17F). Meiosis takes place and four haploid cells are produced. This is followed by four rapid mitoses, giving rise to 64 cells that become sperms. They are discharged into the conceptacle. In the formation of the egg cell, meiosis is followed by a single mitotic division, so 8 haploid cells are formed, which will become mature eggs. They are discharged, unfertilized, from the oogonium and pass with the sperms into the sea water where fertilization takes place. Thus, in *Fucus,* the male gametophyte phase is reduced to four mitotic divisions and 64 sperm cells. Only a single mitotic division occurs in the female gametophyte generation. Compare this with the development of pollen and the embryo sac in higher plants.

SUMMARY OF REPRODUCTION IN THE ALGAE

Reproduction in the prokaryotic Cyanophyta is vegetative insofar as it is known. In all other algae—which are eukaryotes—vegetative, asexual, and sexual reproduction are identical to reproductive processes of other eukaryotes. Reproduction serves the same purpose of either maintaining similarity of generations or of introducing variation between generations. There are broad variations across the algal phyla in the time interval between meiosis and fertilization. In many of the more primitive families, meiosis accompanies the germination of the zygote. In the Bacilliariophyta, meiosis gives rise directly to gametes. In some Chlorophyta, most Rhodophyta and all Phaeophyta, vegetative stages intervene between meiosis and fertilization on the one hand, and between fertilization and meiosis on the other. These intervening vegetative thalli may be similar in appearance or differ greatly in size and form. In *Fucus* the gametophyte, or haploid phase, has been reduced to single haploid cells which, instead of growing into gametophyte plants, become gametes and fuse to produce new diploid plants.

THE PROKARYOTIC ALGAL CELL

The cells of the Cyanophyta resemble those of the bacteria in that they lack nuclei, mitochondria, dictyosomes, and endoplasmic reticulum. They are different from bacteria in that mesosomes (Fig. 21.11) have not yet been recognized. Of much greater significance, the blue-green algae carry on photosynthesis with the liberation of oxygen. The blue-green algal cell thus may be classified as a prokaryotic cell with regard to internal structure, but photosynthesis and respiration are aerobic, making it definitely algal-like.

Each individual protoplast is surrounded by a cell

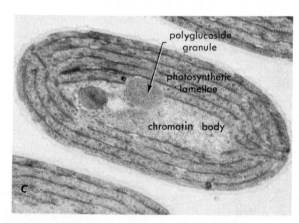

Figure 22.19. Ultrastructure of the blue-green algae. A, *Nostoc*, ×15,000; B, *Oscillatoria*, ×37,000; C, *Spirulina*, ×40,000. (A, B, courtesy of W. Menke; C, courtesy of N. J. Lang.)

Figure 22.20. A vegetative cell from the red alga *Batrachospermum moniliforme.* Chloroplast membranes do not form grana, and the carbohydrate is stored free in the cytoplasm. ×20,000. (Courtesy of David Brown.)

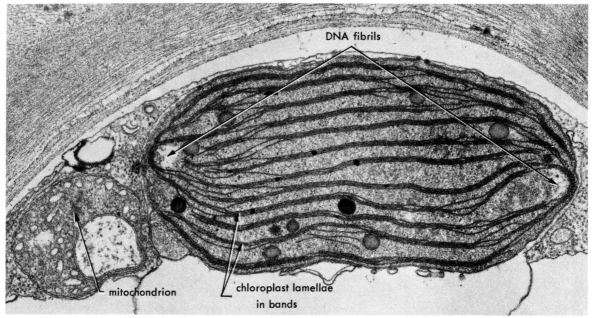

Figure 22.21. A chloroplast and a mitochondrion from a cell of the brown alga *Egregia menziesii*. Note clear areas at opposite ends with DNA fibrils and the chloroplast membranes in parallel bands. ×26,000. (Courtesy of T. Bisalputra and A. Bisalputra, *J. Cell Biol.*, **33**, 511.)

wall that probably contains cellulose and pectin. Under the electron microscope it may be resolved into at least four layers, but details of its ultrastructure have not been agreed upon (Fig. 22.19). A mucilaginous sheath surrounds individual cells, colonies, and trichomes. It varies from species to species in thickness, consistency, and pigmentation. This sheath pigment frequently is responsible for the color of the algal mass. Cells and trichomes with their sheaths may be embedded in this mucilaginous matrix.

Light microscopists early established that the photosynthetic pigment of the blue-green algal cell was confined to the periphery of the cell, and fluorescent studies showed it to be chlorophyll *a*. Figure 22.2*A* shows individual living cells of *Synechococcus* photographed by phase contrast. The mucilaginous sheath, the cell wall, the pigmented peripheral region, and the clear central region are evident. Feulgen staining of *Synechococcus* not only demonstrates the presence of DNA in the center of these cells but also shows that it must extend outward into the peripheral pigmented region. The presence of the chloroplast pigments in the outer region and DNA in the interior region has led to the designation of these regions as the chromatoplasm and the nucleoplasm.

Electron microscopy shows the photosynthetic membranes, or thylakoids, to be confined to the peripheral regions of the cell (Fig. 22.19). Note that the thylakoids are indeed individual flattened vesicles, or sacs, each composed of membranous sheets enclosing a flattened space. The thylakoids are always closed at their margins. Those thylakoids adjacent to the cell wall generally parallel the wall; the more internal thylakoids, while still parallel to each other, range internally toward the center of the cell. The thylakoids, as a group, are not delimited by any membrane from the remainder of the cell.

The nucleoplasm occupies the center of the cell and also extends outward into spaces between the thylakoids (Fig. 22.19). Mitochondria and dictyosomes are lacking. Numerous other particulates may be observed in these cells, but their nature and functions are not understood.

THE EUKARYOTIC ALGAL CELL

All other algae have eukaryotic, compartmented cells similar to those of vascular plants. The cells contain one or more nuclei which divide by mitosis; they contain mitochondria, dictyosomes, endoplasmic reticulum, ribosomes, and chloroplasts (Figs. 4.3, 22.20, 22.23, 22.26). The filaments of the siphonous line lack septa, thus eliminating definite boundaries between otherwise typical eukaryotic cells. Unequal distribution of the other organelles occurs as in other phyla and is probably related to variations in metabolism. For instance, in some unicellular forms the mitochondria have a pe-

division of DNA areas

Figure 22.22. Chloroplast division in *Egregia menziesii* is always preceded by the division and separation of the terminal DNA regions. ×10,000. (Courtesy of T. Bisalputra.)

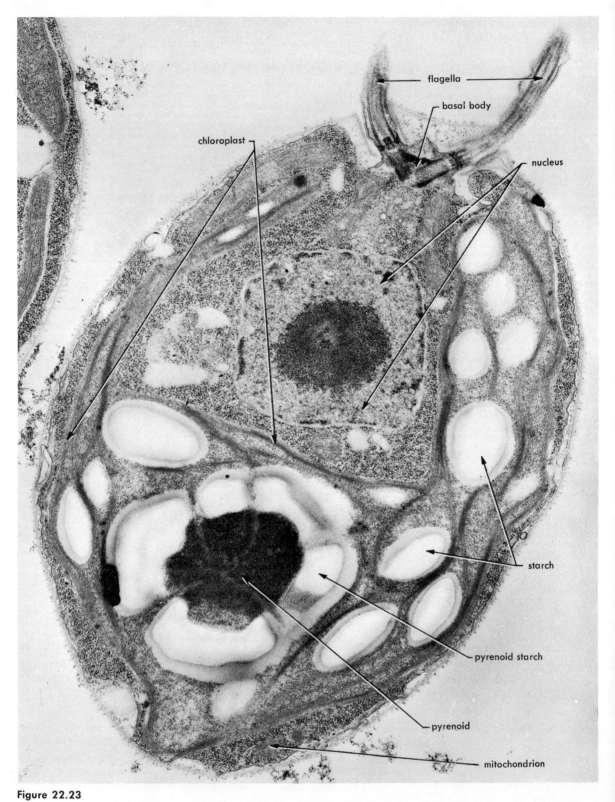

flagella

basal body

chloroplast

nucleus

starch

pyrenoid starch

pyrenoid

mitochondrion

Figure 22.23

ripheral location just under the plasmalemma. In some cases they apparently form a very open reticulum.

Variations in flagellation are important in algal classification, but details will not be considered here. Of interest, however, is the remarkable similarity of the ultrastructure of flagella. The motile orga-

nelles of bacteria are supplied with a single axial filament (Fig. 21.7). The flagella of all other organisms from *Chlamydomonas* to man have a uniform internal arrangement of microtubules.

The flagella of *Chlamydomonas* will serve as an example. Here there are two anterior flagella, entering the cell at an angle. They end in **basal bodies**

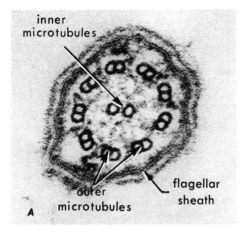

Figure 22.24. Flagella of *Chlamydomonas reinhardi*. A, (*left*) cross section, taken near tip of flagellum. There is an outer circle of nine paired microtubules (doublets) and two central single microtubules. The microtubules are embedded in a matrix which is surrounded by a flagellar membrane and a flagellar sheath. ×100,000; B, (*below*) a longitudinal section of a flagellum at its tip end; the central microtubules are longer than those composing the outer circle of doublets, ×68,000. (Courtesy of D. Ringo, *J. Cell Biol.*, **33**, 543.)

Figure 22.23 (*left*). *Chlamydomonas reinhardi*. At the anterior end two flagella are attached to a basal body. The single cup-shaped chloroplast occupies the periphery of the cell. Starch grains are distributed throughout the chloroplast and starch is prominent in the lower half of the cell, where it surrounds a darkly-staining pyrenoid. The chloroplast envelope separates the nucleus and surrounding cytoplasm from the pyrenoid and inner plastid membranes. Ribosomes are prominent, mitochondria stain diffusely, and dictysomes are not apparent. ×19,000. (Courtesy of D. Ringo, *J. Cell Biol.*, **33**, 543.)

Figure 22.25. The eyespot in *Chlamydomonas* consists of several rows of granules containing carotene. It lies wholly within the chloroplast. ×25,000. (Courtesy Carole Lembi.)

Figure 22.26. The ultrastructure of a portion of a cell of *Scenedesmus quadricauda.* ×15,000. (Courtesy of T. Bisalputra.)

that are connected by a bridge of fibers (Fig. 22.23). A longitudinal section near the tip of the flagellum shows a unit plasma membrane enclosed by a flagellar sheath (Fig. 22.24*B*). Within the flagellum there are three bands of microtubules parallel to the long axis of the flagellum. The central band is longer. The tubules are separated by a matrix of some sort. An array of indefinite cross connections may also be seen. A cross section of the flagellum at this level shows the flagellar sheath and plasmalemma. It also reveals that the outer circle of microtubules is composed of nine sets of double microtubules and that a pair of single microtubules forms the central band (Fig. 22.24*A*). Other cross sections taken at different levels show changes in the relationship of the microtubules as the sections approach the basal body.

Motility is gained by a flagellar motion similar to the breast swimming stroke. At the beginning of the stroke the flagella are extended straight forward. They sweep backwards without bending. The return stroke is an upward wave motion and again resembles the return of the arms in the breast swimming stroke.

Most motile forms, including adult unicells, gametes, sperm cells, and zoospores are provided with a red **eye spot.** This generally appears as two rows of globules of carotene separated by a thylakoid at the periphery of the chloroplast (Fig. 22.25).

There is a close relationship between pigments and the ultrastructure of the plastids. The Cyanophyta and the Rhodophyta have only chlorophyll *a.* The photosynthetic membranes of the Rhodophyta resemble those of the Cyanophyta (Figs. 22.19, 22.20) in being single flattened sacs or lamellae. These lamellae may be more or less parallel, but they never appear as appressed membranes forming grana as occurs in the higher plants. In all Rhodophyta the photosynthetic lamellae are embedded in a stroma and are separated from the cytoplasm of the cell by a double plastid envelope. The **floridean** starch (page 445) characteristically found in the large and more advanced group of the Rhodophyta always occurs outside of the chloroplast and in the cytoplasm (Fig. 22.20). The accessory phycobiliproteins may be present as granules aligned along the photosynthetic lamellae. These pigments are able to trap energy from light of low intensity and to pass the trapped energy on to chlorophyll, where an electron becomes excited in the usual manner (page 243).

Most chloroplasts of the Rhodophyta show denser, roughly spherical areas embedded in stroma, but not delimited from it by a membrane. These are **pyrenoids** and they are thought to play some part in the formation of the starch (Fig. 22.20).

The chloroplasts of the Phaeophyta contain chlorophyll *c* in addition to chlorophyll *a.* Here too the lamellae occur as bands of single parallel thylakoids (Fig. 22.21). They are never appressed to form grana, but they do always occur in groups of three to seven. They are parallel to each other. The storage carbohydrate elaborated in these brown algal cells occurs as granules lying outside the chloroplast in the cytoplasm.

Pyrenoids in the Phaeophyta are limited to definite orders. They project from either the ends or sides of the chloroplast and the photosynthetic lamellae do not enter them. The pyrenoids are frequently separated from the cytoplasm by a distinct cap.

The chloroplast DNA in the brown alga, *Egregia menziesii,* is present in the usual slender strands in two electron-translucent areas located at opposite ends of the plastid. The lamellae converge toward these areas (Fig. 22.21). These DNA areas may divide and separate. The separation of these DNA areas is followed by a longitudinal splitting of the chloroplast (Fig. 22.22). The chloroplasts of *Egregia menziessi* may burst, as do bacterial cells on the surface of distilled water, thus forming a thin film of plastid membranes, stroma, and DNA fibrils. Some of this material may be picked up on a grid and shadowed for observation with the electron microscope. The DNA is present in long slender fibrils similar to those seen in bacteria cytoplasm prepared in a similar fashion (Fig. 21.10). In the *Egregia* chloroplasts the DNA fibrils are distinctly connected to the thylakoids.

The Chlorophyta and Euglenophyta possess both chlorophylls *a* and *b* in a ratio similar to that found in the higher plants. The thylakoids in these algae are appressed into grana just as they are in the chloroplasts of higher plants. However, in many cases the grana in the algal chloroplasts are flat and platelike, not cylindrical as they are in the chloroplasts of higher plants (Figs. 22.23, 22.26). Pyrenoids are present in many species of the Chlorophyta. The pyrenoids are always embedded in the stroma and seem to have a close relationship with the starch grains, which are always within the chloroplast (Figs. 22.23, 22.26).

SUMMARY OF THE ALGAL CELL

The cyanophyte cell is related to the prokaryotic forms by the similarity of its genetic mechanism to that found in bacteria. It contains membrane-associated chlorophyll *a,* photosynthesizes a carbohydrate from carbon dioxide and water, and evolves oxygen. This relates its metabolism to that of higher plants. The cells of all other algae are typical eukaryotic cells. The ultrastructure of their chloroplasts is related to the photosynthetic pigment present. Grana, similar to those observed in the chloroplasts of higher plants, are found only in the Chlorophyta and Euglenophyta, which also have chlorophyll *a* and *b* in ratios close to those of higher plants. In cross section, flagella show the typical arrangement of microtubules found in the motor organelles of all organisms exclusive of the bacteria.

23

Lower Fungi

It is well known that the ravages of human diseases have changed or modified the history of peoples. Similarly, epidemics of diseases that lay waste important food plants may influence greatly the course of events. From 1843 to 1847 a fungus, known scientifically as *Phytophthora infestans,* spread rapidly through the potato fields of Ireland. This fungus causes a severe disease of potatoes known as **late blight.** It not only kills the foliage but also infects the tubers, causing them to rot rapidly. Moreover, when stored, one diseased tuber is a source of infection for all others in the storage bin.

In the middle of the nineteenth century, Irish peasants depended chiefly on the potato as a source of food. Consequently, from 1843 to 1847, when weather conditions were just right for the rapid development of *Phytophthora infestans,* and when the potato disease that it caused attained epidemic proportions, the Irish experienced a disastrous food famine. During these years over a million Irishmen moved across the ocean to New York City. They and their descendants have left an imprint upon American life. The potato famine in Ireland, produced by a fungus, was responsible for the death of a quarter of a million people and the mass movement of Irish people to the United States, chiefly in 1846 and 1847.

Some 80,000 different species of fungi have been recognized. They play an indispensable role in the life of the world. They help to maintain soil fertility; they decompose organic matter of both plants and animals that would soon make the surface of the earth uninhabitable were it not disposed of; they cause diseases of plants and animals, including man; and they are of great importance in various industries, such as in the making of cheese, the manufacture of alcohol, and the retting of flax.

Common examples of parasitic fungi are those that cause mildew of roses, late blight of potatoes, and rust of cereals. Among the saprophytic fungi may be cited those that form mold on bread and cheese and those that cause rotting of wood. A parasite may or may not be "disease-producing," that is, **pathogenic.** The fungus *Phytophthora infestans* is pathogenic; it produces a disease of the potato. On the other hand, certain bacteria subsist (are parasitic) on the roots of alfalfa, beans, and other members of the legume family but do not cause a disease. The organism upon which the parasite lives is the **host.** For example, the potato plant is the host of the parasitic fungus *Phytophthora infestans,* and man is host to the fungi that are responsible for athlete's foot.

Some parasites, like those that cause rusts of cereals, are obliged to secure their nourishment from living tissues. They are known as **obligate parasites.** And certain saprophytes, like some mushrooms, thrive only on nonliving organic materials. They are known as **obligate saprophytes.** On the other hand, some fungi have the faculty of growing as either a parasite or a saprophyte and are referred to as **facultative parasites.**

CLASSIFICATION OF FUNGI

The various ranks of classification in the fungi may be recognized by the terminal syllable or syllables of the name of the rank. The phylum name Mycota may be broken into two syllables; *Myc-* comes from the Greek *myketos,* which means fungus, and *ota* is New Latin meaning "to have." So Mycota literally means "fungus having." The names of the subdivisions end in *-mycotina,* which is a combination of Greek and Old Latin and means "having a fungal likeness." Classes end in -mycetes, which is simply the classical Greek word for fungus. Thus the name Ascomycetes may be broken down into *ascus* (in Greek *ascus* means sac) and *mycetes* or literally, the sac fungi.

There is a single fungal phylum, the Mycota, which may be divided into two subphyla, the Myxomycotina and the Eumycotina. The first of these subphyla comprises the slime molds. It is a small subphylum whose members have a vegetative plant body consisting of a naked mass of protoplasm. The prefix *myxo-* is a Greek combining form meaning slime. Thus Myxomycotina would be "slime having a fungal likeness." The other subphylum, the Eumycotina, is a large group containing thousands of species of diverse forms. Here, the prefix *eu* is also from Greek and means "true." The Eumycotina is divided into five classes, one being a **form class.** We shall consider the following:

Class 1. The Oomycetes (egg fungi). The common water molds and the fungus that causes late blight of potatoes are examples.

Class 2. The Zygomycetes (zygote fungi). Bread mold is a widespread fungus belonging to this class.

Class 3. The Ascomycetes (sac fungi). Representatives in this group include a common mold that grows on jellies and jams. Many sac fungi cause severe diseases of certain orchard trees, and one produces a destructive disease of elms.

Class 4. The Basidiomycetes (club fungi). Well-known members of this group are mushrooms and the fungi that cause rusts and smuts of the cereals.

Class 5. The Fungi Imperfecti (imperfect fungi). The fungi causing athlete's foot are members of this class. Also, many diseases of crop plants result from the attacks of representatives of this group.

Because of the very large numbers of fungi and their great importance in medicine, industry, and agriculture, several branches of science are concerned with them. For instance, one group of workers has studied the structure and activities of the true fungi. These scientists are not greatly concerned with the economic importance of fungi; rather, they are interested in the life histories, in modes of reproduction, and food requirements. They are called **mycologists,** and their science is **mycology. Pathology** deals with the diseases of man, other animals, and plants. A **plant pathologist** is one who specializes in a study of plant diseases. He studies the life histories of the organisms that cause disease, the environmental factors that influence the growth of the organisms, and methods of combating the diseases. He is also interested in plant diseases due to other causes. The annual reduction in yield of orchard, garden, and field crops due to diseases in the United States amounts to many millions of dollars. On the other hand, the development of control methods by plant pathologists has reduced these losses by millions of dollars.

Subphylum Myxomycotina

General Characteristics and Distribution

The slime fungi or slime molds, of which there are some 300 species, are of scientific interest because they seem to combine the characteristics of both plants and animals. The vegetative body consists of a slimy mass of naked protoplasm in which there are many nuclei without separating walls. The vegetative body is called a **plasmodium** (plural, **plasmodia**) (Fig. 23.1). This plasmodium has no definite shape; it creeps slowly by amoeboid movement, usually over shaded, rotting tree trunks, across leaves, or in crevices, engulfing solid particles of food as it goes. Most slime fungi are saprophytes.

The absence of cell walls, and amoeboid movement, and the ability to take solid food particles into the protoplasm are characteristics that are usually associated with animals. When slime fungi reproduce, however, they form spores with cellulose walls; thus they have reproductive characteristics that are definitely those of plants.

Reproduction. Prior to the reproductive stage, the plasmodium moves to a drier substratum. After a time, the plasmodium ceases moving and forms one or more spore-producing structures (**sporangia**) (Fig. 23.1). There is great variation in the form and color of the sporangia among the different species. Some have a very delicate structure and brilliant coloring. The spores formed within the sporangia are uninucleate and are surrounded by a cellulose wall. They are discharged from the sporangia and spread by wind. In the presence of water they germinate; the wall is ruptured, and the contents escape in the form of a flagellated or an amoeboid naked mass of protoplasm that may later multiply by fission. In some species of slime fungi, pairs of these bodies may fuse to form zygotes. Nuclear fusion occurs in the zygote. Each zygote may grow into a plasmodium, or zygotes may coalesce to form a plasmodium.

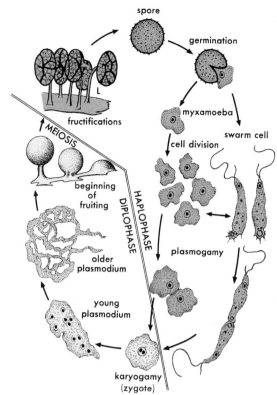

Figure 23.1. Life cycle of a myxomycete. (Redrawn from C. J. Alexopoulous, *Introductory Mycology*, John Wiley and Sons.)

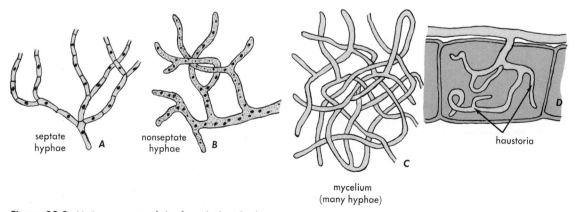

Figure 23.2. Various aspects of the fungal plant body.

Figure 23.3. Portion of hypha of *Rhizoctonia solani.* ×24,000. (Courtesy of C. E. Bracker.)

Subphylum Eumycotina

General Characteristics

With the exception of the simplest of the Eumycophyta, the vegetative thallus, or plant body of the members of this phylum, is composed of a mass of threads of filaments called **hyphae (hypha,** singular) (Fig. 23.2*A, B*). Three of the four different classes of this phylum have distinctive types of hyphae. The vegetative body may be feathery and delicate, as in common bread mold (Fig. 23.17) and other molds (Fig. 23.4), or it may be quite hard and leathery, as in species that cause wood rot. The mass of hyphae forming the vegetative body is called the **mycelium** (Fig. 23.2*C*).

The cell walls of the hyphae may contain cellulose, and also in some species, **chitin,** a substance identical with that found in the cuticle of insects and crustaceans. The nuclei divide normally by mitosis. Cell division does not always follow immediately after nuclear division. In certain instances a half dozen or more nuclear divisions will occur before walls form between the daughter nuclei. In the Oomycetes and Zygomycetes, the vegetative hyphae usually have no cross walls (**septa,** singular **septum**). They are multinucleate, and there is complete continuity of the cytoplasm throughout the whole vegetative thallus. Hyphae of this sort are said to be **coenocytic,** and since no cross walls or septa are formed in such filaments (Fig. 23.2*B*), they are also called **nonseptate.** When cross walls are present, the filaments are said to be **septate** (Fig. 23.2*A*).

The cytoplasm of fungal hyphae resembles that of other plants in having mitochondria and an endoplasmic reticulum. The mitochondria are small rings as may be seen in Fig. 23.3. The endoplasmic reticulum is in the form of slender paired membranes. The small profiles (*S*) may be similar to bodies seen in the cytoplasm of higher plants and which have no generally accepted name, but are sometimes called **microbodies.**

Function of the Mycelium

The vegetative mycelium carries on the general activities of plant cells, such as absorption, digestion, respiration, and secretion, but not photosynthesis. Since it is incapable of synthesizing its own foods it must obtain nourishment either from non-living organic matter or from living plants or animals. The food must be rendered soluble so that it may diffuse through the walls of the hyphae and reach the protoplast. Some fungi obtain food from even the hardest of woods and from solids of many other sorts. These solid substances are liquefied or otherwise rendered diffusible by enzymes that are secreted by the hyphae. For example, certain enzymes are secreted by fungi that break down the complex carbohydrates of wood to simple soluble sugars. Many fungi actually carry on very active digestion of solid materials. Some fungi have specialized hyphae known as **haustoria** that can penetrate living cells and obtain nourishment therefrom (Fig. 23.2D).

Since the principal function of the mycelium is to obtain nutritive materials, it is usually found in close association with a source of food. It may be growing inside a living tree, in a dead stump, on or in a leaf, in aging cheese, in a manure pile, or in many other kinds of organic substances. The mycelium of most fungi is not adapted to withstand much drying; hence it seldom grows openly exposed to the atmosphere, unless the relative humidity is high.

Reproduction. The sporangia, spores, gametangia, and gametes of the fungi show a great range of structure and recall in some respects the reproductive structures found in algae. Simple cell division and fragmentation occur in some fungi. Both motile and nonmotile spores are produced in various types of sporangia. Isogamy is characteristic of several genera. Heterogamy, involving gametangia resembling those of *Vaucheria,* occurs in several families. In two classes of fungi, the Ascomycetes and Basidiomycetes, the sexual cycle is complicated by the failure of the two haploid nuclei to fuse immediately after the gametes unite. This characteristic results in two distinct types of hyphae: (1) those in which the cells are haploid, with one haploid nucleus in each cell and (2) those in which the cells contain two haploid nuclei in each cell. In the latter type the two haploid nuclei will ultimately fuse just before meiosis (Fig. 24.3).

Although the vegetative mycelium of most fungi is the actively destructive portion, the reproductive structures are of more importance from the control standpoint and as a basis of classification. Usually, fungi multiply and are disseminated by abundant spores; it is therefore important to prevent the dis-semination of spores, or to stop their germination, or to kill the young hyphae after germination. It is seldom possible to rid the infected plant of a vigorous mycelium.

Sexual reproduction. As in the algae, sexual reproduction involves two fundamental steps: (1) the union of gametes, during which the chromosome number per nucleus is doubled; and (2) meiosis, during which the chromosome number is halved.

Asexual reproduction. As usual, asexual reproduction does not involve changes in the number of chromosomes. Asexual spores are frequently formed by one of the following methods.

1. Spores may be borne in specialized cells called **sporangia** (spore cases). The sporangia (sporangium, singular) may occur either on typical filaments on special upright hyphae, or on highly branched hyphae. Aerial sporangia are formed as in bread mold (Fig. 23.18). In other species the sporangia are submerged or germinate in thin films of water. The spores produced by such sporangia are motile and are called **zoospores.** The hyphae that bear sporangia are called **sporangiophores** (Figs. 23.9, 23.18).

2. In other cases most of the cells of a septate filament separate from each other. Before separating, heavy walls are secreted around them. Reproductive bodies formed in this manner are known as **chlamydospores.** They are formed in enormous numbers by the various types of fungi.

3. End cells of a hypha may round up and be cut off from the remaining portion of the hypha. These rounded cells are light in weight, a condition that favors rapid dissemination of the fungus. Such cells are called **conidia.** If they germinate to form hyphae or new fungous plants, these conidia may be called **conidiospores.** In many other species the conidia will become sporangia and produce zoospores, in which event they may be called **conidiosporangia.** The hyphae that bear conidia are called **conidiophores** (Figs. 23.9, 24.2).

Class Oomycetes

Many Oomycetes are severe plant pathogens. Some, however, may infect fish or insects or cause diseases in man. Many members of this class, like the common water mold, are normally saprophytes or weak parasites. Some are minute forms of one to several cells, and a few of these smaller sorts are reported to attack small aquatic animals. Most representatives of the Oomycetes develop a more or less extensive mycelium of indefinite form (Figs. 23.4, 23.5, 23.12).

The hyphae of actively growing Oomycetes are

generally **coenocytic** and **nonseptate.** Septa may form occasionally, however, in old hyphae of some species, and normally the reproductive cells are cut off from the vegetative hyphae by cross walls.

The Oomycetes are divided into several orders. Specialists do not agree completely as to the best manner of classifying the plants in this class, and therefore, the number of recognized orders depends upon the authority one is consulting. We shall consider several forms distributed in the orders Saprolegniales and Peronosporales. Note that the names of orders end in *-ales.*

Saprolegniales. Fungi of this order live in fresh or salt water and are generally saprophytic. Even the species that attack the gills of fish grow on tissues that have been weakened or suffered injury. Because of the aquatic habitat of many members of this order they are frequently called **water molds.** They may be easily cultivated by placing small pieces of meat, egg albumin, radish seeds, or dead flies in a dish of pond water (Fig. 23.5). After the mycelium has ramified through the dead fly or other substratum, hyphae grow outward into the water. A small ball of white hyphae, from one-half to an inch in diameter, may thus be formed. Reproductive cells are produced by these hyphae.

Asexual reproduction. With ample food supply the mycelium increases in size with but little tendency to produce reproductive cells. If, however, a well-developed mycelium is transferred to distilled water in which a food supply is lacking, sporangia will usually appear. Sporangia are formed at the ends of hyphae by a cross wall cutting off the tip of a hypha from the rest of the mycelium as in *Saprolegnia* (Fig. 23.6*A*). The sporangium is a multinucleate structure. After a time the protoplasmic contents of the sporangium divide into a large number of spores, each with one nucleus (Fig. 23.6*B, C*). Upon maturity the spores are discharged from the sporangium. In *Saprolegnia* each zoospore has two flagella attached to its anterior end which enable it to swim actively (Fig. 23.6*D*). After a time, the zoospores settle down, lose their flagella, develop a cellulose wall, and pass through a resting period. Upon resuming activity they escape from the wall; the two newly developed flagella are now attached laterally, and the zoospores, thus equipped, swim about for a period. If they come to rest on a suitable substance, each sends out a tubular outgrowth that penetrates and infects this substance.

In a few species of water molds, one or even both zoospore stages are suppressed. When zoospores emerge from the sporangia they may germinate directly into a new mycelium. In certain other species, the spores never leave the sporangia but germinate while still enclosed within it, and the germ tubes pierce the old sporangial wall. In still other species, spores are not even formed; the sporangia germinate directly into a coenocytic mycelium.

Sexual reproduction. The **oogonia** are spherical cells,

Figure 23.4. Fusarium rot on peach. ×3/4. (Courtesy of Plant Pathology Department, University of California.)

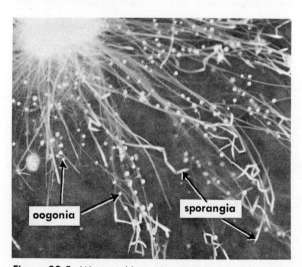

oogonia

sporangia

Figure 23.5. Water mold growing on a hemp seed. ×3.

formed at the tips of short side branches (Figs. 23.7*A*, 23.8*A*). When mature, they are three to four times the diameter of ordinary hyphae. The cytoplasm in the swollen tip becomes denser than that in regular hyphae. From one to twenty eggs may be formed from the protoplasmic contents in each oogonium, depending upon the species. The eggs are spherical, dense bodies of protoplasm, each containing one nucleus.

The **antheridia** are also formed at the tips of branches, in some species near the oogonia (Figs. 23.7*A*, 23.8*B*). Each is separated from the main filament by a cross wall. The antheridia are usually curved and not much greater in diameter than the hyphae from which they arise. The contents of a mature antheridium consist of several nonmotile male gametes. The antheridium comes in contact with an oogonium (Figs. 23.7*A*, 23.8*B*). In some instances, one oogonium may have several antheridia attached to it. A short slender hypha, the **fertil-**

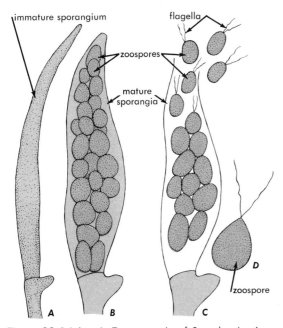

Figure 23.6 (*above*). Zoosporangia of *Saprolegnia*. *A*, immature sporangium; *B*, mature sporangium; *C*, discharge of zoospores; *D*, a zoospore.

Figure 23.7 (*right*). Gametangia in *Saprolegnia*. *A*, oogonia and antheridia with associated hyphae; *B*, antheridium, oogonium with two eggs, fertilization tube in place; *C*, egg cells ready for fertilization. (Redrawn from W. C. Coker, *The Saprolegniaceae*, University of North Carolina Press.)

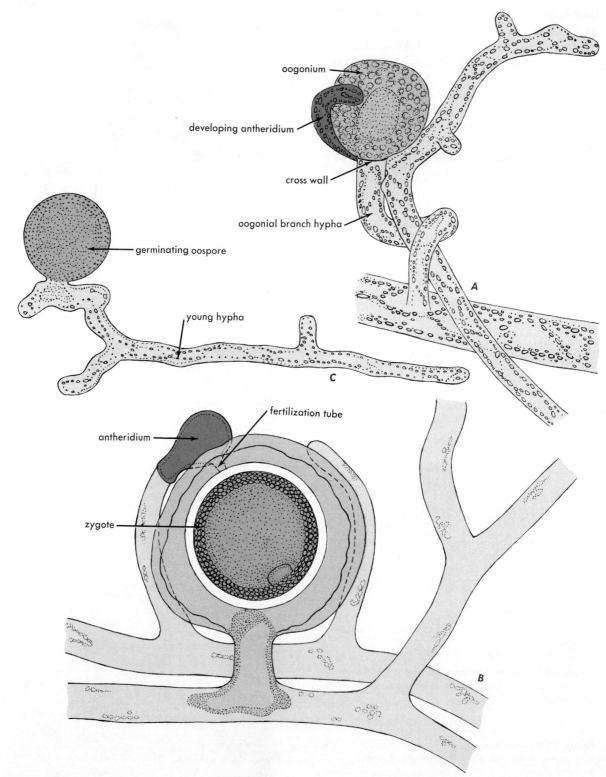

Figure 23.8. Sexual reproduction in beet water mold. *A,* gametangia; *B,* zygote; *C,* germinating oospore. (Redrawn from C. Dreschler.)

ization tube, grows from the side of an antheridium adjacent to an oogonium (Figs. 23.7*B*, 23.8*B*), penetrates the oogonial wall, and comes in contact with one or more eggs. If the oogonium contains several eggs, the fertilization tube usually branches, sending a branch to each egg. Nuclei (male gametes) from the antheridium migrate into the tube and any branches present. One nucleus and possibly some cytoplasm pass into each egg, and fertilization ensues. The fertilized egg or zygote develops a heavy wall, becoming an oospore (Fig. 23.8*B*), and usually will not germinate for several months, even under favorable conditions; hence, it is well adapted to survive unfavorable conditions. Upon germination, the oospore sends out new hyphae (Fig. 23.8*C*), which, if they find a source of food, rapidly grow into a typical mycelium. If food is scarce, the formation of zoospores follows soon after germination. Not uncommonly, eggs may develop into new hyphae without fertilization.

Peronosporales. Nearly all the Peronosporales are parasites. In general, heavy-walled oospores carry them over unfavorable periods (Fig. 23.16), and various sorts of asexual spores (Figs. 23.9, 23.10) bring about rapid multiplication under suitable conditions. Because many of them form a downy growth on the surface of the host or substrate they are sometimes called **downy mildews.**

Asexual reproduction. The asexual reproductive structures of the different genera are adapted to various habitats and show more variation than do the sexual reproductive bodies.

A few Peronosporales, for example, *Pythium,* may live in water or on moist soil as saprophytes; they may infect aquatic plants and are frequently responsible for the "damping off" of the seedlings of many farm and hothouse plants. In keeping with their moist habitat, these fungi are propagated by zoospores that strongly resemble those of the water molds. Sporangia are formed at the tips of the hyphae (Fig. 23.9*A*). When proper conditions prevail, the sporangia open and the zoospores emerge. The zoospores are small and are able to swim in the water films surrounding particles of moist soil.

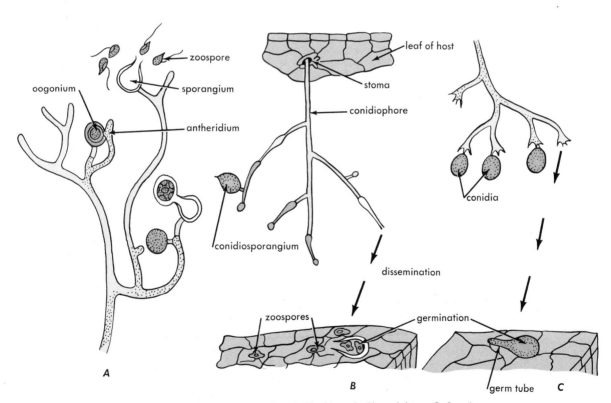

Figure 23.9. Asexual reproduction in the Peronosporales. *A, Phythium; B, Phytophthora; C, Bremia.*

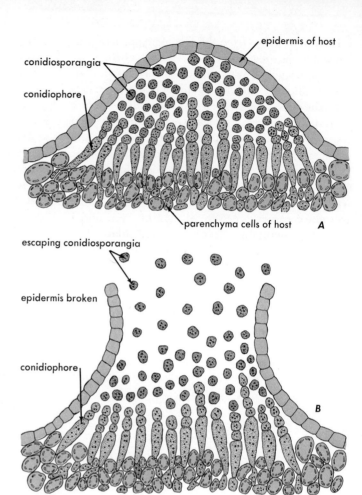

Figure 23.10. Pustules on shepherd's purse caused by conidiosporangia of *Albugo*.

Eventually they germinate and develop a new mycelium. Seedlings may be infected at or just below the soil surface and are quickly killed.

Other Peronosporales (*Phytophthora, Plasmorpara, Albugo*) require a moist habitat and grow best in rainy or humid weather, but they are strictly terrestrial. Many of them are parasites on flowering plants. The mycelium of downy mildews grows within a leaf or stem of a plant (Fig. 23.11) between the cells. Haustoria penetrate the cells.

Aerial sporangia are formed in one of two ways: (1) on long sporangiophores that extend out through the epidermis of the infected plant (Figs. 23.9B, 23.11) (*Phytophthora, Plasmopara*, and others) or (2) in compact pustules beneath the epidermis of infected plants (*Albugo*) (Fig. 23.10). In both structures the sporangia are disjoined from the special hyphae producing them and are disseminated by air currents. Upon germination they usually produce zoospores. Sporangia of this sort may also be called **conidiosporangia** (since they are pinched off), and the branched hyphae that bear them are known as either **sporangiophores** or **conidiophores.**

Some conidiosporangia eventually come to rest on the leaves of susceptible plants. When moisture on the leaf surface is sufficient, zoospores escape from them. The zoospores are very small and are able to move about in the film of water on the leaf surface. They soon send out small hyphae called **germ tubes** that penetrate the host tissues and bring about new infections (Fig. 23.9B, C). *Phytophthora infestans* reproduces asexually in this manner. One fact that the potato growers of England and Ireland learned from the great potato famine of 1846 was that the disease was most severe in the dampest areas and almost nonexistent in dry areas.

Downy mildew is produced by a fungus having asexual reproduction similar to that of *Phytophthora*. Downy mildews are very common infections on many cultivated and wild plants (Fig. 23.12). They are frequently found on grapes, beans, grasses, melons, alfalfa, peas, sugar beets, and other plants. They are easily identified by the conidiophores that may, in severe cases, nearly cover the leaf. Brought into the laboratory, the sporangia can sometimes be induced to germinate by floating them on cold water.

A small group of the Peronosporales, known as the "white rusts," develop sporangia beneath the epidermis of such crop plants as mustards and spinach. There is but one genus (*Albugo*) in this group. The conidiophores do not grow out of the stomata as in the downy mildews. Instead, they collect in pustules under the epidermis of the stem or leaf. Conidiosporangia are cut off in chains from the tips of the conidiophores. They accumulate in large numbers and finally rupture the epidermis, forming creamy-white pustules (Figs. 23.10, 23.13).

In some of the Peronosporales (*Peronospora*) the conidia are formed at the tips of hyphae and dispersed by air currents, but never give rise to zoospores. Instead they germinate directly by sending out one to several germ tubes, which infect the host plant (Fig. 23.9*C*). A downy mildew causing a severe disease of sugar beets behaves in this manner.

Summarizing, the types of asexual reproduction in Peronosporales may be characterized as follows. (1) Sporangia germinate while still attached to the parent hyphae, giving rise to zoospores (*Pythium*). (2) Conidiosporangia are dispersed by wind to other hosts, where the zoospores are liberated (*Phytophthora*). (3) Conidia are identical in appearance, manner of formation, and method of dispersal to conidiosporangia, but, upon germination, give rise to germ tubes (*Peronospora*).

Sexual reproduction. Sexual reproduction in the Peronosporales is similar to that observed in the Saprolegniales. In many forms it immediately precedes death of the host plant. Gametangia (oogonia and antheridia) are formed on short side branches,

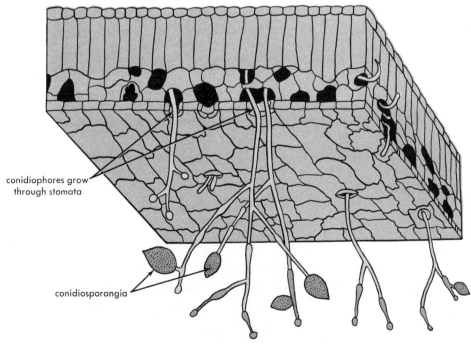

conidiophores grow through stomata

conidiosporangia

Figure 23.11. Conidiosporangia of *Phytophthora infestans* on potato leaves. (Redrawn from L. R. Jones, N. J. Giddings, and B. F. Lutman, "Investigations of the Potato Fungus," *Vermont Exp. Sta. Bull.,* **168,** 9, 1912.)

Figure 23.12. Downy mildew on lettuce leaf. (Courtesy of Plant Pathology Department, University of California.)

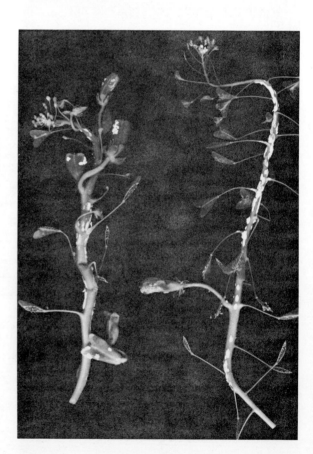

Figure 23.13. *Albugo* on shepherd's purse. (From R. M. Holman and W. W. Robbins, *A Textbook of General Botany*, John Wiley and Sons.)

Figure 23.14. Fertilization in *Pythium*. (Redrawn from K. Miyake, *Ann. Botany* (*London*), **15,** 653.)

the ends of which have been cut off by cross walls (Figs. 23.8, 23.15). Some end cells swell to form spherical oogonia, each of which contains one egg. Close by, the antheridia are formed from end cells of other side branches. One antheridium presses closely to the wall of an oogonium. A fertilization tube forms and penetrates the oogonial wall until it contacts the egg. Fertilization results by the fusing of one sperm nucleus with an egg nucleus. The resulting fertilized egg becomes an oospore by developing a thick cell wall that protects the protoplasm against adverse conditions (Fig. 23.16). When conditions are favorable for growth, the oospore germinates, forming, either immediately or after the development of a short hypha, a large number of zoospores, each of which may develop into a hypha.

Class Zygomycetes

The name of this class refers to the production of a sexual resting spore which typically results from the fusion of two gametangia. No motile cells of any sort are found in this class; all its members produce typical aerial sporangia or conidia, and they regularly have a coenocytic mycelium of indefinite form. Most species are saprophytic, some are weak parasites of plants, a few are specialized parasites of animals and some are actually obligate parasites on other Zygomycetes. We shall consider only the common bread mold.

The mycelium at first grows chiefly within the substrate, which may be composed of various kinds of organic matter. Eventually, aerial hyphae develop so that the surface of the substrate may become covered with a mass of hyphae (Fig. 23.17).

Common bread mold, *Rhizopus stolonifer,* is a member of the order Mucorales and is of worldwide distribution. It grows well on a large variety of organic substances. Although it is mainly a saprophyte, it does attack and considerably damage sweet potatoes, berries, and fruit while in transit or in storage. It grows very luxuriantly on sweet potatoes and bread, and these foods may be utilized as media for the growth of laboratory cultures.

Asexual reproduction. After the mycelium has become well established upon and in a substrate, certain aerial hyphae, usually of larger diameter than

Figure 23.15. Fertilization in *Albugo.* (From F. L. Stevens, *Botan. Gaz.,* **32,** 77. Copyright The University of Chicago Press.)

those in the substrate, grow just above its surface for a short distance and then come in contact again with the substrate. They are known as **stolons.** At the point of contact with the substrate new hyphae form. Hyphae are of three types: (1) stolons; (2) short, branched hyphae, called **rhizoids,** that penetrate the substrate and serve to anchor the mycelium and to absorb nutrients; and (3) hyphae that grow upright and produce sporangia at their tips and therefore are **sporangiophores** (Fig. 23.18).

The development of a sporangium is as follows: The tip of the sporangiophore swells, and a bulging wall cuts off an apical cell, which becomes the sporangium. The dome-shaped wall separating

Figure 23.16. Oospores of *Albugo.* ×150.

Figure 23.17. Hyphae, mature sporangia (black), and immature sporangia (white) of *Rhizopus stolonifer.* ×4.

the sporangium from the parent hypha is called the **columella.** Numerous spores are formed from the protoplasm within the sporangium, and when the spores are ripe the outer wall of the sporangium falls away and the spores are dispersed by air currents and insects. On a suitable substrate, the spores germinate and develop new hyphae. Immature sporangia are white, mature ones black (Fig. 23.17).

Sexual reproduction. The characteristic type of sexual reproduction in the order Mucorales is conjugation. The first step is the chance contact of the tips of short club-shaped hyphal branches (Fig. 23.19). Once in contact, the ends of these two short hyphae swell and elongate slightly. A cross wall forms back from the tip of each hypha, separating

a terminal cell from the parent hypha. The two tip cells thus formed are gametangia. The walls of the gametangia that are in contact dissolve, permitting the two protoplasts to fuse. The zygote resulting from this union develops into a **zygospore,** which has a thick wall and is quite resistant to unfavorable conditions.

There is some variation in the way different species of the Mucorales develop gametangia and zygospores. Let us assume that in a given community no bread mold has ever been found. By chance one bread mold spore is brought in and grows, producing many millions of spores that spread the fungus throughout the community. We now have a strain of bread mold growing in this community

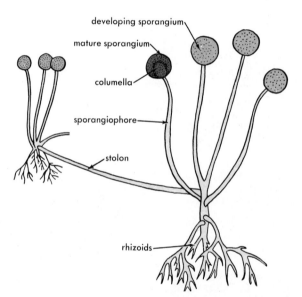

Figure 23.18 (*above*). Asexual reproduction in *Rhizopus stolonifer.*

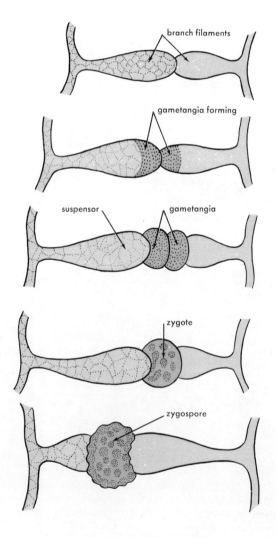

Figure 23.19 (*right*). Sexual reproduction in *Rhizopus stolonifer.*

that has arisen from a single spore. The probabilities are that sexual union will never occur, because many Mucorales, although morphologically similar, are physiologically differentiated into **sexual strains.** Their similar appearance and behavior make it impossible to designate them as male and female. Instead they are called, for convenience, **plus (+)** and **minus (−) strains.** The bread mold spore introduced under these conditions into this community would be either plus or minus. A plus strain will not conjugate with a plus strain, nor a minus strain with a minus strain, and hence no sexual reproduction will occur. Asexual reproduction, however, will be normal. If the opposite strain were now introduced and established, isogametes and zygospores would form. Species of fungi that are differentiated into plus and minus strains are said to be **heterothallic.**

The zygospores of some Mucorales germinate only with difficulty. This is true of common bread mold, *Rhizopus stolonifer.* In these plants little is known about segregation of the plus and minus characters. In other species the zygospores do germinate readily and the segregation of the two sexual strains has been observed.

SIGNIFICANT FEATURES OF THE OOMYCETES AND THE ZYGOMYCETES

A. Vegetative mycelium.
   Saprolegniales, Peronosporales, Mucorales: a coenocytic mycelium; saprophytes, parasites and intermediate types; parasites mainly plant pathogens.

B. Reproduction.
   Asexual
      Saprolegniales: zoospores form in elongated sporangia.
      Peronosporales: zoospores from attached sporangia and conidiosporangia; germination of conidia by germ tubes. Mucorales: aerial sporangia; spores dispursed by air currents.
   Sexual
      Saprolegniales: heterogamy, one to several eggs in each oogonium.
      Peronosporales: heterogamy, one egg in each oogonium.
      Mucorales: isogamy with coenogametes; heterothallic.

CLASSIFICATION

Phylum	Mycota
Subphylum	Myxomycotina
Class	Myxomycetes
Subphylum	Eumycotina
Class	Oomycetes
Order	Saprolegniales
Genus	*Saprolegnia*
Order	Peronosporales
Genera	*Pythum*
	Phytophthora
	Plasmopara
	Albugo
	Bremia
	Peronospora
Class	Zygomycetes
Order	Mucorales
Genus and species	*Rhizopus stolonifer*

24

The Higher Fungi

The higher fungi comprise three classes of the phylum Mycota: the Ascomycetes, the Basidiomycetes and the Fungi Imperfecti. They are of great
importance, being parasitic upon such crops as wheat, corn, stone fruits,
and forest trees. Certain species are required for baking bread, for brewing
and for producing antibiotics.

CLASS ASCOMYCETES

The fine flavors of some cheeses are due to Ascomycetes, and some of the
edible mushrooms are members of this class. Ascomycetes include several
severe, though uncommon, human pathogens, and certain other representatives are known to the medical profession mainly for the beneficial drugs
derived from them. Ergot, a drug widely used to control bleeding, is derived
from an Ascomycete that infects grasses. Yeasts, also members of this class,
are a source of many vitamins and are of great importance in the production
of alcohol and in bread making.

GENERAL CHARACTERISTICS

The hyphae of the sac fungi are septate, and for the greater portion of their
lives the cells are mononucleate. The septa are perforated by a central opening enabling the protoplasm to flow from cell to cell.

The hyphae of the Basidiomycetes, like those of the Ascomycetes, are septate
with pores through the septa. Figure 24.1 shows a septum in *Rhizoctonia solani*.
The pore has a thickened doughnut-like margin much broader than the
adjacent septum. Note the presence of the endoplasmic reticulum paralleling
the wall and turning abruptly away from the pore, which contains a mitochondrion apparently on its way from one cell to another. There is reason
to think that nuclei, too, may pass through these pores from one cell to
another. This mycelium is the active vegetative portion of the plant ramifying
through the host or substrate, causing disease or decay.

Figure 24.1. A small portion of a hypha of *Rhizoctonia solani* showing a septum, with pore, through which a mitochondrion appears to be passing. ×24,000. (Courtesy of C. E. Bracker and E. E. Butler.)

Reproduction

Asexual Reproduction

Various types of asexual spores are chiefly responsible for the dissemination of Ascomycetes during their period of active growth. In mild climates asexual spores may survive the unfavorable seasons, but they are usually killed by cold or by very hot and dry weather.

In many Ascomycetes the spores are formed from the end cells of specialized hyphae. In the process, the end cells round up, are cut off, and are subsequently carried away by wind (Fig. 24.2). The spores so formed are called **conidiospores,** or frequently, simply **conidia,** and the specialized hyphae are called conidiophores (see page 485). Frequently the **conidiospores** remain attached to each other for some time, forming long chains. In many species the conidiophores are associated in a characteristic manner.

Sexual Reproduction

Meiosis followed by mitosis in most Ascomycetes results in eight meiospores in a terminal cell (Fig. 24.3). The appearance of these spores suggests beans or marbles within a cellophane sac, thus the common name **sac fungi.** The Greek word for sac is *ascus* (plural, *asci*), from which is derived the class name, Ascomycetes. The meiospores are generally called **ascospores.** The **asci** are always the terminal cells of special hyphae. They are usually located in a reproductive structure, the **ascocarp** (Fig. 24.4), and have developed by a specialized modification of fertilization from the female gametangium, the oogonium, or, as it is known in this group, the **ascogonium** (Fig. 24.3). The male gametangium is called an **antheridium,** as in preceding groups.

The ascocarp, composed of both vegetative and ascus-bearing hyphae, is characteristic of the species. It may be microscopic or as much as 6 inches in diameter. There are three general types of asco-carps:

1. **Perithecium:** hollow, flask-shaped body with narrow opening (Fig. 24.5*A*).
2. **Cleistothecium:** hollow, completely closed sphere (Fig. 24.5*B*).
3. **Apothecium:** open, cup-shaped body (Fig. 24.5*C*).

The end cells (asci) of the ascus-bearing hyphae, in many forms, line the inner surface of the ascocarp. This surface layer is the **hymenium** or fertile layer. Sterile cells, called **paraphyses,** also arise in the hymenium (Fig. 24.6) and are more numerous and generally longer than asci.

In most Ascomycetes the ascocarps are the direct result of cellular fusion. The description that follows is a generalized account of cellular fusion as it occurs in many Ascomycetes. The gametangia develop from the haploid mycelium growing within the host or substrate. The female gametangium, or oogonium, is a single cell with many nuclei (Fig. 24.3). Antheridia are elongated cells borne on short side branches of adjacent filaments. Both gametangia may have special accessory cells. After an antheridium establishes contact with an oogonium, the male nuclei pass into it. Male and female nuclei pair, but do not fuse as is normally the case. This process now stimulates the growth of hyphae from the oogonium and from the surrounding haploid mycelium. The cells forming from the oogonium are binucleate, each having one male and one female nucleus. This is not a true diploid condition; it may be designated by $n + n$ rather than $2n$ and the paired nuclei are frequently referred to as a **dikaryon.** Since it is from these hyphae that the asci will eventually develop, they may be called **ascogenous hyphae.** The coordinated growth of the haploid and ascogenous hyphae results in the formation of the **ascocarp.** The terminal cells (Fig. 24.6) of the very much branched ascogenous hyphae, together with the ends of some haploid hyphae, form the **hymenial layer.** The hymenial layer is more or less surrounded on the outside and protected by the vegetative haploid hyphae which form a layer called the **peridium.** It is in the young ascus that the nuclei of the dikaryon finally fuse to form a true diploid cell. Meiosis occurs immediately after the nuclear fusion and is followed, in most species, by a mitotic division giving rise to eight ascospores.

The sexual phase of most Ascomycetes is thus

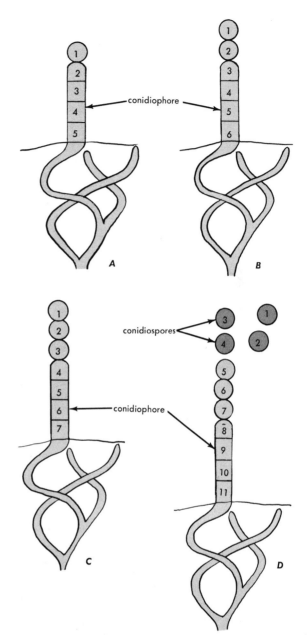

Figure 24.2. Diagram showing formation of conidiophores. *A,* tip of a filament, with the conidiophore divided into cells of similar sizes; first cell (1) has rounded up; *B,* second cell has rounded up and another cell (6) has been added to the series of tip cells; *C,* cell (3) has rounded up and cell (7) has been added to the series of the tip of the conidiophore; *D,* cells (1), (2), (3), and (4) have cut off as conidiophores, (5), (6), and (7) have rounded up, and (8), (9), (10), and (11) have been added to the series.

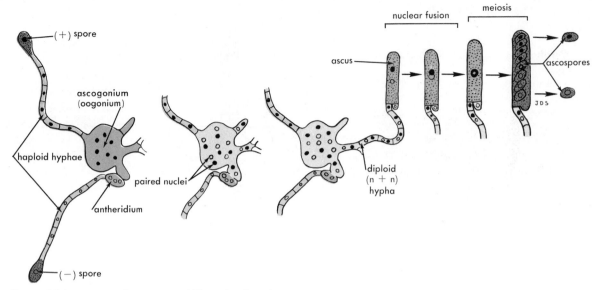

Figure 24.3. Diagram showing sexual life cycle of an Ascomycete.

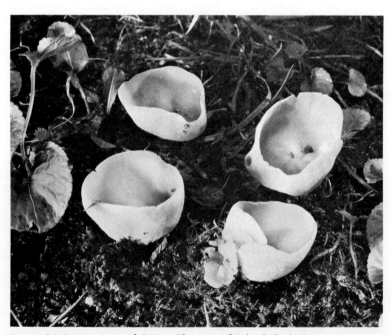

Figure 24.4. Ascocarps of *Peziza*. (Courtesy of John P. Roche.)

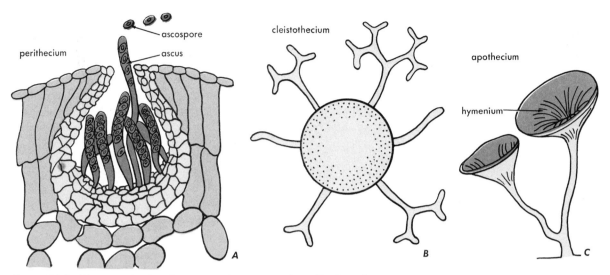

Figure 24.5. Diagrams showing three types of ascocarps. *A*, perithecium; *B*, cleistothecium; *C*, apothecium.

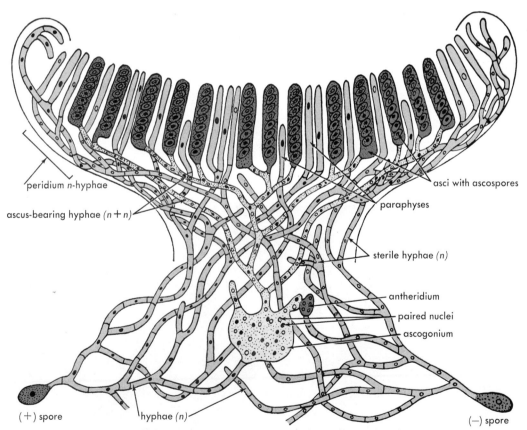

Figure 24.6. Diagram of a cross section of an apothecium. (Redrawn from L. W. Sharp, *Fundamentals of Cytology.* Copyright 1943 by McGraw-Hill Book Company. Used with permission of McGraw-Hill Book Company.)

complicated by the failure of the male and female nuclei to fuse immediately after the union of protoplasts from the antheridium and oogonium. This gives rise to a prolonged stage between the union of the sex protoplasts and the fusion of the sperm and egg nuclei. Fertilization in the Ascomycetes may then be said to involve two steps: (1) the union of the two sex protoplasts, or **plasmogamy,** and (2) the union of the two nuclei, or **karyogamy.** These stages are shown diagrammatically in Figs. 24.3 and 24.6.

The gametangia of the Ascomycetes vary in structure and development. For instance, the simplest type of structure and development occurs in the yeasts, which are single-celled fungi. In some yeasts two cells unite. Meiosis occurs immediately

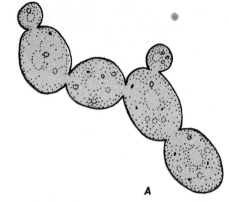

Figure 24.7. Yeast Plants (*Saccharomyces*). A, (*right*) budding cells as seen with the light microscope; B, (*below*) budding cell as seen with the electron microscope. (B, courtesy of E. Vitols, R. J. North, and A. W. Linnane, *J. Cell Biol.,* **9,** 689.)

after conjugation, and ascospores develop (Fig. 24.9).

At the other extreme are filamentous forms in which the oogonium develops a beak and no antheridia are formed. Instead, bodies called **spermatia** are produced at some distance from the oogonium. The spermatia come in contact with the beak of the oogonium. A nucleus from the spermatium passes into the beak and migrates downward to the female nucleus in the inflated base of the oogonium, thus effecting fertilization.

In the Ascomycetes so far described, the development of the ascocarp has been preceded by cellular fusion. In some forms an ascocarp may develop without cellular fusion.

Classification of the Ascomycetes

The fact that about 25,000 species of Ascomycetes exist, many of them severe pathogens, presents a formidable problem to the taxonomist. To group all of these species into a system that will show their relationships and be convenient to use is no mean task. They are commonly grouped into four subclasses and fifteen orders.

Saccharomyces (Yeasts)

Yeast cells may be spherical, ellipsoid, more or less rectangular, or, in vigorously growing cultures, sometimes hypha-like. Their rate of growth seems to influence their shape to some extent. Under ordinary microscopic magnification, living cells appear lacking in much structural detail: only an outer membrane, a protoplast with cytoplasm, vacuoles, and granules of reserve food may generally be seen. A nucleus is present but difficult to observe unless the slides are specially prepared (Fig. 24.7*B*).

Reproduction. Yeasts are normally single-celled Ascomycetes. In old cultures the cells may remain attached, forming short, branched chains. Some yeasts divide by fission, as do bacteria. In the majority of yeasts, however, new cells grow out from the mother cell, much as a small bubble would form if a piece of thin rubber were made to expand through a small opening in some heavier material. The small "bubbles" formed from the mother yeast cell are called **buds**. They enlarge and finally separate from the parent cell. This process of vegetative reproduction is termed **budding** (Fig. 24.7*A*).

Most yeasts form asci, each containing from one to eight ascospores (Fig. 23.8), the number being constant for a given species. As in other Ascomycetes, the production of asci usually is associated with a sexual cycle. In the formation of ascospores, the nucleus of a diploid cell divides into several nuclei, and each, with some associated cytoplasm, becomes delimited as a spore. The spores lie within the parent cell, which is essentially an **ascus** (Fig. 24.8). During the formation of ascospores, a reduction division occurs. Thus, in yeast, the parent cell

Figure 24.8. Ascospores of yeast (*Schizosaccharomyces octosporus*). A and B, as seen in ordinary light; C, as seen by phase contrast. (Material courtesy of Priscilla Stoner.)

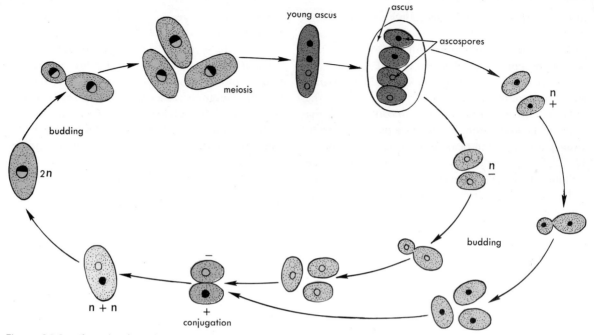

Figure 24.9. Life cycle of *Saccharomyces cerevisiae*.

is diploid, and the ascospores are haploid. In some yeasts the nuclei of adjoining haploid vegetative cells fuse, the resulting zygote dividing to produce ascospores, which multiply by vegetative division. In certain other yeasts a fusion of haploid ascospores takes place.

The life cycle of *Saccharomyces cerevisiae,* the common yeast of commerce, is of considerable interest because it may reproduce asexually by budding in both haploid and diploid phases. Its life history is shown in Fig. 24.9 and is of such a nature as to make this species particularly suited for studies of fundamental biological significance. The life cycle may be written simply as shown below.

Economic importance of the yeasts. The main sources of energy for many yeasts are sugars, which are oxidized in respiratory processes within the yeast cell. Oxygen is not required for the first steps of the oxidation. The sugar, nevertheless, is broken down to a simple organic acid (see page 259), and energy is released in the process. If the oxygen supply is ample this acid is oxidized, with the release of relatively large amounts of energy, to carbon dioxide and water. If the oxygen supply is deficient, the organic acid will be changed to carbon dioxide and *alcohol.* Thus, yeasts growing in a sugar solution well supplied with oxygen will produce carbon dioxide and water, and the yeast plants will multiply rapidly. On the other hand, yeasts growing in a sugar solution poorly supplied with oxygen will form carbon dioxide and alcohol and will multiply slowly. The oxidation of sugars to carbon dioxide

$$\text{Plant } (2n) \rightarrow \text{plant } (2n) \rightarrow \text{ascospores } (n) \nearrow \begin{array}{c} \text{plant } (n) \rightarrow \text{plant } (n) \\ \\ \text{plant } (n) \rightarrow \text{plant } (n) \end{array} \searrow \text{plant } (2n)$$

| Asexual reproduction | Meiosis | Asexual reproduction | Conjugation |

and alcohol by yeasts without the presence of oxygen is known as **alcoholic fermentation.** This process is utilized in the production of industrial alcohol, in wine making, in brewing, and in bread making.

Yeasts have played a very important part in vitamin research. Several vitamins are synthesized by yeast plants, which are an important commercial source of these highly valued substances.

Taphrina

All members of this genus are highly pathogenic to plants, especially fruit trees. They infect leaves, flowers, fruits, and young shoots. They induce unequal growth in leaf cells, causing diseases frequently referred to as **leaf curls.** The fungus causing peach leaf curl is well known. The mycelium of this fungus is intercellular in the palisade tissue of leaves. Here, it stimulates some cells to more rapid growth, which results in wrinkling of the leaf. Eventually the hyphae push their way between the epidermis and the cuticle, where the individual cells swell, round off and thicken their walls, and form a compact layer. These cells then elongate, rupturing the cuticle, and become asci, each one of which contains ascospores. Thus the asci are produced at the surface of the infected organ. The opening buds as they start to grow in the spring are infected by spores that have overwintered between the bud scales. Lime sulfur applied just before the buds open is a very effective control. Typical gametangia are not formed.

Form Genera Penicillium and Aspergillus

Since classification is based on the perfect stages, fungal forms in which the sexual stages have not been found cannot be accurately classified. These forms are grouped together in form genera based upon similarities of the asexual spores.

Asexual reproduction in *Penicillium* and *Aspergillus* is effected by means of conidiospores. In *Penicillium* the spores are formed on profusely branched conidiophores (Fig. 24.10F). In *Aspergillus* the tip of the conidiophore swells and conidiospores form in long chains radiating from this swollen tip (Fig. 24.10G).

In only a few cases are sexual stages known for these forms, but even so they may be divided into five genera in the family Eurotiaceae. One of these may be briefly described. Ascogonia have been found to occur in the form species *Penicillium vermiculatum* which would place it in the true Ascomycete genus *Talaromyces*. *Penicillium vermiculatum* is thus also *Talaromyces vermiculatum* (Fig. 24.10).

In the yeasts the ascospores are produced in typical vegetative cells; in *Taphrina* special asci are formed, but they are not grouped in ascocarps. In *Talaromyces* (*Penicillium*), the asci are grouped in an ascocarp but distributed at random within it rather than being oriented in a hymenial layer. These ascocarps are of the cleistothecial type.

The ascogonium of *Talaromyces vermiculatum* is an elongated slender multinucleate cell, and the antheridium is simply a club-shaped, swollen hypha. Details of fertilization and meiosis are obscure, but eventually cleistothecia develop, containing scattered asci. The sexual cycle is shown in Fig. 24.10A to E. Asexual reproduction is by means of conidia. It must be emphasized that the asexual stage is the usual and greatly predominant mode of reproduction, the sexual stage being rare indeed. The life cycle diagram thus gives much greater emphasis to the sexual stage than is warranted.

Penicillium and *Aspergillus* are probably the most widespread fungi. They are the common blue and green molds and occur on citrus fruits, jellies, and preserves. Their conidia are everywhere in the air and soil, and in the biological laboratory they are frequent contaminations in culture media. Figure 24.11A shows a growth of *Penicillium* on an apple, and Fig. 24.11B shows it on an orange. Enzymes secreted by these fungi are particularly active in digesting starch and other carbohydrates. When purified, these enzymes are important industrial preparations. *Aspergillus oryzae* is used in the preparation of rice wine and soybean sauces. Several species of *Aspergillus* are important in cheese manufacture. A disease resembling tuberculosis is caused by *Aspergillus fumigatus,* and a number of other *Aspergillus* species cause diseases in plants.

Erysiphe

General characteristics. The powdery appearance of the surface of leaves infected with many members of *Erysiphe* and related genera suggests the common name, **powdery mildews.** All the powdery mildews are obligate plant parasites (Fig. 24.12C; see color plate 11. Their food requirements are closely integrated with the metabolism of the host plants, and frequently the host plant is not killed. This relationship insures the fungus a continued food supply.

Many can live only on a special host, whereas others have a wide host range. They grow poorly or not at all in artificial culture media. These characters indicate a high degree of specialization and a long-standing relationship between parasite and host.

The mycelium is generally confined to the surface of the leaves, flowers, or fruits (Fig. 24.12*B, C*). Haustoria (Fig. 24.13) penetrate epidermal and parenchyma cells, from which they secure nourish-ment. At first the mycelium on the surface of the leaf appears like a delicate cobweb. Eventually it assumes a white powdery or dusty appearance owing to the development of numerous conidiospores.

Asexual reproduction. This type of reproduction is effected by conidiospores and usually accounts for the rapid propagation of the fungus during the growing season. The conidiophores are short fila-

Figure 24.10. *A–E,* sexual stages seen in *Talaromyces vermiculatum* (*Penicillium vermiculatum*); *F,* conidia of *Penicillium* sp.; *G,* conidia of *Aspergillus* sp. (*A–E* redrawn from C. J. Alexopoulos, *Introductory Mycology,* John Wiley and Sons.)

Figure 24.12. *A*, cleistothecium with asci of powdery mildew (*Erysiphe*), ×5; *B*, whitish areas on grape leaf are mycelium of *Erysiphe*, ×⅓; *C*, grapes heavily infected with powdery mildew, ×½. (Courtesy of Joe Ogawa.)

A

Color Plate 11

Figure 24.16. *A*, apothecia of the brown rot fungus (*Monolinia fruticola*), ×½; *B*, brown rot on cherry compared with a healthy fruit, ×½; *C*, brown rot on peach, ×⅕. (Courtesy of Joe Ogawa.)

B

C

A

B

C

Figure 24.43. Various types of lichens. *A,* the conspicuous green foliose lichen is *Peltegera aphthosa,* and it is surrounded by the white ascocarps of the fructicose lichen *Cladonia* sp.; *B,* crustose lichen *Acarospora flava* colors a rock on the sky-line of a Sierra Nevada ridge; *C, Ramalina reliculata,* a fructicose lichen, clothes the branches of an oak in the foothills of the Coast Ranges along the Pacific Coast.

ments that stand outward from the mycelium on the surface of the host (Fig. 24.14). The spores themselves may remain attached to each other and form long characteristic chains. Sulfur dusted on the host plants at this stage of the fungal life cycle is an effective control.

Sexual reproduction. The ascogonium is always located at the end of a hypha and may be slightly swollen. The antheridia are small cells, also occurring at the tips of hyphae. An antheridium becomes closely appressed to an ascogonium; an opening appears in the wall separating the gametangia; and the male nucleus migrates through it into the ascogonium. Binucleate hyphae develop from the binucleate zygote, and simultaneously adjacent haploid vegetative hyphae also develop. A small **cleistothecium** (an ascocarp without an opening) is formed and encloses one of several asci (Fig. 24.12*A*). Nuclear fusion and reduction division take place in the ascus, and ascospores develop. The ascospores are usually discharged by force from the cleistothecium.

The cleistothecia are readily seen as small black specks on the surface of infected leaves. **Appendages,** characteristic of the genera, extend outward from the cleistothecia (Fig. 24.12). Appendages may aid in the dispersal of the cleistothecia and in attaching them to a new host.

While the powdery mildews may not kill their host, they weaken it and greatly reduce the crop yield. Powdery mildews cause diseases of apples (Fig. 24.11*A*), grasses, grains, grapes, cherries, and many other plants.

Monilinia

The ascocarps of the order to which *Monilinia fructicola* belongs are cup or disk-shaped (apothecia) (Fig. 24.16*A*; see color plate 11). They may be as much as six inches in diameter, depending on the species, and are sometimes brilliantly colored. Members of one genus, *Morchella* (the morels), are edible. A saprophyte, *Peziza* (Fig. 24.15), is common and probably the best known. Brown rot of stone fruits, a very severe disease of peaches, cherries, plums, apricots, and nectarines, is caused by a disk fungus (*Monilinia,* Fig. 24.16; see color plate 11), which infects mainly blossoms and fruits (Figs. 24.16). Because of these infections the orchardist may experience lower yields or spoilage of the ripe fruit on the tree or during shipment.

Life History of Monilinia fructicola

1. Ascospores formed during early spring infect blossoms (Fig. 24.17*A*, *B*, *C*).

2. A mycelium develops within the ovary (Fig. 24.17*D*), resulting in its complete destruction. At this stage conidiospores may appear.

3. Conidiospores produced in blossoms rapidly infect healthy fruits (Fig. 24.17*E*), under favorable weather conditions.

Figure 24.11. A, *Penicillium* on apple; B, *Penicillium* on orange. (A, courtesy of Plant Pathology Department, University of California.)

4. An extensive mycelium develops within the fruit, causing it to rot (Figs. 24.16*B*, *C*, 24.17*F*, *G*).

5. The rotted fruit dries, becoming a mummy. It may drop to the ground or remain on the tree (Fig. 24.17 *G*).

6. In early spring, gametangia form in the mummied fruits that have fallen to the ground.

7. Fertilization ensues.

8. Apothecia develop from ascogenous and vegetative hyphae (Figs. 24.16*A*, 24.17*H*).

9. The resulting asci shoot ascospores several inches into the air (Fig. 24.15) and re-infection occurs.

10. Bordeaux mixture or lime sulfur applied as the buds swell, together with a destruction of the mummies, will control brown rot.

Claviceps

The genus *Claviceps* is parasitic on grasses, including grains. A dormant mycelium that replaces the mature grain is known as **ergot.** It possesses several alkaloids that have medicinal properties. Ergot constricts the blood vessels, particularly those that pass

Figure 24.13. An epidermal cell of a leaf of atlas barley with haustoria of *Erysiphe graminis nordei.* Note that the haustoria are separated from the hyaloplasm of the host cell by a membrane and that the haustoria are thus in a vacuole-like space. Cellular organelles are prominent in the haustoria and absent from the host cell protoplast. ×11,200. (Courtesy of C. E. Bracker.)

Figure 24.14. Powdery mildew on leaf surface.

conidiospore
conidiophore
haustoria
palisade parenchyma
stoma

into the hands and feet, thus depriving the extremi- ties of a normal blood supply. In humid summers in Central Europe, *Claviceps* may infect rye heavily. In centuries past, before the nature of the fungus was understood, ergot would be milled along with the grains of rye. The contaminated flour, which might contain as much as 10 percent of powdered mycelium, would be baked in bread. A continued diet of bread from this flour resulted in much misery. Because of the contraction of the blood vessels and the limited supply of blood reaching feet or hands, gangrene set in. Hands, arms, and legs would die and finally drop off. The disease was known as "Holy Fire." Today, ergot is a valued drug used to control hemorrhage, particularly during childbirth.

With the knowledge of the poisonous nature of the ergot, diseased grain is no longer milled and *Claviceps* is not of concern in human diet. However, cattle may occasionally feed on infected grasses having ergots in place of seeds (Fig. 24.19) and thus be poisoned.

Life history of Claviceps

1. Ascospores are mature when the first flowers of rye, wheat, or other host grasses open. A given race of *Claviceps* infects certain species of cereals or grasses. The mature ascospores infect the young flowers (Fig. 24.18).

2. A mycelium develops throughout the ovary of the infected flower. Finally it completely replaces the ovary, assuming its general shape (Fig. 24.18*B*).

3. Conidia appear on the surface of this mycelium, together with a sticky, sweet secretion. Insects collect the secretion and, in doing so, disperse the conidia.

4. The compact mycelium replacing the ovary grows upward and becomes a hard and horny body. It is some- what longer than the mature grain and is purple in color. This dormant, elongated mycelium is the **ergot** (Fig. 24.19*B*). It is composed of a compact mass of tough hyphae that, in a dormant condition, survive periods unfavorable to growth (Fig. 24.18*C*).

5. When conditions become suitable, the hyphae of the ergot produce ascogonia and antheridia. After fertiliza- tion the ascogenous hyphae, accompanied by adjacent vegetative hyphae, develop a short, upright stalk. The stalk supports a number of small flask-shaped perithecia (Fig. 24.18*D, E*).

Class Ascomycetes | 511

Figure 24.15. *Peziza.* *A,* apothecia quiescent; *B,* cloud of ascospores discharged upon stimulation of apothecia. × ⅓.

Figure 24.15. *B*

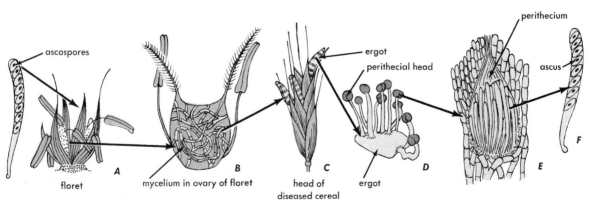

Figure 24.18 (*above*). Life cycle of *Claviceps*. *A*, floret; *B*, mycelium in ovary of floret; *C*, head of diseased wheat; *D*, perithecial head; *E*, perithecium; *F*, ascus.

Figure 24.17 (*left*). Diagram of life cycle of the brown rot fungus (*Monilinia fructicola*).

Figure 24.19. Ergoted head of a grass; A, whole head showing untidy appearance of head; B, enlargement of A, showing several spikelets with a large ergot protruding from the base of one. ×20.

6. Meiosis takes place; asci and ascospores are formed within the perithecia (Fig. 24.18F).

7. As the asci mature, they extend out of the perithecial opening. Pressure develops in the asci, shooting the ascospores into the air.

Significant Features of the Ascomycetes

General Characteristics

1. The mycelium is septate. The cells of the main vegetative hyphae have a single haploid nucleus. The ascogenous hyphae have two haploid nuclei ($n + n$) in each cell.

2. Many are severe plant pathogens, causing diseases of fruit and nut trees and of grains.

3. Some saprophytic forms, such as the yeast, are of considerable economic importance.

4. Yeasts are simple Ascomycetes, consisting of a single cell.

Reproduction

1. Asexual
  (a) Usually by means of conidiospores.
  (b) By budding (in the yeasts).
2. Sexual
  (a) Definite gametangia are not developed in the more primitive species; typical ascogonia and antheridia form in the more advanced types.
  (b) Fertilization generally involves:

(1) The union of two protoplasts without the fusion of the nuclei (plasmogamy).

(2) Stimulation to growth of ascogenous hyphae ($n + n$) and of haploid vegetative hyphae.

(3) Union of two haploid nuclei in ascus (karyogamy), ($n + n \rightarrow \mathbf{2n}$).

(c) An ascocarp develops as a result of the growth of ascogenous and haploid vegetative hyphae.

(1) The inner layer (hymenium) of the ascocarp gives rise to asci.

(2) The outer layer (peridium) of the ascocarp is composed of haploid vegetative filaments.

(d) There are three types of ascocarps, (1) cleistothecia, (2) perithecia, and (3) apothecia.

(e) Meiosis occurs in the formation of the ascospores in the ascus. Eight ascospores are generally produced in each ascus.

CLASS BASIDIOMYCETES

Representatives of the Basidiomycetes include mushrooms, toadstools, bracket fungi, puffballs, and species causing such diseases as wheat rust and corn smut. Many of them are saprophytes, being particularly important in the decay of dead forest trees. Others are parasitic and cause considerable damage to forest and orchard trees, wheat, corn, onions, snapdragons, roses, and many other plants.

Reproduction and Classification

No specialized sex organs are formed by the Basidiomycetes; nearly all, however, reproduce sexually. Conjugation can occur (1) between ordinary hyphal cells, (2) between two special cells, or (3) between special spermlike bodies and receptive hyphae. Some species are **heterothallic;** that is, only plus and minus strains conjugate. Other species are **homothallic,** conjugation taking place between any two cells of any two hyphae, or even between cells of the same hyphae. In both types, prior to conjugation the cells of the vegetative mycelium contain a single nucleus; after conjugation the resulting cell contains two haploid nuclei ($n + n$). The two nuclei in each cell do not fuse until just before meiosis (Fig. 24.22*A, B*). Fusion of the two nuclei

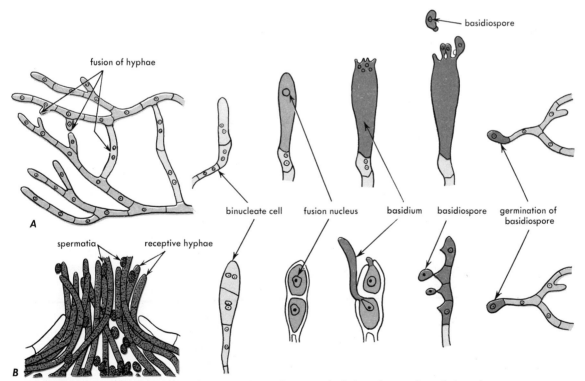

Figure 24.20. Generalized life cycles of Basidiomycetes showing only fusion of protoplasts, fusion of nuclei, and meiosis (production of basidiospores). *A,* common mushroom; *B,* a fungus causing wheat rust.

Figure 24.21. Stages in the development of a basidiocarp. A small button just appearing from the substrate shows at A; B, enlarging basidiocarp, gills still enclosed; C, cap open showing mature basidiocarp; D, additional basidiocarps grow around the first-formed one to give a cluster of fruiting structures. × ½.

and subsequent meiosis take place in special cells known as **basidia** (Fig. 24.22C, D). The meiospores formed by these basidia are called **basidiospores.** The basidia may be either (1) a single club-shaped cell (Fig. 24.20), (2) a single short, filamentous cell, or (3) a short four-celled filament (Fig. 24.20). The club-shaped basidia occur on special spore-bearing structures, as in mushrooms, bracket fungi, and puffballs. This type of basidium characterizes the subclass Homobasidiomycetidae. When the basidium is a single short, filamentous cell or a short four-celled filament, it results from the germination of special resistant spores (Fig. 24.29, 24.34). Basidiomycetes with this type of basidia constitute the subclass Heterobasidiomycetidae. In both types the basidiospores are attached to the basidia by short stalks, the **sterigmata** (singular, **sterigma**).

A generalized life history of a Basidiomycete is shown diagrammatically in Fig. 24.20.

Subclass Homobasidiomycetidae

General Characteristics and Distribution

The mushrooms, bracket fungi, and puffballs are Homobasidiomycetidae. Only a few are parasitic, but these few do extensive damage to forest trees (Figs. 24.27, 24.28), to some orchard trees, and to a small number of garden crops. Several Homobasidiomycetidae are always found growing in association with certain species of trees. It is believed that some of them may form **mycorrhizae:** a symbiotic relationship with tree roots. The fungus is able to make nitrogen and other nutrients more available to the roots. Certain Homobasidiomycetidae obtain their carbohydrates from wood and may even cause extensive decay of timbers in mines and buildings. Specialized tissues and organs may be formed, some of the $n + n$ hyphae becoming specialized to transport food and water, and others becoming tough and woody.

Although both haploid and $n + n$ mycelia develop extensively, the $n + n$ phase is of special interest because it gives rise to the spore-bearing body, or **basidiocarp** (sporophore) (Fig. 24.21), in which the basidia are developed. The hyphae form a tangled mass in the substrate or host and emerge to form a basidiocarp—the mushroom, puffball, or bracket fungus. As the basidiocarp grows, the ends of certain hyphae generally become aligned in a layer that will bear spores when mature. This is the

hymenium layer, the cells of which gradually enlarge to become basidia.

The development of the basidiospore is shown in Fig. 24.22. The young basidium contains the $n + n$ nuclei, or the dikaryon (Fig. 24.22A). These two nuclei fuse to give the true diploid nucleus shown in Fig. 24.22B. Meiosis ensues and the four haploid nuclei destined to pass into the basidiospores appear (Fig. 24.22C). When this stage is reached the sterigmata have extended outward from the tip of the basidium and are ready to receive the basidiospore nuclei. One of the nuclei present in the basidium of Fig. 24.22C would have passed through the sterigma shown in the figure. The basidiospore of Fig. 24.22D is mature and ready for dispersal as the haploid nucleus has moved into it. Note that one nuclear division has been completed here.

There are five orders of Homobasidiomycetidae. We shall consider mainly the order Agaricales, of which common mushrooms and bracket fungi are members. Many of them are edible, and a few are violently poisonous. The common edible mushroom of commerce, *Agaricus campestris,* is a species of the order (Fig. 24.23). The mycelium, or, as it is known to the mushroom growers, the **spawn,** grows extensively in the substrate. It is common in many pastures and is cultivated in well-rotted horse manure.

The Basidiocarp

The basidiocarp of a typical mushroom such as *Agaricus campestris* (Fig. 24.23) consists of a short upright stalk or **stipe,** attached at its base to a mass of mycelium and expanding on top into a broad cap or **pileus.** Many mushrooms have a ring of tissue around the stipe rather close to the pileus. This ring occurs because when young the pileus was attached to the stipe at this point. The underside of the pileus of *Agaricus* is formed by thin **gills** radiating outward from the stipe. These gills are lined with a hymenium or spore-bearing layer. If a young gill is sectioned, the majority of basidiospores will still be attached to the basidia. Four basidiospores are attached to each basidium by short stalks, the **sterigmata.** If the gill is very young, developing basidia on which basidiospores have not yet been formed may be observed (Fig. 24.24). Figure 24.25 shows a section through three gills. Note two stages of development of basidia and the loose tangle of hyphae forming the center of the gill. When a basidiospore is discharged it is shot horizontally from its position on the basidium straight into the space between two gills. A basidiospore is somewhat like a toy rubber balloon, having a large volume for its weight. It is shot from the basidium with a force sufficient to carry it about midway between the two gills. It then falls straight downward. It has been estimated that some basidiocarps may discharge as many as a million spores a minute for several days.

If the cap of a mushroom is placed on a piece of paper and carefully protected from wind, the spores will fall straight downward and form a **spore print.** Spore prints are important in the identification of mushrooms, since they indicate the color of the spores and the pattern of the gills.

The spores of *Agaricus campestris* are brown, whereas those of other Agaricales may be white, black, red, yellow, or ocher. The spores of *Amanita muscaria,* the "Fly-Mushroom," are white. *Amanita muscaria* has other distinguishing features: when the basidiocarp is in the button stage it is completely surrounded by a thin tissue known as the **universal veil** (Fig. 24.26A, B, C). As the basidiocarp expands, this veil is torn (Fig. 24.26D, E). Part of it remains attached to the cap (Fig. 24.26F, G), where ruptured fragments of it may be seen in the mature mushroom; the remaining part stays underground, forming a cup-shaped structure out of which the stalk grows (Fig. 24.26F, G). The enlarged basal structure is referred to as the **volva** or "death cup." The fragments of white on the reddish yellow cap are remnants of the universal veil. The white gills and ring are visible. The volva of *A. muscaria* tightly clasps the stipe; it may be seen here as the swollen, scaly base (Fig. 2.7A; see color plate 2).

Amanita phalloides is one of the most deadly poisonous of all mushrooms. It may be recognized by (1) its white spores; (2) the volva or "death cup" (one must be sure to dig deeply); (3) the ring on the stipe below the cap; (4) the color of cap, which is reddish in the center and fades to yellow toward the edges; and (5) the torn remains of the universal veil on top of the cap. As a general precaution, one should avoid all white-spored mushrooms with a death cup and a ring on the stipe. Many species of Agaricales, however, are excellent eating.

Not all Agaricales have gills. In some the lower surface of the cap or bracket has many small pores that extend upward to form small tubules (Fig. 24.27). The hymenium lines the tubules. Many

Figure 24.22. Basidiospore formation in *Schizophyllum commune*. A, (above) young basidium with *n + n* nuclei ×3000; B, (right) basidium with 2*n* nucleus ×5000. C, (upper right) four *n*-nuclei after meiosis ×3000. Note development of sterigma. D, (lower right) basidiospore attached to sterigma; the *n*-nucleus is completing its first division. (Courtesy of K. Wells.)

bracket fungi found on the trunks of forest trees are **pore fungi.**

Some bracket fungi cause serious damage to forest trees and lumber in mines and wooden buildings. During World War II much green lumber had to be used, and destruction by fungi of improperly cured lumber resulted in much loss. One of the best-known fungi in this group is *Fomes applanatus* (Fig. 24.28). It causes a disease of forest trees known as *white-mottled rot,* to which many hardwoods, such as poplar, birch, maple, basswood, oak, and elm are susceptible. True firs, Douglas fir, spruce, and hemlock may also suffer from this disease.

The basidiocarp of *Fomes applanatus* is a very elegant shelf-shaped bracket or **conk.** The conks are perennial and grow to a large size. They are gray on top, and the lower surface has millions of pores lined with a creamy-white hymenium.

Most of the bracket fungi of the forest are facultative parasites. They normally live in the dead heartwood, to which they gain entrance by wounds. They are, however, able to invade the living cells of sapwood, which they may destroy.

Other types of basidiocarps also occur, such as those in puffballs and bird's-nest fungi.

Subclass Heterobasidiomycetidae

General Characteristics

The basidium of the Heterobasidiomycetidae is a short, swollen filament developing directly from a specialized resistant spore (Figs. 24.29, 24.34). Most members of the subclass are parasites on vascular plants. Like the Homobasidiomycetidae, they possess, as far as is known, both haploid and $n + n$ phases, and reduction division takes place in the development of the basidia or basidiospores. Asexual reproduction may occur in either the haploid or $n + n$ phase or in both. Conjugation may take place between two nonspecialized cells, as in the Homobasidiomycetidae. In some forms, however, special cells conjugate or a spermlike cell may conjugate with a receptive hypha.

Some of the Heterobasidiomycetidae may have two hosts; that is, separate phases of their life history are passed on different plants. Such forms are **heteroecious.** If only one host is required to complete their life history, they are said to be **autoecious.**

Figure 24.23. *Agaricus campestris*, the common edible mushroom. ×1.

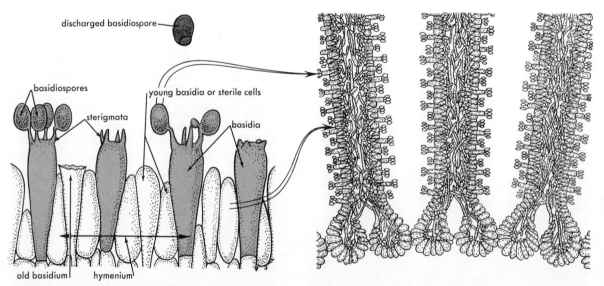

Figure 24.24. Section of gill cut parallel with plane of basidia. It shows the production of basidiospores. (Redrawn from A. H. R. Buller, *Researches on Fungi*. Used with permission of Longman Group Ltd.)

Figure 24.25. Section showing three gills of a mushroom. (Redrawn from A. H. R. Buller, *Researches on Fungi*. Used with permission of Longman Group Ltd.)

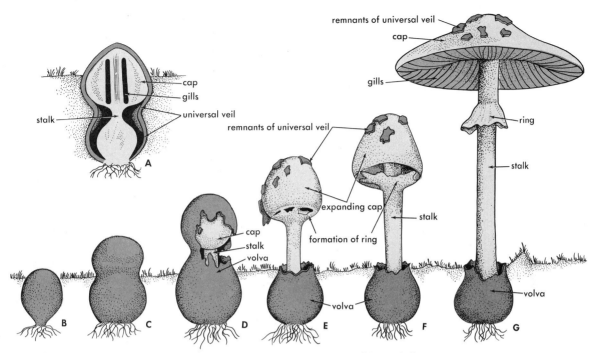

Figure 24.26. Stages in the development of the basidiocarp of death angel (*Amanita*). (Redrawn from Gibson, *Our Edible Mushrooms and Toadstools,* Harper and Bros., 1899.)

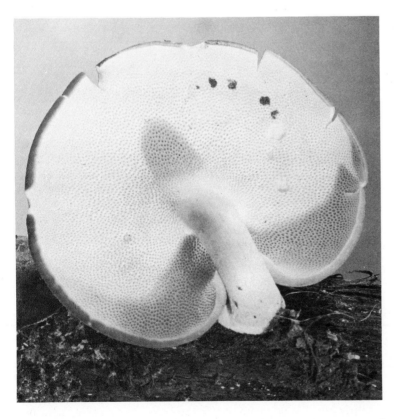

Figure 24.27. The underside of the pore fungus *Polyporus brumalis.* × ⅓. (Courtesy of Roche.)

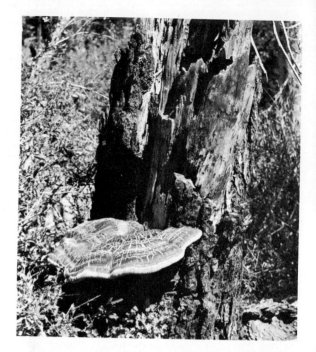

Figure 24.28. Basidiocarps of *Fomes applanatus*. × ⅓. (Courtesy of Brownell.)

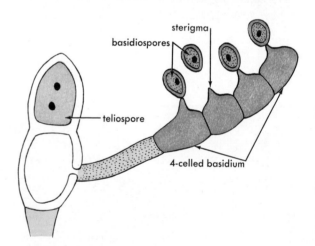

Figure 24.29. Stages in the life cycle of *Puccinia graminis*. Teliospore with basidium and basidiospores.

Uredinales

The Uredinales, or rust fungi, are of world-wide distribution. They are known to infect a very large number of higher plants, and all are obligate parasites. Black or red spores develop in pustules (**sori**) on the leaves of many grasses, roses, and mallows, and they (particularly the red spores), suggested the name of **rust.**

There are many types of rust fungi and they show marked variation in their life histories. In nearly all of them, however, basidiospores are produced on short four-celled basidia that usually develop from an overwintering spore, known as a **teliospore** (Fig. 24.29). Fusion of the two haploid nuclei and reduction division occur during the development of the basidia and basidiospores. The details of conjugation are obscure in many species.

Common wheat rust, caused by *Puccinia graminis,* is of great economic importance, and its life history is well known. It produces all the different types of spores common to the order. We shall follow its life history in detail.

Life history of Puccinia graminis. In the early spring, the overwintering teliospores germinate on the soil or stems of wheat, in fact any place where there is moisture.

Teliospores of *Puccinia graminis* are two-celled spores. A short four-celled basidium develops from each cell; only one is shown in Fig. 24.29. A single haploid basidiospore is formed from each cell, so that four basidiospores eventually develop from each cell of the teliospore. Two of these four basidiospores are plus strain (+), two are minus strain (−).

The basidiospores are carried by air currents and can infect only the young leaves, fruit, or twigs of common barberry bushes. They germinate chiefly on barberry leaf surfaces and produce a germ tube, which penetrates the epidermis. Soon a mycelium develops in the tissues of the host. Then, after a time, pustules, called **spermagonia,** appear on the upper surface of leaves. In section a spermagonium is pear-shaped and has a small pore opening through the upper epidermis to the exterior of the leaf (Fig. 24.30*B, C*). Some hyphae extend upward

Figure 24.30. Stages in life cycle of *Puccinia graminis*. *A,* aecia as they appear on lower surface of barberry leaf; *B,* section through barberry leaf showing spermagonia and aecia; *C,* spermagonium; *D,* aecium. (*A, B,* courtesy of Triarch Products.)

through the pore; others, lining the interior of the spermagonium, cut off large numbers of spermlike cells, called **spermatia.** These are forced out of the spermagonia through the pores.

Puccinia graminis is heterothallic; both plus and minus spermagonia and spermatia appear on barberry leaves. These develop from plus or minus basidiospores, both of which are haploid. Spermatia are transferred by wind or insects to adjacent spermagonia, where they come in contact with the **receptive hyphae,** which are essentially female gametangia. A plus (+) spermatium fuses with a cell of a minus (−) receptive hypha, and vice versa. An $n + n$ diploid mycelium results; its cells each contain two haploid nuclei.

Shortly after fusion, a mass of $n + n$ hyphae ap-

pears in the spongy parenchyma of the barberry leaf close to the lower epidermis. Here a second series of pustules called **aecia** are formed (Fig. 24.30*B*). The aecia face downward, and the upper side toward the leaf mesophyll becomes lined with $n + n$ hyphae. The growth of the aecia eventually brings about the rupture of the lower epidermis, and binucleate **aeciospores,** cut off from the ends of the numerous hyphae lining the base of the aecium, are exposed to the atmosphere (Fig. 24.30*B*, *D*). The aeciospores thus released are disseminated by wind and they can infect only *wheat* plants. In surface view the aecia present the appearance of a group of small cups and are frequently called **cluster cups** (Fig. 24.30*A*).

The hyphae of germinating aeciospores gain en-

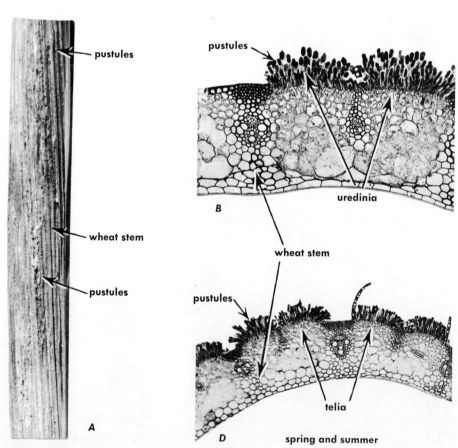

Figure 24.31. Stages in life cycle of *Puccinia graminis* on wheat. *A*, sori on wheat stem; *B*, *C*, (*upper right*) uredinia; *D*, *E*, (*lower right*) telia. (*A*, *B*, *D*, courtesy of Triarch Products.)

trance to wheat plants by growing through the stomata. A delicate mycelium develops and the hyphae form haustoria that draw nourishment from surrounding cells of the wheat plant but do not kill them. About ten days after infection, red binucleate spores, **uredospores,** are produced by the mycelium. The epidermis of the wheat plant is ruptured by the mass of spores, forming open sori, the **uredinia** (Fig. 24.31*A, B, C*). The uredospores thus produced are carried by wind currents to other wheat plants. The fungus is spread throughout a grain field during the growing season by the red uredospores. The progress of the disease throughout a wheat-growing area may be so rapid as to attain epidemic proportions.

As the wheat begins to mature, the production of red uredospores by the infecting mycelium gives way to the production by the same mycelium of heavy-walled two-celled overwintering spores—**teliospores.** The sori bearing teliospores are called **telia** (Fig. 24.31*D, E*).

Each cell of the teliospore is, at the outset, binucleate; the two haploid nuclei, one (+) and one (−) are still present and distinct (Figs. 24.31*E,* 24.32*A*). These two haploid nuclei now fuse (**karyogamy**), the true diploid condition resulting. As in the Ascomycetes and Homobasidiomycetidae, meiosis follows directly upon nuclear fusion. In the rusts, a short basidium grows from each cell of the teliospore, the four haploid nuclei resulting from meiosis migrate into it, and cell walls develop. Basidiospores now form, one from each cell of the basidium (Fig. 24.32), and the cycle starts again.

Puccinia graminis is **heteroecious;** it requires two hosts to complete its life cycle. Certain other rusts are **autoecious,** needing but one host to complete their life cycle. Not all rusts develop all the spore types found in *Puccinia graminis,* nor do they all have an identical life history.

Wheat rust may be controlled in cold climates by eradicating all barberry bushes (the alternate host) near wheat fields. Barberry will not be present to carry the rust over the winter, and if uredospores are killed by the winter cold, the rust will die. In regions with warm winters the uredospores may overwinter, wheat may be infected in the spring, and wheat rust may therefore be more difficult to control.

Different varieties of *Puccinia graminis* are able to attack other grains. *P-g-var. avenae* attacks oats; *P-g-var. tritici* attacks wheat. Breeding work with wheat, however, has shown that many **biological strains** exist even within the variety *tritici.* Indeed, new rust strains seem to appear from time to time, and so it becomes necessary to continue the breeding of wheat varieties that will be relatively resistant to them.

Significant steps in life history of Puccinia graminis

1. Two hosts, wheat and barberry, are necessary to complete the life history. In other words, *Puccinia graminis* is heteroecious.

2. Uredospores (one-celled, red) and teliospores (two-celled, black) are produced on the wheat plant. Uredospores can infect other wheat plants; they are killed by vigorous winters. Teliospores cannot infect any plant;

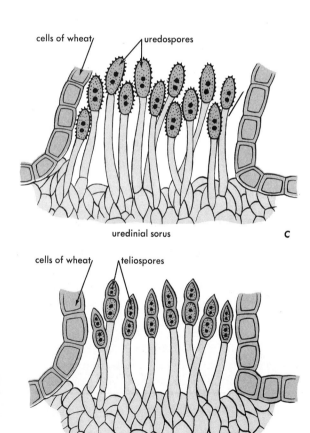

cells of wheat   uredospores

uredinial sorus   C

cells of wheat   teliospores

telial sorus   E

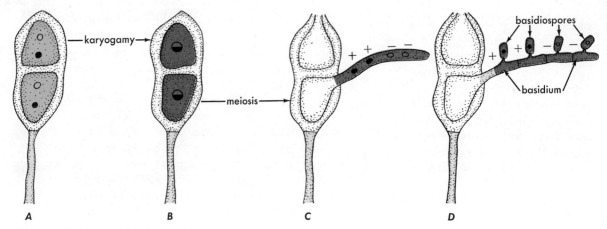

Figure 24.32. Stages in the life cycle of *Puccinia graminis*. Teliospore; *A*, binucleate cells; *B*, 2n cells; *C*, step in formation of basidium and basidiospores; *D*, basidium with basidiospores.

they are overwintering spores, germinating in the spring on the soil or wherever else they find moisture.

3. A short four-celled basidium develops from each of the two cells of a teliospore. Meiosis takes place in this development. One basidiospore (haploid) forms from each of the four cells of the basidium. Basidiospores, either plus or minus, infect young tissues of the barberry.

4. Plus or minus spermagonia (both kinds, haploid) develop in the upper portion of the barberry leaf. The spermagonia produce spermatia and receptive hyphae.

5. Fertilization results from union of a spermatium with a receptive hypha of the opposite strain.

6. The resulting $n + n$ mycelium forms an aecium that discharges binucleate aeciospores from openings in the lower surface of the leaf.

7. Aeciospores infect young wheat plants, producing an $n + n$ mycelium which eventually produces uredospores and teliospores.

Ustilaginales

All members of the Ustilaginales are plant parasites causing a group of diseases known generally as the **smuts.** Two families are generally recognized: Ustilaginaceae and Tilletiaceae. The life histories of the fungi causing cereal smuts vary, particularly in the manner of infection of the hosts: there may be **local infection, blossom infection,** or **seedling infection.** We shall discuss several genera under these headings.

Local infection. *Ustilago zeae,* a member of the Ustilaginaceae, causes common corn smut (Fig. 24.33*A, B*). Infection by spores of U. *zeae* is said to be **local,** meaning that infections remain localized, producing pustules or large tumors. Perhaps the most noticeable tumors are those that occur in the corn ear. The kernels become much enlarged due to the development within them of an extensive mycelium that eventually gives rise to a mass of black, heavy-walled, overwintering spores. Since haploid nuclei fuse and meiosis occurs within them, they are teliospores. These spores may lie dormant in the soil, on old corn stalks, or in manure.

A four-celled basidium develops when the teliospore germinates. Two plus and two minus basidiospores are formed, one from each of the four cells of the basidium (Fig. 24.34). The basidiospores may infect corn plants directly or they may produce **sporidia,** which may multiply by budding and which are able to bring about infection. At any rate, a small haploid mycelium, either plus or minus, develops soon after infection (Fig. 24.35). Plus and minus mycelia develop closely together, and conjugation occurs by the fusion of cells from small haploid mycelia of opposite strains. After conjugation, the $n + n$ mycelium, depending upon its location in the corn plant, may form a small pustule or develop into a tumor as large as a baseball. Most of the hyphal cells in the pustule or tumor change into teliospores.

Blossom Infection. Spores of *Ustilago tritici* and

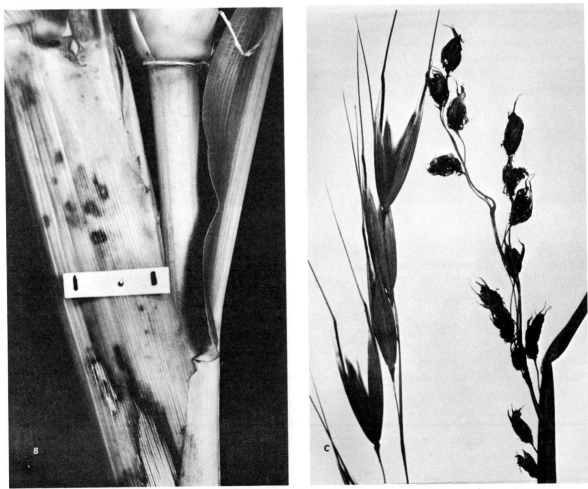

Figure 24.33. Smuts (*Ustilago*). A, ear of corn with smutted kernels; B, smut of corn leaf; C, smut of oats.

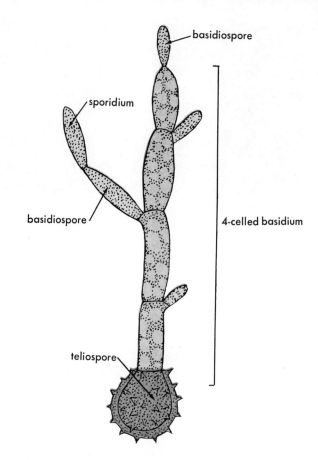

Figure 24.34. Teliospore of *Ustilago* zeae with basidium.

Figure 24.35. Undersurface of upper epidermis of a corn leaf showing diagrammatically the formation of an *n* + *n* mycelium of *Ustilago* zeae.

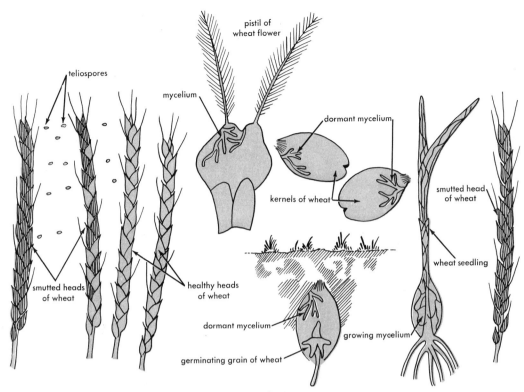

Figure 24.36. Life cycle of *Ustilago tritici*, a blossom-infecting smut.

Ustilago avenae can infect *only the pistil* of wheat and oat flowers. They are known as **blossom-infecting smuts,** and the disease they produce is called the loose smut of wheat or oats. Teliospores that mature in heads of diseased plants at the time healthy plants are in blossom are transferred to healthy blossoms by air currents (Fig. 24.36). The young pistil is directly invaded and a small mycelium eventually develops in the embryo. The infection has no injurious effect on the developing grain, which matures normally. The mycelium within the embryo remains small and dormant, thus carrying the fungus over seasons unfavorable to growth. When the grain germinates, the mycelium within resumes activity. It grows best in or near meristematic tissue, and thus keeps pace with the development of the wheat or barley plant. The presence of the mycelium close to the meristematic tissue has an accelerating influence on the growth of the host, which matures rapidly, producing flower heads. The mycelium invades the developing ovary, eventually replacing the host tissue completely. The entire head of grain becomes a black, loose mass of teliospores

(Fig. 24.33*C*), and when the spores blow away nothing is left but the axis of the spike.

Meanwhile, healthy plants have been growing more slowly and developing normal florets. Teliospores, and florets ready for pollination, occur at the same time. Teliospores are carried by wind to the healthy florets, where the ovaries are eventually invaded. (Fig. 24.36).

Seedling infection. *Tilletia tritici,* belonging to the family Tilletiaceae, causes **bunt** or **stinking smut** of wheat. Only young seedlings can be infected. The smuts of this family are consequently referred to as the **seedling-infecting smuts.**

As in the blossom-infecting smuts, masses of smut spores (teliospores) replace the mature grain of wheat (Fig. 24.37). In contrast to loose smuts, however, the grain coats remain intact, and infected grains resemble normal ones. At harvest time the infected grains are broken and the teliospores are thoroughly mixed with the healthy grains, to which they firmly adhere. As the name stinking smut indicates, the teliospores have a very unpleasant odor. Grain mixed with them is unfit for flour or for

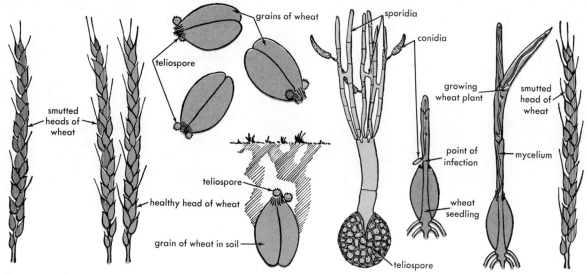

Figure 24.37. Life cycle of *Tilletia tritici*, a seedling-infecting smut.

animal feed. The presence of large numbers of smut spores as dust during threshing greatly increases the fire hazard; explosions and serious fires are not infrequent when heavily smutted grain is threshed.

The teliospores carry the fungus over the season unfavorable for growth. Adhering to the grain, they are sown with it. Meiosis occurs in the teliospore, and frequently several mitotic divisions follow, producing several haploid nuclei. Upon germination of the teliospore, a short haploid mycelium is formed that gives rise to numerous spores called **sporidia.** The sporidia conjugate by means of conjugation tubes, either while still attached to the haploid mycelium or after falling away. After conjugation a short $n + n$ mycelium is formed which produces conidia. Seedlings may be infected by these conidia, or a second short mycelium may form that produces another crop of conidia. At any rate, infection of seedling wheat plants is brought about by conidia. The young mycelium grows best in, or adjacent to, meristematic tissue. It keeps pace with the developing wheat plant. When the grain starts to head out, the mycelium enters the young ovaries of developing flowers. Hyphal filaments may eventually replace all normal cells of the grain, except the coats. Teliospores are formed by these hyphal cells, and the cycle is repeated (Fig. 24.37).

Comparison of Seedling- and Blossom-infecting Smuts. In both types, teliospores are formed in the inflores-cence of cereals. In seedling infection, the teliospores overwinter in the soil or *on the grain.* Upon germination they produce sporidia that conjugate and develop a small mycelium from which conidia are cut off. Conidia infect only seedling grain plants. In blossom-infecting smuts, teliospores lead to the infection of the pistils of nearby flowers, and the rudimentary mycelium remains dormant *in the grain* over winter. In both types of infection the young seedling wheat plant has within its cells an actively growing smut mycelium that keeps pace with the development of the cereal. This mycelium invades young ovaries and fills the ovary with hyphal cells, which become teliospores.

Significant Features of the Basidiomycetes

A. *Vegetative characteristics.*

1. Hyphae septate; extensive haploid mycelium, one haploid nucleus in each cell; extensive $n + n$ mycelium, two haploid nuclei in each cell.

2. Homobasidiomycetidae: mostly saprophytic forms, a few species parasitic on trees.

3. Heterobasidiomycetidae: parasitic on plants; some, such as the rusts, heteroecious with complicated life cycles; in a few species true basidiospores not produced.

B. *Reproductive characteristics.*

1. Asexual: asexual spores not formed by the Homo-basidiomycetidae; various sorts formed by different spe-

cies of Heterobasidiomycetidae; uredospores and aecio-spores formed by the rusts, sporidia formed by some smuts.

2. Sexual: no motile sex cells or gametangia.

(*a*) Homobasidiomycetidae: (1) union of two haploid hyphal cells; (2) *n* + *n* mycelium with two haploid nuclei in each cell developed from the union of the two haploid hyphal cells; (3) basidiocarp (mushroom, puffballs, bracket fungi) produced by the *n* + *n* mycelium; (4) hymenium consisting of club-shaped basidia developed on the basidiocarp; (5) two haploid nuclei fuse in the basidia; (6) meiosis results in four basidiospores.

(*b*) Heterobasidiomycetidae: (1) basidia develop from resistant teliospores; may be short filamentous cells or short four-celled filaments; meiosis occurs in formation of basidiospores; (2) fertilization with eventual fusion of two haploid nuclei varies considerably from species to species.

CLASS FUNGI IMPERFECTI

About 24,000 species of fungi, in some 1200 genera, are known only by their asexual stages. The class name, Fungi Imperfecti, arises from the custom of calling the sexual stages of fungi **perfect stages** and the asexual stages **imperfect stages.** Since only the imperfect stages of this large group of fungi are known, they are called the Fungi Imperfecti. Obviously, the classification is an artificial one.

In general, the structure of the hyphae, which are septate, and of spores suggests that many imperfect fungi may be Ascomycetes; others may be Basidiomycetes. The Fungi Imperfecti may be considered to be Ascomycetes or Basidiomycetes whose sexual stages have not been observed or no longer exist. The classification of these fungi is fraught with great difficulties largely because a natural classification is based upon sexual stages and the morphology of sexual and asexual stages is by no means coordinated. For instance two forms may have very similar conidial stages but very different sexual stages. The various categories of classification in this class are designated as **form genera** or **form families** to show that the members of the genera or families do not necessarily have a natural or family relationship. For instance some species have been named simply on the basis of the host upon which they were found, a procedure that resulted in naming hundreds of literally nonexistent "species."

Only a single form group will be mentioned, the form order Moniliales; it is the largest in the class and has over 10,000 form species. Most of the fungal

pathogens of man belong here. Some of them have recently been demonstrated to be causal agents of various types of allergy. *Penicillium* and *Aspergillus* are genera frequently placed in this order because there are many species belonging to these genera whose sexual stages are unknown. This means that the *Penicillium* of Roquefort cheese and the forms from which penicillin is obtained are Fungi Imperfecti because the formation of asci has never been

Figure 24.38. Testing a bacterium for its ability to prevent or retard the growth of other bacteria. A, seven different cultures have grown from parallel streaks made on an agar plate. A culture of the bacterium to be tested is streaked across the top of the plate; B, the upper streak has grown and an antibiotic from this colony has inhibited the growth of the bacteria in the parallel streaks.

Figure 24.39. Commercial production of an antibiotic. A, fermentation tanks in which *Penicillium* or another organism is grown; B, equipment for the extraction of the antibiotic from the fungal mycelium. (Courtesy of Charles Pfizer and Company.)

observed, even though tons of mycelium of these species have been grown.

Penicillium notatum has won deserved fame because **penicillin,** a drug derived from it, will inhibit the growth of bacteria without injuring human tissue. Substances such as penicillin, formed by one organism which inhibits the growth of other organisms, are called **antibiotics** (Fig. 24.38). Our knowledge of them dates back to 1870, but their significance was not fully appreciated until the discovery of penicillin.

In 1928, Alexander Fleming investigated extracts of *Penicillium* and after demonstrating their antibiotic properties gave the antibiotic the name penicillin. It was not until after 1940 that a concentrated effort was made to produce penicillin on a commercial scale.

Many thousands of pounds of the mycelium of *Penicillium notatum* have been grown, and the species has been subjected to intensive study, yet no sexual stages have ever been observed. The absence of sexual reproduction presents the geneticists who are studying *Penicillium notatum* with a difficult breeding problem. Nevertheless, they have succeeded in growing strains that produce many times more penicillin per pound than do the original strains. Figure 24.39 illustrates several steps in the commercial production of penicillin.

Several forms in the Moniliales are adapted to capture and destroy microscopic animals. A few actually trap nematode worms by forming small rings which quickly constrict when stimulated by the contact of a nematode crawling through them. A nematode-trapping form and several types of conidiospores are shown in Fig. 24.40.

The Parasexual Cycle

We have discussed the sexual cycle of the Ascomycetes and Basidiomycetes and we have further said that no sexual reproduction has been observed in the Fungi Imperfecti. The parasexual cycle has been observed in this group and it provides a means of gene recombination different from, and possibly in addition to, the ordinary type of gene recombination due to the commonly occurring sexual process. The parasexual cycle, as the prefix para indicates, is a sexual cycle that occurs beside or parallel with a normal sexual cycle. At present it is known only for certain fungi.

The first step in the parasexual cycle is the fusion of two ordinary hyphal cells (Fig. 24.41*A*). Recall

that this was the normal step in plasmogamy in the Homobasidiomycetidae but not in the Ascomycetes, where gametangia are normally formed. This fusion results in the formation of a dikaryon (Fig. 24.41*B*). If the fusing cells were parts of the same mycelium, both nuclei of the dikaryon would be identical; if the fusion results from the union of cells from different mycelia, the two members of the dikaryon could carry different genes and the resulting paired nuclei would be called a **heterokaryon.** Since only the combination of different nuclei is of importance in producing variation, we will consider only this condition. By a special process known as **diploidization** this heterokaryotic state is passed from cell to cell in the haplophase mycelium (Fig. 24.41*C*). At some isolated places in the heterokaryotic cells the $n + n$ nuclei may fuse to produce a true heterozygous

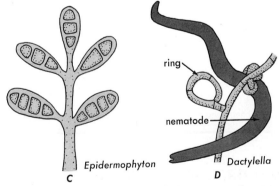

Figure 24.40. *A, B, C,* conidia of form-genera of *Moniliales; D,* nematode-capturing rings of *Dactylella.* (*A, B, C,* redrawn from C. J. Alexopoulos, *Introductory Mycology,* John Wiley and Sons; *D,* redrawn from C. Dreschler.)

diploid nucleus (Fig. 24.41D). This may divide by mitosis and once, in many divisions, crossing over may take place between one pair of chromosomes (Fig. 24.41F). This differs from meiosis in that this mitotic crossing over is rare and does not occur in all homologous pairs of the chromosome complement, but only in one pair. But, as usual, this crossing over results in a gene recombination in the chromosome of the pair involved in the crossover. Such recombination is, however, of no genetic importance as long as the diploid state remains; separation of the members of the chromosome pair must as always occur.

Such a separation does take place, but in a very irregular fashion. For instance, suppose mitotic crossing over has occurred between the alleles *ABCD* and *abcd*, to give us *ABcd* and *abCD*. These recombinates will separate and divide in a normal fashion. But suppose in a certain division both chromatids of *ABcd* go to the same pole, but *abCD* divides as usual (Fig. 24.41E). The resulting daughter cells will differ; one will have two *ABcd* chromosomes and one *abCD* chromosome; while the other will have no *ABcd* chromosome and one *abCD* chromosome. Thus the number of homologous chromosomes bearing the *ABCD* alleles will have been reduced

Figure 24.41. Diagram showing possible steps in the parasexual cycle. A, union of two vegetative cells; B, heterokaryon; C, diploidization; D, diploid nucleus; E, heterozygous chromosomes; F, mitosis of one pair of chromosomes in diploid nucleus. All other divisions are normal; G, haploidization through nonseparation of a crossover pair.

from two to one. If this process is repeated for all other pairs of chromosomes in the complement a haploid cell will eventually be formed. Whereas the end result is similar to that attained in meiosis, the process is very different in that it is slowly achieved and in a haphazard fashion.

Thus the parasexual cycle parallels the steps of fertilization and meiosis, but no definite sexual cells are formed, crossing over is at random, and the haploidization of individual pairs of chromosomes is probably achieved through the random failure of daughter chromosomes to separate rather than through the orderly step of meiosis. The whole process is irregular and there does not seem to be a precise time or place for any of the steps. Once a haploid cell is formed, it multiplies to give a segment of haploid mycelium from which conidia may eventually be formed. The great importance of the process lies in the possibility it gives for variation in forms that do not possess a normal sexual cycle.

THE LICHENS

General Characteristics and Distribution

The lichens are composite plants composed of algae and fungi. The algal components are generally single-celled forms belonging to either the Chlorophyta or the Cyanophyta. When free from the fungus, the algae may exist normally.

The fungal component is generally an Ascomycete, but algal associations with Basidiomycetes are also known. The fungal component of the lichen cannot live separated from the alga unless supplied with a special nutritive medium.

The association of a fungus and an alga in the lichen is generally believed to be **symbiotic;** that is, both the fungus and the alga derive benefit from the association. The alga furnishes food for the fungus, which supplies moisture and shelter from high light intensity for the alga.

A section through a lichen thallus reveals four distinct layers. The top and bottom layers consist of a compact mass of intertwining fungal filaments. The algal cells form a green layer beneath the top mass of fungal filaments, and a loose layer of hyphae lies directly below the algal cells (Fig. 24.42).

Lichens are widespread. They form luxuriant growths on the frozen, northern tundras, where they supply feed ("reindeer moss") for deer. In our own country they frequently cover rocks, trees, and boards exposed to sun and wind (Fig. 24.43, see color plate 12). They are slow but efficient soil formers; rocks are disintegrated slowly by their action. Lichens are pioneer plants in plant succession which begins on bare rock. Lichens and mosses accumulate soil and organic matter sufficient to allow herbs and later shrubs and trees to become established.

For convenience in classifying, lichens may be divided into three groups simply on the basis of their vegetative body. This is a highly artificial grouping. Probably the most conspicuous lichens are members of the **fruticose** type. They are formed by small tubules or branches. Fruticose lichens may make extensive and attractive growths on oaks (Fig. 24.43*C*) or the dead branches of certain pines (Fig. 24.43). Some lichens have a leaf-like plant body. They constitute the **foliose** type and are common in moister climates, where they make extensive growths on tree trunks or on fallen logs (Fig. 24.43*A*). However, the **crustose** lichens are the most common, if less conspicuous, type as they grow extensively on rocks, bark, poles, boards, and other hard surfaces. Indeed any bright coloring on the dark rocks (Fig. 24.43*B*) and gray granites of our northern mountains, on unpainted wood, or on bark of trees is quite likely to be due to a growth of crustose lichens.

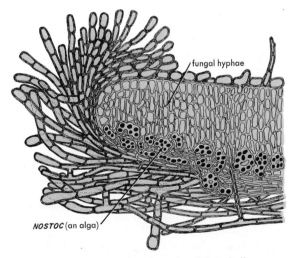

Figure 24.42. Cross section through a lichen thallus. (Redrawn from Darbishire.)

Table 24.1 *Classification of ascomycete genera discussed*

SUBCLASS	ORDER	GENUS	COMMON NAME
Hemiascomycetidae	Endomycetales	*Saccharomyces*	Yeasts
	Taphrinales	*Taphrina*	Leaf-curl fungi
Euascomycetidae	Eurotiales	*Eurotium* (*Aspergillus*)	Green mold
		Talaromyces (*Penicillium*)	Blue mold
	Erysiphales	*Erysiphe*	Powdery mildews
	Clavicipitales	*Claviceps*	Ergot
	Helotiales	*Monilinia*	Brown rot fungus
	Pezizales	*Peziza*	Cup fungus

Asexual Reproduction

Lichens may multiply simply by the distribution of small pieces of the vegetative thallus. Many lichens produce special disk-shaped bodies, composed of both a fungus and an alga, that serve as a means of vegetative reproduction. These bodies are called **soredia** (singular, **soredium**).

Both the algal and the fungal components may produce spores. The algal may grow independently, but if, upon germination of the fungal spores, the young hyphae do not find the required algal cells, they soon die.

Sexual Reproduction

Sexual reproduction is characteristic of the type of fungus present in the lichen. Since the fungal component is, in the great majority of lichens, an Ascomycete, ascocarps with asci and ascospores are formed. Upon the germination of the ascospore the germ tube must find its proper algal associate in order to survive.

MYCORRHIZAE

Mycorrhizae (sing., mycorrhiza) are fungal associations with the roots of higher plants. In some cases the fungi are unicellular and live within individual root cells, but in many cases the fungi have typical hyphae that cover the root tips in a thick mat and penetrate between cortical cells. These hyphae do not form haustoria, but contact with root cells is nevertheless very close and metabolites are transferred in both directions. Radioactive carbon in carbon dioxide, for example, can be fixed in photosynthesis by the higher plant and later be detected in the fungus. Radioactive isotopes of phosphorus, calcium, and potassium can also be shown to be taken up in greater amounts by plants with mycorrhizae than by plants without them.

Mycorrhizal associations occur throughout the plant kingdom. They have been found on roots of many trees, shrubs, and herbs of flowering plants; on conifers, ferns, mosses, and lower vascular plants. Possibly more than 80% of all angiosperms exhibit them. In some cases the presence of mycorrhizae is essential to normal plant development. Seedlings of orchids and heaths fail to survive if the fungi are absent, and plantings of pines in areas where they do not normally grow (such as Australia) do poorly unless the mycorrhizal fungi are introduced to the soil. These fungi may also contribute to plant development by secreting certain hormones.

Fungi of all types are involved in these associations. Those which can not be identified are given a trinomial name beginning with the binomial *Mycelium radicans*. The mycorrhizal fungus of scotch pine (*Pinus sylvestris*), for example, has been isolated but never identified with a known, free-living fungus; it has been given the temporary name *Mycelium radicans sylvestris*.

CLASSIFICATION

Phylum   Mycota
  Class   Ascomycetes
    Subclass   Hemiascomycetidae
      Order   Endomycetales
       Genus and species   *Saccharomyces cervisiae*
      Order   Taphrinales
       Genus   *Taphrina*
    Subclass   Euascomycetidae
      Order   Eurotiales
       Genera   *Eurotium* (*Aspergillus*)
          Talaromyces
          (*Penicillium*)
      Order   Erysiphales
       Genus   *Erysiphe*
      Order   Clavicipitales
       Genus   *Claviceps*
      Order   Helotiales
       Genus   *Monilinia*
      Order   Pezizales
       Genus   *Peziza*
  Class   Basidiomycetes

    Subclass   Homobasidiomycetidae
      Genera and species   *Agaricus campestris*
          Amanita muscaria
          Amanita phalloides
          Fomes applanatus
    Subclass   Heterobasidiomycetidae
      Order   Uredinales
       Genus, species, and strain   *Puccinia graminis tritici*
      Order   Ustilaginales
       Family   Ustilaginaceae
        Genera and species   *Ustilago zeae*
           Ustilago tritici
           Ustilago avenae
       Family   Tilletiaceae
        Genus and species   *Tilletia tritici*
  Class   Fungi Imperfecti
    Form order   Moniliales
      Form genera and species   *Penicillium notatum*
         Aspergillus
         Dactylaria brachophaga

25

The Viruses

Toward the close of the nineteenth century, Iwanowski, a Russian plant scientist, carefully pressed some sap from a stunted and diseased tobacco plant. With a small amount of this sap he inoculated a healthy tobacco plant. The healthy plant soon developed signs of a disease similar to that of the plant from which the sap was taken. The experiment was repeated, but this time the sap from the diseased plant was forced through a filter with pores so small that bacteria could not pass through them. The healthy plants inoculated with this filtered sap soon contracted the disease of the original plant. This was a new discovery—a disease could be transmitted from one plant to another by a filtered sap that was free of visible living bodies even when viewed by the most powerful light microscopes. The sap contained something (a **virus**) that caused an infectious disease.

CHARACTERISTICS AND DISTRIBUTION

This experiment of Iwanowski's has been confirmed many times. The infective principle has been isolated and studied. We have gained from these studies much information about viruses that infect plants, animals and even bacteria. A healthy tobacco plant compared with a plant of the same age but infected with tobacco virus is shown in Fig. 25.1. Diseases transmissible from one plant to another by an agent too small to be distinguished even when enlarged some 3000 times are called **virus diseases.** Such diseases occur in both plants and animals. The name **virus** is given to the agent or substance that causes the disease. The nature of viruses is still an interesting scientific puzzle.

Smallpox, poliomyelitis, and measles are examples of virus diseases of man. Dog distemper is caused by a virus. The so-called mosaic diseases of plants, characterized by a mottling of the leaves, are also virus diseases. Even though the actual agent causing these virus diseases was long unknown, much has been learned of methods to combat the diseases. For instance, in 1796, an English country doctor named Jenner transferred material from the pustules of cows suffering from cowpox to a fresh scratch in the skin of man and found that this treatment would successfully prevent smallpox in man. It was general knowledge among country people that milkers never contracted smallpox. Jenner purposely rubbed dried material from cowpox pustules on scratches on the arm of his small son and later exposed him to virulent smallpox. Application of this principle, now known as **vaccination,** has reduced the incidence of the once-dreaded smallpox almost to zero.

Culture of Viruses

The study of viruses is hampered because it is only possible to culture them in living cells. Rabbit skin, tobacco leaves, bacteria, or cultures of living tissue are necessary for the multiplication of viruses. Tobacco virus is, for instance, increased in quantity by growing it in tobacco leaves; cowpox virus may be grown on rabbit skin or obtained from cattle.

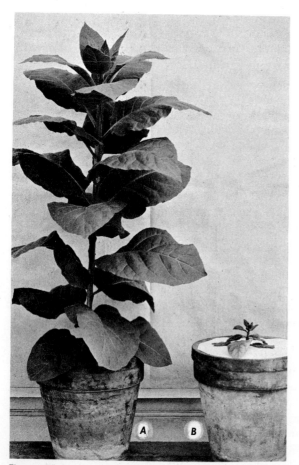

Figure 25.1. A, healthy tobacco plant; B, tobacco plant infected with curly top virus. (Courtesy of K. Esau.)

The virus of poliomyelitis is grown on a culture of HeLa cells derived from human cancer or from living cells obtained from the kidneys of the rhesus monkey.

The Nature of a Virus

Viruses are small particles, less than 0.2 of a micron in one dimension and thus below the resolving power of the light microscope. They are all, as far as known, unable to multiply outside of a host. While these characteristics serve to distinguish the viruses as a group, they do not separate all bacteria from all viruses. The smallpox virus is large enough to be seen with the light microscope, and there are some very small parasitic bacteria. The metabolic relationship between the virus and the host cell, however, differs from the relationship between a bacterium and its host cell.

The bacterium is a true cell comprised of a cell wall and an internal organized protoplast, and supplied with the enzyme systems responsible for energy conversion; these factors make possible the synthesis of cellular products needed for growth and cell division. A bacterium lives as an individual living within the host cell; it synthesizes its own characteristic proteins, catalyzed by its own enzyme system and utilizing energy produced through respiration in its own prototoplast. Its raw materials are absorbed from the host cell. Bacteria give off waste products which may be harmful to the host cell. In so doing, bacteria may change the metabolism of the host cell, but so far as is known, the host cell never elaborates bacterial components.

On the other hand, viruses appear to become a part of the host cell. They do not possess energy-yielding, or synthetic enzyme, systems. Viral constituents are similar to those of their hosts and induce enzyme systems of the host to elaborate viral, rather than host, materials. Therefore viruses cannot multiply outside of cells. Due to this close association with living cells, we know viruses only because of their direct effect upon either bacteria, animal, or plant cells. Indeed, viruses may be more logically compared to cellular constituents than to whole cells. Viruses, as cell constituents, offer a logical explanation of their infectious nature. The similar cell and viral constituents are the nucleic acids. In the cell the nucleic acids have a genetic continuity passing from cell generation to cell generation by mitosis or from one individual generation to another by way of the gametes. However, in virus-infected cells, the nucleic acid metabolism of both host and virus is closely linked.

Viruses, as particles, exist outside of cells, where they may be considered nonliving chemicals. They gain entrance into cells in one way or another from the outside, and once inside they multiply with the assistance of the host cells, doing harm to the host cells in the process. Viruses, then, are true **infectious agents.** Inside the host cell they may become incorporated in its metabolism and be **transmitted** from one cell generation to the next. While thus reproducing, they usually protect the host cell against additional infection by more of the same type of virus particles. They are said to be in a **vegetative stage.** In addition to infection, viruses are also transmitted from cell to cell.

Viral and host cell constituents differ in that host cell constituents never exist as such outside of the cell, while the viral constituents may so exist. Host cell constituents are active strictly in host cell metabolism, while the viral constituents induce the host cell to elaborate viral products. Indeed, it has been reported that viruses may induce mutation-like changes in seedlings from virus-infected fruit trees. Such changes are inheritable, and resemble the symptoms produced in the parent tree originally infected by the virus.

All viruses are small infectious agents. They can be seen only with the electron microscope, and the study of them with it has only begun. We may catalog them as rods. (Fig. 25.7), as spherical particles (Fig. 25.7), or as more complicated particles, but we cannot classify them by their morphology as we do many living organisms. Table 25.1 gives a list of several plant viruses with shapes and sizes. For convenience, the present procedure is to group them as bacterial, plant, or animal viruses. This is an artificial classification, and we may find that some viruses are infectious for a wide range of host organisms.

Bacterial Viruses

Bacterial viruses were first known as **bacteriophages,** a term which has been shortened to **phage** and is now used to designate a bacterial virus. Phages were discovered independently by two investigators, Twort (1915) and d'Herelle (1917), both of whom were curious over the sudden dissolution or **lysis** of colonies or portions of colonies of the bacteria they were studying. These spots of lysed bacteria are known as **plaques.** They were able to

Table 25.1 *Some plant viruses with their dimensions and vectors*

VIRUS	SIZE IN MILLIMICRONS	VECTOR
Spherical viruses		
Tobacco necrosis	20	Unknown
Bean mosaic	30	Unknown
Squash mosaic	30	Beetle
Aster yellows	40	Leafhopper
Potato yellow dwarf	110	Leafhopper
Short-rod viruses		
Hydrangia ringspot	44×16	Unknown
Maize stunt	240×50	Unknown
Long-rod viruses		
Tobacco mosaic	280×15	Many vectors
Bean yellow mosaic	750×12	Aphids
Beet yellows	1250×10	Aphids

show that the destruction of the bacteria was due to an infectious agent (1) too small to be seen with the light microscope, (2) capable of passing through a fine filter, and (3) capable of being transmitted in bacterial cells. Its characteristics thus conformed to the definition of a virus.

The discovery of an infectious agent that could rapidly destroy the bacterial cell explained the failure of transmission of human diseases in such places as the religious rivers of India where large numbers of sick bathe. The pathogenic bacteria discharged in such places are rapidly destroyed by phages.

Perhaps the best known phages are those infecting strains of *Escherichia coli,* for which several phage strains are known. These phage strains are designated as T1, T2, T7, T-even and so on.

THE REPRODUCTIVE CYCLE OF VIRUSES

Life Cycle of a Virulent Phage

When phage particles are added to a thinly spread culture of bacteria on an agar plate, they become absorbed by their **tails** on the cell walls of the susceptible bacteria (Fig. 25.2). Note that the phage particle consists of the aforementioned tail and a **head** which in side view has six sides. The tail ends in an attachment region. In Fig. 25.3 this attachment can be seen to consist of spread protein

Figure 25.2. Phage T2 attached to the surface of an *E. coli* cell. *A*, before discharge of DNA from the head; *B*, after discharge of DNA from head. *A*, ×20,000; *B*, ×22,000. (Courtesy of E. H. Cota-Robles and M. D. Coffman, *J. Bacteriol.*, **63**, 266.)

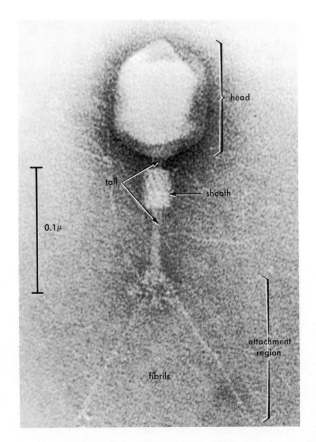

Figure 25.3. An electron micrograph of a T-even phage particle. ×60,000. (Courtesy of S. Brenner, R. W. Horne, et al., *J. Mol. Biol.*, 1959. Copyright Academic Press.)

fibrils. The tail is a hollow protein rod with, it is thought, a **pin** at the bottom end, and a **sheath** which surrounds the rod just below the head. The head consists of a protein coat and a DNA core. The head is about 700A in diameter. The viral DNA does not contain the base cytosine (page 102), as does bacterial DNA, but another base, hydroxy-methylcytosine (HMC). This is a convenient circumstance, as it makes it possible to distinguish the viral and bacterial DNAs. Shortly after the attachment of the phage particle to the bacterium, the phage DNA moves down the hollow core of the tail and is injected into the bacterial protoplast. The protein coat of the head remains outside the bacterium, and may be seen in Fig. 25.2*B*. Immediately upon the entry of the viral DNA, the replication of bacterial DNA stops. The synthesis of new bacterial enzymes also stops, but those present continue to function for a time. Under the influence of viral DNA the bacterial enzyme systems either synthesize, or are themselves converted into systems capable of elaborating the necessary phage constituents. Some type of new RNA synthesis may also be necessary. Phage DNA is now replicated in the bacterial cell. This activity results in the elaboration of 100 or more new phage particles, complete with protein fibrils, tails, and head coating, the latter enclosing a viral DNA core. The bacterium now disintegrates or undergoes lysis and the viral particles are released to infect more bacteria (Fig. 25.4). The whole process takes about one hour and is shown diagrammatically in Fig. 25.5. Phages which bring about such a rapid lysis of the host cell are called **virulent phages.**

Life Cycle of a Temperate Phage

There are many phages known that do not immediately kill the bacteria they invade. The bacterial hosts are called **lysogenic** bacteria and the phages that infect them are known as **temperate** phages. The metabolism of these temperate phages is so subtly linked to that of the host bacterium that both may exist through many generations. In this state the phage is known as a **prophage** and it confers immunity upon its host against invasion by other similar phage particles. The prophage changes to an active phage when the growth of phage and bacterium get out of step. At this juncture the phage DNA directs the bacterial cell to elaborate phage constituents; the phage multiplies and eventually matures; infectious, temperate phage particles are

released from the lysing bacterial host.

During replication of the phage, bacterial DNA may be incorporated in the phage DNA. When the temperate phage particle carrying bacterial DNA infects another bacterium, the new host may be influenced by the bacterial DNA carried by the phage. This process is known as **transduction** and occurs as described below.

Transduction

A strain of *Salmonella typhinurium* is known which is resistant to streptomycin; it may be designated as str-r. Phages may be reproduced in bacteria resist-

Figure 25.4. Disintegration of a bacterium and multiplication of virus particles. (Courtesy of S. E. Luria, M. Delbrück, and T. F. Anderson, *J. Bacteriol.*, **46,** 57.)

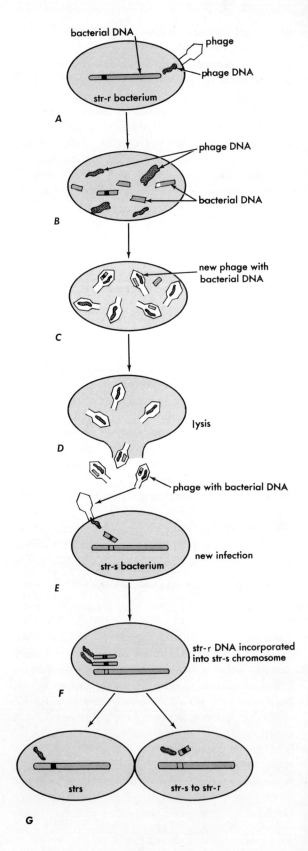

Figure 25.5 (*above*). Diagram of a life cycle of a virulent phage.

Figure 25.6 (*right*). Diagram to show probable steps in the transduction of genetic information from one bacterium to another by way of a transducing virus. (Redrawn after R. Sager and F. J. Ryan, *Cell Heredity*, John Wiley and Sons.)

ant to streptomycin. A phage becomes attached to the bacterial cell wall and phage DNA enters the bacterial protoplast (Fig. 25.6A). The bacterial DNA disintegrates and there is a multiplication of the vegetative virus (Fig. 25.6B). The disintegration of bacterial DNA does not destroy the individual bits of genetic information but seems to distribute it at random in the bacterial cell (Fig. 25.6C). When the phage DNA becomes incorporated in the protein heads of the newly forming phage particles, the replicated phage DNA carries with it some of the original bacterial DNA (Fig. 25.6D). This original bacterial DNA is known as the **transducing fragment.** Upon lysis, the phage particles are released into the medium, where they may be collected and transferred to a culture of streptomycin-sensitive bacteria (str-s). If this culture is now grown on a medium containing streptomycin, about one in a million bacterial cells will be str-r. What has happened is this: When the phage DNA is injected into the bacterium the transducing fragment enters also (Fig. 25.6E), and is replicated along with the phage DNA. Since this is a temperate phage, lysis will not occur immediately, and upon division of the bac-

terial cell the str-r gene in the transducing fragment somehow takes the place of the str-s gene in the bacterial DNA (Fig. 25.6F). Note that in this case the transduced fragment merely goes along with the viral DNA. There is also evidence that the host str-r gene may become incorporated in the viral DNA by gene recombination. **Transduction** is the transfer of host DNA by means of a phage to another host where it is incorporated into the DNA of the new host. It might be compared to transformation; in transduction, bacterial DNA is transferred by a virus, while in transformation, bacterial DNA is transferred by agents other than phage.

Lysogenic Conversion

When lysogenic bacteria are infected with a phage, lysis does not occur. The phage becomes a **prophage;** its replication is synchronized with the replication of bacterial DNA so that bacteria and phage may exist together in a type of symbiosis. Under these conditions the phage DNA may directly influence the bacterial DNA. This is known as **lysogenic conversion.**

SUMMARY OF PHAGE

At this point, to summarize, we might mention that it has been rigorously demonstrated that the DNA of bacterial cells may be changed by the incorporation of DNA from other outside sources. In other words, DNA in these forms does not have a strict genetic continuity; it may fail to pass unchanged from one cell generation to another, and instead be influenced by DNA introduced from sources outside of strictly hereditary lines. We do not know whether this happens in higher forms. There have been, for instance, suggestions that cancer might be due to a virus.

PLANT VIRUSES

Plant viruses occur as either spheres or elongated rods (Fig. 25.7). In both cases there is some sort of RNA core surrounded by a protein sheath. In the case of the spherical bushy stunt virus, the center of the sphere is the RNA core, which is surrounded by a protein sheath (Fig. 25.8). The rod forms, as illustrated by tobacco mosaic, seem to be more complicated; for (although there is still a central core of RNA) this elongated rod is twisted into a uniform spiral to form what appears, in gross form, to be a hollow rod (Fig. 25.9). The presence of virus

Figure 25.7. Rod-shaped particles of tobacco mosaic virus and isodiametric particles of grape fan leaf virus. TMV virus particle is 180 A wide. (Courtesy of Robert Taylor.)

Figure 25.8. Bushy stunt virus stained to show the central core, which is composed of RNA. The clear area around the core is a layer of protein, and the dark circles separating the virus particles are formed by the precipitation of uranyl nitrate absorbed on the protein sheath. ×150,000. (Courtesy of H. E. Huxley and G. Zubay, *J. Biophys. Biochem. Cytol.*, **11**, 273.)

RNA core

Figure 25.9. A diagrammatic interpretation of the structure of a tabacco mosaic rod.

particles may be detected in the chloroplasts of tobacco plants infected with tobacco mosaic virus (Fig. 25.10). Under certain conditions these particles may become arranged in a precise crystalline array (Fig. 25.11).

Classification of Viruses and Virus Diseases

Many angiosperms are susceptible to virus infection. It is also known that bacteria may be attacked by a viruslike agent. So far, no virus diseases of

Figure 25.10. Chloroplast from a tobacco plant infected with tobacco mosaic virus (U-5 strain). Note array of crystalline-like particles. ×25,000. (Courtesy of T. A. Shalla.)

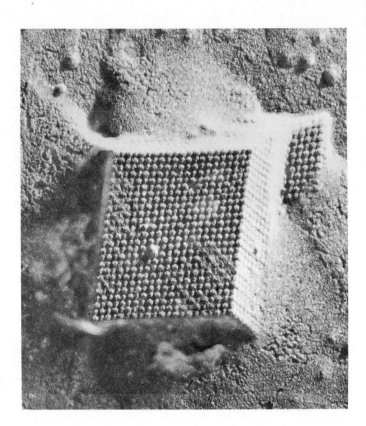

Figure 25.11. A crystalline particle of the tobacco necrotic virus, ×84,000. (Courtesy of L. W. Lablaw, *J. Ultrastruct. Res.,* **2,** 177, Academic Press.)

Figure 25.12. A, healthy sugar beet leaf; B, sugar beet leaf infected with mosaic virus; C, cells from healthy leaf; D, cells from diseased leaf. (Courtesy of K. Esau and C. W. Bennett.)

gymnosperms, mosses, or ferns, have been reported. Tobacco itself may suffer from 15 different viruses, and potatoes from 18. As early as 1935, virus diseases were known for 1100 species of angiosperms. Furthermore, a single virus may break down into distinct strains. Fifty-four strains have been described for the common tobacco mosaic. No satisfactory system for the classification of this large number of viruses has thus far been devised. Although many factors contribute to the difficulty of classifying viruses, the most important ones are: (1) only a few viruses have been isolated in a pure form; (2) viruses may be cultured only in living cells; (3) plant viruses are not specific; one virus usually infects several plants, and one plant may be infected with several viruses. Viruses may be compared with cell constituents rather than with complete cells.

Symptoms of Virus Diseases of Plants

One of the most common symptoms of virus infection in plants is the occurrence of light-green or yellow areas on leaves (Fig. 25.12B). These areas range from small circles to large irregular patches. Uniform stripes or irregular bands of the normal green color may appear in the leaf. In some instances the entire diseased plant may be a much lighter green than the healthy one. Many variegated horticultural varieties of plants undoubtedly are infected with some mild-acting virus. The diseases characterized by mottling or variegation of leaves are referred to as "mosaics," and almost all are regarded as virus diseases. We have, for instance, tobacco mosaic, potato mosaic, and sugar cane mosaic.

The yellow or lighter areas of diseased leaves are usually thinner in cross section than are those of normal healthy leaves. The chloroplasts may be absent or very reduced in size (Fig. 25.12C, D) and in number, or they may be yellow in color. This reduction in the green pigment, and therefore in the color of the leaf, is called **chlorosis.** A chlorotic leaf is one that has light green or yellow areas. Not all chlorotic leaves, however, harbor a mosaic virus; chlorosis may be due to other factors, including mineral deficiencies.

In addition to being chlorotic, leaves of virus-infected plants may be irregular in shape with a rough or puckered surface, may be rolled in along the edges, may be reduced in width, or may turn yellow and drop prematurely (Fig. 25.13).

Figure 25.13. Healthy plants compared with those infected with a virus. A, healthy bean; B, diseased bean; C, healthy grape; D, diseased grape. (Courtesy of Plant Pathology Department, University of California.)

Figure 25.14. Tulip flower infected with a virus disease.

Flowers of infected plants are frequently dwarfed, mottled, or streaked. The streaking of flowers occasionally produces a highly prized horticultural variety; many hundreds of dollars have been paid for tulip bulbs infected with a virus that causes a streaking of the flowers (Fig. 25.14). Fruits may be small, distorted, and mottled. Tomatoes, peppers, mellons, and beans infected with certain viruses show a distinct mottling; tomatoes and cucumbers may be distorted and much reduced in size. Stems may show discoloration of various sorts; frequently they are dwarfed and contain cankers or dead areas.

Viruses belonging to the mosaic group induce changes primarily in parenchyma tissue (see page 120). Vascular tissues, however, also may be injured as the result of a primary injury to the parenchyma tissue of leaves and stems. Other viruses primarily harm the conducting tissue, particularly the phloem. The sieve tubes are injured and frequently killed, forming a mass of dead cells on the outer edge of the vascular bundle. This mass impairs the movements of food in the plant and results in injury to other plant tissues.

In many plants, viruses seem to interfere with those tissues responsible for the production, as well as the translocation, of food (Fig. 25.15). This damage may result in a stunting of all organs of the plant (Fig. 25.1). To the plant grower it means a greatly reduced crop yield or even a loss of plants. For instance, peach yellows (a virus disease) causes an estimated loss of 450,000 trees annually in New Jersey, Pennsylvania, and Maryland. In a single area the spindle tuber virus disease of potatoes decreased the yield 88 percent over a 4-year period.

Transmission of Viruses

Any organism that carries a disease-causing agent from one living thing to another is called a **vector.** Insects, particularly those that feed on plant juices, are the most important vectors of viruses of plants. Certain types of aphids, thrips, and leafhoppers account for the transmission of viruses from one plant to another. The aphids feed by injecting a long stylet into plant leaves or stems and sucking out plant juices. They obtain sap mostly from phloem tissues, from parenchyma cells of the cortex of stems, and from the mesophyll of leaves. The stylet may wind its way between cells of the cortex until it reaches the phloem, or it may pass directly through all cells in its path (Fig. 25.16). Thrips have large mandibles with which they chisel their way into epidermal cells. They then suck out the cell contents.

When insects feed on plants infected with virus

Figure 25.15. Healthy and diseased vascular tissues of tobacco. *A*, healthy; *B*, infected with curly top mosaic. (Courtesy of K. Esau.)

Figure 25.16. Leafhopper feeding on beet petiole. (Courtesy of C. W. Bennett.)

diseases they ingest the virus along with the cell sap. The virus passes into the intestine of the aphid or thrip, is absorbed into the blood stream, and is carried to the salivary glands. The saliva secreted by the feeding insect then carries the virus, and any plant upon which the aphid or thrip feeds may become infected with the virus. There is some reason to believe that the virus may undergo development in the insect. With certain diseases, the insect cannot infect a healthy plant immediately after feeding upon a diseased plant; an **incubation period** (within the insect) of several days must intervene before the virus is capable of causing infection when given off with the saliva.

Some viruses can be transmitted from one host plant to another only through specific insects. Often, however, the relationship between insect, virus, and plant is not specific. One insect may carry several virus diseases, several different insects may carry the same virus, and several species of plants may be infected with the same virus. The virus that causes Pierce's disease of grapes, for instance, may be carried by 15 different species of insects and can also affect 37 other species of plants. Some viruses are so highly infectious that they may be carried from one plant to another by almost any sort of contact. Workers in the tobacco field may carry tobacco virus from one plant to another on their hands or clothing. The tobacco virus is not destroyed in the manufacture of smoking tobacco. Thus, should cigarettes made from virus-infected leaves be smoked by a field worker, he may easily carry the virus on his hands to healthy plants. Tobacco virus may also cause a disease of tomato plants.

A few of the more common virus diseases of plants are: aster yellows, which has been transmitted by leafhoppers to more than 120 species in 30 different families, including such crop plants as celery, lettuce, carrot, and parsley; beet curly top, transmitted by leafhopper to numerous species in at least three families; sugar cane mosaic; mosaic of legumes; phony disease of peach; potato mosaic; tobacco mosaic; tomato spotted wilt; and wheat mosaic, transmitted in some unknown manner through the soil (Table 25.1).

The Life Cycle of a Plant Virus

The precise changes occurring in a plant virus particle as it passes through the body of an insect and a plant, or is transmitted from one plant to another, cannot yet be related. There are still many gaps in our knowledge, and speculation may be permitted. There is, as has been related, an incubation period in the body of an insect after it picks up virus particles from an infected plant. There also appears to be an **incubation** or **eclipse** period in the plant cell after its infection with virus particles. Infection may be with viral RNA or with whole virus particles, but it appears that viral movement within the plant body is always in the form of RNA. All attempts to modify the course of infection by modifying the protein moiety have ended in failure. On the other hand, changes of viral RNA reduce or change the course of viral infection. There is some evidence that RNA appears in the host nucleus before it appears in the host cytoplasm and that viral RNA is elaborated by the host cytoplasm before the viral protein is formed. The replication of the viral RNA is indeed somewhat of an enigma because current ideas of RNA replication and protein synthesis are generally related to information carried in the DNA molecule. It is difficult, at present, to understand how the host plant DNA uses viral RNA to build more viral RNA and viral protein rather than continuing in the normal pathway of plant DNA to plant RNA and plant protein. One must also account for the genetic specificity of the viral RNA. The viral RNA has been related to RNA particles present in the cytoplasm and known as **ribosomes.** Since these particles have not been regarded as having genetic continuity, it is difficult to associate the replication of specific RNA with them. However, plant viruses are not as specific as animal viruses and phages; we have already noted the wide range of hosts attributed to certain plant viruses. This lack of specificity may be related to the absence of DNA from the plant viruses. Some of the problems of the replication of viral RNA may be solved if the association of plant viruses with chloroplasts should prove to be widespread. RNA and DNA have been demonstrated in chloroplasts. The replication of viral RNA could thus occur in infected chloroplasts without the aid of the nucleus. At the present time, this account of the RNA replication of plant viruses is highly speculative.

The Bryophytes

The phylum Bryophyta includes the mosses and liverworts. Although all bryophytes are morphologically distinct from the thallophytes, it is difficult to separate them from the algae in general on the basis of any single character. Perhaps the two greatest differences between the algae and bryophytes are the general structure of the plant bodies and gametangia. It will be recalled (page 447) that the plant body of the thallophytes is generally composed either of single cells (or filaments of cells), or of intertwining filaments, resulting in a more or less complex body. Layers of parenchyma tissue occur in some forms. With the exception of one stage in the life history of the mosses, the bryophyte plant body is never filamentous. It is composed of blocks or sheets of cells forming a parenchymatous tissue (see page 120).

GENERAL CHARACTERISTICS

The gametangia of the thallophytes are unicellular or, if multicellular as in *Ectocarpus,* lack a protective jacket of sterile cells. The gametangia of the bryophytes, on the other hand, are multicellular and always have a protective jacket of sterile cells. Each and every cell of the multicellular structure of *Ectocarpus* produces gametes. However, the multicellular female gametangium characteristic of all members of the Bryophyta produces but a single gamete, the egg, which is surrounded by a layer of protecting cells. The male gametangia of both Ectocarpus and the bryophytes produce numerous sperms, but the sperm-producing cells of the bryophytes are always surrounded by a protective layer of parenchyma cells. These differences are shown in Table 26.1 (see page 575).

Water is required by most algae and bryophytes to effect fertilization. A definite alternation of generations occurs in all members of the Bryophyta, and the sporophyte (diploid) generation is always more or less dependent on the gametophyte (haploid) generation. In all Bryophyta an **embryo sporophyte,** consisting of a spherical or elongated mass of tissue, is formed directly after the germination of the zygote (Figs. 26.2D, 26.3C).

Asexual spores are not formed in Bryophyta. Members of this phylum reproduce asexually either by fragmentation of the gametophyte or by special

$$\text{Gametophyte } (n) \text{ (haploid)} \xrightarrow{\text{egg } (n)}_{\text{sperm } (n)} \text{zygote } (2n) \rightarrow \text{sporophyte } (2n) \text{ (diploid)} \rightarrow \text{meiospores } (n) \rightarrow \text{gametophyte } (n) \text{ (haploid)}$$

bodies known as **gemmae** (Fig. 26.8*B*).

The bryophytes occur on soil, on the trunks of trees, on rocks, and many are truly aquatic.

Alternation of Generations

As just mentioned, all Bryophyta have a distinct alternation of generations. A haploid plant, producing gametes, alternates with a diploid plant; meiosis occurs in the diploid plant, and haploid spores containing *n* chromosomes in each nucleus are formed. Thus a haploid gamete-producing plant—the **gametophyte**—regularly alternates with a diploid spore-producing plant—the **sporophyte,** as shown in the above diagram.

Formerly, it was customary to call the gametophyte a sexual plant and the sporophyte an asexual plant. We have seen, however, that fertilization and meiosis are but two phases of a single sexual cycle. It is believed that the sporophyte plant may have originated when unknown factors delayed meiosis so that this process occurred, not immediately upon the germination of the zygote, as it does in many algae, but only after many mitotic cell divisions had intervened. These cell divisions intervening between fertilization and meiosis have resulted in the sporophyte plant. We shall see in our study of the phyla of vascular plants (page 577) how the interval between fertilization and meiosis lengthens, resulting in larger and more complex sporophytes. Because of this intimate relationship between fertilization and meiosis, the spores formed by the sporophyte as a result of meiosis are one stage of the sexual cycle. They do not, as do the asexual zoospores of *Saprolegnia* or *Vaucheria* or the uredospores of wheat rust, reproduce an exact replica of the plant that produced them. Thus we should consider both the sporophyte and gametophyte as two phases in a complete sexual life history.

Classification

The Bryophyta are divided into three classes, the Hepaticae (liverworts), the Anthocerotae (*Anthoceros*), and the Musci (mosses). The Hepaticae are the most primitive bryophytes. The simplest liverworts consist of a flat, ribbonlike, green thallus which produces gametes and a meiospore-producing

capsule or sporangium embedded within the gametophyte thallus. The name liverwort is very old, having been used in the 9th century. It was probably applied to these plants because of their fancied resemblance to the liver and the belief that plants resembling human organs would cure diseases of the organs they resembled. At any rate, a prescription for a liver complaint in the 1500's called for "liverworts soaked in wine."

Less primitive Hepaticae have simple stems and leaves and the sporophytes are borne above the green gametophyte. The mosses are probably the best known class of the Bryophyta. They have either an upright or a prostrate leafy gametophyte, and may form extensive mats on moist, shady soil. The sporophyte is raised above the leafy gametophyte. *Anthoceros* has the simplest gametophyte of the group but its sporophyte has a meristematic region, which is a type of tissue more characteristic of higher forms than of the Bryophyta.

CLASS HEPATICAE

General Characteristics and Distribution

The Hepaticae are commonly known as the liverworts. There are some 8500 species of liverworts. The great majority of them grow in moist, shady localities. The gametophyte is the prominent plant, and the sporophyte is usually partially parasitic on the gametophyte. The gametophyte is green and may grow either as a flat ribbon or as a leafy shoot. In either event the plant body is frequently called a thallus, even though its internal structure does not correspond to that of the thallophytes. Of the four orders in the Hepaticae, we shall consider briefly the characteristics of two genera, *Riccia* and *Marchantia*, both belonging to the order Marchantiales.

The gametophytes of the Marchantiales are small, green, ribbon-shaped plants (Figs. 26.1, 26.4*A*, 26.5*A*). They branch regularly by a simple forking at the growing tip, resulting in a number of Y-shaped branches. In some species a rosette may be formed. Air chambers that open by a pore and that are formed by, or contain, special cells adapted

Figure 26.1. Habit of, A, *Ricciocarpus*, and B, *Riccia*, ×1. (Courtesy of W. Russell.)

for photosynthesis, form the upper portion of the thallus (Figs. 26.2*A*, 26.4*B*). Several types of storage cells generally make up the lower half. **Rhizoids,** specialized elongated cells performing the functions of roots (anchorage, and absorption of water and mineral nutrients), extend downward from the lowermost layer of cells. The thallus is thus differentiated into distinct upper and lower portions. Scales, frequently brown or red, are formed on the lower surface (Fig. 26.2*A*).

Riccia

Riccia is a widely distributed genus, and, although it requires water for active growth, most species will withstand considerable drought. Several species are aquatic, growing either on mud or on the surface of small ponds.

Gametophyte. The gametophyte is a small green thallus, frequently forming a rosette (Fig. 26.1). The tissue on the lower side is composed of colorless cells that may contain starch. The tissue on the upper side consists of vertical rows of chlorophyll-bearing cells, between which are air chambers (Fig. 26.2*A*). The gametangia are embedded in deep, lengthwise depressions or furrows, in the dorsal surface of the thallus. Antheridia and archegonia are usually found on the same gametophyte.

Antheridia. The antheridia of different representatives of the Bryophyta are similar in structure, though they may vary somewhat in shape. In *Riccia* they are pear-shaped and composed of two types of cells, (1) fertile and (2) sterile (Figs. 26.3*A*, 26.2*B*). Fertile cells give rise to sperms, which are relatively numerous, small, and dense with protoplasm. The sterile cells form a protective jacket, one cell in thickness, around the fertile cells. The antheridia of *Riccia*, and of other Bryophyta, are thus composed of a mass of fertile cells, all of which develop into sperms, and a protective jacket of sterile cells. The antheridia are connected to the gametophyte by a short stalk.

Mature sperms consist mainly of an elongated nucleus with two long flagella. They may be shot from the mature antheridium with considerable force, or they may be extruded slowly in a single gelatinous mass. In any event they do not leave the antheridium until enough moisture is present to allow them to swim about.

Archegonia. The archegonium of *Riccia* is a flask-shaped structure consisting of two parts, (1) an expanded basal portion, the **venter,** and (2) an elongated **neck** (Fig. 26.2*C*). Four **cover cells** are located at the top of the neck. Each archegonium contains a single egg cell, which is located in the

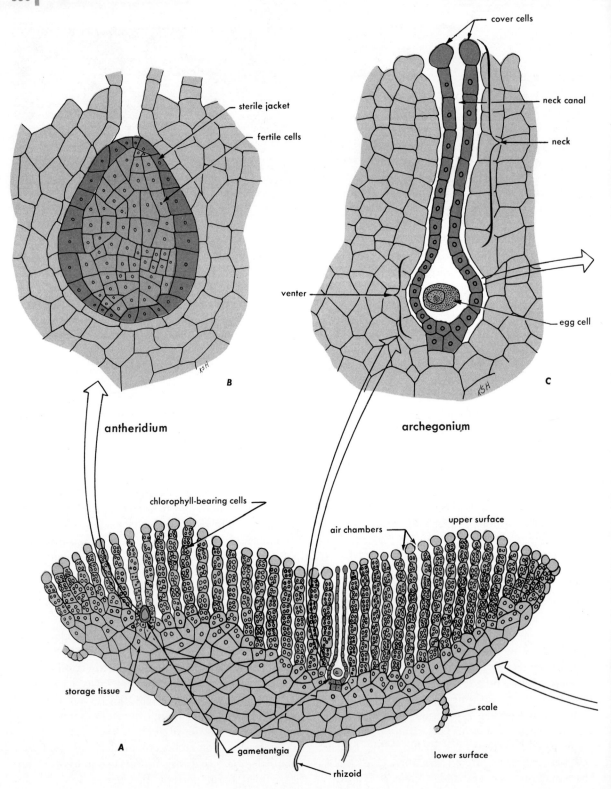

Figure 26.2. *Riccia, A,* cross section through a thallus; gametangia are on the upper surface of the thallus between the photosynthetic filaments or in a notch in the thallus, ✕8; *B,* an antheridium; note jacket of sterile cells surrounding the developing sperm cells; *C,* archegonium on thallus, neck canal open,

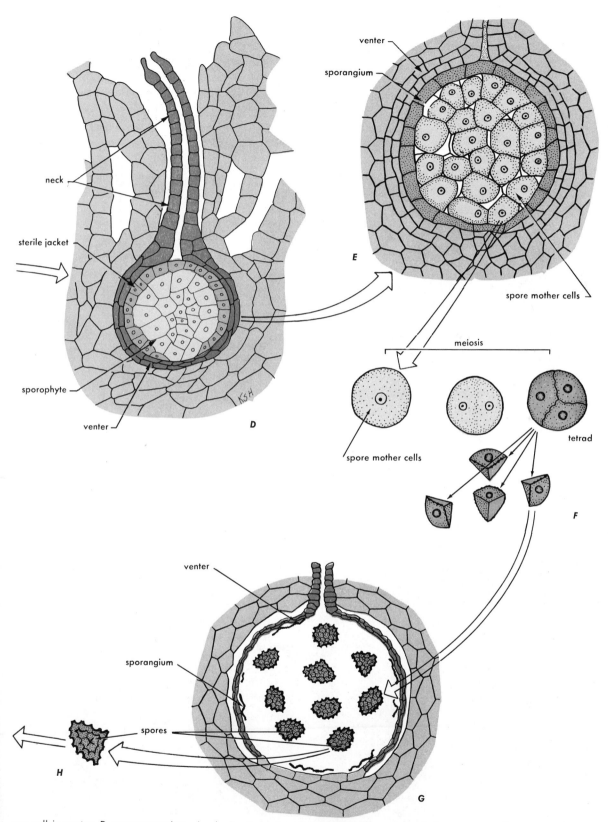

egg cell in venter; *D*, young sporphyte developing in ventor of old archegonium; *E*, sporophyte with spore mother cells still enclosed in venter of old archegonium; *F*, meiosis occurs within the sporophyte; *G*, the venter now contains fragments of the old spore case (sterile cells of the sporophyte) and spores; *H*, a spore; *B*, through *G* about ×50; *H*, ×100.

venter. A short stalk attaches the archegonium to the gametophyte. Archegonia of a similar structure are found in other Bryophyta.

Shortly before the egg cell is mature, the cover cells separate. At the same time the cells in the center of the neck dissolve, so that an open canal connects the venter with the moisture outside the archegonium (Fig. 26.5). The free-swimming sperms move toward certain chemical substances formed by the archegonium, for they swim to it and then down

Figure 26.3

560   26 *The Bryophytes*

Figure 26.3. *Riccia*, *A*, antheridium, note jacket of sterile cells; *B*, archegonium in notch; *C*, developing sporophyte within venter of old archegonium; *D*, nearly mature sporophyte, meiosis has occurred but spores have not separated; *E*, mature spores in venter of old archegonium; all about ×100.

the canal opened by the dissolution of the neck canal cells. This response to chemical stimuli is known as a **chemotactic response.** Several sperms may enter the archegonium, but only one fertilizes the egg in the venter.

Sporophyte. As a result of the fusion of sperm and egg nuclei, the zygote contains the diploid, or 2*n*, number of chromosomes. Mitosis now proceeds in a more or less orderly manner until a spherical mass of some 30 or more similar cells is formed (Figs. 26.2*D*, 26.3*C*). These cells are partially dependent upon the gametophyte. This mass of undifferentiated, parasitic cells comprising the young sporophyte is an **embryo.** It is located within the venter of the archegonium.

Further development of the embryo involves differentiation of an outer layer of cells to form a protective jacket of sterile tissue surrounding a mass

of cells that are capable of forming spores. This fertile or spore-forming tissue is called **sporogenous tissue.** The sporogenous cells continue to divide by ordinary cell division until many have formed. All these cells are **spore mother cells,** and will undergo meiosis, producing spores with the haploid number of chromosomes (Fig. 26.2*E* to *H*).

Meiosis consists of two cell divisions, during which the number of chromosomes is halved. Since two divisions are involved, four spores are formed from each spore mother cell. Each group of four spores, formed from a single spore mother cell, is usually referred to as a **tetrad** or quartet of spores.

The mature sporophyte of *Riccia* (Figs. 26.2*D*, 26.3*E*) is called a **sporangium** or **capsule.** It is composed of a jacket of sterile cells enclosing a mass of spores. The sporophyte wall breaks down when the spores are mature. These spores remain em-

$$\text{Gametophyte } (n) \begin{array}{c} \nearrow \text{egg } (n) \searrow \\ \\ \searrow \text{sperm } (n) \nearrow \end{array} \text{zygote } (2n) \rightarrow \text{embryo sporophyte } (2n) \rightarrow$$

mature sporophyte $(2n) \rightarrow$ spore mother cells $(2n) \rightarrow$ spores $(n) \rightarrow$ gametophyte (n)

bedded in the gametophyte thallus and are not released until after the death and decay of the gametophyte. Furthermore, the sporophyte is dependent on the gametophyte for its nutrients. When the spores germinate, each grows into a typical gametophyte plant.

The complete life history of *Riccia* may be diagrammed as shown above.

Significant steps in the life history of Riccia

1. The gametophyte, a small, green, flat plant, absorbs water and mineral salts from the soil and carbon dioxide from the air. It contains chlorophyll and can synthesize food. All gametophytic nuclei are haploid (contain n chromosomes).

2. The gametangia (antheridia and archegonia) develop in a furrow on the upper surface of the gametophyte.

3. Each antheridium produces thousands of sperms.

4. A single egg is formed in the venter of each archegonium.

5. The gametes (eggs and sperms) contain n chromosomes.

6. When sufficient moisture is present, the mature sperms are extruded from the antheridium. The neck of the mature archegonium is opened and sperms swim down to the egg in the venter.

7. Each egg is fertilized by a single sperm.

8. The zygote, as a result of the union of two haploid gametic nuclei, contains $2n$ chromosomes.

9. An embryo sporophyte, consisting of a spherical mass of undifferentiated cells, partially parasitic upon the gametophyte, develops from the zygote. Each nucleus of the embryo sporophyte contains $2n$ chromosomes.

10. The cells of the embryo differentiate into a sporangium consisting of a jacket of sterile cells enclosing a mass of fertile cells.

11. Spore mother cells, each with $2n$ chromosomes, form in the sporangium.

12. The spore mother cells undergo meiosis. A quartet of spores forms from each spore mother cell. All the spore mother cells contain $2n$ chromosomes, and the spores n chromosomes.

13. Upon the death and decay of the old gametophyte, spores are liberated and new gametophytes are developed from them.

Marchantia

Marchantia may grow in large mats on moist rocks and soil in shady locations. It is widely distributed and fairly common on the banks of cool streams. It is somewhat better adapted to growing on land than is *Riccia*, but considerable moisture is still required for active growth and for fertilization.

Gametophyte. The gametophyte of *Marchantia*, and of other members of the family to which it belongs, differs from that of *Riccia* in that the gametangia are on special disks raised some distance above the vegetative thallus (Figs. 26.4*A*, 26.5*A*). The thallus, although having the same general appearance as that of *Riccia*, is coarser and larger. On its upper surface are polygonal areas, with a small but conspicuous pore in the center. These areas, with their air pores, mark the outlines of air chambers, each filled with short filaments of cells containing chloroplasts. The thallus has a midrib (Fig. 26.4*A*). As in *Riccia*, the lower surface of the thallus is composed of colorless cells, some of which are modified for storage. Rhizoids, which anchor the thallus and serve as organs for absorption, grow from the cells covering the lower surface. Several rows of scales are also attached to this lower surface.

Asexual reproduction. *Marchantia* reproduces asexually in two ways. (1) Older parts of the thallus die and the younger portions, no longer attached, develop into new individual plants; and (2) small cups, known as **gemmae cups,** form on the upper surface, and small disks of green tissue, called **gemmae,** grow from the bottom of these cups (Fig. 26.8*B*). The gemmae, when mature, break off from the thallus and are distributed into nearby areas. New gametophyte plants grow from the gemmae.

Sexual reproduction. The gametangia are very similar in structure to those of *Riccia*. The antheridia are pear-shaped bodies composed of a jacket of sterile tissue surrounding the sperms (Fig. 26.4*B*). They are borne on disks raised above the thallus on slender stalks, called **antheridiophores** (Figs. 26.4*A*, *B*). The antheridiophores are modified portions of the thallus having furrows, rhizoids, and

Figure 26.4. *Marchantia, A,* male plants, note strap-shaped thallus, with midrib. The antheridial discs are an upward extension of the midrib, antheridia are borne in ridges on top of antheridial discs, ×1; *B,* section through an antheridial disc showing a nearly mature antheridium, note the sterile jacket and the maturing sperm cells. The channel to the surface of the disc is just visible. Also note the air spaces with photosynthetic filaments and pore opening to the atmosphere, ×100.

air chambers. The antheridia develop in cavities on the upper surface of the disk, the youngest antheridia being close to the outer margin of the disk. The mature sperms are extruded in a mucilagenous mass.

The archegonia are flask-shaped and have the same structures as those of *Riccia*—the venter, the neck, and the cover cells (Fig. 26.5*C*).

The antheridia and archegonia of *Marchantia* are found on separate plants; that is, the gametophytes are either male or female. The archegonia are borne on specialized branches called **archegoniophores.** The disk at the top of the archegoniophore is frequently an eight-lobed structure. In development, the lobes bend downward and then grow inward toward the stalk. Thus the youngest portion of an older disk bearing archegonia is on the underside and close to the stalk (Fig. 26.5*B*). The archegonia develop in the lower surface of the disk but in tissue similar to that found on the upper surface of the thallus. Striking fingerlike processes grow out from

between the lobes. The archegonia mature, and fertilization takes place when the disk of the archegoniophore is but slightly elevated above the thallus.

Sporophyte. The zygote develops, as in *Riccia,* into the embryo, which is a spherical mass of undifferentiated, colorless, dependent tissue (Fig. 26.6*A*). It is diploid, the nuclei of all its cells containing 2*n* chromosomes. Subsequent growth, however, is more complicated than in *Riccia*. Some cells of the embryo divide to form a mushroom-like growth that becomes embedded in the gametophyte tissue. This growth is called the **foot.** The cells in the central portion of the embryo form a stalk or **seta.** As the seta elongates, the lower cells develop into a **sporangium** (Fig. 26.6*B*). The lengthening of the seta suspends the sporangium below the disk. While this development is taking place, the stalk of the archegoniophore elongates, lifting the disk with its archegonia well above the thallus (Figs. 26.5*A,* 26.7).

Fertilization stimulates the enlargement of the old

archegonium, which keeps pace with the enlargement of the developing sporophyte. As a result, the sporophyte is continually enclosed within the archegonium, which, because of its increase in size and change in function and shape, is now known as the **calyptra.** In addition, the surrounding gametophyte tissue produces two other envelopes that protect the sporophyte.

The mature sporophyte (Fig. 26.7A) is composed of three parts, the **foot,** the **seta,** and the **sporangium** (Figs. 26.6B, 26.7A). The foot is an absorbing organ. The seta serves to lower the sporangium away from the archegonial disk and thus to facilitate the distribution of the spores. Before meiosis the sporangium or capsule is composed of a jacket of sterile cells surrounding a mass of spore mother cells, among which are a number of sterile elongated cells (Fig. 26.7A). Four spores, each containing n

Figure 26.5. *Marchantia,* A, female plants, note the thallus, the midrib is not so apparent. The light specks arrayed in rows on the thallus are the pores shown in Fig. 26.4B. The archegonia are on the lower surface of the disc, in the white perianth, between the radiating fingers, ×1; B, section through a disc and portion of supporting archegoniophore. Sections are shown through several archegonia, the youngest archegonium is closest to the archegoniophore and, in this section, is just visible, the second oldest is distinct and probably holds a fertilized egg cell, ×50; C, two argechongia, one, very young and undifferentiated, the second is mature and shows the venter with an egg cell enclosed within it, a ventral canal cell, the neck and neck canal cells, ×200.

Figure 26.6. *Marchantia.* A, the developing sporophyte assumes a spherical shape within the venter, the old archegonium is evident; B, the developing sporophyte still enclosed within the archegonium shows three specialized regions; the foot, the seta, and the capsule, ×50.

Figure 26.7. *Marchantia.* A, (*left*) section through a nearly mature sporophyte showing the absorbing organ, the foot, and the elongating seta, and the capsule, the latter with a jacket of sterile cells, meiospores, and elaters; B, a view of an archegonial disc with two sporophytes, the spores are being shed from the capsule of one, while the capsule of the other has not yet opened, ×2.

Class Hepaticae

Figure 26.8. *Marchantia.* A, section through a small portion of the lower end of a capsule. Spores may be seen, both within, and outside of the capsule, elaters are also present, ×200; B, spores landing in a favorable location will germinate to form a flat thallus which as shown here, may reproduce asexually through germmae formed in germmae cups, ×1/20.

chromosomes, develop from each spore mother cell. The elongated sterile cells are transformed into spiral elements that change shape under the influence of varying moisture conditions. These cells are called **elaters** and they aid in the dispersal of the spores (Fig. 26.8A).

The dissemination of spores is aided by two structural features not found in *Riccia:* (1) the sporangium of the sporophyte hangs from the lower side of the raised archegonial disk (Fig. 26.7B), and (2) elaters help to empty the spore case. The spores develop immediately into new gametophytes.

The life history of *Marchantia* can be represented as shown.

Significant Steps in the
Life History of Marchantia

1. The gametophyte thallus absorbs water and mineral salts from the soil and carbon dioxide from the air. It is green and can synthesize foods. All nuclei of the gametophyte have *n* chromosomes.

2. The gametangia (antheridia and archegonia) are borne on upright branches called antheridiophores and archegoniophores.

3. The zygote is diploid, containing 2*n* chromosomes.

4. The embryo sporophyte, consisting of a spherical mass of undifferentiated cells, is formed from the zygote.

5. The cells of the embryo sporophyte differentiate into a foot, a seta, and a sporangium.

6. The foot is an absorbing organ. It is embedded in

$$\text{Gametophyte } (n) \; {\nearrow \text{ egg } (n) \searrow \atop \searrow \text{ sperm } (n) \nearrow} \; \text{zygote } (2n) \rightarrow \text{embryo sporophyte } (2n) \rightarrow$$

mature sporophyte (2*n*) → spore mother cells (2*n*) → spores (*n*) → gametophyte (*n*) the gametophyte, from which it receives nourishment.

7. The sporangium consists of a jacket of sterile cells surrounding a mass of sporogenous tissue interspersed with elongated sterile cells.

8. The seta is a stalk that, in lengthening, lowers the sporangium below the surface of the archegonial disk.

9. Repeated mitotic cell divisions in the sporogenous tissue result in spore mother cells, each containing $2n$ chromosomes.

10. The spore mother cells divide by meiosis into quartets of spores, each containing n chromosomes.

11. The elongated sterile cells are transformed into spiral elaters.

12. The presence of the elaters and the hanging position of the sporophyte aid in disseminating the spores.

13. The spores germinate immediately into new gametophyte plants.

CLASS ANTHOCEROTAE

The Anthocerotae have the simplest gametophytes of the Bryophyta. They are small, green thallus plants with little internal differentiation of vegetative tissues. They are slightly lobed with numerous rhizoids growing from the lower surface (Fig. 26.9). The antheridia are similar in structure to those encountered among the Hepaticae. They are located in roofed chambers in the upper portion of the thallus. The archegonia are embedded within the thallus and in *direct contact* with the vegetative cells surrounding them (compare Figs. 26.2*C* and 26.10).

The sporophyte (Figs. 26.9, 26.11) of the Anthcerotae is in striking contrast to those of the Hepaticae, The subepidermal cells contain chloroplasts, and typical stomata are found in the epidermis. A foot embedded in the thallus serves as an absorbing organ. The sporangium is an upright elongated structure. The sporogenous tissue forms a cylinder parallel with the elongated axis of the sporangium. The spores mature not all at once but in progression from the top down. A meristematic region lies just above the foot and continually adds new cells to the base of the sporangium. Its presence means that spores may be produced over long periods.

Under exceptionally favorable growing conditions, the sporophyte may lengthen greatly. Some sporogenous tissue at the base of the sporangium may be replaced by a conspicuous conducting strand. The foot enlarges and, through decay of the gametophyte, comes into more or less direct contact with the soil. Such sporophytes are capable of

Figure 26.9. *Anthoceros* gametophyte with upright, dependent sporophyte. ×3.

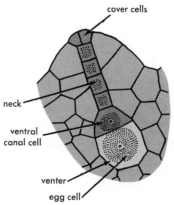

Figure 26.10. Section through an archegonium of *Anthoceros*, ×50.

maintaining themselves independently for some time.

CLASS MUSCI

General Characteristics and Distribution

We have seen that the liverworts are small and inconspicuous plants; they are not generally noticed where they grow on some stream bank or moist roadside. Although the mosses are small plants, they are nevertheless conspicuous. They frequently cover rather large areas of stream banks. They grow on rocks, on trees, and sometimes submerged in streams. Some are able to resist considerable drought, but, like the liverworts, all require moisture for active growth and reproduction.

The mosses as a class show great structural uni-

Figure 26.11. (*right*) Longitudinal median section through sporophyte of *Anthoceros*. (Redrawn from R. M. Holman and W. W. Robbins, *A Textbook of General Botany*, John Wiley and Sons.)

Figure 26.12. *Mnium*, showing location of gametangia, ×1.

formity. The gametophytes of nearly all species have two growth stages: (1) a creeping, filamentous stage (the **protonema**), from which is developed (2) the moss plant with an upright or horizontal stem bearing small, spirally arranged green leaves (Figs. 26.12, 26.13). Rhizoids are found at the base of the stem, whereas gametangia occur at the tips of either the main or lateral branches (Fig. 26.12).

On the protonema, leafy shoots arise from buds (Fig. 26.14), and a protonema derived from a single spore may give rise to many moss shoots, each with numerous rhizoids at its base. The rhizoids, unlike those found in most of the liverworts, are composed of filaments of cells. They absorb water and solutes, anchor the leafy shoot to the soil and may give rise to protonemata under favorable conditions.

The mature sporophyte, like that of some of the Hepaticae, is composed of a foot, seta, and sporangium. Unlike the Hepaticae sporophyte, it has a greater amount of sterile tissue and the capsule is considerably more complex (Fig. 26.20).

Although there are three orders of mosses, their structural uniformity is such that a single life history is sufficient to illustrate the general features of development in this class.

Gametophyte

The germinating spores of mosses do not develop directly into a leafy gametophyte but first become a filamentous structure. This early stage of gametophyte is called the **protonema** (Fig. 26.14). The protonema is not a permanent structure although it may branch considerably under favorable conditions and cover a rather large area of soil, sometimes forming a green coating resembling algal growth. The cells composing the protonema contain numerous chloroplasts.

The mature stem of the leafy gametophyte is differentiated into a **central cylinder,** a **cortex,** and an **epidermis.** Recent observations indicate that certain cells of the central cylinder (Fig. 26.15*A*) may function in the upward movement of water and solutes. There are, however, no vessels or tracheids such as occur in the vascular plants. Some cells are strikingly similar to sieve elements. (Fig. 26.15*B*).

Sexual reproduction. The gametophytes of many mosses are **monoecious;** that is, both antheridia and archegonia are produced by the same gametophyte. (Fig. 26.12). Some mosses are **dioecious;** in other

Figure 26.13. *Mnium,* showing gametophytes with attached sporophytes.

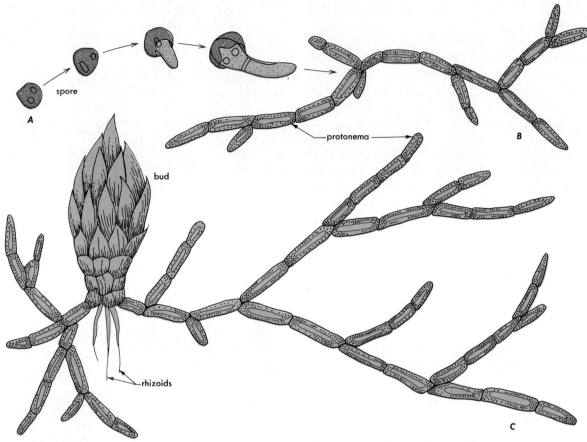

Figure 26.14. *Funaria.* A, germination of spores; B, protonema; C, protonema with bud. (Redrawn from W. P. Schimper, *Recherches sur les mousses*, Strasbourg, 1848.)

words, antheridia and archegonia are produced on separate gametophytes. The gametangia of *Funaria* are formed at the summit of the leafy shoot. First antheridia develop, and then the gametophyte may branch and the anchegonia form at the tips of the branches. In *Mnium* and *Polytrichum* the shoots bearing antheridia are easily recognized, for the leaves surrounding them are spread somewhat like the petals of a flower. The group of antheridia appears as an orange spot in the center of the terminal cluster of leaves (Figs. 26.12, 26.16).

The antheridia of most true mosses (Fig. 26.17) are essentially similar to those of liverworts. The cells forming the sterile jacket of the antheridium contain chloroplasts that become orange-red when the antheridium ripens. As in liverworts, the sperms consist mainly of an elongated nucleus, and each has two long cilia. The antheridia are surrounded by club-shaped, multicellular, sterile hairs with conspicuous chloroplasts. They are called **paraphyses.**

The archegonia differ from those of the Hepaticae in having a longer neck and a longer stalk between the venter and the tip of the gametophyte (Fig. 26.18).

Mosses, like liverworts, are dependent upon the presence of water to effect fertilization as well as the opening of the antheridia and the archegonia.

Sporophyte

Soon after fertilization the zygote begins to develop into a spindle-shaped embryo that differentiates into a sporophyte consisting of a foot, seta, and sporangium or capsule. The foot penetrates the base of the venter and grows into the apex of the leafy

Figure 26.16. Antheridial heads of *Polytrichum*, ×1.

Figure 26.15. Longitudinal sections through gametophyte stems of mosses. *A*, stem of *Mnium* showing central strand of sclerenchyma-like cells, ×100; *B*, stem of *Polytrichum* with phloem-like cells, ×100. (*B*, courtesy of E. M. Gifford, Jr.)

Figure 26.17. Antheridial head of *Mnium*, ×35.

Figure 26.18. Anchegonium surrounded by paraphyses (*Mnium*), × 50.

shoot. It absorbs water and nourishment from the gametophyte for the growth and development of the sporophyte. The seta elongates rapidly, raising the sporangium one-half inch or more above the top of the leafy gametophyte (Figs. 26.13, 26.19). The old archegonium increases in size as the sporophyte enlarges. When the seta elongates, the top of the expanded archegonium is torn from its point of attachment to the gametophyte and elevated with the sporangium. The old archegonium, now known as the calyptra, remains for some time as a covering for the sporangium.

The sporangium of mosses is larger and more complex than that of the Hepaticae. It may measure from a sixteenth to an eighth of an inch in diameter and twice that in length. It is surrounded by an epidermal layer composed of cells similar to those in the epidermis of higher plants. Stomata occur in the epidermis covering the lower half of the sporangium. The sterile tissue forming the inner portion of the sporangium may conveniently be divided into three regions, each of which may be recognized by the type of cells comprising it. A fourth region is formed from the fertile sporogenous cells (Fig. 26.20). The sterile regions may be briefly characterized as follows. (1) The cells comprising the base of the sporangium contain chloroplasts and are fairly compact; (2) the cells forming the outer region

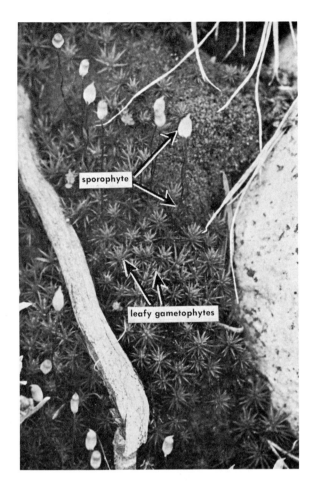

Figure 26.19. *Polytrichum*, showing leafy gametophytes with sporophytes.

of the upper half of the sporangium also contain chloroplasts, but these cells are rather loosely associated, large air chambers being regularly formed; and (3) the cells forming a core in this upper portion are devoid of chloroplasts and are compact, forming a mass of thin-walled cells referred to as the **columella.** The fourth region, composed of sporogenous cells, forms a layer around the columella. As in the liverworts, the cells of the sporogenous tissue may increase in number by mitotic cell division. Spore mother cells, each containing the diploid number of chromosomes, develop from the sporogenous cells. Meiosis takes place in spore mother cells, and spores containing n chromosomes result. They are haploid,

the first cells of the gametophyte generation.

The columella projects upward, forming a small dome above the main mass of the sporangium. The four or five outer layers of cells of this dome differentiate into a dry, brittle cap called the **operculum** (Fig. 26.20). The cells immediately beneath the operculum form a double row of long, triangular teeth called the **peristome.** The broad bases of the teeth are attached to the thick-walled deciduous cells forming the **annulus** around the upper end of the sporangium. When the sporangium matures and becomes dry, thin-walled cells holding the operculum in place break down, allowing the operculum to fall away and thus expose the teeth of the peri-

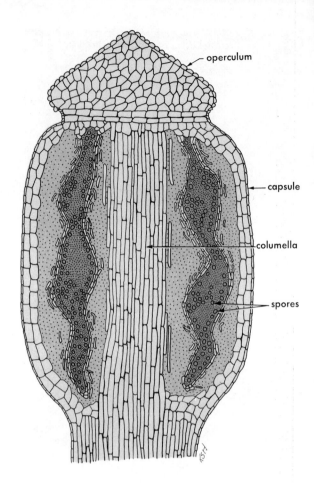

operculum

capsule

columella

spores

Figure 26.20. Median section through a mature but unopened capsule of *Mnium*, ×25.

stome. By this time most of the thin-walled cells within the columella have collapsed and the cavity thus formed is filled with a loose mass of spores.

The peristome teeth are rough and are very sensitive to the amount of moisture in the air. When they are wet or the atmospheric humidity is very high, they bend into the cavity of the sporangium; when dry, they straighten and lift out some of the spores, which are then disseminated by air movements. If a spore comes to rest on moist soil and if illumination and temperature are favorable, it germinates and grows into a protonema.

The life history of *Funaria* may be represented as shown:

$$\text{Gametophyte } (n) \rightarrow \text{leafy plant } (n) \nearrow \overset{\text{egg } (n)}{\searrow} \underset{\text{sperm } (n)}{\nearrow} \text{zygote } (2n) \rightarrow$$

embryo sporophyte $(2n) \rightarrow$ mature sporophyte $(2n) \rightarrow$

spore mother cells $(2n) \rightarrow$ spores $(n) \rightarrow$ gametophyte (n)
(protonema)

Table 26.1 *Tabulation of differences between thallophytes and bryophytes*

THALLOPHYTES	BRYOPHYTES
1. Mostly aquatic	1. Mostly terrestrial
2. A few of the most advanced have sieve-like elements	2. Some mosses have cells suggesting sieve cells; otherwise, food conducted in relatively undifferentiated cells
3. None have water-conducting elements	3. Simple water-conducting cells
4. In general, no specialized water- and nutrient-absorbing tissue. Rhizoids and haustoria may occur in some fungi	4. Rhizoids anchor and absorb water and mineral nutrients
5. Mostly filamentous or a lacework of intertwining filaments; parenchyma in a few	5. Only one stage of mosses is filamentous; all others are formed of parenchyma cells
6. A definite alternation of generations in many forms	6. All have an alternation of heteromorphic generations
7. Both sporophytes and gametophytes independent	7. Gametophyte independent; sporophyte dependent
8. Sporophyte frequently large and complex	8. Sporophyte small and relatively simple
9. Gametangia are either single cells or groups of single cells not accompanied by a jacket of sterile vegetative cells	9. Gametangia always composed of the gamete-producing cells protected by a jacket of sterile vegetative cells

Significant Steps in the Life History of a Moss

1. The gametophyte consists of (*a*) a filamentous, branched, algal-like growth called a protonema and (*b*) leafy shoots that develop from buds on the protonema. The shoots consist of a stalk bearing rhizoids at its lower end and leaves throughout its length. The gametophyte is green and able to synthesize food. All gametophyte nuclei are haploid.

2. The antheridia and archegonia develop at the apex of the leafy gametophyte.

3. Each antheridium produces hundreds of sperms.

4. A single egg is formed in the venter of each archegonium.

5. The gametes (eggs and sperms) each contain n chromosomes.

6. When sufficient moisture is present, the mature sperms are extruded from the antheridium. The neck of the mature archegonium is open and the sperms swim down it to the egg in the venter of the archegonium.

7. The egg is fertilized by one sperm.

8. The zygote, as a result of the union of gametic nuclei, contains $2n$ chromosomes.

9. An embryo sporophyte, consisting of a spindle-shaped mass of undifferentiated cells, parasitic on the gametophyte, develops from the zygote.

10. The embryo develops into a sporophyte consisting of a foot, a seta, and a sporangium. The sporangium of most mosses is more complex than those found in the Hepaticae and the Anthocerotae. In addition to the sporogenous tissue, it contains several types of sterile tissues, among which may be mentioned the operculum, annulus, and peristome.

11. Spore mother cells, each containing $2n$ chromosomes, form from the sporogenous cells.

12. The spore mother cells undergo meiosis. Four spores result from each mother cell.

13. When spores germinate they form the protonema.

CLASSIFICATION

Phylum	Bryophyta
Class	Hepaticae
Order	Marchantiales
Genera	*Riccia*
	Ricciocarpus
	Marchantia
Class	Anthocerotae
Genus	*Anthoceros*
Class	Musci
Genera	*Funaria*
	Mnium
	Polytrichum
	Sphagnum

27

The Lower Vascular Plants

In contrast to the algae, fungi, liverworts, and mosses, the vascular plants possess a well-developed vascular system that serves for the conduction of water, mineral salts, and foods. Most of them are land plants, but some require free water for fertilization. The sporophyte is the dominant generation; except for the youngest stages in the formation of an embryo, it is independent of the gametophyte. The gametophytes are small, and in the conifers and flowering plants they are dependent upon the sporophyte. The plant body of the sporophyte is, in its simplest form, an axis, with meristematic tissue terminating the opposite ends, which form shoots or roots.

CLASSIFICATION

There are ten phyla of vascular plants, six of which are seed-bearing and four of which do not bear seeds. These latter four phyla, the lower vascular plants, are the **Psilophyta,** and **Lycophyta,** and **Sphenophyta** and the **Pterophyta.** The number of families and genera in these phyla is small and their members do not form a very conspicuous part of the current land flora. The Pterophyta is currently the largest of the four phyla. It comprises the ferns, which are prominent in certain cool moist habitats, However, plant fossils indicate that there was a period in the earth's history when the members of the Lycophyta and Sphenophyta formed a large and dominant flora.

That changes have occurred in the vegetation of the earth's surface is a well-established fact; beautifully preserved portions or imprints of plants are found buried in many types of sedimentary rocks, in coal deposits, and in peat bogs. Such plant remains are called **fossils** (Fig. 2.6). Sometimes fossils appear to be remains of plants almost identical with some living today. For instance, although the Coast redwoods of California are now confined to a very narrow fog belt along the coast of northern California and southern Oregon, fossil redwoods show that they were once prevalent in the Northern Hemisphere (Fig. 27.1). Fossil vascular plants differ in greater or lesser degree from present-day forms. All of them, however, can be classified, and it is possible to draw conclusions regarding their relationship with existing plants.

Figure 27.1. Distribution of fossil and living redwoods, *Metasequoia* and *Sequoia*. (Courtesy of R. W. Chaney.)

Some of these fossils will be discussed under the headings of their appropriate phyla.

PHYLUM PSILOPHYTA

This phylum is represented in the existing flora by a single family, the Psilotaceae, comprised of two genera, *Psilotum* and *Tmesipteris.* They are rare plants found mainly in the tropics, one form of *Psilotum* growing as far north as Florida. Figure 27.2 shows the simple plant body of *Psilotum nudum.* Roots are not present and the leaves are small. The branching, upright stem is slightly flattened and contains chlorophyll. A fungus is always associated with the branched rhizome, which is clothed with many rhizoids. The vascular tissue is simple (Fig. 27.3), consisting of poorly developed phloem and

of xylem in a radial arrangement. An endodermis, with a conspicuous Casparian strip, separates the central vascular tissue from the outer cortex. The xylem is composed of spiral and scalariform tracheids. Sporangia are borne in the axils of some of the scale-like leaves (Fig. 27.2*B*).

Fossil Psilophyta are quite well known. They are found in the older sedimentary rocks, indicating that the plants flourished well over 300 million years ago. In spite of their great age their structure may be beautifully preserved. The presumed habit of one species, *Rhynia major,* and a cross section of its stem, are shown in Figs. 27.4, 27.5. Note the remarkable preservation of cellular structure. The plant consisted of a slender rhizome upon which were borne erect, cylindrical stems that branched sparingly. True leaves were absent. Sporangia developed at the tips

Table 27.1 *Comparison of Psilophyta, Lycophyta, Sphenophyta, and Pterophyta*

GENERATION	PSILOTUM	LYCOPODIUM	SELAGINELLA	EQUISETUM	FERNS
Sporophyte	A branching stem, scale leaves, or leaves, no roots	Prostrate branching stem with upright branches, roots and leaves	Prostrate branching stem with upright branches, roots and leaves, also scales	Rhizome, upright, jointed stem, small leaves, stems carry on photosynthesis	Rhizome, roots, and leaves Herbs (mostly) Stem: several strands of vascular tissue with xylem of each surrounded by phloem
	Simple vascular system	Simple vascular system	Simple vascular system but vessels are present	Simple vascular system	Vessels occur in two genera Sieve elements but no companion cells in phloem
	One type of sporangium Homospory No sporophylls	One type of sporangium Homospory Sporophylls present	Megasporangia Microsporangia Heterospory Sporophylls all similar	One type of sporangium Homospory Sporangiophores	One type of sporangium Homospory (mostly) Leaves bear spores in sporangia
	No cone	A cone in some species	Cone	Cone composed of sporangiophores	No cones
		Suspensor	Embryo develops within female gametophyte on sporophyll Suspensor	No suspensor	Embryo attached to gametophyte No suspensor
Gametophyte	Irregular subterranean structure associated with fungus	Irregular to tapered subterranean structure associated with fungus	Male gametophyte, one prothallial cell, and an antheridium within microspore	A single green thallus resembling *Anthoceros*	Heart-shaped, small, green, completely independent thallus
	Gametangia embedded in thallus	Gametangia embedded in thallus	Female gametophyte, small amount of tissue within megaspore, cushion protrudes in which gametangia are embedded	Gametangia embedded in thallus, neck of archegonium protruding	Antheridium of approximately 6–10 cells Archegonia partially embedded in gametophyte
	Motile sperms	Motile sperms	Motile sperms	Motile sperms	Motile sperms

of some branches, and typical quartets of spores suggesting normal meiosis have been found. Rhizoids growing from the rhizome absorbed moisture and nutrients and anchored the plant to the soil.

PHYLUM LYCOPHYTA

The Lycophyta are represented by three families (Lycopodiaceae, Selaginellaceae, and Isoetaceae). There is one living genus from each family in our modern flora: *Lycopodium, Selaginella,* and *Isoetes.* We shall consider two genera, *Lycopodium* and *Selaginella.*

There are many fossil Lycophyta belonging to several orders, Lepidodendrales being one of the best known. Representatives of this order were forest trees, some of which bore seeds. The leaves of the living members of this phylum are generally small and usually arranged in a spiral. In flowering plants leaf gaps occur at the juncture of leaf and stem (Fig. 8.16). Pay particular attention to Fig. 9.13*E*, noting the gaps formed by the departure of the leaves as compared with the solid core of xylem in the root. Leaf gaps are never formed at the junction of stem and leaf in the Lycophyta, so that a rotted stem of a member of this phylum would have, like many angiosperm roots, a core of solid xylem (Fig. 9.13*E*).

Lycopodium

Approximately 400 species are in this genus. Most are trailing plants, many forming short upright branches that sometimes recall pine seedlings (Figs.

Figure 27.2. *Psilotum nudum.* A, plant. Notice sporangia on branches at left, ×1/6; B, end of branch showing scale leaves and sporangia at node-like swellings, ×1.

2.1, 27.6). They are frequently called "ground pine" or "clubmoss." Widely distributed, they are found in largest numbers in subtropical and tropical forests. They cannot grow in arid habitats. Several species grow in the eastern and northwestern United States but none occur in the more arid states of the Southwest. Some eastern species are in danger of extinction because of their ready sale in florist shops for Christmas decorations.

Mature Sporophyte

The main stem branches freely and is prostrate. Upright stems, approximately eight inches in height, grow from the horizontal stem (Fig. 27.7*A*, *B*). Both types of stems are sheathed with small green leaves. Small but well-developed adventitious

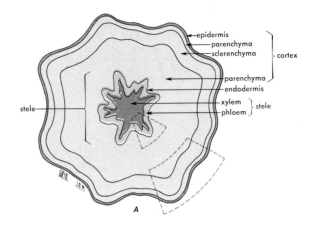

Figure 27.3. Cross section of *Psilotum* stem. *A*, (*above*) diagram to show arrangement of tissues, ×40; *B*, (*below*) enlarged sector showing cellular detail, ×900.

sporangia

upright stems

rhizome

rhizoids

Figure 27.4. Presumed habit of *Rhynia*, a plant that lived over 300 million years ago. (Redrawn from R. Kidston and W. H. Lang, *Trans. Roy. Soc. Edinburgh*, **52,** 831.)

roots rise irregularly from the underside of the horizontal stem (Fig. 27.10*A*). As in many higher plants, the primary root, which grows from the embryo, is not long-lived.

The xylem in a *Lycopodium* stem occurs as a complex system of anastomosing strands, so that its distribution in the stem varies greatly in different transections of the stem. The phloem is present between the strands of xylem and thus, too, forms a complex system of anastomosing strands. The irregular arrangement of the xylem and phloem is shown in Fig. 27.8. The xylem is composed of tracheids, whereas the phloem contains sieve cells and some parenchyma cells. An endodermis encircles the vascular cylinder. *Lycopodium* stems also lack pith, as do the roots of most higher vascular plants.

Reproduction

Lycopodium may reproduce either asexually or through a typical alternation of generations involving, as usual, gametophytic and sporophytic generations. Like some of the more advanced algae, and unlike the bryophytes, both generations are independent of each other. Furthermore, the sporophyte generation is the dominant generation. Meiosporangia are borne in, or close to, the axils of the small leaves (Figs. 27.9, 27.10*B*), sheathing the stem. The spores (Fig. 27.10*C*) and sporangia of a given species are all alike. The leaves bearing the sporangia are called spore-bearing leaves or **sporophylls** (Fig. 27.9). In some species they closely resemble ordinary non-spore-bearing leaves in their structure and appearance (Fig. 27.7*A*). In other species they differ in size, shape, position, and color from the sterile leaves. Such modified sporophylls are grouped together closely at the ends of stems, forming a cone or **strobilus** (Figs. 2.1, 27.6, 27.7*B*, 27.10*A*).

Gametophyte. The spores do not germinate readily. The gametophytes in some species grow above ground and are green; in other species the gametophytes are subterranean and lack chlorophyll. The gametophytes are always associated with a fungus. The underground gametophytes (Fig. 27.10*D*) are hard to find and, being difficult to grow in culture, are not well known. Those that have been studied are monoecious, and the gametangia resemble those of *Anthoceros*. The gametangia are borne on the upper portion of the gametophyte (Fig. 27.10*E*); the cells of this portion are free of fungal filaments. The archegonia consist of a neck projecting slightly from the gametophyte surface, and a venter with one enclosed egg (Fig. 27.10*H*). The antheridia are composed of a mass of fertile cells buried shallowly in the gametophyte (Fig. 27.10*F*) and surrounded by a jacket of sterile cells. The sperms have two cilia, and fertilization occurs when sufficient free water is present to allow the sperms to swim to the mature archegonia.

Embryo sporophyte. The embryo sporophyte possesses (1) a well-developed foot, (2) the rudiments of a short primary root, (3) leaf primordia, and (4) a short shoot apex (Fig. 27.10*I*). The embryo grows directly into the mature sporophyte plant (Fig. 27.10*J, A*).

In some species small masses of tissue, called bulbils, are formed, drop from the parent plant, and grow directly into new young plants.

Selaginella

Selaginella species resemble those of *Lycopodium* in their general appearance but are smaller. They number more than 500 and, although they are widely distributed, most of them are tropical; a few grow in the temperate zones. Some species are adapted to withstand periods of drought and hence may grow in relatively dry localities.

Mature Sporophyte

As in *Lycopodium,* the sporophyte of *Selaginella* generally consists of a branched, prostrate stem with short, upright branches, usually only a few inches

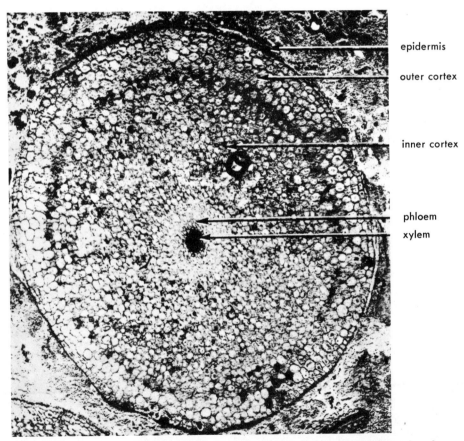

epidermis

outer cortex

inner cortex

phloem
xylem

Figure 27.5. Cross section of *Rhynia* stem. (From R. Kidston and W. H. Lang, *Trans. Roy. Soc. Edinburgh,* **52,** 603.)

Figure 27.6 (*right*). *Lycopodium clavatum* in its natural habitat in the deep shade of a cool forest.

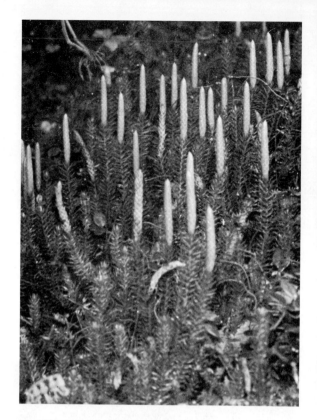

Figure 27.7 (*below*). Sporophylls in *Lycopodium*. A, sporophylls similar to vegetative leaves, *Lycopodium selago*; B, sporophylls different from vegetative leaves and grouped in strobili, *Lycopodium complanatum*. ×1.

sporophylls

A

strobilus

B

high. In some species the stem is upright and slender and may climb 50 feet in height. Both horizontal and upright stems are sheathed with small leaves in four longitudinal rows or ranks (Figs. 27.11, 27.13).

Two species of *Selaginella* are shown in Fig. 27.11. *Selaginella watsonii* (Fig. 27.11*A;* see color plate 13) grows in exposed rock crevices in the higher Sierra Nevada and is able to withstand periods of drought. The more delicate *Selaginella emmeliana* (Fig. 27.11*B*) grows best in a humid environment; it is frequently grown as an ornamental plant.

The vascular system in stems of *Selaginella* consists of from one to several separate strands or **steles,** each having a central core of xylem surrounded by phloem. The cross sections of Fig. 27.12 show a single strand in a stem of *Selaginella.* It occurs in a large air space and is supported by strands of radially arranged endodermal cells. Vessels are present in the xylem of several species of *Selaginella.* Parenchyma and sclerenchyma tissue form a cortex, which is bounded externally by an epidermis.

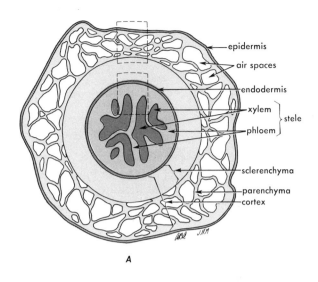

A

Figure 27.8. Cross section of *Lycopodium* stem. A, (above) showing irregular distribution of xylem and phloem, ×65; B, (below) enlarged sector showing cellular detail, ×260.

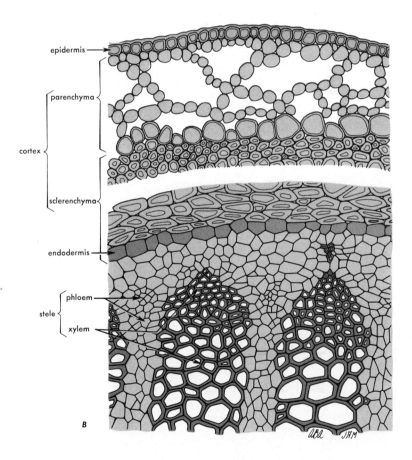

B

Reproduction

As in *Lycopodium,* the spores are borne in sporangia, which grow in or near the axils of sporophylls. Although the sporophylls do not differ greatly in appearance from the sterile leaves, they are always grouped to form cones, or strobili, at the ends of upright branches (Fig. 27.13; see color plate 13 and Fig. 27.14*A*).

Two types of sporangia are formed: **megasporangia** (Gr., *mega,* large) and **microsporangia** (Gr., *micro,* small). As the names indicate, the megasporangia produce larger spores. A single strobilus usually contains both types of sporangia, in some species the microsporangia being borne above the megasporangia (Fig. 27.14*B, C, D*).

Within a developing megasporangium, all but one of the spore mother cells degenerate. This remaining spore mother cell, nourished in part by the fluid resulting from the degeneration of the others and in part by the **tapetal** cells (the layer of cells surrounding the spore cavity) increases greatly in size. During meiosis, four large spores, called **megaspores,** are formed (Fig. 27.14*C, F*). Each megaspore upon germination will give rise to a female gametophyte (Fig. 27.14*M*).

Only a few spore mother cells within the developing microsporangium degenerate. The 250 or so that remain undergo meiosis, each forming four small spores, the **microspores** (Figs. 27.14*E, H*).

The production by a given species of two distinct types of meiospores—megaspores and microspores—is called **heterospory.** Thus *Selaginella* is heterosporous; *Lycopodium,* which produces but one spore type, is said to be **homosporous.**

Female gametophyte. Repeated cell divisions within the megaspore result in the female gametophyte, which is contained within the megaspore until it nears maturity. Its increase in size eventually ruptures the megaspore wall, and a small cushion of colorless gametophytic tissue protrudes from the megaspore along the lines of rupture (Fig. 27.14*M*). Archegonia develop on the protruding cushion. Although similar in structure to the archegonia of the liverworts, they are much more reduced in size (Fig. 26.2*C*). They are sunken within the gametophyte tissue; only two short cells of the neck protrude (Fig. 27.14*N*).

The megaspore (containing the female gametophyte) may be shed from the cone or strobilus at almost any stage of the development of the gametophyte. In some species it may be retained in the strobilus until well after fertilization. In any event, fertilization occurs only when sufficient water, either rain or dew, is present, allowing the sperms to swim to the archegonia.

Male gametophyte. Upon germination, the microspore divides into two cells. One of them, the **prothallial cell,** is small and does not divide further; it represents the vegetative portion of the male gametophyte. The other cell, by repeated divisions, develops into an antheridium composed of a jacket of sterile cells enclosing 128 or 256 biciliated sperms (Fig. 27.14*I, J, K, L*). This development occurs

Figure 27.9. The sporangia of *Lycopodium selago* are borne in the axils of sporophylls resembling vegetative leaves. ×3.

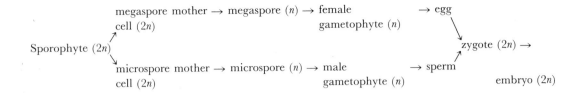

within the microspore wall. Microspores are shed from the microsporangium midway in the development of the male gametophyte and grow to maturity without direct connection with the parent sporophyte or the soil. Usually the microspores sift down to the bases of the megasporophylls (Fig. 27.14*D*). In this position they are close to the developing female gametophytes. The sperms escape when the microspore wall ruptures, and fertilization is made possible.

The sperms swim to the archegonia, which grow on that portion of the female gametophyte protruding from the megaspore (Fig. 27.14*N*). Fertilization ensues, and the resulting zygote initiates the diploid or sporophyte generation.

Embryo sporophyte. Of the two cells formed by the first division of the zygote, only one develops into an embryo. The other cell grows into an elongated structure, the **suspensor,** which forces the developing embryo into the gametophyte, where there is a food supply (Fig. 27.14*N*).

The embryo is a structure with (1) a foot, (2) a root, (3) two embryonic leaves, and (4) a shoot (Fig. 27.14*O*). In certain species of *Sellaginella* the embryo is held by the megaspore and retained within the strobilus. It does not pass into a dormant state, as do seeds, but continues to grow. Young sporophytes may be found extending from the strobilus of parent sporophytes. Should the developing embryo pass into a period of dormancy while being held by the mother sporophyte, a condition would arise that approaches the seed habit.

The life history of *Selaginella* is diagrammed above.

Lepidodendrales

The Lepidodendrales is an important fossil order of the phylum Lycophyta. Many representatives were heterosporous, and true seeds were formed by some species. No members of the order are in existence today. It was, however, a dominant order at the period when the great deposits of coal were being formed. If one can imagine a lush forest of *Selaginella*-like trees, 150 feet tall and six feet in diameter, a fair picture of this forest may be obtained. There were also smaller, shrubby and herbaceous Lepidodendrales. Since two spore types, and seeds, were formed by these ancient plants, we must believe that both heterospory and the seed habit are very old plant characters and not characters confined to present-day plants.

SIGNIFICANT FEATURES OF THE PHYLUM LYCOPHYTA

1. Spores are borne on or near sporophylls, grouped at the ends of upright branches to form cones or strobili.
2. Both homospory and heterospory occur.
3. The gametophytes of *Lycopodium* may be colorless and subterranean, and are associated with a fungus; in other species they are autotrophic.
4. The gametophytes of *Selaginella* develop within the spores.
5. The spores of some species of *Selaginella* are retained in the strobili until the gametophytes have matured and fertilization has occurred.
6. Members of the fossil order Lepidodendrales were heterosporous and produced seeds.
7. The vascular system in the stems of the Lycophyta resembles that found in the roots of Angiospermae.

PHYLUM SPHENOPHYTA

Members of this phylum once grew very abundantly, as is evidenced by the fact that they are well represented in the fossil record. Today, the phylum is represented by a single family (Equisetaceae) with but one genus (*Equisetum*) of about 25 species (Fig. 27.15). Many species inhabit cool, moist places, but *Equisetum arvense* grows in dry habitats. Most species are characterized by the presence of silica in the walls of some of the outer cells. Because of this trait, they were used in colonial days to scour pots and pans and hence were called scouring rushes. The genus is commonly known as the horsetails.

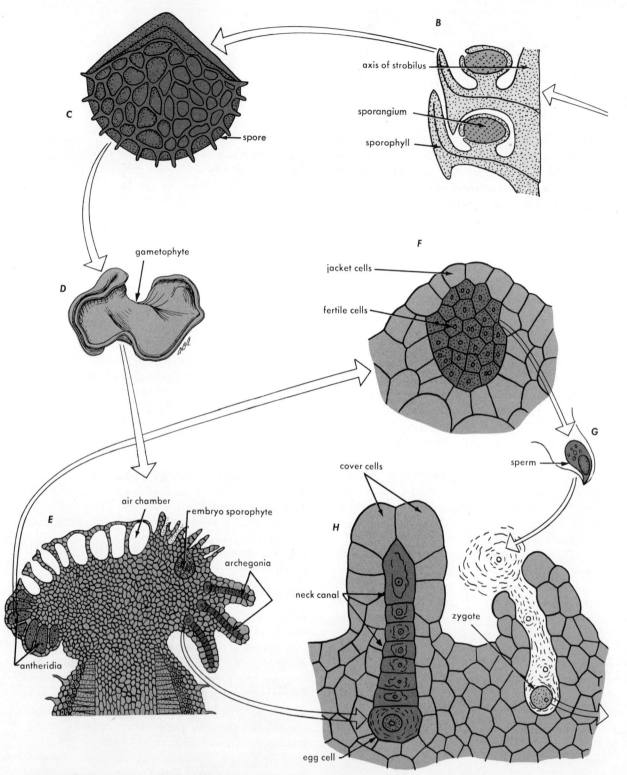

Figure 27.10. Stages in the life cycle of *Lycopodium*. *A, Lycopodium clavatum*, ×1; *B*, section of sporophyll showing location of sporangium, ×3; *C*, spore, ×50; *D*, gametophyte, ×1; *E*, section of gametophyte showing location of gametangia, ×25; *F*, antheridium, ×100; *G*, sperm, ×200; *H*,

mature sporophyte

A

J

young sporophyte

stem

leaf

sporophyte

embryo

root

I

archegonium, ×100; *I*, section of embryo, ×50; *J*, underground gametophyte with sporeling, ×3. (*B*, after M. G. Sykes, *Ann. Botany* (*London*), **22,** 63; *C*, after E. Pritzel in A. Engel and K. Prantl, *Die näturlichen Pflanzinfamilien; E, G, H*, after H. Bruchmann, *Flora*, **101,** 220; *J*, material supplied by A. J. Eames.)

Mature Sporophyte

The sporophyte of *Equisetum* is the dominant phase of its life history. In one tropical species the sporophyte is vinelike and may reach 25 feet in length. Usually, however, five feet represents the maximum upright growth. All species are perennials and have a branched rhizome from which the upright stems arise. The stems, depending upon the species, may branch either profusely or sparingly. In either event, they are straight and marked by ridges and distinct nodes (Figs. 27.15, 27.16). The tissue just above the nodes remains meristematic and structurally weak so that the stems may be easily pulled apart at these points. The bases of the nodes are sheathed by whorls of small simple leaves that are fused laterally (Figs. 27.15, 27.16). When branching occurs, the branches arise at the nodes, immediately below the leaves and, since there are frequently many leaves in a whorl, many branches may form at each node. The leaves are much reduced in size, nongreen and, in many species, short-lived. The stems are green and are, therefore, the organs of food manufacture. Roots occur only at the nodes of the rhizomes or bases of upright stems. The leaves of plants of this phylum are generally arranged in a whorl, and alternate with the branches at nodes. Leaf and branch gaps are present as in the flowering plants but their arrangement is different. The ribbed stem of the Sphenophyta is such a characteristic feature that fossil stems showing ribs are regularly placed in this phylum.

Except at the nodes, the stems of Equisetum are hollow and the ribs make prominent markings on their outer circumferences (Figs. 27.16, 27.17). The ridges, formed of sclerenchyma tissue, not only project outward, but extend inward almost to the small vascular bundles (Fig. 27.17). There are air spaces, or canals, between the vascular bundles. Photosynthetic tissue lies between these air canals and the thin outer layer of sclerenchyma tissue and epidermis. The vascular strands are also marked by a canal (Fig. 27.17). Arms of xylem extend outward from this canal, and phloem tissue lying outward from the canal is present between the radial arms of xylem. An endodermis is present but its location varies from species to species; it may surround the vascular bundles, or encicle the ring of vascular bundles. In some species there is also an inner endodermis.

Reproduction

In all species of *Equisetum* the sporangium-bearing organs, the **sporangiophores,** are specialized structures very different from ordinary leaves. They are grouped together in **strobili** (Fig. 27.18) at the summit of the main upright branches and occasionally on lateral branches (Figs. 27.15, 27.18*A*). In most species the cones or strobili are borne on

A

Figure 27.12. Cross section of a stem of *Selaginella*. *A,* (*above*) diagram showing distribution of stem tissues; a single flattened stele is supported in an air space by filaments of endodermal cells. The cortex is composed of parenchyma and sclerenchyma cells; *B,* (*opp. right*) detail showing cellular structure.

the ordinary vegetative shoots; in a few species they are formed only on special fertile shoots (Figs. 27.15, 27.18A).

The sporangiophores are stalked, shield-shaped structures borne at right angles to the main axis of the cone. The cone may be compared to a pole to which open umbrellas have been fastened, the handle of the umbrella being at right angles to the pole (Fig. 27.18C, D). The sporangia are attached to the underside of the shield, close to its edge. They extend horizontally inward, toward the axis of the cone.

Meiosis is normal, and four meiospores with the haploid number of chromosomes are formed from each spore mother cell. All meiospores are morphologically alike. *Equisetum,* therefore, is homosporous.

Figure 27.12. (continued)

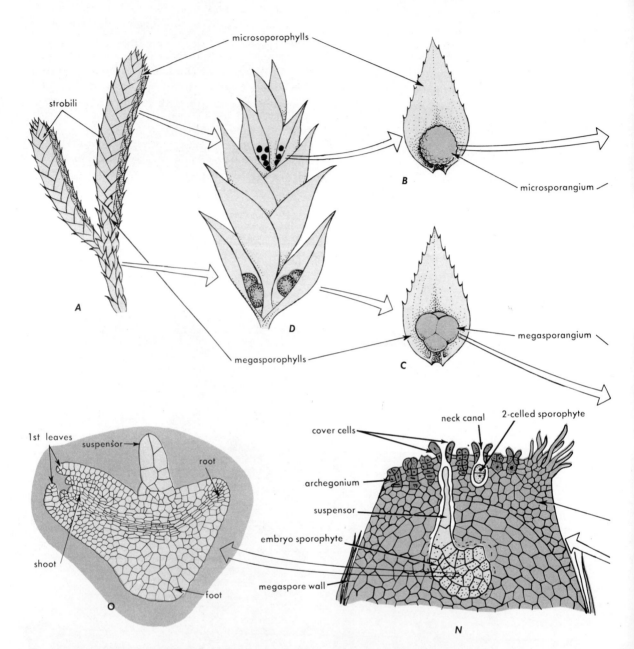

Figure 27.14. Stages in the life cycle of *Selaginella*. A, shoot of *Selaginella watsonii* showing two strobili, ×2; B, microsporophyll with microsporangium, ×5; C, megasporophyll with megasporangium, ×5; D, strobili showing sporophylls with sporangia, ×2; E, microsporangium discharging microspores, ×50; F, megasporangium discharging megaspores, ×50; G, microspores sift downward and lodge in axils of megasporophylls close to megaspores; H, a microspore, ×100; I, section of a microspore, ×150; J, male gametophyte within microspore, ×150; K, antheridium with developing sperm cells; L, mature sperm cells, ×100; M, female gametophyte within megaspore, ×100; N, section of female gametophyte showing location of archegonia in a young embryo pushed, by its suspensor, into the vegetative tissue of the female gametophyte, ×100; O, the young embryo sporophyte generally remains embedded in vegetative gametophyte tissue, ×100. (E, H, I, J, K, L, after R. A. Slagg, *Am. J. Botany*, **19**, 106; I, J, O, after H. Bruchmann, *Flora*, **104**, 180.)

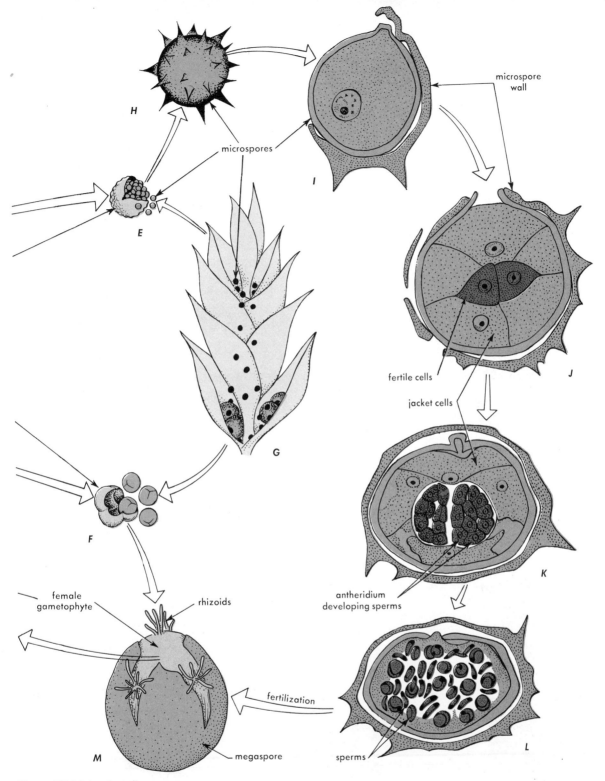

microspores

microspore wall

fertile cells

jacket cells

antheridium
developing sperms

female gametophyte

rhizoids

fertilization

sperms

megaspore

H

I

J

K

L

M

E

F

G

Figure 27.14 (*continued*)

Figure 27.15 (*right*). *Equisetum telmateia,* showing much-branched vegetative shoots and straight unbranched fertile shoots with terminal strobili. ×1/4.

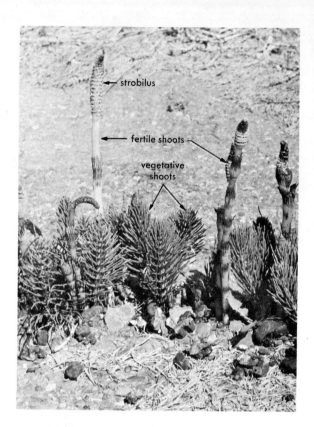

Figure 27.16 (*below*). *Equisetum hyemale, vegetative shoot.* A, upper six nodes showing the telescoping of the youngest internodes and their rapid elongation; ridges and whorls of leaves are evident, ×2; B, enlarged view of stem showing three nodes. Note that the leaves match the ridges below them, but alternate with the ridges above them, ×6; C, the reproductive shoot ends in a strobilus, ×1.

A

Color Plate 13

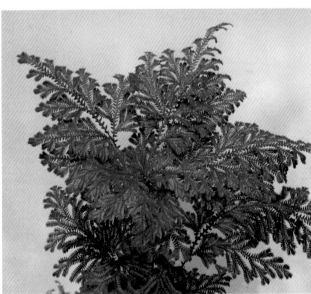

B

Figure 27.11. Two species of *Selaginella*. A, *Selaginella watsonii*, a montane species growing in crevices in granite, ×1; B, *Selaginella emmeliana*, a species native to tropical America and requiring a moist warm climate, ×½.

Figure 27.13. Four strobili of *Selaginella watsonii*.

A

B

C

D

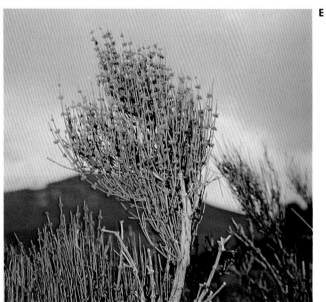

E

Figure 28.1. Gymnosperms. A through D, Coniferophyta. A, cones and branches of *Cedrus deodar*; B, C, young and mature carpellate cones of *Pinus ponderosa*; D, staminate cones of *Juniperus occidentalis*; E, *Ephedra nevadensis*, an evolutionary advanced gymnosperm belonging to the phylum Gnetophyta.

Upon maturity the meiospores are discharged from the sporangia (Fig. 27.18*E*). The spores have a thick, double wall, and, when they are mature, the outer half of the wall unfolds to form narrow hygroscopic ribbons that move under the influence of moisture changes and aid in the distribution of spores (Fig. 27.19). The spores are fragile and normally live but a few days.

Gametophytes. The gametophytes are small green bodies about the size of a pinhead and consist of a cushionlike base with many erect, delicate lobes (Fig. 27.20). They may be easily cultured.

The gametangia are similar to those of *Lycopodium.* Fertilization takes place only when there is free water. The sperms are spiral, multiciliate cells.

The zygote develops directly into an embryo. No suspensor is formed. The embryo is similar to those already described, except that the foot is small or lacking.

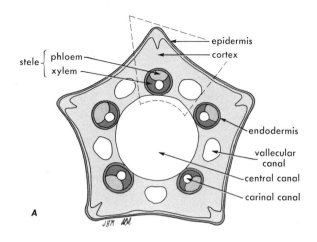

Figure 27.17. Cross section of *Equisetum* stem. A, (*above*) diagram showing arrangement of stem tissues, ×70; B, (*below*) enlarged view showing details of cellular structure, ×400.

Figure 27.18. Strobili of *Equisetum telmateia*. A, (above) fertile stems with strobili; also note much-branched sterile stems, ×¼; B, (above right) strobilus with sporangiophores, ×2; C, (below right) enlarged view of several sporangiophores, ×10; D, (opp. above) cross section of strobilus showing arrangement of sporangiophores around central stem, ×2; E, (opp. center) several sporangiophores showing dehiscence of sporangia and discharge of spores, ×2.

D

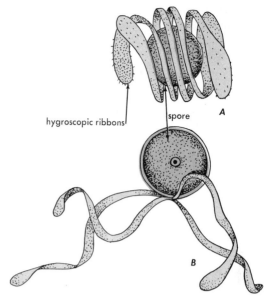

hygroscopic ribbons   spore   A

B

Figure 27.19. Spores of *Equisetum* showing the hygroscopic ribbons, ×100. (Redrawn from R. von Wettstein.)

E

Figure 27.18 (continued)

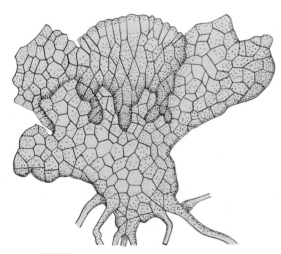

Figure 27.20. Gametophyte of *Equisetum*, ×20. (Redrawn from E. R. Walker, *Botan. Gaz.*, **92,** 1. Copyright The University of Chicago Press.)

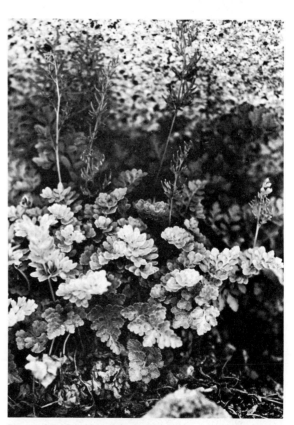

Figure 27.21. A small fern, the rock brake (*Cryptogramma acrostichoides*), growing on a crevice on open granite in the partial shade of a granite boulder.

SIGNIFICANT FEATURES OF THE
PHYLUM SPHENOPHYTA

1. The genus *Equisetum,* consisting of about 25 species in the family Equisetaceae, is the only living group of this once widely distributed phylum.

2. The stem of *Equisetum* has distinct nodes and internodes. Meristematic tissue occurs just above the nodes, which are sheathed by whorls of small, brownish, short-lived leaves. Photosynthesis is carried on by the green stems.

3. The strobili are composed of specialized structures, sporangiophores consisting of shield-shaped disks, supported by short stalks growing at right angles to the stem.

4. The autophytic gametophyte is a single thallus with many lobes.

5. Present-day species of *Equisetum* are homosporus; certain extinct species of Sphenophyta are heterosporous.

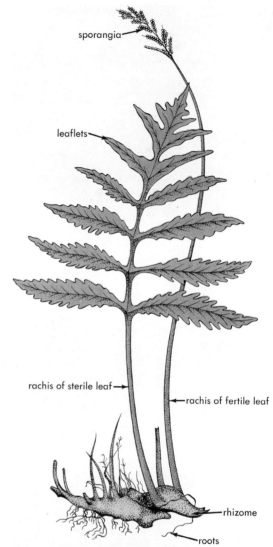

Figure 27.23 (*above*). The sensitive fern (*Onoclea sensibilis*). ×1/20. (Redrawn from L. Diels in A. Engler and K. Prantl, *Die näturlichen Pflanzenfamilien.*)

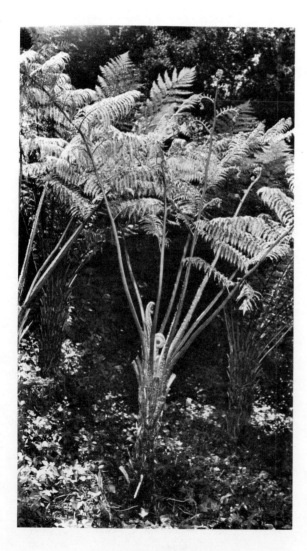

Figure 27.22 (*left*). A tree fern (*Cyathea*) growing in Golden Gate Park, San Francisco. ×1/50.

Figure 27.24 (*opp. right*). Fern fronds retain meristematic tissue at their tightly rolled tips. As differentiation proceeds the tips uncoil. A, tightly rolled tip of a young frond of the tree fern, *Cyathea*, ×1/4; B, maturing frond of the bracken fern (*Pteris aquilina*); the tips are still tightly rolled, ×1/2; C, uncoiling tip of a frond of the sword fern (*Nephrolepis exaltata*), ×2.

PHYLUM PTEROPHYTA

The phylum Pterophyta comprises the ferns, most of which are shade-loving plants of small size (Fig. 27.21), their upright leaves, or **fronds,** generally being their most prominent feature. All species native to temperate zones have an underground rhizome with leaves and roots at the nodes. In this respect they resemble the growth form of many monocotyledons (Fig. 7.32). Some ferns of the tropical regions do grow into fairly large trees (Fig. 27.22). All Pterophyta have a definite alternation of generations, with both the sporophyte and gametophyte being autotrophic plants.

The Pterophyta comprise four orders, the largest of which, the Filicales, includes the **true ferns.** The Filicales are divided into eleven or more families, of which the largest and best known in the United States is the Polypodiaceae. Most of the following discussion will deal with representatives of this family. There are also several fossil orders. One of them, the Coenopteridales, will be briefly discussed because of its characteristic leaf strucutre.

Figure 27.25. Fern fronds. A, simple entire frond of hart's-tongue fern (*Phyllitis scolopendrium*) ×1/10; B, pinnate frond of *Polypodium*, ×1/2; C, bipinnate frond of *Adiantum*; D, bipinnate frond, with serrate leaflets, of the rock brake fern, *Cryptogramma acrostichoides*, ×1; E, fertile leaf of *Cryptogramma*, ×1; F, much-dissected frond of *Asplenium*; young plants form at meristematic leaf tips, ×1/2.

Figure 27.26 (opp. right). Cross sections of a rhizome of *Polypodium*. A, showing relationship of stem tissues, ×45; B, a higher magnification showing cellular detail, ×450.

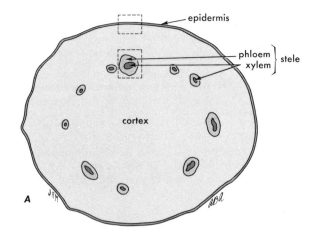

A

Mature Sporophyte

The sporophyte, which is the dominant generation of all ferns, possesses an underground stem or rhizome from which leaves and adventitious roots arise (Fig. 27.23). The leaves are the most prominent part of the fern plant and vary greatly in size and form. They differ from the leaves of the flowering plants in two important respects: (1) they have an apical meristem that usually continues active for some time (Fig. 27.24B), and (2) spores frequently are borne on their lower surface (Fig. 27.28). Young fern leaves are rolled in tight spirals and consist chiefly of meristematic tissue (Fig. 27.24). As the leaf

B

Figure 27.26

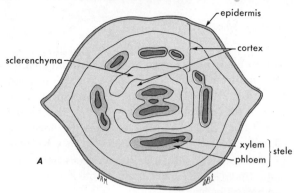

epidermis

cortex

sclerenchyma

xylem
phloem
} stele

A

Figure 27.28 (*opp. right*). Sori. *A*, sporangia in a band close to margin of leaflet, *Pellaea*; *B*, sporangia along margin of leaflet, *Adiantum*; *C*, sporangia in elongated sori parallel to midrib of leaflet with indusia, *Woodwardia*, *D*, sporangia in round sori, *Polypodium*; *E*, sporangia grouped in round sori; centrally placed indusia are present, *Polystichum*; *F*, sporangia in long rows on veins, *Asplenum nidus*. All about ×2.

Figure 27.27 (*left & below*). Cross sections of rhizome of *Pteris*. *A*, showing relationship of stem tissues, ×37; *B*, higher magnification showing cellular detail, ×550.

epidermis

sclerenchyma

cortex

parenchyma

phloem
xylem

stele

pericycle

endodermis

sclerynchyma

cortex

endodermis

pericycle

B

Figure 27.29. Fern gametophytes growing on a flower pot, ×2. (Courtesy of E. M. Gifford.)

Figure 27.28. *A–F*

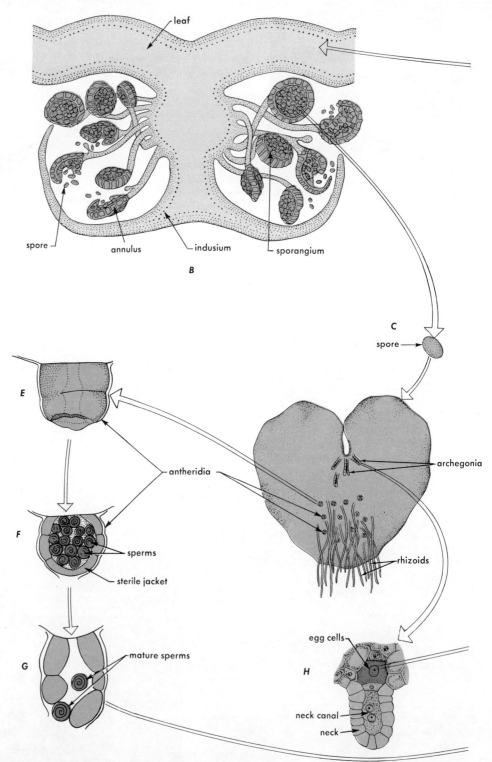

Figure 27.30. Stages in the life cycle of a fern. *A*, a frond of a sword fern, ×1; *B*, section through a sorus, showing indusium, sporangia, and spores, ×50; *C*, spore, ×100; *D*, gametophyte, ×5; *E*, antheridium, ×100; *F*, section of antheridium showing sperm, ×100; *G*, open antheridium discharging sperm, ×100; *H*, archegonium, ×100; *I*, archegonium receptive to sperm, ×100; *J*, section of gametophyte showing zygote within the venter of the archegonium, ×100;

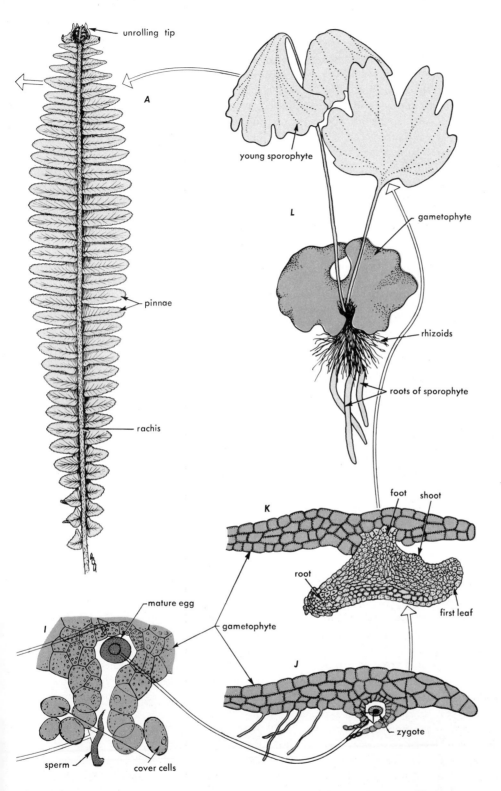

unrolling tip

A

young sporophyte

pinnae

gametophyte

rachis

L

rhizoids

roots of sporophyte

foot *shoot*

K

root

mature egg

first leaf

I

gametophyte

gametophyte

J

sperm *cover cells*

zygote

K, section of gametophyte and embryo, ×100; *L*, gametophyte with attached young sporophyte, ×10. (*E, F, G*, after M. E. Hartmann, *Botan. Gaz.*, **91**, 252. Copyright The University of Chicago Press. *H, I*, after D. H. Campbell, *Mosses and Ferns*, The MacMillan Company, 1905. *J, K*, after R. M. Holman and W. W. Robbins, *A Textbook of General Botany*, John Wiley and Sons.)

matures, it unwinds from the base upwards. The upright, expanded portion is mature, but the leaf continues to grow by cell division at its coiled tip. The uncoiled, fully expanded fern leaf lacks meristematic tissue, as the cells of all such tissue have differentiated into permanent cells. In certain species the apical meristem may remain active for years, resulting in leaves nearly 10 feet long.

Most fern leaves are compound, although simple types exist in all groups. The most common type of fern leaf has a stout or rigid petiole, which is prolonged to form the rachis from which leaflets arise (Figs. 27.23, 27.25).

The vascular tissue of the ferns is organized into one or more vascular strands each having a core of xylem, surrounded by phloem and separated from the ground parenchyma by an endodermis. These vascular bundles are interconnecting, so, as in *Lycopodium*, the precise arrangement will vary from section to section. However, there are arrangements characteristic of various genera. For instance, in *Polypodium* there is a ring of small bundles placed well out from the center of the rhizome (Fig. 27.26A). Each strand is formed of a central core of xylem, surrounded by phloem, and the whole having a bounding endodermis (Fig. 27.26B). In *Pteris* there is an outer circle of somewhat flattened vascular bundles with a central set of inter-connected bundles. Each vascular bundle has a bounding endodermal layer (Fig. 27.27A, B). The xylem consists largely of tracheids, vessels being known in only two genera. Sieve cells occur in the phloem.

Reproduction

Vegetative reproduction. This type of reproduction may occur in one of two ways: (1) by death and decay of the older portions of the rhizome and the subsequent separation of the younger growing ends; and (2) by the formation of deciduous leaf-borne buds, which become detached and grow into new plants. Such buds occur in only a few genera.

Sexual reproduction. In the sexual life cycle, independent sporophyte and gametophyte generations alternate with each other. The vegetative structure of the sporophyte has been described above. The spores are borne in sporangia, which ordinarily develop on the lower surface or margins of the fronds (Fig. 27.28). Not all leaves are fertile, that is, spore-producing; and the fertile leaves may not always be similar to sterile (non-spore-producing) ones (Figs. 27.21, 27.23, 27.25D, E). The distribu-

tion of sporangia on the leaf surface varies considerably in the different genera and species (Fig. 27.28). They may (1) cover much of the lower surface, (2) be grouped in **sori** (singular, **sorus**) and grow in a definite relationship with the veins, or (3) grow only along the margins or edges of the leaf.

When the sporangia are grouped together in sori, a structure called the **indusium** (Fig. 27.28C, E), which is sometimes umbrella-like, may be present, thereby protecting the young and developing sporangia. Frequently, marginal sporangia are protected by the curled edge of the leaf, which forms a so-called **false indusium** (Fig. 27.28B).

The sporangium is a delicate watch-shaped case, consisting of a single layer of epidermal cells, only one row of which possesses heavy walls. This row, which nearly encircles the sporangium (in the Polypodiaceae), is the **annulus**; it functions in opening the dried mature sporangium and aids in the dispersal of the ripe spores (Fig. 27.30B).

The young sporangium is filled with sporogenous cells, which eventually give rise to spore mother cells. Meiosis in the spore mother cell results, as always, in spores with a reduced number of chromosomes. Ferns, with the exception of two families, the Marsileaceae and the Salviniaceae, are homosporous. These two families, though widely distributed, are small, uncommon and unfernlike in appearance.

Gametophyte. The gametophytes, or **prothallia,** of the Polypodiaceae are small, flat, green, heart-shaped structures (Fig. 27.29) with rhizoids on their lower surface (Fig. 27.30D). In most species they apparently mature rapidly and are not long-lived. Antheridia and archegonia are borne on the same prothallium. Antheridia are formed when the prothallium is very young and are scattered over its lower surface. The antheridia are small and have a jacket formed from only three or four sterile cells (Fig. 27.30E, F). Antheridia project only slightly from the surface of the gametophyte. Normally, 32 sperms develop within each antheridium (Fig. 27.30F).

The archegonia form later than the antheridia and are usually clustered close to the notch, also on the undersurface of the gametophyte (Figs. 27.30D, H). They, too, are small but are typical, consisting of a short neck and a venter that encloses the egg cell. The neck cells protrude slightly from the lower surface of the gametophyte.

Fertilization occurs when moisture is present and

the sperms are thus able to swim to the neck of the archegonium. The resulting zygote is diploid and rapidly develops into an embryo sporophyte comprising a foot, root, stem, and leaf (Fig. 27.30*I*, *J*, *K*). The embryo develops directly into a young sporophyte (Fig. 27.30*L*).

SIGNIFICANT FEATURES OF THE PHYLUM PTEROPHYTA

1. The sporophyte plant, in most species, consists of a rhizome bearing adventitious roots and upright leaves.

2. A well-developed vascular system is present.

3. Usually the leaves are dissected into leaflets, arranged pinnately on a stout midrib or rachis. They are condensed branch systems and, when young, have meristematic tissue at their tips.

4. Sporangia are borne on the undersurface of the leaves. They may be grouped into sori, which, in some species, are protected by an indusium.

5. All but two families of the Pterophyta are homosporous.

6. The gametophytes (prothallia) are independent of the sporophyte. They are small, green, heart-shaped, and short-lived plants. Both antheridia and archegonia are borne on the same gametophyte.

Order Coenopteridales (Fossil Ferns)

The fossil ferns as represented by some of the Coenopteridales, though different in external appearance from present-day ferns, possessed sporangia, spores, and a vascular system that resembled the corresponding structures of present-day ferns. They were small plants, either erect, creeping, or climbing. They are of interest because of their primitive leaves, which differed only slightly from the stem. As the shoot developed, it branched; one division continued to grow, whereas the other soon ceased growth. This short, determinate branch of the main axis was the leaf. Structurally it closely resembled the stem. This primitive type of fern leaf forms one basis for the belief that the leaves of flowering plants are in reality condensed branch systems.

A SUMMARY OF THE MORPHOLOGICAL TRAITS OF THE LOWER VASCULAR PLANTS

These phyla of more primitive vascular plants have striking similarities, as well as differences. They can be arranged in a sequence suggesting the evolution of traits better adapting them to survive in an environment becoming increasingly drier. These traits are compared in Table 27.1.

CLASSIFICATION

Phylum	Psilophyta
Family	Psilotaceae
Genera	*Tmesipteris*
	Psilotum nudum
Family	Rhyniaceae
Genus	*Rhynia*
Phylum	Lycophyta
Family	Lycopodiaceae
Genus	*Lycopodium*
Family	Selaginellaceae
Genus	*Selaginella*
Phylum	Sphenophyta
Family	Equisetaceae
Genus	*Equisetum*
Phylum	Pterophyta
Order	Coenopteridales
Order	Filicales
Family	Polypodiaceae
Genus	*Polypodium*
Genus	*Pteris*
Family	Cyatheaceae
Genus	*Cyathea*

28

The Gymnosperms

This group of plants is represented by such common trees as the pines (*Pinus*), spruces (*Picea*), firs (*Abies*), and cedars (*Cedrus*), all of which possess well-developed cones in which the seeds are borne (Fig. 28.1; see color plate 14). Other gymnosperms, such as the *Ginkgo,* yews (*Taxus*), and Mexican tea or joint pine (*Ephedra*), do not bear cones. However, a trait common to all gymnospermous plants is the absence of a protecting case, like the ovary wall, around the seeds. In the yews, seeds are partially surrounded by a red berry-like structure (Fig. 28.19). In *Ephedra* the ovules are borne in the axils of short bracts (Fig. 28.1*E*) and in *Ginkgo* the naked ovules are attached to the ends of short branches (Fig. 28.7*B*). By contrast, in the angiosperms, the seed of the peach, for example, is surrounded by the matured ovary wall, which is in part fleshy and in part hard and stony. In the pines and other cone-bearers the seeds are borne on the surface of the scales that comprise the cone, and, though well protected by the scales, they are not surrounded by floral parts. Such seeds, lacking the protection of an ovary wall, are said to be "naked" (Fig. 28.2)—thus the name gymnosperm, derived from the two Greek words *gymnos* (naked) and *sperma* (seed). It will be recalled that the fossil Lepidodendrales, belonging to the Lycophyta phylum, bore seeds, thus indicating that the seed habit is not of geologically recent origin.

GENERAL CHARACTERISTICS AND CLASSIFICATION

The external morphology of gymnosperms shows considerable variation, as may be seen from the following examples. (1) Fossil gymnosperms so resemble ferns that they are commonly called "seed ferns." (2) Some modern gymnosperms of Mexico and other tropical areas are palmlike in appearance. (3) Most north temperate gymnosperms have needle or scale leaves and are mainly evergreen. (4) Some gymnosperms of the temperate regions, particularly of Australia, are broad-leaved. (5) Certain gymnosperms are shrubby and thrive under very arid conditions.

The vascular tissue of the class has been discussed at some length in the chapter on the stem. It should be recalled that vessels are absent from the

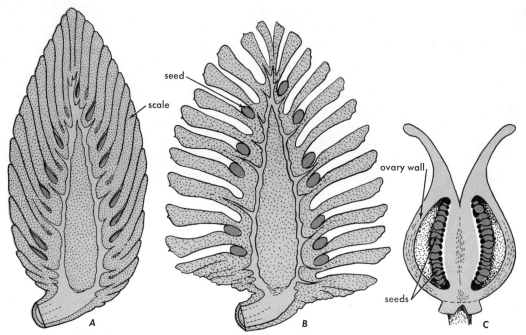

Figure 28.2. *A, B,* median longitudinal sections of cones of a gymnosperm. *A,* immature cone tightly closed to protect ovules; *B,* open to disseminate seeds; *C,* angiosperm fruit showing seeds enclosed in an ovary.

Figure 28.3. Probable appearance of a seed fern. (From H. N. Andrews, Jr.)

xylem in all but a few species, and that companion cells do not occur in the phloem, where there are sieve cells rather than sieve tubes. Pith is present in the stems but not in roots.

There are 5 phyla of gymnosperms, encompassing some 650 species grouped in many genera and families. We shall briefly consider three phyla, the Pteridospermophyta or seed ferns; the Cycadophyta or cycads; and the Ginkgophyta, *Ginkgo,* or the maiden-hair tree. A fourth phylum, the Coniferophyta or conifers, will be considered in more detail.

PHYLUM PTERIDOSPERMOPHYTA

The Pteridospermophyta are extinct, but they are well known as fossil plants. They grew over much of the earth when coal deposits were being formed. Their fossils, consisting of beautifully preserved stems, roots, leaves, seeds, and even pollen grains, have been found encased in rock mined as waste material in coal mines (Fig. 28.4). They have a distinct fernlike appearance (Fig. 28.3) and for many years were regarded as ferns. When, however, seeds were found attached to the leaves, it became necessary to revise their classification. Since the seeds were naked, these plants were regarded as gymnosperms. The structure of the seeds (Fig. 28.4*D*) resembles that of such modern gymnosperms (see page 615) as the cycads and *Ginkgo.* Furthermore, pollen grains have been found associated with the fossil seeds in such a manner as to suggest strongly a life history similar in its essential steps to that of living gymnosperms. Because of these findings, the Pteridospermophyta are now considered to be the most primitive gymnosperms.

PHYLUM CYCADOPHYTA

Some 200 million years ago members of the phylum Cycadophyta formed an extensive portion of the earth's flora, probably constituting the food supply for at least some of the herbivorous dinosaurs. Today there remain some nine well-defined genera of cycads with about 100 species growing in widely separated areas of the earth's surface and largely confined to the tropics. Only one genus, *Zamia* (Fig. 28.6), occurs naturally in the continental United States; it is found in southern Florida. Various genera are, however, under cultivation, out of doors in the warmer regions, and in large glass houses in the colder regions. Two cycads, *Cycas revoluta* and

Dioön spinulosum, are shown in Fig. 28.5*A* and *B.* Both were growing in the conservatory at Golden Gate Park. These trees are palm-like in appearance; in fact, *Cycas revoluta* is known as the Sago palm. In *Zamia,* the genus native to Florida, the trunk is largely subterranean (Fig. 28.6), but in *Cycas* it forms a single straight bole. These trees are very slow growing, a six foot high specimen being perhaps as much as a 1000 years of age.

That these trees are really gymnosperms and not palms is readily evident from the forms of the cones. The large, upright cone shown in Fig. 28.5*A* bears pollen, while that shown in Fig. 28.5*B* is a seed-bearing, or ovulate, cone. While we shall not discuss the nature of these seeds, they are not like those of angiosperms but consist of erect unprotected ovules with their micropyles turned outwards, forming striking examples of naked ovules. An additional primitive feature is the presence of motile sperm. Since *Ginkgo* is cultivated widely in the United States and, like the cycads, possesses both naked ovules and motile sperm, these features are described in somewhat more detail in the following section.

PHYLUM GINKGOPHYTA

Only one living representative, the maiden-hair tree (*Ginkgo biloba*), remains of this very ancient phylum of plants. It has been reported as growing wild today in the forests of remote western China. It has, however, been grown for centuries on Chinese and Japanese temple grounds and is now cultivated in many countries. It is a large tree with characteristic small, fan-shaped leaves (Fig. 28.7), which are divided into two lobes. The trees are dioecious.

Reproduction

With the exception of a few algae and some fungi, all living plants so far studied have been characterized by motile sperms and free water has been necessary for fertilization. Moreover, the gametophytes have been mostly independent of the sporophyte. It will be recalled that *Selaginella* has these characteristics, except that the female gametophytes are occasionally retained by the sporophyte until well after fertilization. The sexual cycle in *Ginkgo* is somewhat comparable to that of *Selaginella;* however, the female gametophytes are almost completely surrounded by sporophyte tissue, and, though motile sperms are formed, the fluid in which

Figure 28.4. Fossil seed ferns. *A*, a section through a mass of coal showing embedded plant material; *B*, cross section of fossilized stem of *Lyginopteris*, an ancient plant that grew in England; *C*, seeds of a large seed fern that grew in Illinois; *D*, median longitudinal section of the seed of a seed fern; *E*, impressions in rock of leaves of *Lyginopteris*, from Scotland. (*A, B, D, E*, courtesy of H. N. Andrews, Jr.; *C*, courtesy of Field Museum of Natural History.)

they swim is produced by the parent sporophyte.

The female gametophytes of *Ginkgo* are buried in the ovule (see page 191), where small, reduced archegonia are formed. The archegonia have two neck cells and no neck-canal cells except the ventral canal cell. Entrance to the ovule is through a small opening, the micropyle, at one end of the ovule. Microspores develop in catkin-like strobili on male trees. When mature, the microspores are shed and are carried by air currents to the mature ovules on female trees. The microspores are drawn into the micropyle and come to lie in contact with the nucellus. In this position they develop into male gametophytes, forked pollen tubes, that grow parisitically into the upper portion of the nucellus. When the end of a pollen tube nears an archegonium, two motile sperms are released into a cavity just above the archegonium. One sperm eventually passes into an archegonium and fuses with the egg nucleus. An embryo develops from the zygote. Integuments surround the dormant embryo thus forming, with the enveloping female gametophyte, a true seed.

The important feature of this life history is the development of a method to *insure fertilization without the necessity of free water*. Such a development has made possible the evolution of an extensive dry land flora.

PHYLUM CONIFEROPHYTA

Classification

Without exception, the well-known and economically important gymnosperms of the temperate zones belong to the Coniferophyta. There are nine families and numerous genera. They comprise the pines, hemlocks, firs, spruces, junipers, yews, redwoods, and many others. Not all of them bear cones, yet the cone is such a conspicuous feature of a large number of them that the phylum has been named the Coniferophyta or the cone bearers. True cones are formed by the majority of them (Figs. 28.1, 28.8). Juniper berries are in reality cones with fleshy adhering scales (Fig. 28.16). In the yews, on the other hand, the seeds are surrounded at the base by a more or less pulpy, berry-like body and are not borne in cones (Fig. 28.19). In either event the seeds are naked, not being surrounded by an ovary wall.

Figure 28.5. *A,* (above) *Cycas revoluta* showing a microsporangiate cone; *B,* (below) *Dioön spinulosum* showing ovulate cone.

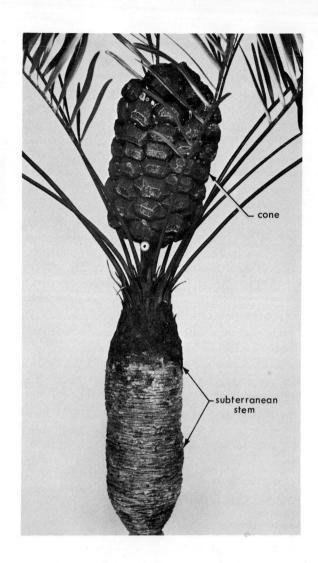

cone

subterranean
stem

Figure 28.6. *Zamia,* showing ovulate cone and subterranean stem. (Courtesy of Field Museum of Natural History.)

The following brief descriptions of several common genera will serve as illustrations of the more important conifers.

Family Pinaceae

Pinus (*pines*). The pines are usually large trees. The trees may be either pyramidal (Fig. 28.9) or flat-topped. The leaves are needlelike, two or more growing together (except in *Pinus monophylla*) in a **fascicle** or group, which is sheathed at the base (Fig. 28.10). The cones vary greatly in size and shape and are very characteristic of the species to which they belong (Fig. 28.8*F, G, H*).

Abies (*firs*). The firs are stately trees of a symmetrical pyramidal shape. The leaves are flat and linear; in cross section they are relatively broad, without marked angles. The cones are erect on the branches and shatter at maturity (Fig. 28.11).

Picea (*spruces*). These trees closely resemble the firs, from which they can be distinguished by the position of the leaves on the branchlets (Fig. 28.12) and by the angular appearance of the leaves in cross

Figure 28.7. *Ginkgo biloba.* A, growing on the campus of the University of California at Berkeley; B, branch showing mature seeds. (B, courtesy of John H. Gerard from National Audubon Society.)

section. The cones are pendent and do not shatter at maturity (Figs. 28.8*B*, 28.13).

Tsuga (*hemlocks*). The trees of this genus are pyramidal with slender horizontal branches. The leaves are usually two-ranked, linear, flat, and with a short petiole. They resemble the leaves of firs but are much shorter. The cones are small (Fig. 28.14).

Pseudotsuga (*Douglas fir*). Only two species occur in the United States. One of them, the Douglas fir (*Pseudotsuga menziesii*), is the most heavily cut timber tree of the United States. It may grow to a height

of 200 feet with a diameter of 12 feet. Its leaves are flat, like those of the true firs, but have white lines on either margin and a groove along the upper surface. The cones are from 2 to 4½ inches long, pendulous, and easily recognized by the bracts that extend outward below each scale (Fig. 28.15).

Juniperus (*junipers*). This genus is composed of both trees and shrubs. The trees are inclined to be somewhat irregular in shape and frequently have more than one trunk. The leaves are of two sorts, (1) in some species, spreading and needlelike and

Figure 28.8. Mature ovulate cones. A, *Cedrus deodara*; B, *Picea sitchensis*; C, *Sequoia sempervirens*; E, *Sequoiadendron gigantea*; D, *Metasequoia glptostroboides*; F, *Pinus monophylla*; G, *Pinus contorta*; H, *Pinus jeffreyi*; I, base of cone of *Pinus jeffreyi*.

Figure 28.8 (*continued*)

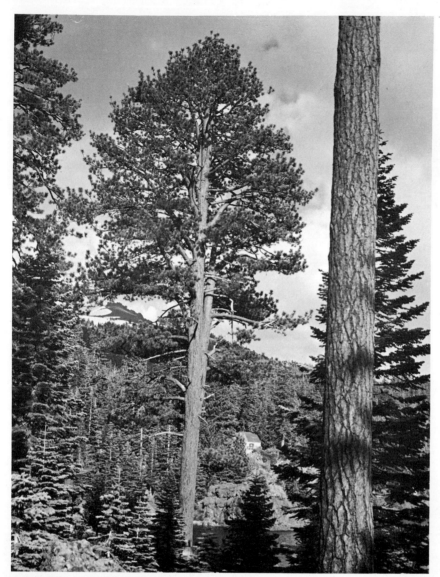

Figure 28.9. *Pinus jeffreyi,* showing the trunk of one tree and the crown of a second tree.

Figure 28.10. Fascicle of yellow pine needles. ×1.

Figure 28.11 (*opp. right*). *Abies.* A, a mature tree; B, top of a young tree showing ovulate cones; C, ovulate cone and branch. Note arrangement of needles; D, staminate cones.

(2) in other species, scalelike and closely pressed to the twigs. The cone of the junipers is usually called a "berry"; it is a fleshy structure enclosing several seeds. Morphologically, it is a modified cone, composed of fleshy adhering scales that completely enclose the seeds (Fig. 28.16).

Larix (larches). The trees belonging to this genus grow in the cooler portions of the Northern Hemisphere. They differ from most other members of the Pinaceae in that they are deciduous. The American larch, or tamarack, is a tall tree frequently found in bogs. The needles are short, linear, and grouped in crowded clusters on short spurs. On the leading shoots, however, the needles are arranged spirally. The cones are small and persistent (Fig. 28.17).

Family Taxaceae (yews)

With two exceptions, ovulate cones are not borne by members of the yew family. The seed of this family so strongly resembles a drupe or a nut that

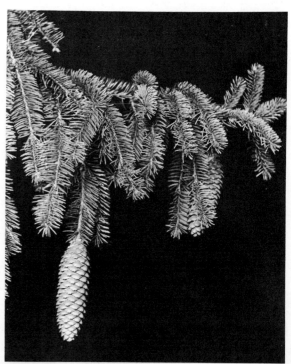

Figure 28.13. Branch of *Picea alba* with ovulate cones. ×1/2. (Courtesy of American Museum of Natural History.)

Figure 28.12. (*left*) Leaves of *Picea*. (Redrawn from L. H. Bailey, *The Cultivated Conifers in North America*, The Macmillan Company.)

it is usually referred to as a fruit. Figure 28.18 shows the seed of *Torreya Californica* or the California nutmeg, a rare forest tree. The embryo is protected by an outer flesh called an **aril**, and an inner hard pit. Both these tissues may be derived from the integuments of the ovule; hence, the structure is morphologically a naked seed. The seed of the ground pine (*Taxus canadensis*), which forms a shrubby ground cover in the forests of the North Atlantic states, is similar (Fig. 28.19). The outer red aril almost en-

Figure 28.14 (*left & above*). *Tsuga mertensiana*. A, mature tree X ½; B, ovulate cones. Note arrangement of needles on branch.

Figure 28.14 A

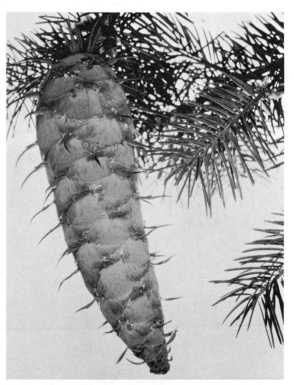

Figure 28.15. Branch of *Pseudotsuga menziesü* (Douglas fir) with ovulate cone X ½.

closes the seed and drops away when the seed is mature.

Yews are not common in the United States. Two examples have been mentioned; two other species occur; the western yew and the stinking cedar of the southern states. Possibly the best-known yew is the English yew, which, because of the excellent bows that were made from its wood, is closely linked with English history and folklore.

The leaves of the yews, as illustrated in Fig. 28.19, are flat and linear, not unlike those of the firs.

Life History of a Conifer

The following outline of a gymnosperm life cycle is based largely on the genus *Pinus*. It differs from the life cycles of other genera, mainly in requiring three years for completion. The significant stages

Figure 28.16. *A, Juniperus occidentalis* growing at 8000 feet in the Sierra Nevada; *B*, branch with staminate cones, *C*, branch with ovulate cones. *B*, *C* ×1.

of pollination, fertilization, and the maturation of the seed occur within a few inches of each other on a given branch, thus making it an ideal subject for class study.

All conifers produce two kinds of spores—**microspores** and **megaspores**—borne in cones that are morphologically distinct. The two types of cones are known, respectively, as **staminate** (Figs. 28.20, 28.22D) and **ovulate cones** (Figs. 28.1B, C, 28.8).

Staminate Cone

Staminate cones average a half-inch, or less, in length by a quarter-inch in diameter. They are borne in groups, usually on the lower branches of the trees (Fig. 28.20). Each cone is composed of a large number of small scales (microsporophylls) arranged spirally on the axis of the cone. Two microsporangia develop on the under surface of each scale (Figs. 28.22E, F).

Figure 28.17. *Larix lyalli* (larch). A, mature trees; B, branch with needles; C, mature ovulate cones. B, C ×1.

Figure 28.18. A branch of *Torreya californica* with seed. ×1.

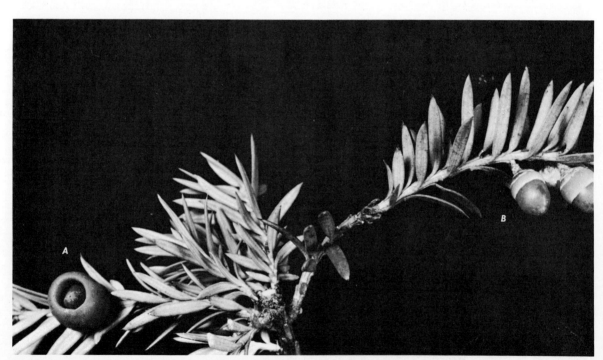

Figure 28.19. Branches of *Taxus* with seeds; note berry-like aril surrounding seed in *A*, and seed without aril in *B*. ×1.

28 *The Gymnosperms*

The stages of microspore development are quite similar to those of *Selaginella* and to the spores of mosses and ferns. Sporogenous cells give rise to spore mother cells, which are surrounded by a nutritive cell layer, the **tapetum.** Each spore mother cell undergoes meiosis and four microspores result. As usual, the microspore contains the haploid number of chromosomes. The nucleus within the newly formed microspore divides several times by mitosis, forming a **pollen grain** that contains two viable haploid nuclei and vestiges of several vegetative cells (Fig. 28.22*G, H*). The pollen grain is finally shed from the microsporangium. Enormous numbers of pollen grains are formed. They are light in weight and bear two **wings** which facilitate dispersal by wind.

Ovulate Cone

The ovulate cone, when mature, is the well-known cone of the pines, firs and other genera of conifers (Fig. 28.8). Each cone is composed of an axis, upon which are borne, in a spiral fashion (Figs. 28.8*H*, 28.22*A*), a number of woody scales. Young ovulate cones appear at the tips of young branches in the early spring (Figs. 28.1*B*, 28.22*A*). Two ovules, each enclosing a single megasporangium, develop on the upper surface of the **ovuliferous scales** (Figs. 28.21, 28.22*B, C*). A longitudinal section through a young cone is shown in Fig. 28.21.

The ovules first appear as small protuberances on the upper surface of this scale, close to the axis of the cone. A protective layer of cells, the **integument,** develops early on the outer surface of the ovule. In the end of the ovule, nearest the axis of the cone, there is a small opening, the **micropyle,** through which pollen grains may enter.

In pine, one megaspore mother cell lies in the center of each ovule (Fig. 28.22*J*); several are contained in the ovules of the redwoods and cypresses. Like the microspore mother cells, the megaspore mother cell is surrounded by a nutritive tissue, here called the **nucellus.** (Fig. 28.22*J*). It is in reality a megasporangium because it is within this tissue that the megaspore arises.

Female Gametophyte

The megaspore mother cell soon divides by meiosis. Four megaspores, usually arranged in a single row of four cells, or a linear tetrad, (Fig. 28.22*J*), result from the meiotic division of each spore mother cell. The nucleus within each mega-

Figure 28.20. End of a branch of *Pinus* with a cluster of staminate cones holding ripe pollen. ×1.

Figure 28.21. Median longitudinal section of young ovulate cones of *Pinus* at time of pollination. ×6.

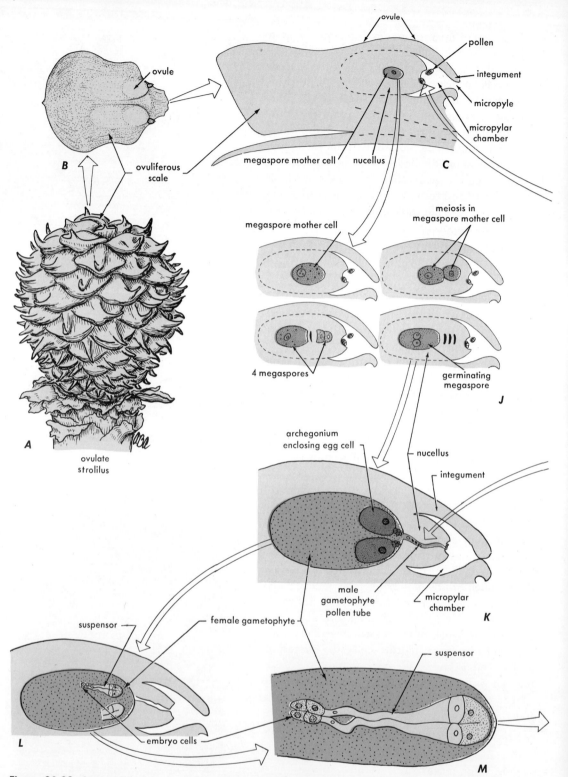

Figure 28.22. Stages in the life cycle of a pine. *A*, young ovulate cone, ×2; *B*, scale from ovulate cone showing two ovules, ×2; *C*, section of ovulate cone showing megaspore mother cell and pollen in pollen chamber, ×50; *D*, staminate cone, ×2; *E*, scale from staminate cone, ×3; *F*, cross section of staminate scale, ×50; *G*, winged pollen grain, ×75; *H*, development of male gametophyte in pollen grain; *I*, pollen tube is penetrating female gametophyte, ×50; *J*, formation of linear tetrad of megaspores, ×75;

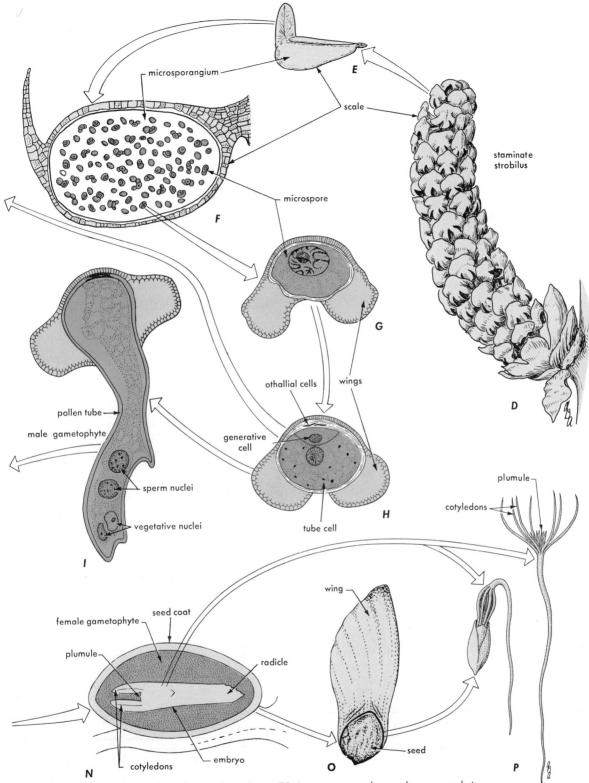

microsporangium

scale

E

staminate
strobilus

microspore

F

D

microspore

G

wings

pollen tube

male gametophyte

othallial cells

generative
cell

sperm nuclei

vegetative nuclei

tube cell

H

I

plumule

cotyledons

wing

female gametophyte   seed coat

plumule

radicle

seed

embryo

cotyledons

N

O

P

K, female gametophyte with egg cell and pollen tube, ×75; *L,* suspensor and proembryo sporophyte
within female gametophyte, ×75; *M,* proembryo, ×100; *N,* section of seed showing seed coats, female
gametophyte, and embryo sporophyte, ×50; *O,* winged seed, ×2; *P,* seedlings ×1. (*G, H, I,* after
Coulter and Chamberlain, *Morphology of the Gymnosperms,* University of Chicago Press.)

Phylum Coniferophyta   627

spore has the haploid number of chromosomes. Generally, only one of the four megaspores develops into a female gametophyte; the other three degenerate. The germination of the megaspore and growth of the female gametophyte progress very slowly. Several months are required in most conifers, and 13 months are required in pine.

The development of the female gametophyte takes place entirely within the ovule. There are approximately eleven mitotic divisions of the megaspore before cell walls begin to appear between the newly formed nuclei. Walls gradually form, however, resulting in a small mass of gametophytic tissue, completely enclosed by the diploid cells of the ovule. The sporophytic tissue of the ovule adjacent to the female gametophyte is the nucellus. While the cell walls are being laid down in the developing female gametophyte, two or more archegonia differentiate at its micropylar end. Reference to Figs. 28.22K, 28.23B shows that the ovule at this stage consists of the integuments, nucellus, and the female gametophyte, which contains several archegonia, each with its enclosed egg. Directly beneath the micropyle is a space, the **micropylar chamber.** Note the nucellar tissue lying between the micropylar chamber and the archegonia.

Pollination

It will be recalled that pollination, in typical flowers, is the transfer of pollen from the anther to the stigma; in conifers, it is the transfer of pollen from the staminate cone to the ovulate cone. Conifer pollen is wind-blown. Since, in many species, the ovulate cones are borne on the higher branches of the tree and the staminate cones are concentrated on the lower branches, and since the pollen does not blow vertically upward, cross-pollination is usual. Pollination occurs in most conifers in the early spring soon after the ovulate cone emerges from the dormant bud (Fig. 28.1B)—at about the time of meiosis. At this age the scales of the young cones turn slightly away from the axis so that the pollen grains sift down to the axis of the cone. Here, they come in contact with a sticky substance secreted by the ovule. As this material dries it draws some pollen grains through the micropyle into the micropylar chamber where they lodge close to the developing female gametophyte (Fig. 28.22C). The pollen grain develops slowly into a male gametophyte (pollen tube) (Figs. 28.22I, 28.22K). Thus the

male and female gametophytes of the conifers develop to maturity in close proximity within the ovule. They are both dependent upon nucellar tissue for nourishment and protection.

Male Gametophyte

A layer of nucellar tissue separates the female gametophyte from the micropylar chamber. The pollen grain lodged in the micropylar chamber sends out a tube, the **pollen tube,** which grows into the nucellar tissue. The tube may branch slightly, and it apparently secretes enzymes that digest the nucellar tissue (Figs. 28.22K, 28.23B). Several nuclear divisions occur in the tube but no cell walls are formed. Two of the last-formed nuclei are **sperm nuclei.** This branched pollen tube, containing two sperm nuclei and several vegetative nuclei, is the **male gametophyte** (Fig. 28.22I).

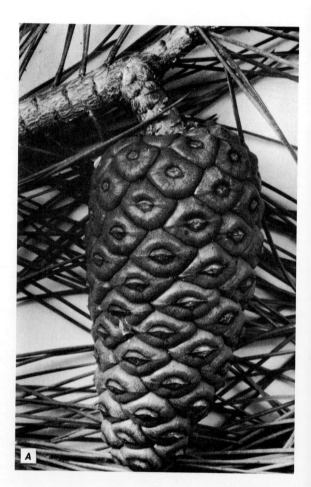

Fertilization

The development of the male and female gametophytes is so coordinated that the egg is formed and ready for fertilization when the pollen tube, containing two sperm nuclei, has reached the archegonium. In pine this development requires about a year. At the time of fertilization pine cones are generally green and the scales are tightly closed, showing the spiral pattern of arrangement (Fig. 28.23). The sperm nuclei, together with other protoplasmic contents of the pollen tube, are discharged directly into the egg cell. Sperm nuclei do not possess cilia and hence are not actively motile. One sperm nucleus comes in contact with the egg nucleus and unites with it. The non-functioning sperm nucleus and the other protoplasmic material discharged into the egg cells soon undergo disorganization.

Embryo

The formation of the embryo is preceded by the development of a relatively elaborate **proembryo.** This structure, in pines, consists of four tiers of four cells each (Fig. 28.22L, M). The four cells farthest from the micropylar end of the proembryo may each develop into an embryo. The intermediate cells are the suspensor cells; they elongate greatly and push the embryo cells deep into the female gametophyte (Fig. 28.22M). While this development is taking place, the female gametophyte continues to grow. It enlarges and becomes packed with food to be used not only for the growth of the embryo but also as a reserve in the seed.

It will be recalled that the female gametophytes usually contain two or more archegonia apiece. Since the egg in each archegonium may be fertilized and since each of the four embryo-forming cells may

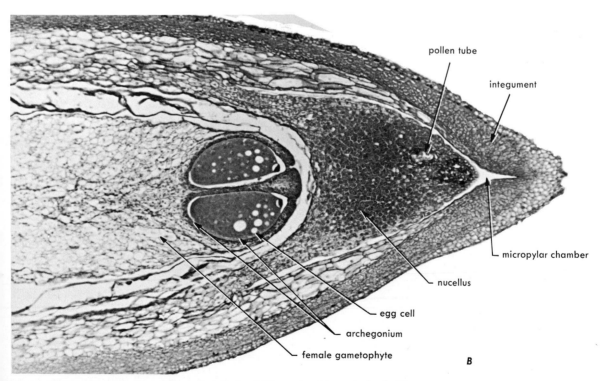

Figure 28.23. Ovulate cones of *Pinus.* A, left about 13 months old, ready for fertilization; B, section through tip of ovule. (B, slide courtesy of Triarch Products.)

Table 28.1 *Comparison of ovule and seed of pine*

TIME OF POLLINATION	TIME OF FERTILIZATION	MATURE SEED
Ovule	Ovule	Seed
Megaspore mother cell	Female gametophyte	Endosperm
	Egg cell → Zygote	Embryo
Pollen grain	Pollen tube sperm nucleus	
Megasporangium	Megasporangium or nucellus	Perisperm
Integument	Integument	Seed coat
Micropyle	Micropyle	Micropyle

Table 28.2 *A generalized life history of a pine*

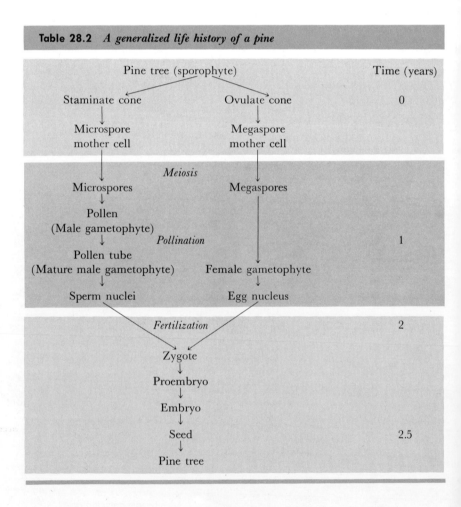

give rise to an embryo, a number of embryos may develop in every seed. Normally, however, only one embryo survives, but seeds with two well-formed embryos are not rare.

The mature embryo consists of several **cotyledons** or seed leaves, the **shoot tip,** and **hypocotyl,** and a **radicle** or rudimentary root (Fig. 28.22N). The embryo is embedded, as previously mentioned, in the enlarged female gametophyte. Both embryo and female gametophyte are surrounded by (1) the dried nucellus, or **perisperm** and (2) a hard protective **seed coat,** formed from the integument of the ovule. The whole structure is the **seed** (Fig. 28.22O). In many pines the seeds are winged. The development of the seed is summarized in Table 28.1.

Normally, in pines, the seed matures some twelve months after fertilization, and since fertilization is effected about thirteen months after pollination, two years intervene between the initiation of the ovule and the formation of the seed. The young pine seedling (Fig. 28.22P) has several cotyledons. A generalized life history of a pine is shown in Table 28.2.

Gymnosperm seeds may remain dormant for many years, and some may remain embedded in the old mature cones for six years or more. Heat causes the cones of some species to open and release the seeds. This behavior has considerable survival value since large numbers of seeds are released after fires, and injured trees are thus replaced by the young ones. In many species, however, the seeds are shed soon after they are mature.

SIGNIFICANT FEATURES OF THE GYMNOSPERMS

1. All forms in this group bear seeds which are not surrounded, or protected, by an ovary wall; a true fruit is never formed.

2. All are trees or shrubs of varying form; many have needle or scale leaves, and most of these forms are evergreen. Some have broad leaves and in others the leaves are palmlike.

3. With few exceptions, vessels are lacking in the xylem, and companion cells are absent from the phloem.

4. The best known gymnosperms of the Northern Hemisphere bear cones of one sort or another. In the yews the seed is surrounded by a fleshy aril.

5. Sexual reproduction involves:

(a) The production of microspores (through meiosis) in staminate cones and their transformation, by several nuclear divisions in the microspore, into pollen grains.

(b) Pollination by wind.

(c) The development of an ovule on the upper surface of the scales of the ovulate cone.

(d) The production of megaspores through meiosis within the ovule.

(e) The development of a female gametophyte within the ovule from one megaspore. Several reduced archegonia form within the female gametophyte.

(f) The penetration of a pollen tube or male gametophyte through the nucellus and discharge of the sperm cells into the egg cells. A few gymnosperms have motile sperms cells.

(g) The formation of a proembryo from the zygote, one end of which gives rise to the embryo.

(h) The enlargement of the female gametophyte, which stores food.

(i) The seed, which consists of a hard seed coat, the perisperm, the food-storing female gametophyte, and the embryo.

CLASSIFICATION

Phylum	Pteridospermophyta
Phylum	Cycadophyta
Genus	*Cycas*
Genus	*Zamia*
Genus	*Dioön*
Phylum	Ginkgophyta
Genus and species	*Ginkgo biloba*
Phylum	Coniferophyta
Family	Pinaceae
Genus and species	*Pinus monophylla*
Genus	*Abies*
Genus	*Picea*
Genus	*Tsuga*
Genus and species	*Pseudotsuga menziesii*
Genus	*Juniperus*
Genus	*Larix*
Family	Taxaceae
Genus and species	*Taxus brevifolia*
species	*canadensis*
Genus and species	*Torreya Californica*
species	*taxifolia*
Phylum	Gnetophyta
Genus	*Ephedra*

The Angiosperms

The angiosperms, or flowering plants, are the dominant plants of the world today. They include nearly all the crop plants of orchard, garden, and field. Hardwood forests, shrublands, grasslands, and deserts are composed chiefly of flowering plants. In fact, except for the coniferous forests and the waters of the world, the conspicuous and dominating vegetation wherever we go is mainly flowering plants. They have bodies showing great variation, from the simple stemless, free-floating duckweed (*Lemna*) through a whole series of herbaceous types to shrubs of varying sizes and finally to trees such as the oaks and beeches. Structurally they are adapted to a land habitat and certain forms, as exemplified by the cacti (Fig. 18.33), are able to live and grow in very dry deserts. We have already studied the structure and physiology of the angiosperm plant body in considerable detail. The angiosperms constitute the phylum Anthophyta which is divided in two classes, the Monocotyledonae and the Dicotyledonae.

THE FLOWER, FRUIT, AND REPRODUCTION

The outstanding and unique structure of the angiosperms is the *flower* (see color plate 5, 15.2, 15.5, 29.1) and the *fruit* (Figs. 16.1, 16.5) (which develops from the ovary of the flower) (Figs. 15.5, 16.1, 16.2) with its enclosed *seeds* (Figs. 15.7, 16.8, 16.2). The flower is a shoot bearing floral leaves. In a complete flower (Figs. 15.1, 29.1) the floral leaves are sepals, petals, stamens, and carpels. Some flowers have only the essential reproductive structures (Fig. 15.11*A, B*), stamens and carpels. Other flowers are unisexual (Fig. 15.12), that is, have either stamens or carpels, not both. Other variations in flower structure are described on pages 280–287.

Free water is not needed for fertilization in the angiosperms; motile, ciliated sperms are not produced by any members of this phylum. Sexual reproduction is made possible by the production of pollen (Figs. 15.4, 15.27, 29.1*C*), by pollination, and by the growth of a pollen tube (Fig. 29.1*D*), phenomena which make possible the eventual union of the egg and sperm (Fig. 29.1*A, B, C*).

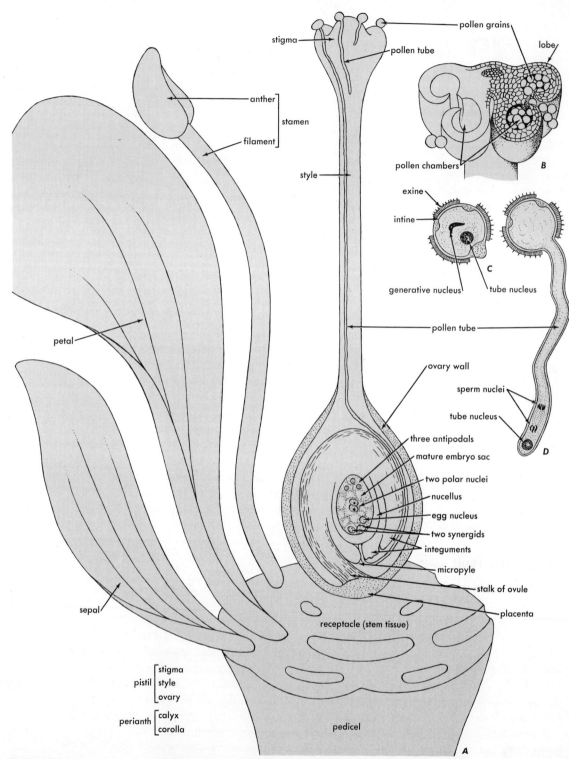

Figure 29.1. *A,* diagram of a flower; *B,* cross section of an anther; *C,* mature pollen grain; *D,* germinating pollen grain. (*C* and *D* redrawn from Bonnier and Sablon, *Cours de Bontanique,* Librairie Generale de l'Enseignement.)

Table 29.1	*Comparative terminology of reproductive parts*
Anthophyta	**other phyla**
Stamen	Microsporophyll
Pollen sac	Microsporangium
Microspore	Microspore
Pollen grain	Germinated microspore or young male gametophyte
Pollen tube	Mature male gametophyte
Carpel	Megasporophyll
Nucellus	Megasporangium
Megaspore	Megaspore
Embryo sac	Female gametophyte
Ovule	Ovule

All flowering plants bear seeds (Figs. 2.14, 16.5, 16.21). In this respect they resemble the gymnosperms. In the angiosperms the seeds are borne within a closed structure, the ovary, which eventually becomes the fruit (Fig. 16.1). In the gymnosperms, on the other hand, the seeds are not surrounded completely by any structure, but are borne on the upper surface of a scale (Fig. 28.2).

Flower Structure Compared with Reproductive Structures of More Primitive Forms

Since the flower is the structure of the angiosperms adapted for sexual reproduction, its parts should, to some degree, be comparable to the reproductive structures of more primitive forms. Let us examine the flower with this in mind. The petals and sepals, the showy and protective whorls of floral leaves, are not directly represented in any of the lower forms. Meiosis occurs in the anthers of stamens and microspores are produced in pollen sacs (Figs. 15.4, 29.1*B*). The stamen, then, is a microsporophyll and a pollen sac is a microsporangium. The pollen grain itself, containing, as it does, two nuclei or cells (Fig. 29.1*C*), would be the first stage in the development of the male gametophyte. The pollen tube would constitute the mature male gametophyte. Meiosis takes place within the ovule. The tissue immediately surrounding the developing megaspore is the nucellus, which is borne within the ovule (Figs. 15.10, 29.1*A*). It will be recalled that the ovules, in many instances, are attached to the margins of the carpels, which are floral leaves (Fig. 15.7). Thus the carpel, since it bears megaspores, is a megasporophyll and

the nucellus is the megasporangium. Generally one of the megaspores gives rise to an embryo sac or female gametophyte (Fig. 29.1*A*). These two sets of terms are summarized in Table 29.1.

Life Cycle

With these comparisons in mind, let us proceed with an outline of the essential steps in the life cycle of an angiosperm. The dominant vegetative plant is the sporophyte and, as already emphasized, the flower is the organ concerned with sexual reproduction.

The Male Gametophyte

The anther is the particular part of the stamen responsible for the production of microspores, which are produced by meiosis from a microspore mother cell within the pollen sac (Fig. 15.4). Each microspore contains one haploid nucleus, which soon divides into generative and tube nuclei (Fig. 29.1*C*). While this mitosis is taking place, a heavy, sculptured wall forms about the microspore and a mature pollen grain results. The pollen is now shed and conveyed in one fashion or another (page 297) to a stigma where it germinates; a pollen tube penetrates the stigma (Fig. 29.1*A* and *D*) and grows through the style, down to the ovule within the ovary. The tube nucleus, depending upon the species of plant, may degenerate early or may persist until the pollen tube is well developed. In any event, the generative nucleus usually divides within the pollen tube into two sperm nuclei (Fig. 29.1*D*). Each sperm nucleus, with its associated cytoplasm, is a sperm cell. The pollen tube with the sperm cells and the tube nucleus, if present, constitute the mature male gametophyte.

The Female Gametophyte

An enlarged cell within nucellar tissue of a young ovule undergoes meiosis and forms four megaspores arranged in a row (page 277). While this process is occurring, integuments form about the nucellar tissue, resulting in the formation of a typical ovule (Fig. 15.10). The three megaspores closest to the micropyle generally disintegrate. The remaining megaspore undergoes three successive mitotic divisions, resulting in eight nuclei of the embryo sac or female gametophyte: one egg cell, two synergid cells, one endosperm mother cell with two polar nuclei, and three antipodal cells (Fig. 29.1*A*).

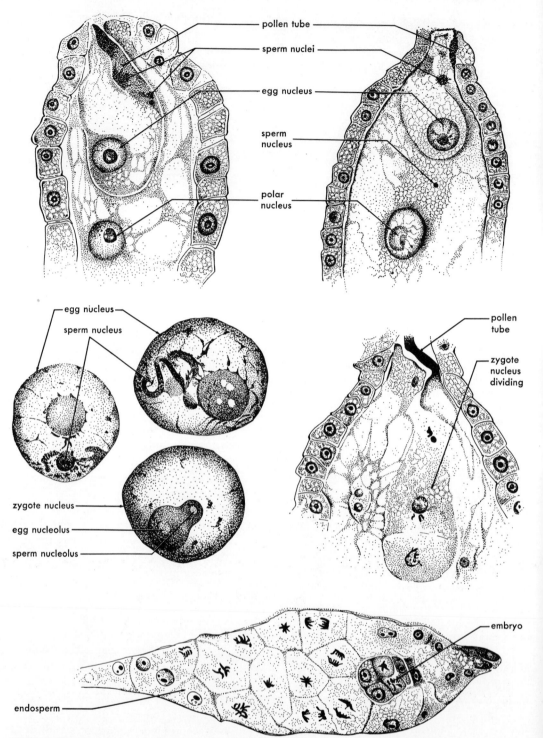

Figure 29.2. Fertilization and embryo development in *Crepis.* (Redrawn from H. Gerassimova, *La Cellule,* **42,** 103.)

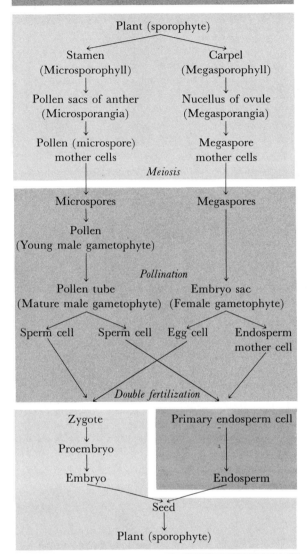

Table 29.2 *Generalized life history of an angiosperm*

Plant (sporophyte)

Stamen (Microsporophyll) — Carpel (Megasporophyll)

Pollen sacs of anther (Microsporangia) — Nucellus of ovule (Megasporangia)

Pollen (microspore) mother cells — Megaspore mother cells

Meiosis

Microspores — Megaspores

Pollen (Young male gametophyte)

Pollination

Pollen tube (Mature male gametophyte) — Embryo sac (Female gametophyte)

Sperm cell — Sperm cell — Egg cell — Endosperm mother cell

Double fertilization

Zygote — Primary endosperm cell

Proembryo

Embryo — Endosperm

Seed

Plant (sporophyte)

Fertilization and Seed Development

With the penetration of the pollen tube to the mature embryo sac, the stage is set for fertilization. In all previous forms we have studied, this involves only the union of the egg cell with a sperm. In the angiosperms, fertilization involves not only the union of the egg cell with the sperm cell but, in addition, the union of a second sperm cell with the endosperm mother cell to form the primary endosperm cell (Fig. 29.2). Since there are two nuclei in

the endosperm mother cell the nucleus of the primary endosperm cell will have three sets of chromosomes (page 277). The union of the two sperms, one with the egg, the second with the endosperm mother cell, is **double fertilization.**

The zygote generally develops a small proembryo, from one end of which the typical embryo of the seed develops. The fate of the primary endosperm cell depends upon the species. Probably in all forms it develops into an endosperm which plays some role in nourishing the developing embryo (Fig. 16.21). In a few genera, notably the grains (page 314, Fig. 16.14), the endosperm enlarges and persists in the seed to form the main source of nourishment for the young seedling. There are three general types of seeds, depending upon the various combinations of cotyledon number and the location of stored food (Figs. 16.22, 16.23, and 16.14). A seed always includes a dormant embryo and a food supply to enable the young seedling to establish itself. The food is generally stored in the cotyledons of the embryo or in the endosperm.

With the germination of a seed and the development of the seedling into a flowering plant, the life cycle of an angiosperm is completed. The essential steps are shown in Table 29.2.

Comparative Reproductive Structures

Certain comparative details between the structures and life histories of members of the gymnosperms and angiosperms are given in Table 29.3. The pine seed and the corn grain are compared in Table 29.4. Details concerning the Psilophyta, Lycophyta, Sphenophyta and Pterophyta are given in Table 27.1.

In all Pterophyta the sporophyte is the dominant generation and both gametophyte and sporophyte generations are independent. In the gymnosperms and the angiosperms there occurs a further reduction in the size and complexity of the gametophyte to a condition of complete dependency on the parent sporophyte. This is accompanied by an overall increase in the general complexity of the sporophyte. It can be stated that, as a guiding principle, the more advanced forms in these phyla are more highly adapted to grow and flourish in a dry land habitat than are the more primitive forms. The most obvious developments designed to accomplish this adaptation are (a) those which remove any dependence of the plants on free moisture for fertilization, and (b) adaptations, both vegetative and

Table 29.3 *Comparison of the conifers and the seed plants*

GENERATION	GYMNOSPERMS	ANGIOSPERMS
Sporophyte	Roots, stems, and leaves	Roots, stems, and leaves
	Trees and shrubs, no herbs	Trees, shrubs, and herbs
	Stem: one core of xylem surrounded by phloem, vessels in one shrubby genus only	Stem: one core of xylem surrounded by phloem, or several vascular strands with phloem exterior
	No companion cells	Companion cells and sieve tube elements
	Sieve cells	
	Pith present in stems, not in roots	Pith in stems, not in roots, of dicots; in some monocot roots
	Cambium develops in all species	Cambium in perennial forms; absent generally in annuals
	Staminate and ovulate cones, generally, with staminate and ovulate scales	Flowers, stamens, and carpels
	Heterospory	Heterospory
	Microspores in microsporangia	Microspores in anther sacs
	Megaspores in megasporangia or nucellus	Megaspores in nucellus
	Ovules present, exposed on scales	Ovules present, enclosed within carpel
	Embryo develops within a seed	Embryo develops within a seed
Gametophyte	Pollen tube: male gametophyte	Pollen tube: male gametophyte
	Female gametophyte within ovule	Female gametophyte (embryo sac) within ovule
	Antheridium practically absent	Antheridia absent
	Sperm cells formed by pollen tube	Sperm cells formed by pollen tube
	Motile sperms in several primitive genera, but free water not needed for fertilization	Nonmotile sperms only formed in pollen tube
	Nonmotile sperms in all other forms	
	A very greatly reduced archegonium completely embedded in female gametophyte	Archegonium only suggested by synergid cells of the embryo sac
	Embryo within female gametophyte	Embryo within seed coats; female gametophyte absent

Table 29.4 *Comparison of pine seed and corn grain*

PINE SEED	CORN GRAIN
Seed lacks pericarp	Seed coats fused with pericarp
Numerous cotyledons	One cotyledon
Food stored in female gametophyte	Food stored in endosperm
Radicle	Radicle protected by coleorhiza
Plumule protected by cotyledons	First leaves protected by coleoptile

reproductive, which enable the plant to grow and to reproduce on dry land.

Protection of the Female Gametophyte

The gynoecium of the angiosperm flower is essentially a modified leaf, or leaves, enclosing an ovule, in which the female gametophyte is to be found. The ovary, as we know, eventually develops into a fruit. This envelopment of the ovules, and subsequently the seed, by the ovary wall is a new development which sets the angiosperms apart from all other groups of plants. Recall that in ferns the egg cells have little protection from their immediate environment. The gametophyte thallus in ferns is a delicate structure, and the venter of the archegonium opens directly, by way of the ventral canal, to external moisture (Fig. 27.30*I*, *J*). The female gametophyte of *Selaginella* (Fig. 27.14*M, N*) may be afforded some protection by the parent sporophyte, but the motile sperms must reach the egg under their own power and the archegonium opens to the external environment by a very short neck canal. In pine the female gametophyte has gained the protection of the integuments of the ovule and the scales of the ovulate cone, as well as being retained within the megasporangium. Pollen must still, however, be deposited very close to the female gametophyte in order to effect fertilization. Pine no longer requires free water for movement of sperms and, except for a few primitive gymnosperms, motile sperms are absent from this phylum.

The female gametophyte (embryo sac) of the angiosperms has gained, in the ovary wall, an added protective barrier. However, this has necessitated

further adaptations to enable the sperms to pass this added protection: (1) the development of the receptive stigma (*frontispiece*) and (2) the growth of the pollen tube through the style (Fig. 29.1).

Size of Gametophytes

Reduction in the size of the gametophytes has thus proceeded to a point in which the male gametophyte, the pollen tube, consists of one vegetative nucleus and two sperm cells (Fig. 29.1*D*). The female gametophyte is a seven-celled structure, one of whose cells is an egg cell. A second cell, the endosperm mother cell, is an innovation in that it too is receptive to a sperm cell. The remaining five cells are vegetative cells (Fig. 29.1*A*). It should be pointed out that the embryo sacs of some angiosperms have more cells, while a few have less. Should the gametophyte stages be completely eliminated, gametes, instead of meiospores, would result from meiosis. This actually occurs in animals and in some diatoms (page 466). The haploid phase in these forms is reduced to the sperm and egg cells.

Nourishment for the Developing Embryo

Double fertilization is still another difference between the life cycle of the angiosperms and more primitive forms. In the latter forms, nourishment for the young developing sporophyte is generally provided by the female gametophyte (Figs. 27.14*N*, *M*, 28.22*L*) or, in the case of many ferns, by the simple monoecious gametophyte (Fig. 27.30*J*). In the angiosperms, food for the developing seedling is stored either in the endosperm (Fig. 16.14) or in the cotyledons of the embryo itself (Fig. 16.22).

The Fruit

Fertilization in the angiosperms, as in all other forms, stimulates the zygote and certain surrounding tissue to further development. In the angiosperms the ovary wall is stimulated to produce a fruit, and the fruit is a development originating with, and characteristic of, the angiosperms. The protection of the seed from its surroundings by the ovary wall is accompanied by various devices for releasing the seed from its enclosure or otherwise bringing it under conditions favorable for germination. Among these adaptations may be mentioned the different sorts of dehiscence (Figs. 16.10, 16.11), various kinds of fleshy fruits (Figs. 16.7, 16.20), and the reduction of the ovary wall so that it becomes indistinguishable from the seed coats (Fig. 16.14).

Evolution in the Angiosperms

If, as assumed, the angiosperms represent the culmination of an extensive evolutionary development, it would appear likely that, within the phylum itself, evolution has been, and probably still is, active. In other words, families of angiosperms must be related and it should be possible to arrange the families in an order that would give some indication of their relationship. Some families will be found to be more primitive than others; and from the more primitive types the more advanced and specialized forms have presumably arisen (see chapter 19). What is the primitive floral type? What have been the trends in the evolution of the flower?

Figure 29.3. Ranunculaceae. Flower and fruit of buttercup (*Ranunculus*). A, flower stalk; B, lengthwise section of flower; C, floral diagram; D, cluster of fruits; E, a single fruit. (A, B, and D, redrawn from Korsmo.)

What system of classifying flowering plants will be most truly natural?

The Besseyan System

Many systems of classifying plants have been proposed. Two are in general use in America today. One of these was proposed by Engler and forms the basis for the arrangement of plants in many books on classification. A second system that has found much favor with American botanists is described here. This system (Besseyan) regards the order Ranales as being the most primitive. In this order the Magnoliaceae is one of the more primitive families, with the Ranunculaceae somewhat more advanced. The Christmas rose (Fig. 15.2), magnolias (Figs. 15.3*A*, 16.10*A*), and buttercups (Fig. 29.3) are representatives of these two families, and while both have flowers with many characteristics in common, the fact that magnolias are trees or shrubs and most Ranunculaceae are herbaceous forms, and geologically trees appeared before herbs (Table 30.1), indicates that the Magnoliaceae is the more primitive family. In both magnolias and buttercups the flower parts are arranged as follows.

1. The floral axis is elongated.
2. All, or at least some, of the flower parts are spirally arranged.
3. Stamens and carpels are numerous and separate.
4. Sepals and petals are numerous and separate.
5. The flowers are perfect and complete.
6. The floral symmetry is regular.
7. The perianth segments and stamens are attached to the receptacle *below* the points of attachment of the carpels (i.e., the flowers are hypogynous).

These, then, are a group of characteristics indicative of primitive flowers; changes in any of them indicate more advanced forms. The principal tendencies in the evolution of the flower, according to the Besseyan system, are changes in these characteristics and they may be enumerated as follows:

1. From an elongated to a shortened floral axis.
2. From a spiral to a whorled condition of floral parts.
3. From numerous and separate stamens and carpels to few and coalesced stamens and carpels.
4. From numerous and separate sepals and petals to few and coalesced sepals and petals.
5. From complete and perfect flowers to incomplete and imperfect flowers.
6. From regular to irregular flowers.
7. From hypogyny to epigyny.

According to the Besseyan view, there were at least three main lines of advance from the primitive Ranalian type of flower as exemplified by magnolias and buttercups. These lines culminated in (1) the mints (Labiatae), (2) the asters (Compositae), and (3) the orchids (Orchidaceae). In comparing plants with evolution in mind, it must be emphasized that not only do certain characteristics have more weight than others, but groups of characteristics must be given thoughtful consideration. For example, epigyny is thought to be an advanced condition. However, it should be noted that all flowers in one of the three main lines of ascent are hypogynous. This line culminates in the mints and places this family at a level of evolutionary advance comparable to that held by the Compositae and Orchidaceae.

SELECTED FAMILIES OF ANGIOSPERMS

Let us now proceed to examine selected families of angiosperms with two objectives in mind: (1) to learn the characteristics of some important angiosperm families and (2) to discover how these families fit into the Besseyan system of angiosperm classification. In our discussion of these matters so far, floral characteristics have been emphasized. The single exception is that of the tree form of the magnolias versus the herbaceous habit of the buttercups. The emphasis is placed on floral parts because their form and structure are little influenced by the external environment, and varies much less from generation to generation than do vegetative parts. Because of this constancy in structure, floral parts have a much greater value in both plant identification and in tracing evolutionary relationships than do vegetative characters. However, the latter cannot be neglected altogether, and many of the characteristics discussed in detail in the chapters on the stem (page 107), leaf (page 185), and fruit (page 305), are used in identification and in evolutionary studies. It might be well to review briefly and independently the morphological details presented in these chapters.

Since the magnolia, tulip tree, and buttercups are thought to be primitive forms, we shall begin our discussion with these families, then proceed to follow through a line of ascent ending in the mint family. We shall consider the line culminating in the composites next and complete our discussion with the line of monocotyledonous families. It should be

further noted that the lines of development are not straight but branched and tree-like. The arrangement of the families discussed here as they occur in the Besseyan system is shown in Fig. 29.4.

From Ranales to Labiatae

In this line of ascent from the Ranales, no change has occurred from the condition of hypogyny; all flowers, in all families in the line are hypogynous. Syncarpy occurred very early, as did a reduction in the number of floral parts. Other types of coalescence and the change to irregular flowers appeared somewhat later but still early in the line of ascent. The families involved here are the Magnoliaceae, the Ranunculaceae, the Malvaceae, the Cruciferae, the Solanaceae, the Salicaceae and the Labiatae.

The Malvaceae is a clearly marked family with relatively primitive characters as shown by its regular, perfect, and hypogynous flowers. There are numerous stamens with coalesced anthers, and the five or more carpels are more or less firmly united. In the Cruciferae, flower parts are reduced in numbers and only the carpels are coalesced. In the Solanaceae, the flowers, while perfect and regular, show coalescence of sepals, petals, and carpels, with the stamens adnate to the corolla tube. The willows (Salicaceae) are an example of a family in which the reduction in both number and size of floral parts has resulted in an advanced condition. The flowers are imperfect and grouped in catkins, and both calyx and corolla are lacking. The pistillate flowers are hypogynous and, in the staminate flowers, the numbers of stamens has been reduced to one or two. The willows are a part of a branch from the main line in which wind, rather than insect, pollination is the rule. The Labiatae, or mints, represent the most advanced family in this line of ascent. The flowers, even in this advanced family, are hypogynous; they are, however, irregular; there is a reduction in the number of parts to four or two; and the sepals, petals, and carpels are coalescent. The stamens are not only adnate to the corolla tube but are also highly specialized for insect pollination. Let us consider each family in more detail.

Magnoliaceae

The Magnoliaceae shares with the Ranunculaceae the distinction of being a very ancient group of plants. The tulip tree, *Liriodendron,* as shown by fossil remains, once had a very widespread distribution. Most species of the Magnoliaceae are trees

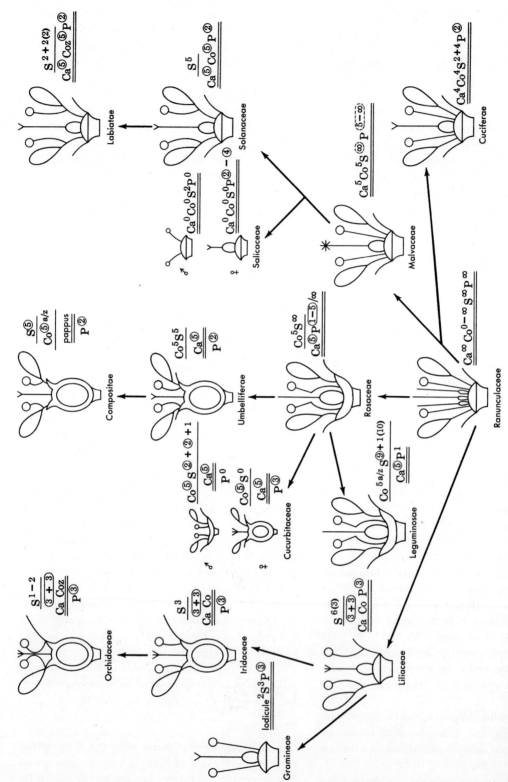

Figure 29.4. Diagram showing the evolutionary tendencies in the Angiospermae according to the Besseyan system. (Courtesy of McMinn.) Key to symbols is on page 000.

or fairly large shrubs. The magnolias themselves are magnificent trees, *Magnolia grandiflora* growing to 100 feet or more with its stiff, large, evergreen leaves and large white blossoms seven to eight inches across. The magnolias are largely warm climate species, but the tulip trees will stand northern winters. There are some 10 genera and 80 species in the family. The leaves are alternate, mostly entire, and pinnately veined with large stipules generally enclosing the buds.

Flowers are perfect, regular, and with distinct, separate parts. There are six or more petals. There are generally numerous stamens and carpels all arranged spirally on the receptacle (Fig. 15.3*A*). The fruits (Fig. 16.10*A*) are follicles or samaras and are frequently grouped to form a cone-like body.

Ranunculaceae

Most members of the family are herbs, although there are a few small shrubs and woody climbers. The leaves are characteristically very much divided, a feature which has led to the common name of Crowfoot Family. There are about 300 genera and approximately 1200 species growing largely in the temperate and Arctic regions. Many of our common garden flowers belong to this family. Among these are *Delphinium, Aquilegia, Paeonia, Ranunculus, Anemone* and *Clematis*. The marsh marigold and hepatica, common spring flowers, are members of this family. One species, *Aconitum Napellus,* yields an important drug, and some of the wild *Delphinium* are poisonous to livestock.

The floral parts are quite similar in structure and arrangement to those of the Magnoliaceae. They indicate a primitive angiosperm family (Fig. 29.3). There are numerous separate parts to each whorl, and they are arranged in a spiral order on the receptacle. The flowers are generally regular and hypogynous. The fruit is a follicle, an achene, or sometimes a berry. There are exceptions to this arrangement of parts; for instance, the flowers of *Delphinium* are very irregular.

Cruciferae

The Cruciferae is a large, distinct family containing many cultivated forms, as well as some that are noxious weeds. Cabbage, cauliflower, broccoli, Brussel sprouts, kohlrabi, and kale are all varieties of a single species, *Brassica oleracea*. There is reason to think that kale was the original form. Wild

mustard, another *Brassica* species, is a noxious weed in grain fields, though sometimes used as cover crop in orchards. Radishes and stocks are other members of this family, as are many garden herbs. The family has a worldwide distribution in the temperate and subarctic zones; all are herbs. There are about 200 genera and 1800 species.

The flowers and fruits are very characteristic of the family. The flowers (Fig. 29.5*A*) are small but conspicuous because of their large numbers. They are complete, regular, and hypogynous. There are four sepals and four petals, usually somewhat sharply angled close to their midline and bending backward. There are six stamens, four longer than the other two. The single pistil is composed of four carpels and has two locules. The arrangement of the petals generally suggests a cross, hence the family name, Cruciferae or "cross-bearing." The fruit is a silique (Fig. 16.12), generally dehiscing in the typical fashion. The silique of the radish (*Raphanus*) does not dehisce.

Malvaceae

This is another large family of some 50 genera and 900 species. There are no important food-producing species, and the ornamentals in the family, such as *Hibiscus* and *Abutilon,* are not as conspicuous as roses, irises, and orchids. The cotton, however, taken from the seed coats of the various members of the genus *Gossypium,* make this family of plants as important from the viewpoint of politics, agriculture, and industry as the grasses, leg-

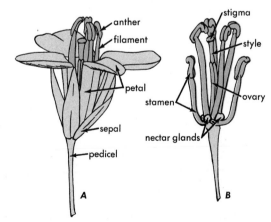

Figure 29.5. Cruciferae. Mustard flower (*Brassica*). A, all flower parts intact; B, sepals and petals removed. Note four long and two short stamens and the single pistil.

umes, and roses. There are herbs, shrubs, and trees in the family; their leaves are mostly palmately veined and lobed, with small deciduous stipules.

The flowers (Figs. 15.15*B*, 29.6) are regular, generally perfect, with five sepals usually united, and a tubular, five-lobed corolla. The many stamens are united by their filaments to form a tube surrounding the pistil and frequently adnate at its base to the five petals. The ovary is superior with generally two to many locules, each containing one to

Figure 29.6. Malvaceae. *A* and *B*, regular flowers of *Anoda* and *Malope*; *C*, double staminal column of *Sidalcia*; *D*, ovary and staminal column of *Anoda*; *E*, gynoecium of *Malope*; *F*, dehisced capsule or boll of *Gossypium* with cotton fibers; *G*, seed of *Gossypium* with attached cotton fibers. (Redrawn from L. H. Bailey, *Manual of Cultivated Plants*, The Macmillan Company.)

several seeds. In most species the number of carpels can be told from the number of styles and stigmas. The fruit of most species is a capsule, although in a few it is a berry. The cotton of commerce occurs as long hair or fuzz on the seeds, which are borne in large capsules.

Solanaceae

This is a large family with many tropical forms, but it is also well represented in the temperate regions. There are about 75 genera and over 2000 species, some 1200 of the latter being in a single genus, *Solanum*. To this family belong tomatoes, potatoes, tobaccos, eggplant, peppers, *Petunias,* and *Salpiglossis.* While the family is of world-wide distribution, most of the cultivated forms were brought under domestication in the Western Hemisphere. There are many poisonous and drug plants, such as belladonna, in the family. Even such a common plant as the tomato was long supposed to be poisonous, as appears to be indicated by its scientific name *Lycopersicon,* which means "wolf peach." There are many erect and climbing herbs in the family, also some shrubs and small trees. The alternate leaves are mostly entire, although in a few species they may be variously dissected.

The perfect flowers (Fig. 29.7) are mostly regular; the calyx and corolla are coalesced and generally five-lobed. The number of adnate, hypogynous stamens (five) corresponds to the number of corolla lobes with which they alternate. Sometimes one or more of the stamens may be sterile. The superior ovary normally has two locules (five in tomato), with a single style and a two-lobed stigma. The fruit is a berry or, less frequently, a capsule with many seeds.

Salicaceae

The willow family is comprised of only two genera (*Salix,* willows, and *Populus,* poplars), with about 200 species. They are mostly trees, but some shrubby forms are known. They are very abundant in the Northern Hemisphere, mostly in the temperate zones. Baskets are woven from branches of the basket willow and paper pulp is obtained from the trunks of one species. Willows are common along water courses, frequently overhanging or even choking mountain streams, to the ill comfort of fishermen. All species have alternate simple stipulate leaves.

The trees are generally dioecious and flowers are borne in catkins in the spring in advance of the leaves. The perianth is lacking; the one or more stamens and single pistil are found in the axils of bracts. There are mostly two stigmas (Fig. 29.8). The fruits are small capsules having in a single locule many seeds with long down or hairs. Willows have one or two stamens per flower while the poplar flowers have between eight and ten. Pussy willow is a familiar example of this family.

Labiatae

The mints, like the orchids and composite families, represent the supposed highest advance of one of the three lines of angiosperm evolution. It is a large family of considerable economic importance, largely because of the volatile oils produced by certain of its members. Peppermint, spearmint, thyme, sage, and lavender are examples. There are about 160 genera with 3000 species well distributed over the surface of the earth. There are herbs and shrubs in the family, generally with characteristic square stems. The leaves are simple and opposite or whorled.

The flowers (Fig. 29.9) may be showy, are generally irregular, and are usually complete, although some may be imperfect. The calyx is regular or two-lipped, partly coalesced and commonly five-toothed. The four or five parts of the corolla are coalesced characteristically to form two distinct lips, from which the name of the family is derived. There are four stamens, of two sizes, two shorter than the other two. The superior ovary comprised of four carpels is four-parted, each carpel containing a seed. The mature fruit consists of four single-seeded nutlets.

From Ranales to Compositae

The second line of ascent from the Ranales involves the Leguminosae, the Rosaceae, the Umbelliferae and the Compositae, with an offshoot family, the Cucurbitaceae. This large group of families is characterized by an early change from hypogyny to epigyny. Following this there is a division into two sublines, one being marked by a lack of coalescence and a retention of regular flowers. In the other subline coalescence and irregular flowers both occur in advanced forms.

The Rosaceae (rose family) is one of the more primitive families in this line of evolutionary devel-

Figure 29.7. Solanaceae. *A*, branch of *Lycopersicon showing leaf, flower, and buds; B*, longitudinal section of flower of *Lycopersicon; C*, cross section of tomato fruit; *D*, unopened flower of *Nicotiana; E*, longitudinal section of *Nicotiana* flower; *F*, capsule of *Nictotiana*, calyx is persistent. (*A, B, C, D, F* redrawn from L. H. Bailey, *Manual of Cultivated Plants*, The Macmillan Company.)

opment. There exists a considerable variety of flower structure within the family. The more primitive members of the family are regular, perfect, with numerous parts and with such a slight degree of perigyny as to be recognizable only with careful observation. This is exemplified by the strawberry (Figs. 16.6, 29.10). Changes within the family involve a reduction in the number of floral parts, a shift from slight perigyny to true perigyny (Figs. 15.16B, 29.10) and to epigyny (Fig. 16.20). Coalescence also occurs (Fig. 16.20). Members of the family generally have regular flowers.

The Leguminosae (pea family) are considered to be a more advanced family than the Rosaceae, even though their flowers are hypogynous. There is a reduction in number of parts, the gynoecium having only one carpel; coalescence of stamens and of some petals occurs and the corolla parts are irregular. The Umbelliferae represent a further advance over the Rosaceae in that the flowers are epigynous and exhibit syncarpy (Fig. 29.13). This line of ascent finds its climax in the Compositae, whose flowers all possess advanced characters, such as reduction in number of parts, modification of parts (pappus), irregular flowers, some imperfect flowers, the presence of coalescence in each whorl, and adnation.

The Cucurbitaceae culminate a branch line from the rose order. The flowers of the Cucurbitaceae have such advanced characters as loss of floral parts, imperfect flowers, epigyny, and coalescence.

Rosaceae

Whereas grasses and the legumes supply man with bread and vegetables of basic importance, the rose family supplies dessert, fruits, and almonds and roses for decoration. There are better than 2500 species in this family and over 100 genera, not counting the almost numberless cultivated forms of roses, peaches, apples, cherries, strawberries, almonds, and so on. The family is of world-wide distribution and somewhat heterogeneous. Many of the principles of angiosperm evolution can be demonstrated with its members. There are trees, shrubs, and herbs in the family. The leaves are usually alternate and bear stipules.

While the flowers of different genera show much variation, they are generally perfect and regular. The parts are borne on a floral disk which, as it develops after fertilization, becomes responsible for certain of the fruits characteristic of the family. In the rose the disk or calyx tube is hollow and sur-

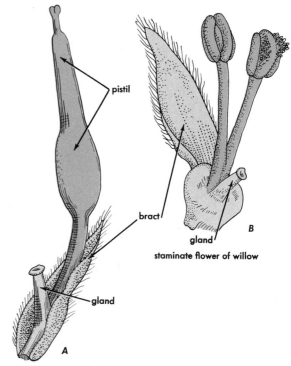

Figure 29.8. Salicaceae. A, a pistillate flower of willow (*Salix*); B, a staminate flower. (Redrawn from A. M. Johnson, *Taxonomy of the Flowering Plants.* Copyright 1931, 1959 by Meredith Corporation. Used with permission of Appleton-Century-Crofts.)

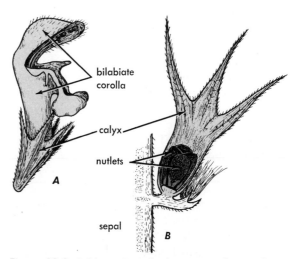

Figure 29.9. Labiatae. Irregular hypogynous flower of mint, showing union of sepals and petals.

rounds numerous pistils; the numerous stamens and corolla are adnate to this tube, the flower being perigynous. Almonds (Fig. 15.16*B*), plums, peaches, and apricots have a similar structure except that the number of parts of the gynoecium is reduced to one or occasionally two. In the strawberry (Figs. 16.6, 29.10), the reverse situation obtains, in that instead of a hollow structure bearing the parts of the gynoecium, the carpels are borne on the outside of a raised disk. In the apple (Fig. 16.20) and quince, the calyx tube or floral disk grows attached to the ovary and expands to become the edible portion of the fruit. A simple classification of the flowers of the rose family is not possible. There is much variation in the type of fruits. In the strawberry the matured ovary is an achene; in the apples and pears it is a pome; in the peaches and apricots, blackberries and cherries, it is a drupe; capsules occur in some genera and follicles in others.

Cucurbitaceae

The gourd or melon family is an interesting one, in that it is of world-wide distribution in at least the warmer regions of the world, and various peoples have selected different forms for domestication. It is also probably the only group in which the fruits are highly prized for ornamental purposes and for use as vessels of various types. The pumpkins and squashes are thought to be of American origin. Cucumbers and melons have probably come from Africa and central Asia. Other species appear to have been first cultivated in the tropics of Asia, Polynesia, and India. They are frequently tendril climbing, annual, or perennial vines. Most are rapid growing and frost tender. There are about 90 genera with 700 species. The stems are usually soft and hairy or prickly. The generally simple leaves are large and sometimes deeply cut. Lateral tendrils are often present and they may be simple or branched.

The flowers (Figs. 15.12*C*, *D*, 29.11) are imperfect and the plants may be either dioecious or monoecious. In the pistillate flower the five-lobed calyx is adnate to the ovary. There may be five petals or they may be coalesced into a five-lobed corolla. The inferior ovary is composed commonly of three carpels with styles and stigmas also coalesced, but the latter may be lobed. The fruit is the pepo or berry so characteristic of the family. The perianth of the staminate flower is similar. There are five stamens, two of which may be fused. The anthers twist so that the stamens appear to form a solid central column.

Leguminosae

The legume family has such distinctive characteristics that its members can frequently be recognized with but little experience. It is a large family with close to 500 genera and several thousand species. All sorts of plant bodies are represented: herbs, both annuals and perennials, shrubs, vines, and trees. It is of world-wide distribution. While less heterogeneous than the rose family, considerable variation is found among its various members. It is a family of considerable importance in supplying

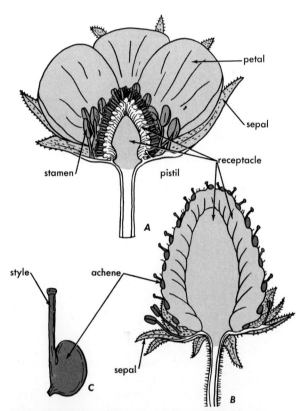

Figure 29.10. Rosaceae. Flower and aggregate fruit of strawberry (*Fragaria*). A, median lengthwise section of a flower; B, median lengthwise section of a fruit; C, single achene. (Redrawn from W. W. Robbins, *Botany of Crop Plants*. Copyright 1931 by Blakiston. Used with permission of McGraw-Hill Book Company.)

Figure 29.11 (*right*). Cucurbitaceae. A, *Cucurbita pepo*, flowering and fruiting stem; B, *Cucurbita pepo*, fruit; C, *C. maxima*, staminal column; D, *C. maxima* stigma, longitudinal section through pistil, corolla tube removed; E, *C. maxima*, cross section through ovary; F, *C. maxima*, style and stigma. (Redrawn from L. H. Bailey, *Manual of Cultivated Plants*, The Macmillan Company.)

food for man and his animals. Many legumes are used for ornaments, from shade trees to cut flowers. Some lumber is obtained from the black locust and some of the tropical species furnish wood for fine cabinet work. The association of the nitrogen-fixing bacteria with the roots of legumes places this family in a unique position relative to the maintenance of soil fertility. Peas, beans, peanuts, clovers, and lupines are common herbaceous legumes; wisteria is a vine, brooms and redbuds are shrubs or low trees, and the locusts and acacias are tree types. The *Mimosa* genus alone has some 300 species, including the sensitive plant of the florist shop; there are others of varying habit from tall trees to low herbs. The leaves of the legumes are prevailingly pinnately compound and quite generally with stipules. Sometimes a leaflet will be modified to a tendril.

The flowers of most genera are of the well-known sweet pea or "butterfly" type (Figs. 15.5*A*, 29.12). They are thus irregular, the five-parted corolla forming a standard, two wings, and a keel. Sepals are present. Nine of the ten stamens are united to

anthers

pistil
separate stamen

united stamens

C

E

standard

wing

ovules

locule

keel

B

D

stamens   pistil

corolla

calyx

G

inflorescence

H

F

I

form a tube around the ovary; the tenth stamen is free. The gynoecium is composed of a single carpel. In some genera the flowers, while quite irregular, are not of the typical sweet pea type and in *Mimosa* and allied genera the flowers are regular and small. In this group some species have numerous, free stamens. The legume (Figs. 16.2, 16.9) or pod is the characteristic fruit of the family. It forms from a single carpel, and has one locule and numerous seeds. In many genera it dehisces at maturity along both sutures; in a few genera it is indehiscent.

Umbelliferae

The parsley family forms a very distinctive group of plants, many of which can be recognized with little experience as belonging to the family. There are about 250 genera and 2000 species growing in all regions of the world, though confined to the mountains in the tropic zone. The name is derived from the typical umbel type of inflorescence (Fig. 15.26). The family contains many crop plants, as well as some that produce drugs and a number of kitchen herbs. Poison hemlock, which is famous because of its use by the Greeks as a means of carrying out the death sentence and which was given as such to Socrates, is a member of this family. The genus is widespread, constituting a hazard to livestock not only in the pastures of Greece, but on the open ranges of California. Carrots and parsnips as well as parsley are members of this family. They are mostly herbs, rarely small shrubs with generally hollow stems. The leaves are alternate, mostly compound, and frequently much dissected. The petioles expand at the base and may somewhat sheath the stem.

The flowers are small, being borne in umbels, which are in turn grouped in umbels, so a large inflorescence is formed. Frequently, the umbels of all levels are subtended by bracts, which form a characteristic involucre. The flowers (Fig. 29.13) are regular or the outer flowers irregular, always epigynous and perfect, although they may not always be complete. There are five stamens alternating with the petals. The gynoecium is composed of two carpels, each with but a single seed. The fruits are indehiscent, although the carpels separate from each other when mature.

Compositae

The composite family forms an extremely large group of plants, only a few of which are woody and except for rare instances are not designated as trees or shrubs. In this family of herbaceous plants there are around 800 genera and about 12,000 known species. The family is not only of world-wide distribution but in most places is relatively abundant. The family is not noted for its food plants; endive, artichokes, lettuce, and sunflower, form most of its contribution in this respect. Neither is it famous as a producer of drugs or other commercial products. One species, safflower, is now being grown for oil and another has been considered as a possible source of latex for rubber. But paging through a seed catalog or a visit to a florist shop will show members of this family in their full grandeur. *Dahlia, Chrysanthemum, Aster, Zinnia, Tagetes, Gaillardia,* ageratum, and many others are representatives of this family. Dandelions color lawns, and various wild species add to the fall color of meadows and fields. There are 200 species of the sagebrush, genus *Artemisia,* that cover arid portions of the world and supply, among other things, tarragon for fancy vinegar and nectar for desert honey. Leaves are of various shapes.

The flowers and fruits of the family have been described in detail on pages 288 and 313 and are shown in Fig. 29.14. They will not be discussed further here. They are, however, so characteristic of the family that any composite should be recognizable with a minimum of experience.

Figure 29.12 (*left*). Leguminosae. *A,* inflorescence and leaf of *Lathyrus; B,* exploded view of *Lathyrus* corolla; *C,* essential organs of *Lathyrus* flower; *D,* cross section of ovary; *E,* pod of *Lathyrus; F,* branch showing inflorescences and leaves of *Acacia; G,* a single *Acacia* flower; *H,* flower of *Bauhinia; I,* dehisced pod of *Bauhinia.* (Redrawn from L. H. Bailey, *Manual of Cultivated Plants.* The Macmillan Company.)

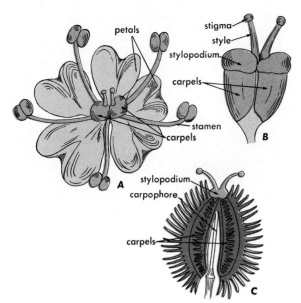

Figure 29.13. Umbelliferae. Flower and fruit of carrot (*Daucus carota*). A, flower; B, two carpels; C, mature fruit.

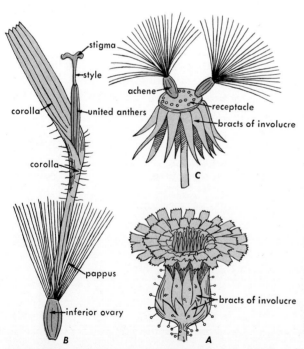

Figure 29.14. Composite flower and fruit of sow thistle (*Sonchus*). A, flowering head; B, single ray flower; C, mature achenes. (Redrawn from Korsmo.)

The Monocotyledonous Line from Ranales to the Orchids

There are many obvious differences between the Ranales and the monocotyledonous line of evolution. These have been discussed and may be briefly reviewed in Table 29.5.

However, some primitive monocotyledons, particularly certain water weeds such as the arrowheads and water plantains, have much in common with the buttercups and marsh marigolds. The families selected to represent this line of ascent are the Liliaceae, the Gramineae, the Iridaceae, and the Orchidaceae. The lilies differ from the Ranales in all of the distinctive monocot characters. In addition they show a reduction in the number of floral parts and a coalescence of carpels. The irises show a single advance over the lilies in that they are epigynous. Many are regular, perfect, and with separate perianth parts. The stamens show no coalescence. Syncarpy occurs as it does in the lilies. The orchids show evolutionary advance in that they have irregular flowers very highly specialized for insect pollination. The stamens have been reduced in number to one or two.

The order to which the grasses belong is thought to have arisen from the order to which the lilies belong. The primitive characters of the grasses include hypogyny, regular flowers, and only coalescence of the carpels. The loss of perianth parts and reduction in the number of stamens and carpels marks them as a more advanced family than the Liliaceae.

Liliaceae

Unlike the grasses, lilies are grown largely for ornamental purposes, although two genera, the onions (*Allium*) and *Asparagus*, are grown extensively for food. Some species yield drugs and others have poisonous properties that may cause trouble in pastures or on the ranges of the western states. There are about 2000 species in some 200 genera. Most of the lilies grow from a bulb or bulblike organ, and flower in a single growing season after which the shoot dies down. A few, such as the Joshua trees, are woody perennials. Tulips, hyacinths, day lilies, and aloes are other examples of the lily genera.

Members of the lily family may be generally quite easily recognized because of the typical showy flowers (Fig. 15.1*C*). The regular perianth has six parts, usually separate; but if united, they are lobed,

Monocotyledons

1. One cotyledon or seed leaf.
2. Generally marked parallel leaf venation.
3. Flower parts typically in groups of three or multiples of three.
4. Vascular bundles of stems scattered throughout a cylindrical mass of ground tissue.
5. Vascular cambium lacking in most forms.

Dicotyledons

1. Two cotyledons or seed leaves.
2. Generally marked netted venation of leaves.
3. Flower parts typically in groups of four or five.
4. Vascular bundles of stems usually arranged in the form of a cylinder.
5. Vascular cambium present in those forms having secondary growth.

and the corolla and calyx are generally not distinct from each other. There are six stamens and one pistil. The superior ovary is single and three-celled; there are three stigmas. The fruit is mostly many-seeded and may be a capsule (Fig. 16.5) or a berry.

Gramineae

Not only man, but mammals in general, have been associated with the grass family for far more years than those of recorded history. It is probably not an overstatement to say that without the grass family and man's ability to learn to exploit it, civilization, as we know it, could not have developed. Different grasses were domesticated by the three great civilizations: the wheats, rye, barley, and oats by the Mediterranean people; corn (*Zea mays*) by the Aztecs, Mayans, and Incas and other natives of the Western Hemisphere; and rice and millet by the Chinese. Grasses are grown for human and animal consumption, for ornament and for uses in the arts and industry. There are over 400 genera and about 4500 species. Most grasses are herbaceous, either annuals or perennials; a few bamboos are climbers and some bamboos are woody. Grasses of one kind or another grow in all kinds of soil and situations. Grasses may be conveniently divided into five

groups according to their use. (1) The bamboos, mostly evergreen, stout perennials. They are used for construction in many parts of the Orient and some have been introduced into the warmer parts of the United States for ornamental purposes. (2) The cereals, supplying grain and forage for both man and animals. (3) The sugar-producing kinds, such as sugar cane. These are strong upright perennials growing only in the tropics. (4) The sod-forming grasses, perennials that cover many square miles of the earth's surface. They are used in lawns and meadows. (5) There is a large group of grasses grown for ornamental purposes, such as pampas grass. The flowers and vegetative characteristics of the grasses have been described in some detail on pages 188 and 287 and shown in Figs. 15.18 and 15.19. The leaf is characteristically divided into a sheath, which surrounds the stem and expands to a linear blade (Fig. 10.4).

The flowers (Figs. 15.18 and 15.19), known as florets, are small and usually grouped in spikelets. The latter may be arranged in a variety of ways, frequently in spikes or heads. The spikelet normally has two glumes at its base and a number of florets attached to the rachilla. The floret itself has two bracts, the lemma and palea, from which may extend an awn, protecting the single pistil, with its feathery stigmas, and the three stamens. Lodicles may also be present. The fruit is the typical grain or caryopsis, with a single seed whose seed coats have become attached to the ovary wall.

Iridaceae

The Iridaceae or iris family contains about 60 genera and some 1000 species. They are all herbaceous, largely perennial forms, usually with rhizomes, bulbs, or corms, as in the lilies. The leaves and flowering stalks last for only one season. Some of the choicest florists' plants occur in this family; the *Iris, Gladiolus, Freesia, Crocus, Watsonia,* and so forth.

The regular perianth, like that of the lily, has six parts, united in some genera and separated in others (Fig. 29.15). In some genera the flowers are irregular. Three parts of the perianth may differ from the other three parts, or all may be similar. The flowers are showy and usually emerge from a spathe consisting of a pair, sometimes more, of herbaceous or scaly bracts. The single ovary is inferior and sets this family definitely apart from the Liliaceae. The

Figure 29.15. Iridaceae. *A*, *Gladiolus* inflorescence; *B*, longitudinal section of a gladiolus flower; *C*, *Iris* inflorescence; *D*, *Iris germanica*; *E*, capsule of *Iris sibirica*; *F*, cross section of ovary of *Freesia*. (Redrawn from L. H. Bailey, *Manual of Cultivated Plants*, The Macmillan Company.)

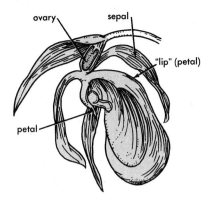

Figure 29.16. Orchidaceae. Irregular epigynous flower of orchid, longitudinal section.

ovary is three-celled and has three stigmas. The fruit is always a many-seeded capsule.

Orchidaceae

The orchids are agreed by all to represent the most advanced family in the Monocotyledoneae. They are all herbaceous plants, circling the world largely in the tropics, but with a few genera extending in the colder temperate regions. There are some 15,000 recognized species in several hundred genera. All forms are perennial, with tuberous, bulbous, or otherwise thickened roots. They may be erect, prostrate, or climbing. A few are saprophytic and lack chlorophyll; others are epiphytes growing on trees without benefit of soil.

The flowers are very showy and highly specialized for insect pollination. A typical flower is shown in Fig. 29.16. Sepals are present, the corolla is irregular, and the petals are united. The ovary is epigynous, one- or three-celled. The stamens unite with the pistil to form a structure sometimes called the **column.** There may be only one fertile stamen, sometimes with a number of staminodes or infertile stamen-like structures. The pollen hangs together in masses known as **pollinia.** Insects lift such a pollinium and place it on the beak or projection of the stigma, which may stand at the base of the anthers. The fruit is a dehiscent capsule having many minute seeds.

SUMMARY OF ANGIOSPERM INNOVATIONS

In our short discussion of the life history of an angiosperm, several references were made to structures which were new to, and characteristic of, the angiosperms. This combination of structures has adapted the angiosperms to the present climatic conditions and resulted in their great variety and extensive vegetative growth. While this discussion has considered only reproductive structures, various vegetative adaptations have also played their part in making the angiosperms such successful competitors for their ecological niche. We shall not discuss these vegetative adaptations here. The reproductive innovations may, however, be summarized as follows:

1. The development of the carpel, involving (*a*) the ovary, (*b*) the style, (*c*) the stigma.
2. The enclosure of the ovule within the ovary, affording added protection to the ovule and the female gametophyte within it.
3. The provision, by the style and stigma, of a precise pathway guiding the sperm cells in the pollen tube to the egg cell in the embryo sac.
4. The reduction of the female gametophyte to a seven-celled embryo sac.
5. Various highly successful systems of pollination, made possible by the stamen, which (although not an innovation of comparable rank to that of the carpel) has proved to be a very adaptable part.

6. The reduction of the male gametophyte to a small pollen tube and the complete elimination of motile sperms.

7. The elimination of the female gametophyte as the source of nourishment for the young embryo and seedling.

8. The presence of double fertilization, resulting in the formation of an endosperm as a source of nutrient, at least for the young embryo and in some cases for the seedling.

9. Food storage within the embryo itself.

10. The development of the fruit from the ovary wall.

SIGNIFICANT FEATURES OF THE ANGIOSPERM LIFE CYCLE

1. Microspores are produced by meiosis in microsporangia or pollen sacs of the anthers; the latter are a part of the microsporophylls or stamens.

2. Each pollen grain, developing from a microspore by mitosis, has a generative and tube nucleus and is provided with a thick, sculptured cell wall.

3. After pollination the pollen tube or male gametophyte grows down the style to the ovule.

4. Meiosis in the megaspore mother cell within the nucellus or megasporangium of the ovule results in four megaspores aligned in a row.

5. One megaspore commonly develops into a seven-celled embryo sac or female gametophyte.

6. The two sperms are discharged into the embryo sac and double fertilization occurs.

7. The zygote develops an embryo.

8. The primary endosperm cell forms an endosperm that may not persist into the seed.

9. After fertilization, the ovule becomes the seed.

SUMMARY OF ANGIOSPERM EVOLUTION

1. Angiosperm families, according to the Besseyan system of classification, are all derived from the primitive Ranales, to which order buttercups and magnolias belong.

2. The important changes in floral evolution, according to the Besseyan system, are as follows: (a) from a spiral to a whorled arrangement of floral parts; (b) from many parts to few or even to a loss of parts; (c) from separate floral parts to coalesced parts; (d) from a regular flower to an irregular flower; (e) from hypogyny to epigyny.

3. According to the Besseyan system of classification there are three branched lines of ascent from the Ranales.

4. The first line of ascent following the Ranales includes the Malvaceae (mallow family), the Cruciferae (mustard family), the Solanaceae (potato family), the Salicaceae (willow family), and the Labiatae (mint family).

5. The second line comprises the Rosaceae (rose family), the Leguminosae (pea family), the Umbelliferae (carrot family), and the Compositae, with the Cucurbitaceae (melon family) as an offshoot from the Rosaceae.

6. The third line comprises the Monocotyledoneae in the following order: Liliaceae (lily family), Iridiaceae (iris family), and Orchidaceae (orchid family). The Gramineae (grass family) are an offshoot from the Liliaceae.

CLASSIFICATION

In our survey of the families of flowering plants, we have generally been following the Besseyan system of classification. Many taxonomists, however, follow other systems of classification, the system of Engler being a favorite alternative. For the purpose of comparison, the families studied are listed here according to the systems of both Bessey and Engler. For each system the classes and families considered most primitive are given first, the advanced are given last. Note that Bessey considered the monocotyledons the more advanced while Engler considered them to be more primitive.

Bessey's System	Engler's System	Primitive
DICOTYLEDONAE	MONOCOTYLEDONAE	
Magnoliaceae	Gramineae	
Ranunculaceae	Liliaceae	
Malvaceae	Iridaceae	
Cruciferae	Orchidaceae	
Solanaceae	DICOTYLEDONAE	
Labiatae	Salicaceae	
Rosaceae	Ranunculaceae	
Leguminosae	Magnoliaceae	
Cucurbitaceae	Cruciferae	
Salicaceae	Rosaceae	
Umbelliferae	Leguminosae	
Compositae	Malvaceae	
MONOCOTYLEDONAE	Umbelliferae	
Liliaceae	Labiatae	
Gramineae	Solanaceae	
Iridaceae	Cucurbitaceae	
Orchidaceae	Compositae	Advanced

Evolution

In a broad sense the term evolution refers to a process involving gradual changes. It is well known that neither animals, nor plants, nor cities, nor states remain the same. We may speak of the evolution of means of transportation, the evolution of human clothing, the evolution of mountains and valleys, and the evolution of many other nonorganisms. **Organic evolution** pertains to the gradual changes that have taken place in living things (organisms). Coal, for instance is formed from the remains of plants that were different in appearance from those growing today. The student of organic evolution may be interested in, among other things, accounting for the disappearance from the earth's flora of the Coal Age plants and the appearance of the modern flowering plants.

Organic evolution is closely allied to genetics and plant breeding. We have seen that the factors which determine the characteristics and the activities of plants are associated with the nuclei. Furthermore, these factors or genes are transmitted from generation to generation by the chromosomes during the processes of fertilization and meiosis. If the origin of new plants is to be understood, experiments on genetics will furnish important information. Many such experiments have been carried out and they have achieved two ends: (1) they have made possible clearer insight into some of the mechanisms of evolution, and (2) they have greatly increased the yield and improved the quality of agricultural crops.

THE GEOLOGIC HISTORY OF PLANTS

Since plants cannot be dissociated from their environment, some knowledge of the principal changes in the earth's crust is essential to an understanding of plant evolution. Broadly speaking, two different types of rock are found in the earth's crust. One of them, like lava, was formed from molten material. The oldest known rocks are of this nature; granite is an example. Because of their molten origin no fossils are ever found in these rocks, but there is good reason to believe that unicellular algae inhabited the waters of the earth when these old rocks were being formed. Such rocks were predominant when

659

the earth was young, during the period in the earth's history known as the **Archeozoic Age.** It was about 550 million years in length and it came to an end some 1500 million years ago.

The second type of rocks (sedimentary rocks) are formed by the deposition of large amounts of sediment in bodies of water. As layer upon layer of sediment accumulates, the pressure upon the lower layers becomes enormous and they are turned to rock. Shale, limestone, and sandstone are examples of sedimentary rocks. If plants or animals are buried in the sediment, and if decay does not occur, the buried individuals are preserved as **fossils.** It is important to note that plant fossils consist almost entirely of plant remains that grew in or near water or were carried by streams into lakes or other sites of deposition. Algal limestones are known from the Archeozoic Age. The remains of plants growing in arid or mountainous regions are rarely deposited where conditions are favorable for their preservation as fossils. Either it is too dry or the site is undergoing erosion and the deposits are of very short duration. This means that the geological record of plants must probably remain incomplete and that the records we do have usually represent only the floras of lowlands and moist habitats. Considerable evidence indicates that the great environmental extremes prevalent in arid or semiarid mountainous areas are especially favorable to rapid evolution. This factor, we shall see, may be very important in connection with the problem of the origin of the angiosperms and, together with the rarity of fossils representing plants of arid habitats, may serve to explain peculiarities that surround the first records of the angiosperms.

The best available evidence indicates that living organisms existed at least 2 billion years ago, for sedimentary rocks that old may contain fossils. The 2 billion years since these first fossils were formed in the Archeozoic Age are divided into four ages; the oldest, the **Proterozoic,** is followed in order by the **Paleozoic,** and **Mesozoic,** and the **Cenozoic** (Table 30.1). The present Age is the Cenozoic, and it is already 60 million years old. Man may have been an inhabitant of the earth for over two million years, but civilized man of historical times is a newcomer of some 7000 years. By comparison with the geologic age of his near relatives, the apes, he is hardly three days of age. The relationship of these four ages, their climatic characteristics, and dominant floras are shown in Table 30.1.

Evidences of these geological ages may be seen in the exposed strata of road cuts, canyons, or mines. Sometimes fossiliferous rocks lie exposed in plateau or level areas. Figure 30.1*A* illustrates a road cut along an Oregon highway. The dark strata contain the remains of plants that resembled modern horsetails. Chunks of this material are continually breaking away and rolling down to the shoulder, (Fig. 30.1*B*) where they may easily be examined. In other regions fossils may be strewn around on the surface of the ground. The "stones" of Fig. 30.1*C* are carbonized pieces of stems of *Metasequoia,* the dawn redwood (Figs. 10.5, 27.1 28.8*E*). This tree was formerly abundant in the Western United States, but was thought to be extinct until a few living trees, virtually living fossils, were found some years ago in China. Of course, not all sedimentary rocks contain fossils, but many do; and furthermore, many fossils remain undiscovered. Probably no field of botany can be as easily enriched by the amateur collector as paleobotany.

Relationship of Phyla of Plants to Geological Age

A survey of the plant kingdom shows that certain phyla growing today are but remnants of once extensive floras. A few fossils representative of them have been described. Let us now see how the different phyla are related to each other and to the geological period during which they flourished.

Fossil thallophytes are little known. The type that has been described has been found chiefly in Proterozoic rocks, although fossil algae are known from the early Archeozoic Age, over 1500 million years ago. Some evidence indicates, moreover, that thallophytes dominated the vegetation of the early Paleozoic Age. There is general agreement that the first plants to inhabit the earth were algae. Consequently they are shown as the earliest plants and as those from which all other plants have been derived. (Table 30.1).

It is supposed that the bryophytes and vascular plants have arisen, quite independently of each other, from the algae or algae-like plants. The evidence of their independent origins is not, however, conclusive.

The first vascular plants that made their appearance on the earth are known today only as fossils. They are placed in the order Psilophytales (phylum Psilophyta) and it seems likely that they were the ancestors of the other phyla of vascular plants. The

TODAY

Man and herbaceous plants.
Climate slowly warming.
Atlantic coast eroding.
Pacific coast rising and
moving slowly northward.
Oceans rising.

Fan diagram labels (outer to inner):

- Psilophyta / Psilotum and Tmesipteris
- Bryophyta / Algae
- Bacteria / Fungi
- Modern club mosses
- Modern horsetails
- Modern ferns
- Modern gymnosperms
- AGE OF MAN AND HERBACEOUS PLANTS
- Extinction of many trees / Increase in dominance of herbs such as grass
- Retreat of forests from poles / Reduction of forests / Rise of herbaceous plants
- Appearance of Man
- Modernization of woody angiosperms / Development of extensive forests
- Woody angiosperms dominant / Many modern living genera present / Angiosperms increasing
- First angiosperms
- Gymnosperms dwindling
- AGE OF GYMNOSPERMS
- Gymnosperms dominant
- Cycads and conifers
- Extinction of many forms as climate changes from moist to dry
- AGE OF LYCOPODS, SEED FERNS, AND AMPHIBIANS
- Extensive coal-forming swamps
- Seed ferns and other ferns
- Dominant club mosses
- horsetails
- Primitive Bryophyta
- Primitive gymnosperms
- Seed ferns
- Primitive ferns, lycopods, and horsetails
- First forests
- First land plants
- Psilophytales / Rhynia
- AGE OF MARINE ALGAE
- Bacteria
- Algae

Major Division	Climate and Major Geological Changes
Cenozoic Quaternary, 2 million years ago	Periodic glaciation. Warming and cooling.
Cenozoic Late Tertiary	A general cooling and change to semi-arid conditions. Appearance of temperate zones and great mountain ranges.
Cenozoic Early Tertiary, 60 million years ago	Began with mountain glaciers and semi-arid conditions. Ended warm and humid.
Late Mesozoic, 125 million years ago	Climate generally warm with some fluctuation. Ended cool. Appearance of Rocky Mts., Sierra Nevada, and Andes.
Early Mesozoic, 185 million years ago	Climate becomes drier, with periodic glaciation and much change in the earth's crust. Appalachian and Sierra Nevada Mts. form. Seas drain from continents.
Late Paleozoic, 309 million years ago	Climate becomes warmer and drier.
Middle Paleozoic, 381 million years ago	Climate warm over most of earth. Widespread shallow seas over North America.
Early Paleozoic, 553 million years ago	Uniformly warm shallow seas over North America.
Proterozoic, 1,500 million years ago	Young Laurentian Mts. form. Glaciation.

Table 30.1 *Relation of phyla of plants and geological age*

Psilophyta (Fig. 27.4) were apparently the first true land plants, arising some 550 million years ago (Table 30.1). It is interesting to note that the first air-breathing animals appeared at the time these land plants flourished.

For some hundreds of millions of years after the appearance of the Psilophyta, during the late Paleozoic Age, the horsetails, the lycopods, and the seed ferns dominated the vegetation. True ferns were also present in considerable numbers, but they were quite small in size. Figure 30.2 shows how a Paleozoic forest probably looked.

Four groups of plants, the horsetails, the lycopods, the ferns, and the seeds ferns, are thought to have originated from different representatives of the Psilophyta (Table 30.1). Their relationship is

Figure 30.1. Fossils. *A,* a road cut along a highway showing fossil-bearing strata; *B,* a chunk of fossiliferous rock; *C,* carbonized stems of *Metasequoia.*

more like that of brothers than of direct lineal descendants. As we have learned, the living herbaceous club mosses and horsetails are the relics of a once-large club moss and horsetail flora. Ferns occupy a position today not unlike that which they held in the late Paleozoic Age, some 380 million years ago. They may be somewhat more numerous now, but they are still relatively small in size. The seed ferns, however seem to be related to modern gymnosperms and to angiosperms, each of which, in turn, dominated the earth's vegetation. This shift in vegetation type probably reflects climatic changes. The humid Paleozoic Age, illustrated in Fig. 30.2, slowly gave way to the Mesozoic, which began as a very cool, arid period and ended as a warm, wet period. The Paleozoic plants that demanded much water were at a disadvantage in the early Mesozoic, and the way was opened for the development and spread of new floras and of new vegetation types (Table 30.1).

Origin of the Angiosperms

The angiosperms as we now know them seem to have appeared already highly evolved into many families and genera, without leaving us a record of the stages in their evolution or suggesting the nature of their immediate ancestors. They appeared in the later, warm, moist part of the Mesozoic Age (Table 30.1). Their great diversity at this time can only mean a long period of development, which projects their beginnings back into the early, arid periods of the Mesozoic, or possibly even to the cool, dry period that marked the close of the Paleozoic. Therefore, it is unlikely that any satisfactory fossil evidence as to the origin of the angiosperms will ever be found. In addition to great aridity, much mountain building and subsequent erosion has served to destroy any chance record of the origin of the angiosperms. Studies of present-day angiosperms, however, in which vessels are lacking or in which the carpels are not completely fused at maturity may furnish evidence as to the origin of angiosperms.

CHARLES DARWIN AND EVOLUTION

Many biologists since the time of the Greek philosophers have attempted to explain the mechanism of evolution. Great progress has been made during the twentieth century. It will be seen (page 668) that the origin of new species can in some instances be explained and that new species can still arise. Our greater understanding was made possible chiefly because the work of Charles Darwin and of Gregor Mendel provided a basis that enabled biologists to approach the problem of evolution in an entirely new light.

Darwin, as a result of many years of careful study and observation of a large number of plants and animals, laid emphasis upon the following points.

1. Numbers of plants or animals may increase in a geometric ratio. For example, a given plant may produce 1000 seeds. If each seed grows into a new plant and each

Figure 30.2. Reconstruction of a Paleozoic forest. Note the seeds attached to fernlike leaf in left center. (Courtesy of Field Museum of Natural History.)

Figure 30.3. Environment variation in sorghum. Shade plant, left; sun plant, right.

new plant produces 1000 seeds, 1 million new plants could result. A third generation from the one original plant would result in 1 billion plants. It may be seen from this example that plants, and animals also, potentially, could increase in numbers at a tremendous rate.

2. Actually, the number of individuals in a given plant or animal association remains fairly constant. There is no such tremendous increase in the number of individuals as seed production seemingly makes possible.

3. No two individual plants or animals are identical; there is **variation.**

Reasoning from these observations, Darwin arrived at these conclusions:

1. Any given population is usually able to reproduce many more young individuals than can adequately be raised in the region it occupies. Therefore, a struggle for existence occurs among the individuals.

2. In this struggle for existence, only those individuals survive which, because of their particular variation, are best adapted to their immediate environment. Thus, a natural selection takes place; the unfit do not survive.

3. These selected variations may be inherited, that is passed from one generation to another, and thus may gradually give rise to new species.

Variation

Since variations play an important part in plant breeding and evolution, it will be well to examine the concept a little more closely. Variations are of two kinds, (1) environmental and (2) genetic.

Environmental Variations

Suppose that two plants of the same species are growing close together but that one is in the shade of a tree whereas the second, six feet away, receives full sun (Fig. 30.3). The plant in the shade is small and stunted; it has few flowers and the yield of seeds is low. The plant in full sun is large and robust; it has many flowers and a good yield of seed. Let us collect seeds from each of these plants and divide each collection into halves. If we plant seeds from the sun plant in both shady and sunny locations and do likewise with the seed from the shade plant, we shall again have poor, stunted plants in the shade and healthy plants in the sun. The conditions under which the seed was formed has not influenced, in the least, the type of plant produced. The environment alone has been responsible for the variation. This is an **environmental variation,** that is, variation induced by differences in the environment.

Genetic Variations

If we glance through a seed catalog, we shall find several varieties of beans offered for sale—green string beans, yellow wax beans, navy beans, and so on (Fig. 30.4). No matter where we plant these bean seeds, so long as the conditions permit growth and the production of seeds, we shall obtain bean seeds similar to those that we planted. Such variations are inherited variations; they are **genetic variations,** and it appears probable that they are the more important variations from an evolutionary viewpoint.

The Mechanism of Evolution

The conclusions of Darwin (page 664), together with those of Mendel (page 346), not only laid the foundation for much of the present knowledge of evolution, but they were also a great stimulus to research, and many hundreds of biologists have subsequently carried out experiments that have further increased our knowledge of the mechanism of evolution.

Let us briefly consider the nature of the gene as it has been revealed by some of these experiments. Genes are regarded as, or are related to, complex chemical compounds, and it appears probable that every gene pair differs from every other gene pair —that is, that *TT* is different from *RR*—and even the members of a pair may not be identical, for instance *Tt*. Genes are too small to be seen with the best light microscopes, but we know their approximate size, some of their probable physical properties, and in some plants and animals the locations of many genes within the chromosomes. The salivary gland chromosomes of a fruit fly show dark bands thought to contain genes (Fig. 30.5). Since the majority of plants possess from seven to forty pairs of chromosomes and each chromosome contains many hundreds of genes, individual plants contain several thousand genes. A diagram of a chromosome of Indian corn (*Zea mays*) (Fig. 30.6) shows the location of a few of the known genes of this plant. Obviously, all the associated genes of a single chromosome will move together during meiosis. They are said to be **linked.**

In our discussion of Mendel's work, we described one gene as being responsible for one aspect of a single character; the presence of the dominant gene *T* resulted in a tall pea plant. It is now known that many genes may influence one character. For instance, the color of chrysanthemum flowers is controlled by many genes. On the other hand, a single gene may be active in determining the nature of several characters. Furthermore, genes are not iso-

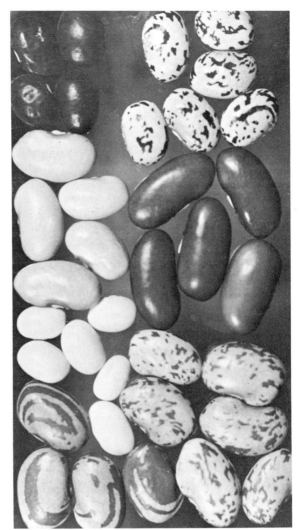

Figure 30.4. Genetic variations shown in the seed coats of beans. (Courtesy of Francis Smith.)

Figure 30.5. Chromosomes from the salivary gland of *Drosophila.*

lated units acting independently of each other. The action of one gene may influence the action of other genes, and all genes react in conjunction with the cytoplasm. If genes are lost through the loss of a piece of a chromosome, some character of the plant will reflect their absence. It is also known that genes may undergo slight alterations, with a resultant change in some character of the plant containing them. Any modification of a chromosome and the accompanying alteration of the plant in question is a **mutation** (Fig. 30.7).

In a normal, healthy plant the genes are in a delicate balance with each other and with the protoplasm containing them, with other cells, and finally with the environment in which the plant is growing. Any disturbance in this equilibrium results in some upset in the metabolism of the plant. Mutations disturb the equilibrium and, if large, are likely to result in the death of the plant containing them. As a rule, only plants with small mutations are able to survive. Furthermore, most small mutations are recessive.

Mutations occur normally in most plants and the rate of their occurrence has been measured in some instances. The rate varies, depending upon the particular mutation and the species involved. An average rate might be one mutation in 100,000 individuals. The mutation rate can be artificially increased by treating pollen grains, anthers, seeds, etc., with x-rays, ultra-violet radiation, and certain chemicals. The mutation of a gene causes some change in the structure of a plant; mutations, therefore, are of great importance in evolution.

It is important to remember that the gene complex is in a delicate equilibrium with the external environment of the plant. Therefore, only those mutations, combinations of mutations, or other changes in the chromosomes which confer some advantage to the plant are of evolutionary significance. In other words, only the gene changes that better adapt an organism to a specific environment are likely to result in new plant or animal forms. For instance, "vestigial" is a mutation in fruit flies that results in small, underdeveloped wings. If a population composed of equal numbers of normal flies with vestigial wings is grown in still air, the normal flies increase rapidly, whereas those with the vestigial wings die and soon disappear from the population. If a similar population is placed in a windy environment, however, the small-winged mutants survive, whereas the flies with normal wings die off.

Any given gene mutates infrequently; for instance, the gene *Su*, sugary, in corn changes 2.4 times in one million gametes.

It may be seen that a given gene is remarkably constant; that is, it is able to reproduce itself exactly many times without a change. Now, several thousand genes are present in any given gamete. If, for example, we assume that 5000 genes exist in a given species and that the average mutation rate is one in 100,000, then one out of every 20 gametes would carry a mutated gene.

A mutation rate of one in 100,000 genes may appear too low to be of evolutionary significance. It is actually not, however. Geological time is sufficiently long to permit the development of new forms

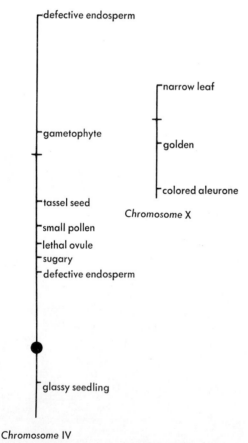

Figure 30.6. Genetic map of fourth and tenth chromosomes of corn (Zea mays).

even at this slow rate. Naturally, those species with a high mutation rate show greater evolutionary development than do those with a low mutation rate. Species with a high mutation rate are **plastic species.** But even in plastic species the formation of new species is a complicated process involving mutations, recombinations of genes, environmental changes, and time.

For example, the effect of a desirable mutated gene, which first appears in one individual, prob- ably as a recessive factor, must become visible; this means that two gametes each containing the mu- tated gene must unite. Furthermore, during suc- ceeding years, through reproduction, the mutated gene must be distributed throughout the population in question.

The occurrence of mutations is well known, but how they have become distributed throughout a population is less well understood. Many factors are involved, four of which are listed below.

1. The mutation rate of the gene in question. A high mutation rate makes possible a more rapid distribution of the gene throughout a population.

2. The usefulness of the mutation. If the new gene confers a great advantage to the individuals possessing it, their chances of survival are enhanced.

3. The number of interbreeding individuals in a pop- ulation. For the most rapid spread of a mutated gene, populations of intermediate sizes are best.

4. Geographical features, such as mountain barriers or broad valleys, which either hinder or aid migration of individuals.

Figure 30.7. Mutations in sunflower (*Helianthus annuus*). A, common wild sunflower; B, mutant known as Sun Gold. × ½. (Courtesy of C. Heiser.)

Hybridization, which serves to reassort the genes into various combinations, occasionally results in the immediate formation of a new species or even genus. Several examples are well known; some have occurred in experimental breeding plots and others in nature. One example concerns two species of grasses. One species, *Spartina stricta,* is a native of England, and the other, *Spartina alterniflora* (smooth cordgrass), grows in the New England states. Commerce between Boston and London resulted in the accidental transportation of seeds of the New England grass across the ocean to localities in southern England. Thus *Spartina alterniflora* became established on the channel coast of England, where it grew in close proximity to the English *Sparina stricta.* Hybrids resulted from the two species, and one of them proved to be very vigorous: it is called *Spartina Townsendii* and is recognized as a distinct species. It rapidly spread over many thousand acres of southern England and crossed the channel into France. Its agricultural properties proved to be desirable and it has since been cultivated in many parts of the world. It is particularly useful in aiding in the reclamation of salt-marsh lands in Holland, France, and England.

Spartina Townsendii is very distinct from its parents. It will not backcross to either of them. It has 126 chromosomes; *Spartina alterniflora* has 70, and *Spartina stricta* has 56. Note that 70 + 56 equals 126. *Spartina Townsendii* received, in some manner, complete sets of chromosomes from each of its parents.

In nature, plants with four or six sets of chromosomes are often larger and more vigorous than those plants having only two sets. A plant with four sets of chromosomes is said to be a **tetraploid.** Any increase in chromosome numbers is known as **polyploidy.** Polyploids of snapdragons, phlox, and marigolds can now be purchased from some seed houses.

Polyploids are especially numerous in harsh, environments, such as tundra or desert.

A modern theory of the mechanism of evolution can be outlined as follows.

1. The genes, contained in the chromosomes, are largely responsible for the development, structure, and metabolism of plants and animals.

2. The complement of genes does not remain absolutely constant. Mutations (changes in chromosomes and in genes) occur, which modify the structure and metabolism of the individuals containing them.

3. Mutations causing considerable phenotypic change are likely to kill or to greatly weaken the plant because they upset the delicate equilibrium existing between the plant and its environment.

4. Hybrids, because of the resulting new combinations of genes, differ from their parents. Hybrids within a species are fertile; hybrids between species are usually sterile or partially so.

5. If the variants produced by mutations or hybridization are better adapted to the environment than the parent plants, the parent plants may be replaced by the new forms. Many complicated factors are involved in this replacement.

6. As mountains are elevated or eroded, as glaciers advance or recede, as the climate of the earth becomes more or less arid, or as the habitats of plants are changed in other ways, those plants best adapted to the new environment replace the forms less well adapted.

7. Evolution thus results from slow changes in the earth's surface, variation in plants and animals, and an adjustment between the changes in the earth and changes in the living organisms.

8. In summarizing evolutionary mechanisms, Dobzhansky, writing in the *Scientific American* has said, "Evolution is a creative response of the living matter to the challenges of the environment. Evolution is due neither to chance nor to design; it is due to a natural creative process."

Glossary

TECHNICAL TERMS INCLUDING THEIR ORIGIN

Abbreviations

A.S. stands for	Anglo-Saxon
D.	Dutch
F.	French
Gr.	Greek
It.	Italian
L.	Latin
Lapp.	Lappland
M.E.	Medieval English
M.L.	Medieval Latin
N.L.	New Latin
O.F.	Old French
R.	Russian
Sp.	Spanish

The glossary serves two purposes, (1) to define terms used in the text and (2) to give their derivation. It should be noted that in carrying out the first purpose, with few exceptions, only those meanings actually used in the text are given. Reference to dictionaries will give additional meanings, while other textbooks may define the same words in slightly different ways.

Scientific language and slang have much in common. They are both vital, growing, and changing phases of modern English. The same words will have different meanings in different parts of the country. Both slang and scientific vocabularies contribute their full share of standard everyday English words. And finally, both vocabularies are largely unintelligible to those not fully initiated into their intricacies.

Scientific language grows in a number of ways. Some botanical terms, such as seed, are used by scientist and layman alike and can be traced back to Anglo-Saxon days, when the word *sed* was used to designate anything sown in the ground, or, in our terms, a seed. In other instances, words from Greek or Latin have been taken into the scientific language with their identical original meaning; for instance, the Greeks called wood *xylon,* and we use this term, changed to xylem, to designate the woody conducting tissue of plants. Other words have a more complex history. *Metabolos* is the greek word *"to change."* When the scientists of Europe wrote in Latin they took this Greek word, made a Medieval Latin word from it, and used it to describe the changes undergone by some insects. It was used in 1639 to mean changes in health, and in 1845 the German scientist who elaborated the cell doctrine made a German word of it and first used it as we use it here, designating the changes taking place within a cell. Perhaps the newest scientific word in your text is *eukaryote.* Look it up.

In using the glossary, pay as much attention to derivation as you do to the definition. To memorize the definition alone is to learn, parrot-fashion, only one word. To learn the derivation is to understand the word and possibly to introduce yourself to a whole family of new words. Pay particular attention to such combining forms as *hetero, auto, micro, phyll, angio, plast* or *plasm, spore,* and the like.

A, abbreviation for an **Angstrom,** 0.0001 of a micron. There are 10,000,000 Angstroms in a millimeter and 10 in a millimicron, mμ.

Abscission zone (L. *abscissus,* cut off), zone of delicate thinwalled cells extending across base of petiole, the breakdown of which disjoins the leaf or fruit from the stem.

Absorb (L. *ab,* away + *sorbere,* to suck in), to suck up, to drink up, or to take in. In plant cells materials are taken in (absorbed) in solution.

Accessory (M. L. *accessorius,* an additional appendage), something aiding or contributing in a secondary way, such as buds in addition to the main axillary bud.

Achene, simple, dry, one-seeded indehiscent fruit, with seed attached to ovary wall at one point only.

Acid (F. *acide,* from L. *acidus,* sharp), a substance belonging to a class of which the commonest and most typical members are sour, and are compounds of hydrogen with another element or elements; oxygen is generally the third element. Examples of inorganic acids are sulfuric acid, (H_2SO_4) and hydrochloric acid (HC1); examples of organic acids are acetic acid (vinegar) and pyruvic acid.

Actinomorphic (Gr. *aktis,* ray + *morphe,* form), said of flowers of a regular or star pattern, capable of bisection in two or more planes into similar halves.

Active solute absorption, the intake of dissolved materials by cells against a concentration gradient and requiring an expenditure of energy.

Adaptation (L. *ad,* to + *aptare,* to fit), adjustment of an organism to the environment.

Adenine, a purine base present in DNA, RNA, and nucleotides such as ADP and ATP.

Adhesion (L. *adhaerere,* to stick to), a sticking together of unlike things or materials.

Adnation (L. *adnasci,* to grow to), in flowers, the growing together of two or more whorls to a greater or less extent. Compare adhesion.

ADP, Adenosinediphosphate.

Adsorption (L. *ad,* to + *sorbere,* to suck in), the concentration of molecules or ions of a substance at a surface or an interface (boundary) between two substances.

Advanced (M.E. *advaunce,* to forward), said of a taxonomic trait thought to have evolved late in time from some more primitive trait.

Adventitious (L. *adventicius,* not properly belonging to), referring to a structure arising from an unusual place: buds at other places than leaf axils, roots growing from stems or leaves.

Aeciospore (Gr. *aikia,* injury + spore), one of the binucleate asexual spores of rust fungi formed as a result of the sexual fusion of cells but not of nuclei.

Aecium (plural, **aecia**) (Gr. *aikia,* injury), in rust, a sorus that produces aeciospores.

Aerate, to supply or impregnate with common air, such as by bubbling air through a culture solution.

Aerobe (Gr. *aer,* air + *bios,* life) an organism living in the presence of oxygen.

Aerobic respiration, respiration involving molecular oxygen.

Agar (Malay, *agaragar*), a gelatinous substance obtained mainly from certain species of red algae.

Aggregate fruit (L. *ad,* to + *gregare,* to collect; to bring together), a fruit developing from the several separate carpels of a single flower.

Alcohol (M.L. from Arabic *al-kuhl,* a powder for painting eyelids; later applied, in Europe, to distilled spirits which were unknown in Arabia), a product of the distillation of wine or malt. Any one of a class of compounds analogous to common alcohol; the ending **ol** designates a member of this class of compounds.

Aleurone (Gr. *aleuron,* flour), proteinaceous and fatty material, usually in the form of small grains, occuring in the outermost cell layer of the endosperm of wheat and other grains.

Alga (plural, **algae**) (L. *alga,* seaweed), a member of the large group of thallus plants containing chlorophyll and thus able to synthesize carbohydrates.

Algin, a long chain polymer of mannuronic acid found in the cell walls of the brown algae.

Alkali (Arabic, *alqili,* the ashes of the plant saltwort), a substance with marked basic properties.

Allele (Gr. *allelon,* of one another, mutually each other), either of a pair of contrasting Mendelian characters.
—multiple-, several alleles at a single locus.
—pseudo-, factors that recombine within a single locus.

Allopolyploidy (Gr. *allos,* other + polyploidy), ploidy in which sets of chromosomes are derived from different parents, generally not closely related.

Alpine (L. *Alpes,* the Alps Mts.), meadowlike vegetation at high elevation, above tree line.

Alternate, referring to bud or leaf arrangement in which there is one bud or one leaf at a node.

Amensalism (L. *a,* not + *mensa,* table), a form of biological interaction in which one organism is inhibited by another, but the other is neither inhibited nor stimulated.

Amino acid. (Gr. *Ammon,* from the Egyptian sun god, in N.L. used in connection with ammonium salts) an acid containing the group NH_2; one of the building blocks of a protein.

Ammonification (*Ammon,* Egyptian sun god, near whose temple ammonium salts were first prepared from camel dung + L. *facere,* to make), decomposition of amino acids, resulting in the production of ammonia.

Anaerobe (Gr. *a,* without + *aer,* air + *bios,* life), an organism able to respire in the absence of free oxygen, or in greatly reduced concentrations of free oxygen.

Anaphase (Gr. *ana,* up + *phais,* appearance), that stage in mitosis in which half chromosomes or sister chromatids move to opposite poles of the cell.

Androecium (Gr. *andros,* man + *oikos,* house), the aggregate of stamens in the flower of a seed plant.

Aneuploid (Gr. *aneu,* without + *ploid*), the condition in which the number of chromosomes differ from *n* times the basic number, i.e., $3n + 1, 2n - 1, 2n + 4.$

Angiosperm (Gr. *angion,* a vessel + *sperma* from *speirein,* to sow, hence a seed or germ), literally a seed borne in a vessel, thus a group of plants whose seeds are borne within a matured ovary.

Angstrom (After A. J. Angstrom, a Swiss physicist, 1814–1874), a unit of length equal to 0.0001 of a micron. $10,000A = 1$ micron $= 0.001$ millimeter.

Anisogamy (Gr. *an,* prefix meaning not, + *isos,* equal, + *gamete,* spouse), the condition in which the gametes, though similar in appearance, are not identical.

Annual (L. *annualis,* within a year), a plant that completes its life cycle within one year's time and then dies.

Annular vessels (L. *annularis,* a ring), vessels with lignified rings.

Annulus (L. *anulus* or *annulus,* a ring), in ferns, a row of specialized cells in a sporangium, of importance in opening of the sporangium; in mosses, thick-walled cells along the rim of the sporangium to which the peristome is attached.

Anther (M.L. *anthera*—from the Greek word *anthros,* meaning flower—a medicine extracted from the internal whorls of flowers by medieval pharmacists; confined to pollen-producing parts by herbalists in 1700), pollen-bearing portion of stamen.

Antheridium (anther + *-idion,* a Gr. dim. ending, thus a little anther), male gametangium or sperm-bearing organ of plants other than seed plants.

Anthocyanin (Gr. *anthros,* a flower + *kyanos,* dark blue), a blue, purple, or red vacuolar pigment.

Antibiotic (Gr. *anti,* against or opposite + *biotikos,* pertaining to life), a natural organic substance which retards or prevents the growth of organisms; generally used to designate substances formed by microorganisms that prevent growth of other microorganisms.

Antibody (Gr. *anti,* against + body), a protein produced in an organism, in response by the organism to a contact with a foreign substance, and having the ability of specifically reacting with the foreign substance.

Antipodal (Gr. *anti,* opposite + *pous,* foot), generally, referring to one object situated opposite another object. Specifically, referring to cells or nuclei at the end of the embryo sac opposite that of the egg apparatus.

Apex (L. *apex,* a tip, point, or extremity), the tip, point, or angular summit of anything: the tip of a leaf; that portion of a root or shoot containing apical and primary meristems.

Apical meristem, a mass of meristematic cells at the very tip of a shoot or root.

Apomixis (Gr. *apo,* away from + *mixis,* a mingling), the production of offspring in the usual sexual structures without the mingling and segregation of chromosomes.

Apothecium (Gr. *apotheke,* a storehouse), a cup-shaped or saucer-shaped open ascocarp.

Archegonium (L. dim. of Gr. *archegonos,* literally a little founder of a race), female gametangium or egg-bearing organ, in which the egg is protected by a jacket of sterile cells.

Aril (M.L. *arillus,* a wrapper for a seed), an accessory seed-covering formed by an outgrowth at the base of the ovule in *Taxus.*

Ascocarp (Gr. *askos,* a bag + *karpos,* fruit), a fruiting body of the Ascomycetes, generally either an open cup, a vessel, or closed sphere lined with special cells called asci (*see* Ascus).

Ascogenous hyphae, hyphae arising from the ascogonium, after the formation of $n + n$ paired nuclei. The hymenial layer of the ascocarp develops from the ascogenous hyphae.

Ascogonium, the oogonium or female gametangium of the Ascomycetes.

Ascomycetes (Gr. *askos,* a bag + *mykes,* fungus), a large group of true fungi with septate hyphae producing large numbers of asexual conidiospores and meiospores called ascospores, the latter in asci.

Ascospore (Gr. *askos,* a bag + spore), meiospore produced within an ascus.

Ascus (plural, **asci**) (Gr. *askos,* a bag), a specialized cell, characteristic of the Ascomycetes, in which two haploid nuclei fuse, immediately after which three (generally) divisions occur, two of which constitute meiosis, resulting in eight ascospores still contained within the ascus.

Asexual (Gr. *a,* without + L. *sexualis,* sexual), any type of reproduction not involving the union of gametes or meiosis.

Aspect (L. *aspectus,* appearance), the direction of slope of a surface, as a hillside with a south-facing aspect.

Assimilation (L. *assimilare,* to make like), the transformation of food into protoplasm.

Atoms (F. *atome,* from the Greek *atomos,* indivisible), the smallest particles in which the elements combine either with themselves, or with other elements, and thus the smallest quantity of matter known to possess the properties of a particular element. Atoms were first postulated by the Greek philosopher-scientists Leucippus and Democritus about 450 B.C.

ATP, adeosinetriphosphate, a high-energy organic phosphate of great importance in energy transfer in cellular reactions.

Auricles (L. *auricula,* dimin. of *auris,* ear), earlike structures; in grasses, small projections that grow out from the opposite side of the sheath at its upper end where it joins the blade.

Autoecious (Gr. *auto,* self + *oikia,* dwelling), having a complete life cycle on the same host.

Autophyte (Gr. *auto,* self + *phyton,* a plant), a plant that can make its own food.

Autoradiograph (Gr. *auto,* self, + L. *radiolus,* a ray + Gr. *graphe,* a painting), a photographic print made by a radioactive substance acting upon a sensitive photographic film.

Autotetraploidy the condition in which the doubling of the chromosome number occurs in one cell or between cells on the same plant.

Autotrophic (Gr. *auto,* self + *trophein,* to nourish with food), pertaining to a plant that is able to manufacture its own food.

Auxin (Gr. *auxein,* to increase), a plant growth-regulating substance regulating cell elongation.

Axil (Gr. *axilla,* armpit), the upper angle between a petiole of a leaf and the stem from which it grows.

Axillary bud, a bud formed in the axil of a leaf.

Bacillus (L. *baculum,* a stick), a rod-shaped bacterium.

Bacteria (Gr. *bakterion,* a stick), common name for the class Schizomycetes.

Bacteriology (bacteria + Gr. *logos,* discourse), the science of bacteria.

Bacteriophage (bacteria + Gr. *phagein,* to eat), literally, an eater of bacteria; a virus that infects specific bacteria, multiplies therein, and usually dissolves the bacterial cells.

Banner (M.L. *bandum,* a standard), large, broad, and conspicuous petal of legume type of flower.

Bark (Swedish *bark,* rind), the external group of tissues, from the cambium outwards, of a woody stem or root.

Base pair, the nitrogen bases that pair in the DNA molecule, adenine with thymine, and guanine with cytosine.

Basidiomycetes (M.L. *basidium,* a little pedestal + Gr. *mykes,* fungus), group of fungi characterized by the production of meiospores on special cells, the basidia.

Basidiospore (M.L. *basidium,* a little pedestal + spore), type of meiospore borne by basidia in the Basidiomycetes.

Basidium (plural, **basidia**) (M.L. *basidium,* a little pedestal), a specialized reproductive cell of the Basidiomycetes in which nuclei fuse and meiosis occurs. It may be a special club-shaped cell, a short filamentous cell, or a short four-celled filament.

Benthon (Gr. *benthos,* the depths of the sea), attached aquatic plants and animals, collectively.

Berry, a simple fleshy fruit, the ovary wall fleshy and including one or more carpels and seeds.

Biennial (L. *biennium,* a period of two years), a plant which completes its life cycle within two years and then dies.

Bifacial leaf (L. *bis,* twice + *facies,* face), a leaf having upper and lower surfaces distinctly different.

Binomial (L. *binominis,* two names), two-named; in biology each species is generally indicated by two names, the genus to which it belongs and its own species name.

Biology (Gr. *bios,* life + *logos,* word, speech, discourse), the science that deals with living things.

Biosystematics (Gr. *bios,* life + *synistanai,* to place together), a field of taxonomy which emphasizes breeding behavior and chromosome characteristics.

Biotic (Gr. *biokitos,* relating to life), relating to life.

Biotin, a vitamin of the B complex.

Bordered pit, a pit in tracheid or other cell of secondary xylem having a distinct rim of the cell wall overarching the pit membrane.

Botany (Gr. *botane,* plant, herb), the science dealing with plant life.

Bract (L. *bractea,* a thin plate of precious metal), a modified leaf, from the axil of which arises a flower or an inflorescence.

Bud (Middle English *budde,* bud), an undeveloped shoot, largely meristematic tissue, generally protected by modified scale-leaves.

Bud-scale, a modified protective leaf of a bud.

Bud-scar, a scar left on a twig when the bud or bud scales fall away.

Bulb (L. *bulbus,* a modified bud, usually underground), a short, flattened, or disk-shaped underground stem, with many fleshy scale-leaves filled with stored food.

Bundle scar, scar left where conducting strands passing out of the stem into the leaf stalk were broken off when the leaf fell.

Cactus (Gr. *kaktos,* a kind of prickly plant), any plant of the cactus family, famous plants of the desert.

Callose (L. *callum,* thick skin + *ose,* a suffix indicating a carbohydrate), an amorphous, hardened carbohydrate constituent of cell walls, commonly developing upon injury.

Callus (L. *callum,* thick skin), mass of large, thin-walled cells, usually developed as the result of wounding.

Calyptra (Gr. *kalyptra,* a veil, covering), in bryophytes, an envelope covering the developing sporophyte, formed by growth of the venter of the archegonium.

Calyx (GR. *kalyx,* a husk, cup), sepals collectively; outermost flower whorl.

Cambium (L. *cambium,* one of the alimentary body fluids supposed to nourish the body organs), a layer, usually regarded as one or two cells thick, of persistently meristematic tissue, giving rise to secondary tissues, resulting in growth in diameter.

Canopy (Gr. *kanopeion,* a cover over a bed to keep off gnats), the leafy portion of a tree or shrub.

Capillaries (L. *capillus,* hair), very small spaces, or very fine bores in a tube.

Caprification, an artificial method of pollinating certain cultivated varieties of figs.

Capsule (L. *capsula,* dimin. of *capsa,* a case), a simple, dry, dehiscent fruit, with two or more carpels.

Carbohydrate (chemical combining forms, *carbo,* carbon + *hydrate,* containing water), a food composed of carbon, hydrogen, and oxygen, with the hydrogen and oxygen frequently in a 2 to 1 ratio, as in water, H_2O.

Carotene (L. *carota,* carrot), a reddish-orange plastid pigment.

Carpel (Gr. *karpos,* fruit), a floral leaf bearing ovules along the margins.

Carpogonium (Gr. *karpos,* fruit + *gonos,* producing), female gametangium (in red algae).

Carpospore (Gr. *karpos,* fruit + spore), one of the spores produced in a carpogonium.

Caruncle (L. *caruncula,* dimin. of *caro,* flesh, wart), a spongy outgrowth of the seed coat, especially prominent in the castor-bean seed.

Caryopsis (Gr. *karyon,* a nut + *opsis,* appearance), a simple, dry, one-seeded, indehiscent fruit, with pericarp firmly united all around to the seed coat.

Catalyst (Gr. *katalyein,* to dissolve), a substance that accelerates a chemical reaction but which is not used up in the reaction.

Catkin (Literally a kitten, apparently first used in 1578 to describe the inflorescence of the pussy willow), a type of inflorescence, really a spike, generally bearing only pistillate flowers or only staminate flowers, which eventually fall from the plant entire.

Cell (L. *cella,* small room), a structural and physiological unit composing living organisms, in which take place the majority of complicated reactions characteristic of life.

Cellulose (cell + *ose,* a suffix indicating a carbohydrate), a complex carbohydrate occurring in the cell walls of the majority of plants. Cotton is largely cellulose. It is composed of hundreds of simple sugar molecules linked together in a characteristic manner.

Chalaza (Gr. *chalaza,* small tubercle), the region on a seed at the upper end of the raphe where the funiculus spreads out and unites with the base of the ovule.

Chernozem (R. *cherny,* black + *zem,* earth), a soil characteristic of some grassland vegetation in warm areas with moderate rainfall; dark in color because of a high content of organic matter.

Chiasma (Gr. *chiasma,* two lines placed crosswise), the cross formed by breaking, during prophase I of meiosis, of two nonsister chromatids of homologous chromosomes and the rejoining of the broken ends of different chromatids.

Chitin (Gr. *chiton,* a coat of mail), a horny substance forming the outer coat of insects and crustaceans and also found in the cell walls of many fungi.

Chlamydospore (Gr. *chlamys,* a horseman's or young man's coat + spore) a heavy-walled resting asexual spore.

Chlorenchyma (Gr. *chloros,* green + *-enchyma,* a suffix meaning tissue), tissue possessing chloroplasts.

Chlorophyll (Gr. *chloros,* green + *phyllon,* leaf), the green pigment found in the chloroplast, important in the absorption of light energy in photosynthesis.

Chloroplast (Gr. *chloros,* green + *plastos,* formed), specialized cytoplasmic body, containing chlorophyll, in which occur important reactions of starch or sugar synthesis.

Chlorosis (Gr. *chloros,* green + *osis,* diseased state), failure of chlorophyll development, because of a nutritional disturbance or because of an infection of virus, bacteria, or fungus.

Chromatid (chromosome + *-id,* L. suffix meaning daughters of), the half-chromosome during prophase and metaphase of mitosis, and between prophase I and anaphase II of meiosis.

Chromatin (Gr. *chroma,* color), substance in the nucleus which readily takes artificial staining; also, that portion which bears the determiners of hereditary characters.

Chromatin bodies, bodies in bacteria that give some of the histochemical reactions that are associated with the chromosomes of higher organisms.

Chromatophores, (Gr. *chromo,* color, + Gr. *phorus,* a bearer), in algae, bodies bearing chlorophyll; in bacteria, small bodies, about 1000 A in diameter, containing chlorophyll, protein, and a carbohydrate.

Chromoplast (Gr. *chroma,* color + *plastos,* formed), specialized plastid containing yellow or orange pigments.

Chromosome (Gr. *chroma,* color + *soma,* body), a group of nuclear bodies containing genes, and largely responsible for the differentiation and activity of a cell, and undergoing characteristic division stages; or, one of the bodies into which the nucleus resolves itself at the beginning of mitosis and from which it is derived at the end of mitosis.

Cilia (singular, **cilium**) (Fr. *cil,* an eyelash), protoplasmic hairs which, by a whiplike motion, propel certain types of unicellular organisms, gametes, and zoospores through water.

Cisterna (plural, **cisternae**) (L. *cistern,* a reservoir), generally referring to sections of the endoplasmic reticulum that appear as parallel membranes, each about 50 A thick bounding a space about 400 A wide.

Cistron (L. preposition *cis,* on this side of, relating to a genetical test of gene location + *tron* or *ton,* from the last syllable of proton, electron and other such words indicating an elementary particle). A complex unit of the gene, probably consisting of from a hundred to thousands of nucleotide pairs.

Cladode (Gr. *kladodes,* having many shoots), a cladophyll.

Cladophyll (Gr. *klados,* a shoot + *phyllon,* leaf), a branch resembling a foliage leaf.

Clay soil particles less than 2 μ in diameter, composed mainly of aluminum (Al), oxygen (O), and silicon (S).

Cleistothecium (plural **cleistothecia**) (Gr. *kleistos,* closed + *thekion,* a small receptacle), the closed, spherical ascocarp of the powdery mildews.

Climax community, the last stage of a natural succession; a community capable of maintaining itself as long as the climate does not change.

Clone (Gr. *klon,* a twig or slip), the aggregate of individual organisms produced asexually from one sexually-produced individual.

Closed bundle, a vascular bundle lacking cambium.

Coalescence (L. *coalescere,* to grow together), a condition in which there is union of separate parts of any one whorl of flower parts.

Coccus (plural, **cocci**) (Gr. *kokkos,* a berry), a spherical bacterium.

Coenobium (Gr. *koinois,* common + Gr. *bios,* life), a colony of unicellular organisms surrounded by a common membrane.

Coenocyte (Gr. *koinos,* shared in common + *kytos,* a vessel), a plant or filament whose protoplasm is continuous and multinucleate and without any division by walls into separate protoplasts.

Coenogamete (Gr. *koinos,* shared in common + gamete), a multinucleate gamete, with no walls separating the many nuclei.

Coenyzme, a substance, usually nonprotein and of low molecular weight, necessary for the action of some enzymes.

Cohesion (L. *cohaerere,* to stick together), union or holding together of parts of the same materials; the union of floral parts of the same whorl, as petals to petals.

Coleoptile (Gr. *koleos,* sheath + *ptilon,* down, feather), the first leaf in germination of monocotyledons, which sheaths the succeeding leaves.

Coleorhiza (Gr. *koleos,* sheath + *rhiza,* root), sheath which surrounds the radicle of the grass embryo and through which the young root bursts.

Collenchyma (Gr. *kolla,* glue + *-enchyma,* a suffix, derived from parenchyma and denoting a type of cell tissue), a stem tissue composed of cells which fit rather closely together and with walls thickened at the angles of the cells.

Colloid (Gr. *kolla,* glue + *eidos,* form), referring to matter composed of particles, ranging in size from 0.0001 to 0.000001 millimeter, dispersed in some medium. Milk and mayonnaise are examples.

Community (L. *communitas,* a fellowship), all the populations within a given habitat; usually the populations are thought of as being interdependent.

Companion cells, cells associated with sieve-tube elements.

Competition (L. *competere,* to strive together), a form of biological interaction in which both organisms (at least initially) decline in growth or success because of the insufficient supply of some necessary factor(s).

Complete flower a flower having four whorls of floral leaves: sepals, petals, stamens, and carpels.

Compound leaf, a leaf whose blade is divided into several distinct leaflets.

Conceptacle (L. *conceptaculum,* a receptacle), a cavity or chamber of a frond (of *Fucus,* for example) in which gametangia are borne.

Cone (Gr. *konos,* a pine cone), a fruiting structure composed of modified leaves or branches, which bear sporangia (microsporangia, megasporangia, pollen sacs, or ovules), and frequently arranged in a spiral or four-ranked order; for example, a pine cone.

Conidia (singular, **conidium**) (Gr. *konis,* dust), asexual reproductive cells of fungi, arising by fragmentation of hyphae, by the cutting off of terminal or lateral cells of special hyphae, or by being pushed out from a flask-shaped cell.

Conidiophore (conidia + Gr. *phoros,* bearing), conidium-bearing branch of hypha.

Conidiosporangium (Gr. *konis,* dust + sporangium),

sporangium formed by being cut off from the end of a terminal or lateral hypha.

Conidiospore (conidia + spore), spore formed as described for conidia.

Conifer (cone + L. *ferre*, to carry), a cone-bearing tree.

Conservative (L. *conservare*, to keep), said of a taxonomic trait whose expression is not modified to any great extent by the external environment; a trait that is constant unless its genetic base is changed.

Convergent evolution, process of successive progeny, originally of quite distinct parents, coming to appear more and more alike through time because of selection pressure.

Conjugation (L. *conjugatus*, united), process of sexual reproduction involving the fusion of isogametes.

Cork (L. *quercus*, oak), an external, secondary tissue impermeable to water and gases.

Cork cambium, the cambium from which cork develops.

Corm (Gr. *kormos*, a trunk), a short, solid, vertical, enlarged underground stem in which food is stored.

Corolla (L. *corolla*, dimin. of *corona*, a wreath, crown), petals, collectively; usually the conspicuous colored flower whorl.

Cortex (L. *cortex*, bark), primary tissue of a stem or root bounded externally by the epidermis and internally by the phloem.

Cotyledon (Gr. *kotyledon*, a cup-shaped hollow), seed leaf; two, generally storing food in dicotyledons; one, generally a digestive organ in the monocotyledons.

Cristae (L. *crista*, a crest), crests or ridges, used here to designate the infoldings of the inner mitochondrial membrane.

Cross-pollination, the transfer of pollen from a stamen to the stigma of a flower on another plant, except in clones.

Crossing over, the exchange of corresponding segments between chromatids of homologous chromosomes.

Cuticle (L. *cuticula*, dimin. of *cutis*, the skin) waxy layer on outer wall of epidermal cells.

Cutin (L. *cutis*, the skin), waxy substance which is but slightly permeable to water, to water vapor, and to gases.

Cutinization, impregnation of cell wall with a substance called cutin.

Cyme (Gr. *kyma*, a wave, a swelling), a type of inflorescence in which the apex of the main stalk or the axis of the inflorescence ceases to grow quite early.

Cystocarp (Gr. *kystos*, bladder + *karpos*, fruit), a peculiar diploid spore-bearing structure formed after fertilization in certain red algae.

Cytokinesis (Gr. *kytos*, a hollow vessel + *kinesis*, motion), division of cytoplasmic constituents at cell division.

Cytology (Gr. *kytos*, a hollow vessel + *logos*, word, speech, discourse), the science dealing with the cell.

Cytoplasm (Gr. *kytos*, a hollow vessel + *plasma*, form),

all the protoplasm of a protoplast outside the nucleus.

Cytosine, a pyrimidine base found in DNA and RNA.

Deciduous (L. *deciduus*, falling), referring to trees and shrubs that lose their leaves in the fall.

Decomposer, (L. *de*, from + *componere*, to put together), an organism which obtains food by breaking down dead organic matter into simpler molecules.

Decomposition (L. *de*, to denote an act undone + *componere*, to put together), a separation or dissolving into simpler compounds; rotting or decaying.

Dehiscent (L. *dehiscere*, to split open), opening spontaneously when ripe, splitting into definite parts.

Deletion (L. *deletus*, to destroy, to wipe out) used here to designate an area, or region, lacking from a chromosome.

Dendrogram (Gr. *dendron*, tree + Gr. *gramme*, what is written or drawn), a graph showing relationship between things at different levels of similarity; the graph resembles the limbs of a tree.

Denitrification (L. *de*, to denote an act undone + *nitrum*, nitro, a combining form indicating the presence of nitrogen + *facare*, to make), conversion of nitrates into nitrites, or into gaseous oxides of nitrogen, or even into free nitrogen.

Deoxyribose nucleic acid (DNA), hereditary material; the DNA molecule carries hereditary information.

Deuterium, or heavy hydrogen; a hydrogen atom, the nucleus of which contains one proton and one neutron. It is written as H^2. The common nucleus of hydrogen consists only of one proton.

Diastase (Gr. *diastasis*, a separation), a complex of enzymes that brings about the hydrolysis of starch with the formation of sugar.

Diatom (Gr. *diatomos*, cut in two), member of a group of golden brown algae with silicious cell walls fitting together much as do the halves of a pill box.

Dictyosome (Gr. *diktyon*, a net, Gr. *soma* body), one of the component parts of the Golgi apparatus. In plant cells a complex of flattened double lamellae.

Dicotyledon (Gr. *dis*, twice + *kotyledon*, a cup-shaped hollow), a plant whose embryo has two cotyledons.

Differentially permeable, referring to a membrane, through which different substances diffuse at different rates; some substances may be unable to diffuse through such a membrane.

Differentiation (L. *differre*, to carry different ways), development from one cell to many cells, accompanied by a modification of the new cells for the performance of particular functions.

Diffusion (L. *diffusus*, spread out), the movement of molecules, and thus a substance, from a region of higher concentration of those molecules to a region of lower concentration.

Digestion (L. *digestio,* dividing, or tearing to pieces, an orderly distribution), the processes of rendering food available for metabolism by breaking it down into simpler compounds, chiefly through actions of enzymes.

Dihybrid cross (Gr. *dis,* twice + *hybrida,* the offspring of a tame sow and a wild boar, a mongrel), a cross between organisms differing in two characters.

Dikaryon (Gr. *di,* two + *karyon,* nut), the *n* + *n,* paired nuclei, each usually derived from a different parent, one male, one female.

Dioecious (Gr. *dis,* twice + *oikos,* house), unisexual; having the male and female elements in different individuals.

Diploid (Gr. *diploos,* double + *oides,* like), having a double set of chromosomes, or referring to an individual containing a double set of chromosomes per cell. Usually a sporophyte generation.

Disease (L. *dis,* a prefix signifying the opposite + M. E. *aise,* comfort, literally the opposite of ease), any alteration from the state of metabolism necessary for the normal development and functioning of an organism.

Distromatic (Gr. *di* a prefix meaning two + *stroma,* a bed, currently meaning a supporting framework), referring to a thallus, two cells in thickness.

Divergent evolution, process of successive progeny, originally of quite similar parents, coming to appear more and more different through time because of isolation and selection pressure.

Division, a major portion of the plant kingdom; equivalent to phylum.

DNA, Deoxyribose nucleic acid.

Dominant (L. *dominari,* to rule), referring, in ecology, to species of a community that receive the full force of the macroenvironment; usually the most abundant of such species, and not all of them; referring, in heredity, to that gene (or the expression of the character it influences) which, when present in a hybrid with a contrasting gene, completely dominates in the development of the character. In peas, tall is dominant over dwarf.

Donor, one who gives.

Dormant (L. *dormire,* to sleep), being in a state of reduced physiological activity such as occurs in seeds, buds, etc.

Dorsiventral (L. *dorsum,* the back + *venter,* the belly), having upper and lower surfaces distinctly different, as a leaf does.

DPN, Diphosphopyridine nucleotide, or more recently, NAD—nicotinamide adenine dinucleotide.

DPNH, Reduced diphosphopyridine nucleotide or, more recently, NADH—reduced nicotinamide adenine dinucleotide.

Drupe (L. *drupa,* an overripe olive), a simple, fleshy fruit, derived from a single carpel, usually one-seeded, in which the exocarp is thin, the mesocarp fleshy, and the endocarp stony.

Ecology (Gr. *oikos,* home + *logos,* discourse), the study of plant life in relation to environment.

Ecosystem (Gr. *oikos,* house + *synistanai,* to place together), an inclusive term for a living community and all the factors of its nonliving environment.

Ecotype (Gr. *oikos,* house + *typos,* the mark of a blow), a genetic variant within a species, which is adapted to a particular environment yet remains interfertile with all other members of the species.

Ectoplast (Gr. *ektos,* outside + Gr., *plastos,* formed), cytoplasmic membrane surrounding the outside of the protoplast.

Edaphic (Gr. *edaphos,* soil), pertaining to soil conditions that influence plant growth.

Egg (A. S. *aeg,* egg), a female gamete.

Elater (Gr. *elater,* driver), an elongated, spindle-shaped, sterile, hygroscopic cell in the sporangium of a liverwort sporophyte.

Electron (Gr. *elektron,* gleaming in the sun, by way of L. *electrum* a bright alloy of gold and silver, and finally amber, from which the first electricity was produced by friction), the elementary charge of negative electricity, equal to 4.80×10^{-10} electrostatic units. Normally, the number of orbital electrons balances the positive charge on the nucleus and renders the atom electrically neutral.

Element (L. *elementa,* the first principles; according to one system of medieval chemistry as recent as 1700, there were four elements composing all material bodies: earth, water, air, and fire), in modern chemistry, a substance that cannot be divided or reduced, by any known chemical means, to a simpler substance. Ninety-two natural elements are known, of which gold, carbon, oxygen, and iron are examples. Several including plutonium have been formed in atomic piles.

Embryo (Gr. *en,* in + *bryein,* to swell), a young sporophytic plant, before the start of a period of rapid growth (germination in seed plants).

Embryo sac, the female gametophyte of the angiosperms; generally a seven-celled structure. The seven cells are two synergids, one egg cell, three antipodal cells (each with a single haploid nucleus) and one endosperm mother cell with two haploid nuclei.

Emulsion (L. *emulgere,* to milk out), a suspension of fine particles of a liquid in a liquid.

Endocarp (Gr. *endon,* within + *karpos,* fruit), inner layer of fruit wall (pericarp).

Endodermis (Gr. *endon,* within + *derma,* skin), the layer of living cells, with various characteristically thickened walls and no intercellular spaces, which surrounds the vascular tissue of certain plants and occurs in nearly all roots and certain stems and leaves.

Endogenous (Gr. *endon*, within, Gr. *genos*, race, kind), developed or added from outside the cell.

Endoplasmic reticulum (Gr. *endon*, within + Gr. *plasma*, anything formed or molded; L. *reticulum*, a small net), originally, a cytoplasmic net adjacent to the nucleus, made visible by the electron microscope. Now, any system of paired membranes within the cytoplasm. Frequently abbreviated to ER.

Endosperm (Gr. *endon*, within + *sperma*, seed), the nutritive tissue formed within the embryo sac of seed plants. It is often consumed as the seed matures, but remains in the seeds of corn and other cereals.

Endosperm mother cell, one of the seven cells of the mature embryo sac, containing the two polar nuclei and, after reception of a sperm cell, giving rise to the primary endosperm cell from which the endosperm develops.

Enzyme (Gr. *en*, in + *zyme*, yeast), a protein of complex chemical constitution produced in living cells, which, even in very low concentration, speeds up certain chemical reactions but is not used up in the reaction.

Epiblast (Gr. *epi*, upon + *blastos*, a sprout or shoot), a small appendage in front of the plumule in the embryo of many grasses.

Epicotyl (Gr. *epi*, upon + *kotyledon*, a cup-shaped hollow), the upper portion of the axis of embryo or seedling, above the cotyledons.

Epidermis (Gr. *epi*, upon + *derma*, skin), a superficial layer of cells occurring on all parts of the primary plant body: stems, leaves, roots, flowers, fruits, and seeds. It is absent from the root cap and not differentiated on the apical meristems.

Epigyny (Gr. *epi*, upon + *gyne*, woman), the arrangement of floral parts in which the ovary is embedded in the receptacle so that the other parts appear to arise from the top of the ovary.

Epiphyte (L. *epi*, upon + *phyton*, a plant), a plant that grows upon another plant, yet is not parasitic.

Equatorial rain forest, vegetation with several tree strata; characteristic of warm, wet regions; synonymous with tropical rain forest of other texts.

ER, endoplasmic reticulum.

Ergot (F. *argot*, a spur), a fungus disease of cereals and wild grasses in which the grain is replaced by dense masses of purplish hyphae, the ergot.

Erosion (L. *e*, out + *rodere*, to gnaw), the wearing away of land, generally by the action of water.

Etiolation (F. *etioler*, to blanch), a condition involving increased stem elongation, poor leaf development, and lack of chlorophyll, found in plants growing in the absence, or in a greatly reduced amount, of light.

Eukaryote (L. *eu*, true + *karyon*, a nut referring in modern biology to the nucleus), any organism characterized by having the cellular organelles, including the nucleus, bounded by membranes.

Evapotranspiration (L. *evaporare, e*, out of, + *vapor*, vapor + F. *transpirer*, to perspire), the process of water loss in vapor form from a unit surface of land both directly and from leaf surfaces.

Evolution (L. *evolutio*, an unrolling), the development of a race, a species, genus, or other larger group of plants or animals.

Exine (L. *exterus*, outside), outer coat of pollen.

Exocarp (Gr. *exo*, without, outside + *karpos*, fruit), outermost layer of fruit wall (pericarp).

Exogenous (Gr. *exe*, out, beyond Gr. *benos*, race, kind), produced outside of, originating from, or due to external causes.

F₁, first filial generation in a cross between any two parents.

F₂, second filial generation, obtained by crossing two members of the F_1, or by self pollinating the F_1.

Facultative (L. *facultas*, capability), referring to an organism having the power to live under a number of certain specific conditions, e.g., a facultative parasite may be either parasitic or saprophytic.

Family, (L. *familia*, family) In plant taxonomy, a group of genera. Families are grouped in orders.

Fascicle (L. *fasciculus*, a small bundle), a bundle of pine or other needle-leaves of gymnosperms.

Fascicular cambium, cambium within vascular bundles.

Fermentation (L. *fermentun*, a drink made from fermented barley, beer), an oxidative process in foods in which molecular oxygen is not involved, such as the production of alcohol from sugar by yeasts.

Ferredoxin, an electron-transferring protein containing iron, involved in photosynthesis and in the biological production and consumption of hydrogen gas. It transfers electrons with the greatest reducing power in cellular metabolism.

Fertilization (L. *fertilis*, capable of producing fruit), that state of a sexual life cycle involving the union of two protoplasts and hence the doubling of chromosome numbers.

Feulgen, a histochemical test for DNA first published in 1932 by R. Feulgen, a German histologist.

Fiber (L. *fibra*, a fiber or filament), an elongated, tapering, thick-walled strengthening cell occurring in various parts of plant bodies.

Fiber-tracheid, xylem elements found in pine that are structurally intermediate between tracheids and fibers.

Filament (L. *filum*, a thread), stalk of stamen bearing the anther at its tip; also, a slender row of cells (certain algae).

Fission (L. *fissilis*, easily split), asexual reproduction involving the division of a single-celled individual into two new single-celled individuals of equal size.

Flagellum (plural, **flagella**) (L. *flagellum*, a whip), a long, slender whip of protoplasm.

Flora (L. *floris*, a flower), an enumeration of all the species that grow in a region; also, the collective term for all the species that grow in a region.

Floret (F. *fleurette*, a dimin. of *fleur*, flower), one of the small flowers that make up the composite flower or the spike of the grasses.

Flower (F. *fleur*, L. *flos*, a flower), floral leaves grouped together on a stem and adapted for sexual reproduction in the angiosperms.

Follicle (L. *folliculus*, dimin. of *follis*, bag), a simple, dry, dehiscent fruit, with one carpel, splitting along one suture.

Food (A. S. *fōda*), any organic substance that furnishes energy and building materials directly for vital processes.

Food chain, the path along which caloric energy is transferred within a community (from producers to consumers to decomposers).

Fossil (L. *fossio*, a digging), any impression, natural or impregnated remains, or other trace of an animal or plant of past geological ages which has been preserved in the earth's crust.

Fruit (L. *fructus*, that which is enjoyed, hence product of the soil, trees, cattle, etc.), a matured ovary; in some seed plants other parts of the flower may be included; also applied, as **fruiting body,** to reproductive structures of other groups of plants.

Fucoxanthin (Gr. *phykos*, seaweed + *xanthos*, yellowish brown), a brown pigment found in brown algae.

Fungus (plural, **fungi**) (L. *fungus*, a mushroom), a thallus plant unable to make its own food, exclusive of the bacteria.

Funiculus (L. *funiculus*, dimin. of *funis*, rope or small cord), a stalk of the ovule, containing vascular tissue.

Galen, Claudius Galenus, 131–210 (?), the last of the great Greek biologists, he probably wrote 256 treatises, of which 131 were of a medical nature. His medical studies, as expressed in his book "On the Natural Faculties," had a very great influence on biological and medical thought during the middle ages.

Gametangium (Gr. *gametes*, a husband, *gamete*, a wife + *angeion*, a vessel), organ bearing gametes.

Gamete (Gr. *gametes*, a husband, *gamete*, a wife), a protoplast that fuses with another protoplast to form the zygote in the process of sexual reproduction.

Gametophyte (gamete + Gr. *phyton*, a plant), the gamete-producing plant.

Gel (L. *gelare*, to freeze), jellylike collidal mass.

Gemma (plural, **gemmae**) (L. *gemma*, a bud), a small mass of vegetative tissue; an outgrowth of the thallus.

Gene (Gr. *genos*, race, offspring), a group of base pairs in the DNA molecule in the chromosome that determines or conditions one or more hereditary characters.

Gene recombination, the appearance of gene combinations in the progeny different from the combinations present in the parents.

Genetics (Gr. *genesis*, origin), the science of heredity.

Genotype (gene + type), the assemblage of genes in an organism.

Genus (plural, **genera**) (Gr. *genos*, race, stock), a group of structurally or phylogenetically related species.

Geotropism (Gr. *ge*, earth + *tropos*, turning), a growth curvature induced by gravity.

Germination (L. *germinare*, to sprout), the beginning of growth of a seed, spore, bud, or other structure.

Gibberellins, a group of growth hormones (not identical with auxins), the most characteristic effect of which is to increase the elongation of stems in a number of kinds of higher plants.

Glucose (Gr. *glykys*, sweet + *-ose*, a suffix indicating a carbohydrate), a simple sugar, grape sugar, $C_6H_{12}O_6$.

Glume (L. *gluma*, husk), an outer and lowermost bract of a grass spikelet.

Glycogen (Gr. *glykys*, sweet + *gen*, of a kind), a carbohydrate related to starch but found generally in the liver of animals.

Glycolysis (Gr. *glykys*, sweet + *lysis*, a loosening), decomposition of sugar to simple compounds without involving free oxygen; early steps of respiration.

Golgi body or zone (Italian cytologist Camillo Golgi 1844–1926 who first described the organelle), in animal cells, a complex perinuclear region thought to be associated with secretion; in plant cells, a series of flattened plates, more properly called dictyosomes.

Grana (singular, **granum**) (L. *granum* a seed), structures within chloroplasts, seen as green granules with the light microscope and as a series of parallel lamellae with the electron microscope.

Ground cover, the area of ground covered by a plant when its canopy edge is projected perendicularly down.

Ground meristem (Gr. *meristos*, divisible), a primary meristem that gives rise to cortex, pit rays, and pith.

Guttation (L. *gutta*, drop, exudation of drops), exudation of water from plants, in liquid form.

Grow (A. S. *growan*, probably from Old Teut. *gro*, from which grass is also derived), of living bodies generally: to increase gradually in size by natural development.

Guanine, a purine base found in DNA and RNA.

Gynoecium (Gr. *gyne*, woman + *oikos*, house), the aggregate of carpels in the flower of a seed plant.

Haploid (Gr. *haploos*, single + *oides*, like), having a single complete set of chromosomes, or referring to an individual or generation containing such a single set of chromosomes per cell. Usually a gametophyte generation.

Haustorium (plural, **haustoria**) (M.L. *haustrum*, a pump), a projection of hyphae that acts as a penetrating and absorbing organ.

Head, in inflorescence, typical of the composite family, in which flowers are grouped closely on a receptacle.

Helix (Gr. *helix,* anything twisted), anything having a spiral form. Here, quite generally refers to the double spiral of the DNA molecule.

Hemicellulose (Gr. *hemi,* half + cellulose), any one of a group of carbohydrates resembling cellulose.

Herb (L. *herba,* grass, green blades), a seed plant that does not develop woody tissues.

Herbaceous (L. *herbaceus,* grassy), referring to plants having the characteristics of herbs.

Herbal (L. *herba,* grass), a book that contains the names and descriptions of plants, especially those which are thought to have medicinal uses.

Herbarium (L. *herba,* grass), a collection of dried and pressed plant specimens.

Heredity (L. *herditas,* being a heir), the transmission of morphological and physiological characters of parents to their offspring.

Hermaphrodite flower (Gr. *hermaphroditos,* a person partaking of the attributes of both sexes, represented by Hermes and Aphrodite), a flower having both stamens and pistils.

Heterobasidiomycetidae (Gr. *heteros,* other + Basidiomycete), a subclass of Basidiomycetes with variable basidia, never club-shaped cells.

Heterocyst (Gr. *heteros,* different + *cystis,* a bag), an enlarged colorless cell that may occur in the filaments of certain blue-green algae.

Heteroecious (Gr. *heteros,* different + *oikos,* house), referring to fungi that cannot carry through their complete life cycle unless two different host species are present.

Heterogametes (Gr. *heteros,* different + gamete), gametes dissimilar from each other in size and behavior, like egg and sperm.

Heterogamy (Gr. *heteros,* different + *gamos,* union or reproduction), reproduction involving two types of gametes.

Heterophyte (Gr. *heteros,* different + *phyton,* a plant), a plant which must secure its food ready made.

Heterosis (Gr. *heteros,* different + -osis, suffix indicating a state of), the state of a genotype having a large degree of heterozygosity.

Heterospory (Gr. *heteros,* different + spore), the condition of producing microspores and megaspores.

Heterothallic (Gr. *heteros,* different + thallus), referring to species in which male gametangia and female gamentangia are produced in different filaments or by different individual plant bodies.

Heterotrichy (Gr. *heteros,* different + *trichos,* a hair), in the algae, the occurrence of two types of filaments, erect and prostrate.

Heterotrophic (Gr. *heteros,* different + *trophein,* to nourish with food), referring to a plant obtaining nourishment from outside sources.

Heterozygous (Gr. *heteros,* different + *zygon,* yoke), having different genes of a Mendelian pair present in the same cell or organism; for instance, a tall pea plant with genes for tallness (T) and dwarfness (t).

Hexose (Gr. *hexa,* six + -ose, a suffix indicating, in this usage, a carbohydrate), a carbohydrate with six carbon atoms.

Hfr gene, gene in *Escherichia coli* including high frequency of gene recombination.

Hilum (L. *hilum,* a trifle), scar on seed, which marks the place where the seed broke from the stalk.

Histology (Gr. *histos,* cloth, tissue + *logos,* discourse), science that deals with the microscopic structure of animal and vegetable tissues.

Homobasidiomycetidae (Gr. *homo,* the same + Basidiomycete), a subclass of Basidiomycetes with a typical club-shaped cell as a basidium.

Homologous chromosomes (Gr. *homologos,* the same), members of a chromosome pair; they may be heterozygous or homozygous.

Homospory (Gr. *homos,* one and the same + spore), the condition of producing one sort of spore only.

Homothallic (Gr. *homos,* one and the same + thallus), referring to species in which male gametangia and female gamentangia are produced in the same filament or by the same individual plant body.

Homozygous (Gr. *homos,* one and the same + *zygon,* yoke), having similar genes of a Mendelian pair present in the same cell or organism; for instance, a tall pea plant with genes for tallness (TT) only.

Hormogonia (singular, **homogonium**) (Gr. *hormos,* necklace + *gonos,* offspring), short filaments, the result of a breaking apart of filaments of certain blue-green algae at the heterocysts.

Hormone (Gr. *hormaein,* to excite), a specific organic product, produced in one part of a plant or animal body, and transported to another part where, effective in small amounts, it controls or stimulates another and different process.

Humidity, relative (L. *humidus,* moist), the ratio of the weight of water vapor in a given quantity of air, to the total weight of water vapor which that quantity of air is capable of holding at the temperature in question, expressed as percent.

Humus (L. *humus,* the ground), decomposing organic matter in the soil.

Hybrid (L. *hybrida,* offspring of a tame sow and a wild boar, a mongrel), the offspring of two plants or animals differing in at least one Mendelian character; or the offspring of plants or animals differing in many characters.

Hydrolysis (Gr. *hydro,* water + *lysis,* loosening), union of a compound with water, attended by decomposition into less complex compounds; usually controlled by enzymes.

Hyaloplasm (Gr. *hyalinos,* glossy + *plasma,* anything formed), the clear background portion of the cytoplasm, forming the continuous substance in which other protoplasmic bodies are embedded.

Hydathode (Gr. *hydor,* water + O.E. *thoden,* stem or *thyddan,* to thrust), a structure, usually on leaves, which releases liquid water during guttation.

Hydrogen acceptor, a substance capable of accepting hydrogen atoms or electrons in the oxidation-reduction reactions of metabolism.

Hymenium (Gr. *hymen,* a membrane), spore-bearing tissue in various fungi.

Hypertrophy (Gr. *hyper,* over + *trophein,* to nourish with food), a condition of overgrowth or excessive development of an organ or part.

Hypha (plural, **hyphae**) (Gr. *hyphe,* a web), a fungal thread or filament.

Hypocotyl (Gr. *hypo,* under + *kotyledon,* a cup-shaped hollow), that portion of an embryo or seedling below the cotyledons.

Hypogyny (Gr. *hypo,* under + *gyne,* female) a condition in which the receptable is convex or conical, and the flower parts are situated one above another in the following order, beginning with the lowest: sepals, petals, stamens, carpels.

Hypothesis (Gr. *hypothesis,* foundation), a tentative theory or supposition provisionally adopted to explain certain facts and to guide in the investigation of other facts.

Hypotrophy (Gr. *hypo,* under + *trophein,* to nourish with food), an underdevelopment of an organ or part.

IAA, indoleacetic acid.

Imbibition (L. *imbibere,* to drink), the absorption of liquids or vapors into the ultramicroscopic spaces or pores found in such materials as cellulose or a block of gelatine. An adsorption phenomenon.

Imperfect flower, a flower lacking either stamens or pistils.

Imperfect fungi, fungi reproducing only by asexual means.

Inclusion body, a body found in cells of organisms with a virus infection.

Incomplete flower, a flower lacking one or more of the four kinds of flower parts.

Indehiscent (L. *in,* not + *dehiscere,* to divide), not opening by valves or along regular lines.

Indicator species, a species that has a narrow range of tolerance for one or more environmental factors so that, from its occurrence at a site, one can predict these factors at that site (e.g., nutrient availability or summer temperatures).

Indoleacetic acid, a naturally-occurring growth regulator, an auxin.

Indusium (plural, **indusia**) (L. *indusium,* a woman's undergarment), membranous growth of the epidermis of a fern leaf that covers a sorus.

Infect, (L. *infectus,* to put into, to taint with morbid matter), specifically to produce disease by such agents as bacteria or viruses.

Inferior ovary, an ovary more or less, even completely, attached to or united with the calyx.

Inflorescence (L. *inflorescere,* to begin to bloom), a flower cluster.

Inheritance (Old F. *enheritance,* inheritance), the reception or acquisition of characters or qualities by transmission of parent to offspring.

Inorganic, referring in chemistry to compounds that do not contain carbon.

Integument (L. *integumentum,* covering), coat of ovule.

Inter, a prefix, from the latin preposition *inter,* meaning between, in between, in the midst of.

Intercalary (L. *intercalare,* to insert), descriptive or meristematic tissue or growth not restricted to the apex of an organ, i.e., growth at nodes.

Intercellular (L. *inter,* between + cells), lying between cells.

Interfascicular cambium (L. *inter,* between + *fasciculus,* small bundle), cambium that develops between vascular bundles.

Internode (L. *inter,* between + *nodus,* a knot), the region of a stem between two successive nodes.

Intine (L. *intus,* within), the innermost coat of a pollen grain.

Intra, a prefix from the Latin preposition *intra* meaning on the inside, within.

Intracellular (L. *intra,* within + cell), lying within cells.

Involucre (L. *involucrum,* a wrapper), a whorl or rosette of bracts surrounding an inflorescence.

Ion (Gr. *ienai,* to go), an electrified particle formed by the breakdown of substances able to conduct an electric current.

Irregular flower, a flower in which one or more members of at least one whorl are of different form from other members of the same whorl; zygomorphic flower.

Isobilateral leaf (Gr. *isos,* equal + L. *bis,* twice, twofold + *lateralis,* pertaining to the side), a leaf having the upper and lower surfaces essentially similar.

Isodiametric (Gr. *isos,* equal + diameter), having diameters equal in all directions, as a ball.

Isogamy (Gr. *isos,* equal, + *gamete,* spouse), the condition in which the gametes are identical.

Isogametes (Gr. *isos,* equal + gametes), gametes similar in size and behavior.

Isomers (Gr. *isos,* equal + Gr. *meros,* part), two or more compounds having the same molecular formula, for example, glucose and fructose are both $C_6H_{12}O_6$.

Jansen, Zacharias, sixteenth-century Dutch spectacle maker who made the first compound microscopes with magnifications of about ten times.

Karyogamy (Gr. *karyon,* nut + *gamos,* marriage), the fusion of two nuclei.

Karyolymph (Gr. *karyon,* nut + *lympha,* water), the ground substance of a nucleus.

Karyon (Gr. *karyon,* a nut), term used in conjunction with the nuclei in cells of Ascomycetes and Basidiomycetes; **dikaryon,** two nuclei per cell, each derived from a different parent, *n + n;* **heterokaryon,** the situation in which members of a dikaryon pair carry different alleles.

Keel (A. S. *ceol,* ship), a structure of the legume type of flower, made up of two petals loosely united along their edges.

Kelp (M.E. *culp,* seaweed), a collective name for any of the large brown algae.

Kinetin (Gr. *kinetikos,* causing motion), one of a group of growth hormones that characteristically promote cell division.

Kinetochore (Gr. *kinein,* to move + *chorein,* to move apart), specialized body of a chromosome, which seems to direct its movement.

Lamella (plural, **lamellae**) (Gr. *lamin,* a thin blade), cellular membranes, frequently those seen in chloroplasts.

Lamina (L. *lamina,* a thin plate), blade or expanded part of a leaf.

Lateral bud, a bud that grows out of the side of a stem.

Laterite (L. *later,* a brick), a soil characteristic of rain forest vegetation; color is red from oxidized iron in the A horizon.

Latex (L. *latex,* juice), a milky secretion.

Leaf axil, angle formed by the leaf stalk and the stem.

Leaflet, separate part of the blade of a compound leaf.

Leaf primordium (L. *primordium,* a beginning), a lateral outgrowth from the apical meristem, which will become a leaf.

Legume (L. *legumen,* any leguminous plant, particularly bean), a simple, dry dehiscent fruit with one carpel, splitting along two sutures.

Lemma (Gr. *lemma,* a husk), lower bract that subtends a grass flower.

Lenticel (M.L. *lenticella,* a small lens), a structure of the bark that permits the passage of gas inward and outward.

Leucine (Gr. *leucos,* white), a white, crystalline amino acid.

Leucoplast (Gr. *leukos,* white + *plastos,* formed), a colorless plastid.

Liana (F. *liane* from *lier,* to bind), a plant that climbs upon other plants, depending upon them for mechanical support; a plant with climbing shoots.

Lichen (Gr. *leichen,* thallus plants growing on rocks and trees), a composite plant consisting of a fungus living symbiotically with an alga.

Lignification (L. *lignum,* wood + *facere,* to make), impregnation of a cell wall with lignin.

Lignin (L. *lignum,* wood), an organic substance or group of substances impregnating the cellulose framework of certain plant cell walls.

Ligule (L. *ligula,* dimin. of *lingua,* tongue), in grass leaves, an outgrowth from the upper and inner side of the leaf blade where it joins the sheath.

Linkage, the grouping of genes on the same chromosome.

Linked characters, characters of a plant or animal controlled by genes grouped together on the same chromosome.

Lipase (Gr. *lipos,* fat + *-ase,* suffix indicating an enzyme), any enzyme that breaks fats into glycerin and fatty acids.

Lipid (Gr. *lipos,* fat + L. *ides,* termination meaning son of; now used in sense of having the quality of), any of a group of fats or fatlike compounds insoluble in water and soluble in fat solvents.

Liverwort (liver + M.E. *wort,* a plant; literally, a liver plant, so named in medieval times because of its fancied resemblance to the lobes of the liver), common name for the Class Hepaticae of the Bryophyta.

Loam (O.E. *lam* or Old Teutonic *lai,* to be sticky, clayey), a particular soil texture class, referring to a soil having 30–50% sand, 30–40% silt, and 10–25% clay.

Lobed leaf (Gr. *lobos,* lower part of the ear), a leaf divided by clefts or sinuses.

Locule (L. *loculus,* dimin. of *locus,* a place), a cavity of the ovary in which ovules occur.

Lodicules (L. *lodicula,* a small coverlet), two scale-like structures that lie at the base of the ovary of a grass flower.

Longevity (L. *longaevus,* long-lived), length of life.

Lumen (L. *lumen,* light, an opening for light), the cavity of the cell within the cell walls.

Lysis (Gr. *lysis,* a lossing), a process of disintegration or cell destruction.

Lysogen (from *lysin,* a substance which brings about lysis, + N.L., *genic,* to give rise to), a substance, virus, or bacteria that stimulates the formation of lysins.

Lysogenic bacteria (Lysis + N.L. *genic,* combining form meaning to give rise to), here used to indicate bacteria carrying phage which eventually disrupt the bacteria.

Lysogenic conversion, the influence of prophage DNA on the host bacterium, including colony morphology, and on the presence of particular enzymes and other compounds.

Macroenvironment (Gr. *makros,* large + O.F. *environ,* about), the environment due to the general, regional climate; traditionally measured some four feet above the ground and away from large obstructions.

Macronutrient (Gr. *makros,* large + L. *nutrire,* to nourish), an essential element required by plants in relatively large quantities.

Map distance on a chromosome, the distance in crossover units between designated genes.

Megasporangium (Gr. *megas,* large + sporangium), sporangium that bears megaspores.

Megaspore (Gr. *megas,* large + spore), the meiospore of vascular plants, which gives rise to a female gametophyte.

Megaspore mother cell, a diploid cell in which meiosis will occur, resulting in four megaspores.

Megasporophyll (Gr. *megas,* large + spore + Gr. *phyllon,* leaf), a leaf bearing megasporangia.

Meiocyte (meiosis + Gr. *kytos,* currently meaning a cell), any cell in which meiosis occurs.

Meiosis (Gr. *meioun,* to make smaller), two special cell divisions occurring once in the life cycle of every sexually reproducing plant and animal, halving the chromosome number and effecting a segregation of genetic determiners.

Meiospore (meiosis + spore), any spore resulting from the meiotic divisions.

Meristem (Gr. *meristos,* divisible), undifferentiated tissue, the cells of which are capable of active cell division and differentiation into specialized tissues.

Meristoderm (meristem + epidermis) the outer meristematic cell layer (epidermis) of some Phaeophyta.

Mesocarp (Gr. *mesos,* middle + *karpos,* fruit), middle layer of fruit wall (pericarp).

Mesophyll (Gr. *mesos,* middle + *phyllon,* leaf), parenchyma tissue of leaf between epidermal layers.

Mesophyte (Gr. *mesos,* middle + *phyton,* a plant), a plant avoiding both extremes of moisture and drought.

Mesosome (Gr. *mesos,* middle + *soma,* body), one of a series of paired membranes occurring in many bacteria.

Metabolism (M.L. from the Gr. *metabolos,* to change), the process, in an organism or a single cell, by which nutritive material is built up into living matter, or aids in building living matter, or by which protoplasm is broken down into simple substances to perform special functions.

Metaphase (Gr. *meta,* after + *phasis,* appearance), stage of mitosis during which the chromosomes, or at least the kinetochores, lie in the central plane of the spindle.

Methionine a sulfur-containing amino acid.

Microbody (Gr. *mikros,* small + body), a cellular organelle, always bound by a single membrane, frequently spherical, from 200 to 600 A in diameter, containing a variety of enzymes.

Microcapillary space, exceedingly small spaces, such as those found between microfibrils of cellulose.

Microenvironment (Gr. *mikros,* small + O.F. *environ,*

about), the environment close enough to the surface of a living or nonliving object to be influenced by it.

Microfibrils (Gr. *mikros,* small + fibrils, dim. of fiber; literally, small little fibers), the translation of the name expresses the concept very well. Microfibrils are exceedingly small fibers visible only with the high magnifications of the electron microscope.

Micron (Gr. *mikros,* small), a unit of distance, 0.001 millimeter or 0.000039 inch. Symbol μ.

Micronutrient (Gr. *mikros,* small + L. *nutrire,* to nourish), an essential element required by plants in relatively small quantities.

Micropyle (Gr. *mikros,* small + *pyle,* orifice, gate), a pore leading from the outer surface of the ovule between the edges of the two integuments down to the surface of the nucellus.

Microsporangium (plural, **microsporangia**) (Gr. *mikros,* little + sporangium), a sporangium that bears microspores.

Microspore (Gr. *mikros,* small + spore), a spore which, in vascular plants, gives rise to a male gametophyte.

Microspore mother cell, a cell in which meiosis will occur, resulting in four microspores.

Microsporophyll (Gr. *mikros,* little + spore + Gr. *phyllon,* leaf), a leaf bearing microsporangia.

Microtubule (Gr. *mikros,* small + *tubule,* dimin. of tube), a tubule 250 A in diameter and of indefinite length, occurring in the cytoplasm of many types of cells.

Micurgical dissections (Gr. *mikros,* small + surgical), surgical experiments done on individual cells or groups of cells, such as a shoot apex.

Middle lamella (L. *lamella,* a thin plate or scale), original thin membrane separating two adjacent protoplasts and remaining as a distinct cementing layer between adjacent cell walls.

Millimeter, the 0.001 part of a meter, equal to 0.0394 inch.

Mitochondrion (plural, **mitochondria**) (Gr. *mitos,* thread + *chondrion,* a grain), a small cytoplasmic particle associated with intracellular respiration.

Mitosis (plural, **mitoses**) (Gr. *mitos,* a thread), nuclear division, involving appearance of chromosomes, their longitudinal duplication, and equal distribution of newly formed parts to daughter nuclei.

Mitospore (mitosis + spore), a spore forming after mitosis.

Mixed bud, a bud containing both rudimentary leaves and flowers.

Molecule (F. *môle,* mass + *cule,* a dim.; literally, a little mass), a unit of matter, the smallest portion of an element or a compound that retains chemical identity with the substance in mass. The molecule usually consists of a union of two or more atoms, some organic molecules containing a very large number of atoms.

Monocotyledon (Gr. *monos,* solitary + *kotyledon,* a cup-shaped hollow), a plant whose embryo has one cotyledon.

Monoecious (Gr. *monos,* solitary + *oikos,* house), having the reproductive organs in separate structures, but borne on the same individual.

Monohybrid (Gr. *monos,* solitary + L. *hybrida,* a mongrel), a cross involving one pair of contrasting characters.

Monophyletic (Gr. *mono,* single + *phyle,* tribe), said of organisms having a common (but sometimes quite ancient) ancestor.

Monostromatic (Gr. *monas,* single, solitary + *stroma,* a bed, currently meaning a supporting framework), referring to a thallus, one cell in thickness.

Morphology (Gr. *morphe,* form + *logos,* discourse), the study of form and its development.

Moss (L. *muscus,* moss), a bryophytic plant.

Mu, μ, Greek symbol used to indicate a micron; there are 1000 microns in 1 millimeter.

Multiciliate (L. *multus,* many + F. *cil,* an eyelash), having many cilia present on a sperm or spore or other type of ciliated cell.

Multiple fruit, a cluster of matured ovaries produced by separate flowers.

Mutation (L. *mutare,* to change), a sudden, heritable change appearing in an individual as the result of a change in genes or chromosomes.

Muton (mutation + *tron,* or *on,* from the last syllable of proton, electron, and other such words indicating an elementary particle), the smallest element, the alteration of which can cause a mutation; about 0.08 map units, possibly a single nucleotide pair.

Mutualism (L. *mutuus,* reciprocal), a form of biological interaction in which both organisms must associate together for continued success of both.

Mycelium (Gr. *mykes,* mushroom), the mass of hyphae forming the body of a fungus.

Mycology (Gr. *mykes,* mushroom + *logos,* discourse), the branch of botany dealing with fungi.

Mycorrhiza (Gr. *mykos,* fungus + *riza,* root), a symbiotic association between a fungus and usually the root of a higher plant.

Myxomycophyta (Gr. *myxa,* mucus + *mykes,* mushroom + *phyton,* plant), a phylum comprising the "slime fungi."

NAD, nicotineamideadenine-dinucleotide, a coenzyme capable of being reduced. Also called DPN, diphosphopyridine nucleotide.

NADH, reduced NAD.

NADP, nicotineamideadenine-dinucleotide phosphate, a coenzyme capable of being reduced. Also called TPN, triphosphopyridine nucleotide.

NADPH, reduced NADP.

Naked bud, a bud not protected by bud scales.

Nanometer (Gr. *nanos,* small), one millionth of a milimeter, similar to a milimicron; equals ten angstroms.

Nectar (Gr. *nektar,* drink of the gods), a fluid rich in sugars secreted by nectaries, which are often located near or in flowers.

Nectar guide, a mark of contrasting color or texture, that may serve to guide pollinators to nectaries within the flower.

Nectary (Gr. *nektar,* the drink of the gods), a nectar-secreting gland.

Net productivity, the arithmetic difference between calories produced in photosynthesis and calories lost in respiration.

Net radiation, the arithmetic difference between incoming solar radiation and outgoing terrestrial radiation.

Net venation, veins of leaf blade visible to the unaided eye, branching frequently and joining again, forming a network.

Neutron (L. *neuter,* neither), an uncharged particle found in the atomic nucleus of all elements except hydrogen. The helium nucleus has two protons and two neutrons. Mass of a neutron is equal to 1.67×10^{-24} gm.

Niche (It. *nicchia,* a recess in a wall), the functional position of an organism in its ecosystem.

Nitrification (L. *nitrum,* nitro, a combining form indicating the presence of nitrogen + *facere,* to make), change of ammonium salts into nitrates through the activities of certain bacteria.

Nitrogen base, the nitrogen containing, positively charged, components of nucleic acids and phospholipids.

Node (L. *nodus,* a knot), slightly enlarged portion of the stem where leaves and buds arise, and where branches originate.

Nonseptate, descriptive of hyphae lacking crosswalls.

Nucellus (L. *nucella,* a small nut), tissue composing the chief part of the young ovule, in which embryo sac develops; megasporangium.

Nucleic acid, an acid found in all nuclei, first isolated as part of a protein complex in 1871 and separated from the protein moiety in 1889. All known nucleic acids fall into two classes, DNA and RNA; they differ from each other in the sugar, in one of the nitrogen bases, in many physical properties, and in function.

Nucleolus (L. *nucleolus,* a small nucleus), dense protoplasmic body in the nucleus.

Nucleosides, components of nucleic acids consisting of a nitrogen base and a sugar. In DNA, the sugar is deoxyribose, and in RNA, ribose; adenine, guanine, and cytosine occur in both DNA and RNA, thymine occurs in DNA, and uracil in RNA.

Nucleotides components of nucleic acid: nucleoside (nitrogen base + sugar) + phosphoric acid.

Nucleus (L. *nucleus,* kernel of a nut), a dense protoplasmic body essential in cellular synthetic and developmental activities; present in all eukaryotic cells except mature sieve-tube elements.

Numerical taxonomy, a field of taxonomy that does not place subjective weight on any particular type of evidence that shows relationships between taxa.

Nut (L. *nux,* nut), a dry, indehiscent, hard, one-seeded fruit, generally produced from a compound ovary.

Obligate anaerobe, an organism obliged to live in the absence of oxygen.

Obligate parasite, an organism obliged to live strictly as a parasite.

Obligate saprophyte, an organism obliged to live strictly as a saprophyte.

Oogamy (Gr. *oion,* egg + *gamete,* spouse), the condition in which the gametes are different in form and activity, i.e., sperms and eggs.

Oogonium (L. dim. of Gr. *oogonos,* literally, a little egg-layer), female gametangium or egg-bearing organ not protected by a jacket of sterile cells, characteristic of the thallophytes.

Oospore (Gr. *oion,* an egg + spore), a resistant spore developing from a zygote resulting from the fusion of heterogametes.

Open bundle, a vascular bundle with cambium.

Operculum (L. *operculum,* a lid), in mosses, cap of sporangium.

Opposite, referring to bud or leaf arrangement in which there are two buds or two leaves at a node.

Organ (L. *organum,* an instrument or engine of any kind, musical, military, etc.), a part or member of an animal or plant body or cell adapted by its structure for a particular function.

Organelle, a membrane bounded specialized region within a cell such as the mitochondrion or dictyosome.

Organic, referring in chemistry to the carbon compounds, many of which have been in some manner associated with living organisms.

Osmosis (Gr. *osmos,* a pushing), diffusion of a solvent through a differentially permeable membrane.

Ovary (L. *ovum,* an egg), enlarged basal portion of the pistil, which becomes the fruit.

Ovulate, referring to a cone, scale, or other structure bearing ovules.

Ovule (F. *ovule,* from L. *ovulum,* dimin. of *ovum,* egg), a rudimentary seed, containing, before fertilization, the female gametophyte, with egg cell, all being surrounded by the nucellus and one or two integuments.

Ovuliferous (ovule + L. *ferre,* to bear), referring to a scale or sporophyll bearing ovules.

Palea (or palet) (L. *palea,* chaff), upper bract which subtends a grass flower.

Paleo-ecology (Gr. *palaios,* ancient), a field of ecology that reconstructs past vegetation and climate from fossil evidence.

Palisade parenchyma, elongated cells, containing many chloroplasts, found just beneath the upper epidermis of leaves.

Palmately veined (L. *palma,* palm of the hand), descriptive of a leaf blade with several principal veins spreading out from the upper end of the petiole.

Panicle (L. *panicula,* a tuft), an inflorescence, the main axis of which is branched, and whose branches bear loose racemose flower clusters.

Pappus (L. *pappus,* woolly, hairy seed or fruit of certain plants), scales or bristles representing a reduced calyx in composite flowers.

Parasexual cycle (Gr. *para,* beside), a sexual cycle involving changes in chromosome number differing in time and place from the usual sexual cycle; occurring in those fungi in which the normal cycle is suppressed or apparently absent.

Parallel venation, type of venation in which veins of a leaf blade that are clearly visible to the unaided eye are parallel to each other.

Paraphysis (plural, **paraphyses**) (Gr. *para,* beside + *physis,* growth), a slender, multicellular hair (*Fucus,* etc.); one of the sterile branches or hyphae growing beside fertile cells in the fruiting body of certain fungi.

Parasite (Gr. *parasitos,* one who eats at the table of another), an organism deriving its food from the living body of another plant or an animal.

Parenchyma (Gr. *parenchein,* an ancient Greek medical term meaning to pour beside and expressing the ancient concept that the liver and other internal organs were formed by blood diffusing through the blood vessels and coagulating, thus designating ground tissue), a tissue composed of cells which usually have thin walls of cellulose, and which often fit rather loosely together, leaving intercellular spaces.

Parent material, the original rock or depositional matter from which the soil of a region has been formed.

Parietal (F. *pariétal,* attached to the wall, from L. *paries,* wall), belonging to, connected with, or attached to the wall of a hollow organ or structure, especially of the ovary or cell.

Parietal placentation, a type of placentation in which placentae are on the ovary wall.

Parthenocarpy (Gr. *parthenos,* virgin + *karpos,* fruit), the development of fruit without fertilization.

Parthenogenesis (Gr. *parthenos,* virgin + *genesis,* origin), the development of a gamete into a new individual without fertilization.

Passive absorption of water, the absorption of water by roots due to forces of transpiration in leaves.

Passive solute absorption, absorption due only to forces of simple diffusion.

Pathogen (Gr. *pathos,* suffering + *genesis,* beginning), an organism that causes a disease.

Pathology (Gr. *pathos,* suffering + *logos,* account), the study of diseases, their effects on plants or animals, and their treatment.

Peat (M.E. *pete,* of Celtic origin, a piece of turf used as fuel), any mass of semicarbonized vegetable tissue formed by partial decomposition in water.

Pectin (Gr. *pektos,* congealed), a white amorphous substance which, when combined with acid and sugar, yields a jelly; the substance cementing cells together; the middle lamella.

Pedicel (L. *pediculus,* a little foot), stalk or stem of the individual flowers of an inflorescence.

Peduncle (L. *pedunculus,* a late form of *pediculus,* a little foot), stalk or stem of a flower that is borne singly; or the main stem of an inflorescence.

Penicillin, an antibiotic derived from the mold *Penicillium.*

Perennial (L. *perennis,* lasting the whole year through), a plant that lives from year to year.

Pentose (Gr. *pente,* five, + -ose a suffix indicating a carbohydrate), a five-carbon sugar.

Perfect flower, a flower having both stamens and pistils; hermaphroditic flower.

Perianth (Gr. *peri,* around + *anthos,* flower), the petals and sepals taken together.

Pericarp (Gr. *peri,* around + *karpos,* fruit), fruit wall, developed from ovary wall.

Pericycle (Gr. *peri,* around + *kyklos,* circle), tissue, generally of root, bounded externally by the endodermis and internally by the phloem.

Peridium (plural, **peridia**) (Gr. *peridion,* a little pouch), external covering of the hymenium of certain fungi; in Myxomycetes, the hardened envelope that covers the sporangium.

Perigyny (Gr. *peri,* about + *gyne,* a female), a condition in which the receptacle is more or less concave, at the margin of which the sepals, petals, and stamens have their origin, so that these parts seem to be attached around the ovary.

Peristome (Gr. *peri,* about + *stoma,* a mouth), in mosses, a fringe of teeth about the opening of the sporangium.

Perithecium (Gr. *peri,* around + *theke,* a box), a spherical or flask-shaped ascocarp having a small opening.

Permafrost (L. *permanere,* to remain + A.S. *freosan,* to freeze), soil that is permanently frozen; usually found some distance below a surface layer that thaws during warm weather.

Permeable (L. *permeabilis,* that which can be penetrated), said of a membrane, cell, or cell system through which substances may diffuse.

Petal (Gr. *petalon,* a flower leaf), one of the flower parts, usually conspicuously colored.

Petiole (L. *petiolus,* a little foot or leg), stalk of leaf.

PGA, phosphoglyceric acid.

Phage (Gr. *phago* to eat), a virus infecting bacteria; originally bacteriophage.

Phelloderm (Gr. *phellos,* cork + *derma,* skin), a layer of cells formed in the stems of some plants from the inner cells of the cork cambium.

Phellogen (Gr. *phellos,* cork + *genesis,* birth), cork cambium, a cambium giving rise externally to cork and in some plants internally to phelloderm.

Phenotype (Gr. *phaneros,* showing + type), the external visible appearance of an organism.

Phloem (Gr. *phloos,* bark), food-conducting tissue, consisting of sieve tubes, companion cells, phloem parenchyma, and fibers.

Phosphoglyceric acid, (PGA), a three-carbon compound formed by the interaction of carbon dioxide (CO_2) and a five-carbon compound, ribulose diphosphate; the reaction yields two molecules of PGA for each molecule of the ribulose diphosphate. The first step in the carbon cycle of photosynthesis.

Phosphorylation, a reaction in which phosphate is added to a compound, e.g., the formation of ATP from ADP and inorganic phosphate.

Photophosphorylation, a reaction in which light energy is converted into chemical energy in the form of ATP produced from ADP and inorganic phosphate.

Photon, a quantum of light; the energy of a photon is proportional to its frequency; $E = h\nu$, where E is energy; h, Planck's constant, 6.62×10^{-27} erg-second; and ν is the frequency.

Photoperiod (Gr. *photos,* light + period), the optimum length of day or period of daily illumination required for the normal growth and maturity of a plant.

Photosynthesis (Gr. *photos,* light + *syn,* together + *tithenai,* to place), a process in which carbon dioxide and water are brought together chemically to form a carbohydrate, the energy for the process being radiant energy.

Phototropism (Gr. *photos,* light + *tropos,* turning), a growth curvature in which light is the stimulus.

Phycobiliproteins, pigments found in the red and blue-green algae, similar to bile pigments and always associated with proteins.

Phycocyanin (Gr. *phykos,* seaweed + *kyanos,* blue), a blue phycobilin pigment occurring in blue-green algae.

Phycoerythrin (Gr. *phykos,* seaweed + *erythros,* red), a red phycobilin pigment occurring in red algae.

Phycomycetes (Gr. *phykos,* seaweed + *mykes,* mushroom or fungus), a class of fungi which approaches the algae in some characters.

Phylogeny (Gr. *phylon,* race or tribe + *genesis,* beginning), the evolution of a group of related individuals.

Phylum (Gr. *phylon,* race or tribe), a primary division of the animal or plant kingdom.

Physiology (Gr. *physis,* nature + *logos,* discourse), the science of the functions and activities of living organisms.

Phytobenthon (Gr. *phyton,* a plant + *benthos,* depths of the sea), attached aquatic plants, collectively.

Phytochrome, a reversible pigment system of protein nature found in solution in the cytoplasm of green plants. It is associated with the absorption of light that affects growth, development, and differentiation of a plant, independent of photosynthesis, e.g., in the photoperiodic response.

Phytoplankton (Gr. *phyton,* a plant + *planktos,* wandering), free-floating plants, collectively.

Pileus (L. *pileus,* a cap), umbrella-shaped cap of fleshy fungi.

Pinna (plural, **pinnae**) (L. *pinna,* a feather), leaflet or division of a compound leaf (frond).

Pinnately veined (L. *pinna,* a feather + *vena,* a vein), descriptive of a leaf blade with single midrib from which smaller veins branch off, somewhat like the divisions of a feather.

Pioneer community, the first stage of a succession.

Pistil (L. *pistillum,* a pestle), central organ of the flower, typically consisting of ovary, style, and stigma.

Pistillate flower, a flower having pistils but no stamens.

Pit, a minute thin area of a cell wall.

Pith, the parenchymatous tissue occupying the central portion of a stem.

Placenta (plural, **placentae**) (L. *placenta,* a cake), the tissue within the ovary to which the ovules are attached.

Placentation (L. *placenta,* a cake + *-tion,* state of), manner in which the placentae are distributed in the ovary.

Plankton (Gr. *planktos,* wandering), free-floating aquatic plants and animals, collectively.

Plasmodesma (plural, **plasmodesmata**) (Gr. *plasma,* something formed + *desmos,* a bond, a band), fine protoplasmic thread passing through the wall that separates two protoplasts.

Plasmodium (Gr. *plasma,* something formed + mod. Latin *odium,* something of the nature of), in Myxomycetes, a slimy mass of protoplasm, with no surrounding wall and with numerous free nuclei distributed throughout.

Plasmogamy (Gr. *plasma,* anything molded or formed + *gamos,* marriage), the fusion of protoplasts, not accompanied by nuclear fusion.

Plasmolemma (Gr. *plasma,* anything formed + *lemma,* a husk or shell of a fruit), a delicate cytoplasmic membrane found on the outside of the protoplast adjacent to the cell wall.

Plasmolysis (Gr. *plasma,* something formed + *lysis,* a loosening), the separation of the cytoplasm from the cell wall, due to removal of water from the protoplast.

Plastid (Gr. *plastis,* a builder), the cellular organelle in which carbohydrate metabolism is localized.

Plaque (D. *plak,* flat piece of wood), here, the clear area in a plate culture of bacteria caused by the lysis of the bacteria.

Plastoquinone, a quinone, one of a group of compounds involved in the transport of electrons in photosynthesis in chloroplasts.

Plumule (L. *plumula,* a small feather), the first bud of an embryo or that portion of the young shoot above the cotyledons.

Podzol (R. *pod,* under + *zola,* ashes), a soil characteristic of taiga vegetation; color of the A horizon is gray because of excessive leaching.

Polarity (Gr. *pol,* an axis), the observed differentiation of an organism, tissue, or cell into parts having opposed or contrasted properties or form.

Pollen (L. *pollen,* fine flour), the germinated microspores or partially developed male gametophytes of seed plants.

Pollen mother cell, the cell in which meiosis will occur resulting in the formation of four pollen grains.

Pollen profile, a diagrammatic summary of the sequence and abundance of pollen types that have been chronologically trapped in sediments.

Pollination, the transfer of pollen from a stamen or staminate cone to a stigma or ovulate cone.

Polygenes, many genes influencing the development of a single trait; results in continuous variability.

Polymerization, the chemical union of monomers such as glucose, or nucleotides to form starch or nucleic acid.

Polynomial (Gr. *polys,* many + L. *nomen,* name), a scientific name for an organism composed of more than two words (as in binomial).

Polynucleotides (Gr. *polys,* much, many), long chain molecules composed of units (monomers) called nucleotides. Nucleic acid is a polynucleotide.

Polyphyletic (Gr. *polys,* many + *phyle,* tribe), referring to organisms having more than one common ancestor.

Polyploid (Gr. *polys,* many + *ploos,* fold), referring to a plant, tissue, or cell with more than two complete sets of chromosomes.

Polyribosome (Gr. *polys,* many + ribosomes), an aggregation of ribosomes; frequently simply *polysome.*

Polysaccharides (Gr. *polys,* much, many + Gr. *sakcharon,* sugar), long chain molecules composed of units (monomers) of a sugar. Starch and cellulose are polysaccharides.

Pome (F. *pomme,* apple), a simple fleshy fruit, the outer portion of which is formed by the floral parts that surround the ovary.

Population (L. *populus,* people), a group of closely related, interbreeding organisms.

Prairie (L. *pratum,* meadow), grassland vegetation, with trees essentially absent; often considered to have more rainfall than does the steppe.

Predation (L. *predatio,* plundering), a form of biological interaction in which one organism is destroyed (by ingestion); parasitism is a form of predation.

Primary (L. *primus,* first), first in order of time or development.

Primary endosperm cell, a cell of the embryo sac after fertilization, generally containing a nucleus resulting from fusion of the two polar nuclei with a sperm nucleus. The endosperm develops from this cell.

Primary meristems, meristems of the shoot or root tip giving rise to the primary plant body.

Primitive (L. *primus,* first), referring to a taxonomic trait thought to have evolved early in time.

Primordium (L. *primus,* first + *ordiri,* to begin to weave; literally beginning to weave, or to put things in order), the beginning or origin of any part of an organ.

Procambium (L. *pro,* before + cambium), a primary meristem that gives rise to primary vascular tissues and, in most woody plants, to the vascular cambium.

Producer (L. *producere,* to draw forward), an organism which produces organic matter for itself and all other organisms (consumers and decomposers) by photosynthesis.

Proembryo (L. *pro,* before + *embryon,* embryo), a group of cells arising from the division of the fertilized egg cell before cells which are to become the embryo are recognizable.

Prokaryotes (L. *pro,* before + Gr. *karyon,* a nut, referring in modern biology to the nucleus), primitive organisms, bacteria and blue-green algae, which do not have the DNA separated from the cytoplasm by an envelope.

Prophage (Pro, G. *prefix* meaning before), a noninfectious phage unit which multiplies with the growing and dividing bacteria but does not bring about lysis of the bacteria. Prophage is a stage in the life cycle of a temperate phage.

Prophase (Gr. *pro,* before + *phasis,* appearance), an early stage in nuclear division, characterized by the shortening and thickening of the chromosomes and their movement to the metaphase plate.

Proplastid (Gr. *pro,* before + plastid), a type of plastid, occurring generally in meristematic cells, which will develop into a chloroplast.

Protease (protein + *-ase,* a suffix indicating an enzyme), an enzyme breaking down a protein.

Protein (Gr. *proteios,* holding first place), naturally occurring complex organic substances (egg albumen, meat) composed essentially of carbon, hydrogen, oxygen, and nitrogen, plus sulfur or phosphorus, which are so associated as to form submiscroscopic chains, spirals, or plates and to which are attached other atoms and groups of atoms in a variety of manners.

Prothallium (Gr. *pro,* before + *thallos,* a sprout), in ferns, the haploid gametophyte generation.

Protoderm (Gr. *protos,* first + *derma,* skin), a primary meristem that gives rise to epidermis.

Proton (Gr. *proton,* first), the nucleus of a hydrogen atom is a single positively charged particle, the proton; the nucleus of all other elements consists of protons and neutrons. The mass of a proton is 1.67×10^{-24} gm.

Protonema (plural, **protonemata**) (Gr. *protos,* first + *nema,* a thread), an algal-like filamentous growth; an early stage in development of the gametophyte of mosses.

Protoplasm (Gr. *protos,* first + *plasma,* something formed), living substance.

Protoplast (Gr. *protoplastos,* formed first), the organized living unit of a single cell.

Pseudopodium (Gr. *pseudes,* false + *podion,* a foot), in Myxomycetes, an armlike projection from the body by which the plant creeps over the surface.

Purine, a group of nitrogen bases having a double ring structure, one five-carbon, the other six-carbon.

Pyrenoid (Gr. *pyren,* the stone of a fruit + L. *oïdes,* like), a denser body occurring within the chloroplasts of certain algae and liverworts and apparently associated with starch deposition.

Pyrimidine, a nitrogen base having a single ring structural formula.

Quadrat (L. *quadrus,* a square), a frame of any shape which, when placed over vegetation, defines a unit sample area within which the plants may be counted or measured.

Quantum (L. *quantum,* how much), an elemental unit of energy. Its energy value is hv, where h, Planck's constant, is 6.62×10^{-27} erg-second and v is the frequency of the vibrations or waves with which the energy is associated.

Raceme (L. *racemus,* a bunch of grapes), an inflorescence in which the main axis is elongated but the flowers are borne on pedicels that are about equal in length.

Rachilla (Gr. *rhachis,* a backbone + L. dimin. ending *-illa*), shortened axis of spikelet.

Rachis (Gr. *rhachis,* a backbone), main axis of spike; axis of fern leaf (frond) from which pinnae arise; in compound leaves, the extension of the petiole corresponding to the midrib of an entire leaf.

Radicle (L. *radix,* root) that portion of the plant embryo which develops into the primary or seed root.

Random plant distribution, a distribution of a plant species within an area such that the probability of finding an individual at one point is the same for all points.

Raphe (Gr. *rhaphe,* seam), ridge on seeds, formed by the stalk of the ovule, in those seeds in which the funiculus is sharply bent at the base of the ovule.

Raphides (Gr. *rhaphis,* a needle), fine, sharp, needlelike crystals.

Receptacle (L. *receptaculum,* a reservoir), enlarged end of the pedicel or peduncle to which other flower parts are attached.

Recessive character, that member of a pair of Mendelian characters which, when both members of the pair are present, is subordinated or suppressed by the other, dominant character.

Recon (from recombination and the ending *on* currently indicating an elementary particle), the smallest unit of gene recombination, approximately 0.02 units of map length; possible one nucleotide pair.

Reduction (F. *reduction,* L. *reductio,* a bringing back), originally "bringing back" a metal from its oxide, i.e., iron from iron rust or ore; any chemical reaction involving the removal of oxygen from or the addition of hydrogen or an electron to a substance. Energy is required and may be stored in the process as in photosynthesis.

Regular flower, a flower in which the corolla is made up of similarly shaped petals equally spaced and radiating from the center of the flower; star-shaped flower; actinomorphic flower.

Replication, the production of a facsimile or a very close copy. Here used to indicate the production of a second molecule of DNA exactly like the first molecule.

Reproduction (L. *re,* repeatedly + *producere,* to give birth to), the process by which plants and animals give rise to offspring.

Respiration (L. *re,* repeatedly + *spirare,* to breathe), a chemical oxidation controlled and catalyzed by enzymes in which living protoplasm breaks down carbohydrate and fats, thus releasing energy to be used by the organism in doing work.

Reticulum (L. *reticulum,* a small net), a small net.

Rhizoid (Gr. *rhiza,* root + L. *oïdes,* like), one of the cellular filaments that perform the functions of roots.

Rhizome (Gr. *rhiza,* root), an elongated, underground, horizontal stem.

Rhizophores (Gr. *rhiza,* root + *phoros,* bearing), leafless branches that grow downward from the leafy stems of certain Lycopodineae and give rise to roots when they come into contact with the soil.

Ribose, a five-carbon sugar, a component of RNA.

Ribose nucleic acid, a nucleic acid containing the sugar, ribose; phosphorus; and the bases adenine, guanine, cytosine, and uracil. Present in all cells and concerned with protein synthesis in the cell.

Ribosomes, (*Ribo,* from RNA + Gr. *somatos,* body), small particles 100–200 A in diameter, containing RNA.

RNA, ribose nucleic acid.

Root (A.S. *rōt*), the descending axis of a plant, normally below ground, and serving to anchor the plant and to absorb and conduct water and mineral nutrients.

Root cap, a thimblelike mass of cells covering and protecting the apical meristems of a root.

Root pressure, pressure developed in the root as the result of osmosis and inducing bleeding in stem wounds.

Rootstock, an elongated, underground, horizontal stem.

Runner, a stem that grows horizontally along the ground surface.

Samara (L. *samara,* the fruit of the elm), simple, dry, one- or two-seeded indehiscent fruit with pericarp bearing a wing-like outgrowth.

Sand, soil particles between 50 and 2000 μ in diameter.

Saprophyte (Gr. *sapros,* rotten + *phyton,* a plant), an organism deriving its food from the dead body or the nonliving products of another plant or animal.

Savannah (Sp. *sabana,* a large plain), vegetation of scattered trees in a grassland matrix.

Scalariform vessel (L. *scala,* ladder + form), a vessel with secondary thickening resembling a ladder.

Schizocarp (Gr. *schizein,* to split + *karpos,* fruit), dry fruit with two or more united carpels, which split apart at maturity.

Sclereids (Gr. *skleros,* hard), more or less isodiametric cells have heavily lignified cell walls.

Sclerenchyma (Gr. *skleros,* hard + -*echyma,* a suffix denoting tissue), a strengthening tissue composed of cells with heavily lignified cell walls.

Scrub (A.S. *scrob,* a shrub), vegetation dominated by shrubs; described as thorn forest in areas with moderate rainfall, or as chaparral or desert in areas with low rainfall.

Scutellum (L. *scutella,* a dimin. of *scutum,* shield), single cotyledon of grass embryo.

Seed (A.S. *sed,* anything which may be sown), popularly as originally used, anything which may be sown; i.e., "seed" potatoes, "seeds" of corn, sunflower, etc. Botanically, a seed is the matured ovule without accessory parts.

Self-pollination, transfer of pollen from the stamens to the stigma of either the same flower or flowers on the same plant.

Seminal root, the root or roots forming from primordia present in the seed.

Sepals (M.L. *sepalum,* a covering; "sepalum" is a modern word formed by analogy with petalum [Gr. *petalon*], outermost flower structures which usually enclose the other flower parts in the bud.

Septate (L. *septum,* fence), divided by cross walls into cells or compartments.

Septicidal dehiscence (L. *septum,* fence + *caedere,* to cut; *dehiscere,* to split open), the splitting open of a capsule along the line of union carpels.

Septum (L. *septum,* fence), any dividing wall or partition; frequently a cross wall in a fungal or algal filament.

Serpentine (L. *serpens,* a serpent), referring to soil derived from metamorphic parent material characterized (among other things) by low calcium (Ca), high magnesium (Mg), and a greenish-gray color.

Sessile (L. *sessilis,* low, dwarf, from *sedere,* to sit), sitting, referring to a leaf lacking a petiole or a flower or fruit lacking a pedicel.

Seta (plural, **setae**) (L. *seta,* a bristle), in liverworts, a short stalk of the sporophyte, which connects the foot and the capsule.

Sexual reproduction, reproduction that requires meiosis and fertilization for a complete life cycle.

Sheath, part of leaf that wraps around the stem, as in grasses.

Shoot (derivation uncertain, but early referring to new plant growth; 1450, "Take a feyr schoyt of blake thorne"), a young branch that shoots out from the main stock of a tree, or the young main portion of a plant growing aboveground.

Shoot tip, that portion of the shoot containing apical and primary meristems and early stages of differentiation.

Sibling species, species morphologically nearly identical, but incapable of producing fertile hybrids.

Sieve cell, a long and slender sieve element with relatively unspecialized sieve areas and with tapering end walls that lack sieve plates.

Sieve plate, perforated wall area in a sieve-tube element through which pass strands connecting sieve-tube protoplasts.

Sieve tube, a series of sieve-tube elements forming a long cellular tube specialized for the conduction of food materials.

Sieve-tube element, that portion of a sieve tube comprised of a single protoplast and separated from other sieve-tube elements by sieve plates.

Silique (L. *siliqua,* pod), the fruit characteristic of Cruciferae (mustards); two-celled, the valves splitting from the bottom and leaving the placentae with the false partition stretched between.

Silt, soil particles between 2 and 50 μ in diameter.

Simple pit, pit not surrounded by an overarching border.

Siphonous line, a line of evolutionary development in the algae in which mitosis is not followed by cytokinesis; this results in an elongated multinucleate, coenocytic filament.

Soil (L. *solum,* soil, solid), the uppermost stratum of the earth's crust, which has been modified by weathering and organic activity into (typically) three horizons: an upper A horizon which is leached, a middle B horizon in which the leached material accumulates, and a lower C horizon which is unweathered parent material.

Soil texture, refers to the amounts of sand, silt, and clay in a soil, as a sandy-loam, loam, or clay texture.

Solute (L. *solutus,* from *solvere,* to loosen), a dissolved substance.

Solution (M.E. *solucion,* from O.F. *solucion,* to loosen), a homogeneous mixture, the molecules of the dissolved substance (e.g., sugar), the solute, being dispersed between the molecules of the solvent (e.g., water).

Solvent (L. *solvere,* to loosen), a substance, usually a liquid, having the properties of dissolving other substances.

Soredium (plural, **soredia**) (Gr. *soros,* a heap), special reproductive body of lichens, consisting of a few algal cells surrounded by fungous hyphae.

Sorus (plural, **sori**) (Gr. *soros,* a heap), a cluster of sporangia.

Species (L. *species,* appearance, form, kind), a class of individuals usually interbreeding freely and having many characteristics in common.

Sperm (Gr. *sperma,* the generative substance or seed of a male animal), a male gamete.

Spermagonium (plural, **spermagonia**) (Gr. *sperma,* sperm + *gonos,* offspring), flask-shaped structure characteristic of the sexual phase of the rust fungi; bearing receptive hyphae and spermatia.

Spermatium (plural, **spermatia**) (Gr. *sperma,* sperm), in rust fungi, a cell borne at the tip of the hyphae which line the interior of spermagonia (on barberry leaves).

Spermatophyte (Gr. *sperma,* seed + *phyton,* plant), a seed plant.

Spike (L. *spica,* an ear of grain), an inflorescence in which the main axis is elongated and the flowers are sessile.

Spikelet (L. *spica,* an ear of grain + dimin. ending *-let*), the unit of inflorescence in grasses; a small group of grass flowers.

Spindle (A.S. *spinel,* an instrument employed in spinning thread by hand), referring in mitosis and meiosis to the spindle-shaped intracellular structure in which the chromosomes move.

Sporangiophore (sporangium + Gr. *-phore,* a root of *phorein,* to bear), a branch bearing one or more sporangia.

Sporangium (spore + Gr. *angeion,* a vessel), spore case.

Spore (Gr. *spora,* seed), a reproductive cell that develops into a plant without union with other cells; some spores such as meiospores occur at a critical stage in the sexual cycle, but others are asexual in nature.

Sporidium (dimin., meaning a little spore), the basidiospore of smut fungi.

Sporophore (spore + Gr. *phorein,* to bear), the fruiting body of fleshy and woody fungi, which produces spores.

Sporophyll (spore + Gr. *phyllon,* leaf), a spore-bearing leaf.

Sporophyte (spore + Gr. *phyton,* a plant), in alternation of generations, the plant in which meiosis occurs and which thus produces meiospores.

Stamen (L. *stamen,* the standing-up things or a tuft of thready things), flower structure made up of an anther (pollen-bearing portion) and a stalk or filament.

Staminate flower, a flower having stamens but no pistils.

Starch (M.E. *sterchen,* to stiffen), a complex insoluble carbohydrate, the chief food storage substance of plants, which is composed of several hundred hexose sugar units and which easily breaks down on hydrolysis into these separate units.

Stele (Gr. *stele,* a post), the central cylinder, inside the cortex, of roots and stems of vascular plants.

Stem (O.E. *stemn*), the main body of the portion aboveground of tree, shrub, herb, or other plant; the ascending axis, whether above or below ground, of a plant, in contradistinction to the descending axis or root.

Steppe (R. *step,* a lowland), an arid grassland vegetation.

Sterigma (plural, **sterigmata**) (Gr. *sterigma,* a prop), a slender, pointed protuberance at the end of a basidium, which bears a basidiospore.

Stigma (L. *stigma,* a prick, a spot, a mark), receptive portion of the style to which pollen adheres.

Stipule (L. *stipula,* dimin. of *stipes,* a stock or trunk), a leaflike structure from either side of the leaf base.

Stolon (L. *stolo,* a shoot), a stem that grows horizontally along the ground surface.

Stoma (plural, **stomata**) (Gr. *stoma,* mouth), a minute opening, bordered by guard cells in the epidermis of leaves and stems, through which gases pass.

Strobilus (Gr. *strobilos,* a cone), a number of modified leaves (sporophylls) or ovule-bearing scales grouped together on an axis.

Stroma (Gr. *stroma,* a bed or covering), 1, a mass of protecting vegetative filaments; 2, the background substance of chloroplasts, probably the location of the carbon cycle of photosynthesis.

Style (Gr. *stylos,* a column), slender column of tissue that arises from the top of the ovary and through which the pollen tube grows.

Suberin (L. *suber,* the cork oak), a waxy material found in the cell walls of cork tissue.

Succession (L. *successio,* a coming into the place of another), a sequence of changes in time of the species which inhabit an area, from an initial pioneer community to a final climax community.

Succulent (L. *sucus,* juice), a plant having juicy or watery tissues.

Sucrose (F. *sucre,* sugar + *-ose,* termination designating a sugar), cane sugar ($C_{12}H_{22}O_{11}$).

Superior ovary, an ovary completely separate and free from the calyx.

Suspensor (L. *suspendere,* to hang), a cell or chain of cells developed from a zygote whose function is to place the embryo cells in an advantageous position to receive food.

Suture (L. *sutura,* a sewing together; originally the sewing together of flesh or bone wounds), the junction, or line of junction, of contiguous parts.

Symbiosis (Gr. *syn,* with + *bios,* life), an association of two different kinds of living organisms involving benefit to both.

Sympetaly (Gr. *syn,* with + *petalon,* leaf) a condition in which petals are united.

Synandry (Gr. *syn,* with + *andros,* a man), a condition in which stamens are united.

Syncarpy (Gr. *syn,* with + *karpos,* fruit), a condition in which carpels are united.

Synergids (Gr. *synergos,* toiling together), the two nuclei at the upper end of the embryo sac, which, with the third (the egg), constitute the egg apparatus.

Synsepaly (Gr. *syn,* with + sepals), a condition in which sepals are united.

Taiga (Teleut *taiga,* rocky mountainous terrain), a broad northern belt of vegetation dominated by conifers; also, a similar belt in mountains just below alpine vegetation.

Tannin, a substance that has an astringent, bitter taste.

Tapetum (Gr. *tapes,* a carpet), nutritive tissue in the sporangium, particularly an anther.

Taxon (pl. **taxa**) (Gr. *taxis,* order), a general term for any taxonomic rank, from subspecific to divisional.

Taxonomy (Gr. *taxis,* arrangement + *nomos,* law), systematic botany; the science dealing with the describing, naming, and classifying of plants.

Teliospore (Gr. *telos,* completion + spore), resistant spore characteristic of the Heterobasidiomycetidae, in which karyogamy and meiosis occur and from which a basidium develops.

Telium (plural, *telia*) (Gr. *telos,* completion), a sorus of teliospores.

Telophase (Gr. *telos,* completion + phase), the last stage of mitosis, in which daughter nuclei are reorganized.

Temperate phage, a phage that does not necessarily cause lysis of the bacterial cells in which it reproduces.

Tendril (L. *tendere,* to stretch out, to extend), a slender coiling organ that aids in the support of stems.

Terminal bud, a bud at the end of a stem.

Testa (L. *testa,* a brick, shell), the outer coat of the seed.

Tetrad (Gr. *tetradeion,* a set of four), a group of four, usually referring to the meiospores immediately after meiosis.

Tetraploid (Gr. *tetra,* four + *ploos,* fold), having four sets of chromosomes per nucleus.

Tetraspores (Gr. *tetra,* four + spores), four spores formed by division of the spore mother cell, used particularly for meiospores in certain red algae.

Tetrasporine line (tetraspore + L. suffix *-ine,* like), a line of evolutionary development in the algae in which mitosis is directly followed by cytokinesis, resulting in a filament, thallus, or complex plant body of varied form.

Thallophytes (Gr. *thallos,* a sprout + *phyton,* plant), a division of plants whose body is a thallus, i.e., lacking roots, stems, and leaves.

Thallus (Gr. *thallos,* a sprout), plant body without true roots, stems, or leaves.

Threonin, an amino acid.

Thylakoid (Gr. *thylakos,* sack + N.L. *oid,* a thing that is like), one of membranes in the chloroplasts.

Thymidine, a nucleoside incorporated in DNA, but not in RNA.

Thymidine H^3, tritiated or radioactive thymidine.

Thymine, a pyrimidine occurring in DNA, but not in RNA.

Tissue, a group of cells of similar structure which performs a special function.

Tonoplast (Gr. *tonos,* stretching tension + *plastos,* molded, formed), the cytoplasmic membrane bordering the vacuole. So-called by de Vries, as he thought it regulated the pressure exerted by the cell sap.

Toxin (L. *toxicum,* poison), a poisonous secretion of a plant or animal.

TPN, triphosphopyridine nucleotide, or, more recently, NADP, nicotinamide adenine dinucleotide phosphate.

TPNH, reduced triphosphopyridine nucleotide or, more recently, NADPH, reduced nicotinamide adenine dinucleotide phosphate.

Tracheid (Gr. *tracheia,* windpipe), an elongated, tapering xylem cell, with lignified pitted walls, adapted for conduction and support.

Tracheophytes (Gr. *tracheia,* windpipe + *phyton,* plant), vascular plants.

Trait, a distinctive definable characteristic; a mark of individuality.

Transduction, gene transfer in bacteria, in which bacterial genes are carried from one bacteria to another by way a temperate phage.

Transformation, mutation or change in genes of bacteria by the direct intervention of extracellular DNA.

Translocation (L. *trans,* across + *locare,* to place), the transfer of food materials or products of metabolism.

Transmit, to pass or convey something from one person, organism, or place to another person, organism, or place.

Transpiration (F. *transpirer,* to perspire), the giving off of water vapor from the surface of leaves.

Trichogyne (Gr. *trichos,* a hair + *gyne,* female), receptive hairlike extension of the female gametangium in the Rhodophyta and Ascomycetes.

Trichome (Gr. *trichoma,* a growth of hair), a short filament of cells.

Triose (Gr. *treis,* three + . . . *ose* suffix indicating a carbohydrate), any three-carbon sugar.

Trisomic (Gr. *treis,* three + *soma,* body), a plant containing one additional chromosome; $2n + 1$.

Tritium, a hydrogen atom, the nucleus of which contains one proton and two neutrons. It is written as H^3. The more common hydrogen nucleus consists only of a proton.

Tropism (Gr. *trope,* a turning), movement of curvature due to an external stimulus that determines the direction of movement.

Tuber (L. *tuber,* a bump, swelling), a much-enlarged, short, fleshy underground stem.

Tundra (Lapp. *tundar,* hill), meadowlike vegetation at low elevation in cold regions that do not experience a single month with average daily maximum temperatures above 50°F.

Turgid (L. *turgidus,* swollen, inflated), swollen, distended; referring to a cell that is firm due to water uptake.

Turgor pressure (L. *turgor,* a swelling), the pressure within the cell resulting from the absorption of water into the vacuole and the imbibition of water by the protoplasm.

Tylosis (plural, **tyloses**) (Gr. *tylos,* a lump or knot), a growth of one cell into the cavity of another.

Type specimen, the herbarium specimen selected by a taxonomist to serve as a basis for the naming and descriptions of a new species.

Umbel (L. *umbella,* a sunshade), an inflorescence, the individual pedicels of which all arise from the apex of the peduncle.

Unavailable water, water held by the soil so strongly that root hairs cannot readily absorb it.

Unicell (L. *unus,* one + cell), an organism consisting of a single cell; generally used in describing algae.

Uniseriate (L. *unus,* one + M.L. *seriatus,* to arrange in a series), said of a filament having a single row of cells.

Universal veil, a membrane completely surrounding the sporophyte of a homobasidiomycete at the button stage.

Uracil, a pyrimidine found in RNA but not in DNA.

Uredinium (plural, *uredinia*) (L. *uredo,* a blight), a sorus of uredospores.

Uredosorus (L. *uredo,* a blight + *soros,* a heap), a group of uredospores.

Uredospore (L. *uredo,* a blight + spore), a red, one-celled summer spore in the life cycle of the rust fungi.

Van Leeuwenhoek, Antony (1632–1723), a native of Delft, Holland, a dealer in dry goods, and a skilled lens maker; made many observations of biological materials at magnifications of about 100 ✕ . Famous for his early observations on bacteria.

Vaccination, the injection of vaccine.

Vaccine (L. *vacca,* cow), a suspension of weakened or dead bacteria or other pathogens injected into the body to immunize against the same species of pathogen or their toxins.

Vacuole (L. dimin. of *vacuus,* empty), a watery solution of various substances forming a portion of the protoplast distinct from the protoplasm.

Vascular (L. *vasculum,* a small vessel), referring to any plant tissue or region consisting of or giving rise to conducting tissue, e.g., bundle, cambium, ray.

Vascular bundle, a strand of tissue containing primary xylem and primary phloem (and procambium if present) and frequently enclosed by a bundle sheath of parenchyma or fibers.

Vascular cambium, cambium giving rise to secondary phloem and secondary xylem.

Vector (L. *vehere,* to carry), an organism, usually an insect, that carries and transmits disease-causing organisms.

Vegetation (L. *vegetare,* to quicken), the plant cover that clothes a region; it is formed of the species that make up the flora, but is characterized by the abundance and life-form (tree, shrub, herb, evergreen, deciduous plant, etc.) of certain of them.

Venation (L. *vena,* a vein), arrangement of veins in leaf blade.

Venter (L. *venter,* the belly), enlarged basal portion of an archegonium in which the egg cell is borne.

Ventral canal cell, the cell just above the egg cell in the archegonium.

Ventral suture (L. *ventralis,* pertaining to the belly), the line of union of the two edges of a carpel.

Vessel (L. *vasculum,* a small vessel), a series of xylem elements whose function is to conduct water and mineral nutrients.

Vessel element, a portion of a vessel derived from a single cell of the vascular cambium or procambium.

Virulence (L. *virulentia,* a stench), the relative infectiousness of a bacteria or virus, or its ability to overcome the resistance of the host metabolism.

Virus (L. *virus,* a poisonous or slimy liquid), a disease principle that can be cultivated only in living tissues, or in freshly prepared tissue brei.

Vitamins (L. *vita,* life + amine), naturally occurring organic substances, akin to enzymes, necessary in small amounts for the normal metabolism of plants and animals.

Volva (L. *volva,* a wrapper), cup at base of stipe or stalk of fleshy fungi.

Volvocine line (Volvox + L. suffix *-ine,* like), a line of evolutionary development in the algae in which the cells remain separate, as in *Volvox,* and are never connected to form a filament or flattened thallus.

Water potential refers to the difference between the activity of water molecules in pure distilled water at atmospheric pressure at 30° C (standard conditions) and the activity of water molecules in any other system. The activity of these water molecules may be greater (positive) or less (negative) than the activity of the water molecules under standard conditions.

Weed (A.S. *wēod,* used at least since 888 in its present meaning), generally a herbaceous plant or shrub not valued for use or beauty, growing wild and rank, and regarded as using ground or hindering the growth of superior vegetation.

Whorl, a circle of flower parts, or of leaves.

Wild-type, in genetics, the gene normally occurring in the wild population, usually dominant.

Whorled, referring to bud or leaf arrangement in which there are three or more buds or three or more leaves at a node.

Wings, lateral petals of legume type of flower.

Wood (M.E., *wode, wude,* a tree), a dense growth of trees, or a piece of a tree, generally the xylem.

Xanthophyll (Gr. *xanthos,* yellowish brown + *phyllon,* leaf), a yellow chloroplast pigment.

Xerophyte (Gr. *xeros,* dry + *phyton,* a plant), a plant very resistant to drought, or that lives in very dry places.

Xylem (Gr. *xylon,* wood), a plant tissue consisting of tracheids, vessels, parenchyma cells, and fibers; wood.

Zoology (Gr. *zoon,* an animal + *logos,* speech), the science having to do with animal life.

Zoosporangium (Gr. *zoon,* an animal + sporangium), a sporangium bearing zoospores.

Zoospore (Gr. *zoon,* an animal + spore), a motile spore.

Zygomorphic (Gr. *zygo,* yoke, pair + G. *morphe,* form), referring to bilateral symmetry; said of organisms, or a flower, capable of being divided into two symmetrical halves only by a single longitudinal plane passing through the axis.

Zygospore (Gr. *zygon,* a yoke + spore), a thick-walled resistant spore developing from a zygote resulting from the fusion of isogametes.

Zygote (Gr. *zygon,* a yoke), a protoplast resulting from the fusion of gametes (either isogametes or heterogametes).

Zymase (Gr. *zyme,* leaven), an intracellular, sugar-fermenting complex of enzymes.

Subject Index

question
Conidia, 485, 500
Conidiophores, 485, 490
Conidiosporangia, 490
Conidiospores, 485; *see also* plant in question
Conifer, 16
Coniferophyta, 16, 613-631
  anatomy, 149-150, 190
  classification, 613-622
  fertilization, 629
  gametophyte, 625-629
  life history, 622-631
  sporophyte, 623-625, 629-631
Conjugation, 466, 515, 519
Consumers, 378
Copper, 218
Copy error, 344
Cork, 151, 157
Cork cambium, 151, 157
  monocotyledons, 138
  root, 180
  stem, 151, 157
Corm, 136
Corn (*Zea mays*), caryopsis, 326
  commercial hybrid, 353
  flower, 287
  fruit, 326
  leaf, 188
  root, 166
  smutted kernels, 526
  stem, 132
Corolla, 268
Cortex, 119
  bark, 156
  brown algae, 454
  moss, 567
  root, 176
  stem, 116, 119, 578, 583, 585, 595, 601
  summary, 123
Cotton, 66, 643
Cotyledon, 17, 29, 326, 631
Coupled reaction, 80
Crick, 99
Cristae, 46
Crossing over, 342-343
Cruciferae, 643
Crystals, 52, 59
Cucurbitaceae, 648
Curvature, plant, 402
Cuticle, 13
  of leaf, 191, 226, cover
  stem, 119
Cutin, 59, 67, 119, 177
Cuttings, 407, 414
Cyanophyta, 11, 442, 472
Cycadophyta, 16, 611
Cycads, 16
Cycle, carbon, 247
  inductive, 410
  Kreb, 261
  life, 27
    gametic, 466
    sporic, 466
    virus, 541-552
    zygotic, 466
  nitrogen, 378
  organic acid, 261
  parasexual, 533
  photophosphorylation, 244
  sexual, 268
Cyme, 296
Cysteine, 67
Cystocarp, 461, 471
Cystolith, 226
Cytochrome, 69, 241
Cytokinesis, 86, 96-97
Cytology, 4

Cytoplasm, 34, 41
  chemistry of, 63-71
Cytoplasmic membrane, 41
Cytoplasmic streaming, 43
Cytosine, 102

Damping off, 489
Darwin, C., 213, 384, 402, 663-665
Davson-Danielli, 44
DDT, 3, 79
Decomposers, 373
Dehiscence, 309
De Historia Plantarum, 384
Deoxyribose nucleic acid, 2, 85, 98
Desert, 377
Diatoms, 11, 445
  reproduction, 460
Dicotyledonae, 17
  leaf, 185-188
  root, 165
  seed, 323-326
  stem, 107-132
Dictyosome, 34, 50, 474, 201
Differentiation, ascocarp, 501
  cell, 85
  cell wall, 96
  DNA control in bacteria, 432
  flower, 292
  fruit, 307, 415
  Fucus, 454
  hormonal control, 401, 407
  leaf, 202
  light effects, 407
  micurgical experiments, 399
  nuclear control, 63-65
  polarity, 450
  Polysiphonia, 456
  primary tissue
    stems, 128
    roots, 170
  secondary tissues, stems, 143-150
    roots, 166
  seed, 323
  stem of Anthophyta, 116-119
  root of Anthophyta, 171, 174
  temperature effects, 412
  tissue culture, 401
Diffuse porous wood, 148
Diffusion, 73-77
  gradient, 77, 222
  protoplasmic streaming, 232
  in transpiration, 222
Digestion, 79-80
  in seeds, 326
Dihybrid cross, 348
Dikaryon, 501
Diploid, 342, 463
Diploidization, 533
Disaccharide, 65
Disease, 423
  bacterial, 420
  fungal, 481, 507-530
  mineral deficiency, 213
  viral, 539
Disk flower, 288
Dissemination of seeds and fruits, 324
Distromatic, 452
DNA, 2, 85
  algal cell, 474
  bacterial, 427, 433
  chloroplast, 47, 478
  mitochondria, 47
  molecule, 99, 102
  replication, 98-104
  viral, 543
Dormin, 205
Drupe, 317

Ecology, 4, 359-381
  paleo, 371
  plant communities, 370
  plant populations, 369
Ecosystem, 359, 377
Ecotype, 369
Edophic factors, 207
Elater, 566
Electron, 81
  acceptor, 241, 243
  flow, 80, 244
  microscope, 39
  transfer in photosynthesis, 242-245
Elements essential for plant growth, 64
Embryo, 29
  adventitious, 302
  angiosperm, 323, 637-639
  bryophyte, 561, 563, 570
  gymnosperm, 629-631
  lower vascular plants, 583, 587, 595, 607
Embryo sac, 277, 323, 635
Embryoid, 86
Endemic, 365
Endocarp, 305
Endodermis, 173, 176, 578, 582, 585, 590, 606
Endoplasmic reticulum, 34, 45, 201, 471
Endosperm, 277, 323, 326
  mother cell, 277
Energy, 2, 64-65
  absorption, 77
  chemical, 243
  flow in ecosystems, 80, 377
  light, 242
  photosynthesis, 79, 237
  respiration, 80, 257
  rich compounds, 80
  stored, 258
Engelmann, T. W., 239
Engler, A., 385, 657
Environment, 1, 2, 359
  absorption, 230
  algae, 441
  bacteria, 437
  biological factors, 366
  DDT, 339
  energy flow, 377-378
  evolution, 664
  flowering, 408-412
  fossil, 371
  forests, 376
  fungi, 481
  geologic, 659
  Lake Erie, 379
  leaf shape, 199
  light, 363, 412-414
  macro-, 360
  micro-, 360
  moisture, 361
  photosynthesis, 237, 247, 252
  pollution, 375-380
  prairie, 377
  respiration, 264
  scrub, 377
  soil, 207, 365
  stomata, 194
  temperature, 412, 437
  tiaga, 375
  transpiration, 223-225
  tundra, 374
  water, 437
Enzymes, 68
  mineral nutrients, 214
  photosynthesis, 245
  respiration, 260
  seed germination, 326
  stomatal movement, 195

Paraphyses, 471, 501
  mosses, 570
Parasite, 419, 482
Parenchyma, 120, 366
  abscission, 203
  border, 197
  cork cambium, 151
  cortex, 120, 176, 569, 581, 583, 591, 595, 601
  lenticels, 152
  monocotyledon, roots, 177
    stem, 137
  palisade, 196
  parenchyma-like,
    Laminariales, 454
    Rhodophyta, 456
  pheloderm, 151
  phloem, 123
  pith, 137
  pith ray, 150
  ray, 147
  root, 176, 180
    cap, 171
  spongy, 196
  wood, gymnosperm, 149
  xylem, 131, 147
Pathenocarpy, 307
Parthenogensis, 302
Partition, 47
Pasteur, L., 420
Pasturization, 420, 437
Pathogenic, 481
Peat, 210
Pectates, 59
Pedicle, 187
Peltate, 187
Penicillin, 533
Pentose, 65
Pepo, 319
Perianth, 269
Pericarp, 307
Periclinal, 94
Pericycle, 173, 177
Peridium, 501
Perigyny, 286
Perisperm, 302, 325, 631
Peristome, 573
Perithecium, 501
Permafrost, 374
Permanent wilting, 230
Permeability, 71-77
Permeable membrane, 71-72
Peronosporales, 489-493
Peroxisome, 52
Petals, 268
Petiole, 187, 191, 197
PGA, 247
Phaeophyta, 11, 445, 452-454, 471
Phage, 543
  summary, 545
  temperate, 543
Phelloderm, 151
Phenotype, 348
Phloem, 117
  conduction, 233
  fibers, 127
  leaf, 197
  parenchyma, 123
  primary, root, 177-178
  primary, stem, 123-127
  ray, 15
  secondary, root, 180
    stem, 141, 143
Phosphoglyceric acid, 247
Phosphorus, 215
Phosphorylation, cyclic, 244
  noncyclic, 244
  in photosynthesis, 244

in respiration, 259-260, 262
Photon, 81, 242
Photo-oxidation, 48
Photoperiod, 408
Photosynthesis, 78-79, 185, 237-255
  chemical reaction, 79
  chlorophyll, 69
  cytochromes, 69
  leaf adaptation, 185
  photosynthetic bacteria, 428
  pigments, 69
  plastids, 47-49
  sugars, 65
Phototropism, 402
Phycobilin, 69
Phycocyanin, 444
Phycoerthrin, 444
Phyla, 8
  algal, 11, 441-478
  Anthophyta, 17, 633-657
  bacteria, 419-439
  Bryophyta, 14, 555-575
  fungal, 12, 481-537
  gymnosperm, 15, 609-631
  key to, 18
  lower vascular plants, 14, 577-607
Physiology, defined, 5
  cell, 71-83
Phytochrome, 69, 411
Phytokinins, 70
Pi, 244
Pigments, 69
Pileus, 517
Pineapple, 321
Pinaceae, 614
  genera, 614-620
  life cycle, 614-631
Pistil, 269, 270
Pit, 128
Pith, 116, 131
  rays, 131
Placenta, 271, 275
Plant, 3
  body, 27-31
  cell, 33-67
  classification, 7-9, 12-14, 18, 383-395
  community, 359, 370
  competition, 367
  day-neutral, 409
  development, 397-417
  ecosystem, 359
  endemic, 365
  groups, 9
  growth, 397-417
  identification, 383
  long day, 409
  phyla, 9-12, 14-17
  pitcher, 201
  populations, 359-369
  scientists, 3, 5
  short day, 409
  succession, 373
  succulent, 201
Plaque, 541
Plasmalemma, 43
  bacterial, 427
Plasmodesmata, 45
Plasmodium, 482
Plasmogamy, 504
Plasmolysis, 76
Plastid, 34, 47-50
Plastoquinon, 241
Plywood, 159, 160
Pod, 309, 310
Podzol, 365
Polar nuclei, 277
Polarity, cell division, 94, 454, 458
Pollen, 270, 296-303

germination, 301
  pine, 625
  taxonomy, 297
Pollen mother cells, 270
Pollen tube, 270
  pine, 628
Pollination, 270, 297-301
  cross, 299, 300
  fig, 297
  insect, 297
  mutalistic, 369
  pines, 628
  profile, 372
  self, 297, 299
  types, 298-301
  wind, 297
Pollinia, 655
Pollution, 378
Polygenes, 352
Polymerization, 66
Polynomial, 385
Polypeptide, 68
Polyploidy, 342, 352, 354, 668
Polypodiaceae, 599
Polyribosome, 45
Polysaccharide, 66
Pome, 319-320
Population, 8, 359, 369
Potassium, 216
Potential, chemical, 74
  water, 74
Powdery mildews, 507
Prairie, 377
Prantl, K., 385
Predation, 369
Pressure, atmospheric, 75
  osmotic, 75
  turgor, 75
  wall, 74
Priestley, J., 238
Primary endosperm cell, 277
Primary endosperm nucleus, 277
Primary meristems, 116-119
Primary plant body, 108
  monocotyledons, 132
  root, 170-179
  stems, 107-139
Primary wall, 59
Primordia, 31
  flower, 292
  leaf, 116
Procambium, 117
  root, 171
  stem, 116
Producers, 378
Productivity, net, 378
Proembryo, 324
  pine, 629
Prokaryotes, 3, 9
  algae, 472-479
  bacteria, 419-439
Prolamellar body, 48
Propagation by means of roots, 168
  by means of stems, 134-136
Prophage, 543
Prophase, meiosis, 342
  mitosis, 90
Proplastid, 48
Protein, 67-68
Proterozoic Age, 660
Prothallia of ferns, 606
Prothallial cell or male gametophyte
  Selaginella, 586
Protoderm root, 171
  stem, 116
Protonema, 869
Protoplasm, 27
Protoplast, 40

Index to Genera

Illustrations in parentheses.